CAMBRIDGE SPECTRUM MATHS 8

J. Goodman

PUBLISHED BY THE PRESS SYNDICATE OF THE UNIVERSITY OF CAMBRIDGE
The Pitt Building, Trumpington Street, Cambridge, United Kingdom

CAMBRIDGE UNIVERSITY PRESS
The Edinburgh Building, Cambridge CB2 2RU, UK
40 West 20th Street, New York, NY 10011–4211, USA
477 Williamstown Road, Port Melbourne 3207, Australia
Ruiz de Alarcón 13, 28014 Madrid, Spain
Dock House, The Waterfront, Cape Town 8001, South Africa

http://www.cambridge.edu.au

First published in 2003

Printed in Singapore by Craft Print International Ltd

Typeface [New Aster 9.5 pt] pt *System* QuarkXPress® [PH]

National Library of Australia Cataloguing in Publication data
Goodman, Jennifer.
Cambridge Spectrum maths 8.
For year 8 students.
ISBN 0 521 53042 3.
1. Mathematics — Textbooks. Title.
510.76

ISBN 0 521 53042 3 paperback

Photograph Acknowledgments
Cover: Martineau Arts. Pages: 25, 62, 77, 270, 284, 292, 338, 353, 384, 402 courtesy Photodisc Australia; 27, 38 courtesy www.imageafter.com; 37, 252 courtesy Tourism New South Wales; 39 Heinz Australia; 67 USDA Photo by Ken Hammond; 74 Di Jones Real Estate; 75 National Australia Bank; 111 Flags courtesy www.theodora.com; 129 McDonalds Australia; 142 photo courtesy RAS Victoria; 157 Karen Eubanks, Midwest Surveys Inc., Canada; 244 © Amanda Pinches; 255 Ford Australia; 257 Newscast; 282 Australian Sports Commission; 288 courtesy www.photolibrary.com; 367 Library of Congress, Prints and Photographs Division (reproduction number LC-USZ62-25564); 368, 369, 374 image: www.freeimages.com; 375 photograph courtesy Chess World; 402 map © Commonwealth of Australia. All rights reserved. Reproduced by permission of the Chief Executive Officer, Geoscience Australia, Canberra ACT.

Every effort has been made to trace and acknowledge copyright but there may be instances where this has not been possible. Cambridge University Press would welcome any information that would redress this situation.

Contents

Introduction

Whilst the content changes in the 2004 Mathematics Syllabus are small, Cambridge has taken the opportunity to fully revise the original successful *Spectrum Maths 8*. Extensive reviews were conducted with teachers from a wide range of schools in order to to improve on the original text.

As a result, this new edition is able to cater for a wider range of abilities. Revisions were made to the chapter sequence and the level of development within a chapter, and the exercises have been given two levels. Questions in the first level introduce the concepts of the section — they are straight forward in style and closely tied to the Worked Examples to ensure early success. The questions in the second level extend and broaden understanding — they involve more than one step, as well as explanations, applications, new concepts and links to previous topics

The theory uses colour extensively to highlight new ideas and to demonstrate important steps in the solutions to the Worked Examples. The material has been written to meet the requirements of both a classroom lesson and homework. The enhanced CD-ROM version of the text can be taken home, weighty school bags can be avoided and the student has access to a large range of interactive resources that aid understanding.

Central to the new Syllabus is the idea of continuity—'take the student from where they are at'. Cambridge Spectrum Maths links the prior learning to the new work in each chapter through **MathsCheck**. There are exercises throughout the text that address Stage 5 outcomes (see the list of outcomes at the start of each chapter). The last chapter of the book begins to address the outcomes for Stage 5 Algebra.

The use of colour enhances readability and the frequent hint boxes coach the student on anticipated common misunderstandings. **Now try this** provides extension and variety. Working Mathematically is treated as an approach rather than a topic and is reflected in the selection of exercises, the style of the **Investigations**, the inclusion of **Language Links** and the provision of the important **Keeping Mathematically Fit** exercises.

This work can be supplemented by the wealth of additional planning, extending, revising and assessing tasks that can be printed from the Teacher's CD-ROM. The CD also includes a chapter from Spectrum Maths 7 and a chapter from Spectrum Maths 9 5.3. Teachers are permitted to make copies of these for use within their school.

We are confident we have published a text that will enhance the learning of mathematics and appeal to both students and teachers. We welcome any feedback, which may be sent to:

Education Commissioning Editor, Mathematics
Cambridge University Press
477 Williamstown Road
Port Melbourne 3207
Email: educationpublishing@combridge.edu.au

At each level Spectrum offers ...

The full colour text book

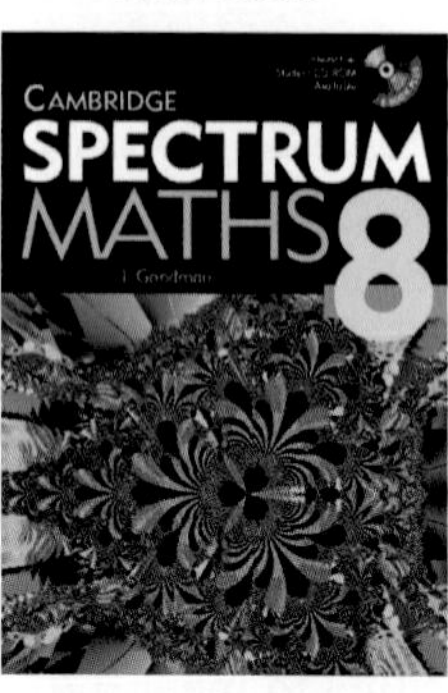

A Student CD-ROM offering the full text PLUS additional interactive study resources

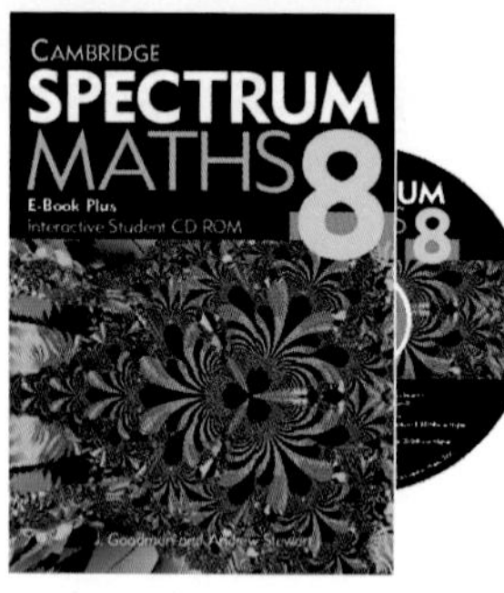

Book and student CD-ROM package

A Resource CD-ROM for your teacher

In each chapter you will find ...

Keeping Mathematically Fit exercises to refresh maths fundamentals

Chapter Review Exercises

Maths Check Revision of existing maths skills required for success in this chapter

In-depth study of the topic

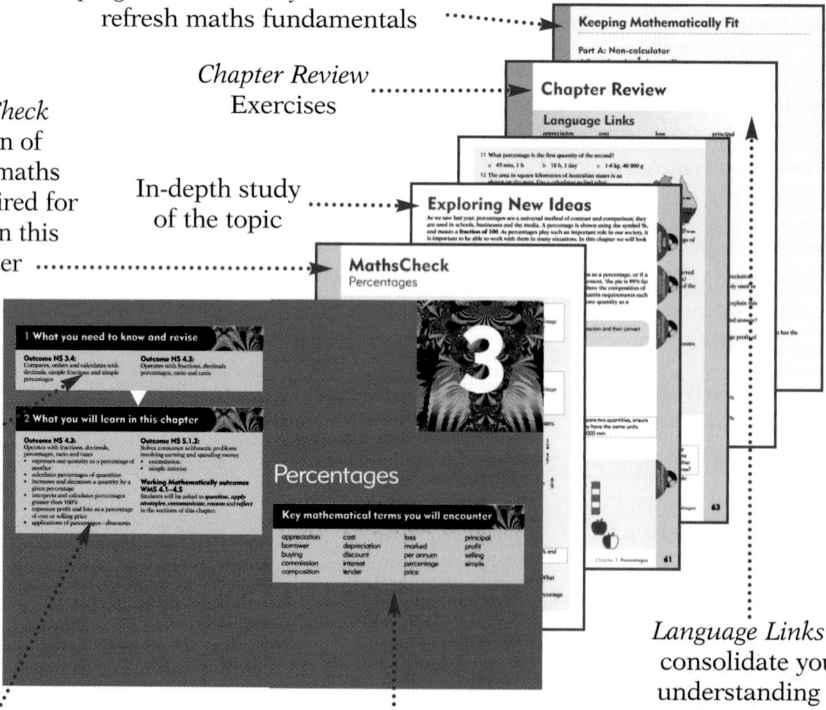

Keeping Mathematically Fit

Part A: Non-calculator

Chapter Review

Language Links

Exploring New Ideas

MathsCheck

Percentages

1 What you need to know and revise

2 What you will learn in this chapter

3

Percentages

Key mathematical terms you will encounter

A list of maths skills and concepts you should know and revise

Language Links to consolidate your understanding of key terms

An outline of what you will learn in this chapter (coded to syllabus Outcomes for the teacher)

Key mathematics terms you will encounter

How to use our navigation tools ...

Theory explanations introduce each new idea with clear and concise explanations that make every topic immediately understandable.

Links to support material on Teacher CD-ROM.

Worked examples demonstrate how to set out your work. Effective use of colour highlights mathematical steps.

Colour-coded links to the student CD-ROM help you to confidently select the appropriate additional interactive resources provided on the CD-ROM.

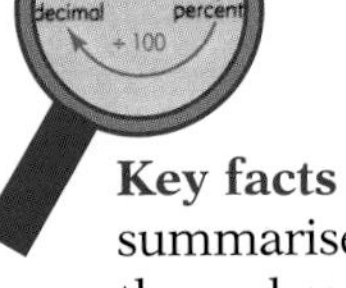

Key facts are summarised throughout the text.

Exploring New Ideas

Investigation

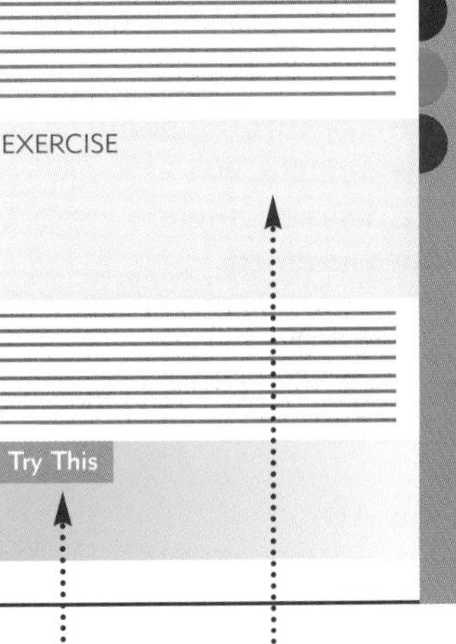

Hint boxes offer invaluable advice on solving mathematical questions.

Levelled exercises
These allow you to gain immediate confidence in the new idea.
Level 2 Exercises require a little more thought.

Investigations strengthen your understanding through further activities and topics for research. These activities may require a little preparation!

Try This activities consider the topic from a different angle to reinforce your understanding of the key concepts.

Symbols and abbreviations

Symbol	Meaning
$=$	is equal to
$\neq$	is not equal to
$\approx$ or $\doteqdot$	is approximately equal to
$<$	is less than
$>$	is more than
$+$	addition operation (plus)
$-$	subtraction (minus)
$\times$	multiplication (times)
$\div$	division (divided by)
$\sqrt{\ }$	square root
$\sqrt[3]{\ }$	cube root
2^3	two to the third power ($2 \times 2 \times 2$) index notation
HCF	highest common factor
LCM	lowest common multiple
$0.\dot{8}$	a decimal in which the 8 recurs forever
$0.\dot{2}3\dot{6}$	a decimal in which the 236 repeats forever
$\frac{3}{4}$	the common fraction three fourths, or $3 \div 4$
LCD	lowest common denominator
$\angle$ABC	angle A, B, C; B is the vertex of A$\hat{\text{B}}$C or turning point
$\llcorner$	right angle; 90°
72°	72 degrees; angle measurement
A B	interval AB
A B	line AB
A B	ray AB
$\perp$	is perpendicular to
$\parallel$	is parallel to
	parallel lines

Symbol	Meaning
	lines equal in length
km	kilometre
m	metre
cm	centimetre
mm	millimetre
kg	kilogram
g	gram
mg	milligram
ML	megalitre
kL	kilolitre (capacity; liquids)
L	litre
mL	millilitre
ha	hectare
m^2	square metre (area)
m^3	cubic metre (volume)
$6t$	$6 \times t$, where t is an algebraic symbol or pronumeral whose value may vary
%	percentage; hundredths
$P = 2(l + b)$	perimeter of a rectangle equals 2 times the sum of length and breadth
$P = 4s$	perimeter of a square equals 4 times the side
$A = lb$	area of a rectangle equals length times breadth
$A = s^2$	area of a square equals side squared
$A = \frac{bh}{2}$	area of a triangle equals base times perpendicular height divided by 2
$V = lbh$	volume of a rectangular prism equals length times breadth times height
3:45	digital clock display for forty-five minutes past three **am** or **pm**
0345 h	24-hour system of expression for the same time **am**
1545 h	24-hour time for 3:45 **pm**

Reviewer details

Dr Irene Abbot, formerly Mathematics Coordinator at Mt St Benedict College, Pennant Hills, is currently teaching part-time at Normanhurst Boys' High School.

Tony Ward has been a Mathematics teacher for over 30 years at Cumberland High School, with a special interest in gifted students.

Joseph Mirabito has been a Mathematics teacher for over 18 years and is currently teaching at St.Andrews College, Marayong.

Graham Fardouley has been a Mathematics teacher for 16 years and is currently teaching at Cherrybrook Technical High School.

Tony Priddle, a key author for the first edition of Spectrum Maths 8, provided encouragement and support in the development of this new edition.

CD icons

Links to web research activities on the *E-Book Plus*
Research various aspects of mathematics – including concepts, personalities and the history of mathematical ideas – using the web. Links to particularly useful web sites are included.

Links to interactive geometry activities on the *E-Book Plus*
Explore geometric concepts independently using current technology applications. Activities for both Cabri Geometry and Geometer's Sketchpad are offered.

Links to spreadsheet activities on the *E-Book Plus*
Enhance your understanding of mathematical concepts using dynamic Excel activities. Many activities are linked to HELP files offering hints for solving the problems.

Links to investigative activities on the *E-Book Plus*
Strengthen your understanding of key concepts using a variety of resources to complete these investigations.

These icons indicate to your teacher that there are additional resources on the *Spectrum Maths 8 Teacher CD-ROM*.

1 What you need to know and revise

Outcome NS 4.1:
Recognises the properties of special groups of whole numbers and applies a range of strategies that aid computation

Outcome NS 4.2:
Compares, orders and calculates with integers

Outcome NS 4.3:
Operates with fractions, decimals, percentages, ratio and rates

2 What you will learn in this chapter

Outcome NS 4.1:
Recognises the properties of special groups of whole numbers and applies a range of strategies that aid computation

Outcome NS 4.2:
Compares, orders and calculates with integers

Outcome NS 4.3:
Operates with fractions, decimals, percentages, ratio and rates

Outcome MS 4.3:
Performs calculations of time that involve mixed units

- uses index notation
- finds square roots and cube roots of numbers using the calculator
- uses the calculator with integers, fractions and decimals
- adds and subtracts time using the calculator
- rounds calculator answers to the nearest minute or hour
- interprets calculator answers for time calculations

Outcome NS 5.1.1:
Applies index laws to simplify and evaluate arithmetic expressions and uses scientific notation to write large and small numbers

- index notation with numerical bases
- four index laws—using numerical bases
  - $a^m \times a^n = a^{m+n}$
  - $a^m \div a^n = a^{m-n}$
  - $(a^m)^n = a^{m \times n}$
  - $a^0 = 1$

Working Mathematically outcomes WMS 4.1–4.5
Students will be asked to ***question***, ***apply strategies***, ***communicate***, ***reason*** and ***reflect*** in the sections of this chapter.

Number

Key mathematical terms you will encounter

approximate
base
calculator
decimal
expanded
exponent
factor
fraction
index
integer
multiple
negative
notation
opposite
positive
power
prime
product
quotient
reciprocal

MathsCheck
Numbers

1 Find the value of:

a 5×12 b $1700 \div 5$ c $\frac{460}{20}$ d 194×85

2 Write in expanded form:

a 74 080 b 0·68 c 605·008

3 Write the place value of the 8 in the numbers given in question **2**.

4 Use <, > or = to make the following true:

a 1001 ☐ 999 b 3079 ☐ 3100 c 808 ☐ 880 d 0·7 ☐ 0·70

5 Evaluate:

Remember the order of operations: ( ), ×/÷, +/–

a $56 - 7 \times 4$ b $96 \div 4 + 3 \times 7$
c $56 - 16 + 10 \times 3$ d $7 \times (10 - 3)$
e $150 - [7 \times (10 - 3 + 2)]$ f $8 + [6 + 10 \times (8 - 3)]$

6 Put in grouping symbols to make each statement true.

a $17 - 4 \times 2 = 26$ b $18 \div 2 + 4 = 3$ c $15 \times 9 - 3 \div 3 = 30$

7 True or false?

a $7 \times 2 + 3 = 7 \times 2 + 7 \times 3$ b $5 \times 8 = 8 \times 5$ c $6 + (7 + 5) = (6 + 7) + 5$
d $16 \div 8 = 8 \div 16$ e $17 \times 0 = 17$ f $6 \div 0 =$ undefined

8 Express a 45 b 75 using prime factors.

9 Using prime factors, find:

A factor tree is a good way of finding a number's prime factors.

a the HCF of 45 and 75
b the LCM of 45 and 75

10 Calculate:

a $^-8 + {}^-6$ b $5 + {}^-9$ c $^-3 + 10$
d $^-27 + 27$ e $7 - 12$ f $^-3 - 8$
g $6 - {}^-9$ h $^-3 - {}^-9$ i $^-6 - 6$
j $^-18 + 7 - {}^-3$ k $5 - 12 + {}^-6$ l $12 + {}^-18 - {}^-20$

11 Calculate the products:

a $2 \times {}^-7$ b $7 \times {}^-2$ c $^-2 \times {}^-7$
d $^-8 \times 9$ e $^-9 \times 5$ f $^-6 \times 1$
g $^-1 \times {}^-3$ h $^-4 \times 0$ i $^-5 \times {}^-5$

12 Calculate the quotients:

a $12 \div {}^-3$ b $^-12 \div 3$ c $^-12 \div {}^-3$
d $^-16 \div 8$ e $^-4 \div 1$ f $36 \div {}^-6$
g $^-333 \div {}^-3$ h $\frac{56}{^-7}$ i $0 \div {}^-5$
j $10 \div 0$ k $^-24 \div 24$ l $8 \div {}^-16$

13 Evaluate:

a $3 \times (1 + {}^{-}5)$ b $3 \times 1 + 3 \times {}^{-}5$
c ${}^{-}5 \times ({}^{-}8 - 2)$ d ${}^{-}5 \times {}^{-}8 - {}^{-}5 \times 2$

14 Calculate:

a $0{\cdot}7 \times 0{\cdot}4$ b $2{\cdot}8 \times 0{\cdot}6$ c $0{\cdot}95 \times 1{\cdot}2$
d $0{\cdot}78 \times 100$ e $1{\cdot}35 \times 2000$ f $6{\cdot}42 \times 10$
g $(0{\cdot}2)^3$ h $(0{\cdot}6)^2$ i $(0{\cdot}12)^2$
j $0{\cdot}08 \div 0{\cdot}4$ k $0{\cdot}0304 \div 0{\cdot}8$ l $81{\cdot}18 \div 0{\cdot}09$
m $2765 \div 1000$ n $9{\cdot}35 \div 1000$ o $0{\cdot}048 \div 10$
p $0{\cdot}085 \div 0{\cdot}04$ q $\sqrt{0{\cdot}04}$ r $\sqrt[3]{0{\cdot}027}$

15 Simplify:

a $\frac{10}{24}$ b $\frac{16}{30}$ c $\frac{140}{150}$ d $\frac{32}{44}$

16 Simplify:

a $\frac{3}{6} + \frac{2}{6}$ b $\frac{7}{8} - \frac{3}{8}$ c $\frac{3}{10} + \frac{2}{5}$ d $\frac{5}{6} + \frac{1}{3}$

e $\frac{1}{2} - \frac{3}{8}$ f $\frac{1}{2} + \frac{2}{3}$ g $\frac{3}{5} - \frac{1}{3}$ h $\frac{3}{4} - \frac{1}{6}$

i $\frac{5}{6} + \frac{5}{8}$ j $2\frac{1}{4} + 1\frac{1}{2}$ k $4\frac{2}{3} - 3\frac{1}{2}$ l $7 + 1\frac{3}{4} + \frac{2}{3}$

17 Perform these multiplications:

a $\frac{1}{4} \times \frac{1}{3}$ b $\frac{4}{5} \times \frac{1}{4}$ c $\frac{3}{7} \times \frac{7}{10}$ d $\frac{5}{8} \times 40$

e $\frac{4}{5} \times \frac{15}{20}$ f $1\frac{1}{3} \times \frac{4}{5}$ g $4\frac{1}{5} \times 6\frac{1}{3}$ h $1\frac{1}{2} \times 1\frac{1}{3} \times 2\frac{1}{5}$

18 Write the reciprocals of:

a $\frac{4}{5}$ b $\frac{8}{7}$ c 6 d $5\frac{1}{3}$

19 Calculate:

a $15 \div \frac{1}{3}$ b $5 \div \frac{1}{2}$ c $10 \div \frac{1}{5}$ d $\frac{1}{5} \div \frac{1}{4}$

e $\frac{7}{4} \div 4$ f $\frac{3}{8} \div \frac{3}{16}$ g $\frac{9}{20} \div \frac{10}{11}$ h $2\frac{1}{4} \div 4$

20 Follow the order rules to simplify:

a $\frac{4}{7} + \frac{1}{2} \times \frac{1}{3}$ b $\frac{1}{2} \times \left(\frac{3}{4} - \frac{5}{8}\right)$ c $\frac{{}^{-}4}{7} \times \frac{14}{20}$

d $\left(\frac{1}{6} + \frac{8}{9}\right) \times \left(\frac{1}{2} - \frac{3}{7}\right)$ e $\frac{4}{9} \div \frac{1}{3} \times \frac{1}{2}$ f $\frac{4}{9} \div \left(\frac{1}{3} \times \frac{1}{2}\right)$

TEACHER

Exploring New Ideas

1.1 Index notation

To quickly show when a number is being multiplied by itself, index notation is used; it is a way of writing products of identical factors.

The base shows the factor that is repeated and the index/exponent/power represents the number of times the multiplication occurs.

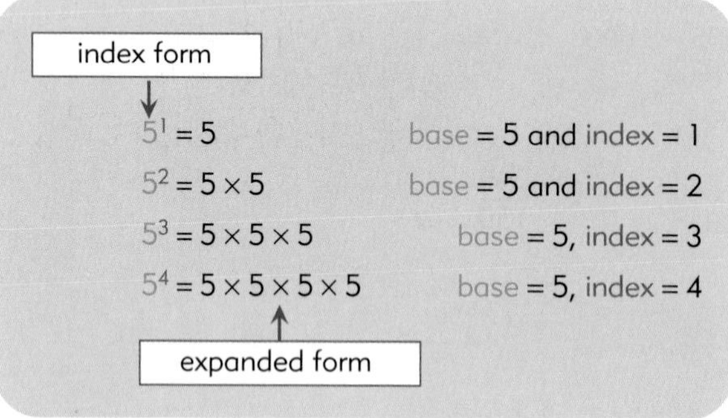

Remember:
- the opposite of squaring is finding the square root
- the opposite of cubing is finding the cube root

EXAMPLE 1
Write the following in index form:

a $3 \times 3 \times 3 \times 4 \times 4 \times 4 \times 3 \times 3$

b $6 \times 7 \times 6 \times 7 \times 6 \times 7$

Solution

a $3 \times 3 \times 3 \times 4 \times 4 \times 4 \times 3 \times 3$
$= \underbrace{3 \times 3 \times 3 \times 3 \times 3}_{\text{5 factors}} \times \underbrace{4 \times 4 \times 4}_{\text{3 factors}}$
$= 3^5 \times 4^3$

b $6 \times 7 \quad \times 6 \times 7 \quad \times 6 \times 7$
$= (6 \times 7)^3$
or $6^3 \times 7^3$

EXAMPLE 2
Write in expanded form:

a 3^4

b $5^4 \times 2^3$

c $(7 \times 9)^5$

Solution

a 3^4
$= 3 \times 3 \times 3 \times 3$

b $5^4 \times 2^3$
$= 5 \times 5 \times 5 \times 5 \times 2 \times 2 \times 2$

c $(7 \times 9)^5$
$= 7 \times 9 \times 7 \times 9 \times 7 \times 9 \times 7 \times 9 \times 7 \times 9$
or $7 \times 7 \times 7 \times 7 \times 7 \times 9 \times 9 \times 9 \times 9 \times 9$

EXERCISE 1A

1 Write each product in index notation:

a $7 \times 7 \times 7$ **b** $10 \times 10 \times 10 \times 10 \times 10$ **c** 8 **d** $6 \times 6 \times 5 \times 5 \times 5 \times 5$

2 Write each power in expanded form:

a 8^3 **b** 1^5 **c** 12^2 **d** $7^4 \times 9^2$

3 Evaluate:

a 0^4 b 10^3 c 2^5 d $2^3 \times 3^2$

4 Express as powers of primes by completing the factor trees:

a 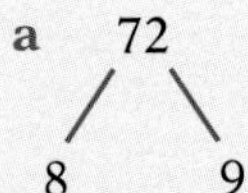

b 2700 → 27, 100

c 144 → 12, 12

d 1701 → 81, 21

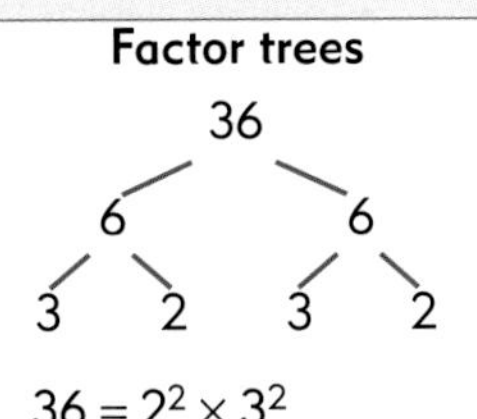

Level 2

5 Simplify:

a $1^4 + 2^3$ b $2^4 - 3^2$ c $(5^1 + 2^4) \times 10^6$

6 Arrange in ascending order: 4^3, 1^{100}, 0^7, 3^4, 10^2.

7 Write each of the following as a product of prime factors:

a 625 b 243 c 2187 d 6125

8 Write each of the following as a power of 10:

a 100 b 10 000 c 100 000 d 10
e 10×100 f $10 \times 100 \times 1000$ g ten thousand h 100^2

9 How many zeros follow the one in each of the following numbers?

a 10^2 b 10^7 c 10^1 d $10^4 \times 10^2$

10 Write the following numbers as powers of prime factors:

a 1225 b 1089 c 16 900 d 74 088

11 By using your answers to question **10**, find the value of:

a $\sqrt{1225}$ b $\sqrt{1089}$ c $\sqrt{16\,900}$ d $\sqrt[3]{74\,088}$

12 Find the numbers missing from the triangles.

a $\triangle^2 = 81$ b $\triangle^3 = 27$ c $\triangle^5 = 1$
d $\triangle^2 = 144$ e $7^{\triangle} = 343$ f $11^{\triangle} = 14\,641$

13 Find the unknown number, x, each time.

a $5^x = 25$ b $5^x = 125$ c $x^3 = 8$
d $6^x = 36$ e $x^3 = 216$ f $x^5 = 243$

1.2 First and second index laws

Let us look at the rules needed when working with numbers written in index notation.

First index law:

$$a^m \times a^n = a^{n+m}$$

When multiplying numbers in index form, if the bases are the same, the powers are added.

If we write 2^3 and 2^4 as products of factors, then we can see what happens to the product of 2^3 and 2^4:

$$
\begin{aligned}
&2^3 \times 2^4 \\
&\quad \text{3 factors} \quad \text{4 factors} \\
&= (2 \times 2 \times 2) \times (2 \times 2 \times 2 \times 2) \\
&\quad (3 + 4) \text{ factors} \\
&= 2 \times 2 \times 2 \times 2 \times 2 \times 2 \times 2 \\
&= 2^{3+4} \\
&= 2^7
\end{aligned}
$$

Second index law:

$$a^m \div a^n = a^{m-n}$$

When dividing numbers in index form, if the bases are the same, the powers are subtracted.

If we write 3^6 and 3^2 as products of factors, then we can see what happens when 3^6 is divided by 3^2:

$$
\begin{aligned}
&3^6 \div 3^2 \\
&= \frac{3 \times 3 \times 3 \times 3 \times \cancel{3} \times \cancel{3}}{\cancel{3} \times \cancel{3}} \quad \begin{matrix}\text{(6 factors)} \\ \text{(2 factors)}\end{matrix} \\
&= 3 \times 3 \times 3 \times 3 \qquad \text{(4 factors)} \\
&= 3^{6-2} \\
&= 3^4
\end{aligned}
$$

Note: $3^6 \times 2^4$ or $3^6 \div 2^4$ cannot be simplified using the rules above, as the *bases are different.*

EXAMPLE 1

Simplify each of the following using the first index law:

a $2^4 \times 2^7$ **b** $6^5 \times 6^3$ **c** $2^7 \times 3^2 \times 2^2$

Solution

a $2^4 \times 2^7 = 2^{4+7} = 2^{11}$

b $6^5 \times 6^3 = 6^{5+3} = 6^8$

c $2^7 \times 3^2 \times 2^2$
$= 2^7 \times 2^2 \times 3^2$
$= 2^{7+2} \times 3^2$
$= 2^9 \times 3^2$

EXAMPLE 2

Simplify each of the following using the second index law:

a $2^7 \div 2^2$ **b** $7^4 \div 7$ **c** $8^3 \div 4^2$

Solution

a $2^7 \div 2^2 = 2^{7-2} = 2^5$

b $7^4 \div 7$
$= 7^4 \div 7^1$
$= 7^{4-1}$
$= 7^3$

c $8^3 \div 4^2$
cannot be simplified as the bases are different.

EXERCISE 1B

1 Write in expanded form, then simplify:

a $2^2 \times 2^3$ b $10^3 \times 10^1$ c $3^4 \div 3^2$ d $7^5 \div 7^4$

2 Complete the following:

a $3^4 \times 3^2 = 3 \times 3 \times 3 \times 3 \times 3 \times 3$
$= 3^{\square}$

b $2^3 \times 2^5 = 2 \times 2 \times 2 \times 2 \times 2 \times 2 \times 2 \times 2$
$= 2^{\square}$

c $4^4 \times 4 = 4 \times 4 \times 4 \times 4 \times 4$
$= 4^{\square}$

d $7^2 \times 7^3 = 7 \times 7 \times 7 \times 7 \times 7$
$= \square^5$

3 Complete the following, using index notation:

a $6^4 \times 6^2 = 6^{\square}$ b $9^6 \times 9^4 = 9^{\square}$ c $2^3 \times 2^5 = 2^{\square}$
d $2^3 \times 2^{\square} = 2^7$ e $4^3 \times 4^5 = 4^{\square}$ f $12^3 \times 12 = 12^{\square}$

4 Complete the following:

a $2^4 \div 2^2 = \dfrac{2 \times 2 \times 2 \times 2}{2 \times 2}$
$= 2^{\square}$

b $8^4 \div 8 = \dfrac{8 \times 8 \times 8 \times 8}{8}$
$= 8^{\square}$

c $3^5 \div 3^4 = \dfrac{3 \times 3 \times 3 \times 3 \times 3}{3 \times 3 \times 3 \times 3}$
$= 3^{\square}$

d $5^6 \div 5^2 = \dfrac{5 \times 5 \times 5 \times 5 \times 5 \times 5}{5 \times 5}$
$= 5^{\square}$

5 Complete the following by applying the second index law:

a $6^4 \div 6^2 = 6^{\square}$ b $8^7 \div 8^3 = 8^{\square}$ c $7^3 \div 7 = 7^{\square}$
d $4^5 \div 4^3 = 4^{\square}$ e $11^{10} \div 11^7 = 11^{\square}$ f $9^6 \div 9^5 = 9^{\square}$
g $7^6 \div 7^4 = \square^2$ h $16^5 \div 16^3 = \square^{\triangle}$ i $2 \div 2 = 2^{\square}$

Level 2

6 Simplify each of the following using the first index law:

a $2^7 \times 2^4$ b $3^7 \times 3^2$ c $6^2 \times 6$
d $10^3 \times 10 \times 10$ e $5^4 \times 5^5 \times 5^2$ f $7^2 \times 7^{20} \times 7$

7 Simplify each of the following using the second index law:

a $2^8 \div 2^4$ b $5^8 \div 5^3$ c $7^9 \div 7^3$
d $17^{26} \div 17^{20}$ e $9^3 \div 9 \div 9$

8 Simplify, leaving your answer in index form:

a $5^4 \div 25$ b $5^6 \times 5^4 \times 5$ c $\dfrac{8^2 \times 8^2}{8^3}$
d $\dfrac{6^4 \times 6^3}{36^2}$ e $3^4 \times 9$ f $16^4 \times 2^2 \div 4$

Investigation Numbers in index notation

Complete the table below by writing each number in index form or as a sum or difference of two or more numbers in index form. (You may not use a number to a power of one!)

1 Can you find more than one answer to some or all of these? Check with your classmates.

Number	Index expression		Index expression
1	1^3	11	$2^3 - 1^2 + 2^2$
2	$1^3 + 1^2$	12	
3	$1^2 + 1^3 + 1^4$	13	
4	2^2	14	
5		15	
6		16	
7		17	
8		18	
9		19	
10		20	

2 Check your answers work with a calculator. Try numbers up to 100.

1.3 Third and fourth index laws

Third index law:

$$(a^m)^n = a^{m \times n}$$

When a number in index notation is raised to a power, the powers are multiplied.

If we write 3^4 as a product of factors, then we can see what happens when 3^4 itself is cubed.

$$\begin{aligned}(3^4)^3 &= (3 \times 3 \times 3 \times 3)^3 \\ &= (3 \times 3 \times 3 \times 3) \times (3 \times 3 \times 3 \times 3) \times (3 \times 3 \times 3 \times 3) \\ &= 3 \times 3 \times 3 \times 3 \times 3 \times 3 \times 3 \times 3 \times 3 \times 3 \times 3 \times 3 \\ &= 3^{12} \qquad \boxed{3^{4 \times 3} = 3^{12}}\end{aligned}$$

Fourth index law:

$$a^0 = 1 \ (a \neq 0)$$

Any number (except zero) raised to a power of 0 is 1.

If we write 2^4 as a product of factors, then we can see what happens when it is divided by itself.

By cancelling

$$2^4 \div 2^4$$

$$\frac{2^4}{2^4} = \frac{2 \times 2 \times 2 \times 2}{2 \times 2 \times 2 \times 2}$$

$$= 1$$

By index law 2

$$2^4 \div 2^4 = 2^{4-4}$$

$$= 2^0$$

$$\therefore 2^0 = 1$$

This process will work for any base except zero ($0^0 \neq 1$). Can you think why?

Could it be that multiplying by zero always produces an answer of zero?

EXAMPLE 1

Simplify each of the following using the third index law:

a $(2^6)^4$

b $(7^2 \times 3^3)^4$

Solution

a $(2^6)^4 = 2^{6 \times 4}$
$= 2^{24}$

b $(7^2 \times 3^3)^4 = (7^2)^4 \times (3^3)^4$
$= 7^{2 \times 4} \times 3^{3 \times 4}$
$= 7^8 \times 3^{12}$

EXAMPLE 2

Simplify each of the following using the fourth index law:

a $3^7 \div 3^7$

b 8^0

c $(5 \times 6)^0$

Solution

a $3^7 \div 3^7 = 3^{7-7}$
$= 3^0$
$= 1$

b $8^0 = 1$

c $(5 \times 6)^0 = 5^0 \times 6^0$
$= 1 \times 1$
$= 1$

EXERCISE 1C

1 Complete the following, using index notation:

a $(2^3)^2 = 2^{\square}$

b $(2^4)^5 = 2^{\square}$

c $(3^7)^2 = 3^{\square}$

d $(7^2)^2 = 7^{\square}$

e $(4^5)^3 = 4^{\square}$

f $(7^5)^2 = \square^{\triangle}$

2 Simplify:

a 5^0

b 10^0

c 15^0

d $7^3 \div 7^3$

e $(2 \times 7)^0$

f 27^0

3 Complete the following using index notation:

a $(2^5 \times 3)^4 = 2^{\square} \times 3^{\triangle}$

b $(7^3 \times 5^2)^2 = 7^{\square} \times 5^{\triangle}$

c $(9 \times 3^3)^4 = 9^{\square} \times 3^{\triangle}$

d $(9 \times 5^2 \times 3^3)^2 = 9^{\square} \times 5^{\triangle} \times 3^{\bigcirc}$

Level 2

4 Simplify each of the following:

a $5^0 + 7$

b $8 - 3^0$

c $10 + 5^0 - 2$

d $(3^2)^4$

e $(7^4)^5$

f $(7^5)^4$

g $(4^6)^0$

h $(10^6)^5$

i $(9)^7$

j $(8^0)^3 \times 8^2$

k $\frac{(6^3)^2}{6^4}$

l $\left(\frac{3}{5}\right)^2$

m $(5^2 \times 8)^3$ **n** $(5^0 \times 4^3)^0$ **o** $(5^2)^0 \times 5^2$

p $5^3 \times 5^2 \div (5^2)^2$ **q** $\frac{3^7 \times 3^3}{(3^4)^2}$ **r** $(5^0 \times 4^3)^2 \div 7^0$

1.4 The calculator

We are lucky to have calculators. They certainly are time-saving devices.

For each question, enter the example supplied and check the answer given. See your teacher if the answers vary.

	Operation	Keys	Example	Answer
1	Addition	+	96·547 + 3·801	100·348
2	Subtraction	–	18·6 – 19·079	$^-0·479$
3	Multiplication	×	15 × 16	240
4	Division	÷	7 ÷ 9	0·7777 …
5	Fraction	$a^b/_c$	$6\frac{1}{2} + 7\frac{1}{4} \times \frac{1}{3}$	$8\frac{11}{12}$
6	Change of direction (integers)	(–)	$^-6 - (^-3) + 9$	6
7	Brackets	[( )]	(8 × 4 – 6) + (9 × 3 – 1)	52
8	Square	x^2	9^2 $(8 - 9 \times 2)^2$	81 100
9	Square root	$\sqrt{\ }$	$\sqrt{144}$ $\sqrt{(15·6 + 7 \times 9·1)}$	12 8·905 054 …
10	Powers	x^y	9^5	59 049
11	nth root	$\sqrt[x]{y}$	$\sqrt[4]{81}$	3
12	Cube root	$\sqrt[3]{x}$	$\sqrt[3]{64}$ $\sqrt[3]{\frac{16·6}{8·1}}$	4 1·270 206 …
13	Rounding	MODE Fix	Round 7·65481328 to 3 dec. pl. $\frac{4·71^2 - 8·54}{\sqrt{4·2}}$ (2 dec. pl.)	7·655 6·66
14	Normal	MODE Norm	Removes the rounding 'fix'.	
15	Time	° ' ' ' or D°M'S	Convert 1·5 h to hours and minutes. 1·5 ° ' ' '	1°30°0° reads 1 hour 30 minutes 0 seconds.

The Casio fx-82TL was used here. If your calculator is different, see your teacher.

Remember: brackets are very important.

EXERCISE 1D

1 Use a calculator for the following; round answers to 2 decimal places where appropriate:

a $56{\cdot}9 \times 0{\cdot}3$ b $\sqrt{12{\cdot}96}$ c $\sqrt[3]{125}$

d $6 - 3\frac{4}{5}$ e $1\frac{1}{2} \times \frac{3}{4}$ f $5{\cdot}6 \div 8{\cdot}9$

g $6{\cdot}3 + {}^{-}9{\cdot}7$ h ${}^{-}8 - 10$ i $[8 \times (4 + 3 - 2)]^2$

j $16{\cdot}971 + 8{\cdot}91 + {}^{-}0{\cdot}63$ k $\frac{4}{7} \times \frac{1}{3} \div \frac{1}{6}$ l $\frac{2}{3}$ of \$100.50

2 Evaluate:

a $6{\cdot}1^2$ b 19^3 c 5^7 d $(3 \times 4^2)^3$

e $({}^{-}2)^2$ f $({}^{-}2)^3$ g $({}^{-}4{\cdot}6)^2$ h $({}^{-}5 \times 6)^2$

3 Find, correct to 3 decimal places, where necessary:

a $\sqrt[4]{100}$ b $\sqrt[5]{2}$ c $\sqrt[4]{7}$ d $\sqrt[5]{9{\cdot}3 \times 4}$

e $\sqrt[3]{8}$ f $\sqrt[3]{{}^{-}8}$ g $\sqrt{0{\cdot}25}$ h $\sqrt{{}^{-}2}$

4 By evaluating the numerator and denominator separately, or by using brackets, evaluate:

a $\frac{4 + 5}{3 - 1}$ b $\frac{12}{2 + 6}$ c $\frac{3 - 2{\cdot}78}{0{\cdot}2}$ d $\frac{\sqrt{16 - 15{\cdot}64}}{2^2 + 1}$

5 Convert the following hours to hours, minutes and seconds on your calculator:

a $5{\cdot}2$ h b $1{\cdot}25$ h c $7{\cdot}4$ h

d $10{\cdot}23$ h e $9{\cdot}05$ h f $1{\cdot}125$ h

g $6\frac{1}{2}$ h h $\frac{3}{4}$ h i $3\frac{5}{8}$ h

Level 2

6 By using brackets, evaluate, correct to 2 decimal places where necessary:

a $\sqrt{\frac{19.1}{5{\cdot}1}}$ b $\frac{18 + [9 - 7 \times 3 \div 2]}{6}$ c $\frac{19^2 + 7{\cdot}3}{8 \times 3}$

d $\frac{5{\cdot}4 - 3^2}{\sqrt{16}}$ e $\frac{9{\cdot}6}{\sqrt[3]{27} + 16 - 4 \times 3}$ f $\sqrt[3]{\frac{9{\cdot}1}{7{\cdot}2 \times 3{\cdot}8}}$

7 120 minutes can be converted into hours by dividing by 60. Convert the following into hours by dividing by 60, then convert using the time key into hours, minutes and seconds.

a 200 min b 78 min c 550 min

d $100{\cdot}8$ min e $245{\cdot}5$ min f $690{\cdot}375$ min

8 When decimals are rounded, we round up at the halfway point; that is, when the following digit is 5 or more. With time we also round up at the halfway point; that is, at or over 30 minutes or 30 seconds.

Write your answers to question **7** to the nearest:

a hour b minute

9 Use your calculator to find:

3 h 12 min 6 s = 3 [° ' "] 12 [° ' "] 6 [° ' "]

a 2 h 48 min + 3 h 57 min   b 10 h 12 min – 6 h 46 min

c 5 h 12 min 8 s + 42 min 33 s   d 5 h 2 min 48 s – 4 h 36 min 51 s

10 Simplify, leaving your answers in index notation:

a $\frac{5^4}{5^4}$ b $9^6 \times 9^2 \div 9$ c $7^4 \div 7^3 \times 7$

d $10^5 \div 100$ e $7^4 \times 7^3 \div 7^2$ f $5^4 \times 5^3 \times 6^2$

g $4^8 \times 4^2 \div 2^6$ h $9^{12} \div (9^3 \times 9^5)$ i $\frac{7^4 \times 7^3}{7^2 \times 7^3}$

Investigation Find a number

TEACHER

Round all decimal answers to the nearest whole number. Highlight the answers in the number grid.

3	6	2	0	4	8	4
2	6	2	1	4	4	8
2	7	1	8	5	0	2
5	5	4	3	2	1	6
8	5	3	1	3	9	8
9	9	3	7	2	8	0
1	8	7	2	6	7	9
1	1	5	7	9	1	6
7	8	0	8	1	0	9
9	9	6	5	5	6	8
7	9	4	5	1	9	5
1	5	9	4	3	2	3

1 64^3

2 $[(1 + 3 \times 6)^2 + 31] \div \sqrt[3]{64}$

3 $\frac{96 + 104}{5}$

4 $(8 \cdot 1)^3$

5 $64 - (^-150)^3$

6 $12^2 + 12^3$

7 $[10 + 6 \times 8] - \sqrt{64} \div 8$

8 $\sqrt{97\,436\,641}$

9 $(6\frac{2}{3})^2 - (\frac{2}{3})^2$

10 $16^4 - (^-3)^3 + \sqrt[3]{125}$

11 $\frac{194 \cdot 8 - (9 \cdot 6 \times 5 \cdot 1 - 6 \cdot 4)}{\sqrt{4 \cdot 8}}$

12 $(13^2)^3$

13 $3^9 \times 3^7 \div 3^3$

14 $\sqrt[5]{4^3} + 65\,566$

15 $6 \times (4 \times 7 - 8) - 3$

1.5 Problems using the calculator

The calculator allows us to work quickly with otherwise difficult and time consuming calculations, like the ones in this exercise below.

EXAMPLE

Melanie buys 45 kilograms of soil to use in her garden at a cost of \$5.78 per 5 kg bag. Find the cost of buying the soil.

Solution

Number of bags: $45 \div 5 = 9$

Cost $= 9 \times 5 \cdot 78$

$= 52 \cdot 02$

∴ cost of the soil is \$52.02.

EXERCISE 1E

1 Margaret makes 112 calls per day as part of her job as a telemarketer. How many calls does she make in a month in which she worked 24 days?

2 Graham buys 24 cans of white paint at $45.90 a can. What is the total cost for the paint?

3 Joshua estimates the crowd at a football game to be 47 695, while Lara estimates the crowd to be 51 419.

- **a** What is the difference between the two estimates?
- **b** What is the average of the two estimates?

4 A syndicate of 84 people each have an equal share of the 5.4 million dollar jackpot lottery prize. What is the value of each individual share?

5 a A Sunday newspaper costs $1.50. If one paper is purchased each Sunday for a year, what is the total spent on newspapers in the year? (Take the number of Sundays in a year to be 52.)

- **b** If a newspaper costing $1.20 was bought every day for a year, what would the total cost be for a year with 365 days?

6 The average human heart beats 36 921 400 times in 1 year (365 days). How many times does it beat in:

- **a** 1 day?
- **b** 1 hour?
- **c** 1 minute?
- **d** a 50-minute maths lesson?

7 As you might in your health or PE class, find your resting pulse rate for 1 minute. How many times does your heart beat in:

- **a** 1 hour?
- **b** 1 day?
- **c** 365 days?

8 Find the average length of wood cut by seven students if their pieces measure

56·7 cm, 48 cm, 50·9 cm, 62 cm, 62·5 cm, 81·6 cm, 49·8 cm.

9 How many books costing $24.95 each can be bought with $1000?

Level 2

10 A fruit shop bought 12 boxes of apples at $9.60 a box, 9 boxes of pears at $11.40 a box and 24 boxes of mangoes at $14.20 a box. Calculate the average price per box for the fruit purchased by the shop.

11 A number multiplied by itself produces an answer of 93·1225. What is the number?

12 The number of bacteria in a beaker in a science lab doubles every 30 seconds. If five bacteria are present in the beaker initially, how many will there be at the end of 10 minutes?

13 Use your calculator alone to find the 15th term in each of the following patterns:

a 3, 6, 9, ...
b 7, 10, 13, ...
c 4, 12, 36, ...

14 a If the carbon-14 in a log halves approximately every 5700 years, how much carbon remains at the end of 100 000 years if there were originally 460 grams of carbon-14 in the log?
b Iodine-131, which is used in the treatment for thyroid cancer, has a half-life of 8 days. If a patient is injected with 12 milligrams of iodine-131, how much iodine-131 is left in the patient's system at the end of 96 days?
c Stromtium-90 is a by-product of nuclear fallout responsible for causing bone cancer and has a half-life of 28 years. How much remains at the end of 300 years if there were initially 2 grams of stromtium-90?

15 The average of nine numbers is 21·5 and the average of 12 other numbers is 19·9. What is the average, correct to 2 decimal places, of the 21 numbers?

16

$^{-}0·64$	0·3	$^{-}1·2$	0·8
$^{-}2$	$^{-}0·04$	$^{-}0·2$	0·32
$^{-}0·1$	1·8	0·64	0·12
8·4	3	0·2	0·4

Using the numbers in the grid, work out the following:

a Find a set of four whose total comes to zero.
b What is the product of your four numbers?
c What is the total of the remaining numbers?

1.6 Problem solving

Remember the strategies for exploring the unknown that we developed last year:

1 Understand the situation
Read the description carefully.
Underline key facts and what you are asked to find.

2 Look for useful strategies
Estimate before you start. Look for upper and lower limits. What kinds of numbers may suit?
Do you know any *rules* or formulas that seem to fit?
Guess, check and refine.
Work systematically. Make lists. Keep a record of everything you try, one thing at a time.
Models, drawings, diagrams or graphs may help you to visualise the situation.
Number patterns may show up in a table.
Break it into parts or *simplify the numbers.*
Work backwards.
Act it out with classmates.
Eliminate possibilities.

3 Follow your best strategy towards a solution
Show all working clearly, even if you can't finalise it.
If the strategy doesn't lead to a solution, *try a different strategy.*
Check that your answer makes sense and answers the question.
Explain your solution in a way that others can follow.

4 EXTEND THE SOLUTION IF POSSIBLE
Ask yourself: 'What if …?' Have you learned anything that might apply elsewhere?
Changing the data or situation slightly would have what effects on the outcome?

EXERCISE 1F

Use the hints above as a checklist for exploring the following:

1 Hamish can paint a room in 10 hours. Sean takes 15 hours to paint an identical room. How long should it take to paint such a room if both work together?

2 A highrise building has as many floors above ground level as it has below (ground floor is at the middle of the building). A security guard randomly checking the floors enters at ground level, goes up 3 floors, down 7, then up 10 to the topmost floor. How many floors does the building have?

3 Each team in a round-robin tennis competition must play every other team once. How many matches are played if there are 12 teams?

> Try simpler versions by reducing the number of teams, and look for a pattern. A diagram should help.

4 Sara and Cristina each have $100 in a savings account. Every month Sara adds $10 to her account, while Cristina deposits $30 in hers. In how many months will Cristina have twice as much as Sara?

> List the balance and deposits in two columns, or model the situation as an algebraic equation.

5 In my pocket I have eight coins with a total value of $4.10. They are made up of $1, 50c and 20c coins. How many of each type do I have?

Level 2

6 If each letter represents a different digit (0–9), find the value of each letter so that the following statement is true:

```
 FOUR +
 FIVE
 ----
 NINE
```

7 Most number plates are made up of three letters and three numbers. How many number plates could be generated if Roslyn requires it to be:

a ROS _ _ _ 6 4
b ROS _ _ _ _ 4

8 A local café has the following menu:

Entree	*Dessert*
• Soup	• Profiteroles
• Calamari	• Ice-cream
• Caesar salad	• Creme caramel
Main	
• Beef stirfry	
• Tofu burger	

For how many days could you have a different three-course meal at the café if:

a you always wanted to start with the caesar salad?
b you always had the tofu burger?
c you picked three different items each time?

9 Postcodes are used for each suburb in Australia, with those belonging to NSW beginning with a 2 and containing 4 digits.

a How many postcodes are possible that:
  i start with 227?
  ii start with 22?
  iii end in a zero in NSW?
b Are all possible postcodes used? Discuss.

10 In the following matchstick arrangement, move three matches to new locations so that the arrangement is of three squares.

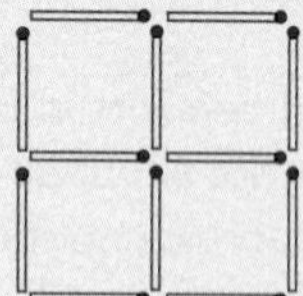

11 Using the digits from the year you were born (for example, 1992) in the order they are written and any mathematical operations you like, including brackets, form as your answers the square numbers less than 100.

12 a How many palindromic numbers are there between 10 and 100?
b List all the two-digit palindromic numbers you found.
c How many palindromic numbers are there between 10 and 1000? You are *not* required to list them all.

Chapter Review

Language Links

approximate, base, calculator, decimal, expanded, exponent, factor, fraction, index, integer, multiple, negative, notation, opposite, positive, power, prime, product, quotient, reciprocal

1 Which word from the list above means:

a invert a fraction? b multiply? c whole number?
d a number with only two factors? e the opposite of positive?

2 Label from the list above:

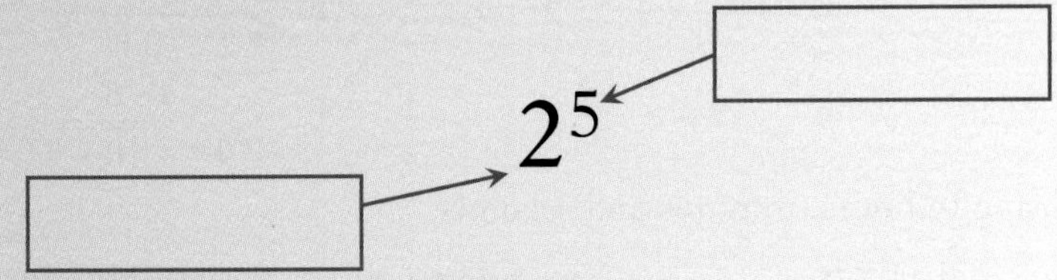

3 $2 \times 2 \times 2 \times 2 \times 2$ is written in _______ form.

4 List any other words you know that mean:

a multiply b divide

5 Write a short paragraph explaining the benefits and disadvantages of using a calculator. (You may like to organise a class debate on the use of technology in the classroom.)

TEACHER

Chapter Review Exercises

MC

1 Without a calculator, find the value of:

a 194×5 b $9540 \div 5$ c $(^-7)^2 \times 2$
d $\sqrt[4]{810\,000}$ e $1\frac{1}{2} \times 4$ f $1{\cdot}2 \times 0{\cdot}6$
g $10 + 7 \times 4 \div 2$ h $\dfrac{15 + 4 \times 3}{3}$ i $5^3 \times 2^2$

2 Write the following in index notation: 1.1

a 8×8 b $6 \times 6 \times 6 \times 6 \times 6 \times 6 \times 6$ c $3 \times 3 \times 5 \times 5 \times 5$

3 Write in expanded form: 1.1

a 5^7 b 2^5 c 11^2 d $5^3 \times 2^4$

1.1 **4** Find the unknown number, x, each time:

a $7^x = 49$ b $5^x = 625$ c $x^6 = 64$ d $x^{50} = 1$

1.2 **5** Simplify by using the first index law, leaving all answers in index notation:

a $9^{10} \times 9^2$ b $7^5 \times 7^4$ c $5^6 \times 5 \times 5$ d $3^4 \times 7^3 \times 3 \times 7^7$

1.2 **6** Simplify by using the second index law, leaving all answers in index notation:

a $9^{10} \div 9^2$ b $7^6 \div 7$ c $2^8 \div 2^4$ d $5^7 \div 5^2$

1.3 **7** Simplify by using the third index law, leaving all answers in index notation:

a $(2^5)^3$ b $(5^4)^2$ c $(7^3)^3$ d $(2^4 \times 3^7)^4$

1.3 **8** Find the value of:

a 7^0 b 15^0 c $(2 \times 4^2)^0$ d $(5^0 \times 2^2)^3$
e 5×4^0 f $7^0 \times 2^0$ g 7×2^2 h $(5 \times 2^4)^0$

1.4 **9** Use a calculator to find the value, correct to 2 decimal places where necessary, of the following:

a $5{\cdot}6^2$ b $3{\cdot}7 + 9{\cdot}4 \times 3{\cdot}9$ c $\dfrac{4{\cdot}09}{1{\cdot}2^2 - 0{\cdot}6}$

d $\sqrt[3]{\dfrac{12{\cdot}4}{0{\cdot}6}}$ e $\sqrt{19{\cdot}6} + \dfrac{4{\cdot}8}{3}$ f $\dfrac{18{\cdot}6^3}{\sqrt{9{\cdot}4 + 10 \times 3^2}}$

1.4 **10** Convert the following to hours, minutes and seconds:

a 406 min b 1·8 h c 5·46 h d 724 min

1.6 **11** The product of two numbers is 512, and one number is half the other. What are the two numbers? Is there more than one solution?

1.5 **12** The first prize in the lottery of $4 567 258 is shared equally between the 14 prize winners. How much does each receive, correct to the nearest dollar?

1.5 **13** Find the average of:

a 10·3, 9·8, 5·1, 7·72, 9·8 and 7·64
b 8·7, 14·2, 19·06, 17·84 and 7·88
c 200, 46, 194, 18 and 172

TEACHER

Keeping Mathematically Fit

Part A: Non-calculator

1 Find 410×500.

2 Find $\frac{2}{3}$ of $6900.

3 If Margaret spends $\frac{1}{5}$ of her money on a new maths book costing $24, how much money does she have left?

4 Find the perimeter of a square with side lengths 7·4 cm.

5 Find $4{\cdot}06 \times 0{\cdot}02$.

6 Find the value of $16 \times 2 + 5^2 - 3 \times 5$.

7 Round 17·965 43 to 1 decimal place.

8 Name a shape that has eight straight lines.

9 Find the perimeter of:

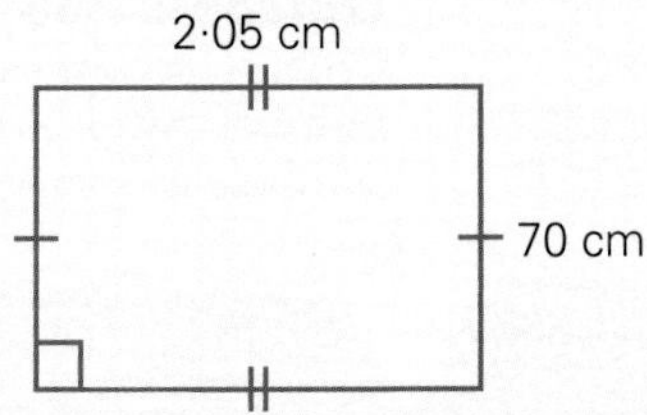

10 Find the value of:

a 15^2 b $1\frac{1}{2} \times \frac{2}{3}$ c $1\frac{1}{2} \div \frac{2}{3}$ d $1{\cdot}4 \div 0{\cdot}07$

Part B: Calculator

1 Find $9{\cdot}6^3 \div (4{\cdot}07 - 3{\cdot}1)$ (answer to 3 decimal places).

2 Evaluate $\sqrt[3]{41{\cdot}069}$ (answer to 2 decimal places).

3 Find $2\frac{4}{5} - 1\frac{3}{4} \times \frac{4}{9}$.

4 Find $\sqrt[3]{91\,125}$.

5 Evaluate $\frac{4{\cdot}6^2 - 3{\cdot}1}{8{\cdot}04}$ correct to 2 decimal places.

6 The perimeter of a rectangle whose length is three more than its width is 34 cm. What are the dimensions of the rectangle?

7 A train leaves Cronulla station at 11:52 am. The train arrives at Central Station 43 min later. What times does the train arrive at Central? If trains leave Cronulla every 20 minutes, what time does the last train of the day leave?

8 Write four questions that result in an answer of 15·02. Use your calculator and be creative.

1 What you need to know and revise

Outcome PAS 4.1:
Uses letters to represent numbers and translates between words and algebraic symbols

Outcome PAS 4.2:
Creates, records, analyses and generalises number patterns using words and algebraic symbols in a variety of ways

2 What you will learn in this chapter

Outcome PAS 4.3:
Uses the algebraic symbol system to simplify, expand and factorise simple algebraic expressions

- recognises like terms and adds and subtracts like terms to simplify algebraic expressions
- recognises the role of grouping symbols
- simplifies algebraic expressions involving multiplication and division
- expands algebraic expressions—the distributive property
- factorises algebraic expressions
- substitutes into algebraic expressions
- generates a number pattern
- translates everyday language to algebraic language and vice versa
- simplifies expressions that involve simple algebraic fractions

Working Mathematically outcomes WMS 4.1–4.5
Students will be asked to ***question***, ***apply strategies***, ***communicate***, ***reason*** and ***reflect*** in the sections of this chapter.

Algebraic expressions

Key mathematical terms you will encounter

algebraic
coefficient
constant
distributive
evaluate
expand
expression
factorise
grouping
pattern
power
pronumeral
sequence
simplify
substitution
term
variable

MathsCheck
Algebraic expressions

1 Translate each rule into an algebraic equation or formula:

- **a** The number of matches (M) equals the number of shapes (S) plus two.
- **b** The number of popsticks (p) equals double the number of hexagons (h).
- **c** The term value (V) equals the term number (n) minus six.
- **d** The number of counters is three-quarters of the number of polygons.

2 Complete the tables using the rule given:

a $M = 2d + 1$

d	1	2	3	4
M				

b $T = \frac{n}{2} + 3$

n	1	2	3	4
T				

3 Write the next two terms of each sequence:

- **a** 1, 3, 6, 10, ______, ______
- **b** $^-2, {}^-4, {}^-6, {}^-8$, ______, ______
- **c** 64, 32, 16, ______, ______
- **d** 1, 1, 2, 3, 5, 8, ______, ______

A **sequence** is a list of numbers built to a rule or pattern. Each number in the list is called a **term**.

A sequence of square numbers is: 1, 4, 9, 16, ...

Term number: 1 2 3 4

4 Copy and complete the rule for each table:

a

x	1	2	3	4
y	5	10	15	20

$y =$ ______ $\times x$

b

n	1	2	3	4
V	2	5	8	11

$V =$ ______ $\times n -$ ______

5 For each sequence:

- **i** Draw up a table showing term number (n) and term value (V).
- **ii** Find a rule and write it as an algebraic equation.

- **a** 4, 8, 12, ...
- **b** 3, 5, 7, ...
- **c** $\frac{1}{4}, \frac{1}{2}, \frac{3}{4}, 1, \ldots$
- **d** 2, 6, 10, ...

6 For the sequence $^-3, {}^-6, {}^-9, \ldots$

- **a** Draw up a table showing term numbers and term values as far as the sixth term.
- **b** Write a formula relating the two variables.
- **c** Use the formula to find the 24th term of the sequence.
- **d** Which term would have the value $^-48$?

7 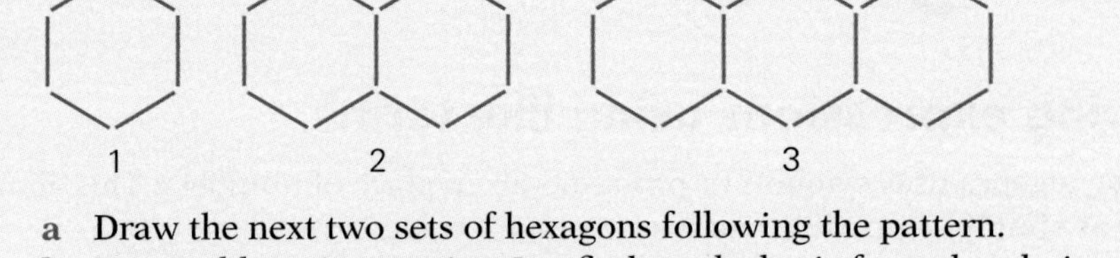

a Draw the next two sets of hexagons following the pattern.

b Use a table as in question **6** to find an algebraic formula relating the number of matches to the number of hexagons.

c Use your formula to find:

  i the number of matches needed for 12 hexagons

  ii the number of hexagons formed from 46 matches

8 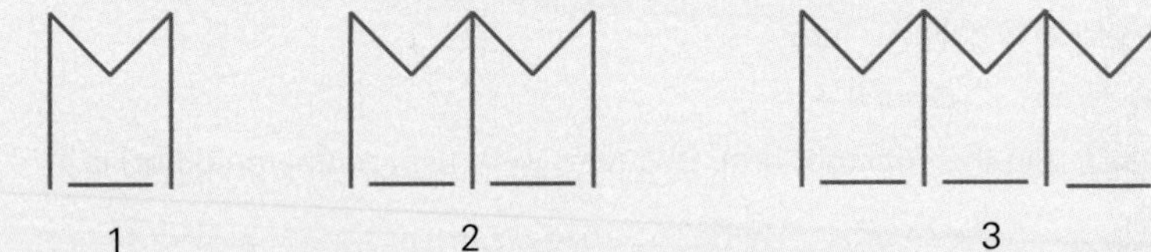

a Write a rule to relate the number of matches needed to build each term and the term number.

b Use your rule to find:

  i the number of matches needed for the 19th term

  ii the term number using 85 matches

c Is it possible for a term to contain exactly 104 matches?

d What is the total number of matches needed to completely build the first 10 terms in this pattern?

Exploring New Ideas

2.1 Simplifying expressions using like terms

As we saw last year, algebra uses symbols or pronumerals in place of numbers. This allows us to write things as simply as possible and devise general rules.

An **algebraic expression** is a group of pronumerals and numbers joined together using the four operations (+, −, ×, ÷).

$x^2 + 2a + 4$ is an algebraic expression.

Each part of an expression is called a *term*.

$x^2 + 2a + 4$ has **three** terms: x^2, $2a$ and 4.

The **coefficient** of a is 2 and the **constant term** (the term with no pronumeral factor) is 4.

In general:

- $m + n = n + m$
- $m \times n = mn$
  $= n \times m$
  $= nm$
- $m - n \neq n - m$
- $\frac{m}{n} \neq \frac{n}{m}$
- $m + 0 = m$
- $m \times 0 = 0$
- $m - 0 = m$
- $m \div 0$ is not defined

To simplify algebraic expressions, you can add or subtract **like terms**.

Like terms contain the same power of the same variable.

$5a$ and $4a$, $7ab$ and ^-3ba, $4p$ and $\frac{p}{2}$ are pairs of like terms. They have the same pronumerals.

$5a$ and a^2, $5x$ and $7xy$ are **not** like terms and cannot be added or subtracted.

EXAMPLE 1

Write an algebraic expression for:

a the sum of x and y

b nine less than the sum of x and y

Solution

a The sum of x and y
$= x + y$

b Nine less than the sum of x and y
$= x + y - 9$

EXAMPLE 2

Simplify:

a $16a - 10a$

b $4p + p^2 - 3p$

c $^-7ab + 5ab$

d $6x + 3 - 5x - 9$

Solution

a $16a - 10a$
$= 6a$

b $4p + p^2 - 3p$
$= 4p - 3p + p^2$
$= p + p^2$

c $^-7ab + 5ab$
$= {^-2ab}$

d $6x + 3 - 5x - 9$
$= 6x - 5x + 3 - 9$
$= x - 6$

Be careful when you use directed numbers, e.g. $^-7 + 5 = {^-2}$
$3 - 9 = {^-6}$

EXERCISE 2A

1 Write an expression for each of the following:

a the sum of 3 and p
b seven more than x
c eight less than x
d the sum of x, y and z
e the sum of m and n is increased by six
f x less y

2 The height of a 40 m building is increased by x m. Write an expression of its new height.

3 A farm has a cows and b chickens. Write an expression for the number of legs on the farm.

4 Simplify each of the following:

a $3x + 2x$ b $3x - 2x$ c $4p + 2p$ d $6m - 2m$
e $7w + 9w$ f $9w + 7w$ g $x - x$ h $12p + 8p$
i $4w + 8w$ j $\frac{1}{2}f + \frac{1}{2}f$ k $4a^2 + 3a^2$ l $12p - p$
m $24m + m + m$ n $24m + 2m$ o $13pq - 9qp$ p $x^3 + x^3$

5 State whether or not each of the following statements is true for **all** values of the pronumerals:

a $x + 3 = 3 + x$ b $5 - a = a - 5$ c $2x - y = 2y - x$ d $x - y + z = x + z - y$
e $a - a - a = ^-a$ f $x + x + x = 3x$ g $a + 0 = a$ h $a + b - c = b - a - c$

6 State whether the following algebraic expressions can be simplified:

a $4x + 5x^2$ b $6ab - ba$ c $p + q - q$ d $x^3 + 2x^3$

7 Simplify:

a $5x + 2x + 3x$ b $4a + 3a + a$ c $4a + 3a - a$
d $7x + x - 4x$ e $12p + p + p$ f $20x - 15x - 2x$
g $p + p - p$ h $x^2 + 6x^2 - 4x^2$ i $6mn + 7mn - 2nm$
j $4x + 5x - 3x$ k $6a - 2a + 3b$ l $7g^2 + 3g + g$
m $2ab + 3ba + ab$ n $xy + xy - 2yx$ o $4w + 7w + w^2$

Level 2

8 Simplify each of the following expressions:

a $5x - 9x$ b $3a - 7a$ c $^-3x - 7x$ d $^-p - p$
e $^-4xy + 8xy$ f $2a^2 - 6a^2$ g $8n - (^-n)$ h $7a - 12a - a$
i $2mn - 8mn + mn$ j $9m + m - 4m$ k $^-8n - (^-3n)$ l $3xy - 6xy$
m $^-4p + 6p - p$ n $5x^2y - 15x^2y$ o $^-a^2 - 4a^2 - 3a^2$ p $^-4xy + xy - 3yx$

9 Simplify each of the following expressions by collecting like terms:

a $5h + 3 + 4h + 7$ b $4 + 8n + 2$ c $x^2 + 3x + 2x^2 + 3x$
d $6x^2 + xy + 4y^2$ e $3a^2b + 7ab^2$ f $2j^3 + j^3 - 5j^3$
g $3 + 4x + 5$ h $20x + 4 - 3x$ i $4p + q + 6$
j $12 - 4x + 3$ k $p + q + p - q$ l $5a + 4b - 6a$
m $16x + 3y - 12x - y$ n $x + 5y - 2x + y$ o $^-4x + 6y - x - 3x$
p $a + 3b + 3a - a^2$ q $a^2 + 5a - 6a^2 - a$ r $^-3q + 8p - 4q - 10p$
s $x^2 - 5x - 3x^2 + 2x$ t $x - 2y + 8x + y$ u $^-3m + 5m - p - 5q$
v $xy + 8x + 4xy - 4x$ w $ab - 7ba - a^2 + a$ x $a^2b - 5ab^2 + 5a^2b + 2ab^2$

10 Find the simplest expression for each perimeter:

a

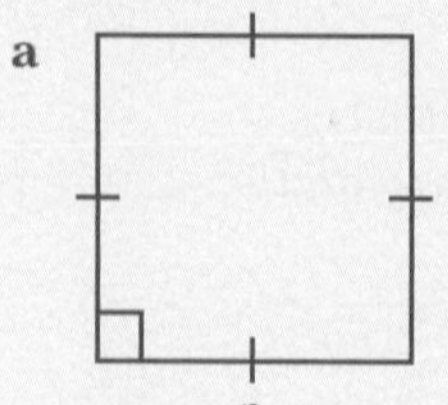

b

x

y y

x

c

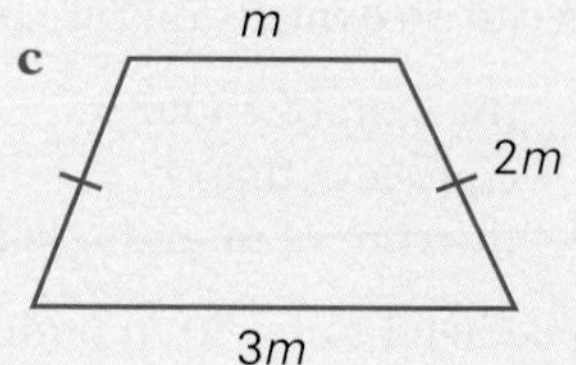

d

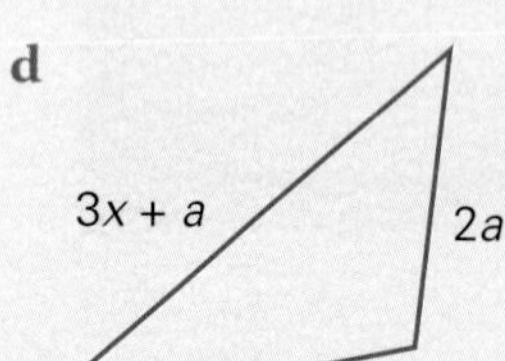

e

$4a - b$

$a - 3b$

f

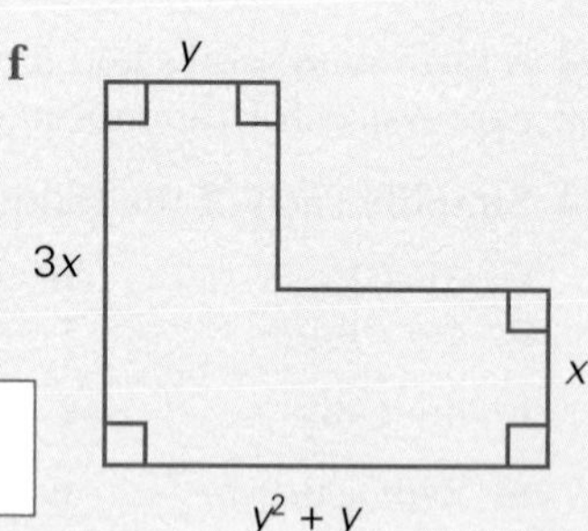

Remember: the perimeter is the distance around the boundary of a closed figure.

2.2 Multiplication and division

WEB RESEARCH

It is possible to write algebraic expressions in a brief and simple way for situations involving multiplication and division. One such situation is to find the number of tiles in a rows of b floor tiles.

$a \times b = ab$

$a \div b = \frac{a}{b}$

$a \times a = a^2$

EXAMPLE 1

Write an algebraic expression for each of the following:

a double the product of x and y

b one-third of a

c 3 is multiplied by p and the result is divided by n

Solution

a $2 \times x \times y = 2xy$ **b** $\frac{1}{3}$ of $a = \frac{a}{3}$ **c** $3 \times p \div n = \frac{3p}{n}$

EXAMPLE 2

Simplify each of the following expressions:

a $4a \times 3b \times c$ **b** $x \times (^-3x)$ **c** $\frac{^-10ab}{8a}$ **d** $^-12 \div (2a + a)$

Solution

a $4a \times 3b \times c = 4 \times 3 \times a \times b \times c$

$= 12abc$

b $x \times (^-3x) = {^-3} \times x \times x$

$= {^-3x^2}$

The rules for ×/÷ integers apply:

Neg × Pos = Neg Neg × Neg = Pos

Neg ÷ Pos = Neg Neg ÷ Neg = Pos

As we know, negative 4 can be written as $^-4$ or -4; likewise $\frac{^-4}{a}$ and $\frac{4}{^-a}$ are often written as $^-\left(\frac{4}{a}\right)$ or $-\frac{4}{a}$

c $\frac{^-10ab}{8a} = \frac{^-10}{8} \times \frac{ab}{a}$

$= \frac{^-5}{4} \times b$

$= \frac{^-5b}{4}$

We use improper fractions when writing algebraic expressions.

d $^-12 \div (2a + a) = {}^-12 \div (3a)$

$= \frac{^-12}{3a}$

$= \frac{^-4}{a}$

As in arithmetic, brackets are performed first, before division—just follow the order of operations.

EXERCISE 2B

1 Write an algebraic expression for each of the following:

a 5 times the value of a
b one-quarter of m
c the product of p and q
d x is multiplied by n and then divided by 5
e the sum of x and y is divided by 5

2 Write an expression for each of the following, if the number is n:

a A number is multiplied by 5 and 3 is added to the result.
b A number is doubled and 8 is added to the result.
c A number is divided by 8 and 2 is subtracted from the result.
d A number is increased by 1 and the result is divided by 4.
e The square of a number is divided by the product of the number and 3.

3 **a** What is the cost of eight cans of soup if each can costs 95 cents?
b What is the cost of eight cans of soup if each can costs p cents?
c What is the cost of n cans of soup if each can costs p cents?

4 Trish earns $\$x$ a week. How much does she earn in:

a 3 weeks? **b** a year? **c** n weeks? **d** a day?

5 Bede and Jean win $\$x$ in a lottery. They plan to share the prize equally between themselves and their three daughters. How much will each member of the family receive?

6 A taxi driver charges a flag fall of \$3 plus \$2 per kilometre travelled. How much does Roslyn owe the driver for a trip of:

a 4 km? **b** 15·5 km? **c** n km?

7 Simplify:

a $2 \times 3a$
b $4a \times 3$
c $5a \times 3b$
d $2n \times 6n$
e $5a \times 2ab$
f $2x \times 3 \times 4y$
g $3 \times 3a \times 4b$
h $t \times 3n \times 2$
i $2a \times 5 \times 3a$

8 Write each of these divisions as a fraction:

a $m \div 4$ b $m \div n$ c $x \div 5$ d $a \div (^-3)$
e $^-x \div 4$ f $2 \div m$ g $13 \div w$ h $^-m \div (^-n)$
i $ab \div c$ j $3 \times m \div n$ k $5 \div (^-x) \times y$ l $6 \times (^-n) \div (^-m)$

Level 2

9 Copy and complete:

a $6a = 2 \times$ ______ b $12ab = 3a \times$ ______ c $8n^2 = 2n \times$ ______
d $10a^2b = 5a \times$ ______ e $15d = 5d \times$ ______ f $9e^2 = 3e \times$ ______

10 Simplify each quotient:

a $\frac{10x}{5}$ b $\frac{18c}{3}$ c $\frac{6n}{12}$ d $\frac{8d}{24}$
e $\frac{7}{14x}$ f $\frac{9x}{3x}$ g $\frac{8xt}{2t}$ h $\frac{6ab}{9ba}$

11 Simplify:

a $^-3a \times 2b$ b $4c \times {}^-5$ c $^-5a \times {}^-2b$
d $^-4t \times 4t$ e $\frac{3}{4}p \times \frac{1}{3}m$ f $0{\cdot}8n \times 0{\cdot}2t$
g $1\frac{1}{2}a \times 1\frac{1}{3}b$ h $ab \times ba \times {}^-3$ i $^-3p + 2 \times 4p$

12 Simplify each of the following expressions:

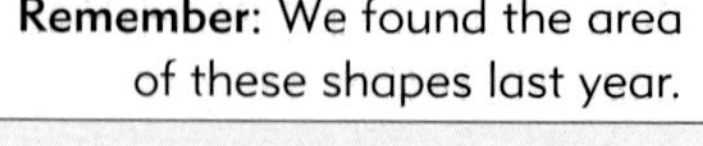
Remember the order of operations.

a $3 \times x + 1$ b $a \times b - 3$ c $5 + 4 \times p$
d $7 + p \times 4$ e $^-6 + 8 \times x$ f $2 - 5 \times x$
g $x \times (^-3) + 6$ h $x \times (^-y) - 3$ i $4 \times p \times 3 - 6$
j $4x \times 3y + 2x$ k $4 \times 5a - 3 \times b$ l $9a \times 3b - 2a \times b$
m $6a \times 3a - 6a \times 3$ n $10 - 3a \times 7a$ o $5a \times (7a + 3a)$

13 Simplify:

a $4xy \div 4$ b $4xy \div 4x$ c $xy \div x$
d $16p \div 4$ e $16p \div 4p$ f $a \div ab$
g $20x \times 5 \div 4$ h $8a \times 4b \div 2a$ i $16a \div 4 \times a$
j $10x^2 \div 2x$ k $5a \div 15a^2$ l $0{\cdot}8mt^2 \div 0{\cdot}4t$
m $^-8cd \div 12cd$ n $\frac{7n}{21n^2}$ o $\frac{^-12at}{18ta}$
p $\frac{14nx}{-35x}$ q $\frac{^-9abc^2}{^-6ca}$

Remember: We found the area of these shapes last year.

14 Find the simplest expression for each area:

a
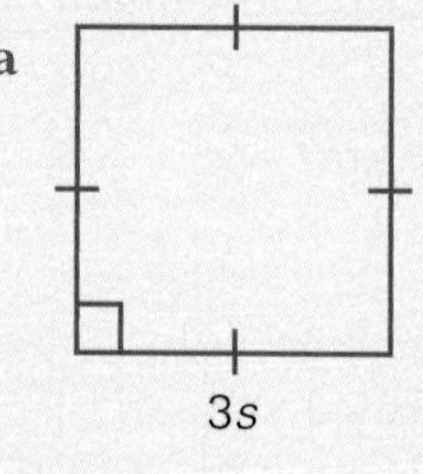

b
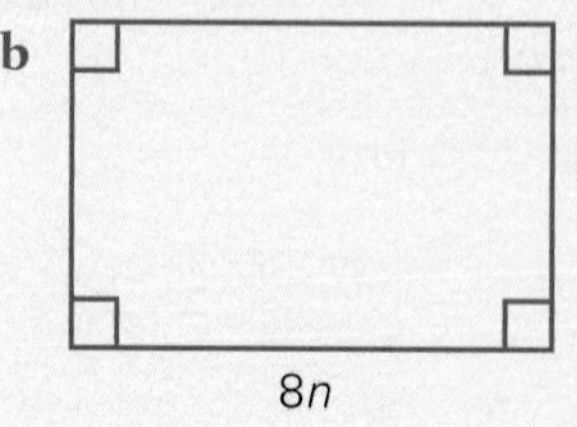

c
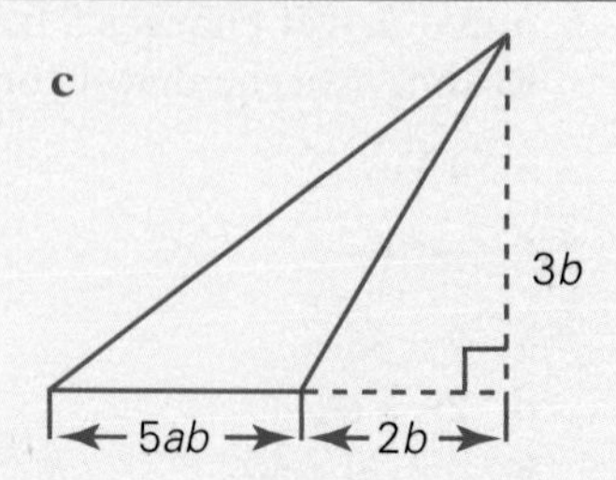

2.3 The distributive law

SPREADSHEET ACTIVITY

Algebraic expressions can often be written more simply using grouping symbols.

We use the **distributive law** to remove the brackets. The terms within the grouping symbols are multiplied by the term immediately outside.

$$\begin{aligned}3(a+b) &= 3\times a + 3\times b\\ &= 3a + 3b\end{aligned}$$

The process of removing the bracket is called expansion.

In general:

$a(b+c) = ab + ac$

$a(b-c) = ab - ac$

Last year we looked at expanding by multiplying by a positive term.

This year we extend our look at expanding to include multiplying by a negative term.

When the number outside the brackets is negative, the sign of each term inside the brackets changes.

In general:

$^-a(b+c) = {}^-ab - ac$ Neg × Pos = Neg

$^-a(b-c) = {}^-ab + ac$ Neg × Neg = Pos

^-a is often written as $-a$.

EXAMPLE 1

Write an algebraic expression for each of the following:

a double the sum of a and 3

b half of x is increased by 3 and the result is multiplied by 5

Solution

a Double the sum of a and three $= 2\times(a+3)$

$= 2(a+3)$

b Half of x is increased by 3 and the result is multiplied by 5 $= (\frac{a}{2}+3\Big)\times 5)$

$= \Big(5(\frac{a}{2}+3)\Big)$

EXAMPLE 2

Expand each of the following:

a $4(a+5)$ **b** $3x(4y-5x)$ **c** $^-6(2x-7)$ **d** $^-(x+5y-3)$

Solution

a $4(a+5)$
$= 4a + 20$

b $3x(4y-5x)$
$= 12xy - 15x^2$

c $^-6(2x-7)$
$= {}^-12x + 42$

d $^-1(x+5y-3)$
$= {}^-x - 5y + 3$

Note the signs change.

EXAMPLE 3

Expand and simplify each of the following:

a $10 + 4(2-3a)$ **b** $2(3x-4) - 8(x+5)$

Solution

a $10 + 4(2 - 3a)$
$= 10 + 8 - 12a$
$= 18 - 12a$

b $2(3x - 4) - 8(x + 5)$
$= 6x - 8 - 8(x + 5)$
$= 6x - 8 - 8x - 40$
$= {}^-2x - 48$

Be careful of your signs.

EXERCISE 2C

1 Write an algebraic expression for each of the following:

- **a** 5 times the sum of x and y
- **b** 5 is added to x and the result is multiplied by 2
- **c** 3 is subtracted from a and the result is multiplied by 4
- **d** n is subtracted from m and the result is doubled
- **e** a is added to $5b$ and the result is multiplied by $^-3$

2 Andre delivers x magazines and y newspapers each day to earn pocket money. How many magazines and newspapers does he deliver in 5 days?

3 In a sidewalk café sit m men and n women. If each person spends $25, how much money will the café receive?

4 Expand each of the following expressions:

a $2(m+n)$	**b** $2(m-n)$	**c** $3(a+4)$	**d** $5(2-p)$
e $4(p+q)$	**f** $(x-y)7$	**g** $5(5+x)$	**h** $4(2-a)$
i $8(3-w)$	**j** $10(a-2)$	**k** $4(n+a)$	**l** $3(x+2)$
m $2(a-t)$	**n** $5(n-2)$	**o** $6(2-p)$	**p** $(c+d)3$

5 Expand each of the following expressions:

a $^-4(x-y)$	**b** $^-3(x+y)$	**c** $^-5(a-b)$	**d** $^-3(2-a)$
e $^-9(a+b)$	**f** $^-10(a-7)$	**g** $^-10(a+7)$	**h** $^-4(3-m)$
i $^-2(2+p)$	**j** $^-6(a-6)$	**k** $^-a(b+2)$	**l** $^-(a+3b)$
m $^-(x+y)$	**n** $^-(x-y)$	**o** $^-(5-2w)$	**p** $^-(x-3y+7)$

6 Expand:

a $5(2x+1)$	**b** $2(3x-1)$	**c** $(3x+2y)4$	**d** $4(3-2a)$
e $3(2d+3)$	**f** $(6+3x)5$	**g** $2(4n-3)$	**h** $4(3x+4)$
i $5(2x-8)$	**j** $7(2a+3)$	**k** $4(6a-3b)$	**l** $3(5a+4b)$
m $^-3(5m-2)$	**n** $^-3a(b+c)$	**o** $3(2a-b+c)$	**p** $^-2(3b-4+a)$

7 Expand each of the following:

a $a(a+1)$	**b** $a(a+b)$	**c** $a(2a-3)$	**d** $m(m+1)$
e $p(p-3)$	**f** $n(n+5)$	**g** $^-a(a-2)$	**h** $^-3p(4-p)$
i $2x(x+5)$	**j** $3p(5-2p)$	**k** $^-5n(m+n)$	**l** $^-2a(3a-7)$
m $a(a+b-c)$	**n** $a(3a-2b)$	**o** $^-3a(^-2a-5)$	**p** $^-6a(3a-5b)$

Level 2

8 Expand and simplify by collecting like terms:

a $2(x+3)+3$ **b** $3(4-a)+5$ **c** $5(x+y)+2x$

d $3(2x-7)+3$ **e** $3(2x-7)-3$ **f** $3(n-2)+4n$

g $5(2+3a)-2$ **h** $7(5-3p)+20p$ **i** $4(a+5x)-15x$

j $^{-}2(p+4)+5$ **k** $^{-}6(2x-3)-3x$ **l** $^{-}4(a+b)+4a$

m $9(a+b)-3a+b$ **n** $x(x+3)-5x$ **o** $^{-}3(2x-5)-4x-2$

9 Expand and simplify:

a $2+3(a+2)$ **b** $6+5(2+a)$ **c** $2+4(a+3)$

d $9+3(a-3)$ **e** $5a+2(3a-5)$ **f** $2m+3(m-4)$

g $10-(x+5)$ **h** $10-(x-5)$ **i** $2a-(a+3)$

j $12a-3(a+4)$ **k** $4a^2-(a^2-2)$ **l** $7x-2(2x-5)$

m $12x+2(x+3)-6$ **n** $7a-2(5-3a)$ **o** $^{-}2x-3(5-6x)$

10 Expand and simplify each of the following:

a $2(a+3)+3(a+1)$ **b** $3(x+4)+2(x-1)$

c $5(2x+1)-2(x-1)$ **d** $4(g-5)+3(g-2)$

e $5(2x+5)-4(x+2)$ **f** $2b(b+5)+b(b-5)$

g $x(x-6)+x(4-x)$ **h** $a(2a+1)-(a+7)$

i $4(5a+8)-6(a-6)$ **j** $2(a-7)-3(a-9)$

k $^{-}a(a+5)+a(3a-1)$ **l** $^{-}3(w-5)+3(w+5)$

m $^{-}5(3h+1)-(2h-5)$ **n** $^{-}(x+1)-4(x-1)$

o $^{-}2a(3a+b)+b(a+4)$ **p** $7(p^2-p+1)-3(2p-1)$

Investigation What am I?

I am housed in the British Museum and am dated from about 1650 BC. I contain Egyptian algebraic problems.

Simplify each of the following algebraic problems to work out what I am.

U $3x-7y+6x$ **P** $8ab \div ab$ **N** $9x-(4-2x)$

H $ab \div 8ab$ **R** $12a+3a \times 7$ **Y** $8a-7b-8a-7b$

A $12ab \div 24a^2$ **D** $3(3x-7)-x+2$ **S** $24ab \div 12a^2b$

I $9x \times 5x - 3x^2$

$33a$	$\frac{1}{8}$	$42x^2$	$11x-4$	$8x-19$

8	$\frac{b}{2a}$	8	$^{-}14b$	$33a$	$9x-7y$	$\frac{2}{a}$

2.4 Factorisation

SPREADSHEET ACTIVITY

Factorising is the opposite of expanding.

To **factorise** a sum or difference is to change it into a *product* by taking the **HCF outside** and placing the remaining factors inside grouping symbols. It is the reverse of *expanding*.

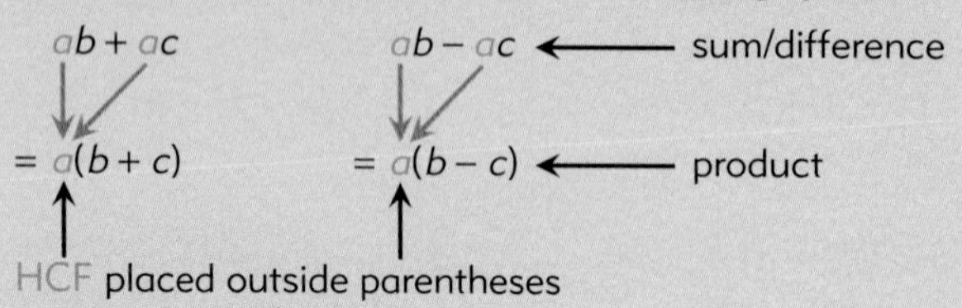

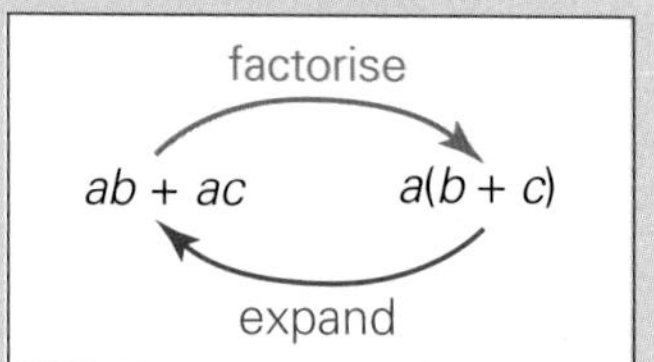

EXAMPLE 1

Find the highest common factor of:

a $12x$ and 8 **b** $8x$ and $24xy$ **c** $4a^2$ and $6ab$

Look at the numbers first, then the pronumerals.

Solution

a $12x = 3 \times 4 \times x$
$8 = 4 \times 2$
HCF is 4

b $8x = 8 \times x$
$24xy = 8 \times 3 \times x \times y$
HCF is $8 \times x$
$= 8x$

c $4a^2 = 2 \times 2 \times a \times a$
$6ab = 2 \times 3 \times a \times b$
HCF is $2 \times a$
$= 2a$

EXAMPLE 2

Factorise each of the following:

a $4x + 8$ **b** $6t - 9$ **c** $x^2 + 3x$ **d** $12a^2b - 16ab^2$

Solution

a $4x + 8$
HCF of $4x$ and 8 is 4
$= 4 \times x + 4 \times 2$
$= 4(x + 2)$

b $6t - 9$
HCF of $6t$ and 9 is 3
$= 3 \times 2t - 3 \times 3$
$= 3(2t - 3)$

c $x^2 + 3x$
HCF of x^2 and $3x$ is x
$= x \times x + x \times 3$
$= x(x + 3)$

d $12a^2b - 16ab^2$
HCF of $12a^2b$ and $16ab^2$ is $4ab$
$= 4ab \times 3a - 4ab \times 4b$
$= 4ab(3a - 4b)$

EXAMPLE 3

Factorise each of the following:

a $^-3x - 9$ **b** $^-x^2 + 5x$

Factorising out a negative changes the sign/s within the brackets.

Solution

a $^-3x - 9$
$= {}^-3(x + 3)$

b $^-x^2 + 5x$
$= {}^-x(x - 5)$
or $5x - x^2$
$= x(5 - x)$

EXERCISE 2D

1 Find the highest common factor of each of the following pairs of terms:

a 6 and 3 **b** 16 and 20 **c** 24 and 36
d $2p$ and p^2 **e** $6x$ and 3 **f** ab and a
g $5x$ and 5 **h** $12a$ and 4 **i** $20p$ and $12q$
j x and x^2 **k** $12ab$ and $24a$ **l** $20mn$ and $5m$
m $10x$ and 6 **n** $10x^2$ and $6x$ **o** $4x^2y$ and $2xy$

2 Copy and complete:

a $2t + 2 \times 3 = 2(\_\_\_\_\_ + \_\_\_\_\_)$
b $ab - ax = a(\_\_\_\_\_ - \_\_\_\_\_)$
c $3d + 6 = 3(\_\_\_\_\_ + \_\_\_\_\_)$
d $4e - 6 = 2(\_\_\_\_\_ - \_\_\_\_\_)$
e $h^2 - 7h = h(\_\_\_\_\_ - \_\_\_\_\_)$
f $6km + 9k = 3k(\_\_\_\_\_ + \_\_\_\_\_)$
g $^-5n + {^-5p} = \_\_\_\_\_(n + p)$
h $6w - 12 = \_\_\_\_\_(w - 2)$
i $9c - 3 = \_\_\_\_\_(3c - 1)$
j $3s + 6s^2 = \_\_\_\_\_(1 + 2s)$

3 Factorise each of the following expressions. Your answers to question **1** should help.

a $2p + p^2$
b $6x + 3$
c $ab - a$
d $5x + 5$
e $12a - 4$
f $20p + 12q$

4 Factorise each of the expressions by taking out the highest common factor:

a $3x + 3$
b $4p - 4$
c $10a - 10$
d $6v - 6$
e $7p + 7$
f $4 + 4x$
g $2x - 4$
h $9 - 9a$
i $4x - 2$
j $10a - 5$
k $10 + 4p$
l $12 + 8w$
m $16 + 12a$
n $18 - 10h$
o $24x + 18$
p $16ab - 12$

5 Factorise each of the following expressions:

a $2a + 2b$
b $3a - 3b$
c $3a - 6b$
d $4x - 4y$
e $5xy - 5$
f $6a - 12b$
g $3p - 9q$
h $7x + 14y$
i $8m - 12n$
j $25p - 5q$
k $4a - 8b$
l $18p - 24q$

6 Factorise each of the following expressions:

a $ab + b$
b $4ab - 3b$
c $12a - 6b$
d $a^2 + a$
e $4ab - 2b$
f $x^2 - 3x$
g $5d - 5d^2$
h $4pq - 16$
i $4q - 16p$
j $m^2 + 4m$
k $3x^2 - 6$
l $w - w^2$
m $p + p^2$
n $c^2 - c$
o $5p - p^2$
p $n + 4n^2$
q $4d^2 + d$
r $15y - 15x^2$
s $w^2 - wx$
t $7p^2 - 6p$

Level 2

7 Factorise fully each of the following expressions:

a $9ab - 12a$
b $24x^2 - 12x$
c $15 - 5p$
d $5y^2 - 20y$
e $12ab + 16b$
f $4xy - 12y$
g $14x^2 - 7x$
h $9t - 12t^2$
i $10a^2 + 4ab$
j $100x - 10xy$
k $20a^2 + 15b^2$
l $16x - 30x^2$
m $12pq - 16q$
n $36 - 36m^2$
o $2p + 2q - 4$
p $x^2 + 3x - xy$

8 Factorise fully each of the following expressions:

a $^-2x + 4$
b $^-5 - 5a$
c $^-6m + 12$
d $^-4 - 12x$
e $^-2p - 8$
f $^-3x + 6$
g $^-3x - 6$
h $^-10h + 15$
i $^-9y - 18$
j $^-7m - 7$
k $^-15 + 20p$
l $^-6 + 18x$
m $^-11 - 11q$
n $^-12x + 16y$
o $^-45 + 15a$
p $^-40a + 30b$

9 Factorise fully each of the following:

a $^-x^2 + x$
b $^-x^2 - x$
c $^-p^2 + p$
d $^-p^2 - p$
e $^-2pq - 10p$
f $^-3x^2 + 6x$
g $^-8ab + 4a$
h $^-12a^2 - 16a$
i $^-8a - 4a^2$
j $^-5x + 5x^2$
k $^-x^2y - xy$
l $^-4pq - 16p^2q$
m $^-15x^2 - 20y$
n $^-xy + 5x^2$
o $^-pq + p^2$
p $^-8ab + 12a - 8b$

10 Factorise where possible:

a $^-4pq - 16p^2$
b $8x + 12y$
c $12b - 7c$
d $r^2 - r + ar$
e $x^2 - 4x + y$
f $^-18ab - 25c$

TEACHER

Try these

Factorise:

a $2(a + 3) + b(a + 3)$ b $w(w + 4) + 2(w + 4)$ c $a(a + b) - b(a + b)$
d $4(x + y) + 4k$ e $p(p - q) + 3p$ f $a^2(a + b) + a(a + b)$

2.5 Evaluating expressions

A pronumeral can be regarded as a **variable** if it can take more than one value. An algebraic expression may have different values depending on the values we substitute for each pronumeral (i.e. when we replace the pronumeral with a number).

EXAMPLE 1

Given $a = 4$, $b = 8$ and $c = {}^-2$, evaluate each of the following expressions:

a ${}^-2ab$ b $a + b - c$ c $\frac{4a}{b}$ d $3(a^2 + c^2)$

Solution

a
$${}^-2ab$$
$$= {}^-2 \times a \times b$$
$$= {}^-2 \times 4 \times 8$$
$$= {}^-64$$

b
$$a + b - c$$
$$= 4 + 8 - ({}^-2)$$
$$= 4 + 8 + 2$$
$$= 14$$

c
$$\frac{4a}{b}$$
$$= 4 \times \frac{a}{b}$$
$$= 4 \times \frac{4}{8}$$
$$= \frac{16}{8}$$
$$= 2$$

d
$$3(a^2 + c^2)$$
$$= 3 \times (4^2 + [{}^-2]^2)$$
$$= 3 \times (16 + 4)$$
$$= 60$$

EXAMPLE 2

Copy and complete the following table:

x	${}^-3$	${}^-2$	${}^-1$	0	1	2
$5 - x$						

Solution

Substitute each x value into the rule $5 - x$,
e.g. if $x = {}^-1$, then $5 - x = 5 - ({}^-1)$
$= 6$

x	${}^-3$	${}^-2$	${}^-1$	0	1	2
$5 - x$	8	7	6	5	4	3

EXERCISE 2E

1 If $a = 4$, find the value of:

a $3a$ b ${}^-a$ c $4 - a$ d $5a - 10$ e $a^2 - 4a + 1$

2 If $x = 3$ and $y = {}^-4$, find the value of:

a $x + y$ b $x - y$ c $2x + 3y$ d $x^2 + 3x - y$ e $3(x + 7y)$

3 If $a = 8$ and $b = \frac{1}{2}$, evaluate:

a ab　　b $a \div b$　　c $4b$　　d $4a - 6b$　　e $\frac{a}{8} - b$

Level 2

4 If $a = 8$, $b = 4$ and $c = 5$, evaluate:

a $6b$　　b $ab - c$　　c $c^2 + a$　　d $3a + 2b$
e abc　　f $^-3 + a$　　g $\sqrt{b} + c$　　h $2(b + c)$
i 2^b　　j $(a - c)b$　　k $\frac{bc}{a + 2}$　　l $\frac{a}{b^2} + \frac{c}{b}$

5 Complete the table:

a	b	$a + b$	$b - a$	ab	$a \div b$	$a^2 + b$	$a(b + 2)$
3	4						
0	9						
0·02	0·04						
$\frac{1}{2}$	$\frac{1}{4}$						

SPREADSHEET ACTIVITY

6 Simplify, then evaluate if $x = 2$ and $y = 5$:

a $2x + 4y + 6x + 3y$　　b $5x + 8 + 3y - 5$　　c $2x^2 - y^2 + 3y^2 - x$

7 Simplify, then evaluate if $a = ^-2$, $b = 3$:

a $4ab - 3ab$　　b $3a - b + 2a + 2b$　　c $a^2 + 2ab - 4ba$

8 If $a = 6$ and $b = ^-\frac{1}{3}$, find the value of:

a $4ab$　　b $4ba$　　c $7a^2b$　　d $(7a)^2$
e $\frac{b}{a}$　　f $\frac{a}{b}$　　g $3ab - 3ba$　　h $\sqrt{a}$

9 Copy and complete:

x	2	3	5	11	$^-5$	$\frac{1}{2}$
$x^2 + 2x + 1$						

TEACHER

Investigation Substitution puzzle

When does the sun always shine?

If $a = 7$, $b = 3$ and $c = ^-1$ find the answers to the questions below and use them to solve the riddle.

S $6a + b$　　**N** $a + b + c$　　**U** a^2b^2
O $b^2 + c^2$　　**D** $7a + 7c$　　**N** $6b^2$
Y abc　　**A** ab　　**A** b^3

__ 10　__ 9　__ 21　__ 45　__ 441　__ 54　__ 42　__ 27　__ $^-21$

2.6 Simplifying algebraic fractions

As you are aware, fractions can be simplified by dividing the numerator and denominator by the highest common factor. This process can also be used to simplify algebraic fractions.

The highest common factor of 12 and 8 is 4.

$12 = 4 \times 3$ and $8 = 4 \times 2$

$$\therefore \frac{12}{8} = \frac{4 \times 3}{4 \times 2}$$
$$= \frac{3}{2}$$

The highest common factor of $2xy$ and $4x$ is $2x$.

$2xy = 2x \times y$ and $4x = 2x \times 2$

$$\therefore \frac{2xy}{4x} = \frac{2x \times y}{2x \times 2}$$
$$= \frac{y}{2}$$

It is sometimes necessary to factorise the numerator, denominator or both before simplifying the algebraic fraction:

$$\frac{ab + a}{a} = \frac{a(b + 1)}{a}$$
$$= b + 1$$

EXAMPLE 1

Simplify:

a $\dfrac{12xy}{18x}$

b $\dfrac{4(a + b)}{16}$

Solution

a $\dfrac{12xy}{18x} = \dfrac{6x \times 2y}{6x \times 3}$

$= \dfrac{2y}{3}$

b $\dfrac{4(a + b)}{16} = \dfrac{4(a + b)}{4 \times 4}$

$= \dfrac{a + b}{4}$

EXAMPLE 2

Factorise and then simplify the following algebraic expressions:

a $\dfrac{3a + 6}{3}$

b $\dfrac{4x + 4y}{6x + 6y}$

Solution

a $\dfrac{3a + 6}{3} = \dfrac{3(a + 2)}{3}$ factorise

$= \dfrac{a + 2}{1}$ cancel

$= a + 2$ simplify

b $\dfrac{4x + 4y}{6x + 6y} = \dfrac{4(x + y)}{6(x + y)}$ factorise

$= \dfrac{4 \div 2}{6 \div 2}$ cancel

$= \dfrac{2}{3}$ simplify

EXERCISE 2F

1 Simplify each of the following:

a $\dfrac{12a}{6}$ b $\dfrac{6a}{12}$ c $\dfrac{ab}{2a}$ d $\dfrac{5a^2}{ab}$

e $\dfrac{xy}{x}$ f $\dfrac{xy}{x^2}$ g $\dfrac{4a}{6}$ h $\dfrac{4a}{6a}$

2 Simplify each of the following algebraic fractions:

a $\dfrac{3(x+2)}{3}$ b $\dfrac{3(x+2)}{6}$ c $\dfrac{3(x+2)}{9}$ d $\dfrac{3(x+2)}{4(x+2)}$

e $\dfrac{a(a+1)}{a}$ f $\dfrac{a(a-b)}{ab}$ g $\dfrac{3(x-y)}{x-y}$ h $\dfrac{a^2(a+b)}{a}$

Level 2

3 Factorise, where possible, and simplify the following algebraic fractions:

a $\dfrac{2a+2b}{4}$ b $\dfrac{3x-6}{6}$ c $\dfrac{4a-2b}{4}$ d $\dfrac{5x+15}{15}$

e $\dfrac{7a-7}{7}$ f $\dfrac{a^2+a}{2a}$ g $\dfrac{3p+3q}{6p+6q}$ h $\dfrac{ab+b}{2b}$

i $\dfrac{8a-16}{16}$ j $\dfrac{4p-8}{16}$ k $\dfrac{x^2+x}{5x+5}$ l $\dfrac{^{-}xy-y}{x^2+x}$

2.7 Adding and subtracting algebraic fractions

As in arithmetic, a common denominator is required to add or subtract algebraic fractions.

- $\dfrac{x}{5}+\dfrac{x}{5}=\dfrac{x+x}{5}$

  $=\dfrac{2x}{5}$

If the denominators are the same, keep the denominator and add or subtract the numerators.

- $\dfrac{x}{5}+\dfrac{x}{2}=\dfrac{2x}{10}+\dfrac{5x}{10}$

  $=\dfrac{2x+5x}{10}$

  $=\dfrac{7x}{10}$

If the denominators are different, change to a common denominator (the lowest common multiple of the denominators is usually used), then add or subtract the numerators.

EXAMPLE 1

Simplify each of the following:

a $\dfrac{5x}{6}+\dfrac{2x}{6}$ b $\dfrac{4a}{7}-\dfrac{a}{7}$

Solution

a $\frac{5x}{6} + \frac{2x}{6} = \frac{5x + 2x}{6}$

$= \frac{7x}{6}$

b $\frac{4a}{7} - \frac{a}{7} = \frac{4a - a}{7}$

$= \frac{3a}{7}$

EXAMPLE 2

Simplify, by creating common denominators:

a $\frac{x}{3} + \frac{x}{4}$

b $\frac{5a}{8} - \frac{a}{4}$

Solution

a $\frac{x}{3} \times \frac{4}{4} + \frac{x}{4} \times \frac{3}{3} = \frac{4x}{12} + \frac{3x}{12}$

$= \frac{4x + 3x}{12}$

$= \frac{7x}{12}$

Lowest common denominator of 3 and 4 is 12.

b $\frac{5a}{8} - \frac{a}{4} \times \frac{2}{2} = \frac{5a}{8} - \frac{2a}{8}$

$= \frac{5a - 2a}{8}$

$= \frac{3a}{8}$

Lowest common denominator of 8 and 4 is 8.

EXERCISE 2G

1 Simplify:

a $\frac{a}{3} + \frac{a}{3}$ b $\frac{2x}{3} - \frac{x}{3}$ c $\frac{5a}{7} + \frac{a}{7}$ d $\frac{6w}{7} - \frac{2w}{7}$

e $\frac{x}{5} + \frac{2x}{5}$ f $\frac{3x}{5} - \frac{x}{5}$ g $\frac{9x}{10} - \frac{2x}{10}$ h $\frac{9x}{10} + \frac{2x}{10}$

i $\frac{9a}{11} - \frac{3a}{11}$ j $\frac{x}{9} + \frac{2x}{9}$ k $\frac{3a}{4} - \frac{a}{4}$ l $\frac{4a}{7} - \frac{4a}{7}$

Simplify **all** answers.

Level 2

2 Simplify:

a $\frac{a}{3} + \frac{a}{2}$ b $\frac{x}{3} + \frac{x}{5}$ c $\frac{x}{3} + \frac{x}{7}$ d $\frac{x}{9} + \frac{x}{5}$

e $\frac{w}{7} - \frac{w}{8}$ f $\frac{m}{9} + \frac{m}{8}$ g $\frac{x}{3} + \frac{x}{15}$ h $\frac{a}{3} - \frac{a}{9}$

i $\frac{x}{4} + \frac{x}{6}$ j $\frac{x}{4} - \frac{x}{6}$ k $\frac{w}{9} + \frac{w}{3}$ l $\frac{m}{10} + \frac{m}{5}$

m $\frac{3a}{4} - \frac{a}{2}$ n $\frac{2w}{3} - \frac{w}{9}$ o $\frac{2p}{7} - \frac{3p}{5}$ p $\frac{4x}{7} - \frac{x}{2}$

3 Fully simplify each of the following additions and subtractions:

a $\frac{x}{4} + \frac{x+1}{4}$ b $\frac{x+1}{2} + \frac{x+3}{5}$ c $\frac{x+1}{4} - \frac{x}{2}$

d $\frac{k+3}{4} - \frac{k}{2}$ e $\frac{(x+3)}{4} - \frac{(x+1)}{3}$ f $\frac{3w}{4} - \frac{w+1}{2}$

g $\frac{x+1}{5} + \frac{x-3}{2}$ h $\frac{x+4}{5} - \frac{x}{6}$ i $\frac{3x+4}{5} - \frac{x+5}{2}$

Brackets may be useful.

2.8 Multiplication and division of algebraic fractions

The rules we use in arithmetic are also the rules we use when multiplying and dividing algebraic fractions.

To multiply algebraic fractions:

- Multiply the numerators together.
- Multiply the denominators together.
- Simplify the final fraction to its lowest form.

$$\frac{a}{b} \times \frac{c}{d} = \frac{a \times c}{b \times d} = \frac{ac}{bd}$$

To divide algebraic fractions:

- Find the reciprocal of the second fraction.
- Multiply the fractions as above.

$$\frac{a}{b} \div \frac{c}{d} = \frac{a}{b} \times \frac{d}{c} = \frac{ad}{bc}$$

The reciprocal of $\frac{x}{y}$ is $\frac{y}{x}$. Just flip.

EXAMPLE 1

Find the reciprocal of:

a $\frac{a}{4}$

b $\frac{5a}{2}$

Solution

a The reciprocal of $\frac{a}{4}$ is $\frac{4}{a}$

b The reciprocal of $\frac{5a}{2}$ is $\frac{2}{5a}$

EXAMPLE 2

Simplify:

a $\frac{x}{4} \times \frac{x}{3}$

b $\frac{a}{4} \times \frac{6}{a}$

Solution

a $\frac{x}{4} \times \frac{x}{3} = \frac{x \times x}{4 \times 3} = \frac{x^2}{12}$

b $\frac{a}{4} \times \frac{6}{a} = \frac{a \times 6}{4 \times a} = \frac{6a}{4a} = \frac{3}{2}$

Simplify.

EXAMPLE 3

Simplify these divisions by multiplying by the reciprocal of the second fraction:

a $\frac{a}{4} \div \frac{1}{3}$

b $\frac{2a}{5} \div \frac{a}{3}$

Solution

a $\frac{a}{4} \div \frac{1}{3} = \frac{a}{4} \times \frac{3}{1} = \frac{a \times 3}{4 \times 1} = \frac{3a}{4}$

b $\frac{2a}{5} \div \frac{a}{3} = \frac{2a}{5} \times \frac{3}{a} = \frac{2a \times 3}{5 \times a} = \frac{6a}{5a} = \frac{6}{5}$

Multiply by the reciprocal.

The reciprocal of $\frac{1}{3}$ is 3.

EXERCISE 2H

1 Write the reciprocal of each of the following fractions:

a $\frac{4}{5}$ **b** $\frac{7}{3}$ **c** 8 **d** $\frac{x}{y}$

e $\frac{a}{3}$ **f** $\frac{a}{b}$ **g** x **h** x^2

Don't forget that whole numbers have a denominator of one: $5 = \frac{5}{1}$

2 Simplify:

a $\frac{a}{5} \times \frac{3}{4}$ **b** $\frac{x}{7} \times 3$ **c** $\frac{3x}{4} \times \frac{1}{3}$ **d** $\frac{a}{4} \times \frac{a}{5}$

e $\frac{x}{4} \times \frac{x}{2}$ **f** $\frac{a}{5} \times \frac{4}{a}$ **g** $\frac{5w}{6} \times \frac{2}{3}$ **h** $\frac{4a}{5} \times \frac{1}{2}$

i $\frac{a}{5} \times \frac{5}{7}$ **j** $\frac{2w}{6} \times \frac{6}{11}$ **k** $\frac{3w}{4} \times \frac{4}{5}$ **l** $\frac{ab}{5} \times \frac{5}{a}$

3 Simplify:

a $\frac{a}{4} \div \frac{1}{4}$ **b** $\frac{w}{3} \div \frac{2}{3}$ **c** $\frac{a}{7} \div \frac{3}{7}$ **d** $\frac{5m}{4} \div 5$

e $\frac{x}{3} \div \frac{x}{2}$ **f** $\frac{m}{4} \div \frac{m}{2}$ **g** $\frac{3m}{5} \div \frac{m}{10}$ **h** $\frac{m}{n} \div \frac{m}{4}$

i $\frac{3w}{8} \div \frac{3w}{7}$ **j** $\frac{5a}{7} \div \frac{2a}{3}$ **k** $\frac{x^2}{y} \div \frac{x}{1}$ **l** $\frac{ab}{4} \div \frac{a}{2}$

Level 2

4 Simplify each of the following:

a $\frac{(x+1)}{4} \times \frac{4}{5}$ **b** $\frac{ab}{5} \times \frac{5}{a}$ **c** $\frac{x}{y} \times \frac{xy}{6}$ **d** $\frac{10a}{7} \div \frac{5a}{b}$

e $\frac{x}{5} \times \frac{x}{3} \times \frac{4}{x}$ **f** $\frac{a}{3} \times \frac{a}{2} \times \frac{2}{5}$ **g** $\frac{1}{3} \times \frac{x^2}{4} \times \frac{4}{x}$ **h** $\frac{ab}{5} \div \frac{a^2}{10}$

i $\frac{x}{4} \times \frac{3x}{5} \div \frac{x}{2}$ **j** $\frac{a^2}{5} \times \frac{5a}{7} \div \frac{1}{7}$ **k** $\frac{a}{3} \div \frac{1}{3} \times a$ **l** $\frac{ab}{5} \div \frac{3a}{10} \div \frac{b}{3}$

TEACHER

Investigation Playing with patterns

You will need: calculators

Each of these investigations leads you into the unknown. Be sure to record any finding that seems significant. Show it in symbols and describe it in words.

1 $\frac{5}{9} = 0{\cdot}555\ldots \quad \frac{7}{90} = 0{\cdot}0777\ldots$ Investigate.

2 $\frac{1}{2}, \frac{2}{3}, \frac{3}{4}, \frac{4}{5}, \ldots$ Investigate.

3 $\frac{1}{2} + \frac{1}{3} + \frac{1}{4} + \frac{1}{5} + \ldots$ Investigate.

Chapter Review

Language Links

algebraic
coefficient
constant
distributive
evaluate
expand
expression
factorise
grouping
pattern
power
pronumeral
sequence
simplify
substitution
term
variable

1 Write a set of instructions for a person to follow so that they are able to factorise an algebraic expression without having completed the task before.

2 What is the opposite of factorising an algebraic expression?

3 What is meant by HCF? How was it used in this chapter?

4 Explain the significance of like terms in algebraic expressions.

5 Write a one page summary of the sections dealt with in this chapter. You can use this to study from later.

6 Copy and complete:

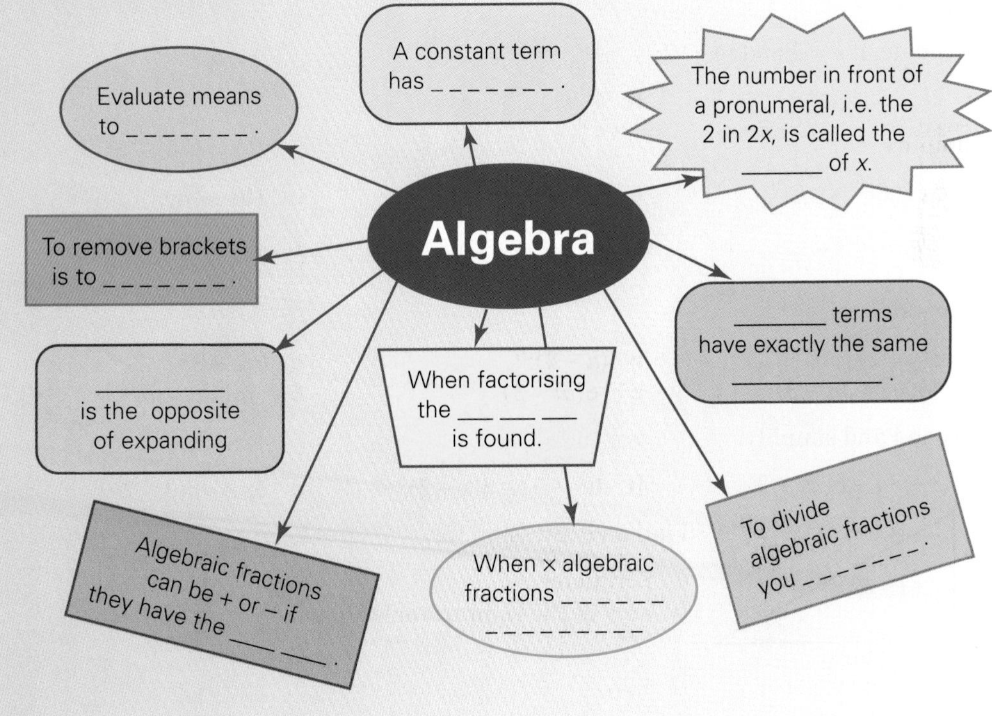

Chapter Review Exercises

2.1 **1** Write an expression for:

a the sum of a and 2
b the product of a and 2
c double m
d half of x
e double the sum of x and y
f m divided by 5 and x

2.1 **2** For the expression $4x^2 + 3x + y + 8$, what is the:

a coefficient of x^2?
b constant term?
c coefficient of y?

2.1 2.2 **3** Simplify each algebraic expression:

a $4 \times 3k$
b $5n \times 6t$
c $a + a$
d $l \div l$
e $b + b + b$
f $x \times x$
g $a \times \frac{1}{a}$
h $7h + 3h$
i $2b \times 3b \times {}^-4$
j $8x - 2x - x$
k $3ab + 2 + 4ab + b$
l $7x + 9 - 3x$

MC **4** Generate the first three terms of a sequence following the rule:

$$V = \frac{n}{4} + 1$$

MC **5** Find an algebraic rule for the sequences:

a $\frac{1}{1}, \frac{1}{2}, \frac{1}{3}, \ldots$
b 2, 5, 10, 17, …

2.5 **6** Evaluate, if $k = 3$ and $m = {}^-2$:

a $2k + 3$
b $4km$
c $2(m^3 - k^2)$

2.1 **7** Simplify:

a $5x - 8x$
b $5n \times {}^-2n$
c $10x \times {}^-6y$
d ${}^-9a - 7a + a$
e $\frac{7p + 3p}{p}$
f $a + 3b + 4a - 7b$

2.3 **8** Expand:

a $3(k + 7)$
b $(a - 4)5$
c $b(2 + 3b)$
d $4n(n^2 - 2n + 3)$
e ${}^-6(m - 3)$
f ${}^-7a(4a - 3b)$

2.3 **9** Expand and simplify:

a $3 + 5n + 6(2n + 3)$
b $4(x + 3) - 3(x + 2)$
c $3p + 4p(3 - 2p)$

2.3 **10**

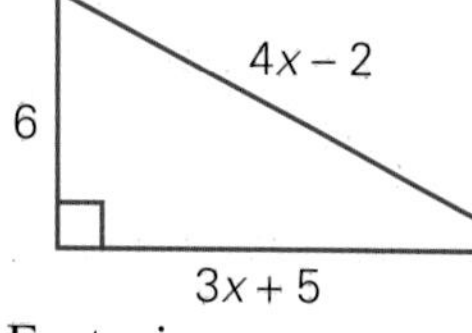

Find an expression for:

a perimeter
b area of the right triangle shown

2.4 **11** Factorise:

a $mt - mp$
b $5x + 10$
c $6c + 9cd$
d $12xy - 16x^2$
e $x^2 - 5x$
f ${}^-7h + 35$
g $16k - 24k^2$
h ${}^-10p^2 - 12qp$

12 Copy and complete:

a

x	$^-2$	$^-1$	0	1	2	3
$4 - x^2$						

b

m	$^-3$	$^-1$	0	1	3	7
$^-m^3$						

c

x	$^-2$	$^-1$	0	1	2	3
$x^2 - x + 2$						

2.5

13 Simplify:

a $\dfrac{5xy}{5}$ b $\dfrac{5x + 5y}{5}$ c $\dfrac{x^2 - xy}{x}$ d $\dfrac{3x + 3y}{4x + 4y}$ e $\dfrac{2x + 6}{6x}$

2.6

14 Simplify:

a $\dfrac{3x}{7} - \dfrac{x}{7}$ b $\dfrac{5a}{6} + \dfrac{3a}{6}$ c $\dfrac{w}{5} + \dfrac{w}{2}$ d $\dfrac{3a}{4} + \dfrac{a}{2}$

2.7

15 Write the reciprocal of:

a $\dfrac{3}{5}$ b 4 c $\dfrac{x}{8}$ d $\dfrac{x}{y}$ e $4a$

2.8

16 Simplify:

a $\dfrac{a}{5} \times \dfrac{5}{6}$ b $\dfrac{x}{4} \times \dfrac{2}{3x}$ c $\dfrac{4a}{3} \times \dfrac{3}{8}$ d $\dfrac{1}{5} \times 20a$

2.8

17 Simplify:

a $\dfrac{x}{4} \div \dfrac{x}{8}$ b $\dfrac{ab}{7} \div \dfrac{1}{7}$ c $\dfrac{a^2}{5} \div \dfrac{ab}{10}$ d $\dfrac{xy}{8} \div \dfrac{x^2}{y}$

2.8

18 Simplify:

a $\dfrac{2x}{5} \times \dfrac{5}{8} \div \dfrac{x}{2}$ b $\dfrac{4p}{6} \div \dfrac{p^2}{12} \times \dfrac{1}{2}$

2.8

TEACHER

Keeping Mathematically Fit

Part A: Non-calculator

1 Copy and complete $8{\cdot}407 \times \_\_\_\_\_\_ = 840{\cdot}7$.

2 Increase $4ab$ by the product of $3a$ and $7b$.

3 Find $2\frac{2}{5}$ of \$1000.

4 Round 346·09761 to 2 decimal places.

5 Find x if $\dfrac{5^7 \times 5^2}{5^3} = 5^x$.

6 Evaluate $94 \times 116 + 6 \times 116$.

7 Simplify $\frac{4{\cdot}6 \times 0{\cdot}2}{0{\cdot}01}$.

8 Convert 5·6 km to cm.

9 Find the value of $15 \times (7 - 4 \times 3)^2$.

10 Simplify $9^3 \times 9^2 \div 9^5$.

Part B: Calculator

1 Find the value of $\sqrt[3]{\frac{9{\cdot}6^2}{3{\cdot}1 - 1{\cdot}2^2}}$ correct to 3 decimal places.

2 Find the value of $7^5 - 2^4$.

3 Find the value of $\frac{9{\cdot}7 - 3 \times 4{\cdot}8^2}{3{\cdot}4}$ (answer to 1 decimal place).

4 Find the value of x in:

a

b
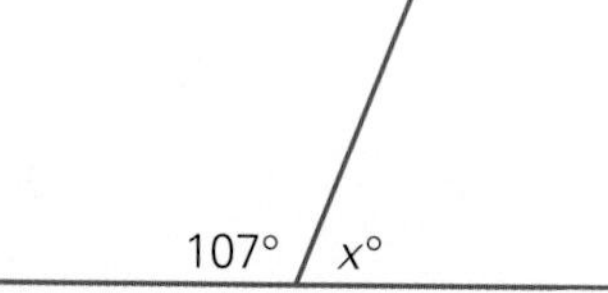

c
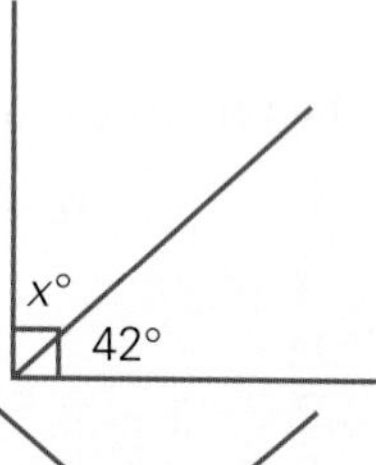

d
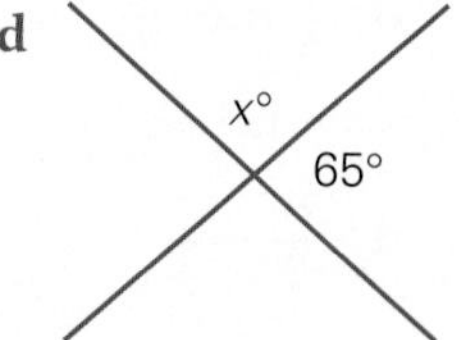

e
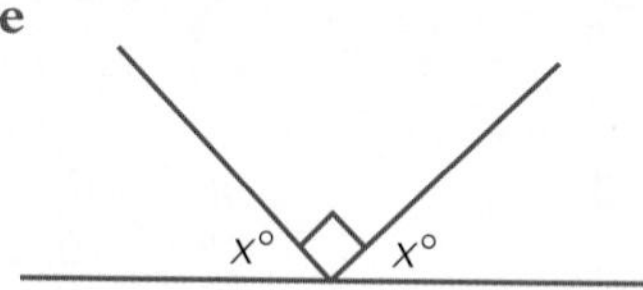

f
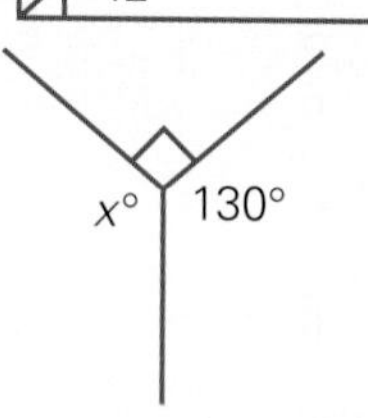

5 Simplify:

a $2\frac{1}{2} + 3\frac{1}{3}$

b $1\frac{1}{4} - \frac{3}{5}$

c $3\frac{4}{7} \times 1\frac{1}{2} \div \frac{2}{3}$

6 Find the perimeter of:

a
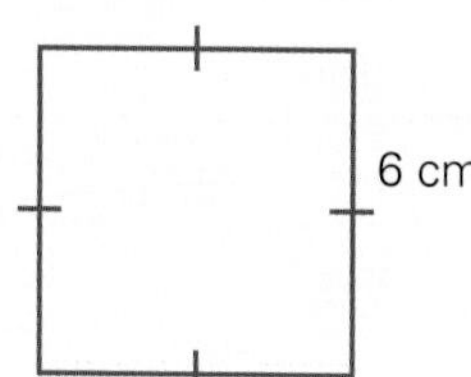

b
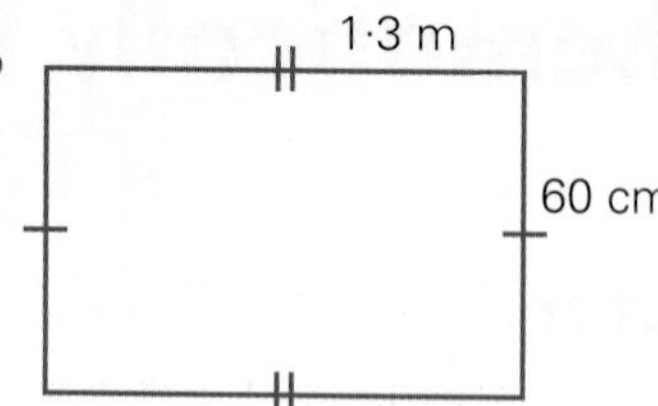

c
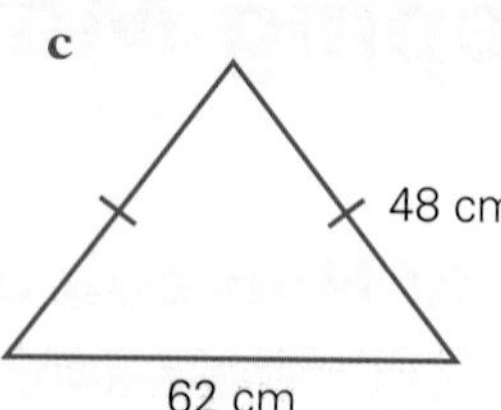

d
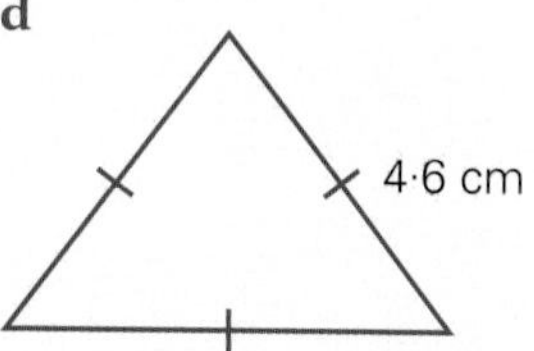

e
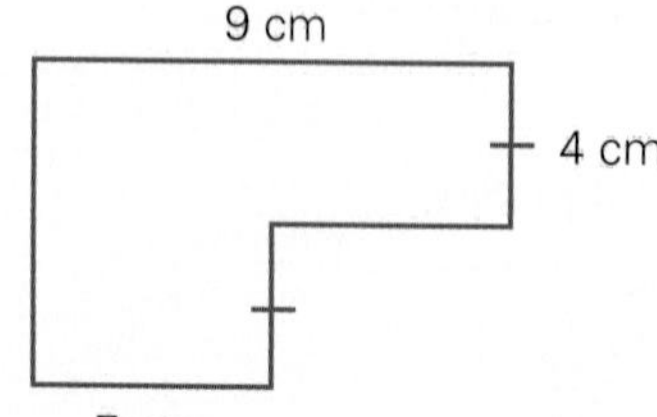

f
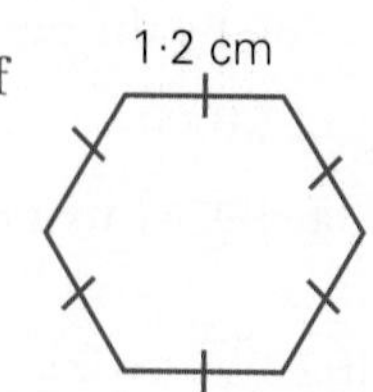

TEACHER

7 Simplify:

a $6a + 3a - a$

b $\frac{4x \times 3y}{24x}$

c $3(2x + 1) - 4x + 5$

d $7ab \times 3ac$

e $4(x + 3) + x$

f $\frac{10a^2}{20a}$

8 The product of two terms is $\frac{x^2}{2}$. What are the two terms?

1 What you need to know and revise

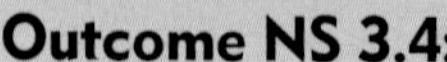

Outcome NS 3.4:
Compares, orders and calculates with decimals, simple fractions and simple percentages

Outcome NS 4.3:
Operates with fractions, decimals percentages, ratio and rates

2 What you will learn in this chapter

Outcome NS 4.3:
Operates with fractions, decimals, percentages, ratio and rates

- expresses one quantity as a percentage of another
- calculates percentages of quantities
- increases and decreases a quantity by a given percentage
- interprets and calculates percentages greater than 100%
- expresses profit and loss as a percentage of cost or selling price
- applications of percentages—discounts

Outcome NS 5.1.2:
Solves consumer arithmetic problems involving earning and spending money

- commission
- simple interest

Working Mathematically outcomes WMS 4.1–4.5
Students will be asked to ***question***, ***apply strategies***, ***communicate***, ***reason*** and ***reflect*** in the sections of this chapter.

Percentages

Key mathematical terms you will encounter

appreciation
borrower
buying
commission
composition
cost
depreciation
discount
interest
lender
loss
marked
per annum
percentage
price
principal
profit
selling
simple

MathsCheck

Percentages

1 Express the following as decimals:

a 15% b 27% c 56% d 19%
e 95% f 82% g 99% h 7%
i 1% j 8% k 10% l 90%

$\times 100$
decimal → percentage
decimal ← percentage
$\div 100$

2 Express the following decimals as percentages:

a 0·75 b 1·4 c 0·95 d 0·56
e 0·27 f 0·03 g 0·99 h 0·05
i 0·3 j 0·03 k 1·04 l 0·1225

3 Express the following percentages as fractions with denominators of 100, simplifying where necessary.

a 19% b 23% c 99%
d 24% e 50% f 25%
g 74% h 60% i 5%

$\times 100$
fraction → percentage
fraction ← percentage
denominator of 100

4 Write these percentages as **i** decimals **ii** fractions:

a $12\frac{1}{2}\%$ b 8·7% c $33\frac{1}{3}\%$ d $66{\cdot}\dot{6}\%$ e 190%

5 Convert the following to: **i** decimals **ii** percentages.

a $\frac{1}{5}$ b $\frac{4}{5}$ c $\frac{8}{10}$ d $\frac{3}{10}$ e 1 f $\frac{1}{4}$

g $\frac{3}{4}$ h $\frac{21}{70}$ i $\frac{1}{8}$ j $\frac{12}{20}$ k $\frac{1}{80}$ l $\frac{5}{7}$

6 Write each of the following fractions as percentages:

a $\frac{1}{9}$ b $\frac{2}{9}$ c $\frac{1}{3}$ d $\frac{4}{9}$ e $\frac{5}{9}$ f $\frac{2}{3}$ g $\frac{7}{9}$ h $\frac{8}{9}$

7 Write each of the fractions in question **7** as a decimal.

8 Convert the following to percentages:

a 0·017 b 8·4 c 8 d $\frac{17}{200}$

e 0·375 f 0·5647 g $\frac{19}{200}$ h $\frac{7}{40}$

9 Determine which in each of the following groups is the greatest:

a 65%, $\frac{3}{5}$ b $\frac{7}{12}$, 72% c $\frac{3}{5}$, $\frac{4}{7}$, 0·66

d $0{\cdot}\dot{6}$, 66%, $\frac{3}{5}$ e $\frac{4}{3}$, 120%

Convert to % and compare.

TEACHER

10 A class in Year 8 at Mathsville High School consists of 13 boys and 17 girls. What percentage of the class are girls?

11 The cost of an airline ticket to Bali was \$890. Kaitlyn had saved \$475. What percentage of the cost had Kaitlyn still to save?

Exploring New Ideas

As we saw last year, percentages are a universal method of contrast and comparison; they are used in schools, businesses and the media. A percentage is shown using the symbol %, and means a **fraction of 100**. As percentages play such an important role in our society, it is important to be able to work with them in many situations. In this chapter we will look at the many uses of percentages.

3.1 Percentage composition

Have you ever wanted to know what your mark of 34 out of 40 was as a percentage, or if a pie's 3 g of fat really means that the company is justified in its statement, 'the pie is 99% fat free'? Many items bought in the supermarket use percentages to show the composition of the ingredients, their fat content and the percentage of the daily vitamin requirements each item contains. In the following section we look at how to express one quantity as a percentage of another.

> To express one quantity as a percentage of another, write them as a fraction and then convert this fraction to a percentage by multiplying by 100%.

EXAMPLE 1

Express the mark of 34 out of 40 in a class test as a percentage.

Solution

$$34 \text{ out of } 40 = \frac{34}{40} \times 100\% = 85\%$$

or 34 ÷ 40 × 100%

EXAMPLE 2

Express 2100 mm as a percentage of 3 m.

Solution

$$\begin{aligned} 2100 \text{ mm of } 3 \text{ m} &= 2100 \text{ mm out of } 3000 \text{ mm} \\ &= \frac{2100}{3000} \times 100\% \\ &= 70\% \end{aligned}$$

To compare two quantities, ensure that they have the same units.
3 m = 3000 mm

EXERCISE 3A

1 What percentage has been shaded?

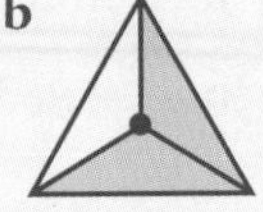
c

d
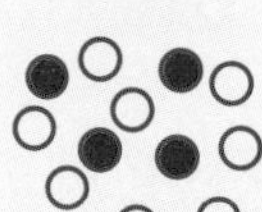
e

f

2 Copy and complete the table of sporting preferences:

Sport	Number preferring	Fraction of total	Percentage
Tennis	40		
Golf	30		
Swimming	70		
Volleyball	60		
TOTAL	200		100%

3 In each case, express the first quantity as a percentage of the second.

a 5 g, 20 g **b** 10 mL, 50 mL **c** 9, 15 **d** 10 km, 20 km
e \$16, \$20 **f** 15, 500 **g** 7c, 50c **h** 5 m, 50 m
i 5 men, 10 men **j** 8 g, 200 g **k** 19, 200 **l** 4 h, 12 h
m 18 L, 40 L **n** 3 s, 60 s **o** $\frac{1}{2}$ km, 3 km **p** \$1.20, \$4

4 Albory obtained 32 marks out of 40 in a test. What percentage was this?

5 Phuong obtained 28 out of 35 in a similar test. Find his percentage.

6 Lydia scored 15 out of 18, while Hayley scored 13 out of 15. Compare percentages to decide who did better in the tests.

7

Year	Girls	Boys	Total
7	72	88	
8	90	85	
9	56	84	
10	75	95	
11	54	56	
12	48	49	
Total			

a Complete the table.
b What percentage of the school's population are girls?
c What percentage of Year 8 are boys?
d What percentage of the school's population are in years 11 and 12?
e What percentage of the boys in the school are in Year 12?
f If two boys leave Year 8 and two girls arrive in their place, what percentage of Year 8 are now boys?

Level 2

8 What percentage is the first quantity of the second (1 decimal place if necessary)?

a 4 cm, 1 m **b** 40 s, 1 min **c** 80 s, 2 min **d** 3 km, 5000 m
e 58c, \$2 **f** 80 m, 4 km **g** 48 min, 1 h **h** 125c, \$3.60
i $\frac{1}{2}$ L, 2100 mL **j** \$4.70, 900c **k** 7 mm, 2 cm **l** 410 mm, 80 cm
m 7 min, 7 h **n** 12c, \$7.40 **o** $\frac{3}{4}$ kg, 3000 g **p** 48 g, 5 kg

9 Express 1500 mm as a percentage of:

a 2000 mm **b** 5 m **c** 90 cm

10 What percentage of \$80 is:

a \$60? **b** \$100? **c** \$70? **d** \$2.80?

11 What percentage is the first quantity of the second?

a 45 min, 1 h **b** 10 h, 1 day **c** 1·8 kg, 40 000 g

12 The area in square kilometres of Australian states is as shown on the map. Use a calculator to find what percentage of the total land area is occupied by each state.

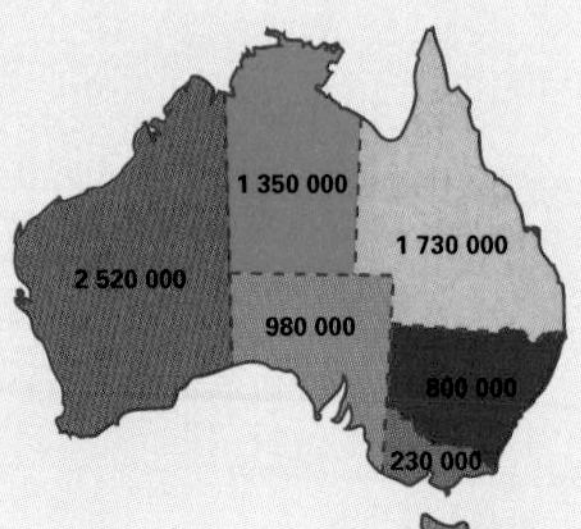

13 **a**

A 250 g apple pie has 7 g of fat. What percentage of the pie's mass is fat?

b Marcus had to jog 3 km in a cross-country event. If the first checkpoint occurred after 1200 m, what percentage of the event had Marcus covered by this point?

c The school day lasts $6\frac{1}{2}$ hours. Lunchtime lasts 40 minutes. What percentage of the school day is spent in the playground at lunchtime?

d Tax is charged at 31.5 cents in the dollar. Express this as a percentage.

14 Two tubs of diet ice-cream claim the following:

Tub A: An 84 g tub advertises only 3 g fat.
Tub B: A 64 g tub advertises only 3% fat.

Which of the two brands contains the lesser amount of fat?

15 A motel needs to average 50% occupancy in a week to break even. There are 36 rooms in the motel.

a Complete the table for the week.

Day	M	T	W	T	F	S	S
Rooms occupied	12	9	4	18	24	30	15
% occupancy							

b Does the motel break even for the week? Explain.

3.2 Percentage of quantities

The media often quotes percentages in their stories and advertisements:

- 52% of the voters surveyed plan to vote Liberal in the next election.
  If 1000 people were surveyed, how many plan to vote Liberal?

> Can you think why people would quote the percentage rather than the true figures?

- 90% of all nutritionists recommend Wheaties.
  If 100 nutritionists were asked for their recommendations, exactly how many people chose to support the product?

- A store requires a 25% deposit.
  If the cost of a TV set is \$1560 exactly, how much is the deposit?

It is often necessary to find the percentage of a particular quantity or amount.

To find a percentage of a quantity, write the percentage as a fraction or decimal, then multiply by the quantity.

$$n\% \text{ of } t = \frac{n}{100} \times t$$

EXAMPLE 1

Find:

a 9% of 150 **b** 7·8% of 30·9 kg

Solution

a 9% of 150
$= 0·09 \times 150$
$= 13·5$

b 7·8% of 30·9 kg
$= 0·078 \times 30·9$ kg
$= 2·4102$ kg

$9\% = \frac{9}{100} = 9 \div 100 = 0·09$

EXAMPLE 2

Sarah plans to save 12% of her weekly income. How much does Sarah save in a week where she earns \$474.60?

Solution

Savings = 12% of 474·60
$= 0·12 \times 474·60$
$= 56·952$

$\therefore$ Sarah plans to save \$56.95 each week.

EXAMPLE 3

How many centimetres in 5% of $\frac{1}{2}$ km?

Solution

5% of $\frac{1}{2}$ km
$= 0·05 \times 50\,000$ cm
$= 2500$ cm

Convert $\frac{1}{2}$ km into cm:
$\frac{1}{2}$ km = 50 000 cm

EXERCISE 3B

1 Find:

a 10% of 20	**b** 5% of 20	**c** 90% of 20	**d** 20% of 20
e 10% of 740	**f** 10% of 74	**g** 20% of 74	**h** 5% of 74
i 50% of 56	**j** 50% of 90	**k** 50% of 84	**l** 25% of 84

2 Find:

a 10% of 200	**b** 15% of 200	**c** 20% of 200	**d** 90% of 200
e 5% of \$20	**f** 12% of 500 cm	**g** 8% of 2000 L	**h** 99% of 1000
i 7% of 30 m	**j** 95% of 44 km	**k** 27% of 56 g	**l** 15% of \$70

3 Evaluate, giving your answers correct to 2 decimal places, where necessary.

a 8% of \$7.20	**b** 15% of 75 m	**c** 24% of 897	**d** 7% of 365
e 15% of \$15.70	**f** 25% of $\frac{3}{8}$	**g** 40% of 79·6	**h** 7% of 0·82
i $12\frac{1}{2}$% of \$9	**j** $7\frac{1}{4}$% of \$45	**k** $33\frac{1}{3}$% of 6 m	**l** $\frac{1}{2}$% of 900

4 At a local hospital 49·5% of all births are female. In a given month, 90 newborns are expected at the hospital. Of these births, how many are expected to be female?

5 A stage show will break even if 70% of the seats are occupied. For a theatre holding 2000 people, how many seats need to be filled?

6 People spend 24% of their day in front of the TV or home computer. Of the 24 hours in a day, how many hours and minutes are spent in front of these appliances?

7 Calculate the number of pages in a magazine dedicated to advertisements if of the 126 pages, 56% is set aside for ads.

8 Peta pays 31·5% of her total income in tax. If Peta earned \$51 257 last financial year, how much of this did Peta lose in tax?

9 An art class of 25 students had 4% of the class absent due to an outbreak of the flu. How many students were present?

Level 2

10 Complete:

a 5% of 1 km = __________ m
b 25% of 1 h = __________ min
c 7% of 3 km = __________ m
d 40% of 3 d = __________ h
e 4% of \$5 = __________ c

11 Buying a \$280 bicycle required Renée to pay a deposit of 15% of the purchase price. After paying the deposit, what balance was left?

12 Elvis scored 93% in a test marked out of 500. How many more marks did Elvis need to end up with a score of 95%?

13 Malcolm plans to save 18% of his weekly income of \$595.60. How much will Malcolm have saved at the end of a year?

14 47·9% of a local government's budget is spent on garbage collection. If a rate payer pays \$107.50 per quarter in total rate charges, how much do they contribute in a year to garbage collection?

15 Banks require a deposit of 10% on the value of the home purchased.

a Michael and Jo have $7\frac{1}{2}$% of the \$280 500 value. How much more do they need to save?

b Michael and Jo are eligible to receive the \$14 000 government grant. With this grant, what percentage of the house value do they have?

16 9·6% of 1 day is the same as x hours and y minutes. Find the value of x and y.

17 If 10% of x equals y, what is x% of 10?

3.3 Increasing and decreasing by a given percentage

To increase a quantity or amount by a percentage is to make it larger by adding to it.

To decrease a quantity or amount by a percentage is to make it smaller by subtracting from it.

To increase by a given percentage:

- add the percentage increase to 100%
- multiply the amount by this new percentage

100% equals the original amount.

To decrease by a given percentage:

- subtract the percentage from 100%
- multiply the amount by this new percentage

That is:

- To **increase** by 20%, find (100% + 20%) or 120% of the amount.
- To **decrease** by 20%, find (100% – 20%) or 80% of the amount.

Applications of increasing and decreasing by a given percentage include:

- discounts—a reduction in the cost of an item
- mark-ups—increases in the cost of an item
- depreciation—a loss in value
- appreciation—an increase in an item's worth

EXAMPLE 1

Raelene's weekly wage of \$450 was increased by 5%. Find the new wage.

Solution

$100\% + 5\% = 105\%$

New wage $= 105\%$ of \$450

$= 1{\cdot}05 \times \$450$

$= \$472.50$

EXAMPLE 2

Decrease 9724 by $8\frac{1}{2}\%$.

Solution

$100\% - 8\frac{1}{2}\% = 91\frac{1}{2}\%$

New amount $= 91\frac{1}{2}\%$ of 9724

$= 0{\cdot}915 \times 9724$

$= 8897{\cdot}46$

EXERCISE 3C

1 What percentage of an amount must be found if you wish to increase it by:

a 5%? **b** 7%? **c** 12%? **d** 1%?
e 10%? **f** 27%? **g** $12\frac{1}{2}\%$? **h** $33\frac{1}{3}\%$?

2 What percentage of an amount must be found if you wish to decrease it by:

a 8%? **b** 12%? **c** 10%? **d** 50%?
e 5%? **f** 25%? **g** $7\frac{1}{2}$%? **h** $33\frac{1}{3}$%?

3 Increase 970 by each of the percentages given in question **1**.

4 Decrease 5000 by each of the percentages given in question **2**.

5 **a** Increase 90 by 7%. **b** Increase 57 by 2%. **c** Increase 98 by 8%.
d Decrease $965 by 5%. **e** Decrease 75 by 8%. **f** Decrease 110 by 10%.
g Increase 110 by 10%. **h** Decrease 18 by 2%. **i** Increase 7·48 by 19%.
j Increase 9·6 by $12\frac{1}{2}$%. **k** Increase 66 by $33\frac{1}{3}$%. **l** Decrease 684 by $7\frac{1}{2}$%.

6 Anne's annual salary of $25 680 is to be increased by 5%. What is her new salary?

7 Inflation is currently running at 3% p.a. That means that all goods and services will cost 3% more in a year's time. What will be the cost in 1 year's time of:

a a $20 CD? **b** a $64 dental checkup? **c** a $49.95 pair of jeans?

8 To set the *retail* price, an appliance shop *marks up* (increases) the cost (wholesale buying price) of all its stock by 35%. Find the retail price of each appliance:

a toaster: cost price $30 **b** hair dryer: cost price $25
c espresso coffee machine: cost price $127.52

9 Giorgio sells perishables such as fruit and vegetables. In his costing, he allows for 12% wastage.

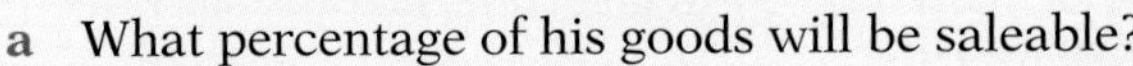

a What percentage of his goods will be saleable?
b Of 50 cartons each containing 18 mangoes, how many mangoes will be saleable?

10 To sell a car, the car salesman offers a 5% discount off the cash price of $18 759. What is the price paid for the car?

11 A pair of sports shoes is discounted by 39% as they are no longer the latest look. The recommended price was $169.95.

a What is the amount of the discount? **b** What is the new price of the shoes?

12 In 2002 the migrant intake of a country increased by 6% from the year before. If the migrant intake of 2001 was 79 000, how many migrants arrived in 2002?

Level 2

13 Two bridesmaid's dresses and wraps totalled $996. If a 10% discount was offered, how much did each bridesmaid pay for her dress and wrap?

14 **a** Increase $56 by 8% and then decrease the result by 8%. Is your answer $56? Explain.
b Increase $97.40 by 5% and then decrease the result by 5%.
c Decrease $97.40 by 5% and then increase the result by 5%. What do you notice?
d Increase 5430 by $12\frac{1}{2}$% and then decrease this result by 10%. Is this the same as increasing the initial amount by $2\frac{1}{2}$%?

15 Find a car's worth at the end of a year if it depreciates at a rate of 13% p.a. and its original price was:

a $20 700 b $34 550 c $7990

16 The value of Thomas' new bike depreciated 22% this year. If the value at the beginning of the year was $888, what is the value of the bike at the end of the year?

17 The price of a new car depreciates in value by 10% as soon as the car leaves the car yard. Calculate the value of a car immediately after it has left the yard if it was purchased for $24 500.

18 If a car worth $38 000 depreciates at 15% p.a., find its value at the end of 1 year.

19 A house appreciates at a rate of $7\frac{1}{2}$% p.a. If a house is now worth $540 000, find its value at the end of:

a 1 year b 2 years c 3 years

TEACHER

Try these

1 What percentage does 950 need to be increased by to give 980? Is it the same as the percentage required to decrease 980 to 950?

2 $\$x$ is increased by 8% and then decreased by 8%. What is the overall percentage change?

3 An amount is increased by 18% and then the result decreased by 10%. What is the overall percentage change? Describe how you can work this out without being told the amount.

3.4 Finding the percentage change

SPREADSHEET ACTIVITY

A car worth $18 000 was sold for $17 200. What was the percentage change? In this section we look at how we can work out percentage change.

$$\text{Percentage increase/decrease} = \frac{\text{amount of increase/decrease}}{\text{original amount}} \times \frac{100}{1}\%$$

EXAMPLE 1

David's train fare increased from $2.50 to $3. What was the percentage increase in the price?

Solution

Increase = $3 – $2.50

= 50c

$$\% \text{ increase} = \frac{50}{250} \times 100\%$$
$$= 20\%$$

or $\frac{0 \cdot 5}{2 \cdot 5} \times 100\%$

EXAMPLE 2

Petrol was 97.4 c/L on Monday. It fell to 89.9 c/L on Tuesday. Calculate the percentage change.

Solution

$$\text{Change} = 97{\cdot}4 - 89{\cdot}9$$
$$= 7{\cdot}5$$

$$\% \text{ decrease} = \frac{7{\cdot}5}{97{\cdot}4} \times 100\%$$

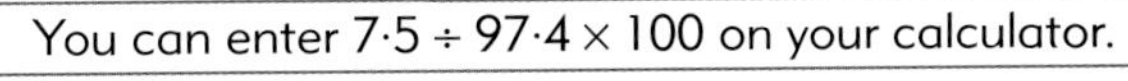

$$= 7{\cdot}700\ 20 \ldots$$
$$= 7{\cdot}7\% \text{ (1 dec. pl.)}$$

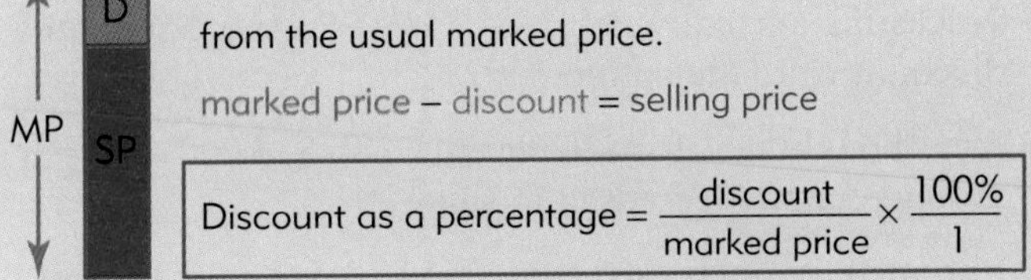

EXAMPLE 3

A netball usually marked at \$39.95 is offered for sale at \$29.95. What percentage discount is this?

Solution

$$\text{Discount} = \$39.95 - \$29.95$$
$$= \$10$$

$$\% \text{ discount} = \frac{10}{39{\cdot}95} \times 100\%$$
$$= 25{\cdot}031\ 28 \ldots$$
$$\approx 25\% \text{ (nearest \%)}$$

EXERCISE 3D

1 Find the percentage increase for each of the following:

- **a** A table was \$600, now \$660.
- **b** A house was \$500 000, now \$550 000.
- **c** A chocolate bar was \$1, now \$1.40.
- **d** A loaf of bread was \$2, now \$2.50.
- **e** A can of oil was \$10, now \$15.

2 Find the percentage decrease for each of the following:

- **a** A photo frame was \$12, now \$6.
- **b** A textbook was \$20, now \$15.
- **c** A cassette player was \$50, now \$30.
- **d** A car was \$20 000, now \$18 000.
- **e** A computer was \$1500, now \$1000.

3 Complete — give answers to the nearest percentage where necessary.

a

Original amount	New amount	Increase	% increase
56	80		
100	150		
24	30		
8·76	10·2		
51 mm	6 cm		

b

Original amount	New amount	Decrease	% decrease
93	90		
150	100		
9	8		
7·564	6·09		
9 km	8.3 km		

4 A computer store offered a $15 discount on a game marked at $69.95. What percentage discount was this?

5 Tracksuits normally ticketed at $49.95 were discounted by $19.95. What percentage discount were they offered at?

6 A car bought new for $30 000 is worth $24 000 one year later. By what percentage has it depreciated (decreased in value)?

7 Simone's pay increased from $532.78 per week to $544.16 per week. What percentage increase was this?

8 A discount of 12 cents in the dollar was offered on a mountain bike worth $1025.

- **a** What is the discount as a percentage?
- **b** What percentage of the marked price does it sell for?
- **c** Calculate the selling price.

9 A calculator wristwatch marked at $69.98 is advertised at a discount of $15. What percentage discount is this, to the nearest whole percentage?

Level 2

10 A greengrocer bought a carton of 24 dozen apples in which 84 were unable to be sold. Calculate the percentage wastage.

11 In 1933 the population of Sydney was 1 235 000. In 2001 the population of Sydney was 4 154 722. Calculate the percentage increase.

12 A plant was measured at the beginning of the week and had a height of 12 cm. At the end of the second week the plant had increased in height by 85 mm. Calculate the percentage increase.

13 A house was bought in 1999 for $189 000. In 2003 it was sold for $235 000.

- **a** Calculate the percentage increase.
- **b** Express this percentage increase as a annual percentage rate.

3.5 Profit and loss

When an item is resold for more than its original price, it is said to be sold for a profit. This can be expressed as a percentage. Obviously it is just as easy to have made a loss on a deal as to have made a profit.

SALE AT A PROFIT

$$\text{selling price} = \left.\begin{matrix}\text{cost}\\ \text{buying}\end{matrix}\right\}\text{price} + \text{profit}$$

$$\text{profit \%} = \frac{\text{profit}}{\text{cost}} \times \frac{100}{1}\,\%$$

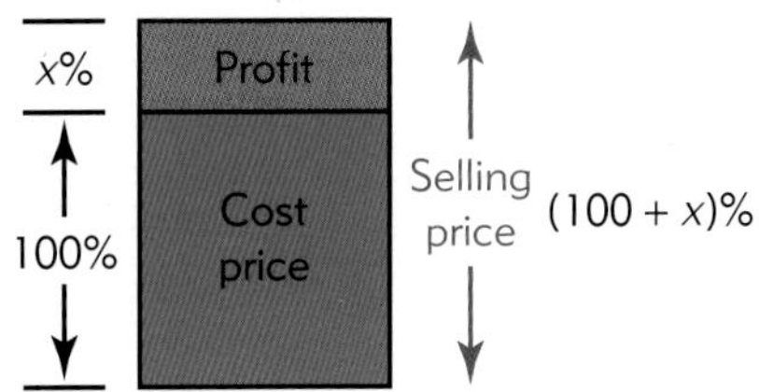

SALE AT A LOSS

$$\text{selling price} = \left.\begin{matrix}\text{cost}\\ \text{buying}\end{matrix}\right\}\text{price} - \text{loss}$$

$$\text{loss \%} = \frac{\text{loss}}{\text{cost}} \times \frac{100}{1}\,\%$$

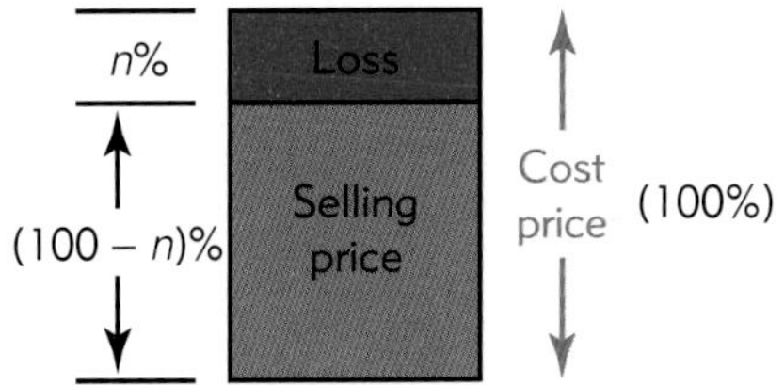

EXAMPLE 1

The cost price for a new car was \$18 950 and its selling price was \$15 400.

a Calculate the loss in dollars. **b** Calculate the percentage loss.

Solution

a 18 950 – 15 400 = 3550

∴ loss = \$3550

b $\%\text{ loss} = \dfrac{\text{loss}}{\text{cost price}} \times 100\%$

$= \dfrac{3550}{18\ 950} \times 100\%$

$= 18\ 7335\ \ldots$

$= 18{\cdot}7\%$ (1 dec. pl.)

EXAMPLE 2

A fruitshop owner buys cartons of peaches for \$5.70. The peaches are then sold to the public at a price of \$8.90 per carton.

a Calculate the profit in dollars. **b** Calculate the percentage profit.

Solution

a 8·9 – 5·7 = 3·2

∴ profit = \$3.20

b $\%\text{ profit} = \dfrac{\text{profit}}{\text{cost price}} \times 100\%$

$= \dfrac{3{\cdot}2}{5{\cdot}7} \times 100\%$

$= 56{\cdot}140\ 35\ \ldots$

$\approx 56\%$ (nearest %)

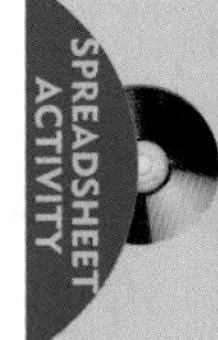

EXERCISE 3E

1 Copy and complete the tables below.

a

Cost price ($)	Selling price ($)	Profit ($)	% profit (1 dec. pl.)
10	15		
24	30		
100	150		
250	255		
17.50	20		

b

Cost price ($)	Selling price ($)	Loss ($)	% loss (1 dec. pl.)
10	5		
16	8		
100	50		
34	19		
94.75	90.50		

2 The cost price of an item is \$15 and the selling price is \$24. Calculate:

a the profit in dollars
b the percentage profit

3 A car was bought for \$24 900 and sold for \$21 000. Calculate:

a the loss in dollars b the percentage loss

4 If the car in question **3** was sold for \$15 000, recalculate the percentage loss.

5 A whitegoods store offers a fridge normally worth \$1320 for \$1200 for a person paying cash. What discount is being offered for a cash sale?

6 The selling price of a secondhand lounge chair was \$572. What was the percentage loss if the lounge chair was originally worth \$950?

Level 2

7 Copy and complete the table below:

Cost price	Selling price	Profit/loss	% profit/loss
\$470	\$400		
	\$470	Loss \$50	
\$50		Profit \$25	
	\$96		10% profit
\$500			17% profit
	\$600		$33\frac{1}{3}$% profit

8 Mei-lee paid \$369 for her formal dress. After graduation, she sold it back to the dressmaker for \$199. Find:

a her loss from the sale
b the loss as a percentage of the cost price
c the loss as a percentage of the selling price

9 A merchant pays \$19.20 for a case of 80 apples. If he allows for 10% wastage, selling the remainder at 3 for a dollar, what is his gain or loss percentage?

10 By selling an article for \$28.50, a trader made a loss of 5% on the cost price. What had the article cost her?

11 A retailer buys a carton of 24 cans of tennis balls for \$78. He then sells the cans individually for \$5.70. Calculate percentage profit.

3.6 Commission

SPREADSHEET ACTIVITY

Car sales people and real estate agents often earn a percentage of the value of the goods they have sold.

A **commission** is a percentage of the selling price of goods sold, offered as an income incentive to salespeople.

Commission = x% of selling price

EXAMPLE 1

Bryce works as a salesman where he earns 3% commission on all sales. Calculate the commission on Bryce's sales if they totalled \$2764.

Solution

$$\begin{aligned}\text{Commission} &= 3\% \text{ of } 2764\\ &= 0{\cdot}03 \times 2764\end{aligned}$$

∴ Bryce earned \$82.92 commission.

EXAMPLE 2

Lachlan earns 5% commission on the first \$20 000 of sales and 2% on the balance of total sales. Calculate Lachlan's commission on sales totalling:

a \$15 000 **b** \$30 000

Solution

a $15\,000 \times 0{\cdot}05$
$= 750$
∴ commission = \$750

b $\begin{aligned}\text{Commission} &= 5\% \text{ of } 20\,000 + 2\% \text{ of } 10\,000\\ &= 0{\cdot}05 \times 20\,000 + 0{\cdot}02 \times 10\,000\\ &= 1000 + 200\end{aligned}$
∴ commission = \$1200

EXERCISE 3F

1 Calculate the 7% commission on sales totalling:

a \$500 **b** \$1000 **c** \$1200 **d** \$4750 **e** \$35 000

2 If the commission in question **1** changes to $7\frac{1}{2}$% on all sales, recalculate the commission.

3 Calculate 4% commission on:

a $24 000 b $356.70

4 A real estate agent received 3% commission on the sale of a house for $280 000.

a How much was her commission?
b How much did the homeowner receive from the sale price after paying the agent's commission?

5 A direct realty firm offers a flat fee commission of $5000 instead of the usual 2·75% of the sale price of the property. It hopes to entice prospective sellers to use their firm instead of others. Copy and complete the table.

Sale price	Fee @ 2·75%	Flat fee	Saving
$400 000		$5000	
$500 000		$5000	
$600 000		$5000	
$700 000		$5000	
$750 000		$5000	

Level 2

6 Christine is offered a sales position in which she is to earn $350 per week plus 2% commission on all sales. How much would Christine receive in a week if she had total sales of:

a $1000? b $1450? c $5500? d $12 000?

7 Calculate Christine's yearly income if her total sales for the year came to $370 000.

8 Christine hopes to earn $40 000 each year. Calculate the total sales Christine is required to make for her goal to be reached.

9 A real estate agent earns commission of 5% on the first $200 000, 3% on the next $200 000 and 1% on any remaining amount. Calculate the commission earned on properties selling for:

a $150 000 b $200 000 c $300 000
d $400 000 e $375 000 f $475 000
g $729 500 h 1 million dollars

10 David earns $50 per day plus 3% commission on the amount he sells over $500 each day.

a Calculate his commission if his sales per day total:
i $300 ii $500 iii $2000
b How much does he need to sell for David's wage for the day to total $100?

11 Dheeraj is given a choice of packages when he joins a local hardware store:
Package A: $600 per week
Package B: $375 per week plus 2% commission on the value of all his sales
a How much will Dheeraj earn for a week where his total sales are $7000 with
i package A? ii package B?
b If Dheeraj chooses package B instead of package A, how much do his sales need to amount to each week for his choice of package to be financially sound?

3.7 Simple interest

Almost every individual and company *borrows* money, from banks, credit unions, building societies, etc. The borrower must expect to pay for the service (the use of someone else's money). This payment is called **interest**.

If you deposit money in a bank then you are *lending* your money to the bank, and it pays interest to you.

The amount borrowed is called the **principal**. Interest is quoted as a percentage of the principal. The *borrower* eventually pays back *principal + interest*. The *investor* receives back his principal + interest.

Simple or **flat rate interest** is charged *each year on the full amount borrowed* at the beginning of the loan.

A \$200 loan for 3 years at 9% simple would mean that each year's interest = 9% of \$200 = 0·09 × 200 = \$18. This calculation would be repeated three times.

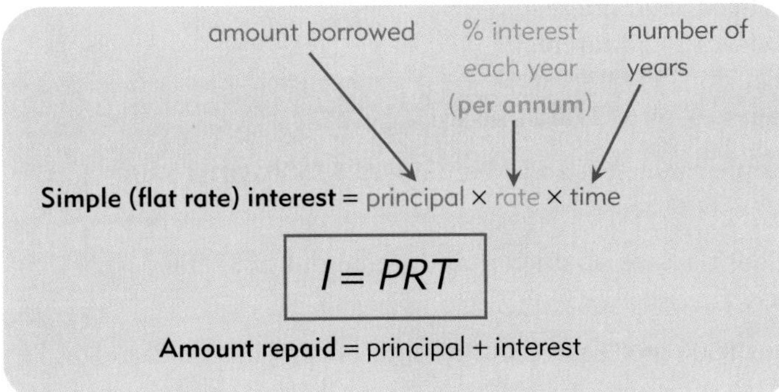

EXAMPLE 1

Mike borrows \$8000 for 3 years at 9% p.a., simple interest, to buy a car. Find:

a the interest charged **b** the total amount repaid

Solution

a $I = PRT$
$= 8000 \times 0{\cdot}09 \times 3$
$= 2160$
Interest = \$2160

b Amount repaid = \$8000 + 2160
= 10 160
Mike repays \$10 160.

EXAMPLE 2

Find the simple interest if \$7000 is invested at $6\frac{1}{4}$% p.a. for 18 months.

Solution

$I = PRT$
$= 7000 \times 0{\cdot}0625 \times \frac{18}{12}$
$= 656{\cdot}25$
Interest = \$656.25

18 months = $\frac{18}{12}$ years

EXAMPLE 3

Calculate simple interest rate if \$9000 is invested for 4 years earning \$1980 interest.

Solution

$I = PRT$
$1980 = 9000 \times R \times 4$
$1980 = 36\,000 \times R$
$= \frac{1980}{36\,000}$
$= 0{\cdot}055$
$\therefore$ interest rate = 5·5%

Remember to multiply by 100% to convert to a percentage.

EXERCISE 3G

1 Find the simple interest earned on:

a $5000 at 6% p.a. for 1 year
b $5000 at 6% p.a. for 3 years
c $120 at 3% p.a. for 3 years
d $4500 at 5% p.a. for 2 years
e $8000 at 4% p.a. for 5 years
f $10 000 at 2% p.a. for 5 years

Level 2

2 Calculate the simple interest earned on:

a $500 at 7% p.a. for 10 months
b $1000 at 3·5% p.a. for 10 months
c $2000 at 4% p.a. for 6 months
d $4500 at 3·75% p.a. for 15 months
e $50 000 at 5·99% p.a. for 30 months
f $1000 at 4% p.a. for 200 days

3 Find the simple interest rate if:

a $7000 invested for 4 years earns $700 interest
b $4000 invested for 5 years earns $500 interest
c $555 invested for 2 years earns $24.50 interest
d $780 invested for $2\frac{1}{2}$ years earns $50 interest
e $100 000 invested for 12 years earns $12 000 interest

4 Find the percentage rate of simple interest if an investment of $2500 earns $300 interest in 1 year.

5 What percentage rate of simple interest is charged if an investment of $7500 earns $1125 in interest over 3 years?

6 For how long must Jenna invest $800 at 4% p.a. flat rate interest in order to earn $64 in interest?

Investigation Percentage puzzle

How do mathematics teachers say goodbye?

Complete the questions below to unlock the puzzle.

R: Find 5% of 60. **L:** Write $33\frac{1}{3}$% as a fraction. **A:** Increase $56.50 by 7%.

A: Calculate $7\frac{1}{2}$% of $276. **O:** Write $\frac{13}{20}$ as a percentage.

C: Mark works 12 hours each day, 4 days a week. What percentage of his week is spent at work?

U: The Australian dollar rose from 39·22 English pence to 43·4. What was the percentage increase?

L: Decrease $27 by 3%.

C: What is $17\frac{1}{2}$% of $1500?

T: Calculate interest on $27 000 at 2% p.a. for $3\frac{1}{2}$ years.

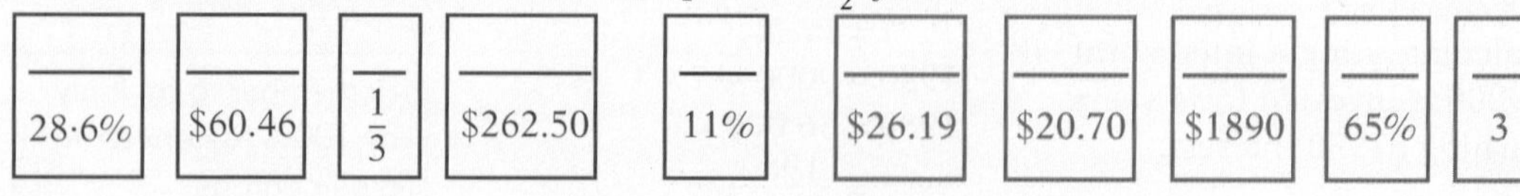

3.8 Finding an amount when given a percentage of it

Often we know the final price after an increase, but not the price before. By setting up a simple equation of the type dealt with last year, it is possible to calculate the price before the increase.

EXAMPLE 1

Find the value of x if:

a 5% of $x = 1970$ **b** $7\frac{1}{2}\%$ of $x = 56{\cdot}7$

Solution

a $0{\cdot}05 \times x = 1970$
$x = 1970 \div 0{\cdot}05$
$= 39\,400$

b $0{\cdot}075 \times x = 57{\cdot}6$
$x = 57{\cdot}6 \div 0{\cdot}075$
$= 768$

EXAMPLE 2

The Rockford family bought their house in Cronulla for \$675 000 from the Milburn family. If the Milburn family made a profit of 27%, what did the Milburn family pay for the house?

Solution

Selling price = 127% of the cost price
$675\,000 = 127\%$ of x
$1{\cdot}27 \times x = 675\,000$
$x = 675\,000 \div 1{\cdot}27$
$= 531\,496{\cdot}063$

$\therefore$ the Milburns bought the house for \$531 496.

EXERCISE 3H

1 Calculate the full amount if:

a 6% equals 27 **b** 2% equals 1 **c** 5% equals 10 **d** 12% equals 60
e 3% equals 2·7 **f** 9% equals 450 **g** 12% equals 96 **h** 50% equals 9
i 8% equals 480 **j** 24% equals 132 **k** $33\frac{1}{3}\%$ equals 96 **l** 15% equals 54

Level 2

2 Find the value of x (answer correct to 1 dec. pl. if necessary) if:

a 8% of $x = 90$ **b** 2% of $x = 9$ **c** 15% of $x = 19$
d 27% of $x = 486$ **e** 5% of $x = 100$ **f** 2% of $x = 4$
g 10% of $x = 134$ **h** 7% of $x = 21$ **i** 15% of $x = 400$
j $33\frac{1}{3}\%$ of $x = 3{\cdot}6$ **k** $66\frac{2}{3}\%$ of $x = 56$ **l** $12\frac{1}{2}\%$ of $x = 89{\cdot}7$

3 A car was sold for \$15 500, representing a loss of 28% on the cost price. Calculate the cost price of the car.

4 After an increase of 10% the cost of a can of soft drink became \$1.80. What was the price before the increase?

5 Edmund bought a second-hand textbook from Stephen for \$12.50. If Stephen lost 80% on the deal, what was the original cost of the textbook?

TEACHER

SPREADSHEET ACTIVITY

Chapter Review

Language Links

appreciation	cost	loss	principal
borrower	depreciation	marked	profit
buying	discount	per annum	selling
commission	interest	percentage	simple
composition	lender	price	

1 Write words from the list which mean:

- **a** the amount borrowed when taking out a loan
- **b** money gained in a transaction
- **c** a reduction in the marked price of an article
- **d** the amount invested each year
- **e** flat rate interest

2 Explain the meaning of:

- **a** per annum
- **b** simple interest
- **c** commission
- **d** percentage

3 Write the *opposite* of:

- **a** increase
- **b** sell
- **c** lend
- **d** profit
- **e** gain
- **f** deposit

4 Write a sentence describing the difference between appreciation and depreciation.

5 Write out a list of places and occupations where percentages are commonly used in society.

TEACHER

6 The working for a question was $0{\cdot}05 \times 4000$. Write a question that could explain this working. Be creative.

7 Present your question to a classmate. Do they obtain the same working and answer? **Discuss** any differences.

8 Write out a **procedure** that a classmate could follow to find the percentage profit of an article costing $\$x$ and sold for $\$y$.

Chapter Review Exercises

MC

1 Convert to a fraction:

- **a** 50%
- **b** 56%
- **c** 75%
- **d** $33\frac{1}{3}\%$

MC

2 Convert to a decimal:

- **a** 77%
- **b** 7%
- **c** 25%
- **d** $12\frac{1}{2}\%$

3 Convert to a percentage:

a $0{\cdot}05$ b $\frac{1}{5}$ c $\frac{3}{8}$ d $\frac{2}{3}$ MC

4 a What percentage of 60 is 40? 3.1
b Express 150 g as a percentage of $\frac{1}{2}$ kg.
c 1750 mm is what percentage of 2 m?
d A mark of 37 out of 45 represents what mark as a percentage?
e Express \$1.40 as a percentage of \$1.25.

5 Find: 3.2

a 7% of 555 b 6% of 900 c 12% of \$1.20 d 25% of 450
e 70% of 1 kg f $66\frac{2}{3}$% of 66 g $12\frac{1}{2}$% of \$180 h $\frac{1}{2}$% of 1 million

6 a Increase 97 by 10%. b Decrease 97 by 10%. 3.3
c Increase 682 by 3%. d Increase 9 km by 12%.
e Decrease \$17.90 by 5%. f Decrease \$900 by $12\frac{1}{2}$%.

7 Increase \$7520 by 8% and then decrease the result by 10%. 3.3

8 A discount of 15% was offered on all clothing. 3.5

a How much was a shirt marked at \$39.95 sold for?
b What was the price reduction?

9 A firm buys TV sets for \$865 each and sells them for \$1395. 3.4

a What profit is made on each sale?
b What is the profit percentage?

10 Ticketworld earns a commission of $1\frac{1}{2}$% of the value of all tickets sold for a rock concert. If 20 000 tickets were sold at \$59 each, what total commission was earned? 3.6

11 Airfares go up by 12% over the Christmas period. How much will a \$258 fare from Sydney to Brisbane return become at the increased rate? 3.3

12 Gina borrowed \$680 for a stereo at $7{\cdot}6$% p.a. simple interest over 15 months. 3.7

a What amount must she repay?
b How much is each of the equal monthly repayments?
c A later model was released at \$699. What percentage increase in price was this?

13 If 2% of x equals 15, find: 3.3

a 1% of x b 20% of x c 80% of x
d 5% of x e 50% of x f 100% of x

14 a If 5% of an amount is 240, find the full amount. 3.8
b If 10% of an amount is $7{\cdot}5$, find the full amount.

15 A French-polisher restores an antique table and sideboard. He sells them both for a total of \$8870. If this represents a profit of 45%, find the cost of the items before the restoration. 3.5

16 A laptop computer depreciates at a rate of 15% p.a. Find the value of a laptop computer purchased for \$4000, at the end of: 3.3

a 1 year b 2 years

TEACHER

Keeping Mathematically Fit

Part A: Non-calculator

1 Copy and complete: $\frac{4}{5}$ of ______ = 30 cm.

2 Find the value of $103 \times 943 - 3 \times 943$.

3 Find the HCF of 125 and 220.

4 Copy and complete: $3x -$ ______ $= 8x$.

5 Find the value of $^-9 + 6 \times {^-7} + (^-2)^2$.

6 Find $\sqrt{6 \times 9 \times 6}$.

7 Convert 4600 mm to metres.

8 Find the value of $\dfrac{16{\cdot}46 - 0{\cdot}42}{0{\cdot}2}$.

9 Find the value of $(1{\cdot}2)^2 - 0{\cdot}4 \times 0{\cdot}6$.

10 Find the value of x if $2^5 \times 2^7 \times 2^x = 2^{14}$.

Part B: Calculator

1 Find the value of $\sqrt{\dfrac{180{\cdot}4 - 6{\cdot}1^2}{4{\cdot}7}}$ (answer to 2 decimal places).

2 Find the value of $\sqrt[3]{^-125\,000}$.

3 Simplify:

a $6x - 2(x + 1)$ b $9xy \times 2x$ c $\dfrac{3ab}{6}$

4 Factorise:

a $12ab - 12$ b $12ab - 24a$ c $^-x^2 - 4x$

5 Find the value of x in:

a 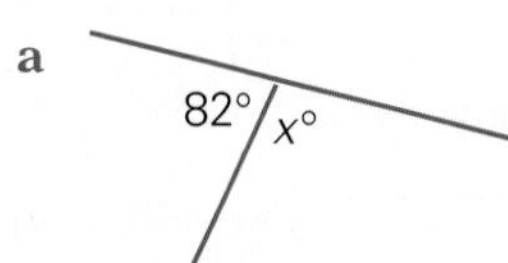

b 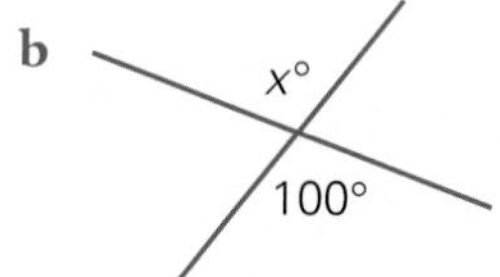

c 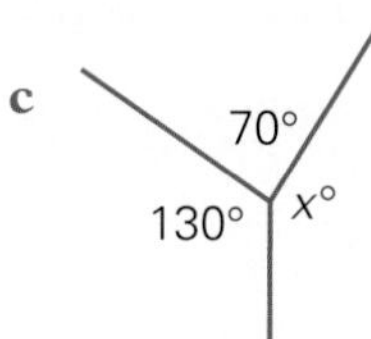

6 Find:

a $12\frac{1}{2}\%$ of 860 b 7% of \$96.40 c $\frac{1}{2}\%$ of 10

7 If 8% of a number is 6·4, what is the number?

8 The perimeter of a rectangle is 28 m. Find the dimensions of the rectangle that has the largest area.

TEACHER

Cumulative Review 1

Part A: Multiple-choice questions

Write the letter that corresponds to the correct answer in each of the following.

1 $16 \times 9 - 4 \div 2 =$

A 70 B 142 C 40 D 112

2 76 528 rounded to the nearest thousand is:

A 77 000 B 76 000 C 76 500 D 80 000

3 The largest of 2, $^-160$, $^-7$ and 4 is:

A $^-160$ B $^-7$ C 2 D 4

4 The temperatures in four cities are: Nagano $^-5$°C, Moscow $^-2$°C, Warsaw 0°C and Anchorage $^-10$°C. The coldest city is:

A Warsaw B Nagano C Anchorage D Moscow

5 $^-3 - 2 \times {}^-5 =$

A $^-13$ B 7 C 25 D $^-25$

6 The number $3\frac{7}{10}$ is:

A a proper fraction B a mixed number
C an improper fraction D an integer

7 The simplest equivalent fraction for $\frac{24}{39}$ is:

A $\frac{8}{13}$ B $\frac{2}{3}$ C $\frac{12}{15}$ D $\frac{4}{5}$

8 Gloria spends $\frac{3}{7}$ of her income of \$280 on clothes and saves the rest. She saves:

A \$120 B \$$\frac{4}{7}$ C \$160 D \$40

9 $36{\cdot}005 \div 1000 =$

A 0·036 005 B 36 005 C 0·360 05 D 36·000 005

10 Hakim wants to divide 1·65 by 0·5 without using a calculator. Which calculation would give him the correct answer?

A 165 ÷ 5 B 16·5 ÷ 5 C 165 ÷ 5 D 1·65 ÷ 0·05

11 $\dfrac{9{\cdot}6}{4{\cdot}7 + 8{\cdot}35}$ is approximately:

A 10·4 B 0·7 C 13·3 D $^-6{\cdot}3$

12 $5^3 \times 5^4$ is the same as:

A 5^{-1} B 25^7 C 5^7 D 1

13 $(3^4)^2$ is equivalent to:

A 9^8 B 9^6 C 3^8 D 3^6

14 $10^1 \times 10^2$ is equal to:

A $10 + 20$ B 10^2 C 1000 D 10 000

15 If $x = 6$ and $y = 2$, the value of $3(x - y)^2$ is:

A 48 B 144 C 38 D 24

16 $4x + 5 + 3x$ simplifies to:

A $7x + 5$ B $12x$ C $12 + x^2$ D $2x + 12$

17 $2a \times 3a$ simplifies to:

A $6a$ B $5a^2$ C $5a$ D $6a^2$

18 $4p(2t - 3)$ expands to:

A $6pt + p$ B $6pt - 7p$ C $8pt - 12p$ D $6pt - 12p$

19 $12x + 18$ factorises to:

A $2(6x - 9)$ B $^-6(2x + ^-3)$ C $^-6(^-2x + 3)$ D $6(2x + 3)$

20 10% discount off the marked price of $\$x$ gives a selling price of:

A $\$(x - 10)$ B $\$0{\cdot}9x$ C $\$\frac{x}{10}$ D $\$0{\cdot}1x$

21 Shirts formerly $\$s$ each increased in price to $\$d$. The percentage price increase was:

A $d - s$ B $\frac{d-s}{s}$ C $\frac{s-d}{100}$ D $\frac{d-s}{s} \times \frac{100}{1}$

22 5 cents from every sale of a \$2.95 hamburger goes to the Olympic team. If \$2 950 000 of hamburgers are sold, the money raised is:

A \$1 000 000 B \$5000 C \$5 000 000 D \$50 000

23 The simple interest earned on \$660 invested at 5% p.a. for 18 months is:

A \$49.50 B \$40.50 C \$594 D \$33

24 A Sydney tollway increased its toll by 200% from its previous \$1.20. The cost of the new toll is:

A \$2.40 B \$3.60 C \$240 D \$1.44

Part B: Short-answer questions

Show full working for each of the following.

1 Evaluate:

a 2^5 b 1^{20} c $3^5 \div 3^5$ d $\frac{2^3 \times 2}{2^6}$

e $\left(\frac{1}{3}\right)^4$ f $(0{\cdot}2)^3$ g $(^-4)^3$ h $\sqrt{81} \times 9^2$

2 Express as percentages:

a $0{\cdot}72$ b $0{\cdot}02$ c $1{\cdot}053$ d $\frac{3}{8}$

3 Express 1200 mm as a percentage of 2 metres.

4 What percentage is paid after a discount of $12\frac{1}{2}\%$ is given?

5 What percentage is paid after a price increase of 8%?

6 How much does a salesperson earn on sales of \$48 000 at the commission rate of 3%?

7 What is John's new salary if he previously earned \$450 per week and was given a rise of 5%?

8 Calculate **a** the interest **b** the amount repaid on a loan of \$15 000 over $2\frac{1}{2}$ years at 8% simple interest.

9 A VCR bought wholesale for \$250 was sold retail for \$489. What percentage profit or loss was the retailer making?

10 If $x = 3$, evaluate $x^2 - 3x$.

11 Simplify:

a $2n + 3 + n$	**b** $2x \times 3y$	**c** $3x + 2y - 9x$
d $3(x + 4) - 2x$	**e** $6a \div 3a$	**f** $p^3 + 4p^3$
g $\dfrac{12ab}{20a}$	**h** $3m^2 + 2m - m^2 + 3m$	**i** $4(x + 3) - 2x$

12 Expand:

a $3(b + 1)$	**b** $6(3 - 2k)$	**c** $(p + 4a)3p$
d $^-(2x - 3y)$	**e** $7(3 - 4a)$	**f** $x(x + 4) + x$

13 Factorise:

a $5x + 10$	**b** $6y - 6$	**c** $8xy - 6x$
d $7p - 14q$	**e** $^-x^2 - x$	**f** $p^2 + pq + 2p$

14 Simplify:

a $\dfrac{x}{7} + \dfrac{3x}{2}$	**b** $\dfrac{4a}{5} - \dfrac{a}{2}$	**c** $\dfrac{7x}{8} - \dfrac{3x}{4}$
d $\dfrac{a}{2} + \dfrac{a}{2} + \dfrac{a}{2}$	**e** $\dfrac{4w}{7} - \dfrac{w}{2}$	**f** $\dfrac{x + 1}{3} + \dfrac{x}{6}$

15 Simplify:

a $\dfrac{x}{y} \times \dfrac{x}{4}$	**b** $\dfrac{a}{5} \times \dfrac{5}{7}$	**c** $\dfrac{x^2}{2} \times \dfrac{4}{x}$
d $\dfrac{3}{4} \div \dfrac{a}{4}$	**e** $\dfrac{10}{a} \div \dfrac{5}{2a}$	**f** $\dfrac{x^2}{5} \div \dfrac{xy}{5}$

1 What you need to know and revise

Outcome DS 3.1:
Displays and interprets data in graphs with scales of many-to-one correspondence

- finds the average/mean for a small set of data
- picture graphs—draws and interprets
- column graphs—draws and interprets
- line graphs—draws and interprets
- divided bar graph and sector (pie) graphs—interprets

2 What you will learn in this chapter

Outcome DS 4.1:
Constructs, reads and interprets graphs, tables, charts and statistical information

- draws and interprets graphs
- chooses appropriate scales
- travel graphs
- uses lines for continuous data only
- reads and interprets tables, charts and graphs
- uses tally to organise data into a frequency distribution table

Working Mathematically outcomes WMS 4.1–4.5
Students will be asked to ***question***, ***apply strategies***, ***communicate***, ***reason*** and ***reflect*** in the sections of this chapter.

Graphs and tables

Key mathematical terms you will encounter

axis	data	score	tally
bar	frequency	sector	title
chart	graph	statistics	
column	picture	survey	
conversion	scale	table	

MathsCheck
Graphs and tables

1

Top of the charts: Mathsville week ending 20/6/2002	⊙ = 1000
Single title	**Sales**
1 Love Triangle	⊙⊙⊙⊙⊙⊙
2 Decagon Dance	⊙⊙⊙⊙◐
3 He's a Square	⊙⊙⊙⊙◐
4 Equilateral Blues	⊙⊙⊙⊙
5 Rhombus Rock	⊙⊙⊙
6 Don't Count on Me	⊙⊙◐

a Which was the most popular single? How many were sold?
b What approximation was used?
c Which had sales of 4000 or less?
d Which had sales of at least 4000?
e Draw the symbols for the 2170 sales of the number 7 single *The Circle of Life*.
f If the total sales of all singles for that week were 32 000, what percentage of sales were made by the top 6?

2

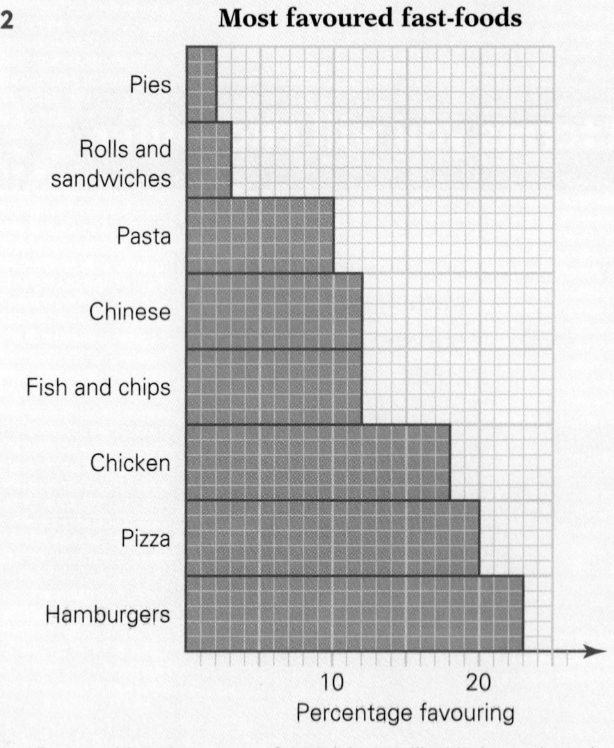

(*Source:* Year 8 survey of 100 Mathsville High students)

a Which food is most favoured? By what percentage of students surveyed?
b Which is the least favoured? What percentage favoured it?
c How much ahead in percentages is pizza over Chinese?
d What is the total of all the columns? Why *must* this be so?
e How many times more popular than sandwiches is chicken?
f What extra column could have been added to round out the survey better?

3 Expenses in staging a show

Theatre rental	Advertis-ing	Production costs	Agent's fees	Insurance	Artists' fees

a Which category makes up the greatest proportion of expenses?
b What section is the smallest?
c Which is greater, production costs or advertising?
d How long is the graph? How long is the theatre rental strip?
e What fraction of the chart is occupied by the theatre rental strip?
f If the total expenses for staging a show were $100 000, calculate the actual amounts spent on theatre rental and insurance.
g If tickets are sold for $25 each, how many must be sold for the promoter to cover all costs?
h What scale was used in constructing the graph?

4 The sector graph below shows the favourite colours in a class of students.

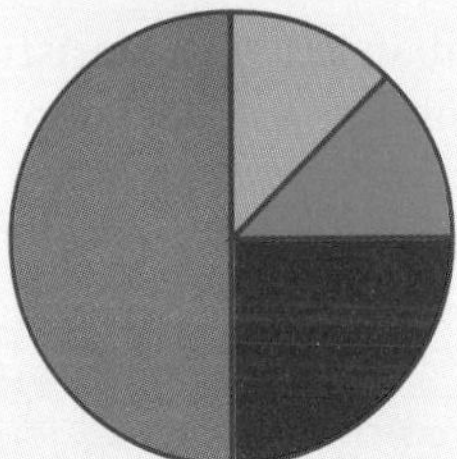

a Which two colours are equal in popularity?
b Which is the most popular colour? How can you tell this by looking at the graph?
c If the classroom is to be painted, which of the above colours would you choose and why?
d If the class has 28 students who voted, how many students choose blue as their favourite colour?

Investigation Graphs in the media

1 Collect advertisements, tables, graphs—any uses of statistics in newspapers, magazines, radio or TV, junk mail, etc. Use them progressively in classroom discussions and displays as you cover this chapter.

2 Find out how data is gathered and processed to come up with TV and radio program ratings, top 10 rankings in music, videos, books, etc.

Investigation Personal data sheet

Fill out a data sheet by listing your personal height, age, weight and so on. You can use this throughout the chapter.

Exploring New Ideas

4.1 Statistics and you

Every day in every way we are all bombarded with statistical information. Understanding the processes involved can help us to make wise decisions, when statistics are used to persuade us to believe something.

Statistics covers the process of *collecting, sorting, analysing* and *presenting* **data** (information). We then draw *conclusions* and make *predictions* and *decisions* based on the data.

COLLECTING, SORTING AND ORGANISING DATA

Before the data can be displayed in a graph, chart or table, it has to be collected. Data can be collected by taking a survey, by experiment, by observation or from records kept by various institutions.

Data is information.

A **survey** is a set of questions designed to discover information. It could be about your favourite colour, chocolate, the number of cars your family owns, how old you are or any other topic the surveyor is interested in.

Once the data is collected, it is often displayed in a frequency distribution table.

A **frequency distribution** table shows each item of data (called a score) with its **frequency** (number of times it occurs).

EXAMPLE

A class of 27 were surveyed about the pets they would most like to own. The data is given below.

D = dog
C = cat
B = bird
R = reptile
G = guinea pig
M = mouse
H = horse

D B R D C D M C D

C C D M C B D D C

B D C D M G D B C

a Organise the data into a frequency distribution table.
b Which pet is the most popular? The least popular?
c What fraction of all pets are dogs or cats?

Solution

a

Score	Tally	Frequency
D	~~||||~~ ~~||||~~	10
C	~~||||~~ |||	8
B	||||	4
R			1
G			1
M	|||	3
H	–	0
	Total	27

b Most popular: dog
Least popular: horse

c $\frac{\text{Dogs/cats}}{\text{Total}} = \frac{18}{27}$

$= \frac{2}{3}$

The highest frequency indicates the most popular. The lowest frequency indicates the least popular.

A 'gatepost' indicates 5 scores and helps us 'count' the frequency. ~~||||~~

EXERCISE 4A

1 a Copy and complete the frequency distribution table.
b What was Holly's most frequent score?
c How many times did she score more than 7?
d How often did she score 9 or better?
e How many scores altogether?

Holly's maths test scores:
6 9 7 9 10 7 7 7 8 7

Score	Tally	Frequency
6			1
7	~~||||~~	5
8		
9		
10		
	Total	

Level 2

2 Jamal surveyed his class on regular bedtimes, to the nearest half-hour. Here is the raw data he collected:

8:30	7:30	9:00	NST	8:00	9:30	8:30
8:00	9:00	8:30	NST	9:30	9:00	7:30
9:30	8:00	10:00	8:30	9:00	9:30	10:00
10:00	8:30	NST	9:00	9:30	8:00	7:00
NST	9:30					

(NST: no set time)

Score	Tally	Frequency
7:00		
7:30		
. .		
. .		
. .		
NST		
	Total	

a Complete a frequency distribution table like the one shown.
b How many children are in the class?
c What is the most common bedtime.
d How many regularly retired *before* 9:00 pm?
e How many regularly went to bed at 9:00 pm or later?
f What fraction of those surveyed had no set bedtime?

3 A random sample of 30 families was surveyed to find the number of school-age children in each. Here is the raw data collected:

2 0 3 1 1 2 1 3 0 4 3 2 1 1 2
2 1 0 1 2 3 2 1 3 1 0 1 2 0 1

a Sort the data into a frequency distribution table (check the total).
b Which score had the highest *frequency*?

c What was the highest *score*?
d Which score had the lowest frequency?
e Which was the lowest score?
f How many families had *fewer than* two school-age children?
g How many had *at least* three school-age children?
h What fraction of the families had two children of school age?
i What percentage of the families had fewer than three children?

Investigation Collecting data

Draw up frequency distribution tables for your class on:

a left-handed or right-handed
b hair colour
c eye colour
d TV sets per household
e favourite chocolate bar
f continent of birth
g favourite sport
h number of children in family

4.2 Column graphs, picture graphs and divided bar graphs

COLUMN GRAPHS

A **column graph** displays data using two axes and columns of equal width.

EXAMPLE 2

Display the data in example 1 about Mathsville College sporting choices in a column graph.

Solution

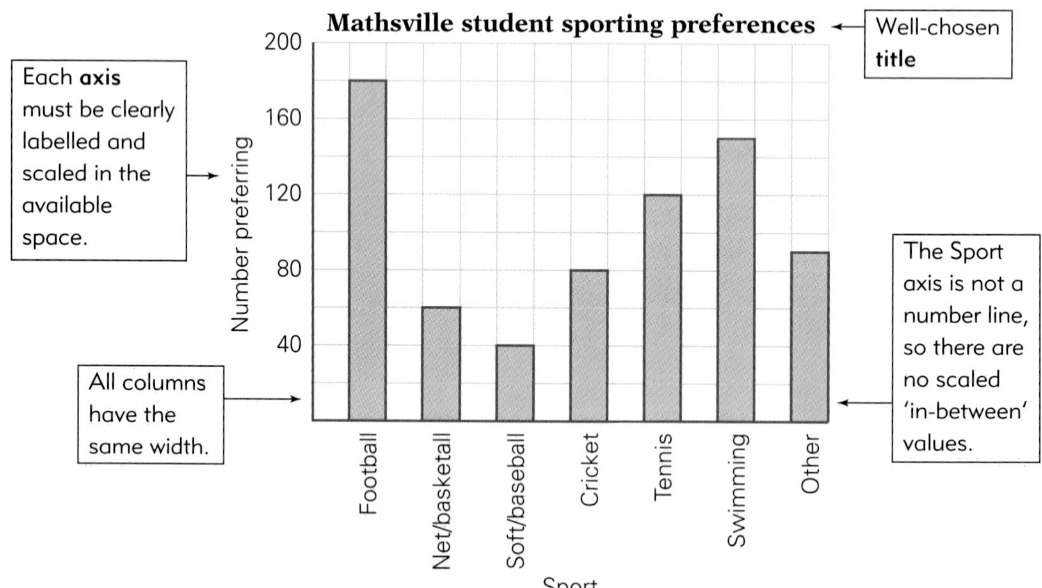

Always check scales carefully to find the *value of each subdivision* before taking readings. Axes could be interchanged to produce horizontal bars.

PICTURE GRAPHS

A **pictogram** or **picture graph** is probably the simplest type of graph. It conveys information by using symbols or pictures to stand for an amount of items.

EXAMPLE 1

Let's suppose we have taken a survey of the total population of our school, producing the following raw data. Display this as a picture graph.

Mathsville College student sport preferences

Sport	Football	Net/ basketball	Soft/ baseball	Cricket	Tennis	Swimming	Other	TOTAL
Number preferring	180	60	40	80	120	150	90	720

Solution

A graph enables us to *visualise* number relationships better than a table.

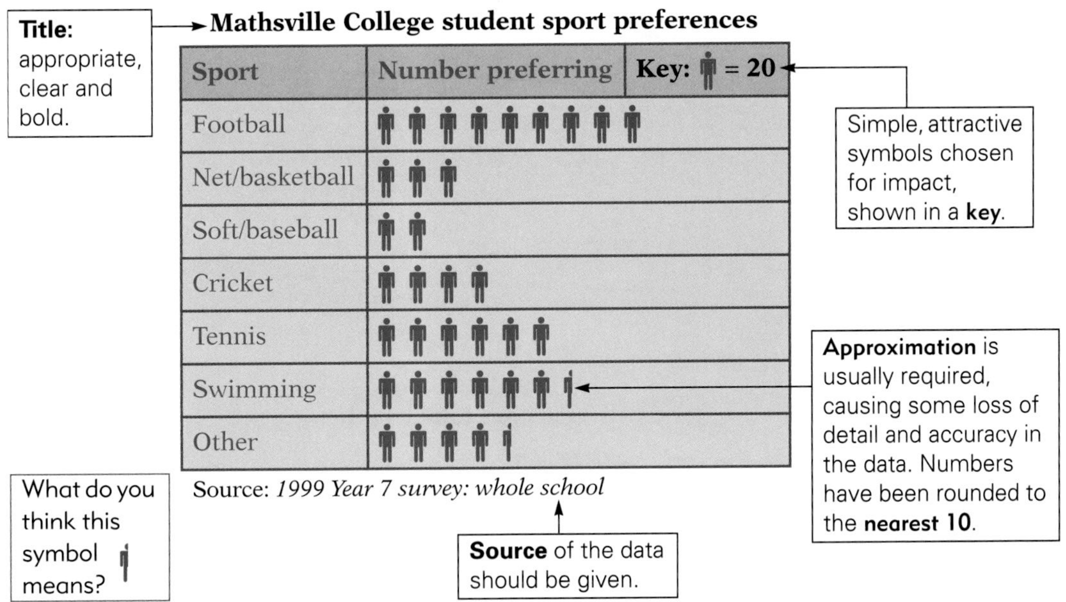

A DIVIDED BAR GRAPH

A **divided bar graph** is created from a rectangle or bar that is divided into parts. The size of each part is determined by what fraction or percentage it makes up of the whole rectangle's length. It has the advantage of comparison without taking up too much space.

EXAMPLE 3

Draw a bar graph for the sporting choices of Mathsville students.

Solution

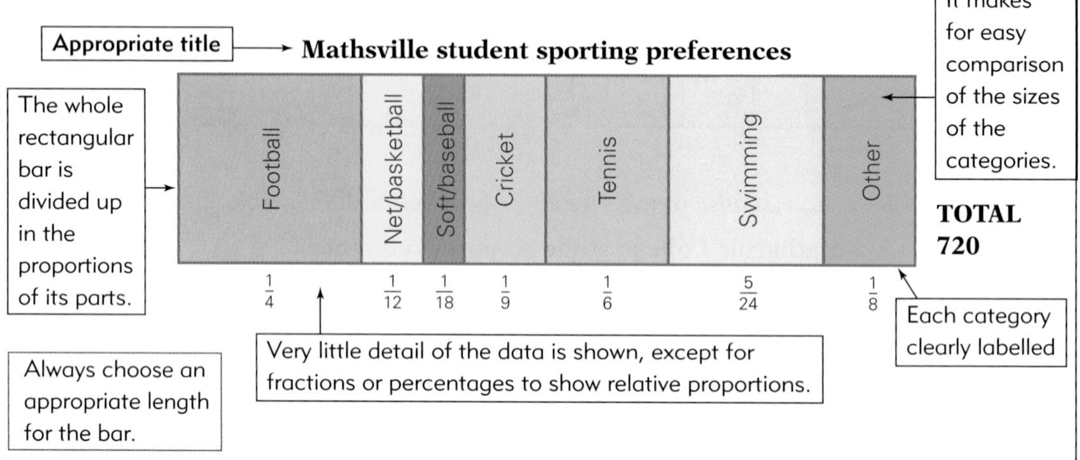

To find the fractions:

$$\text{e.g. football} = \frac{180}{720} = \frac{1}{4}$$

$\therefore$ football is represented by $\frac{1}{4}$ of the divided bar's total length.

EXERCISE 4B

1 a Construct a picture graph for the frequency distribution table supplied below. Use the symbol ⊙ to represent 4 households.

b How many households had more than 4 CD players?

c How many households had fewer than 2 CD players?

d What fraction of households surveyed had no CD players?

e How many households were surveyed?

CD players per household

Score	Tally	Frequency
0	||	
1	~~||||~~ |	
2	~~||||~~ ~~||||~~ ~~||||~~ |	
3	~~||||~~ |||	
4	||||	
Over 4	||	

2 Construct a pictograph to display the data shown in the table. Use a symbol such as 📺 to represent 50 households. Remember to show the essential features of title, key, source, well-spaced symbols. From the *pictograph*, answer the following:

a Which city has the most pay-TV connections per 1000 households? How many?

b Which city has the least connections per 1000 households? How many?

c How many more connections per 1000 has Brisbane than Adelaide?

d If the cities were listed in descending order of connections, what position would Adelaide be in? What position does it occupy in the table?

e Compare the *average* connection figures from the pictograph and the table.

f If Sydney has 1 200 000 households, how many are connected to pay TV?

Pay-TV connections per 1000 households	
Brisbane	175
Sydney	250
Melbourne	225
Hobart	75
Adelaide	100
Perth	150
Darwin	25
Source: *Australian Telecommunications Commission*	

Level 2

3 Estimate the value of each location with a letter name:

a 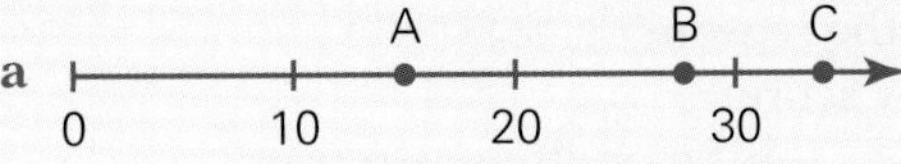

b 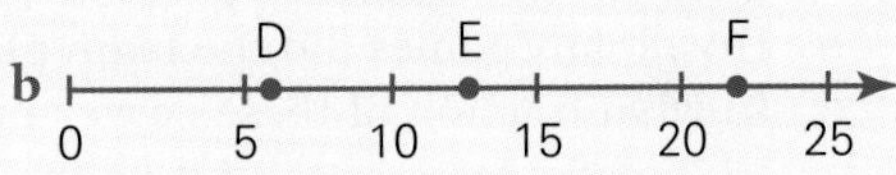

4 a What is the value of each subdivision?

b Give the value represented by each letter.

c Describe where you would locate a dot to show the value **i** 60 **ii** 75.

A B C
0 50 100 150

5 Scale each number line so that the largest number listed will be near the top. Locate each given number on the scale and show its letter.

Example:
A 12
B 39
C 25
D 7

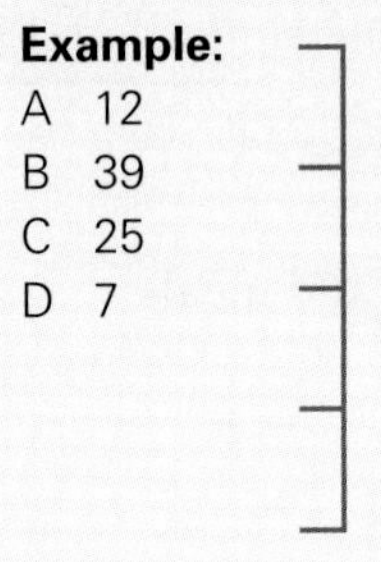

Solution:

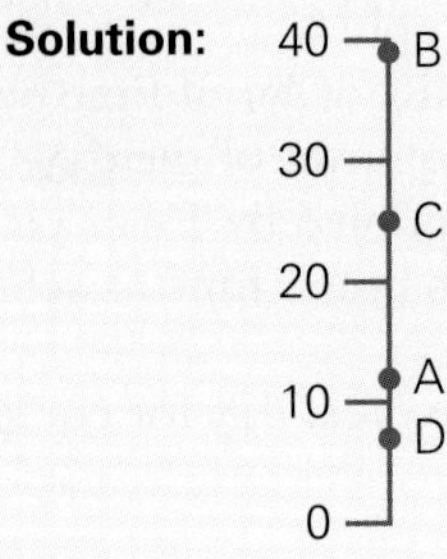

a A 12
B 18
C 14
D 3

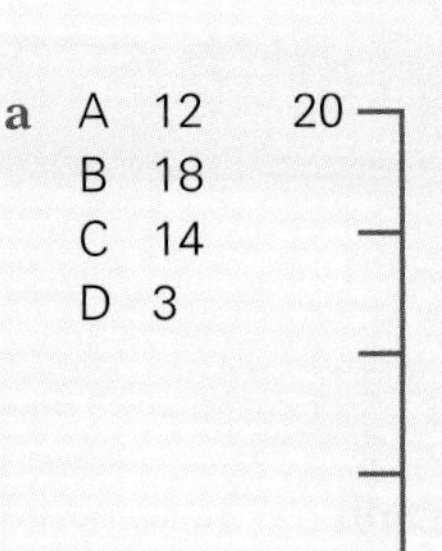

b A 60
B 30
C 75
D 13

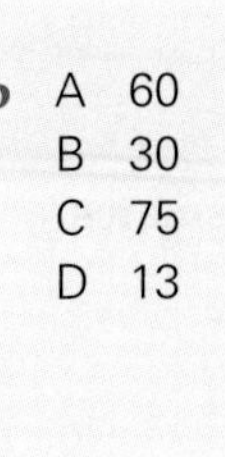

c A 300
B 50
C 450
D 375

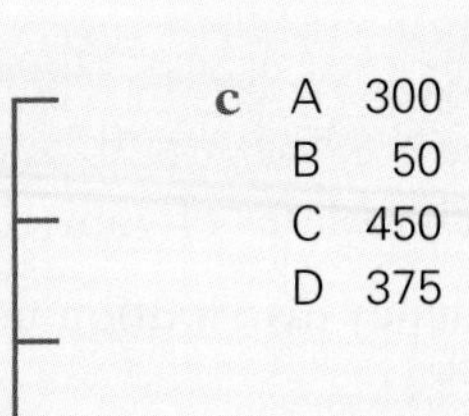

d A 95
B 20
C 55
D 38

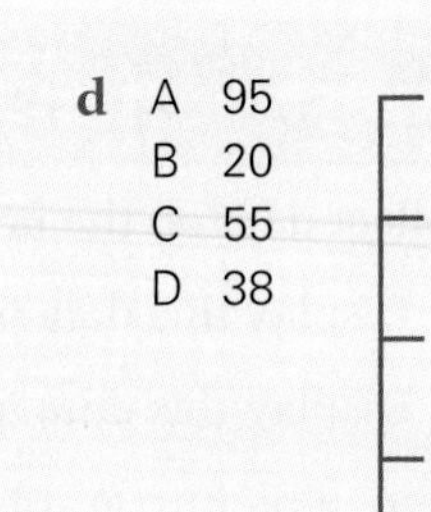

6

Animal gestation periods	
Species	**Days**
Mouse	20
Kangaroo	42
Tiger	105
Human	270
Camel	402
Elephant	645
Source: *Encyclopaedia of Animal Life*	

This table shows the length of time from conception to birth (*gestation* period) in various animal species. Draw a column graph to display the data. Follow this plan:

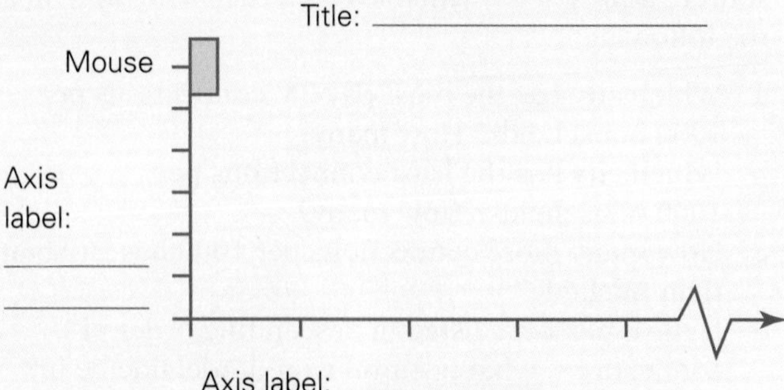

7 The tries scored by Mathsville's mixed Austag team are shown in the table below.

No. of tries	0	1	2	3	4	5
No. of games	1	3	6	6	4	2

a Construct a column graph to display this data.
b How many games did the team play in the season?
c In what fraction of these games did they score:
  i no tries? **ii** two tries? **iii** more than two tries?
d What was the average number of tries scored in the season?

8 Simon's regular weekday time allocation is shown in the table at right.

Sleep	8 h
School	6 h
Travel/meals	2 h
TV/play	3 h
Homework	1 h
Jobs/other	4 h

a Display his data as a bar graph. Give it a title and use a rectangular strip 12 cm by 2 cm to represent the whole day of 24 hours. (Scale is 0·5 cm = 1 h)

Answer the following from the graph:

b List the categories in descending order of time occupied.
c Which category takes up one-quarter of the day?
d What fraction of the day is occupied by TV or play?
e Which single category takes as much time as school, travelling and meals altogether?
f Draw up a bar graph of your typical day for comparison.

Investigation Chocolates

Collect data on the favourite chocolates of your classmates.

1 Display this data in a frequency distribution table.

2 Display this data in:

a a picture graph **b** a column graph **c** a divided bar graph

3 Write a paragraph comparing and contrasting the different graphs.

4.3 Sector graphs

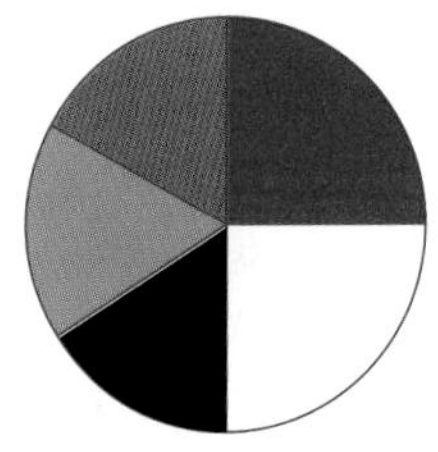

Like the bar graph, **sector** or **pie graphs** have the advantage of displaying data without taking up a lot of space. Also, like the bar graph, the sectors are decided by the fraction of the whole. In this case, each category is a *fraction of 360°*.

EXAMPLE

a Find the required angle for each of the sporting choices for Mathsville.

b Draw a sector graph (radius 2 cm) and display this data.

Mathsville College student sport preferences

Sport	Football	Net/ basketball	Soft/ baseball	Cricket	Tennis	Swimming	Other	TOTAL
Number preferring	180	60	40	80	120	150	90	720

Solution

a Finding the angles:

- football $= \dfrac{180}{720} \times 360°$ $= 90°$
- cricket $= \dfrac{80}{720} \times 360°$ $= 40°$
- swimming $= \dfrac{150}{720} \times 360°$ $= 75°$
- net/basketball $= \dfrac{60}{720} \times 360°$ $= 30°$
- tennis $= \dfrac{120}{720} \times 360°$ $= 60°$
- other $= \dfrac{90}{720} \times 360°$ $= 45°$
- soft/baseball $= \dfrac{40}{720} \times 360°$ $= 20°$

b **Mathsville student sporting preferences** ← Appropriate title

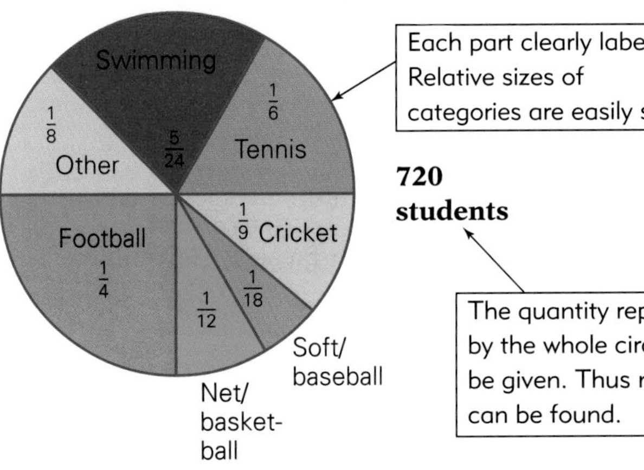

Each part clearly labelled. Relative sizes of categories are easily seen.

720 students

The quantity represented by the whole circle should be given. Thus numbers can be found.

EXERCISE 4C

1 Copy and complete the table:

Sector (slice) central angle	180°	120°	90°	60°	45°	40°	72°	270°
Fraction of 360°	$\frac{180}{360}$							
Simplest fraction	$\frac{1}{2}$							

2 Refer to the TV programs sector graph to answer the following:

TV programs, 5 pm – 11 pm

Police dramas
Movies
News/ current affairs
Comedies
Quiz shows

- **a** Which sector occupies the most time?
- **b** Which slice covers the least time?
- **c** Which parts run for equal times?

Level 2

3 Refer to the sector graph in question **2**.

- **a** How many minutes are represented by the entire circle?
- **b** Measure the central angle of the 'police drama' sector. What fraction of the circle does it represent?
- **c** What fraction of the time is occupied by News/current affairs?
- **d** How many minutes did the movies run for?

4 **Rugby League Football Club Income**

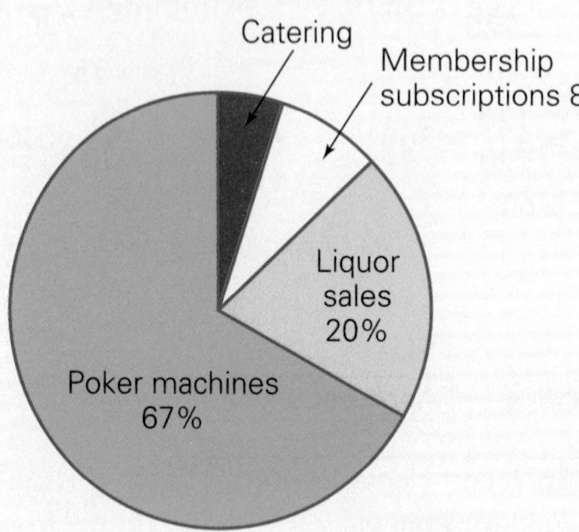

- **a** What is the club's major source of income?
- **b** What does the club make least money from? Why do you think this is?
- **c** Measure the central angle of the 'poker machines' sector. What fraction of the circle does it occupy?
- **d** Similarly, find the fraction of the circle taken up by liquor sales.
- **e** If the annual club income totals \$7 200 000, how much is received from membership subscriptions?
- **f** Sketch a sector graph to show how you think such a club's income slices would change if poker machines were banned.

5 A class of 120 students were asked to choose their favourite flavoured ice-cream. The results were:

chocolate: 60 strawberry: 30 vanilla: 20 banana: 10

- **a** Find the angle required by each sector.
- **b** Draw a sector graph of radius 4 cm to show the data above.

6 Draw a sector graph using the following information:

Survey of different eye colours

Eye colour	Blue	Brown	Green	Hazel
No. of students	15	12	8	5

7 Draw up a similar sector graph for the eye colour of the students in your class.

8 This sector graph shows how a family's annual budget is distributed.

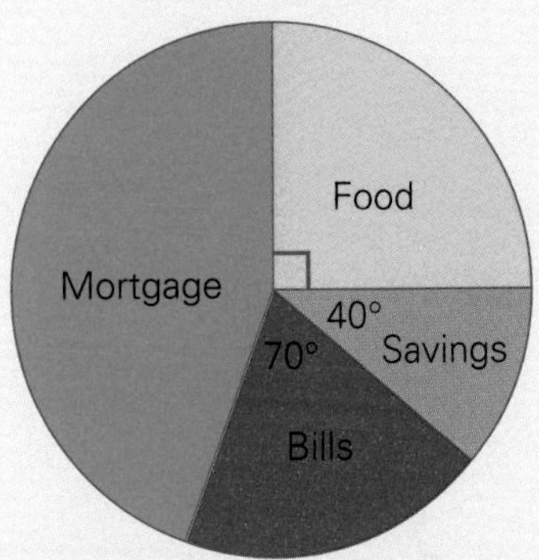

a What is the angle measure for the sector marked Mortgage?

b What fraction of the family's annual budget is spent on food?

c If the family saves $9000 per year, calculate:

i how much is spent on food

ii how much the mortgage is per year

iii the total of the family's annual budget

Investigation The M & M graph

You will need: packets of multicoloured Smarties or M&Ms, cardboard, graph-drawing materials

1 In groups of three or four, tally the different colours in your box of chocolates in a frequency table at the top corner of your cardboard sheet.

2 Use this data to build a display of pictograph, column graph, bar graph and sector graph for the classroom walls.

3 Don't eat your results until you tally!

4.4 Line graphs

Line graphs are commonly used because they are often more accurate and detailed than other graphs looked at so far. They involve straight lines, curves or a section of straight lines. They have a horizontal and vertical axis with different scales and units.

Line graphs are used when the data collected is quantitative and continuous; that is, when the data is numerical with the full range of values being possible, e.g. temperature, mass and height.

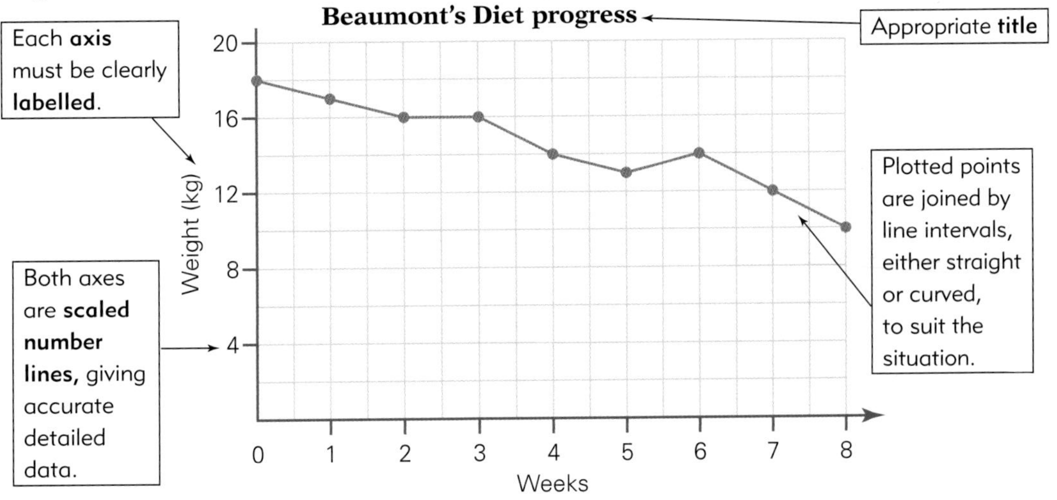

The line is continuous—we can read off information *between* the plotted points.

EXAMPLE

Display this data about the height of a weed over 6 weeks.

Week	0	1	2	3	4	5
Height (mm)	0	4	6	7	9	16

Solution

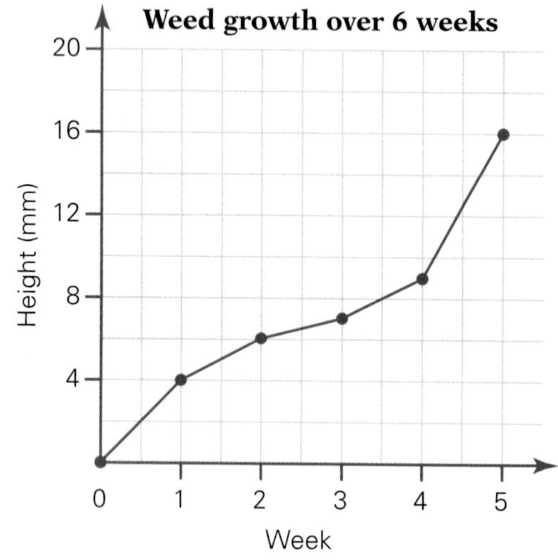

EXERCISE 4D

1 Refer to the graph of Beaumont's diet progress on page 97 to answer the following:

a How heavy was Beaumont at the start of her diet?
b How much weight had Beaumont lost over the 8-week period?
c What was the greatest amount of weight lost in any one week?
d During which week did Beaumont gain weight?
e In which week was Beaumont's weight stable? How does the graph show this?
f What was Beaumont's average weight loss per week over the 8 weeks?
g The line showing weight loss in weeks 7 and 8 is straight rather than broken, as in weeks 4 and 5. What does the straight pattern show?
h What did Beaumont weigh halfway through week 4?
i At the end of which week did Beaumont weigh 13 kg?

2 Use the graph to complete the table below.

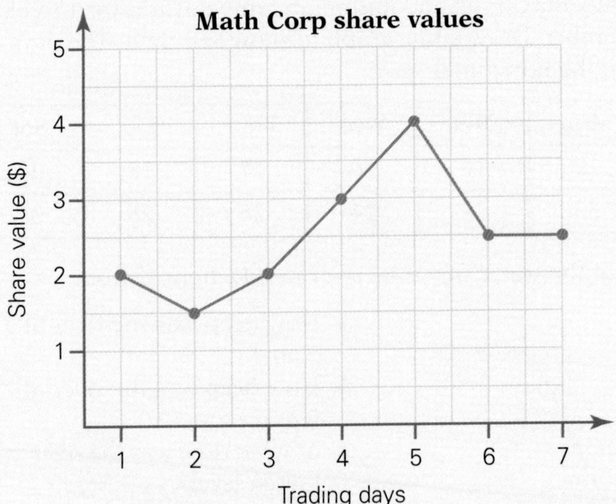

Trading day	Price ($)
1	2
2	
3	
4	
5	
6	
7	2.50

3 a What do the rising intervals *AB* and *BC* show is happening in the bath?

b How many litres of water were in the bath, as shown at point *C*?

c What happened from *C* to *D*? What could have caused this?

d What does the horizontal interval *DE* show about the volume of water in the bath?

e How long did the bath take to fill? to empty?

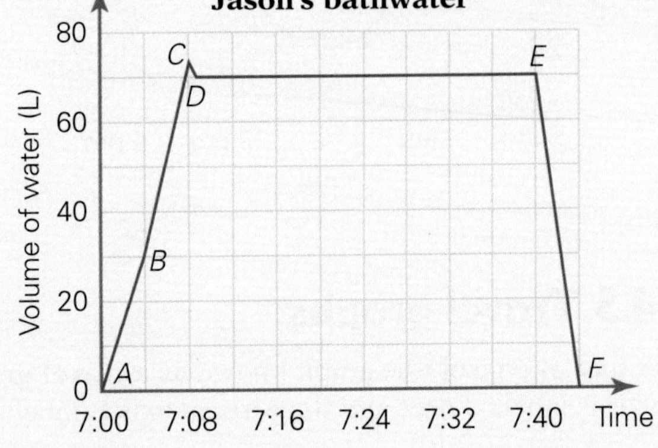

f How much water was in the bath at **i** 7:04 **ii** 7:28 **iii** 7:42?

g How long does it seem Jason was in the full tub for?

h Jason turned on just the hot water at 7:00 pm. At what time did he turn on the cold as well?

i Why is the slope of interval *BC* steeper than that of *AB*?

j What does the even slope *EF* show?

k At what rate in L/min did the water drain out of the tub?

Can you write a story for this graph?

4 Draw line graphs to display the following information.

a

Time	midnight	4 am	8 am	12 pm	4 pm	8 pm	midnight
Temp °C	10	8	16	24	22	16	14

b

Month	Jan	Feb	Mar	Apr	May	Jun	Jul	Aug	Sep	Oct
Profit	50	40	30	30	24	10	50	40	60	20

Level 2

5 The table below shows the daily maximum and minimum temperatures for 1 week in Sydney in the month of November. Draw a line graph to show this data. The days of the week should be drawn on the horizontal axis.

Day of week	Sun	Mon	Tue	Wed	Thur	Fri	Sat
Min temp °C	18	19	16	16	17	18	24
Max temp °C	30	33	28	24	26	26	31

6 This graph shows the depth of the water in a tidal river in a 12-hour period.

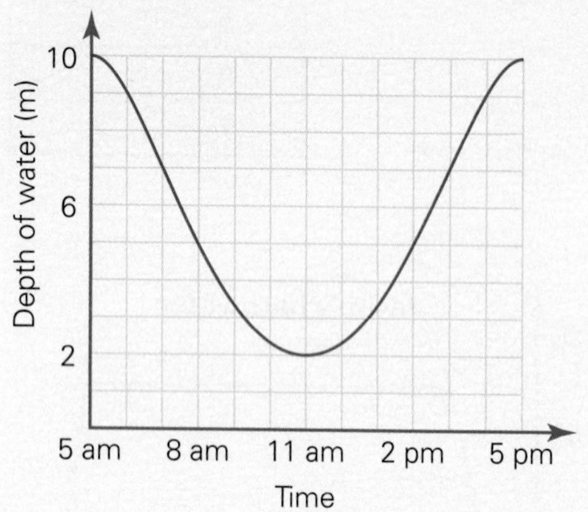

a How deep was the river at 5 am?

b How deep was the river at 5 pm?

c At what time was the river at its lowest level?

d What was the average depth? Explain.

e A boat needing 6 m depth wants to enter the river. Between what times shown is this not possible?

4.5 Travel graphs

A special type of line graph known as a **travel graph** is used for journeys. The horizontal axis is labelled *time* and the vertical axis is labelled *distance*.

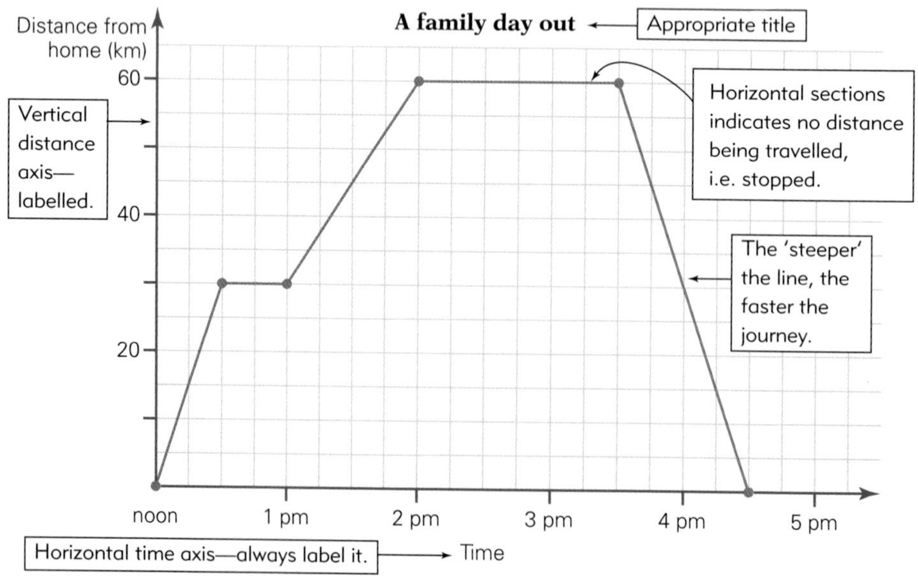

EXAMPLE

Refer to the travel graph 'A Family Day Out' and answer the following questions.

a At what time did the journey begin? **b** Between what times did the family stop?

c How far did the family travel in the day? **d** At what time did the family return home?

Solution

a 12 noon

b 12:30 pm – 1 pm and again at 2:30 pm – 3:30 pm.

c 60 km there and back. Total distance travelled = 120 km.

d 4:30 pm

EXERCISE 4E

1 Three friends, Anna, Billy and Cianne, travel the 5 km from school to the local library. Their journeys are displayed in the three graphs below.

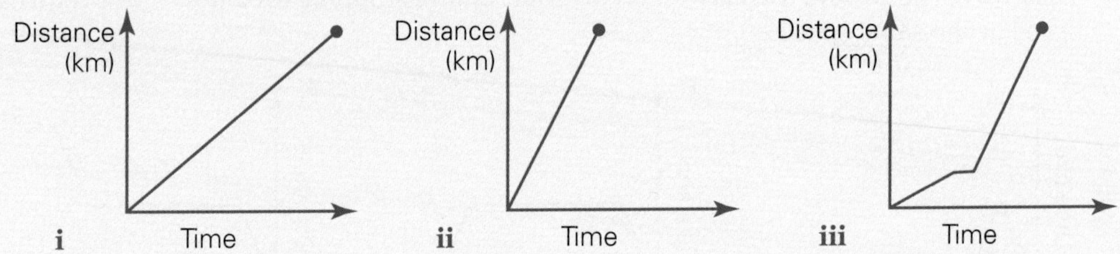

a If Anna walked a short distance before getting picked up by her Mum, which graph represents her trip?

b If Cianne arrived at the library last, which graph best represents her journey?

c Which graph represents the fastest journey? Explain your answer.

2

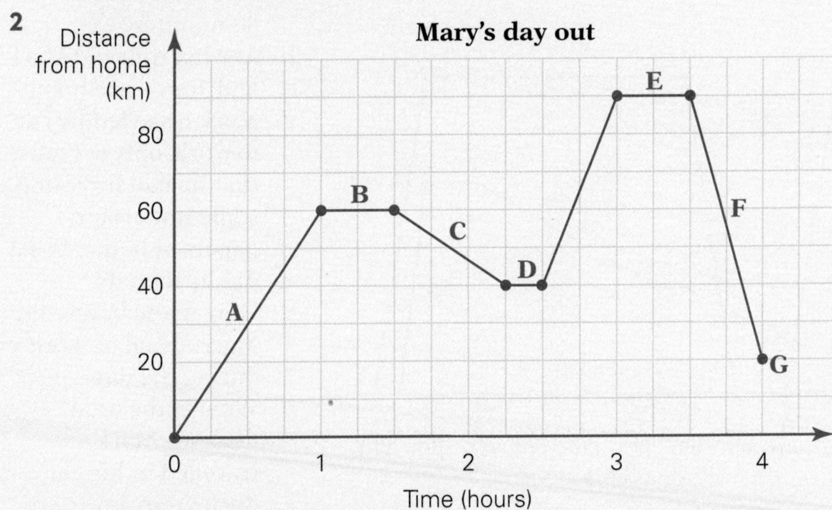

Choose the letter/s from the graph that correspond to the following:

a the fastest part of the journey

b the first time that Mary stops

c a 15-min break

d a speed of 60 km/h

e Mary's stops along the way

f 20 km from home

3 Write a story that could explain the graph of Mary's journey in question **2** (make reference to her speed, the kilometres travelled and times of each stop).

Level 2

4 Choose the graph that best describes the speed of a car as it is driven around a sharp bend in the road.

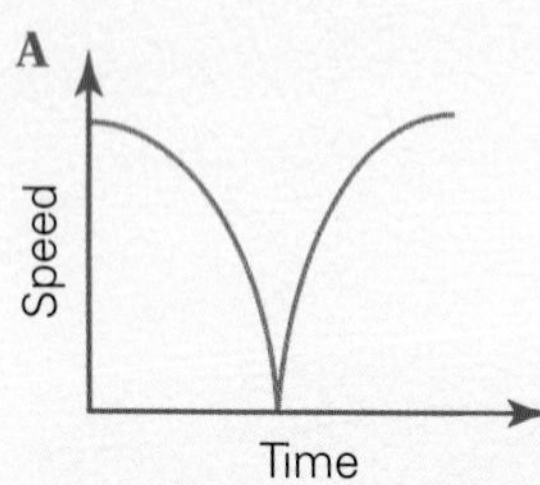

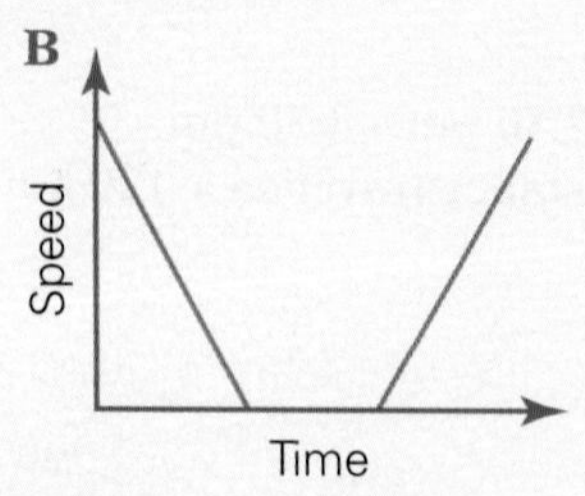

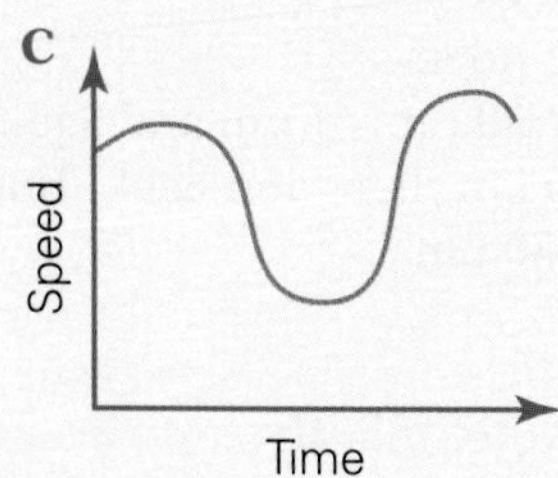

5 Choose the graph that best describes the following:

'Rose leaves home and travels for half an hour. Before stopping for 2 hours, she returns home at the same speed.'

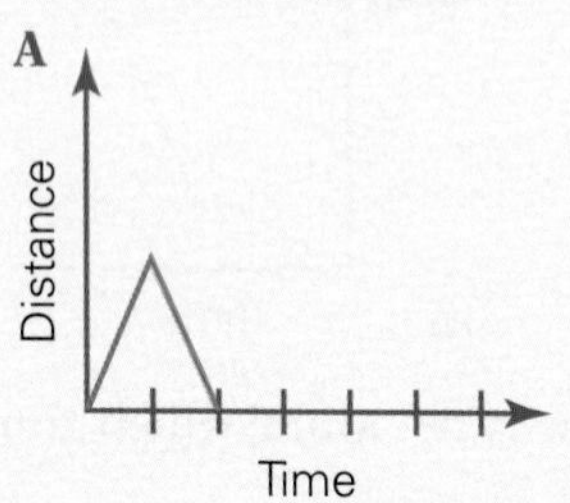

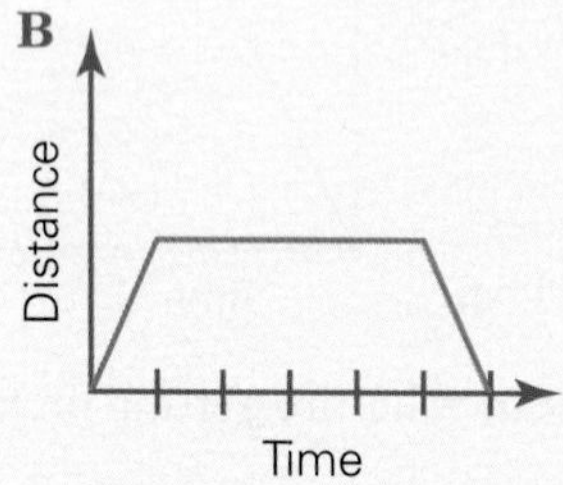

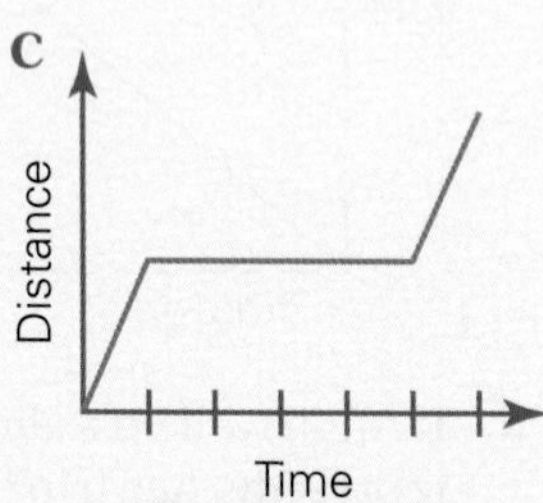

6

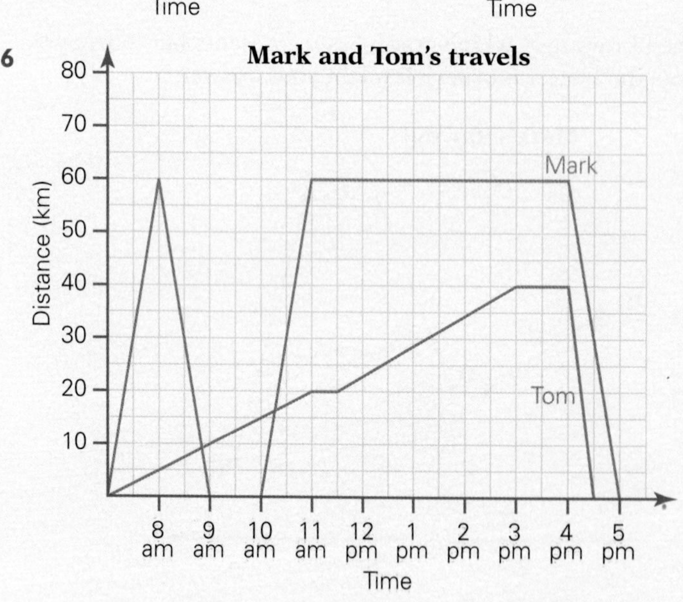

a At what time did Tom take a break for 30 minutes?

b At what time did Mark and Tom leave home?

c Mark travelled by car to work, only to realise that he had forgotten some important papers at home. What did he then do?

d How many hours did Mark spend at work during the day?

e What is the total distance Mark travelled in his car during the day?

f What was the total time Mark spent in his car during the day?

g Tom left home at 7 am and rode his bike 20 km. He then stopped for half an hour. How far did Tom ride his bike before stopping again?

h For how long did Tom stop the second time?

i Between what times did Tom travel the fastest?

j What could explain the speed of Tom's return journey?

k Calculate the speed of Mark's trip to work.

l What does it mean when the lines intersect (cross)? Discuss.

7 a Display the following as a travel graph.

Time	7:30 am	7:45 am	8 am	9:15 am	2 pm	2:30 pm	4 pm
Distance from home in km	0	15	15	75	75	90	0

b Calculate the speed of the return journey from 2:30 pm to 4 pm.
c For how long was this traveller away from home during the day?
d Write a story that could relate to the graph you have just drawn.

4.6 Conversion graphs

As the name suggests, a **conversion graph** is used to show a direct relationship between two variables. For example, it is useful in showing the relationship between currencies.

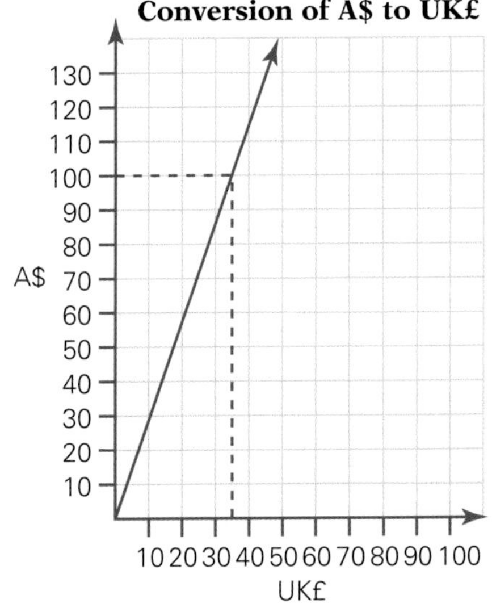

As with all other line graphs, axes are labelled and the graph is suitably titled.

EXAMPLE

a A person wishes to convert A$100 to pounds. How many pounds will this buy?
b Express this conversion as an exchange rate for A$1.
c Convert 10 pounds to Australian dollars.

Solution

a \$100 Australian = 35 pounds
b \$100 Australian = 35 pounds
$\div 100 \qquad \div 100$
\$1 Australian = 0·35 pounds
c $10 \div 0{\cdot}35 = 28{\cdot}571\ \ldots$
10 pounds converts to \$28·57 Australian.

$$\frac{10}{0{\cdot}35} \rightarrow \frac{1000}{35}$$

$$35\overline{)1000{\cdot}00} = 28{\cdot}571$$

EXERCISE 4F

1 This is a **conversion graph** showing the changeover from Australian money to US currency. You can see that \$20 Australian would be exchanged for US\$10.

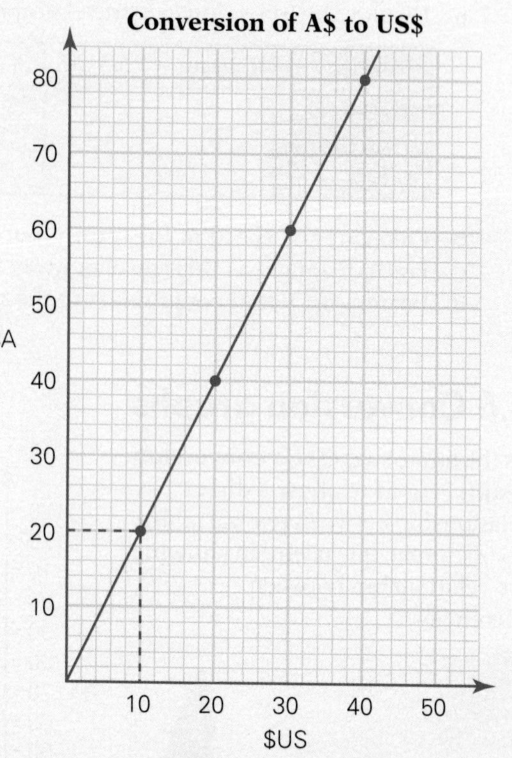

- **a** What is the scale used on each axis?
- **b** What is the value of each subdivision?
- **c** Convert to US\$:
  - **i** A\$40 **ii** A\$60
- **d** Change to \$A:
  - **i** US\$60 **ii** US\$10
- **e** Convert 700 dollars of Australian currency into US dollars.

2 The conversion graph shown converts degrees Celsius to degrees Fahrenheit.

- **a** What temperature in degrees Fahrenheit does 50°C correspond to?
- **b** What temperature in degrees Celsius corresponds to 100° Fahrenheit?
- **c** How many degrees Fahrenheit correspond to the temperature decrease of 35° to 15°C?

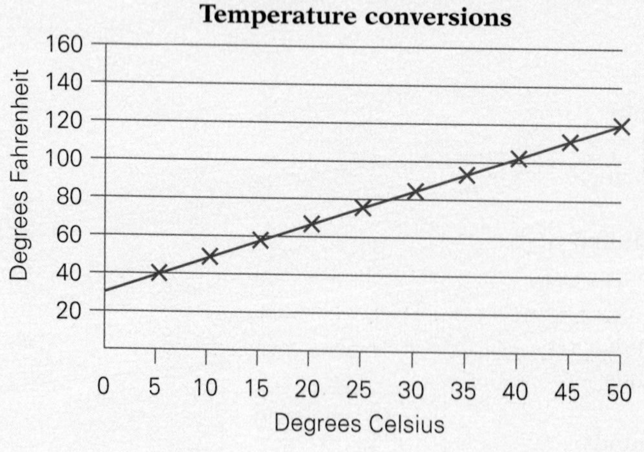

Conversion graphs are sometimes used in subjects such as science and economics.

3 The following graph shows the conversions from kilometres to miles.

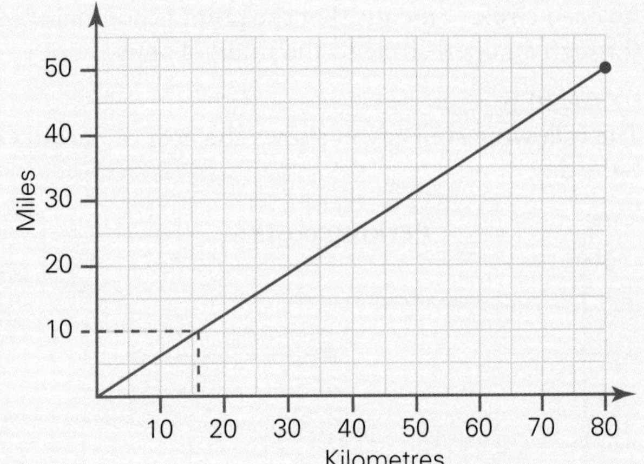

a How many miles correspond to 80 km?

b How many miles correspond to 40 km?

c How many kilometres does a car travel if it goes 20 miles?

d A speed of 60 km/h converts to how many miles per hour?

e How many miles does each kilometre represent?

Level 2

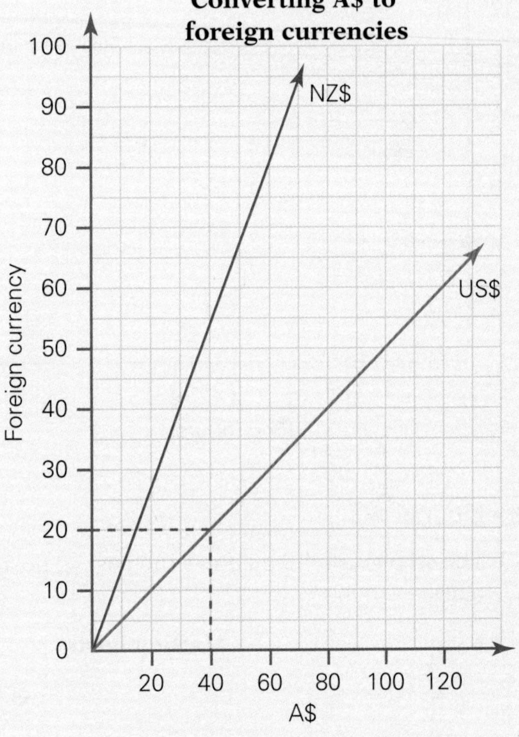

4 a Which axis represents the Australian dollar?

b Convert A$40 to:

i US$ **ii** NZ$

c A New Zealander arrives in Sydney with NZ$80. How many Australian dollars does this convert to?

d How much does NZ$800 convert to in A$?

e Convert US$50 to:

i A$ **ii** NZ$

f Convert US$100 to A$.

g Convert US$30 to NZ dollars.

h What rule can be used to express Australian dollars in:

i US dollars? **ii** NZ dollars?

i Use a graphics calculator to sketch these conversion graphs. Find how much $2.75 Australian converts to.

5 One Australian dollar converts to four South African rands. Display this as a conversion graph like the previous ones seen in this exercise. (A graphics calculator may be used.)

4.7 Step graphs

A **step graph** is another type of line graph that is made up of a set of lines (instead of one continuous line). These lines have definite breaks between them. It makes the overall graph appear like a series of steps. For example, they are useful to show graphing parking costs and mailing costs.

The end of one step is on the same vertical line as the beginning of the next step.

A *coloured* circle ● means that the point (cost) is included.

An *open* circle ○ means that the point is not included. Like an open doorway, you can travel through an open circle to the interval above.

EXAMPLE

The following step graph shows the cost of parking at an inner city parking station on a Saturday.

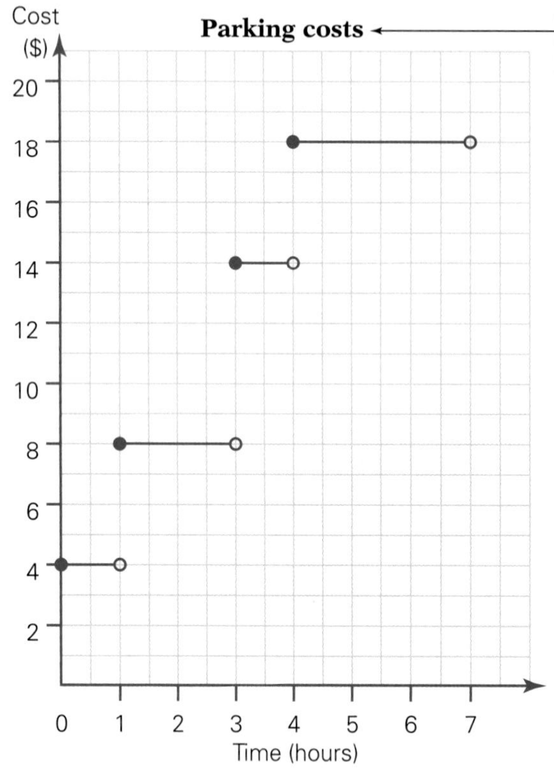

a What is the cost of parking for 30 minutes?

b What is the cost of parking for 3 hours?

c For how long can a person park for $10?

Solution

a The cost for 30 minutes of parking is $4.

b The cost of parking for 3 hours is $14.

c For $10 a person can park for under 3 hours.

EXERCISE 4G

1 The step graph here shows the cost of posting a parcel within the state based on its mass.

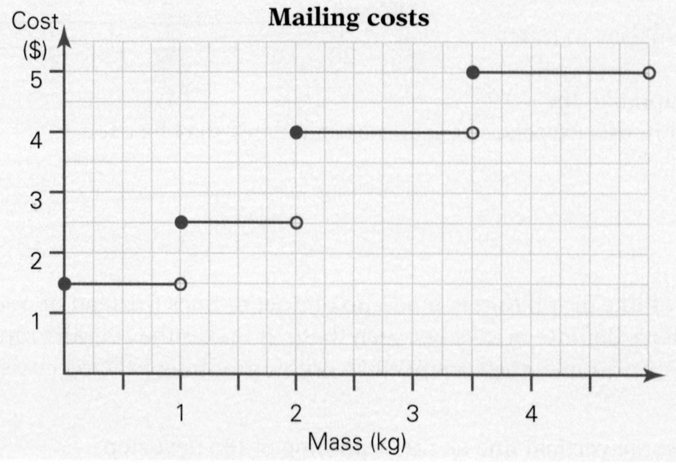

a What is the cost of mailing a package weighing 500 g?

b What is the cost if the package weighs 1 kg?

c What is the cost of sending a parcel with mass **i** 3000 g **ii** 3·5 kg?

d What is the heaviest package that can be sent for a maximum cost of $4?

2 An alternative courier service has the following charges for delivering packages, within the state, based on mass.

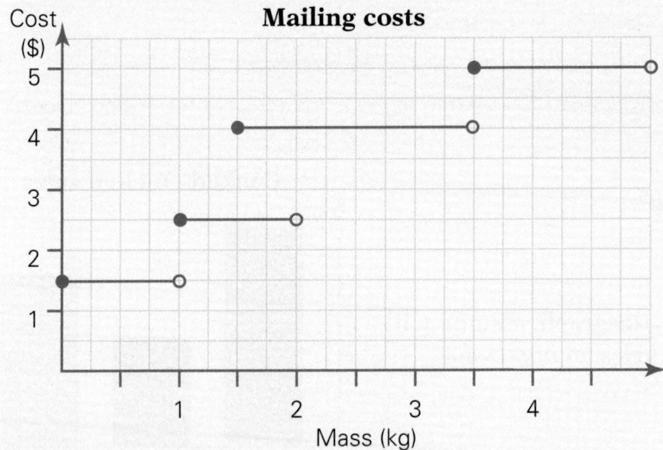

a What is the cost of mailing a package weighing 500 g?

b What is the cost if the package weighs 1 kg?

c What is the cost of sending a parcel with mass **i** 3000 g **ii** 3·5 kg?

d What is the heaviest package that can be sent for a maximum of $4?

e What is the saving for mailing a 2 kg package with the alternative courier rather than the company in question **1**?

f What is the cheapest way to mail two items weighing 1 kg each: separate 1 kg packages or a combined 2 kg package?

Level 2

3 Copy and complete the step graph below for the parking costs shown in the table.

Total number of hours	Cost ($)
1 hour and under	$1.50
2 hours and under	$2.00
3 hours and under	$4.00
4 hours and under	$6.00
Over 4 hours	$10.00

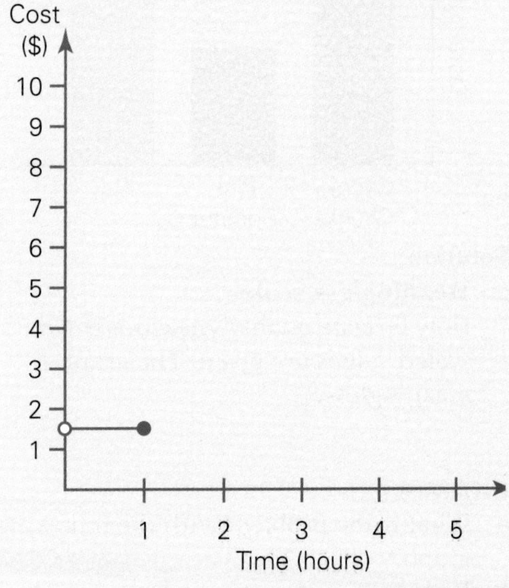

4.8 Graphs that mislead

Graphs can be used to persuade us to accept a certain belief—in a product, a politician's promises, or a service. Advertising is the lifeblood of today's media. You are its target. There may be deliberate or accidental deception in the graphs of the data.

GRAPH EVALUATION CHECKLIST

- Are the scales clearly *marked,* with *values shown?*
- Are the *axes labelled* with words of *clear meaning?*
- Are the scales *uniform,* starting at a suitable value, and *unbroken?*
- Are enlarged areas (2D) or *volumes* (3D) *used to magnify the effect* of the scaled values?

EXAMPLE 1

Why is this graph misleading?

Solution

No scale shown

We don't know the numbers; the graph, assuming it's based on accurate data, just gives an *impression*.

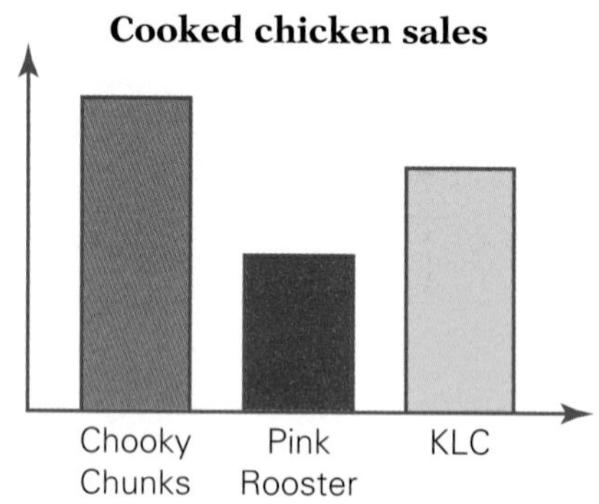

EXAMPLE 2

Why are the following graphs misleading?

a

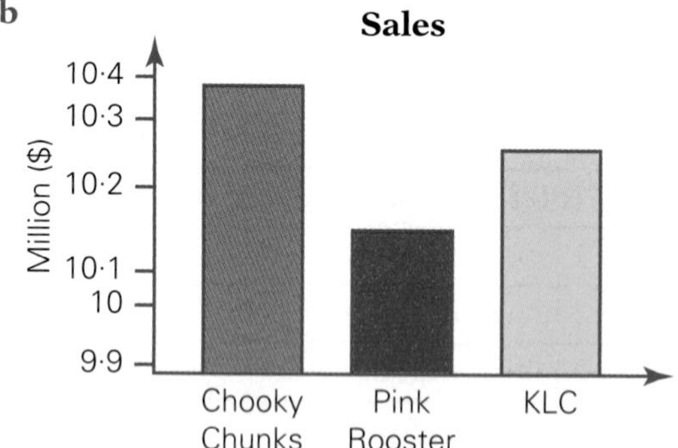

b

Solution

a **Meaningless scale**
How is 'taste quality' measured? No scaled values are given. The graph is meaningless.

b **Irregular or distorted scale**
Notice how the vertical axis does not start at zero, and scale divisions are irregular to distort the upper parts in a way that enhances CC's position.

EXAMPLE 3

a What is the problem with the following graph?

b Redraw it so that it is no longer misleading.

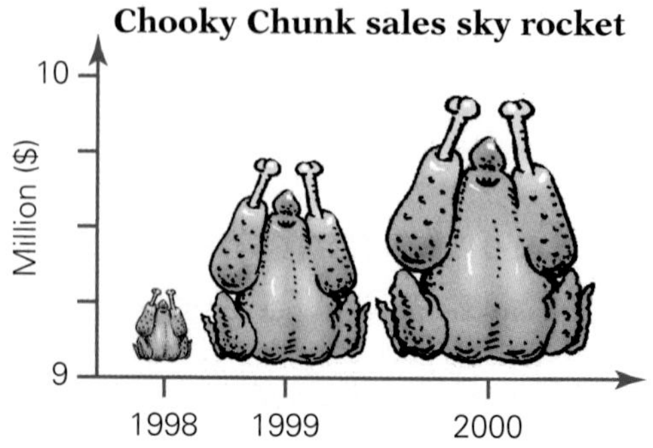

Solution

a **Using volume of a 3D shape**
Though it's the *height* of each bird that gives the true reading, the sheer *volume* increase in the drawings gives the impression of enormous increase in sales.

b

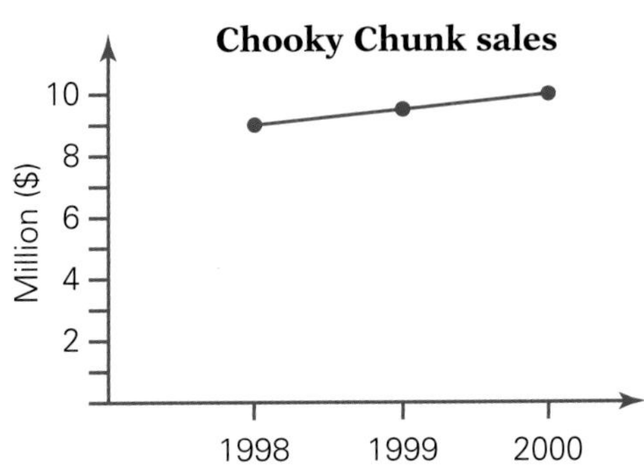

EXERCISE 4H

1 Using the checklist above as a guide, comment on how each graph has been presented in order to mislead. Redraw each to give the correct impression.

a

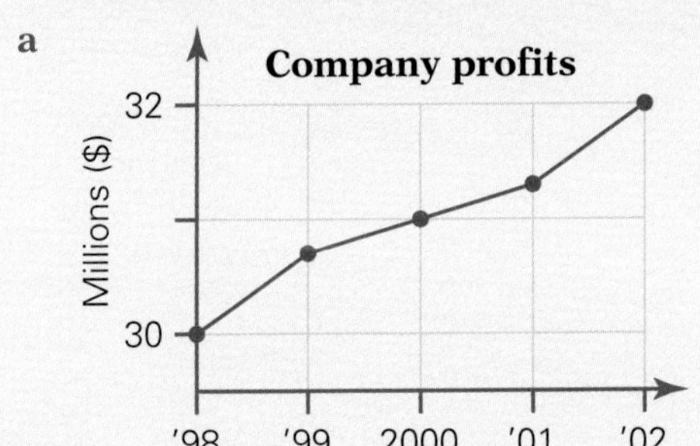

b

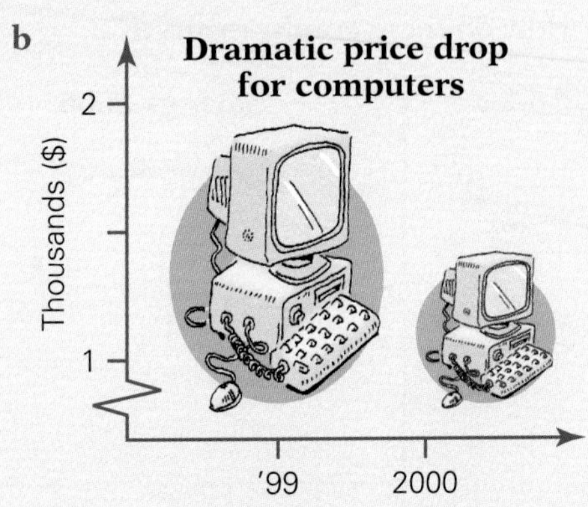

c

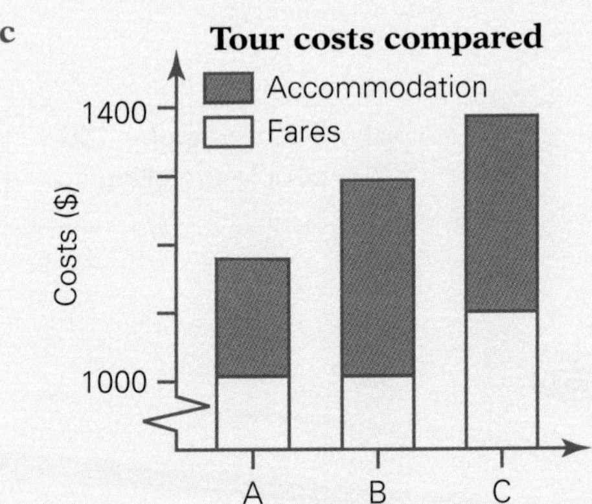

d

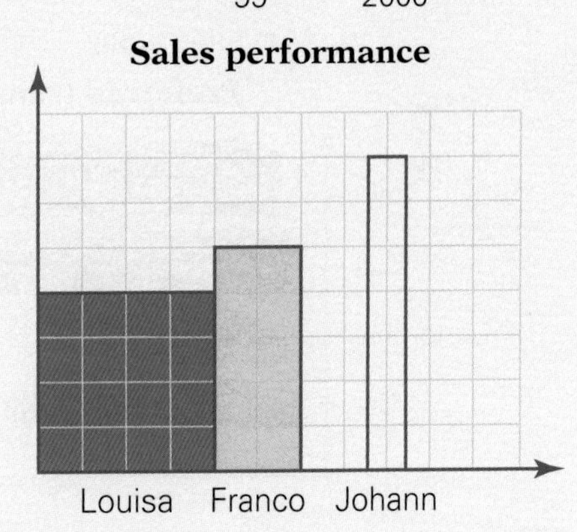

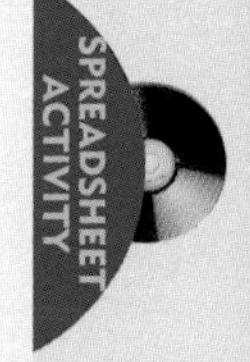

2

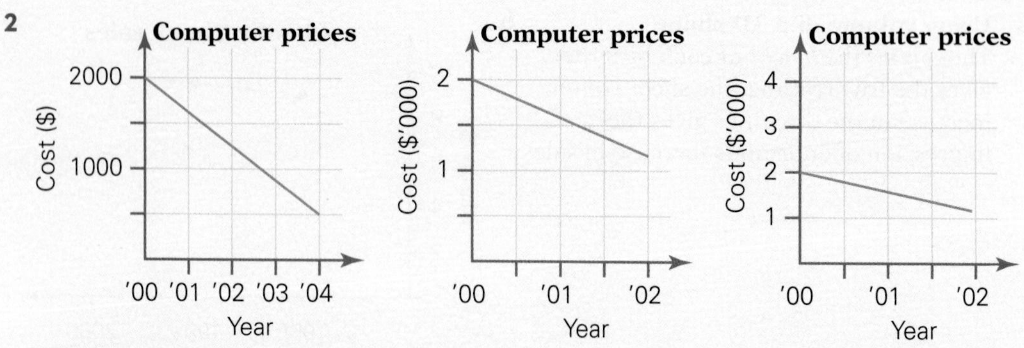

All three graphs show the same data up to the year 2002.

a Find the cost of a computer in year 2001 from each graph. Do all graphs agree?

b **Compare** vertical axis scales and comment.

c **Compare** horizontal axis scales and comment.

d **Comment** on the overall impression about computer prices given by each graph.

3 How do these graphs mislead?

a

b

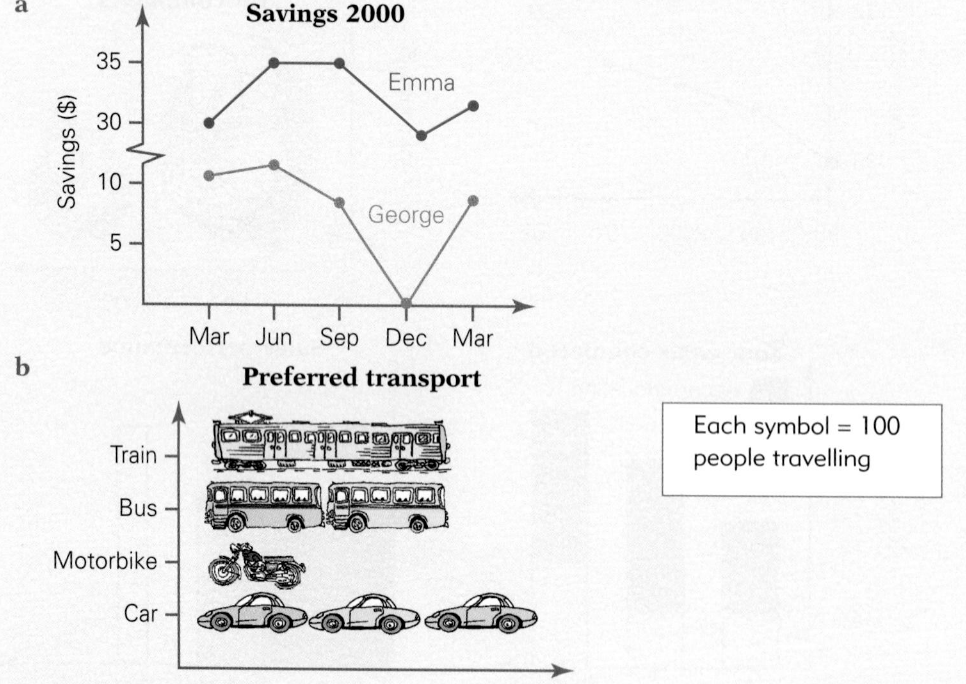

Level 2

4

School	Enrolment		
	2000	2001	2002
Mathsville High	1200	1220	1250

Draw two graphs of this data, one accurate in every way, the other designed to impress the reader with growth in enrolments.

5 Find an example of a misleading graph from any source. Paste it into your workbook and write a paragraph **explaining** its weaknesses.

4.9 Tables and charts

YOUR DOLLAR

	US$	59.16	–1.30
	Jap	69.93	–1.03
	UK£	36.93	–0.71
	Euro	54.42	–0.39
	Sfr	0.800	–0.004
	NZ$	1.084	–0.006
	Ind rph	5275	–92
	HK$	4.621	–0.093

Australia

City	Today	Maximum
Adelaide	Fine	21
Alice Springs	Fine	20
Brisbane	Fine	23
Broome	Fine	34
Cairns	Windy. Showers	27
Canberra	Fine	17
Darwin	Fine	31
Hobart	Fine	16
Newcastle	Fine	21
Perth	Showers	19
Sydney	Mostly sunny	21
Townsville	Brief shower	27
Wollongong	Fine	21

Flags courtesy of www.theodora.com/ flags appear with permission

Everywhere in the community, data is sorted and presented in tables or charts: sports statistics, telephone and postal charges, foreign currency exchange rates, and so on. Reading and drawing information from tables is an essential skill.

- Read headings carefully.
- Use a ruler to guide your eye across rows, to avoid 'skipping' to the wrong row.

Essential features of a well-constructed table are:

- an appropriate, bold **title**
- clear headings for columns and labels for rows
- items **well spaced** for easy reading
- systematic arrangement—
  alphabetical or numerical ordering
- source of the data given, if appropriate

JET CARGO RATES (from Sydney)

Destination	$/kg
Adelaide	6·25
Brisbane	5·80
Canberra	3·69
Darwin	8·76
Hobart	7·58
Melbourne	5·40
Perth	10·49

Source: *Kangaroo Airlines*

Example

From the tables above, answer the following:

a What is the rate ($/kg) of sending a package from Sydney to Perth?

b What is the weather forecast for Newcastle?

Solution

a $10.49 per kg **b** Fine and 21°C

EXERCISE 4I

Refer to the table of cargo rates above to answer the following typical questions:

1 Direct readings

a How much per kilogram does it cost to send cargo by jet from Sydney to Brisbane?

b What is the charge to send a 1-kg parcel to Darwin?

2 Comparison of readings

a Which destination is the most expensive?
b Which is the least expensive?
c What determines the difference in prices?
d List the cities in order of their distance from Sydney, from closest to furthest.

3 Mathematical operations

a How much more per kilogram does it cost to send cargo to Hobart than to Melbourne?
b How much less per kilogram does it cost to Darwin than to Perth?
c What does it cost to send 3 kg of cargo to Adelaide?
d How much change would there be from $50 after paying for 3400 g of cargo to Brisbane?
e How many whole kilograms of cargo could I send to Canberra for $20?

Level 2

4 Reading/estimating between or beyond

a Newcastle is about one-quarter of the way to Brisbane. How much per kilogram would you *expect* the rate to be?
b Alice Springs is roughly half as far as Darwin. Estimate the cargo rate.

> This means that you are charged for a whole minute if you speak for less than 60 seconds. A 2 minute 10 second call would be charged as 3 minutes.

STD call charges per minute (or part thereof)

Scale \ Distance (km)	50–85	85–165	165–745	Over 745
DAY RATE* 8:00 am – 6:00 pm Monday to Friday	24¢	34¢	35¢	50¢
EVENING RATE* 6:00 pm – 8:00 pm Monday to Friday	19¢	25¢	26¢	36¢
ECONOMY RATE* 8:00 pm – 8:00 am Monday to Friday and 8:00 pm Fri – 8:00 am Mon	10¢	14¢	15¢	18¢

* Minimum charge: 20¢ per call   Source: *Teltus*, 2002

5 Use the STD charges table above to answer the following:

a How much per minute does a call over a distance of 200 km cost if it is made at 10:00 am Tuesday?
b What is the cost per minute of a call made on Sunday at 3:00 pm, over a distance of 125 km?
c Which is the *most* expensive rate over all distances? Why is this?
d Which is the *least* expensive rate over all distances? Why is this?

e Which distance category would a call from Sydney to Brisbane fall into?

f How much more per minute over 500 km does it cost to ring at 5:00 pm Wednesday than 7:00 pm Wednesday?

g Calculate the cost of these calls:
  i 2 min, Saturday 2:00 pm, 80 km
  ii 3 min, Thursday 7:00 pm, 940 km
  iii 4 min 20 s, 11:30 am Tuesday, 100 km
  iv 8 min 5 s, 9:45 pm Friday, 623 km

h How much would Rebecca save on her 8 min 42 s call to Perth from Sydney by making it at 9:00 am Saturday rather than at 9:00 am Thursday?

i Why does the distance chart not show distances under 50 km?

Investigation Water usage

1 Carefully review your habits during a typical week, then complete the chart below.

2 If every Australian (18 million people) showered rather than bathed each day, how much water per day would be saved? How much per year?

3 If your water use is typical of Sydney's three million people, what is the city's total water consumption each week?

The water I use

How used	Usual amount in litres	Number of times/week	Total litres used
Bath	80		
Shower	50		
Flushing toilet	5/11		
Washing hands, face	5		
Drinks	1		
Brushing teeth	1		
Washing dishes (one meal)	20		
Cooking (one meal)	8		

Investigation Temperature changes

1 Use a thermometer to measure classroom temperature at hourly intervals from 8:00 am to 3:00 pm. Record the data in a table.

2 Use grid paper and label the vertical axis *Temperature* (° Celsius). Scale the axis from 0 to a suitable maximum, using 1 cm to 5°. Label the horizontal axis *Time* (h). Scale it from 8:00 am to 3:00 pm, using 1 cm to 1 h.

3 Give the graph a suitable title.

4 Plot points with clear dots to represent the data from your table.

5 Join the dots with line intervals.

Chapter Review

Language Links

axis	data	score	tally
bar	frequency	sector	title
chart	graph	statistics	
column	picture	survey	
conversion	scale	table	

1 Write adjectives formed from conversion, frequency, statistics, table.

2 Write sentences with sample, score, sector, survey.

3 Write the plural of axis and tally.

4 What is meant when the word 'data' is used?

5 Write a survey designed to gather data on the bedtimes of a family.

6 Write a story to **explain** this graph, and give the graph a title.

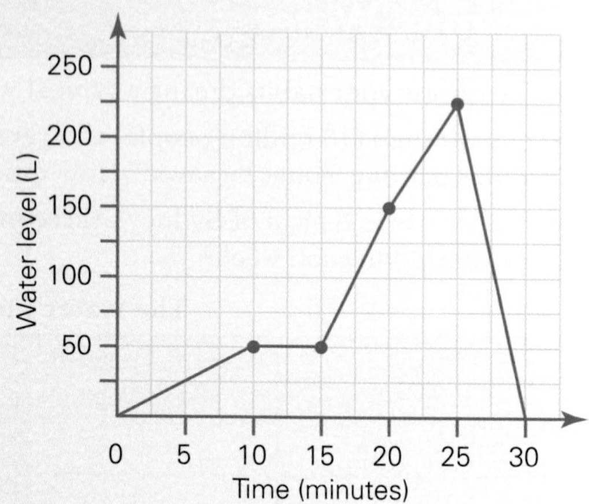

7 By **reflecting** on the graphs in this chapter copy and complete the following table, on the features, advantages and disadvantages of graphs.

Name of graph	Features	Advantages	Disadvantages
Picture graph			
Column graph	• Two axes named with scales marked • Columns of equal width used to show comparisons	• Commonly used and easy to read • Good for discrete or categorical • data • Shows comparisons	• Accuracy in reading scales • Not suitable for all types of data
Line graphs		• Used for continuous data • Shows in-between values • Commonly used	

Name of graph	Features	Advantages	Disadvantages
Sector graph			• Angles not always whole degrees • Difficult if too many sectors required • Not suitable for all types of data
Divided bar graph			• Doesn't show detail
Travel graph			
Step graph			

Discuss your answers other classmates and see if you can add any more detail to your table.

Chapter Review Exercises

1 If a pictograph key shows [car symbol] representing 80 cars: 4.2

a What would [car symbol][car symbol][car symbol] represent?

b Draw the symbols to represent 120 cars.

2 Forty students in a school of 800 participated in a survey of teenagers' favourite drinks. The results were: coffee 3, tea 2, cola 20, other soft drinks 9, milk drinks (all kinds) 6. 4.3

a Write a suitable title for a graph of the data.

b Draw a key symbol and write its number value for a picture graph.

c Show how you would display the cola and other soft drinks data as two parts of a pictograph.

d What fraction of the student population took part in the survey?

3 4.3

18	21	20	20	22
19	18	21	19	20
20	19	20	20	21
19	20	23	22	20

In her science experiment, Simone kept records of the number of days between litters of babies delivered by her female mice.

a Draw up a frequency distribution table of her results.

b What was the most common period between litters?

c How often did the period exceed 20 days?

d How frequently was it 19 days or less?

4.8

4 **a** How many games have Newcastle played?

b How many matches have Manly lost?

c What total points have been scored by St George in all the games played so far?

d How many competition points have been awarded to Parramatta?

e Use the Pts, W and D columns to figure out the points awarded for a win and for a draw.

f Explain why Manly and Parramatta have the same total points awarded, even though Parramatta has won more games than Manly.

g Which teams have won as many games as they've lost?

h When Gold Coast plays its 20th match, what outcome/s would put it ahead of Wests on the table?

Points table

Team	P	W	D	L	F	A	Pts
Manly	19	13	2	4	445	309	28
Parramatta	20	14	—	6	413	321	28
Nth Sydney	20	12	1	7	467	304	25
Newcastle	19	11	1	7	408	284	23
Sydney City	20	11	1	8	438	330	23
Wests	20	10	—	10	327	351	20
Illawarra	19	8	3	8	347	330	19
Gold Coast	19	9	1	9	389	384	19
Balmain	20	9	—	11	303	300	18
St George	19	8	—	11	289	360	16
Sth Sydney	20	4	1	15	297	568	9

P = matches played W = matches won
D = drawn games L = lost
F = total points for A = total points against
Pts = competition points awarded

4.2

5

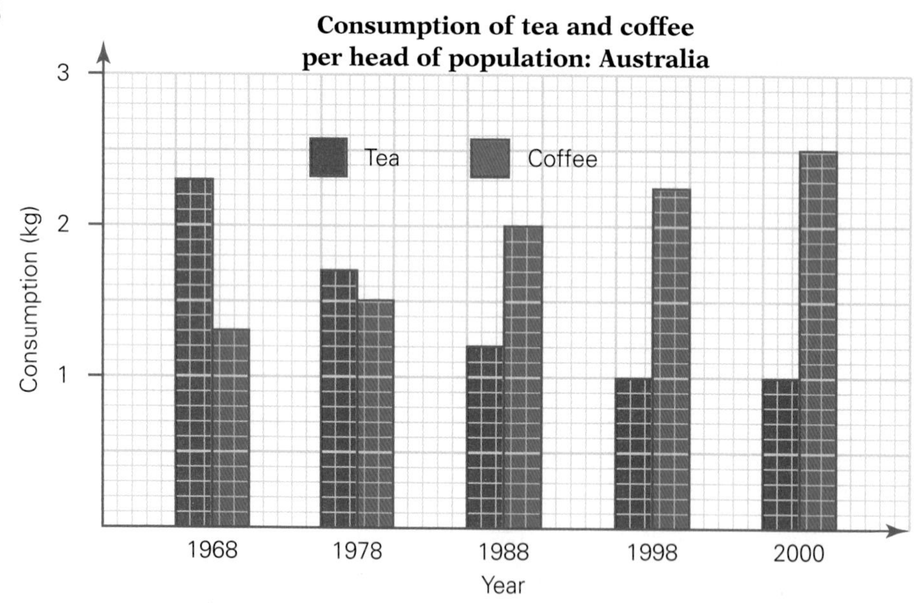

a How much more tea than coffee was consumed per person in 1968?

b When did coffee drinking exceed tea drinking for the first time?

c If the population of Australia was 18 500 000 in 1998, how much coffee was consumed in Australia?

d What is happening in general with the drinking of these two beverages?

e What would you *expect* to be the consumption of each drink by 2008, if current trends continue?

6 **Julio's spending during one week ($24)** 4.2

Food | Movie | Soccer ball | Swap cards | Savings

Scale: 1 cm = $2

a Which category did Julio spend most on?
b Which category did he allocate least money to?
c How much did he pay to go to a movie?
d Which two categories did he spend the same amount on?
e What fraction of his spending went on food?
f At this rate, how many weeks would it take him to pay his dad back $28 for his new soccer ball?

7 **Who gets what from petrol sales** 4.3

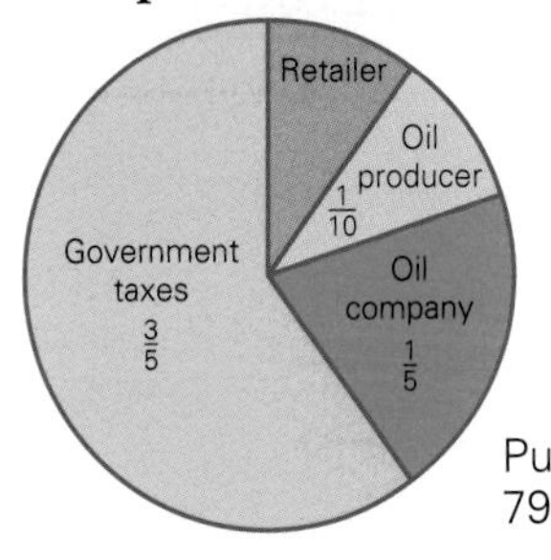

Pump price: 79·9 ¢/L

a Who gets the largest share of petrol money?
b Who gets the least?
c What fraction does the retailer get?
d How much money flows to the oil company from each litre of petrol sold, to the nearest cent?
e How much does the government make out of the sale of 5 billion litres of fuel?
f If the government reduced its share to 30·76 cents, what sized slice (in degrees) would show this in a sector graph?

8 4.4

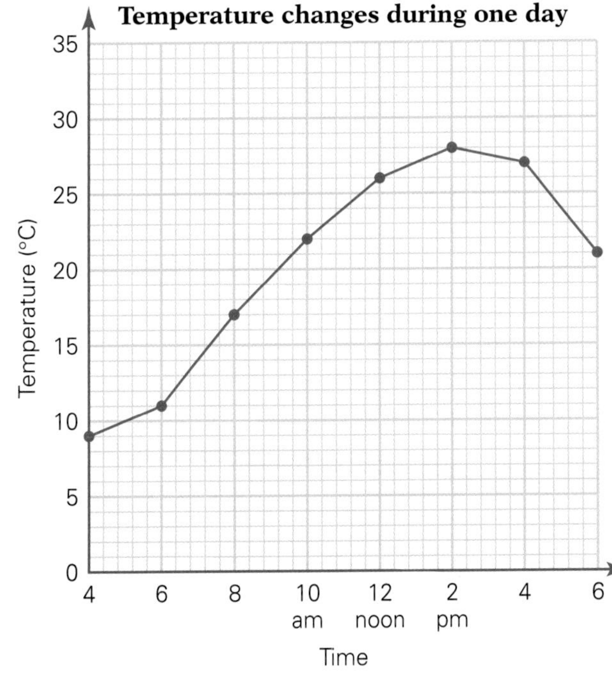

a What is the scale used on the vertical axis?
b What does each 'in-between' grid line represent on the horizontal axis?
c What was the temperature at:
i 6:00 am? ii 4:00 pm?
d At what time/s was the temperature 26°C?
e What is the difference between maximum and minimum temperatures shown?
f What was the temperature at 7:30 am?

4.5 **9** Answer the following questions based on Jim's travel graph.

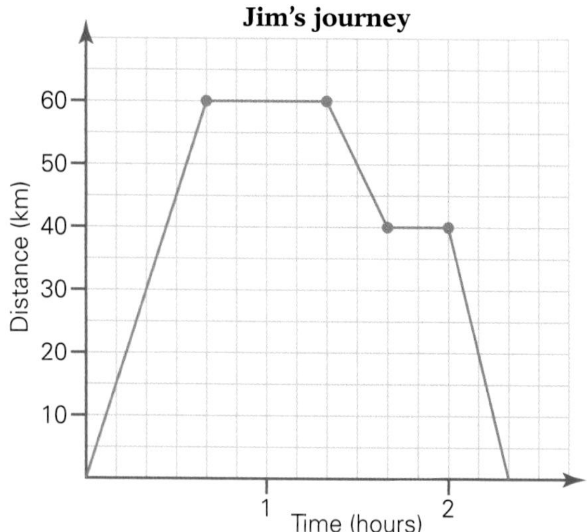

- **a** How far did Jim travel before stopping for the first time?
- **b** How long did Jim stop for the second time?
- **c** How far did Jim travel between stops?
- **d** How far did Jim travel during his journey?
- **e** For how long did Jim travel during the day (excluding his breaks)?

4.7 **10** Parking costs at a local supermarket are shown in this step graph.

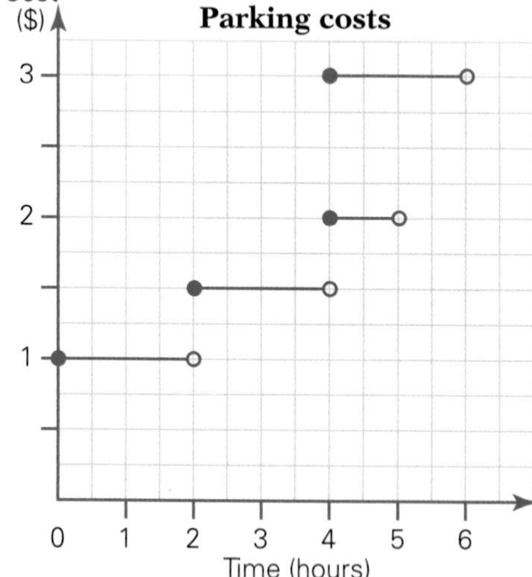

- **a** How much does it cost to park for 1 hour?
- **b** How much does it cost to park for 2 hours?
- **c** How much does it cost to park for 4 hours?
- **d** What is the maximum number of hours shown on this graph?
- **e** What is the maximum length of time a person can park for $2.50?

4.7 **11** Which type of graph would be *most appropriate* to display data on:

a temperature? **b** division of income? **c** colour of cars? **d** number of cars?

Keeping Mathematically Fit

Part A: Non-calculator

1 Copy and complete $3 \times \square \times 4b = 24ab^2$.

2 Simplify $3^4 \div 3^2 \times 3^4$.

3 Find the LCM of 440 and 100.

4 Simplify $3\frac{2}{3} - 1\frac{1}{4}$.

5 Find the value of $^-17 + (^-5 \times 4 - 9)$.

6 Write the fraction that is halfway between $\frac{1}{4}$ and $\frac{1}{5}$.

7 Find the average of 10, 14, 17 and 9.

8 What is the HCF of 240 and 76?

9 Convert $5\frac{1}{4}$ kg into grams.

10 If $a + 7 = 10$, find the value of $4a^2$.

Part B: Calculator

1 Find the cost of $7\frac{1}{2}$ metres at \$4.90 a metre.

2 If $m = \dfrac{N^2 - N}{3}$, find m when $N = 9$.

3 Find the value of $\dfrac{49{\cdot}03 - 19{\cdot}36}{(^-4{\cdot}8)}$ correct to 2 decimal places.

4 An airfare to LA rose from \$1795 to \$1920. What percentage increase was this, to the nearest per cent?

5 If $a = 4$ and $b = {}^-6$, find the value of:

a $3ab$ b $\dfrac{a - b}{2}$ c $8(b - a)$ d ab^2

6 Simplify:

a $7 - (2a + 5)$ b $\dfrac{10x}{5y}$ c $2(x + 5) + 3(x - 1)$

7 Complete the rule

x	3	4	5	6
y	10	11	12	13

$y =$ ______ + ______

8 If $\square$ is an integer and $\frac{\square}{5}$ lies between $\frac{1}{4}$ and $\frac{7}{10}$:

a Find a value for $\square$.

b How many different values can $\square$ take?

TEACHER

1 What you need to know and revise

Outcome PAS 3.1b:
Constructs, verifies and completes number sentences involving the four operations with a variety of numbers

Outcome PAS 4.3:
Uses the algebraic symbol system to simplify, expand and factorise simple algebraic expressions

2 What you will learn in this chapter

Outcome PAS 4.3:
Uses the algebraic system to simplify, expand and factorise simple algebraic expressions

Outcome PAS 4.4:
Uses algebraic techniques to solve linear equations and simple inequalities

- distinguishes between algebraic expressions, where letters are used as variables, and equations, where letters are used as unknowns
- solves linear equations using methods that include backtracking and guess and check
- solves equations using algebraic methods
- checks solutions by substituting
- solves word problems by the use of an equation
- solves equations that arise from substitution into formulae
- finds a range of values that satisfy an inequality
- solves simple inequalities
- represents solutions to simple inequalities on the number line

Working Mathematically outcomes WMS 4.1–4.5
Students will be asked to ***question***, ***apply strategies***, ***communicate***, ***reason*** and ***reflect*** in the sections of this chapter.

Equations

Key mathematical terms you will encounter

backtrack
equation
evaluate
expression
formula
inequality
inspection
inverse
isolate
operation
pronumeral
solution
substitute
translate
unknown

MathsCheck
Equations

1 Simplify these algebraic *expressions*:

a $4 \times 3k$　b $4a \times 5p$　c $2m \times 6m$
d $7n + 3n$　e $6d - d$　f $5ab - 3ab$
g $3x + 6 + 5x + 7$　h $^{-}3 \times 4c + 5 \times 2c$　i $7p + p - 3q^2$

2 Expand using the distributive pattern:

a $3(k + 8)$　b $5(x - 3)$　c $^{-}2(x - 4)$
d $2(l + b)$　e $(2a + 3b)2n$　f $x(x + 3y)$

3 Expand and simplify:

a $4(m + 2) + 3(6 - m)$　b $^{-}2(p - 4) - (3p + 2)$

4 Evaluate, if $a = 3$, $b = {}^{-}2$: $4ab - ab$

5 I think of a number, double it, add 3 and the answer is 23. What is the number?

6 $\boxed{16} \xrightarrow{+9} \square \xrightarrow{\times 3} \square \xrightarrow{\div 3} \square \xrightarrow{-9} \square$　Explain the outcome.

7 Name the *inverse* operation to:

a switch on　b empty
c attack　d add 7
e double　f minus 8

Inverse operations cancel or undo each other.
Addition is the inverse of **subtraction**.
Multiplication is the inverse of **division**.
Inverse can be thought of as **opposite**.

8 Write the value of:

a $6 - 6$　b $^{-}10 + 10$　c $7 \div 7$
d $3 \times \frac{1}{3}$　e $\frac{2}{5} \times \frac{5}{2}$　f $\frac{5}{5}$
g $n - n$　h $2p + {}^{-}2p$　i $^{-}6 \div {}^{-}6$

9 Simplify:

a $a + 7 - 7$　b $\frac{3x}{3}$　c $\frac{3}{5} \times \frac{5p}{3}$
d $8n + {}^{-}3 + 3$　e $\frac{^{-}5t}{^{-}5}$　f $\frac{^{-}7}{10} \times \frac{^{-}10w}{7}$

10 Find the number that when inserted into the $\square$ makes each of the following statements true.

a $6 + \square = 0$　b $8 \times \square = 1$　c $4a \div \square = a$
d $x + \square = x + 4$　e $x + 2 + \square = x$　f $\frac{x}{3} \times \square = x$
g $a - 3 + \square = a$　h $7w \div \square = w$　i $p - 7 + \square = p$
j $^{-}a \times \square = a$　k $3x + 1 - \square = 3x$　l $\frac{x}{7} \times \square = x$

TEACHER

Exploring New Ideas

5.1 Equations

You will recall that an equation is a mathematical statement which says that two things are equal, i.e. the left-hand side *equals* the right-hand side.

LHS = RHS

$2x + 1 = 15$ is an equation where two expressions are equal in value.

$5 + x = x + 5$ is an equation which is true for all values of x.

$ab = ba$ is an equation which is true for all values of a and b.

Often an equation is true for a particular value (or values) of the pronumeral. When we find these values we are *solving* the equation.

- If $\boxed{n} \xrightarrow{+5} \boxed{n+5}$, then $\boxed{n} \xleftarrow[-5]{} \boxed{n+5}$
- If $\boxed{x} \xrightarrow{-3} \boxed{x-3}$, then $\boxed{x} \xleftarrow[+3]{} \boxed{x-3}$
- If $\boxed{a} \xrightarrow{\times 4} \boxed{4a}$, then $\boxed{a} \xleftarrow[\div 4]{} \boxed{4a}$
- If $\boxed{c} \xrightarrow{\div 6} \boxed{\frac{c}{6}}$, then $\boxed{c} \xleftarrow[\times 6]{} \boxed{\frac{c}{6}}$

This process of *undoing* operations in reverse order is called **backtracking** or *reverse flow* technique. It brings us back to just the pronumeral. It *isolates the unknown.*

EXAMPLE 1

Is $x = 3$ a solution to $4x - 1 = 11$?

Solution

Substitute $x = 3$ into the equation.

$$\begin{aligned} \text{LHS} &= 4x - 1 \\ &= 4 \times 3 - 1 \\ &= 11 \\ &= \text{RHS} \end{aligned}$$

$\therefore x = 3$ is a solution to the equation $4x - 1 = 11$

EXAMPLE 2

Solve the following by backtracking:

a $2x + 1 = 15$ **b** $\frac{x}{7} - 1 = 5$

The opposite of + is −.
The opposite of × is ÷.

Solution

a

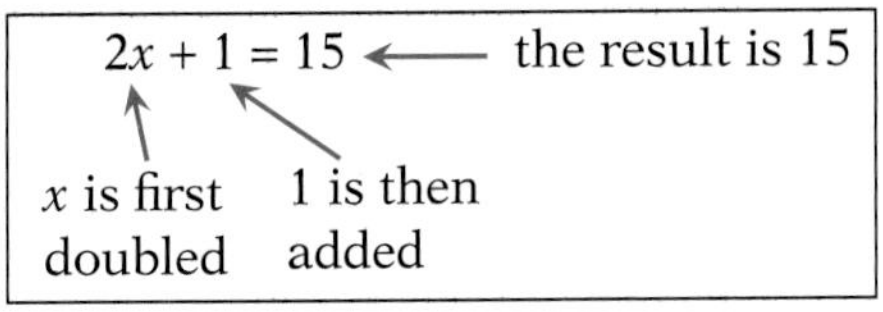

$2x + 1 = 15$ ⟵ the result is 15

x is first doubled; 1 is then added

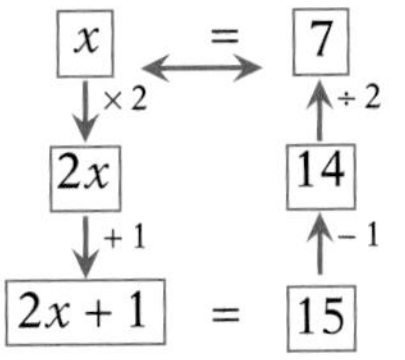

$\boxed{x} \xrightarrow{\times 2} \boxed{2x} \xrightarrow{+1} \boxed{2x+1} = \boxed{15} \xrightarrow{-1} \boxed{14} \xrightarrow{\div 2} \boxed{7}$, so $\boxed{x} = \boxed{7}$

$\therefore$ the solution is $x = 7$

b $\dfrac{x}{7} - 1 = 5$ ← the result is 5

x is first divided by 7; then 1 is subtracted

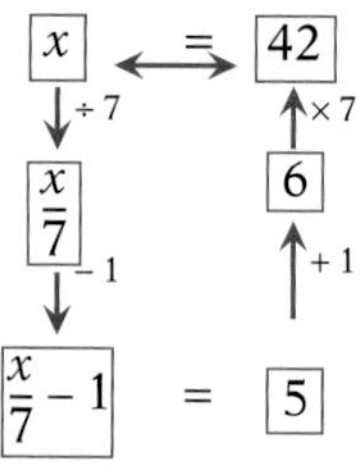

∴ the solution is $x = 42$

EXERCISE 5A

1 Choose the equations from the following:

a $x + 3$ **b** $3x - 6 = 9$ **c** $x^2 - 8$ **d** $3a + b - c$
e $x^2 - x$ **f** $a + 4 = 18$ **g** $3x - y = 12$ **h** $y - 3x$

2 For each equation below check whether the solution given in brackets is correct:

a $p + 5 = 9$ $[p = 4]$ **b** $2x - 4 = 10$ $[x = 7]$
c $3x - 5 = 10$ $[x = 1]$ **d** $w + 1 = 2w$ $[w = 1]$
e $3(a + 2) = 0$ $[a = 2]$ **f** $\frac{x}{5} + 1 = 4$ $[x = 15]$
g $2x + 3 = x + 7$ $[x = {}^-4]$ **h** $2q + 3 = q + 7$ $[q = 4]$

3 Copy and complete each arrow diagram by inserting the operation on top of the arrow:

a $\boxed{n} \to \boxed{n+4}$ **b** $\boxed{a} \to \boxed{a-3}$ **c** $\boxed{b} \to \boxed{5b}$
d $\boxed{c} \to \boxed{c \div 7}$ **e** $\boxed{x} \to \boxed{4x} \to \boxed{4x+8}$ **f** $\boxed{y} \to \boxed{2y} \to \boxed{2y-3}$
g $\boxed{p} \to \boxed{{}^-6p} \to \boxed{\frac{{}^-6p}{5}}$ **h** $\boxed{r} \to \boxed{r+5} \to \boxed{\frac{r+5}{3}}$ **i** $\boxed{m} \to \boxed{m-2} \to \boxed{3(m-2)}$

4 As above, copy and complete these *backtracking* diagrams:

a $\boxed{n} \leftarrow \boxed{n+4}$ **b** $\boxed{a} \leftarrow \boxed{a-3}$ **c** $\boxed{b} \leftarrow \boxed{5b}$
d $\boxed{c} \leftarrow \boxed{c \div 7}$ **e** $\boxed{x} \leftarrow \boxed{4x} \leftarrow \boxed{4x+8}$ **f** $\boxed{y} \leftarrow \boxed{2y} \leftarrow \boxed{2y-3}$
g $\boxed{p} \leftarrow \boxed{{}^-6p} \leftarrow \boxed{\frac{{}^-6p}{5}}$ **h** $\boxed{r} \leftarrow \boxed{r+5} \leftarrow \boxed{\frac{r+5}{3}}$ **i** $\boxed{m} \leftarrow \boxed{m-2} \leftarrow \boxed{3(m-2)}$

5 Use backtracking to find the value of the unknown, in each of the following:

a $\boxed{x} \xrightarrow{\div 4} \boxed{} \xrightarrow{+3} \boxed{5}$

b $\boxed{x} \xrightarrow{-4} \boxed{} \xrightarrow{\times 3} \boxed{21}$

c

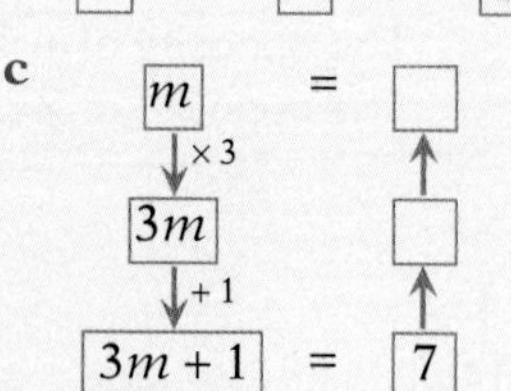

d

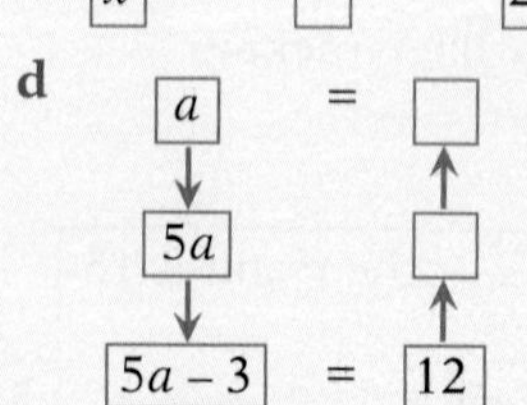

Level 2

6 For each of the following, check whether the unknown number is 6.

- **a** 3 is added to an unknown number; the result is 9.
- **b** Doubling an unknown number gives 13.
- **c** 3 less than an unknown number is 9.
- **d** 30 divided by an unknown number is 5.
- **e** Doubling an unknown number plus 2 is 10.

7 One solution to the equation $a + b = 10$ is $a = 2$ and $b = 8$. Can you think of another solution?

8 For each equation below, check whether the solution given in brackets is correct.

- **a** $a + b = 6$ [$a = 3, b = 3$]
- **b** $a - b = 8$ [$a = 10, b = {}^{-}2$]
- **c** $x \div y = 4$ [$x = 1, y = 4$]
- **d** $xy = 16$ [$x = y = 4$]
- **e** $3(x + y) = 12$ [$x = 1, y = 3$]
- **f** $5x - y = x$ [$x = 1, y = 4$]

9 For each of the equations below state whether it is:

- **A** true for all values of x
- **B** true for a specific value of x only
- **C** never true

- **a** $x + 0 = x$
- **b** $x + 1 = 2$
- **c** $x \times x = x^2$
- **d** $x + x = 2x$
- **e** $x + 4 = 8$
- **f** $x \times x = x + x$
- **g** $3x = 9$
- **h** $x - 1 = {}^{-}1$
- **i** $x + x = x^2$

10 Use backtracking to find the value of x in each of these equations:

- **a** $3x - 1 = 8$
- **b** $4x + 1 = 9$
- **c** $\frac{x}{2} + 3 = 4$
- **d** $\frac{x + 3}{2} = 4$
- **e** $\frac{1}{2}(x + 3) = 4$
- **f** $\frac{x}{4} - 1 = 5$

11 Try out these on your partner:

- **a** I think of a number, subtract 7, then halve the result. The answer is 15. What is the number?
- **b** I think of a number, add 6, then multiply the answer by 5. The answer is 45. What is the number?
- **c** Make up one of your own and try it out.

INVESTIGATION

5.2 Solving equations

Simple equations can often be solved by inspection:

$x + 7 = 10$ $2x = 8$ $\frac{x}{5} = 4$

Other equations such as $4{\cdot}3x + 2{\cdot}8 = 10{\cdot}5$ need to be solved systematically.

In general, an equation remains balanced if you perform the same operation on both sides.

To *solve* an equation, you use inverse (opposite) operations to undo the equation so that the only term remaining on the left-hand side is the required pronumeral.

Balance Beams

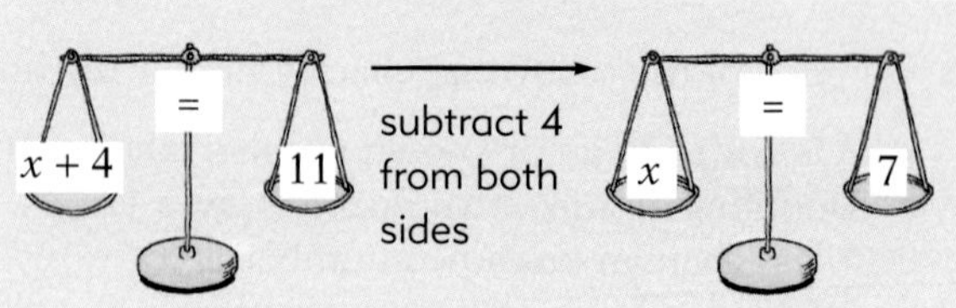

EXAMPLE 1

Solve each of the following equations:

a $x + 7 = 10$ **b** $2x = 8$ **c** $\frac{x}{5} = 4$

Solution

Even though the solution could be found by inspection, we shall apply the algebraic method to each of the equations here.

a
$$x + 7 = 10$$
$$-7 \quad -7$$
$$x = 3$$

b
$$2x = 8$$
$$2x = 8$$
$$\div 2 \quad \div 2$$
$$x = 4$$

c
$$\frac{x}{5} = 4$$
$$5 \times \frac{x}{5} = 4 \times 5$$
$$x = 20$$

EXAMPLE 2

Solve:

a $3a - 4 = 8$ **b** $\frac{x}{8} + 3 = {}^-4$ **c** $\frac{x+3}{8} = {}^-4$

Solution

a
$$3a - 4 = 8$$
$$+4 \quad +4$$

What number less 4 gives 8?

$$3a = 12$$
$$\div 3 \quad \div 3$$

What number times 3 gives 12?

$$a = 4$$

Check: $3 \times 4 - 4$
$= 8$ ✔

b
$$\frac{x}{8} + 3 = {}^-4$$
$$-3 \quad -3$$

What number plus 3 gives $^-4$?

$$\frac{x}{8} \times 8 = {}^-7 \times 8$$

What number divided by 8 gives $^-7$?

$$x = {}^-56$$

Check: $\frac{^-56}{8} + 3$
$= {}^-7 + 3$
$= {}^-4$ ✔

c
$$\frac{x+3}{8} = {}^-4$$

What number divided by 8 gives $^-4$?

$$\left(\frac{x+3}{8}\right) \times 8 = {}^-4 \times 8$$

What number plus 3 gives $^-32$?

$$x + 3 = {}^-32$$
$$-3 \quad -3$$
$$x = {}^-35$$

Check: $\frac{^-35 + 3}{8}$
$= \frac{^-32}{8}$
$= {}^-4$ ✔

EXAMPLE 3

Find the value of x for:

a $^{-}x = 5$ **b** $^{-}x = {^{-}7}$ **c** $4 - 2x = {^{-}6}$

Solution

a $^{-}x = 5$

$\div {^{-}1} \quad \div {^{-}1}$

$x = {^{-}5}$

The opposite of 5 is ⁻5.

b $^{-}x = {^{-}7}$

$\div {^{-}1} \quad \div {^{-}1}$

$x = 7$

The opposite of ⁻7 is 7.

c $4 - 2x = {^{-}6}$

$-4 \quad -4$

$^{-}2x = {^{-}10}$

$\div {^{-}1} \quad \div {^{-}1}$

$2x = 10$

$\div 2 \quad \div 2$

$x = 5$

EXERCISE 5B

1 Copy and complete to solve each equation:

a $t + 5 = 8$
$-5 \quad$ ___
$t =$ ___

b $n - 3 = 12$
$+3 \quad$ ___
$n =$ ___

c $8p = 24$
$\div 8 \quad$ ___
$p =$ ___

d $\frac{x}{5} = 10$
$\times 5 \quad$ ___
$x =$ ___

e $m - 5 = {^{-}7}$
___ ___
$m =$ ___

f $^{-}5c = 30$
___ ___
$c =$ ___

2 Apply the inverse operation on both sides to solve each equation.

a $t + 7 = 12$ **b** $5m = 35$ **c** $x \div 3 = 11$

d $5 + n = 9$ **e** $\frac{a}{6} = 4$ **f** $27 = 3d$

g $x + 6 = 2$ **h** $8 + a = 5$ **i** $l - 9 = {^{-}6}$

j $\frac{r}{7} = {^{-}3}$ **k** $^{-}6p = 18$ **l** $^{-}16 = 7 + c$

m $b + \frac{2}{3} = \frac{5}{6}$ **n** $1\frac{3}{4} + x = 2\frac{1}{2}$ **o** $\frac{3}{5}a = \frac{9}{10}$

p $0{\cdot}04h = 2{\cdot}8$ **q** $^{-}10p = 10$ **r** $^{-}x + 1 = 6$

3 Copy and complete the solution to each equation:

a $2m - 3 = 1$
$+3 \quad +3$
$2m =$ ___
$\div 2 \quad \div 2$
$m =$ ___
Check: ___

b $5 + 8a = 21$
$-5 \quad -5$
$8a =$ ___
$\div 8 \quad \div 8$
$a =$ ___
Check: ___

c $4(x + 3) = 20$
$\div 4 \quad \div 4$
$x + 3 =$ ___
$-3 \quad -3$
$x =$ ___
Check: ___

d $(h - 4)7 = 14$
$\div 7 \quad \div 7$
___ = ___
$+4 \quad +4$
$h =$ ___
Check: ___

e $3 \times \frac{c+2}{3} = 7 \times 3$
___ = ___
$-2 \quad -2$
$c =$ ___
Check: ___

f $7 + 3t + 5t = 15$
$(3t + 5t) + 7 = 15$
$-7 \quad -7$
___ = ___
$\div 8 \quad \div 8$
$t =$ ___
Check: ___

4 Translate each description into an equation, then solve:

- **a** Five times p equals 40.
- **b** When 9 is subtracted from t, the result is 32.
- **c** What number increased by 7 gives 50? (Use x.)
- **d** The product of 8 and what number is 56?
- **e** What number divided by 9 gives 8?
- **f** Two-thirds of a certain number is $^{-}12$.

5 Solve:

a $^{-}x = 7$ **b** $^{-}x = 9$ **c** $^{-}p = {}^{-}5$
d $^{-}x = \frac{1}{2}$ **e** $^{-}2x = 8$ **f** $-\frac{x}{3} = {}^{-}4$

6 Solve each of the following equations. Check your answers.

Your solutions should look like:

$$3p - 7 = 11$$
$$+7 \quad +7$$
$$3p = 11$$
$$\div 3 \quad \div 3$$
$$p = 6$$

a $2x + 1 = 9$ **b** $2x - 1 = 9$ **c** $4x - 5 = 7$
d $3 + 5x = 13$ **e** $2 + 2p = 8$ **f** $3a - 4 = 8$
g $5a - 1 = 9$ **h** $6d + 2 = 8$ **i** $8 + 2x = 10$
j $7p - 1 = 13$ **k** $7p + 1 = 15$ **l** $6 + 6s = 12$
m $4d - 3 = 9$ **n** $4d + 3 = 9$ **o** $3 + 9a = 12$
p $5h - 9 = 6$ **q** $5 + 5h = 20$ **r** $6t - 12 = 0$
s $4 - 3x = 7$ **t** $10 - 3p = 16$ **u** $^{-}5 - 2w = 5$

Level 2

7 Solve each of the following equations:

a $\frac{x}{2} + 1 = 3$ **b** $\frac{a}{5} - 1 = 4$ **c** $4 + \frac{a}{2} = 6$
d $\frac{d}{4} - 1 = 9$ **e** $\frac{s}{9} + 9 = 9$ **f** $\frac{a}{3} - 6 = 5$
g $\frac{w}{5} - 2 = 3$ **h** $\frac{q}{7} + 9 = 10$ **i** $\frac{r}{10} + 5 = 10$
j $\frac{x}{4} - 3 = 3$ **k** $\frac{x}{4} + 3 = 3$ **l** $\frac{2x}{3} - 1 = 5$
m $5 - \frac{x}{2} = 5$ **n** $1 - \frac{a}{2} = {}^{-}3$ **o** $^{-}2 - \frac{w}{3} = 1$

8 Isolate the pronumeral in each of the following equations:

a $\frac{a+1}{7} = 2$ **b** $\frac{a-1}{7} = 2$ **c** $\frac{x+5}{2} = 3$
d $\frac{n-4}{2} = 4$ **e** $\frac{w-8}{3} = {}^{-}2$ **f** $\frac{5-w}{2} = 6$
g $\frac{4+2x}{3} = 4$ **h** $\frac{3x-1}{2} = 2$ **i** $\frac{2a-6}{3} = 4$
j $\frac{2-a}{3} = {}^{-}4$ **k** $\frac{3p+4}{5} = 5$ **l** $\frac{5y-2}{3} = {}^{-}2$

9 Solve each of the following equations. Check your answers.

a $x + 0{\cdot}2 = 0{\cdot}7$ **b** $x - 0{\cdot}2 = 0{\cdot}7$ **c** $h - \frac{1}{2} = 3$

d $x - 3 = 4{\cdot}5$ **e** $p + 3{\cdot}2 = 6$ **f** $2x = \frac{1}{2}$

g $\frac{x}{2} = 0{\cdot}6$ **h** $2x = 0{\cdot}6$ **i** $p - 0{\cdot}9 = 1$

j $p + 0{\cdot}9 = 1$ **k** $w + \frac{1}{2} = 4$ **l** $2w - \frac{1}{2} = \frac{1}{2}$

10 Find the value of x in each of the following equations:

a $^-x = 9$ **b** $^-x = \frac{^-3}{4}$ **c** $^-x = {}^-9$ **d** $^-x = 100$ **e** $^-x = 0$

11 Use your calculator to help you solve each of the following equations (answer to 2 dec. pl. if necessary).

a $\frac{a}{63} = 94$ **b** $\frac{^-x}{6} = 15$ **c** $\frac{k}{46} = {}^-920$

d $9a - 2{\cdot}5 = 7{\cdot}9$ **e** $^-21{\cdot}3x = 961$ **f** $2x + 456 = 9078$

12 Simplify each of the following equations before solving each one:

a $4x + 3x = 21$ **b** $9x - 5x = 16$ **c** $^-3a - 5a = 16$

13 Translate into equations, then solve:

a p is multiplied by 2, then 5 is added. The result is 11.
b k is divided by 3, then 4 is subtracted, giving 1.
c 6 is added to a number, then the result is multiplied by 5. The answer obtained is 40. Find the number.
d John has saved a certain amount. He put in another $10, then split the total into three equal parts, ending up with $15 in each part. What was the original amount saved?

5.3 Using equations in problem solving

You can use equations to help you to solve problems. Just follow these steps:

1 Introduce a pronumeral.

2 Write an equation.

3 Solve the equation.

4 Check your solution.

5 Answer the question in words.

EXAMPLE 1

Rahnee earns $7 an hour for casual work at a fast-food store. At 8 hours per week, how many weeks must she work to earn $500?

Solution

Let w be the number of weeks. — Step 1: Introduce a **pronumeral.**

$7 \times 8 \times w = 500$ — Step 2: Write an **equation.**

$56w = 500$ — Step 3: **Solve.**

$\div 56 \quad \div 56$

$w = 8{\cdot}9$ — Step 4: **Check.**

$\therefore$ she must work for 9 weeks. — Step 5: Answer the question in words.

EXAMPLE 2

Three consecutive odd numbers total 81. What are they?

Solution

Let the first odd number be x. — Step 1: Introduce a **pronumeral.**

$x + x + 2 + x + 4 = 81$ — Step 2: Write an **equation.**

$3x + 6 = 81$ — Step 3: **Solve.**

$-6 \quad -6$

$3x = 75$ — Step 4: **Check.**

$\div 3 \quad \div 3$

$x = 25$ — Step 5: Answer the question in words.

$\therefore$ the odd numbers are 25, 27, 29.

EXAMPLE 3

Find the value of n given the perimeter of the rectangle shown is 50 cm.

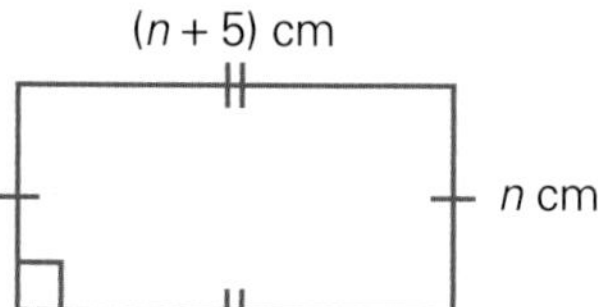

Solution

Perimeter $= 2(l + b)$ — Step 1: Introduce the **pronumeral.**

$\therefore 50 = 2(n + 5 + n)$ — Step 2: Write an **equation.**

$50 = 2(2n + 5)$

i.e. $4n + 10 = 50$ — Step 3: **Solve.**

$-10 \quad -10$

$4n = 40$

$\div 4 \quad \div 4$

$\therefore n = 10$ — Step 4/5: Check, and answer the question.

EXERCISE 5C

1 For each of the following, write an equation and then solve it to find the unknown number.

- **a** If 10 is added to a certain number, the result is 19.
- **b** The product of a certain number and 7 is 35.
- **c** Eight less than a certain number is 25.
- **d** If a certain number is divided by 4, the result is 9.
- **e** If I think of a number, double it and add 7, the result is 27. What is the number?
- **f** If I think of a number, halve it and subtract 4, the result is 12. What is the number?
- **g** If I multiply a number by 5, the result is 17. What is the number?

h Three less than a number is $^{-}3$. What is the number?

i When a certain number is quadrupled (multiplied by 4), then 9 is subtracted, the result is 23. Find the number.

j Tripling a certain number and then increasing the product by 5 gives $^{-}13$. What is the number?

k The sum of two consecutive integers is 69. What are they?

Level 2

2 A shirt is offered on sale for \$10 off. If x represents the original cost of the shirt and the shirt was sold for \$55, which of the following equations represents this situation?

A $x + 10 = 55$ **B** $10x = 55$ **C** $x \div 10 = 55$ **D** $x - 10 = 55$

3 The cost of hiring a taxi is \$3 plus \$2 per kilometre travelled. If n represents the number of kilometres travelled for a cost of \$12, which of the following equations best represents this?

A $3 - 2n = 12$ **B** $3 + 2n = 12$

C $5 + n = 12$ **D** $2 + 3n = 12$

4 Mark earns \$$x$ while his girlfriend Tanya earns \$20 more than Mark. If combined they earn \$290, which of the following equations represents this situation?

A $x + 20 = 290$ **B** $2x + 20 = 290$ **C** $x - 20 = 290$ **D** $2x - 20 = 290$

5 Two television shows are to be recorded on a 3-hour tape. The first show lasts x minutes; the second show is twice as long. If after taping there are 90 minutes left on the tape, which of the following equations represents this situation?

A $2x + 90 = 3$ **B** $3x + 90 = 180$ **C** $3 + x = 90$ **D** $2x + 90 = 180$

6 a Four consecutive even integers total 188. What is the largest of them?

b The sum of two consecutive even integers is 62. What is the smaller of the two numbers?

c The sum of three consecutive integers is 123. What are the numbers?

d The sum of three consecutive even numbers is 126. What are the numbers?

7 Joanne earns double Patrick's wages, since he works part-time. Together they earn \$960 per week. What are the weekly earnings of Joanne?

8 Find the value of x in each of the following:

a

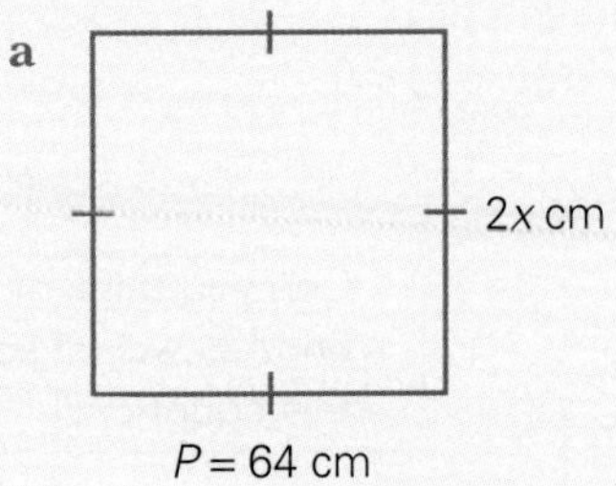

b

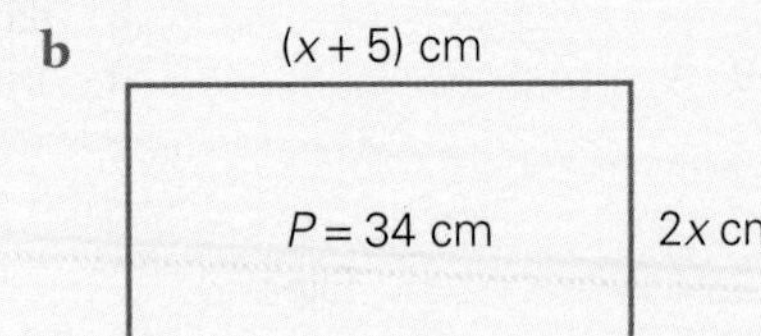

c

d

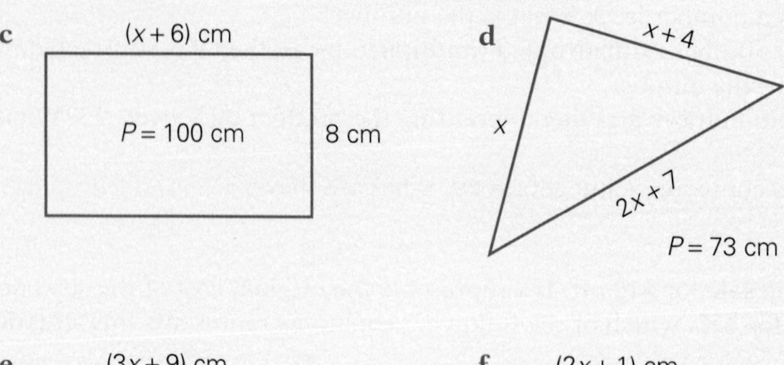

e

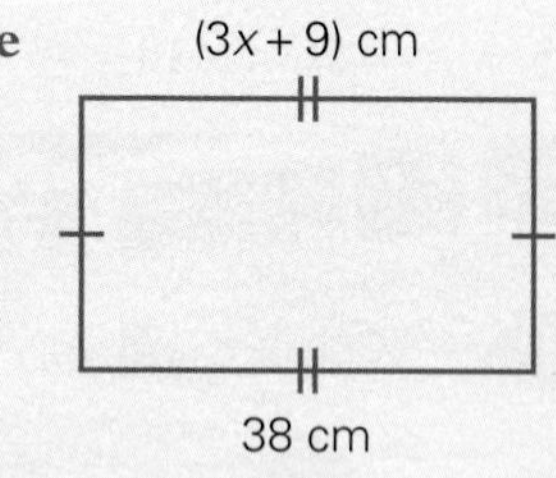

f

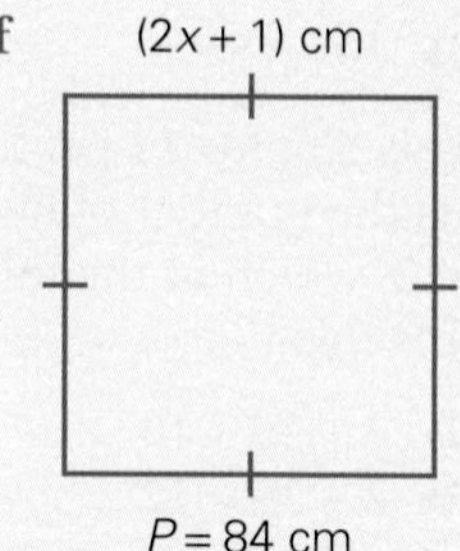

9 An isosceles triangle has each of its equal sides three times as long as its base. Its perimeter is 84 metres. Find its dimensions.

10 A rectangle is 7 metres longer than it is wide. Its perimeter is 46 metres. Find its dimensions, then its area.

11 Gloria is now 5 years older than Constanzia. The total of their ages is 41. How old will each be in 10 years' time?

12 A quantity of 20c and 50c coins has a total value of \$54. There are twice as many 20c coins as 50c coins. How many 20c coins are there?

5.4 Equations with brackets

When dealing with equations involving brackets, it is usually easier to expand the brackets and then solve the equation.

EXAMPLE

Solve each of the following equations by first expanding the brackets.

a $3(2x - 1) = {}^{-}9$ **b** ${}^{-}2(4 - 3x) = 6$

c $2a + 5(a - 1) = 9$ **d** $3(p - 1) - 2(3p + 1) = 0$

Solution

a $3(2x - 1) = {}^{-}9$ (Expand.)

$6x - 3 = {}^{-}9$ (Solve.)

$+3 \quad +3$

$6x = {}^{-}6$

$\div 6 \quad \div 6$

$x = {}^{-}1$

Check: $3(2 \times [{}^{-}1] - 1)$
$= 3({}^{-}3)$
$= {}^{-}9$

Can you think of another way of solving this equation?

b $^-2(4 - 3x) = 6$

$^-8 + 6x = 6$

$+8 \quad +8$

$6x = 14$

$\div 6 \quad \div 6$

$x = \frac{7}{3} = 2\frac{1}{3}$

Expand.

Solve.

Check: $^-2(4 - 3 \times \left[\frac{7}{3}\right])$
$= {}^-2(^-3)$
$= 6$

c $2a + 5(a - 1) = 9$

$2a + 5a - 5 = 9$

$7a - 5 = 9$

$+5 \quad +5$

$7a = 14$

$\div 7 \quad \div 7$

$a = 2$

Expand.

Simplify.

Solve.

Check: $2(2) + 5(2 - 1)$
$= 4 + 5$
$= 9$

d $3(p - 1) - 2(3p + 1) = 0$

$3p - 3 - 6p - 2 = 0$

$^-3p - 5 = 0$

$+5 \quad +5$

$3p = 5$

$\div {}^-3 \quad \div {}^-3$

$p = {}^-1\frac{2}{3}$

Expand—be careful of the signs.

Simplify.

Solve.

Check: $3(^-1\frac{2}{3} - 1) - 2(3 \times {}^-1\frac{2}{3} + 1)$
$= {}^-8 + 8$
$= 0$

EXERCISE 5D

1 Solve each of the following equations:

a $3(x + 1) = 9$ **b** $4(a - 1) = 8$ **c** $3(x + 4) = 12$
d $4(a + 2) = 10$ **e** $5(x - 1) = 5$ **f** $2(x - 9) = 10$
g $6(s + 3) = 6$ **h** $2(5 + a) = 8$ **i** $7(a - 8) = 21$
j $10(a + 2) = 10$ **k** $5(3 + a) = 15$ **l** $2(a - 6) = 0$

2 Solve:

a $2(2x + 1) = 10$ **b** $3(2x - 1) = 12$ **c** $2(5x - 7) = 10$
d $3(2x - 1) = 9$ **e** $5(4x - 8) = 10$ **f** $7(2p + 1) = 7$
g $12(5a - 3) = 24$ **h** $3(3 + 2d) = 9$ **i** $9(3p + 9) = 0$
j $4(3n - 9) = 0$ **k** $3(5k - 10) = 0$ **l** $5(3w - 5) = 35$

3 Solve each of the following equations:

a $3(x - 9) = {}^-21$ **b** $2(a - 3) = {}^-2$ **c** $8(2x + 1) = {}^-16$
d $4(7 - a) = {}^-4$ **e** $^-7(x + 1) = 21$ **f** $^-(2x - 9) = 1$
g $^-5(3 - a) = 10$ **h** $^-(7 - 6x) = 5$ **i** $^-3(2 - w) = 9$
j $^-5(2p + 3) = 15$ **k** $^-5(b - 1) = 20$ **l** $^-7(3 + s) = {}^-28$

4 Solve each of the following equations:

a $3(2 - 3x) = 15$ **b** $^-7(3p - 2) = {}^-2$ **c** $8(3x - 9) = {}^-10$
d $8(2 - 5x) = {}^-5$ **e** $^-2(2 - p) = {}^-7$ **f** $^-5(4 - 3p) = 0$

Level 2

5 Solve each of the following equations:

a $2(2x + 1) + x = 7$ **b** $2(x - 3) + 5x = 1$ **c** $3x + 2(x + 1) = 12$
d $3(2d + 1) - 5d = 6$ **e** $2p + 5(2p - 7) = 1$ **f** $6 + 4(3x - 8) = 12$
g $5(x + 8) - 4x = 9$ **h** $4p - (p - 9) = 3$ **i** $10 - 2(x - 8) = 30$
j $5(8 - 3x) - 10 = 30$ **k** $x - 3(x + 4) = {}^{-}8$ **l** $x - 3(4 - x) = {}^{-}8$

6 Solve:

a $2(x + 1) + 3(x + 4) = 4$ **b** $3(a - 6) + 7(a + 3) = 20$
c $4(3d - 1) + 3(3 + d) = 14$ **d** $8(t - 2) - 6(t + 4) = {}^{-}2$
e $3(x + 2) - 2(x + 1) = 6$ **f** $5(x + 6) - 4(x - 3) = 8$
g $4(2 - p) - (2 + p) = 9$ **h** $5(2w + 1) - 2(1 - w) = {}^{-}2$

7 For each of the following statements, write an equation and then solve it.

a Four is added to a number and the sum is multiplied by 3. If the result is 12, find the number.
b I think of a number, and subtract 5 before multiplying the result by 7. If this gives an answer of 19, what is the number?
c Twice a certain number less 1 is multiplied by 3 to give a result of 9. What is the number?

8 **a** For each of the following rectangles, set up an equation to represent its perimeter.
b If the perimeter of each rectangle is 14 cm, solve each of the equations:

i

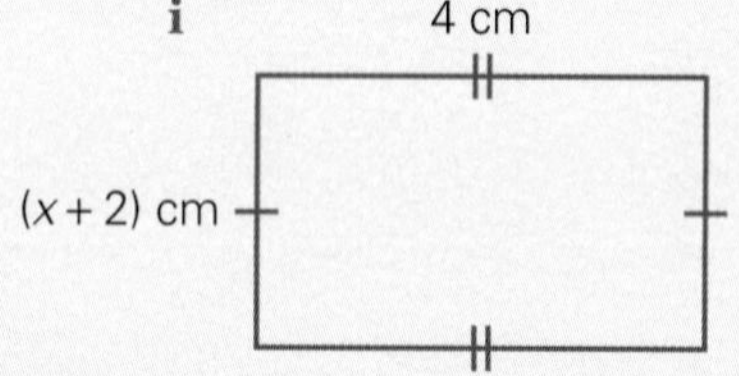

ii

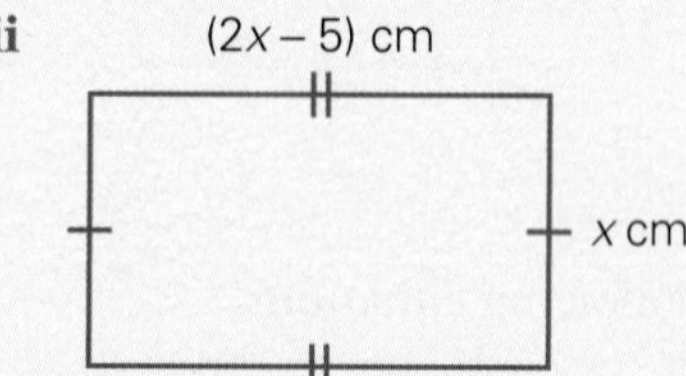

iii

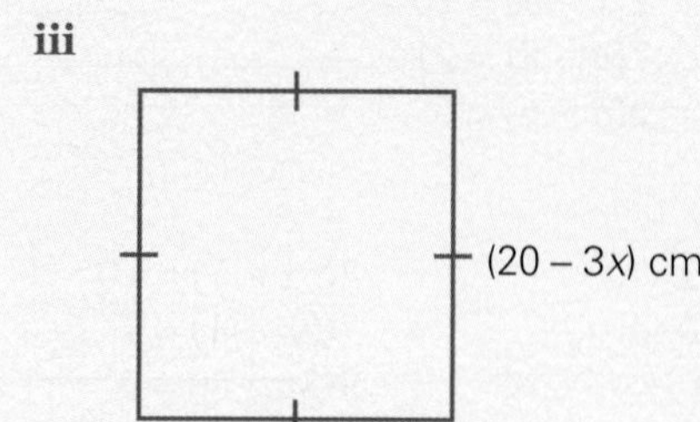

9 During the after-Christmas sale, the price of a DVD decreased by \$5. If Amanda bought six DVDs for \$96 during the sale, what was the cost of each DVD if they were all originally the same price?

10 The cost of a set price meal at a local restaurant increased by \$7 per person. A group of five friends dined at the restaurant and their bill came to a total of \$135. What would the price of the meal have been before the increase occurred?

5.5 Equations with pronumerals on both sides

If an equation has pronumerals on both sides of the equation, you need to collect all the terms involving the pronumeral on one side of the 'equal to' sign and the remaining terms on the other side.

It is usually easier to move the smaller number.

EXAMPLE 1

Solve $5x + 3 = 9 + 2x$

Solution

Strategy: eliminate pronumeral terms from one side—usually the side having the pronumeral with the smaller coefficient/s.

$5x + 3 = 9 + 2x$
$-2x \quad -2x$
$3x + 3 = 9$
$-3 \quad -3$
$3x = 6$
$\div 3 \quad \div 3$
$x = 2$

Check: $5 \times 2 + 3 = 9 + 2 \times 2$
$13 = 13$ ✔

EXAMPLE 2

Solve $26 - 3t = 5t + 2$

Solution

$26 - 3t = 5t + 2$
$+3t \quad +3t$
$26 = 8t + 2$
$-2 \quad -2$
$24 = 8t$
$\div 8 \quad \div 8$
$3 = t$
$\therefore t = 3$

EXAMPLE 3

Solve $2(c + 5) = 3(c - 2)$

Solution

$2(c + 5) = 3(c - 2)$
$2c + 10 = 3c - 6$
$-2c \quad -2c$
$10 = c - 6$
$+6 \quad +6$
$16 = c$
$\therefore c = 16$

EXERCISE 5E

1 Solve, using the strategy suggested in brackets:

a $2n + 5 = 6 + n$ (subtract n from each side, then take 5)
b $8p - 9 = 5p + 12$ (subtract $5p$, add 9, divide by 3)
c $2 + 4a = 16 - 3a$ (add $3a$, take 2, $\div$ 8)

2 Solve the following. Check your solutions.

a $7x = 2x + 15$ b $5t = 12 - t$ c $3p - 8 = p$
d $6n + 7 = 2n + 9$ e $3 + 10c = 8c + 11$ f $5n + 3 = 4n - 5$
g $3x + 5 = 2x + 7$ h $4p - 8 = 2p + 2$ i $5k + 10 = 3k + 4$
j $5x - 1 = x + 3$ k $8 + 2t = 10 + t$ l $3x - 9 = 3 - x$
m $6y - 10 = 12 - 5y$ n $2x - 9 = x - 10$ o $2w + 3 = w + 6$

3 Solve:

a $2h + 32 = 8 + 6h$ b $7l - 20 = 15 + 2l$ c $2 + 7a + 6 = 20 + 3a$
d $5b + 4 + 2b = 3b + 20$ e $5 - 3y + 2y = 4y$ f $5 - 3t = t + 5$
g $5 - 2t = 10 - t$ h $3v + 7 = 11 - v$ i $14 - 5x = 2x$
j $7y - 9 = 9 - 2y$ k $5 - 3p = {}^{-}9 - p$ l $2 - 9x = {}^{-}10x$
m $12 - 5a = 10 - a$ n ${}^{-}p - 1 = p + 3$ o $4 - a = a - 6$

TEACHER

Level 2

4 Find the value of the pronumeral in each equation:

a $5 + 2(z - 4) = 9$
b $5(2p + 3) - 4 = 31$
c $2(3n - 4) + 1 = 17$
d $3(5n - 1) - 5n = 7$
e $3(x + 4) = 2(7 + x)$
f $5(k - 1) = 4(k + 2)$
g $3(2m + 1) = 3(m + 2)$
h $3(2 + p) = 5(2 - p) + 6p$

5 Vladimir found that six fixed-cost restaurant meals plus \$50 in drinks came to the same amount as did Tania's two fixed-cost meals plus \$130 in drinks. What was the cost of each fixed-cost meal?

6 I think of a number and add 9. The result is twice the first number. What is the first number?

7 Given ΔABC is isosceles, find the value of x.

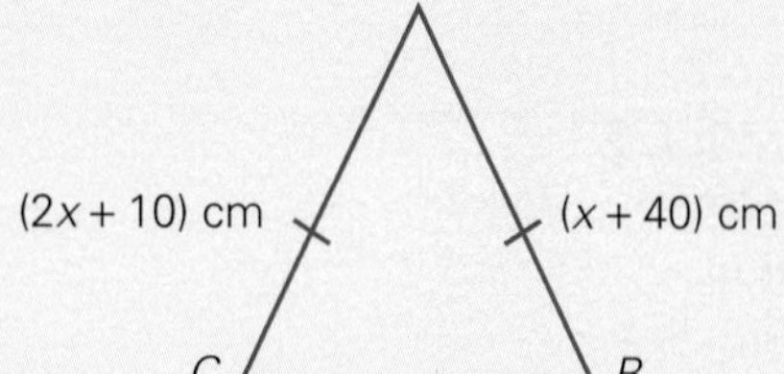

8 Sarah earns \$80 more than Tom and Barbara earns double Tom's income. Find the value of Tom's income if Barbara and Sarah both earn the same amount.

9 Find the value of a and b in each of the following:

a

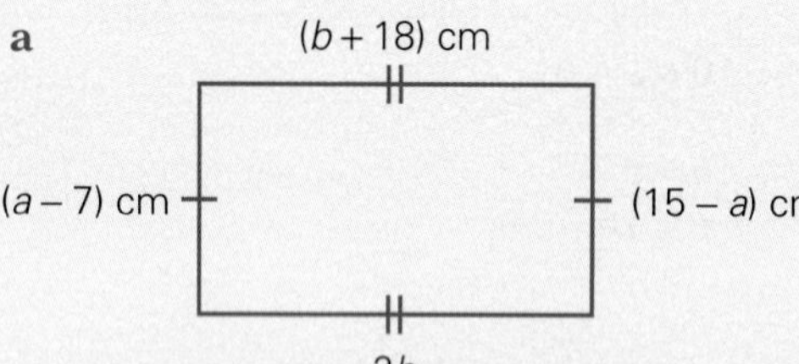

b

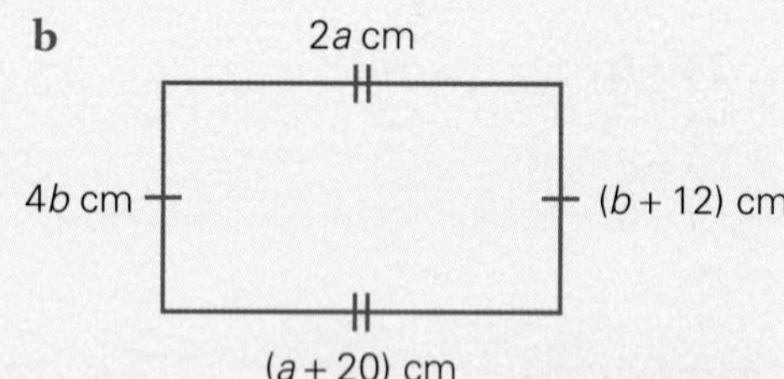

c

4b
2a – 12
a + 28

d

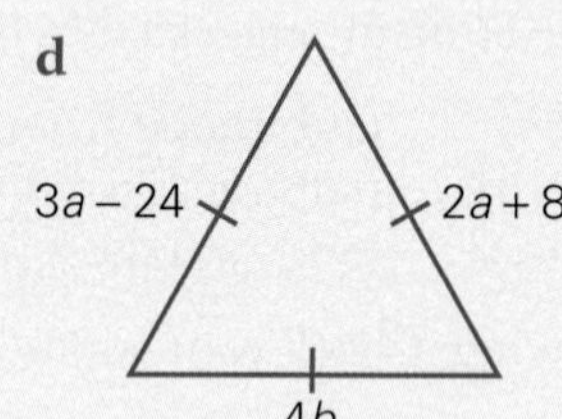

10 a If twice a number is subtracted from 12 and the result is the same as if 4 was added to the number, what is the number?

b Four times a number less 5 is the same as double the number plus 3. Find the number.

c The sum of three consecutive even numbers is the same as six times the smallest of the even numbers in the group. Find the three numbers.

d A number x is increased by 3, then this amount is multiplied by 5. The result is twice x plus 6. Find the number.

e Two sisters, Jodi and Katie, are given the same amount of pocket money. Jodi buys a magazine and has \$1.50 left. Katie buys two magazines and has 50 cents left. What amount of money did each receive?

TEACHER

Investigation Mix and match equations

- Make a set of cards like those shown.
- Give one card to each member of the group and have them organise themselves in the correct order to show the steps in solving each equation.
- Make up your own set of cards.

$2a + 1 = 9$	$a = 3\frac{1}{2}$	$a + 7 = 5$	$4a = 15 - 1$
$3a + 9 = 0$	$2a = 9 - 1$	$a = 4$	$a = 5 - 7$
$5a - 2a = 9 - 3$	$2a = 8$	$a = {}^{-}3$	$3(a + 3) = 0$
$\frac{a+7}{2} = 6$	$3a = {}^{-}9$	$a = \frac{6}{3}$	$5a + 3 = 2a + 9$
$\frac{4a}{3} = 7$	$3a = 6$	$a = 12 - 7$	$a = 2$
$5 - 9 = 2a + 2a$	$a + 7 = 6 \times 2$	$a = \frac{21}{4}$	$a = 5$
$\frac{4a+1}{3} = 5$	$a + 7 = 12$	$a = 5\frac{1}{4}$	$5 - 2a = 2a + 9$
$a = \frac{14}{4}$	$4a = 21$	$a = \frac{{}^{-}4}{4}$	$a = {}^{-}1$
${}^{-}4 = 4a$	$4a + 1 = 15$	$4a = 14$	$a = {}^{-}2$

5.6 Equations and formulae

A formula (or rule) is an equation representing a special relationship.

$A = 1 \times b$, $P = 2l + 2b$ and $E = mc^2$ are all examples of formulae.

The value of one of the pronumerals can be found if you are given the values of the other pronumerals.

> $E = mc^2$ is the formula for energy.

EXAMPLE

Given $A = \frac{(a+b)h}{2}$, find the value of:

a A if $a = 4$, $b = 8$ and $h = 6$

b h if $A = 12$, $a = 4$ and $b = 2$

Solution

a
$$A = \frac{(a+b)h}{2}$$
$$A = \frac{(4+8)\times 6}{2}$$
$$= \frac{12 \times 6}{2}$$
$$= 36$$

b
$$A = \frac{(a+b)h}{2}$$
$$12 = \frac{(4+2)h}{2}$$
$$24 = 6h$$
$$4 = h$$
$$\therefore h = 4$$

EXERCISE 5F

1 The formula for the perimeter of a rectangle is $P = 2(l + b)$. Find the value of P if:

a $l = 3, b = 8$ **b** $l = 8, b = 3$ **c** $l = 9, b = 6$ **d** $l = 10, b = 12$
e $l = 7, b = 9$ **f** $l = 0{\cdot}5, b = 0{\cdot}4$ **g** $l = 9{\cdot}8, b = 5{\cdot}6$ **h** $l = b = 15$

2 The formula for the area of a rectangle is $A = lb$. Find the value of A for the values of l and b given in question **1**.

3 If $S = \frac{D}{T}$, find the value of S given:

a $D = 100, T = 2$ **b** $D = 45, T = \frac{1}{2}$ **c** $D = 90, T = 2{\cdot}5$

4 Given the formula $v = u + at$, find the value of v when:

a $u = 3, a = 1, t = 9$ **b** $u = 5, a = 2, t = 10$ **c** $u = 7, a = 4, t = 12$
d $u = 18, a = 9, t = 15$ **e** $u = 1{\cdot}2, a = \frac{1}{2}, t = 8$ **f** $u = 12{\cdot}8, a = {}^{-}8, t = 0{\cdot}5$
g $u = 100, a = \frac{3}{4}, t = 12$ **h** $u = 3{\cdot}4, a = 4{\cdot}5, t = 6{\cdot}5$ **i** $u = \frac{1}{2}, a = 4, t = 8$

5 For the formula $y = mx + b$, find the value of y if:

a $m = 2, x = 5, b = 3$ **b** $m = {}^{-}2, x = 5, b = 3$ **c** $m = 4, x = 1, b = 9$
d $m = 1, x = 4, b = {}^{-}4$ **e** $m = \frac{1}{2}, x = 6, b = 2$ **f** $m = \frac{1}{2}, x = 0, b = {}^{-}5$
g $m = {}^{-}3, x = {}^{-}2, b = 5$ **h** $m = 4{\cdot}5, x = 6, b = 0$ **i** $m = \frac{{}^{-}3}{2}, x = 6, b = 12$

6 For the formula for energy, $E = mc^2$, find the value of E if:

a $m = 9, c = 2$ **b** $m = 10, c = 1$ **c** $m = 2{\cdot}64, c = 0{\cdot}2$ **d** $m = \frac{1}{2}, c = \frac{1}{4}$
e $m = 15, c = 2$ **f** $m = 4, c = \frac{1}{2}$ **g** $m = 19{\cdot}8, c = 1{\cdot}4$ **h** $m = 3{\cdot}4, c = 0{\cdot}9$

SPREADSHEET ACTIVITY

7 To convert degrees Fahrenheit to degrees Celsius, the formula $C = \frac{5}{9}(F - 32)$ can be used. Convert the following degrees Fahrenheit to degrees Celsius:

a $F = 0$ **b** $F = 10$ **c** $F = 32$ **d** $F = 40$ **e** $F = 100$

8 If $I = PRT$, find I if:

a $P = 1200, R = 0{\cdot}5, T = 4$ **b** $P = 450, R = 0{\cdot}05, T = 12$
c $P = 4500, R = 0{\cdot}065, T = 3{\cdot}5$ **d** $P = 80, R = 0{\cdot}075, T = 5\frac{1}{2}$

9 $c^2 = a^2 + b^2$. Find the value of c given c is positive and:

a $a = 3, b = 4$ **b** $a = 6, b = 9$ **c** $a = 0{\cdot}8, b = 0{\cdot}6$ **d** $a = 8, b = 15$

Level 2

10 a Given $y = 2x + 1$, find the value of x if $y = 9$.
b Given $y = a - 9$, find the value of a if $y = {}^{-}6$.
c Given $y = 9 - 2x$, find the value of x if $y = \frac{1}{2}$.

11 If $A = lb$, find the value of l if:

a $A = 18, b = 6$ **b** $A = 25, b = 5$ **c** $A = 60, b = 3$ **d** $A = 4{\cdot}6, b = \frac{1}{2}$

12 If $A = lb$, find the value of b if:

a $A = 90, l = 9$ **b** $A = 24, l = 4$ **c** $A = 12{\cdot}4, l = 0{\cdot}6$ **d** $A = 100, l = 2{\cdot}5$

13 If $P = 2(l + b)$, find the value of b given:

a $P = 12, l = 3$ b $P = 15, l = 0{\cdot}5$ c $P = 2{\cdot}4, l = 0{\cdot}01$ d $P = 56, l = \frac{3}{4}$

14 If $v = u + at$, find the value of t given:

a $v = 12, u = 9, a = 1$ b $v = 15, u = 5, a = 2$

c $v = 200, u = 8, a = \frac{1}{2}$ d $v = 9{\cdot}6, u = 4{\cdot}3, a = 5$

15 Given the formula $A = \frac{a+b}{2}$, find the value of a if:

a $A = 4{\cdot}6$ and $b = 0{\cdot}2$ b $A = 9{\cdot}36$ and $b = 0{\cdot}3$

16 If x and y are both positive numbers and $y = x^2$, find the value of x given y has the following values:

a 16 b 25 c 64 d 0·49 e 10 000

17 By using the formula (question **7**) for converting degrees Fahrenheit to degrees Celsius, find the degrees Fahrenheit that converts to the following degrees Celsius:

a $C = 0$ b $C = 32$ c $C = 100$ d $C = 25$ e $C = 10$

18 Given $A = \frac{(a+b)h}{2}$, find the value of h given $a = 4$, $b = 9$ and $A = 34{\cdot}6$.

19 If $S = \frac{a}{1-r}$, find the value of r given $a = 10$ and $S = 20$.

20 If $y = ax^2$, find the value of x given $y = 400$, $a = 4$ and x is positive.

21 Given $c^2 = a^2 + b^2$, find the two values of a, correct to 1 decimal place, if $c = 12$ and $b = 9$.

22 Given $v^2 = u^2 + 2aS$, find the two values of u, correct to 1 decimal place, if $v = 10$, $a = 2$ and $S = 3$.

SPREADSHEET ACTIVITY

SPREADSHEET ACTIVITY

SPREADSHEET ACTIVITY

SPREADSHEET ACTIVITY

5.7 Inequations

An inequation (or inequality) is a mathematical statement that indicates that one quantity is greater than or less than another.

Remember:

$<$ less than $>$ greater than

$\leq$ less than or equal to $\geq$ greater than or equal to

When using the inequality symbols, it is important to remember that the narrow end points towards the smaller number and the wider end points towards the larger number.

We can show a value on a number line: $x = 2$

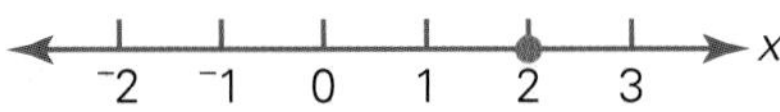

An inequation can also be shown on a number line:

Remember: on number lines, use an open circle for < or > and a closed circle for ≤ or ≥.

$x \geq 2$

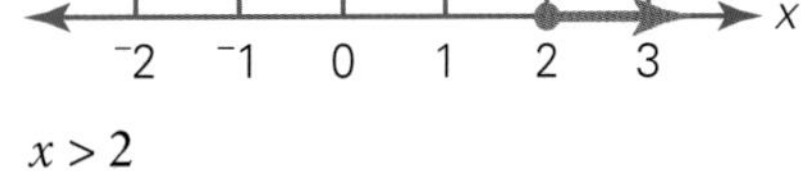

$x > 2$

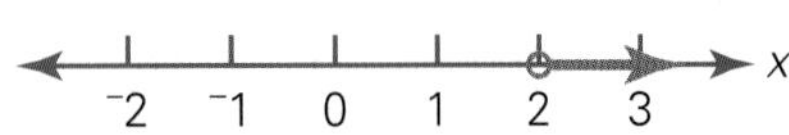

EXAMPLE

Write each of the following as an inequation and then show each solution on a number line:

a x is less than or equal to 3. b x is greater than 1.

c x is less than 0. d x is greater than or equal to $^{-}2$.

Solution

a $x \leq 3$

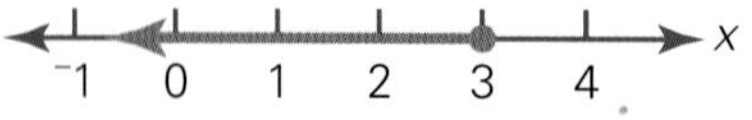

b $x > 1$

c $x < 0$

d $x \geq {}^-2$

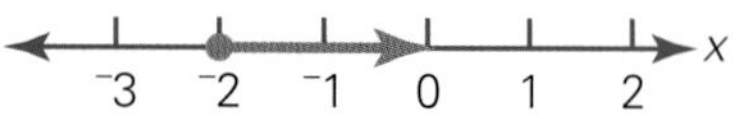

EXERCISE 5G

1 Match each inequation given with the correct number line.

a $x > 4$

b $x < 4$

c $x \geq 4$

d $x > {}^-4$

e $x \leq {}^-4$

1 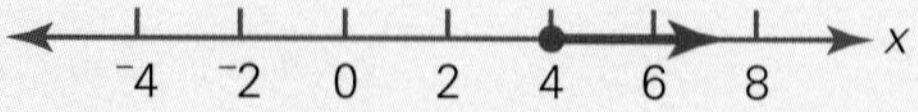

2 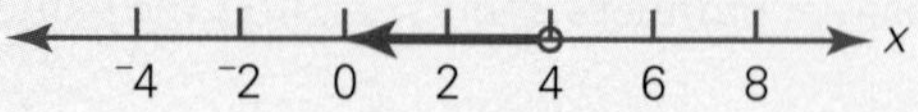

3 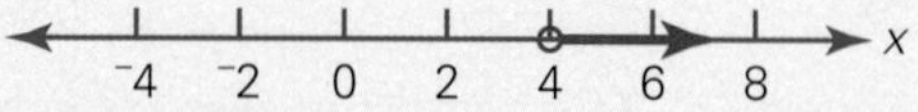

4 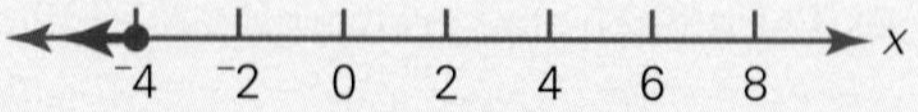

5 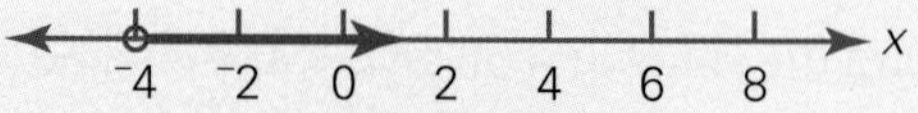

2 Write each of the following as an inequality:

- **a** The number of people who visit the Sydney Opera House each year is over 100 000.
- **b** The number of lollies in a bag should be at least 50.
- **c** A factory worker must pack more than three boxes a minute.
- **d** More than 100 penguins take part in the nightly parade on Phillip Island.
- **e** The weight of a suitcase is less than or equal to 30 kg.

3 Write each of the following statements as an inequation and determine which of the numbers below make each inequation true:

$${}^-6, {}^-2, {}^-\tfrac{1}{2}, 0, 2, 5, 7, 10, 15, 24$$

- **a** x is less than zero.
- **b** x is greater than 10.
- **c** x is greater than or equal to 10.
- **d** x is less than or equal to zero.
- **e** x is greater than or equal to ⁻1.
- **f** x is less than 10.

Level 2

4 Show each of the following on separate number lines:

- **a** $x \geq 7$
- **b** $x > 1$
- **c** $x < 1$
- **d** $x \leq 1$
- **e** $x \geq {}^-1$
- **f** $a \geq 0$
- **g** $p \geq {}^-2$
- **h** $a > {}^-15$
- **i** $h < 5$

5 Write an inequation to describe what is shown on each of the following number lines:

a 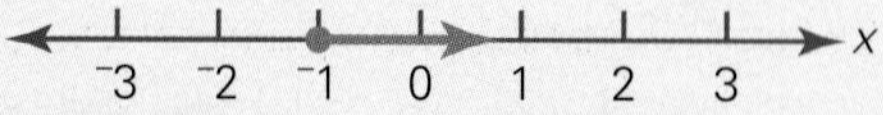

b 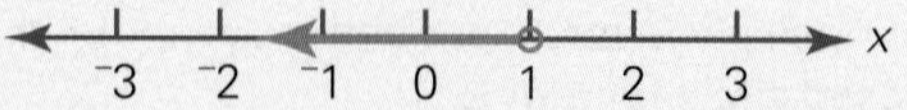

c 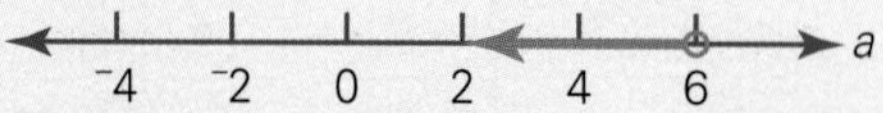

d 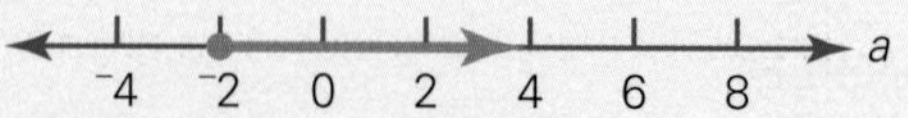

e 

f 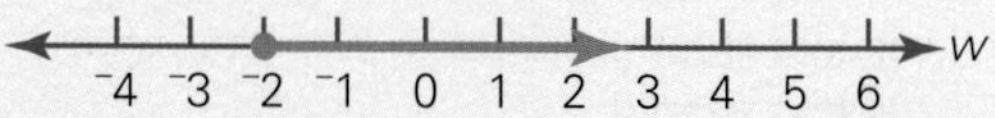

Investigation Inequalities

Let us start with the numbers 4 and 6 and the true relationship $4 < 6$.

Copy and complete the following table:

	4	6	$4 < 6$
Add 3	$4 + 3$	$6 + 3$	$7 < 9$
Subtract 3	$4 - 3$		1
Multiply by 2			
Divide by 2			
Multiply by $^{-}2$			$^{-}8 \nless {}^{-}12$ $\therefore {}^{-}8 > {}^{-}12$
Divide by $^{-}2$	$4 \div (^{-}2)$	$6 \div (^{-}2)$	

What **conclusions** can be drawn from the table on the left if an inequality is multiplied or divided by a negative number?

5.8 Solving inequalities

You can solve inequations in a similar way to solving equations.

An inequation (inequality) remains *true* when we:

- *add* the same number to or *subtract* the same number from both sides
- *multiply* or *divide* both sides by the same *positive* number

but the inequality sign is *reversed* if we *multiply* or *divide* both sides by the same *negative* number.

EXAMPLE 1

Solve each of the following inequations, and show your solutions on a number line:

a $2x - 1 > 17$ **b** $\frac{x}{3} + 4 \le 5$

Solution

a $2x - 1 > 17$

$\quad {+1} \quad {+1}$

$2x > 18$

$\quad \div 2 \quad \div 2$

$x > 9$

0 9 x

b $\frac{x}{3} + 4 \le 5$

$\quad -4 \quad -4$

$\frac{x}{3} \times 3 \le 1 \times 3$

$x \le 3$

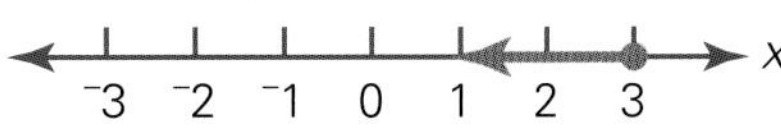

EXAMPLE 2

Solve $4 - x \ge 6$.

Solution

$4 - x \ge 6$ | Move the negative pronumeral.

$\quad +x \quad +x$

$4 \ge 6 + x$

$\quad -6 \quad -6$

$^{-}2 \ge x$ | Reverse.

$\therefore x \le {}^{-}2$ | Note that the inequality sign still 'points' to the x.

Alternatively

$$4 - x \geq 6$$
$$\quad -4 \quad\; -4$$

The opposite of ^{-}x is x.
The opposite of $\geq$ is $\leq$.
The opposite of 2 is $^{-}2$.

$$^{-}x \geq 2$$
$$\downarrow \; \downarrow \; \downarrow$$
$$x \leq {}^{-}2$$

When we divide both sides by a *negative* number, the sign changes.

EXERCISE 5H

1 Solve each of the following inequalities:

a $2x > 10$ b $x + 2 < 7$ c $3x > 15$
d $\frac{x}{2} \geq 8$ e $x - 3 > 4$ f $x - 3 < 4$
g $p + 8 \leq 0$ h $3a > 0$ i $x - 7 < 0$
j $2x \leq 14$ k $5m > {}^{-}15$ l $d - 3 > 2{\cdot}4$
m $\frac{x}{7} \leq 0{\cdot}1$ n $\frac{1}{2}x \leq 6$ o $5 + x > 9$

2 Display each of your answers to question **1** on separate number lines.

3 Solve:

a $2 + 4a \leq 10$ b $5 + 2y > 11$ c $3p - 1 > 14$
d $3x - 2 \geq 10$ e $3x - 2 < 1$ f $5 + 2w \geq 8$
g $5x + 5 < 10$ h $5x - 5 \geq 0$ i $10p - 2 < 8$

Level 2

4 Give the solution set for each of the following:

a $\frac{x + 2}{4} \leq 1$ b $\frac{a - 3}{2} \leq {}^{-}1$ c $\frac{x}{4} - 1 \geq 6$
d $\frac{x}{3} + 7 > 2$ e $5 + \frac{x}{2} < 7$ f $\frac{x + 2}{4} < 8$
g $\frac{2x - 7}{3} > 4$ h $\frac{2x + 1}{5} < 0$ i $\frac{3x}{2} + 1 \geq {}^{-}3$
j $5x - 4 > 2 - x$ k $4(2x + 1) \geq 16$ l $3x + 7 < x - 2$

5 For each of the following, write an inequality and give the solution set:

a If a number is multiplied by 3, the result is less than 9.
b If a number is multiplied by 3 and the result divided by 4, it creates an answer less than 6.
c If a number is doubled and then 15 is added, the result is greater than 20.
d Thomas is x years old and Gary is 4 years older. The sum of their ages is less than 24.
e Kaitlyn has x rides on the ferris wheel at \$4 a ride and spends \$7 on food. The total amount she spends is less than or equal to \$27.

6 Choose an *appropriate strategy* to solve the following:

a $5 - x < 6$ b $7 - x \geq 10$ c $^{-}p \leq 7$
d $9 - a < {}^{-}10$ e $^{-}w \geq 6$ f $^{-}3 - 2p < 9$
g $4w + 5 < 3w$ h $4p + 2 \geq {}^{-}8 - p$ i $10 - 3a > 5 + a$

Chapter Review

Language Links

backtrack	formula	isolate	substitute
equation	inequality	operation	translate
evaluate	inspection	pronumeral	unknown
expression	inverse	solution	

1 Explain, and give an example of, the difference between an algebraic expression and an equation.

2 What does it mean to evaluate an expression?

3 Give a mathematical as well as a non-mathematical definition of the word 'substitute'. Where might you use the word 'substitute' in your everyday life?

4 What does it mean to 'isolate the pronumeral' when solving an equation?

5 Write a word question that could account for the following equations:

a $2x = 16$ b $x - 9 = 11$ c $3p + 9 = 2p + 4$

6 What is the difference between an equation and an inequality?

Chapter Review Exercises

1 Which of the following are equations? 5.1

a $3p$ b $x - 7 > 10$ c $a + 1 = 7$ d $4(c - 3)$

2 Which of the following are true for *all* values of x? 5.1

a $x + 5 = 7$ b $x - x = 0$ c $x \div (^-x) = {}^-1$
d $x(x + 1) = x^2$ e $x^2 \geq 0$

3 In which of the following equations is $w = 3$ a solution? 5.1

a $10 + w = 4w + 1$ b $2w \div 3 = 2$ c $5(w - 1) = 10$
d $4w - 2 = 10$ e $2w = {}^-6$ f $^-w = {}^-6$

4 Solve the following equations: 5.2

a $3p = 27$ b $12 = \frac{m}{6}$ c $4 - x = 3$
d $8w = {}^-16$ e $^-q = {}^-12$ f $^-9 - x = 0$

5.2 **5** Solve each of the following equations:

a $5 + 7t = 40$ b $\frac{c}{2} + 4 = 9$ c $\frac{d-2}{6} = 3$

d $4 + 5w = 12$ e $8 - 2x = 10$ f $\frac{w}{8} - 1 = 3$

g $2 + 2w = {}^{-}6$ h $2t - 7 = 10$ i $\frac{4-x}{3} = 4$

5.4 **6** Find the value of the pronumeral in each of the following equations:

a $6 = 4 - 2n$ b $8(h + 3) = 56$ c $27 = (5k - 1)3$
d $3v + 2 - v = 6$ e $9 - 4p = {}^{-}3$ f ${}^{-}9(2w - 1) = 27$

5.5 **7** Find the value of the pronumeral:

a $5x + 3 = 2x + 6$ b $3w + 4 = 7w + 8$ c $3(b + 5) - 7 = 14$
d $9c + 1 = 7(c + 3)$ e $2(5 - d) = 3(4d - 2)$ f $8(2x - 3) - (x - 1) = 0$

5.6 **8** a If $P = 2l + 2b$, find the value of P given $l = 9$ and $b = 11$.
b If $A = lb$, find the value of l given $A = 124$ and $b = 4$.
c If $A = \frac{(a+b)h}{2}$, find the value of a given $h = 8$, $b = 7$ and $A = 45$.
d If $E = mc^2$, find the value of c if $E = 90$ and $m = 0{\cdot}1$.
e Given $c = \sqrt{a^2 + b^2}$, find the value of a if $c = 5$ and $b = 3$.

5.3 **9** Translate into an algebraic equation and solve:

a Hans is paid a base wage of \$300/week plus \$50 for every appliance he sells. How many appliances must he sell to earn \$750 for the week?
b Two consecutive multiples of 6 add to 150. What is the larger of the multiples?
c The sum of three consecutive odd numbers is 177. Find the value of the three numbers.
d The cost of hiring a hall for Josie's 21st birthday party is \$140 plus \$45 per hour the hall is required. For how many hours can the hall be hired if the budget allows for a cost of \$500?
e The sum of a number and 14 is the same as double the number less 4. What is the number?

5.7 **10** Show on separate number lines:

a $x \geq 1$ b $x < 5$ c $x = {}^{-}1\frac{1}{2}$

5.8 **11** Solve each of the following inequations:

a $3x > 9$ b $3x + 1 < 10$ c $\frac{x}{3} - 6 > 12$
d $5(x + 2) \leq 20$ e $\frac{x-8}{2} \geq 1$ f $\frac{2x}{3} + 5 < 4$

5.8 **12** Give three integer values that satisfy:

a $3x - 5 < 34$ b $1 - \frac{z}{4} \geq 6$ c $3p - 1 < 0$

5.8 **13** Solve:

a ${}^{-}w < 4$ b $1 - 3p > 4$ c $2 - a \leq 4$

5.8 **14** Solve $4a + 12 \leq 26 - 3a$.

Keeping Mathematically Fit

Part A: Non-calculator

1 Find the missing number: $(20 - 4) \times \square = 32 \div 4$

2 How many days from 15 May to 1 July?

3 Insert an operation sign (+, –, ×, ÷) to make the following true: $40 \square 10 = 32 \square 8$

4 Find $2\frac{1}{2}\%$ of 4840.

5 If eight doughnuts cost \$3.20, what would a dozen doughnuts cost?

6 Simplify $\frac{32 + 24}{70}$.

7 If $4x = {}^{-}16$, find $x + 6$.

8 Simplify $\frac{2^5 \div 2^4}{2}$.

9 Find an expression for the nth term in 16, 12, 8, ...

10 Given $624 \times 19 = 11\,856$, what is the value of $0{\cdot}624 \times 1{\cdot}9$?

Part B: Calculator

1 Evaluate $\sqrt[3]{47{\cdot}8 - 2{\cdot}3^2}$, correct to 1 decimal place.

2 \$690 is invested at 6% p.a. for 5 years. Calculate the simple interest earned on this investment.

3 Find the value of $6\frac{4}{7} - 3\frac{1}{2}$.

4 Find 8·7% of \$96.45.

5 Find $\frac{2}{3} \div \frac{5}{12}$.

6 Simplify:

a $3^4 \times 3^7 \times 3^2$ b $6x + 4(2 - x) + 5$ c $\frac{24xy}{6x + 4x}$

7 A 28 cm piece of wire is bent to form a rectangle whose length is three less than its width. Calculate the area of the rectangle.

8 Form a 3 × 3 magic square using the first nine counting numbers. Can this be done in more than one way?

TEACHER

1 What you need to know and revise

Outcome MS 4.1:

Uses formulae and Pythagoras' theorem in calculating perimeter and area of circles and figures composed of rectangles and triangles

Outcome NS 4.1:

Recognises the properties of special groups of whole numbers and applies a range of strategies to aid computation

- $\sqrt{}$ and $\sqrt[3]{}$

2 What you will learn in this chapter

Outcome MS 4.1:

Uses formulae and Pythagoras' theorem in calculating perimeter and area of circles and figures composed of rectangles and triangles

- identifies the hypotenuse
- establishes the relationship between the lengths of sides of a right-angled triangle in practical ways
- uses Pythagoras' theorem to find the lengths of sides in right-angled triangles
- solves problems involving Pythagoras' theorem
- gives answers to questions involving Pythagoras' theorem in exact form as a surd and also as an approximation
- Pythagorean triads
- the converse of Pythagoras' theorem as a test for right-angled triangles

Working Mathematically outcomes WMS 4.1–4.5

Students will be asked to ***question***, ***apply strategies***, ***communicate***, ***reason*** and ***reflect*** in the sections of this chapter.

Pythagoras' theorem

Key mathematical terms you will encounter

diagonal	hypotenuse	right	square
exact	Pythagoras	root	triad

MathsCheck
Pythagoras' theorem

1 i Name each of the triangles below. **ii** What types of triangles are they?

a

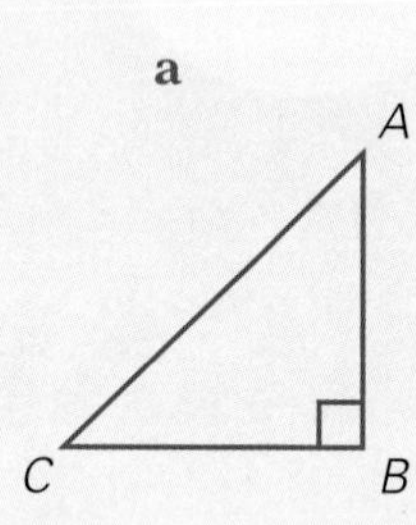

b

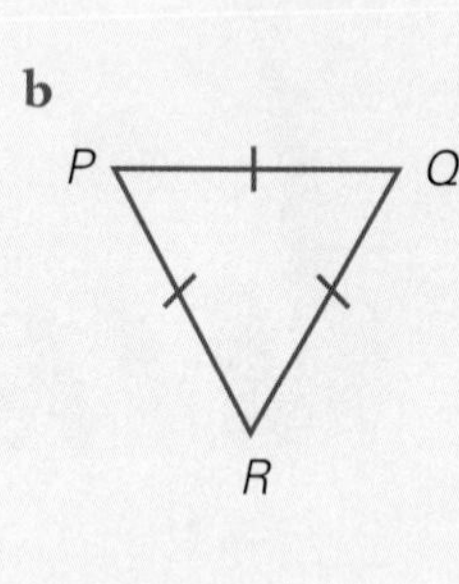

c

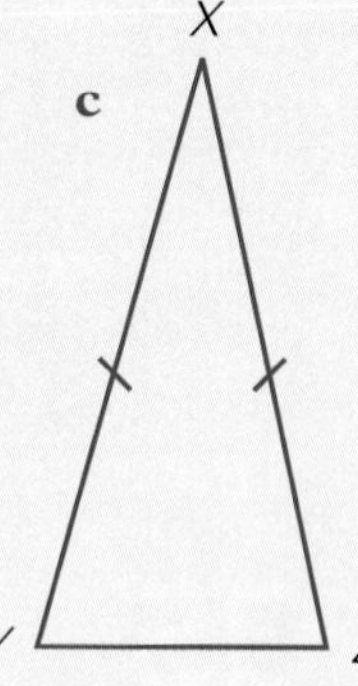

d

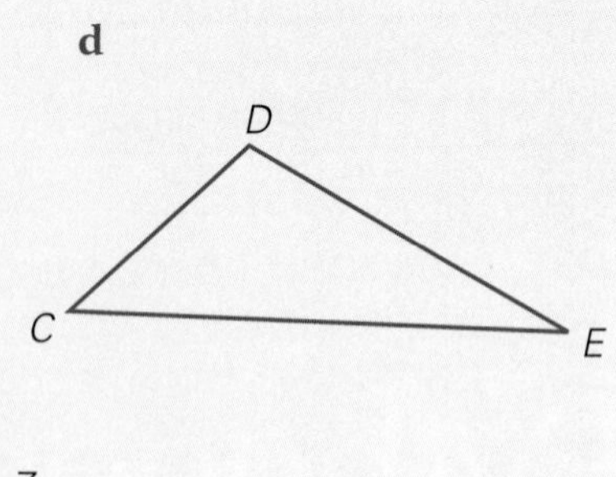

2 Name the side opposite the right angle for each of the triangles below.

a

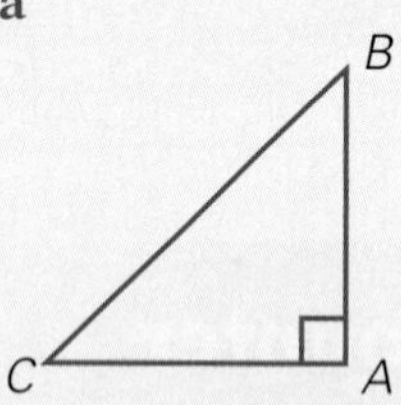

b

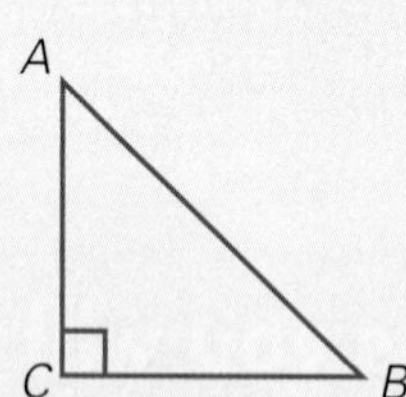

c

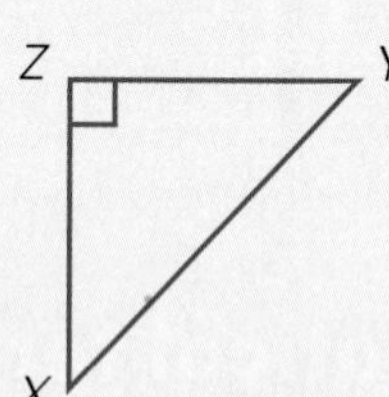

d

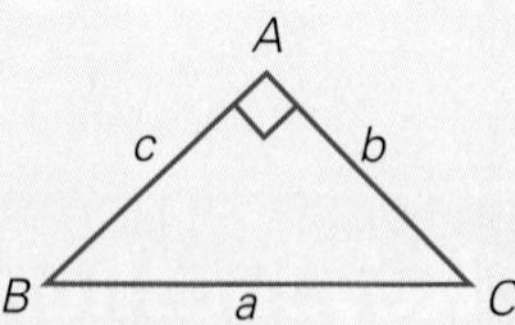

3 Find the perimeter of:

a

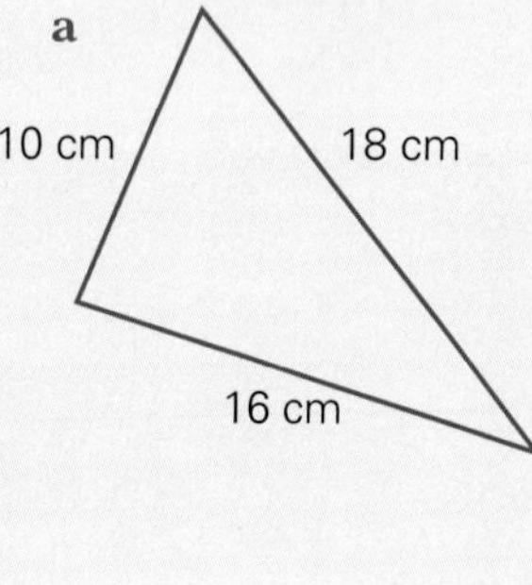

b

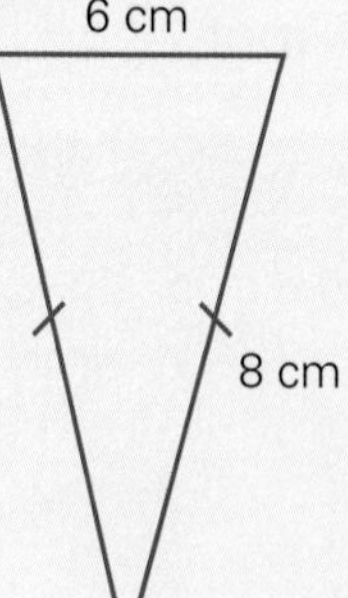

c

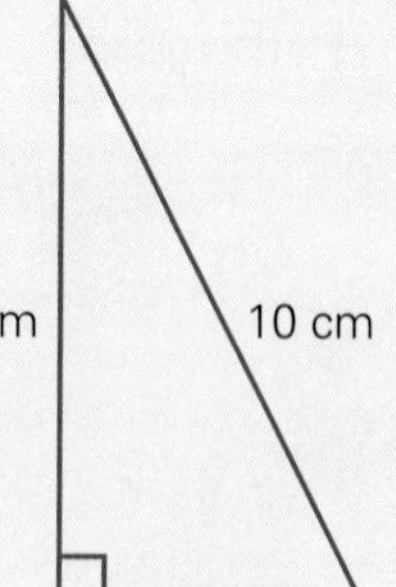

d

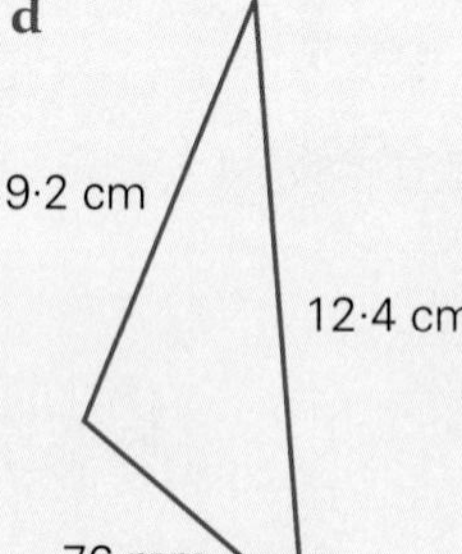

4 Find the value of:

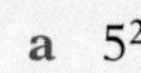

a 5^2 **b** $3{\cdot}4^2$ **c** 15^2 **d** 41^2 **e** $3^2 + 4^2$ **f** $9^2 - 5^2$

5 Evaluate, correct to 2 decimal places, where necessary:

a $\sqrt{164}$ **b** $\sqrt{41}$ **c** $\sqrt{93}$ **d** $\sqrt{81}$ **e** $\sqrt{8^2 - 3^2}$ **f** $\sqrt{6^2 + 8^2}$

TEACHER

Exploring New Ideas

6.1 Discovering Pythagoras' theorem

A Greek philosopher and mathematician born c. 582 BC, Pythagoras spent time studying in Egypt and Babylon before settling in the south of Italy.

Pythagoras was the founder of a secret school or brotherhood. They were concerned with solids, music and the relationships between whole numbers.

In the following exercise, you will be able to discover what the secret society of Pythagoras found out about triangles.

INTERACTIVE GEOMETRY

EXERCISE 6A

To explore triangles, you will need compass, ruler and protractor.

1 Draw a base line of length 8 cm. With compass and ruler, construct the other two sides of ΔABC as shown. What type of triangle is ΔABC? Measure its angle sizes.

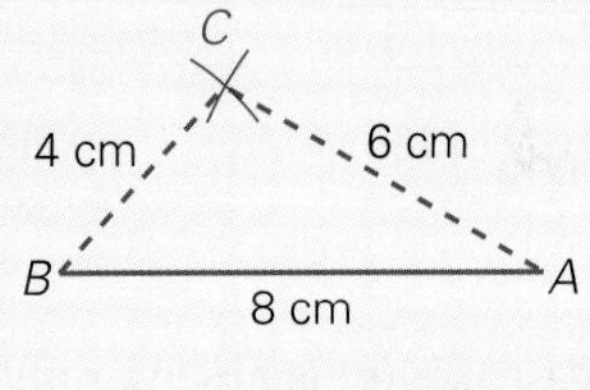

2 Construct ΔXYZ as shown. What type of triangle is it? Measure its angle sizes.

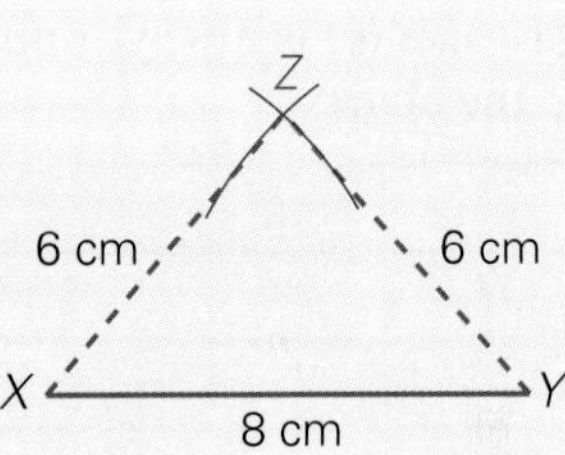

3 Now draw ΔMPQ. Name the type and measure the angles.

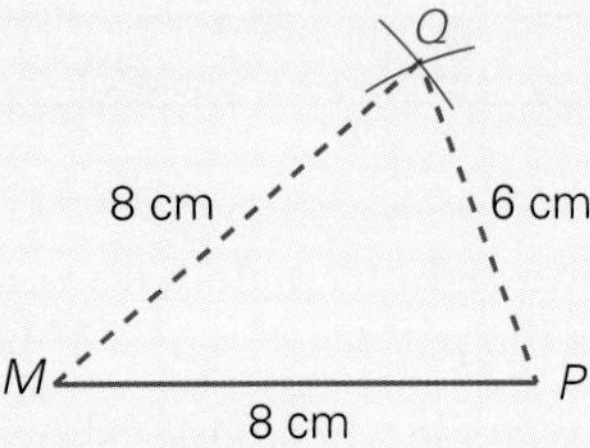

4 Draw ΔTHG. Measure the angles and find what sort of triangle it is.

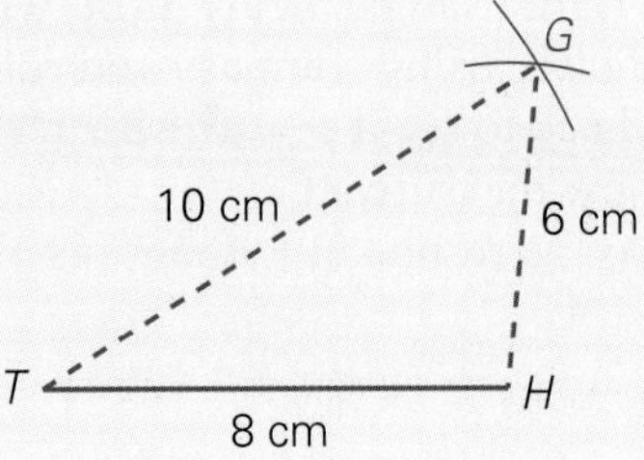

5 Copy and complete the chart:

	Shortest side squared (a^2)	**Middle side squared (b^2)**	**Longest side squared (c^2)**	$a^2 + b^2$
a ΔABC	$4 \times 4 = 16$		$6 \times 6 = 36$	52
b ΔXYZ				
c ΔMPQ				
d ΔTHG				

Which is the only triangle where $c^2 = a^2 + b^2$?

6 Using ruler and protractor, construct right triangles as shown:

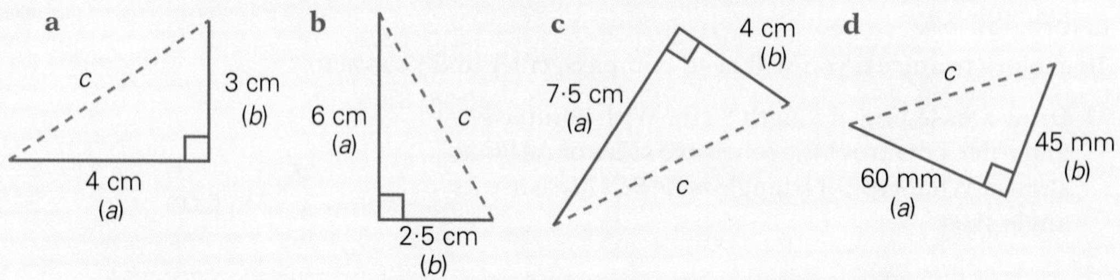

The *longest side* (c) is always opposite the right angle. Measure the side marked c, then complete the chart:

	a^2	b^2	c^2	$a^2 + b^2$
a	16	9		25
b				
c				
d				

What number pattern is true for all four right triangles?

Level 2

7 Draw a right triangle with your choice of side lengths, following question **6** above. Does the pattern $c^2 = a^2 + b^2$ hold true?

8 Any triangle whose sides show this pattern must be a right triangle. Draw ΔABC with measurements as shown, using compass and ruler.
Check to see if $c^2 = a^2 + b^2$.
Now measure $\angle C$.
Do the results fit Pythagoras' rule?

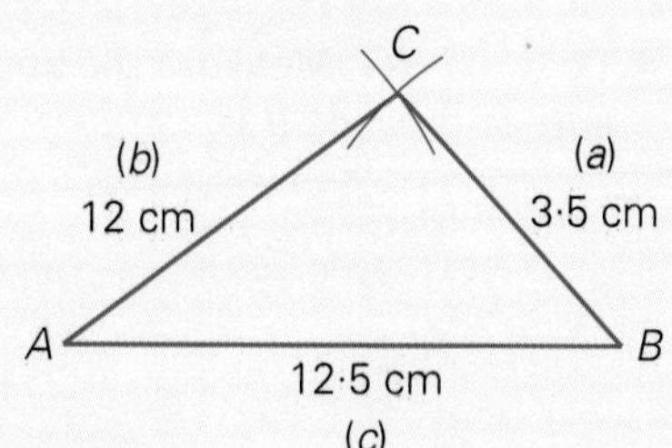

Investigation Pythagoras' proof

You will need: paper, ruler, scissors

As Pythagoras himself never wrote anything down, it is hard to know how he proved his theorem. His students learnt everything in their heads! However, it may have gone something like this.

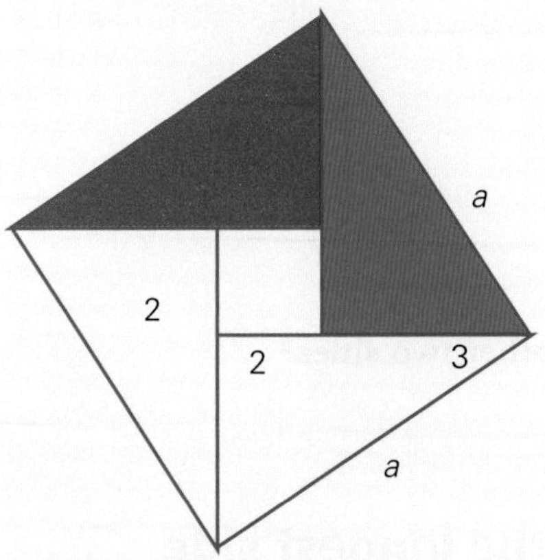

Start with a small square, say 2 cm by 2 cm, extend each side by 3 cm and join to form a larger square as shown.

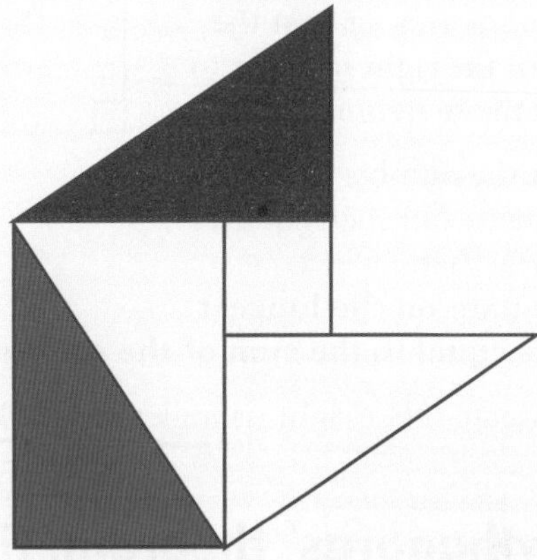

Move the green triangle as shown, then the pink.

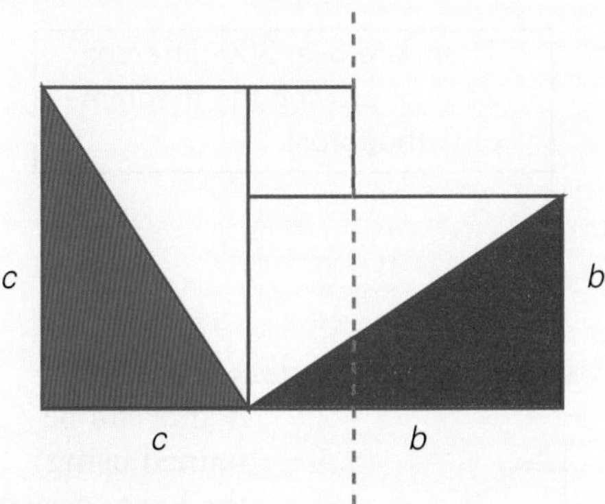

If we divide this shape with a line as shown, we obtain two smaller squares equal in area to the larger.

i.e. $a^2 = b^2 + c^2$

Investigate other ways of proving Pythagoras' theorem.

Suggestions include:

- the Chinese proof
- the President's proof
- Leonardo Da Vinci's proof

Design a poster of your choice, clearly showing the proof of Pythagoras' theorem you decided upon.

Compare this proof to the one given above.

Investigation Pythagoras' theorem by squares

You will need: centimetre square grid paper

1 Copy these shapes onto the grid paper and cut them out.

2 Fit the squares against the sides of the right triangle to check the matching.

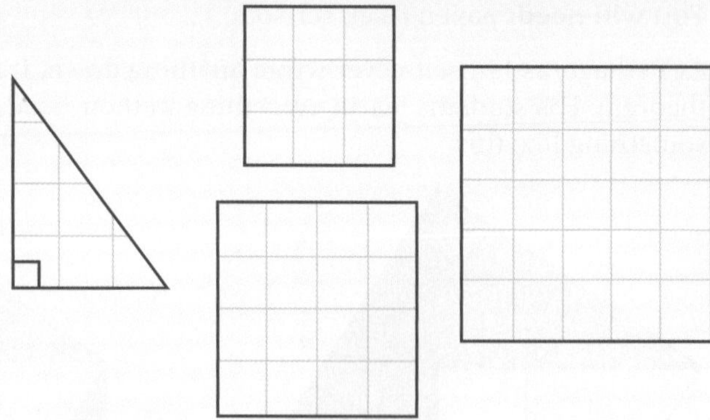

3 Count the number of square centimetres in each square.
Is it true that:
the square on the longest side is equal to the sum of the squares on the other two sides?

4 Investigate other right triangles similarly.

6.2 Pythagoras' theorem: finding the longest side

The longest side of any right-angled triangle is *always* opposite the right angle and is given the name **hypotenuse** (hyp).

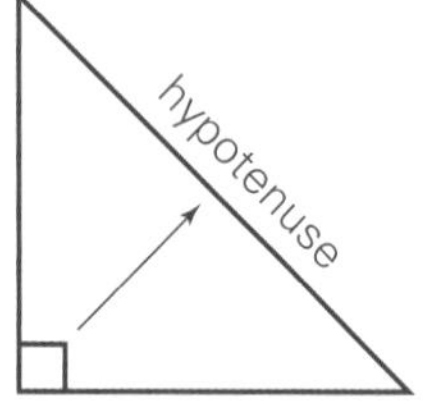

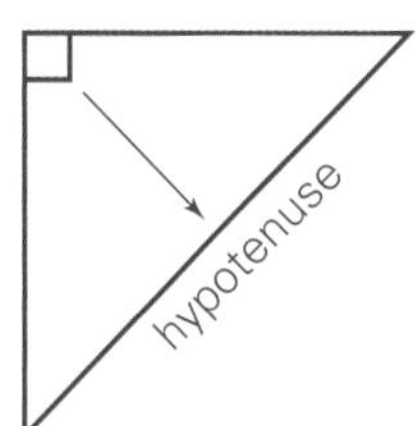

There are over 300 different ways of proving the theorem of Pythagoras!

Naming sides in triangles:

1 Using vertices

hyp: PQ

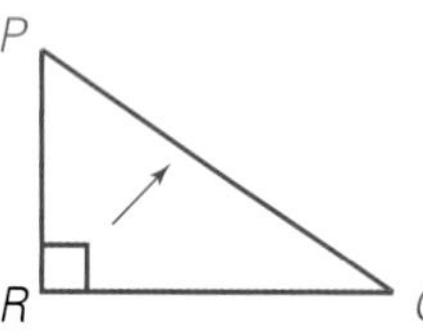

2 Using lower case letters

P
r
R
Q

A side can be named using the lower case of its opposite vertex.
hyp: r

Pythagoras' theorem states that:

The square of the hypotenuse is equal to the sum of the squares of the other two sides.

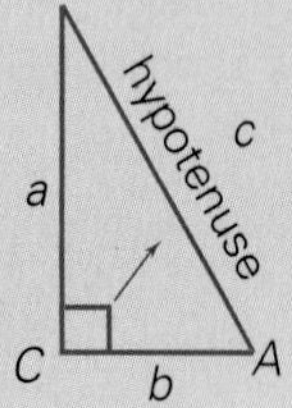

For the triangle shown, the theorem can be written as:

$$c^2 = a^2 + b^2$$

By drawing an arrow across from the right angle, it is easy to find the hypotenuse.

EXAMPLE 1

Write Pythagoras' theorem for each of the triangles given:

a

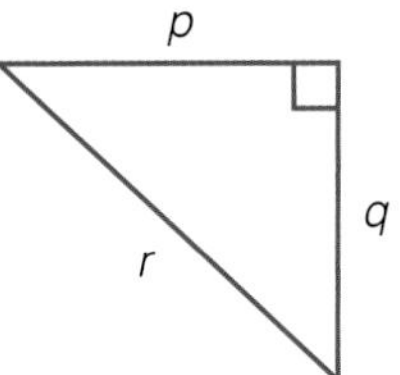

b

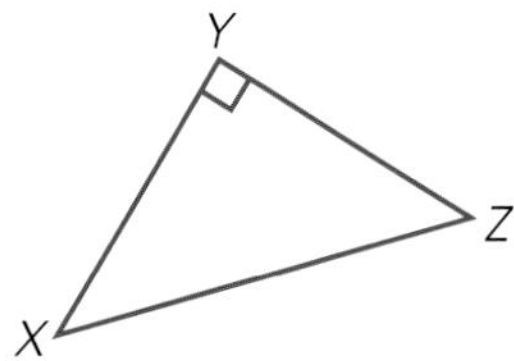

Solution

a

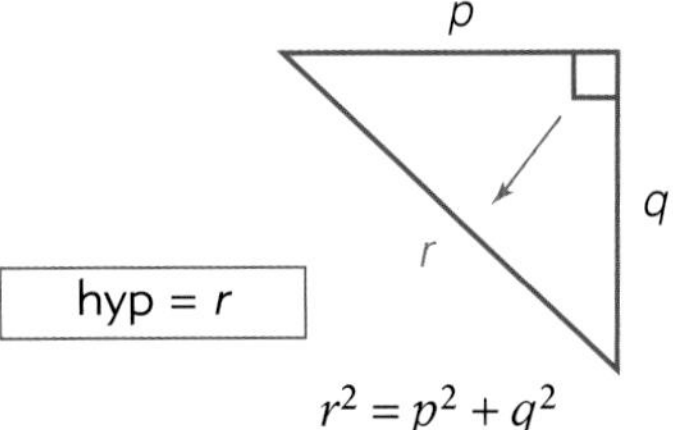

$$r^2 = p^2 + q^2$$

b

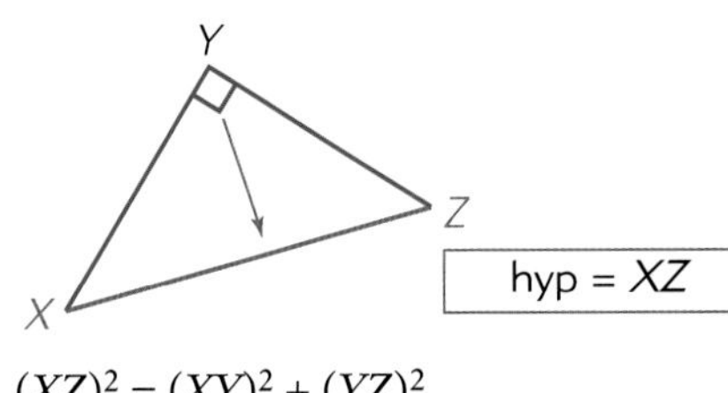

$$(XZ)^2 = (XY)^2 + (YZ)^2$$

or $\quad y^2 = z^2 + x^2$

EXAMPLE 2

Use Pythagoras' theorem to calculate the length of the hypotenuse in each triangle below:

a

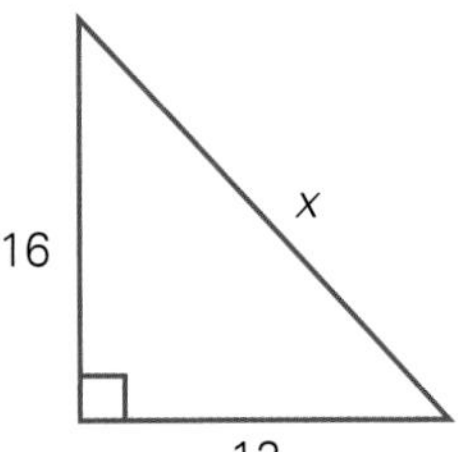

b

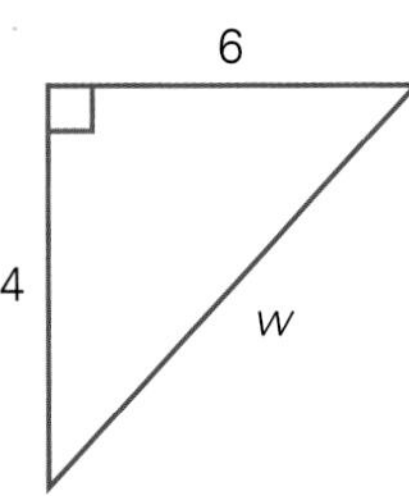

Solution

a

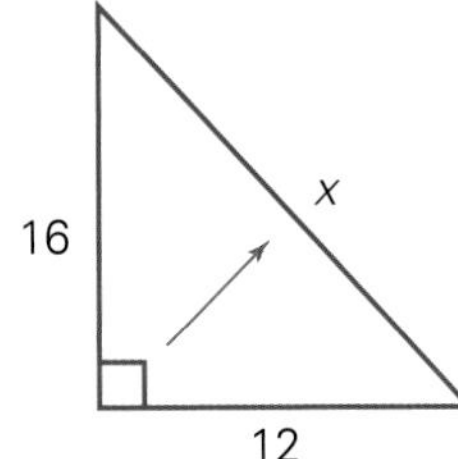

$x^2 = 16^2 + 12^2$
$\quad = 400$
$x = \sqrt{400}$
$\quad = 20$
$\therefore$ the hypotenuse is 20 m.

> All sides are positive.
> $\therefore$ instead of $x^2 = 400$
> $x = \pm 20$
> We know that:
> $x = 20$

b

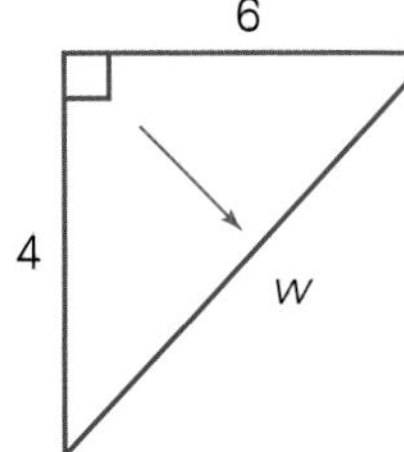

$w^2 = 6^2 + 4^2$
$\quad = 52$
$w = \sqrt{52}$
$\quad = 7{\cdot}211\ 102\ \ldots$
$\quad = 7{\cdot}2$ (1 decimal place)
$\therefore$ the hypotenuse is 7·2 m, correct to 1 decimal place.

> Leaving your answer as $\sqrt{52}$ is known as *exact* form or *surd* form.

> When dealing with measurements and rounding it is usual to give your answers to 1 decimal place more than the measures given in the question, unless the question gives specific instructions.

EXERCISE 6B

1 Give the name of the hypotenuse for each of the triangles given:

a 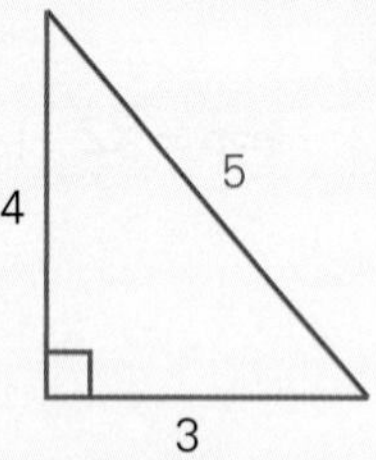

b 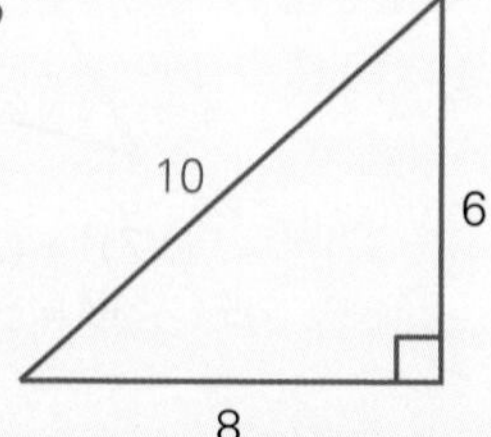

c **d** 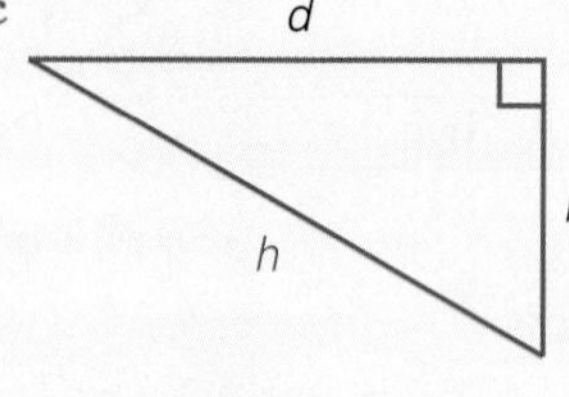

2 For each triangle, select the correct statement of Pythagoras' theorem:

a

4, 5, 3

A $3^2 = 4^2 + 5^2$
B $4^2 = 5^2 + 3^2$
C $5^2 = 3^2 + 4^2$

b

10, 6, 8

A $10^2 = 6^2 + 8^2$
B $8^2 = 6^2 + 10^2$
C $6^2 = 8^2 + 10^2$

c

d, n, h

A $d^2 = h^2 + n^2$
B $h^2 = n^2 + d^2$
C $h = d + n$

3 Write Pythagoras' theorem for each of these triangles:

a 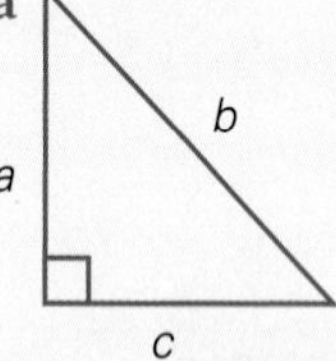

b 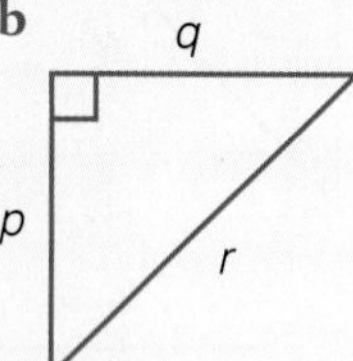

c 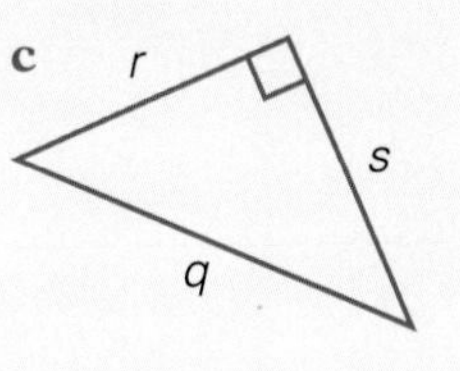

d 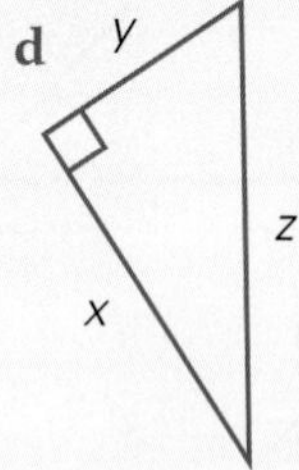

4 Write Pythagoras' theorem for:

a 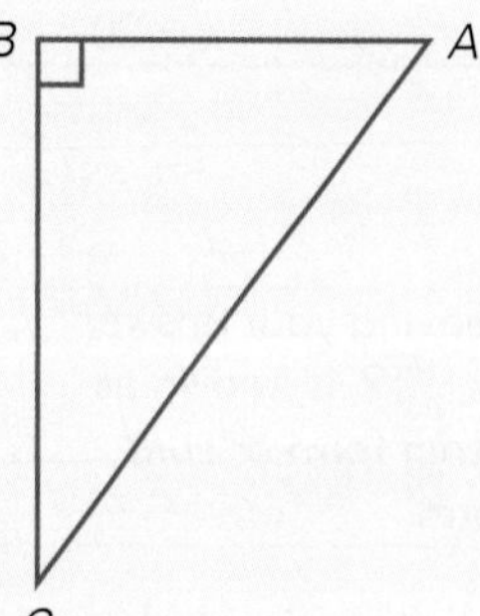

b 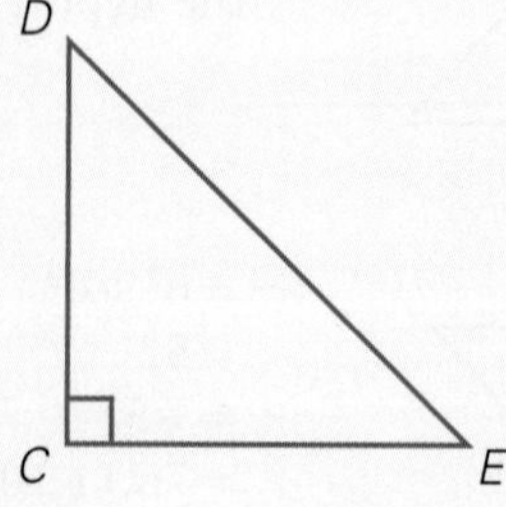

5 Write a statement of Pythagoras' theorem to suit each triangle.

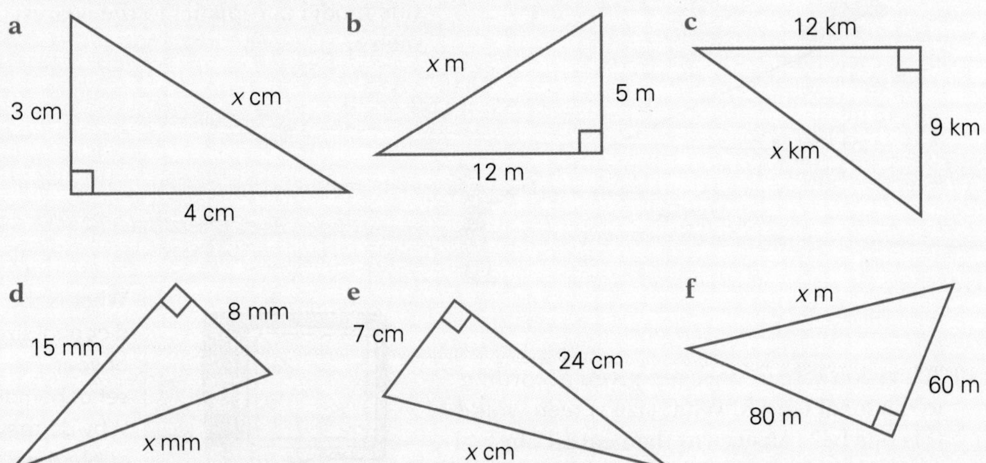

6 Find the length of the hypotenuse for each triangle in question **5**. Your solution should follow the setting out given in the examples.

7 Find the length of the hypotenuse, correct to 2 decimal places where necessary:

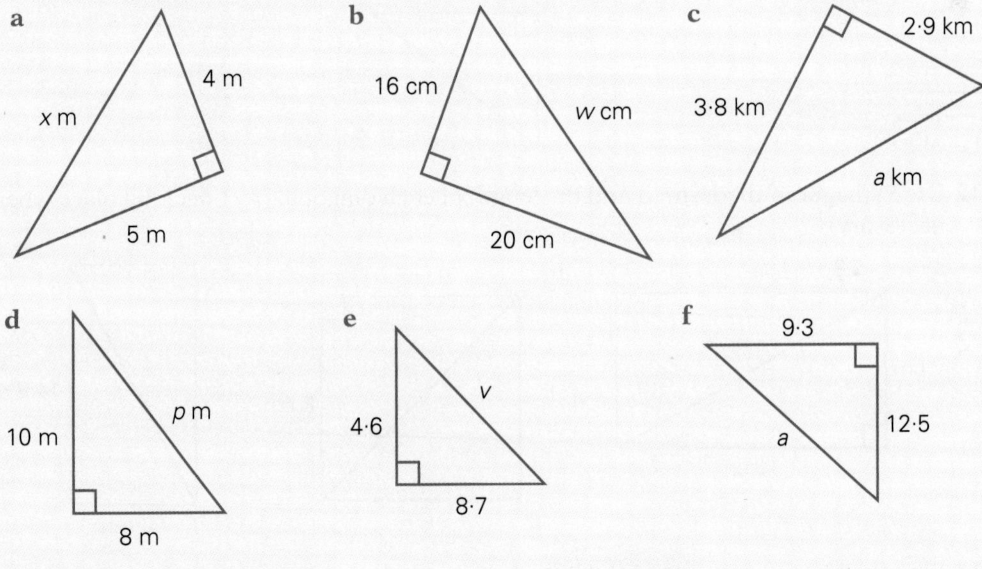

8 What length of cord must be let out to allow the kite to reach a height of 40 metres?

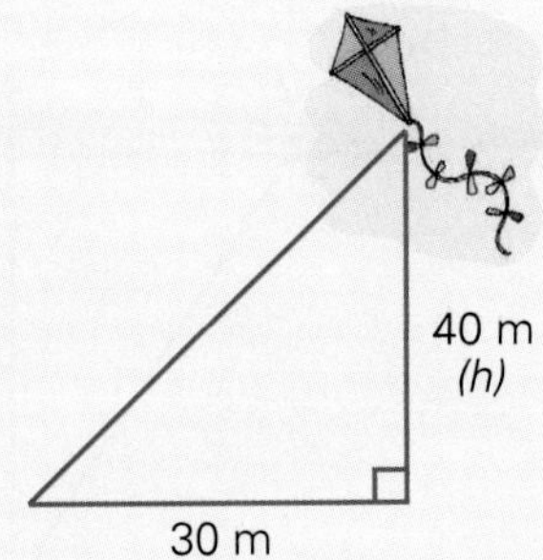

9 Find the length of the control wire for this model aeroplane, to the nearest metre.

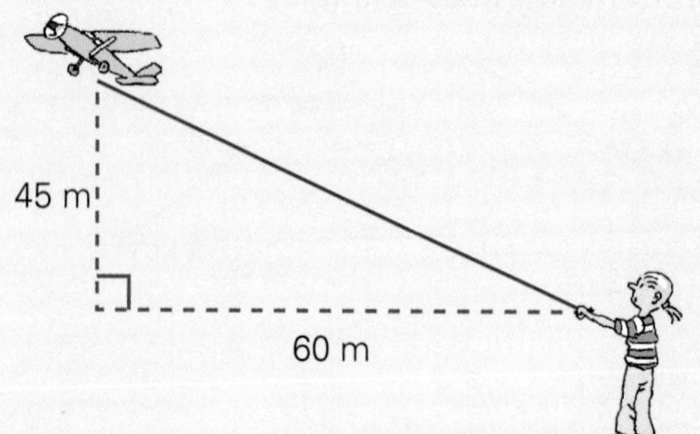

10 Television screen sizes are given according to diagonal length. What size screen would this one be? (Answer to the nearest cm.)

What is the diagonal size of your TV set at home? Why do they quote the diagonal size rather than area?

11 Find the exact value of x.

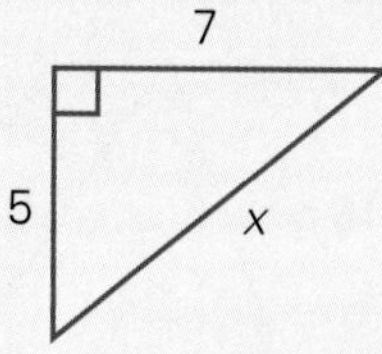

Level 2

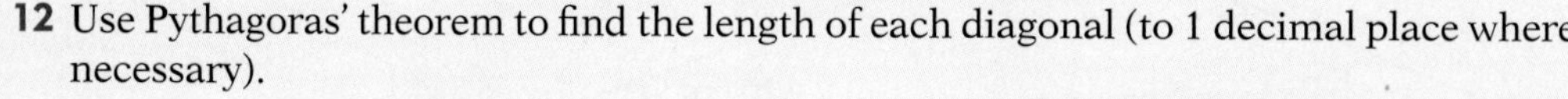

12 Use Pythagoras' theorem to find the length of each diagonal (to 1 decimal place where necessary).

SPREADSHEET ACTIVITY

a

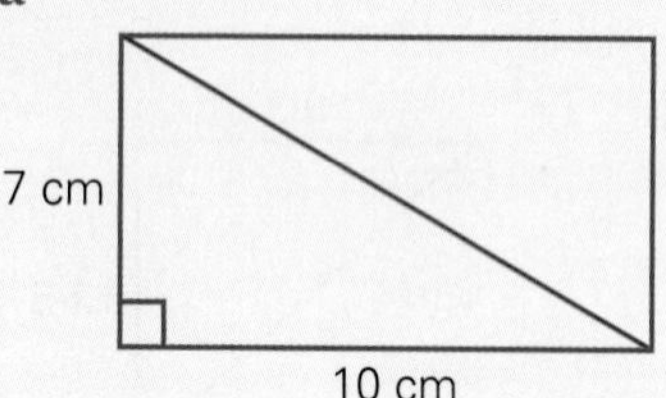

b

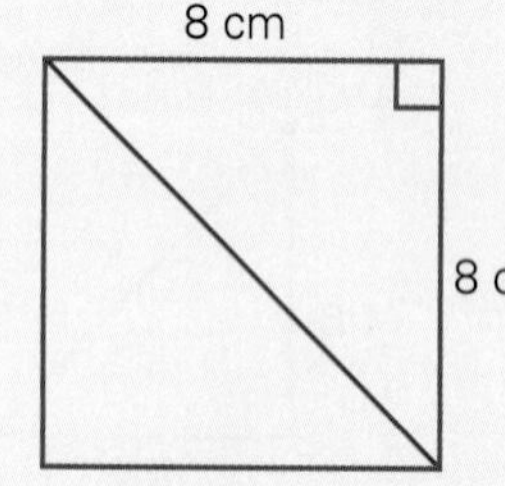

c

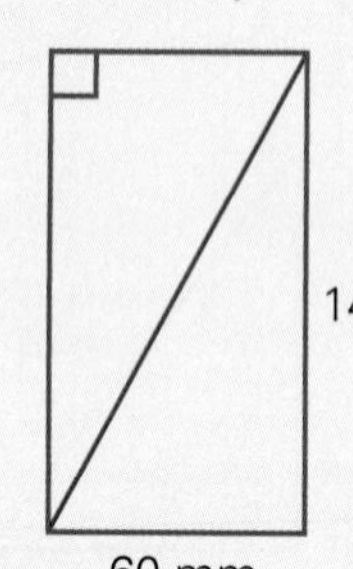

13 Find the length of the interval AB (to 1 decimal place where necessary):

a

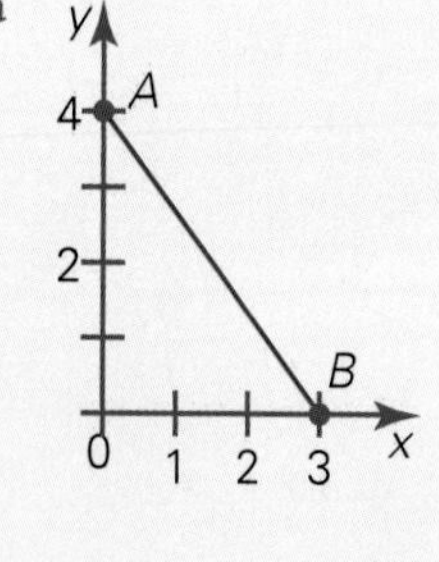

b

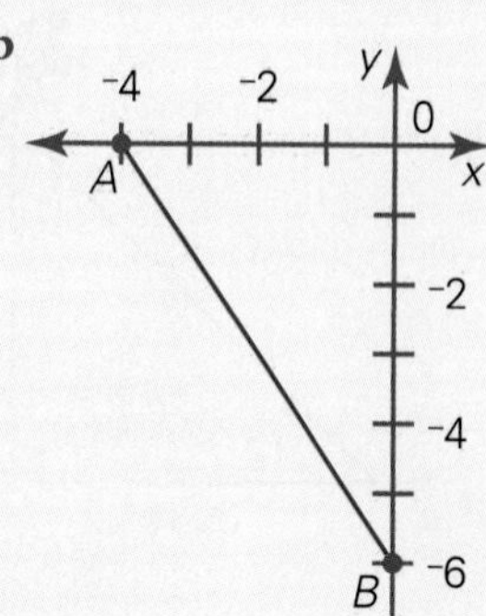

c

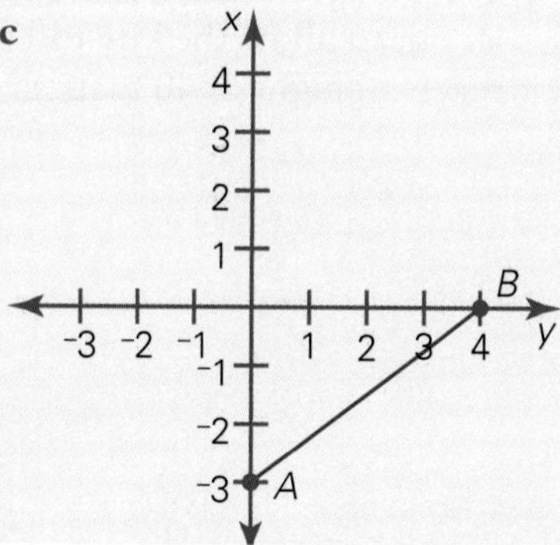

14 A ladder is needed to reach 2 m up the side of a building. If the ladder is to be placed 70 cm from the base of the building, how long does the ladder need to be? *Hint:* draw a diagram.

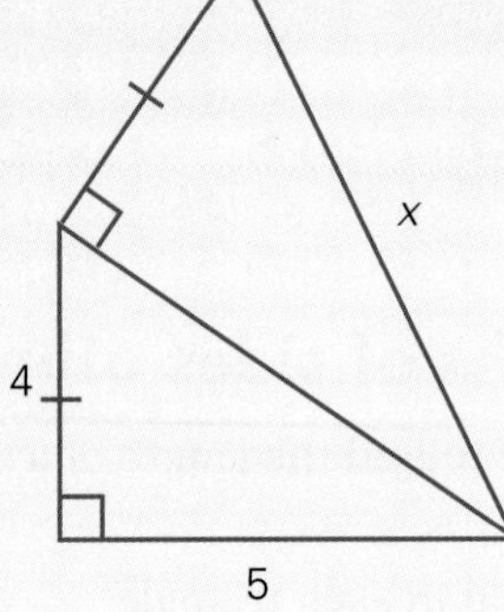

15 Find the exact value of x.

You will need to use Pythagoras' theorem twice.

16 Find the perimeter of triangle *ABC*, correct to the nearest centimetre.

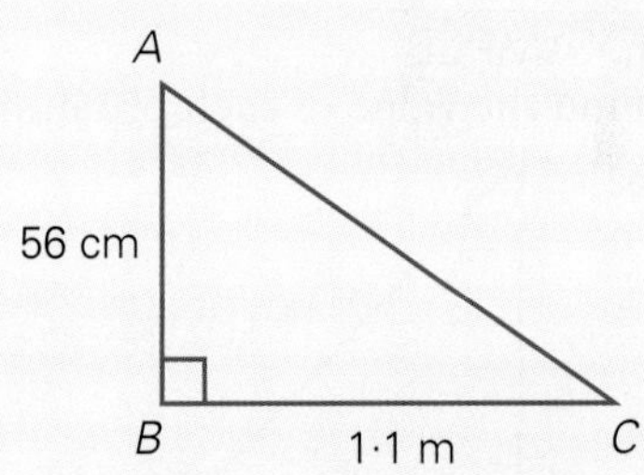

17 Use Pythagoras' theorem to find the value of the pronumerals in each of the following (answer to 1 decimal place), and then find the perimeter of each shape:

a

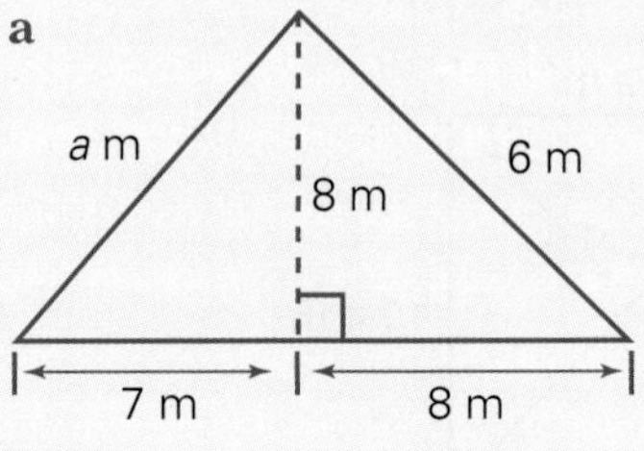

b

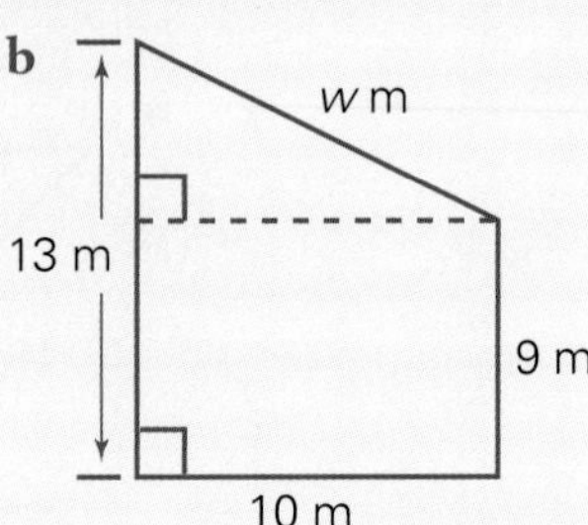

c

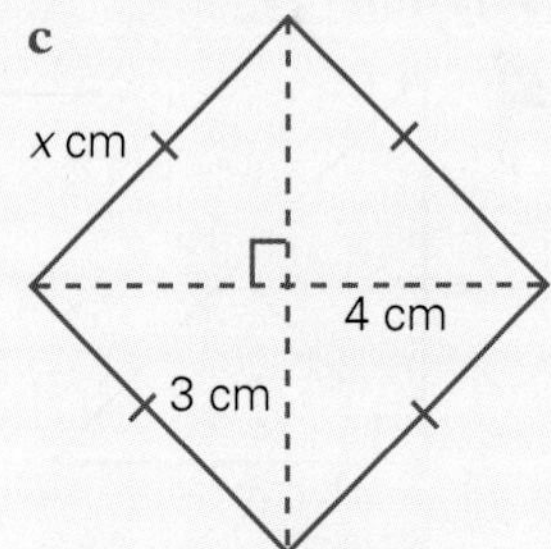

18

B 20 m F
90 m
70 m
D
C 60 m E
80 m 50 m
A

This is an example of an offset or traverse survey. Surveyors can use this when surveying land.

a Find the length of *EF*, correct to the nearest centimetre.
b Find the length of *AE*, correct to the nearest centimetre.
c Calculate the perimeter of the field *AEFBD*.
d If the field is to be fenced, find the cost if the fencing can be purchased only in 1-metre lengths at $24.95 each metre.

You may like to carry out your own traverse survey in groups outside.

Try this

The hypotenuse of a triangle is 7. If the two other sides are a and b with $a > b$, find a possible value for both a and b.

6.3 Finding the lengths of sides other than the hypotenuse

Pythagoras' theorem can be used to find the length of a side other than the hypotenuse.

EXAMPLE

Find the value of each pronumeral in each triangle:

a

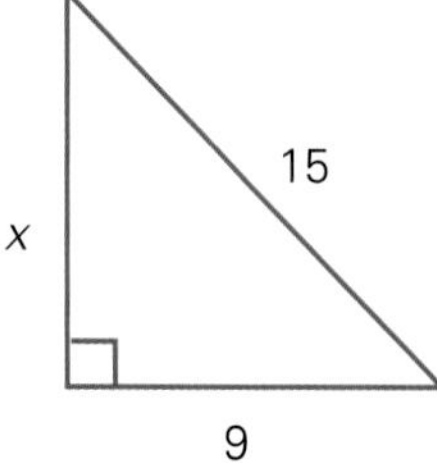

b

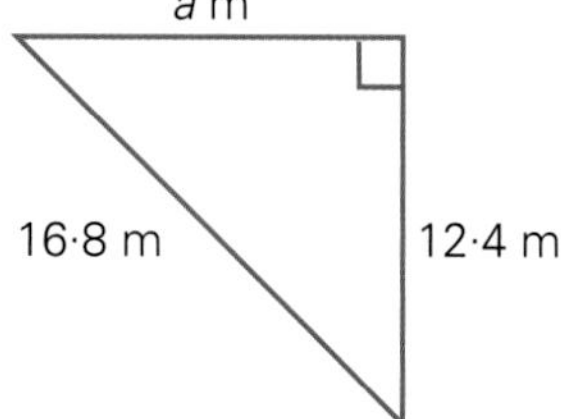

Solution

a

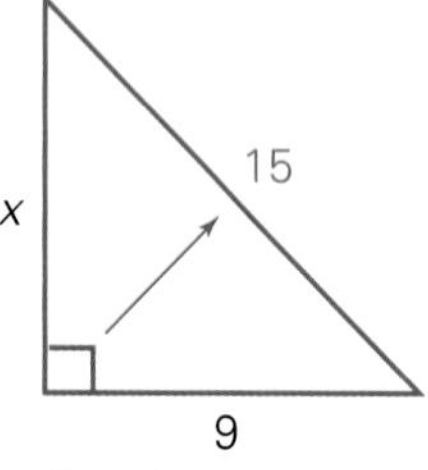

$$15^2 = x^2 + 9^2$$
$$x^2 = 15^2 - 9^2$$
$$= 144$$
$$x = \sqrt{144}$$
$$= 12$$

b

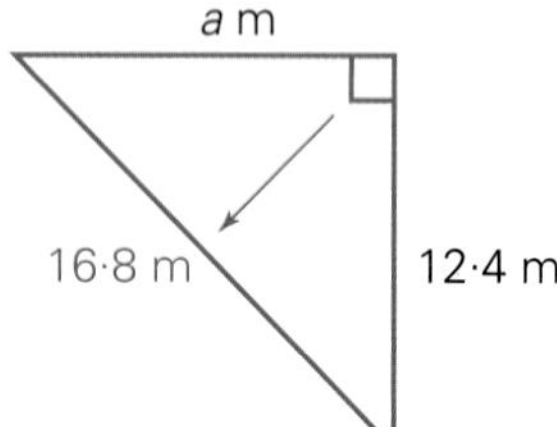

$$16{\cdot}8^2 = a^2 + 12{\cdot}4^2$$
$$a^2 = 16{\cdot}8^2 - 12{\cdot}4^2$$
$$a^2 = 128{\cdot}48$$
$$a = \sqrt{128{\cdot}48}$$
$$a = 11{\cdot}33 \text{ (2 decimal places)}$$

EXERCISE 6C

1 Copy and complete the following (answers to 2 decimal places, where necessary):

a

b cm
10 cm
8 cm

______ $= b^2 + 8^2$

$b^2 =$ ______ $- {}^-8^2$

$=$ ______

$b = \sqrt{\quad}$

$b =$ ______

b

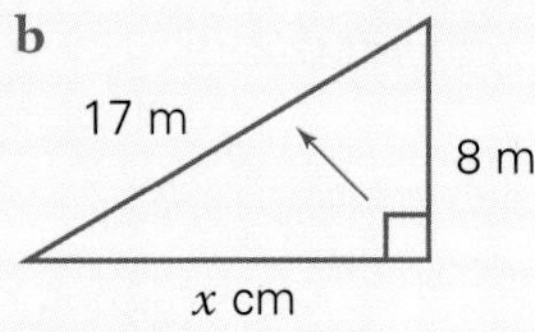

$17^2 = x^2 +$ ______

$x^2 =$ ______ $-$ ______

$=$ ______

$x = \sqrt{\quad}$

$x =$ ______

c

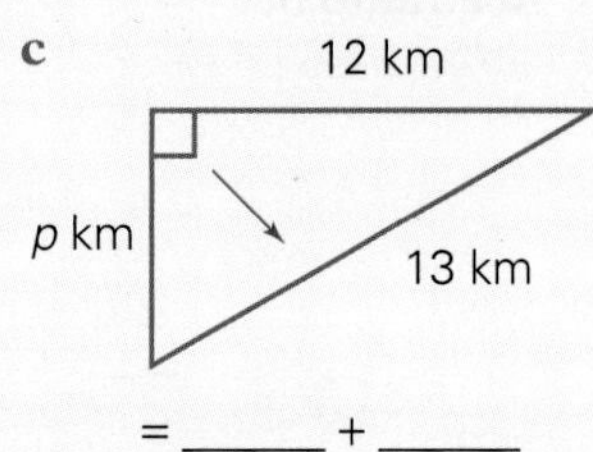

$=$ ______ $+$ ______

$p^2 =$ ______ $-$ ______

$=$ ______

$p = \sqrt{\quad}$

$p =$ ______

2 Find the value of x in the following:

a

x
15
12

b

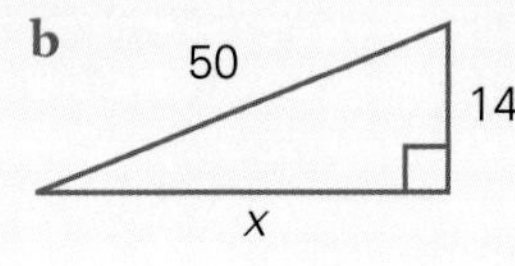

c

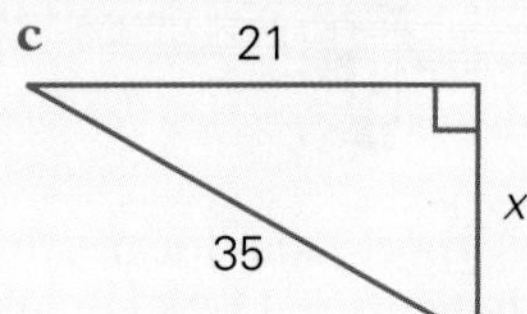

d

x
50
30

e

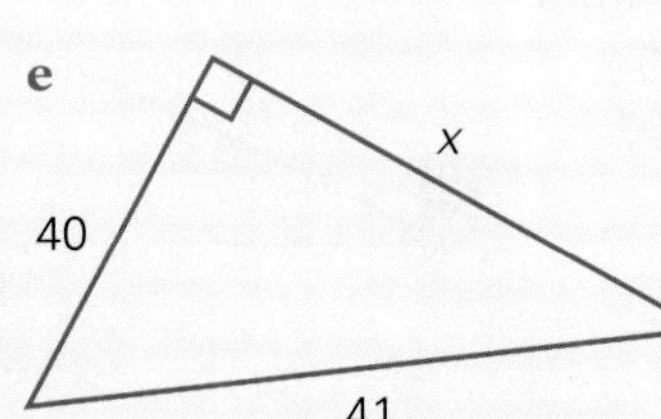

f

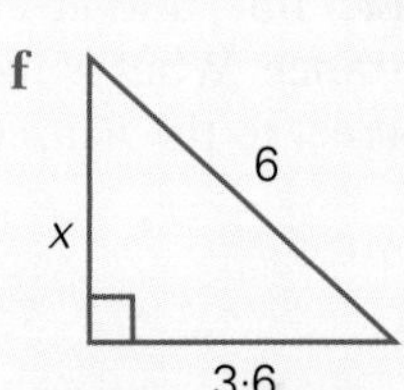

3 Find the length of the side marked p in each of the following, correct to 1 decimal place where necessary:

a

p m
2 m
4 m

b

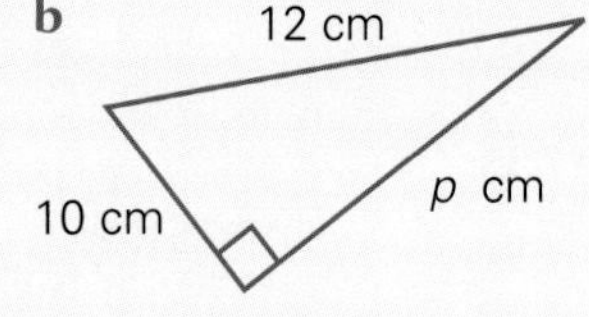

c

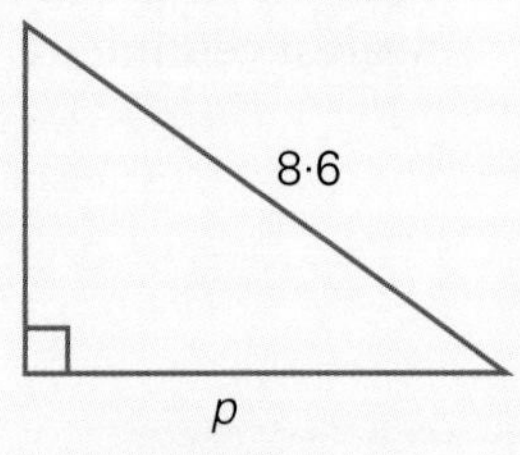

d

p
6·4
10·3

e

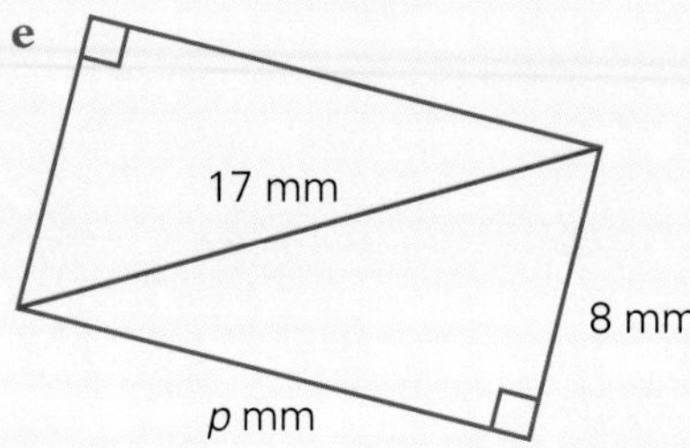

f

p cm
40 cm
56 cm

4 The city council wants to build a skateboard ramp of dimensions shown. How long should the base of the ramp be?

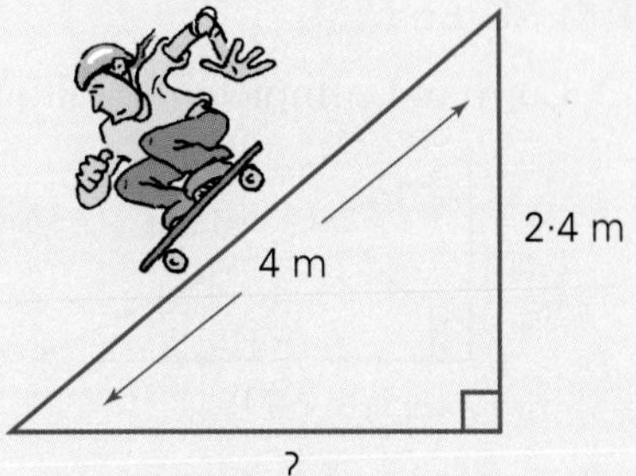

?

120 m

150 m

5 Trung's climbing path across the face of the cliff is 150 metres long, while Angela's direct route up the face is only 120 metres. How far apart will they be when they reach the top?

6 Airport radar locates an incoming helicopter 11·28 km away in a direct line, and at 11 km ground distance. What is its altitude, correct to the nearest metre?

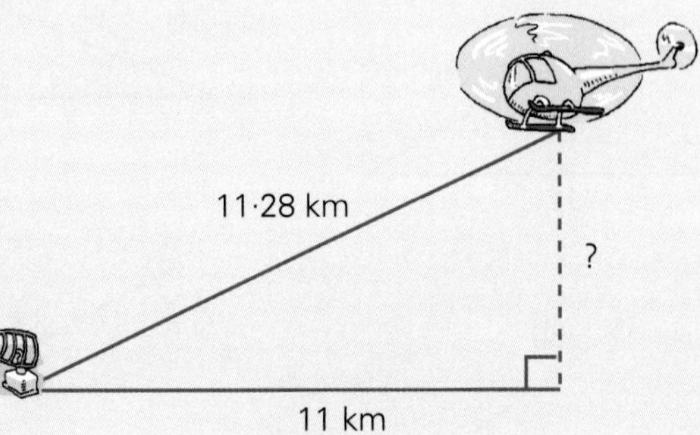

Level 2

7 The diagonal of a square is 7·4 metres. Find the length of each side, correct to the nearest centimetre.

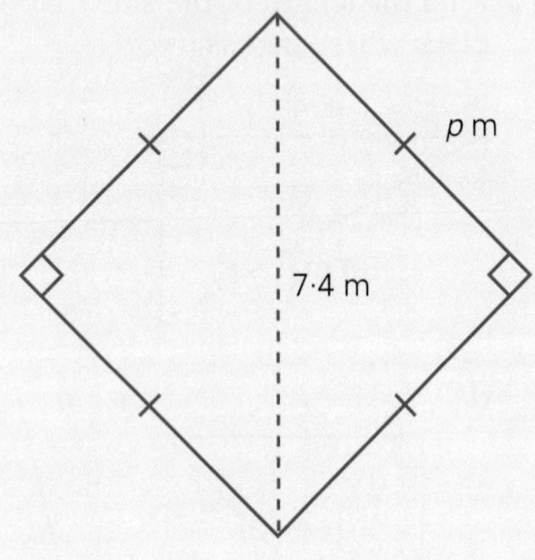

8 On this sailing boat, the wire is 12·6 m long and the boom is 8·6 m long. How high is the tip of the mast above the deck?

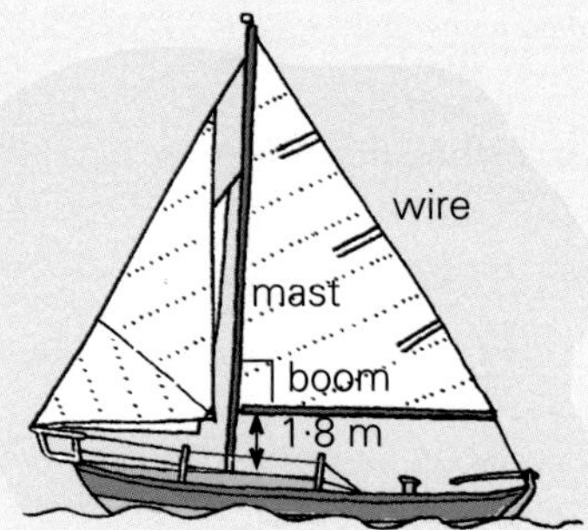

9 Three ships, A, B and C, are in such positions as to form a right angle at B when lines are drawn connecting their positions. If A to C is 8·4 km and A to B is 2·5 km:

a How far is it from B to C (to 1 decimal place)?
b How many minutes would B take to reach C, sailing at 12 km/h?

10 Find the length of each pronumeral and then calculate the perimeter of each figure, correct to 1 decimal place.

a

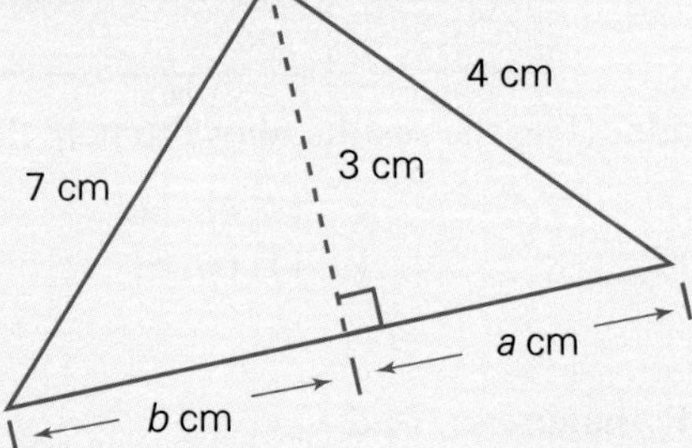

b

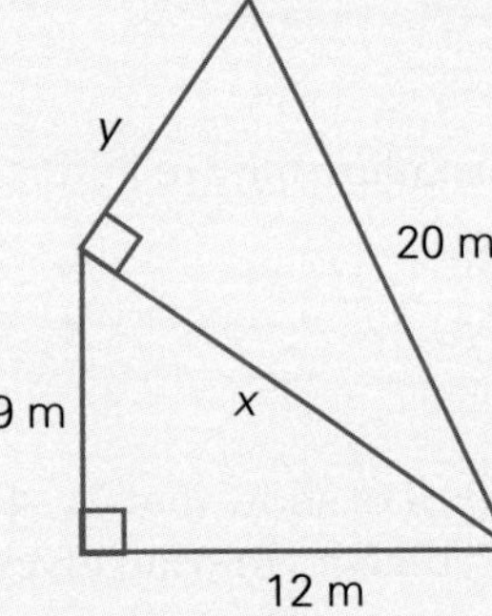

c

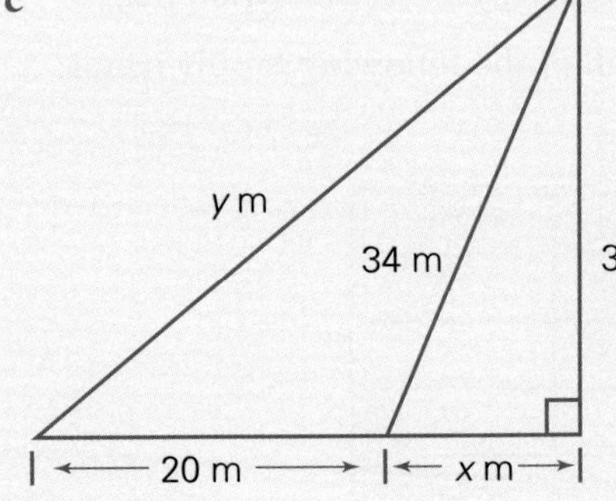

d

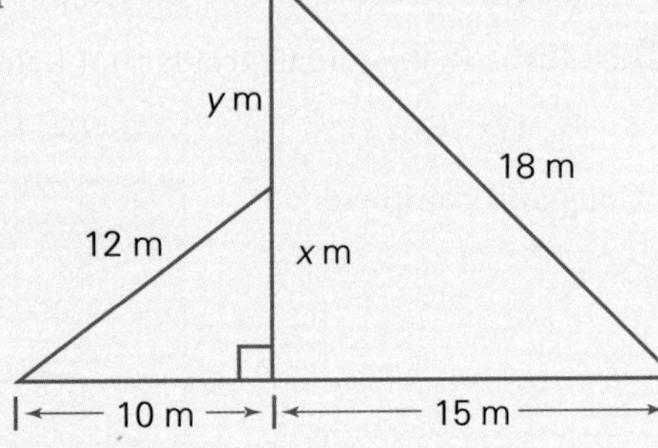

6.4 Pythagorean triads

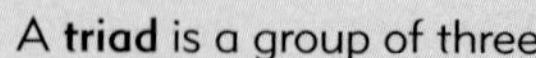

A **triad** is a group of three.

A **Pythagorean triad** is a group of three numbers that obey Pythagoras' theorem.

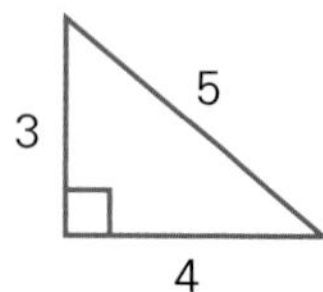

3, 4, 5 is a Pythagorean triad, since $5^2 = 3^2 + 4^2$
$25 = 9 + 16$

EXERCISE 6D

1 Copy and complete this table of Pythagorean triads:

a	b	c
3	4	5
6	8	
9		15
	16	20
1·5	2·0	

2 Which of the following are Pythagorean triads? (Use the results above to help decide.)

a 10, 24, 26 **b** 2·0, 4·8, 5·2 **c** 14, 48, 50
d 24, 45, 51 **e** 4·5, 20, 20·5 **f** 11, 60, 61

Level 2

3 a Use algebra to show that $3x$, $4x$, $5x$ is a Pythagorean triad.
b Find a Pythagorean triad given the value of x is:
i 1 **ii** 7 **iii** 9 **iv** 35
c If 57, 76, x is a Pythagoran triad, find the value of x, if x is the hypotenuse.

4 List three Pythagorean triads that belong to each of the following family triads:

a $5x$, $12x$, $13x$ **b** $8x$, $15x$, $17x$ **c** $7x$, $24x$, $25x$

5 Copy and complete:

a	b	c
3	4	5
5	12	
7		25
	15	17
9	40	

6 The triads in question **5** all belong to the family of triads with the rule

$2x + 1$, $2x^2 + 2x$, $2x^2 + 2x + 1$

a Find a Pythagorean triad given **i** $x = 1$ **ii** $x = 7$
b Find the triads given the shortest side $(2x + 1)$ has the value:
i 27 **ii** 31

7 24, x and 18 are three integers forming a Pythagorean triad. Find the value of x.

Investigation Pythagoras puzzle

What snake has 40% in common with Pythagoras?

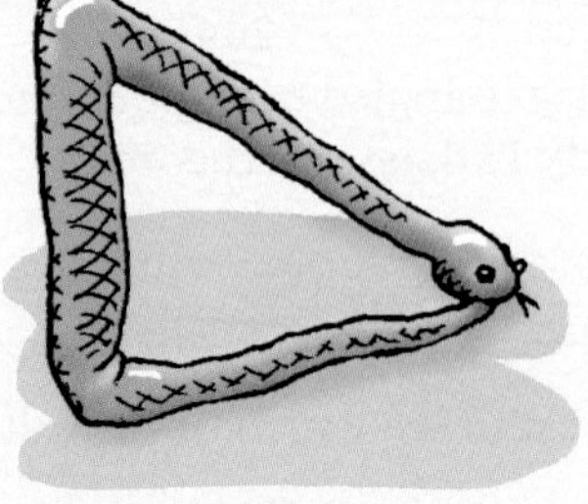

O $12^2 + 13^2$

T $\sqrt{12001}$, correct to the nearest whole

P 55, 300, ?, a Pythagorean triad

N Hypotenuse?

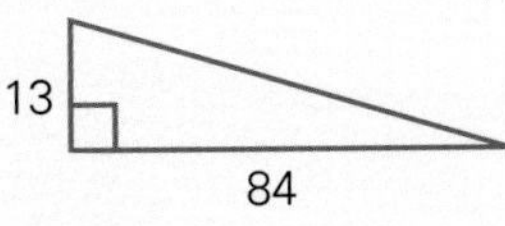

H n?

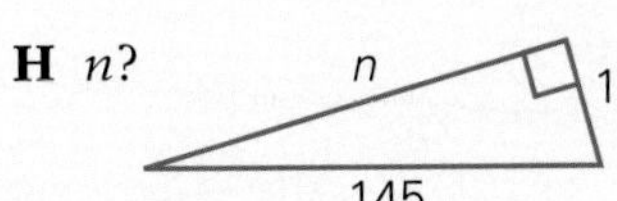

Y Find x.

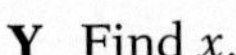

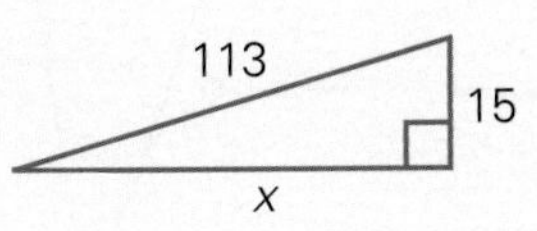

305	112	110	144	313	85

6.5 Testing for right triangles

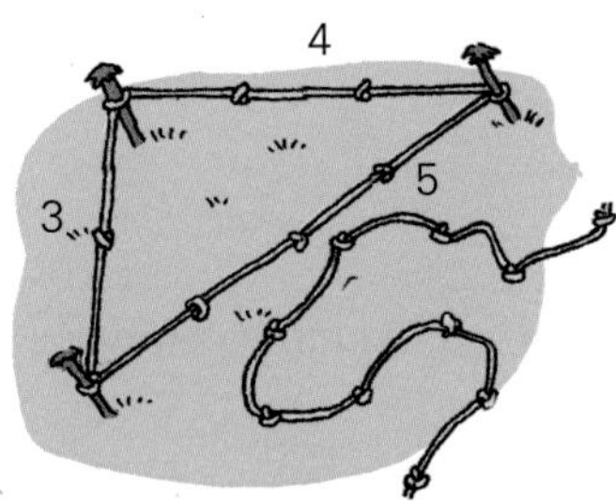

Ancient Babylonians and Egyptians used the special 3, 4, 5 triad in knotted ropes to form right triangles. It is probable that Pythagoras learnt of this in his travels there.

Try it in your classroom to test right angles.

The right angle is the one opposite the longest side.

Discuss how this could be done with a tape measure.

The *converse* of Pythagoras' theorem

If the square on the longest side of a triangle is equal to the sum of the squares on the other two sides, the triangle has a right angle opposite the longest side.

EXAMPLE 1

Is triangle ABC a right-angled triangle?

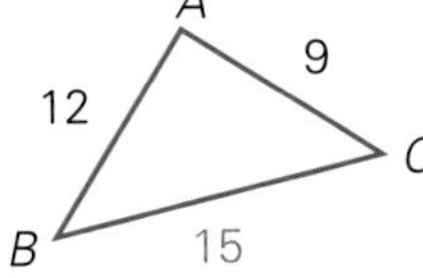

Solution

Longest side $15^2 = 225$

$12^2 + 9^2 = 225$

$\therefore 15^2 = 12^2 + 9^2$ and ΔABC is a right-angled triangle as it satisfies Pythagoras' theorem.

$\therefore \angle BAC = 90°$

EXAMPLE 2

Can a right-angled triangle be drawn with sides 10 cm, 12 cm and 18 cm?

Solution

Longest side	$18^2 = 324$
	$10^2 + 12^2 = 244$
	$244 \neq 324$

$\therefore$ a right-angled triangle *cannot* be drawn with sides 10 cm, 12 cm and 18 cm as it does *not* satisfy Pythagoras' theorem.

EXERCISE 6E

1 Test to see which of the following triangles are right-angled:

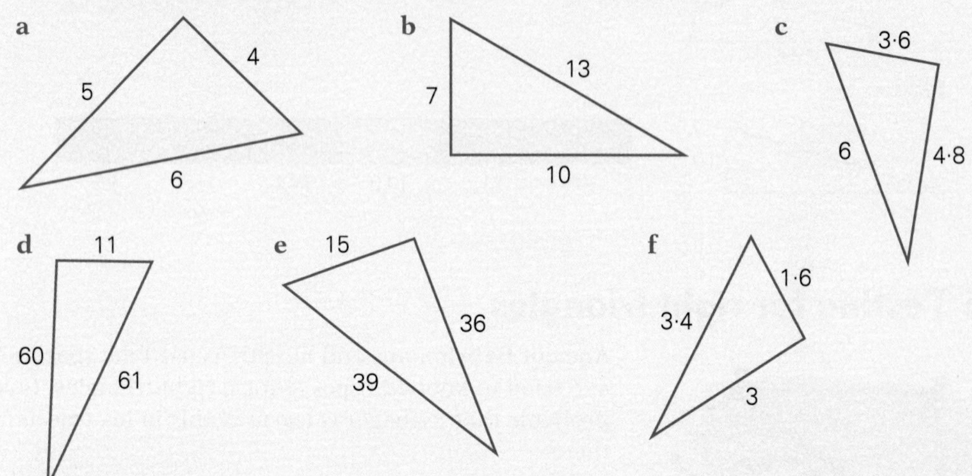

2 Can a right triangle be drawn with the following dimensions? If so, sketch it, showing the right angle.

a 12, 16, 20 **b** 9, 24, 25 **c** 1·4, 4·8, 5

Level 2

3 A builder marking out the foundations for a house put stakes in the ground and nailed timber across as shown. What should the dotted measurement be if the timber forms a correct right angle?

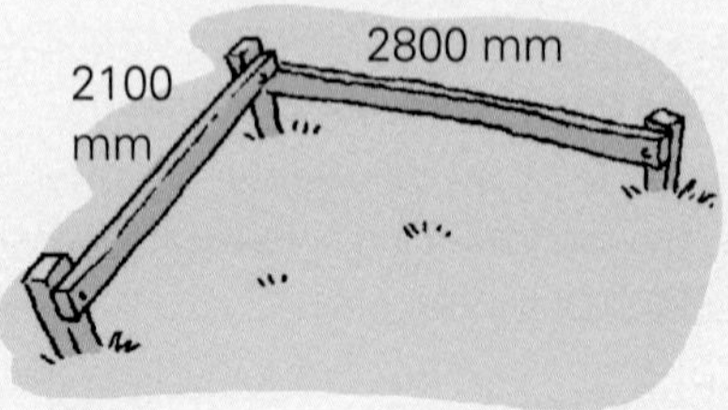

4 From each of the following sets of measures, find three that form the sides of a right-angled triangle:

a 29 m, 20 m, 30 m, 21 m, 36 m
b 26 m, 25 m, 5 m, 13 m, 12 m
c 45 cm, 40 cm, 27 cm, 25 cm, 36 cm
d 3·6 m, 36 m, 4·8 m, 8·4 m, 6 m
e 41 mm, 30 mm, 9 mm, 50 mm, 40 mm
f 3·6 m, 3·3 m, 5·6 m, 5·6 m, 6·5 m, 3·7 m

Investigation Pythagoras on the number line

You will need: graph paper, pencil, set square or protractor, and a pair of compasses

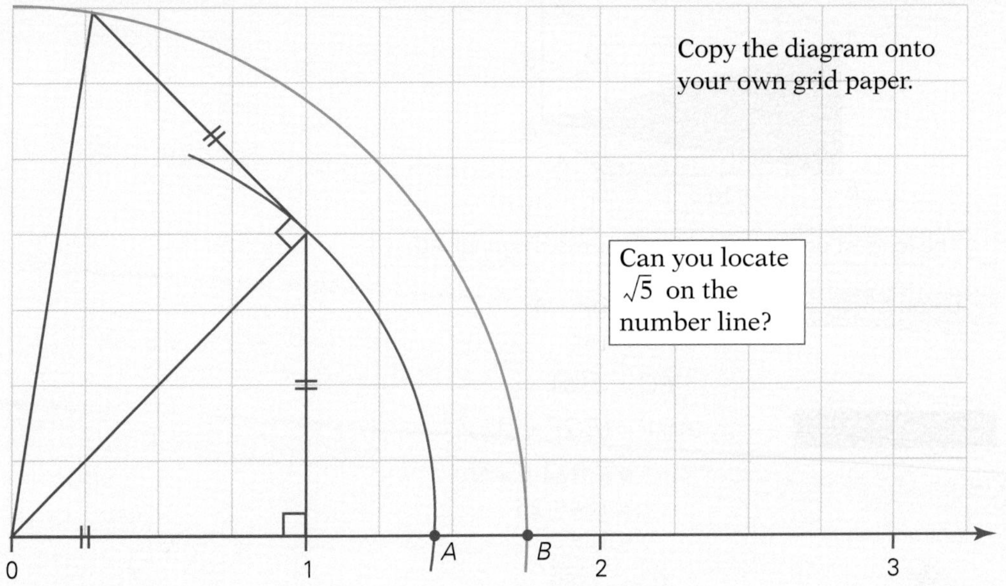

What is the value of A and B on the number line above?

6.6 Miscellaneous exercises involving Pythagoras' theorem

Pythagoras' theorem has many applications, some of which you may have noticed already in the chapter.

Some areas where Pythagoras' theorem is useful include drafting, building and navigation.

EXAMPLE 1

Two hikers leave their camp (P) at the same time. One walks due east for 9 km; the other walks due south for 9·5 km. How far apart are the two hikers at this point (answer to 1 decimal place)?

Solution

Draw a diagram.

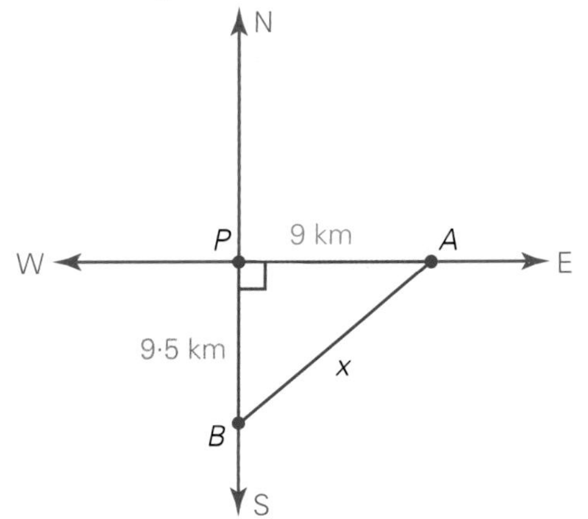

This becomes ΔPAB.

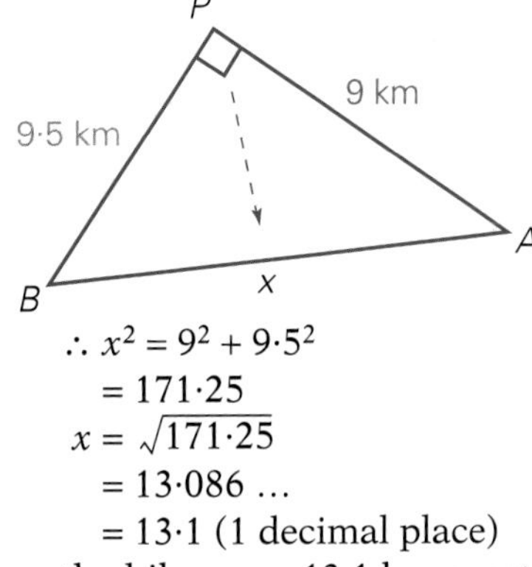

$$\therefore x^2 = 9^2 + 9{\cdot}5^2$$
$$= 171{\cdot}25$$
$$x = \sqrt{171{\cdot}25}$$
$$= 13{\cdot}086 \ldots$$
$$= 13{\cdot}1 \text{ (1 decimal place)}$$

$\therefore$ the hikers are 13·1 km apart.

EXAMPLE 2

A pencil box with internal dimensions 10 cm × 4 cm × 5 cm is built in a woodwork class. What is the length of the longest pencil that can fit into the box?

Solution

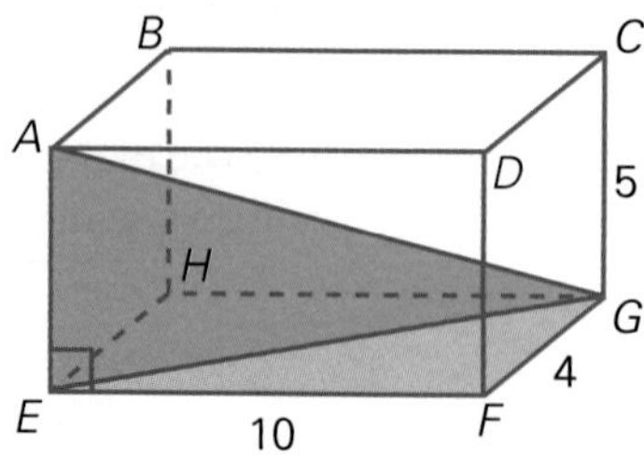

Draw and label a diagram.

The longest pencil is the length of the diagonal AG.

Using ΔEFG

$$(EG)^2 = 10^2 + 4^2$$
$$= 164$$
$$EG = \sqrt{164}$$

Using ΔAEG

$$\therefore (AG)^2 = (EG)^2 + (AE)^2$$
$$= (\sqrt{164})^2 + 5^2$$
$$= 164 + 25$$
$$= 189$$
$$AG = \sqrt{189}$$

$\therefore$ the longest pencil is $\sqrt{189} \doteq 13{\cdot}7$ cm long (to 1 decimal place).

EXERCISE 6F

1 Tom walks 4·5 km east while Zara walks 5·2 km south. How far from Tom is Zara?

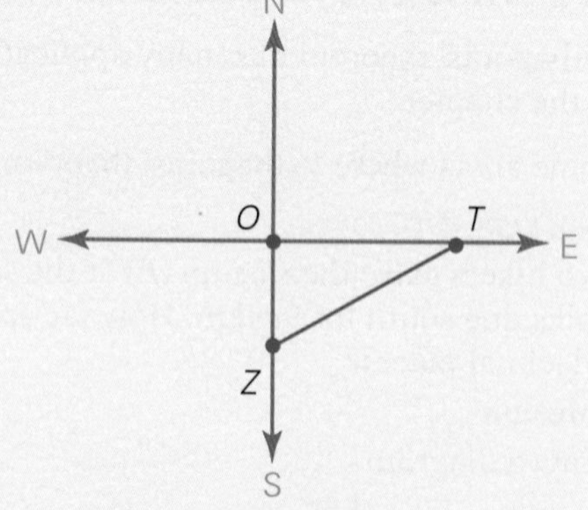

2 This rectangular gate needs a diagonal piece of timber to brace it. How long should the brace be?

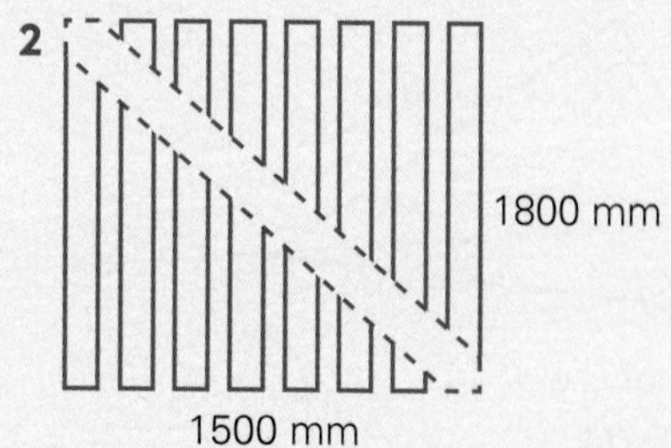

3 The mainsail of this yacht has a staywire 8·2 m long and a vertical wire 5·7 m long. How long is the boom?

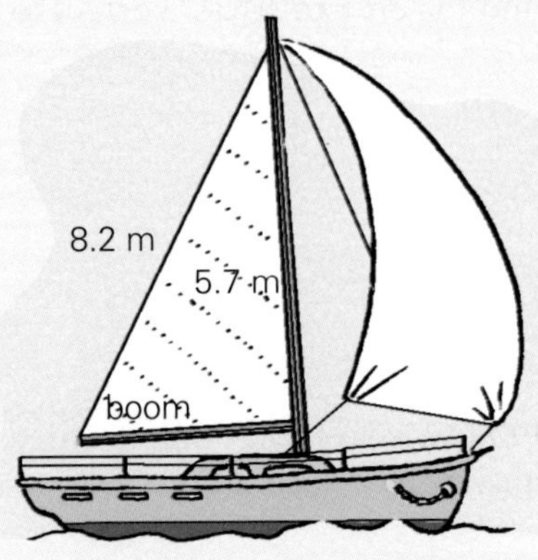

4 This mobile phone signal tower is 12·6 m high, with a staywire anchored to the ground $7\frac{1}{2}$ metres from its base. How long is the wire?

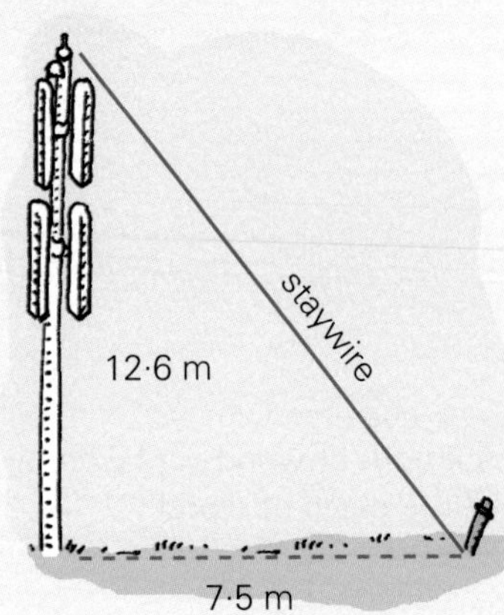

5 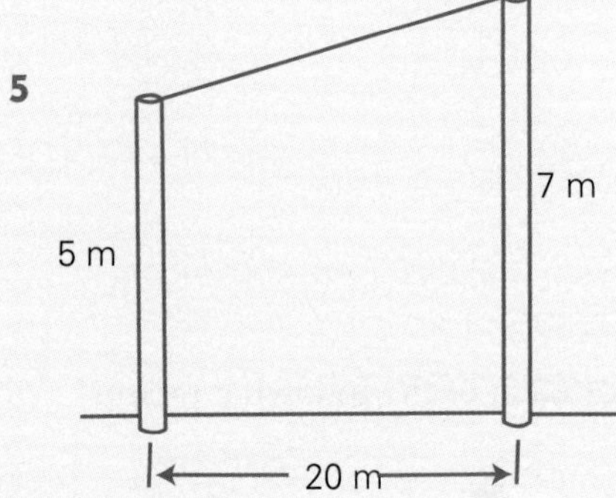

A telephone wire is stretched between two poles, one 7 m high and the other 5 m high. The posts are 20 m apart on level ground. How long is the wire, to the nearest centimetre?

Level 2

6 A boat sets out from port O and sails north-west for 24 km, then changes its course and sails for 10 km in a south westerly direction. How many kilometres is the boat from the port O?

7 A house roof truss has rafters 4·8 m long forming an isosceles triangle with the cross-beam. The height is 3·2 m. How long is the cross-beam, to the nearest millimetre?

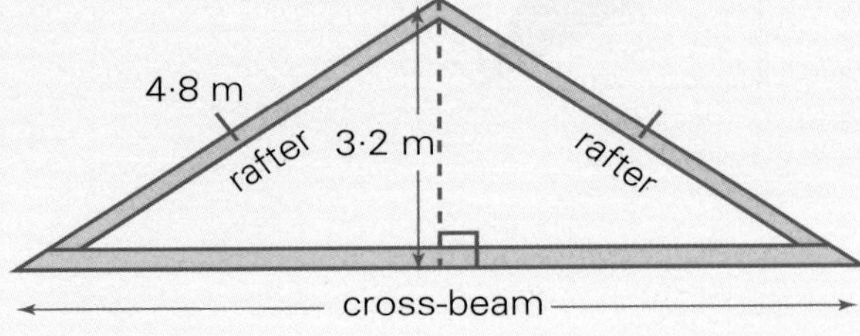

8 Find the area of ΔABC, correct to the nearest whole number.

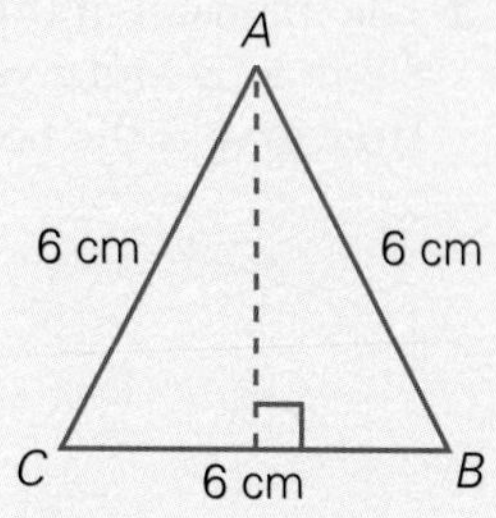

9 If ΔPQR is isosceles, find the area of ΔPQR (answer to 2 decimal places).

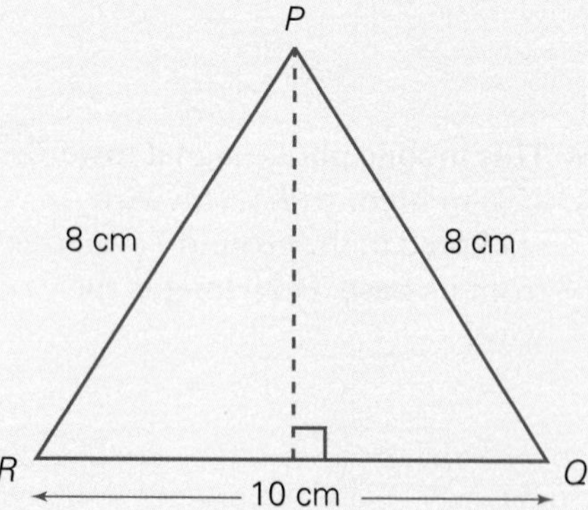

10 A length of wood is placed over a set of stairs to form a ramp. What is the total length of the piece of wood?

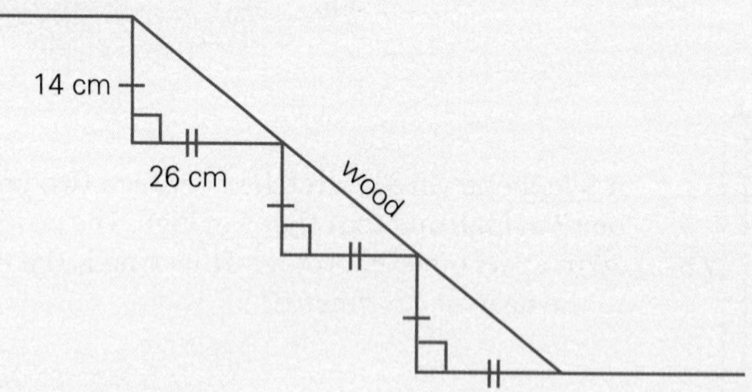

11 Find the value of x, correct to 2 decimal places, for each of the following prisms:

a

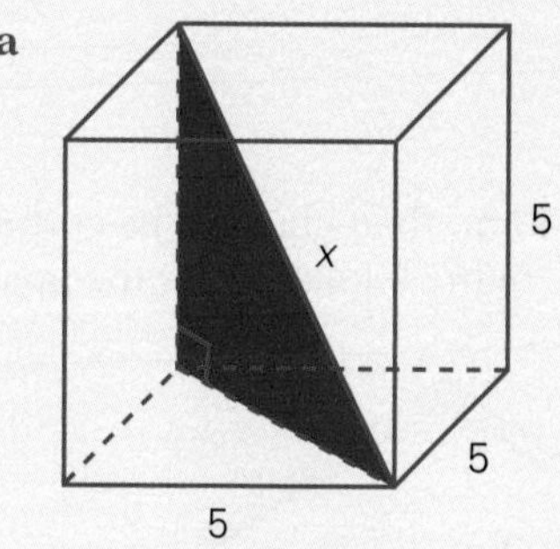

b

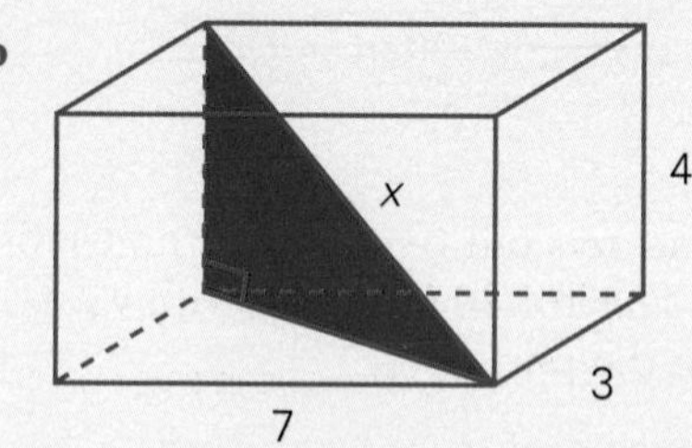

c

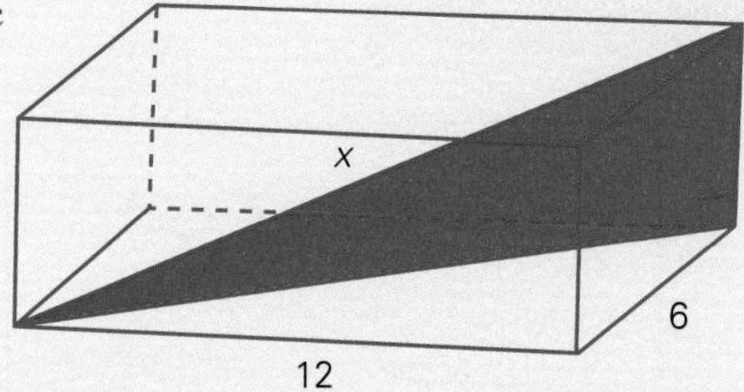

d

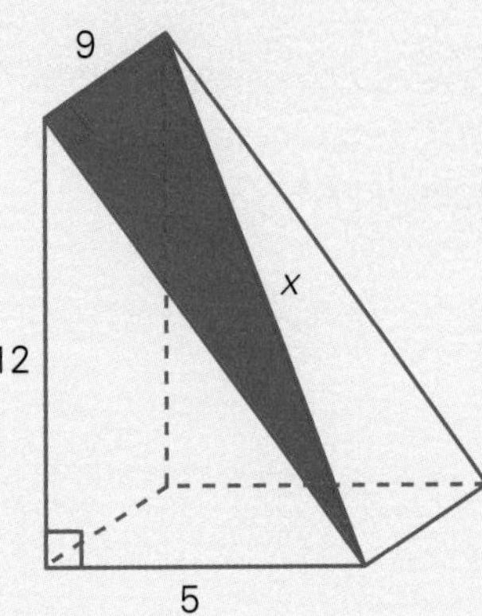

12 Find the value of x in each of the following:

a

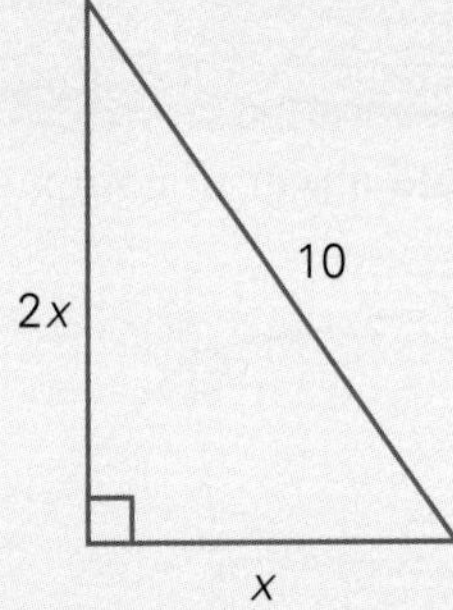

b

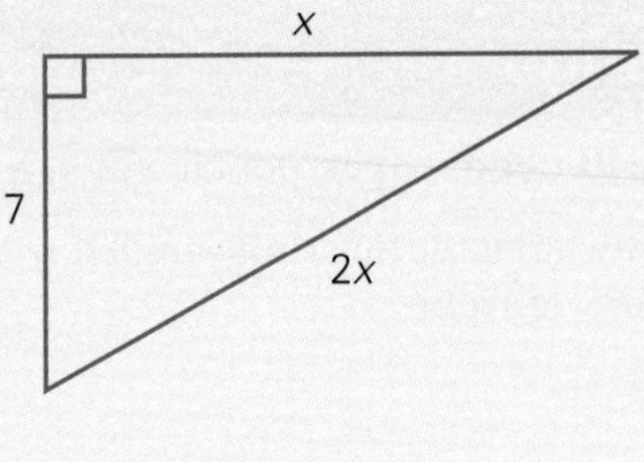

13 A rectangle has a length which is twice its width. If the length of its diagonal is $\sqrt{80}$ cm, find the dimensions of the rectangle.

14

A 25 m ladder is placed against a wall with its foot 20 m from the wall. If the ladder slips 4 m down the wall, how far is the foot of the ladder, to the nearest cm, from the base of the wall?

15 Write down everything you know about ΔPQR.

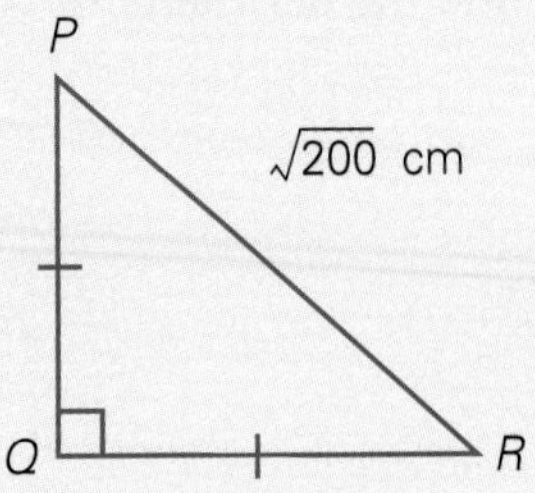

Try this

Given *ABCD* is a rectangle as shown, find the length of *XY*.

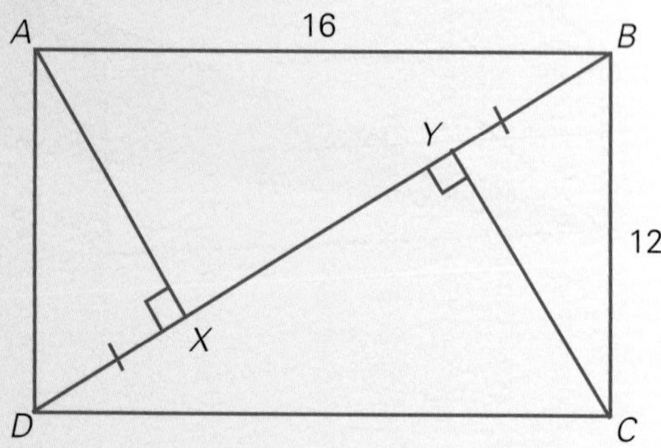

Investigation A Pythagoras spiral

You will need: paper, pencil, ruler, a set square or protractor and a pair of compasses

1 Draw an isosceles right-angled triangle such as ΔABC at right.

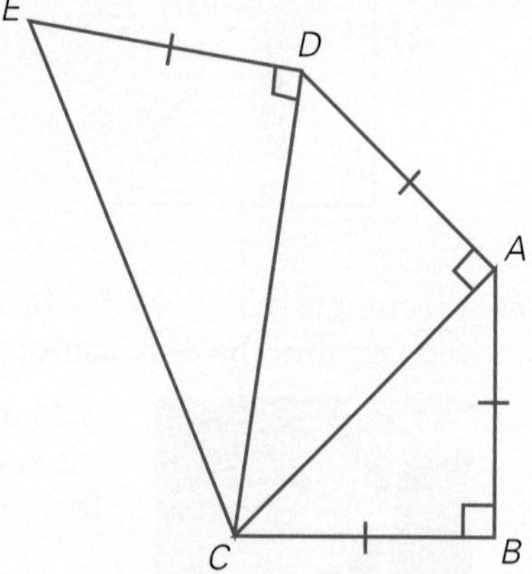

2 Construct another right-angled triangle, ΔCAD, so that $CB = AB = AD$.

3 Draw ΔCDE so that $ED = AD$.

4 Continue to draw triangles following this pattern to create a spiral. How many triangles can you draw?

5 Can you find the perimeter of your spiral?

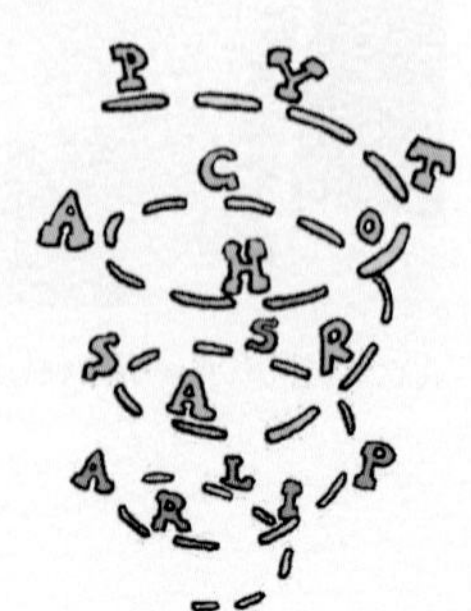

Chapter Review

Language Links

diagonal	hypotenuse	right	triad
exact	Pythagoras	root	square

Design a simple crossword for the terms in the list above. Write a clue for each word. You might like to try your crossword out on a friend.

Pythagoras

Pythagoras, or Pythagoras of Samos, was born in about 580 BC. He was a Greek philosopher and mathematician who spent his youth travelling with his merchant father to the islands around Samos.

Although historians do not know much about Pythagoras' early life, it is believed that he spent time travelling to the East. He was certainly well educated and played the lyre skilfully. He studied the works of Homer (a Greek poet and writer of the time) and studied under some of the most famous mathematicians of the time, Tales and Anaximander to name but two. He was encouraged to spent time in Egypt and Babylon before settling down in Crotona, Italy. He brought back with him knowledge of many areas, including knowledge of regular solids, and he set up a secret society or school among the aristocrats of the city. The symbol of this secret society was the pentagram.

People were suspicious of the Pythagorean brotherhood and many of them were killed in a political uprising. It is not known whether Pythagoras himself was killed or left the city before the uprising took place. The brotherhood itself was finally destroyed in about 400 BC.

Little is known about Pythagoras' work as not much of it was recorded.

Pythagoras is credited with many innovations and ideas including the theorem mentioned in this chapter. It is thought that he was the first to teach that the sun was round and that the earth and the stars moved around in space. Pythagoras and his followers linked music to mathematics; they found out that when a vibrating string is halved in length it sounds an octave higher.

Pythagoras was a teacher, student, musician and mathematician. He believed that all things could be related to numbers.

By referring to the passage above, answer the following questions:

1 Where was Pythagoras born?

2 What is the name of the musical instrument Pythagoras played?

3 What was the profession of Pythagoras' father?

4 Where did Pythagoras travel to in his life?

5 Give the name of at least one of Pythagoras' teachers.

6 Why is not much known about Pythagoras' early life and achievements?

7 Give two ideas that Pythagoras is credited with.

8 What was the name and symbol of the school that Pythagoras set up in Crotona?

9 By carrying out your own research, design a poster on the life and times of Pythagoras.

Chapter Review Exercises

6.2 **1** Identify the hypotenuse of each right triangle:

a

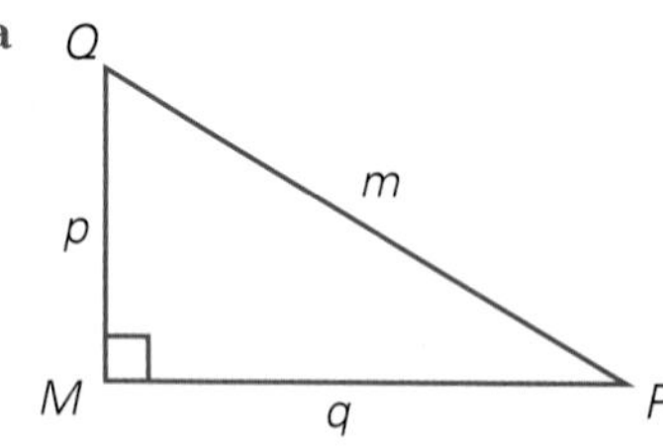

b

A
C
B

c

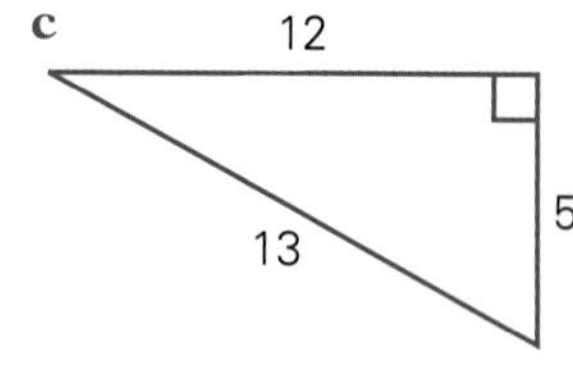

6.2 **2** Write a statement of Pythagoras' theorem for each triangle:

a

b
m
a

b

24
25
7

c

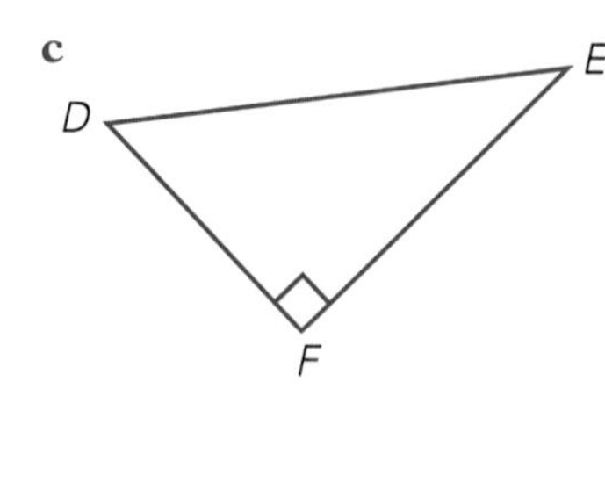

6.3 **3** Find the length of the unknown side in each triangle (correct to 2 decimal places where necessary):

a **b**

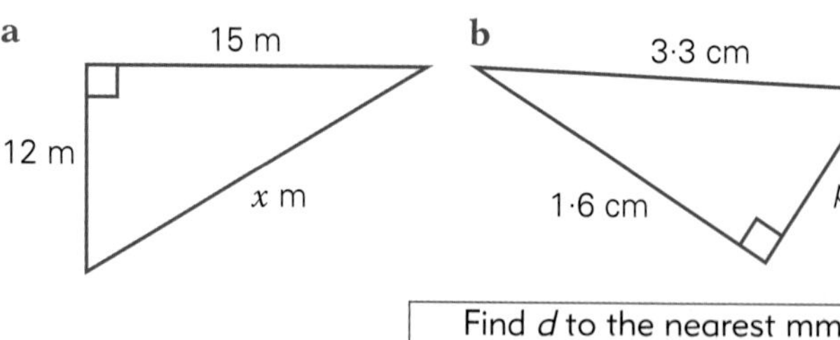

c

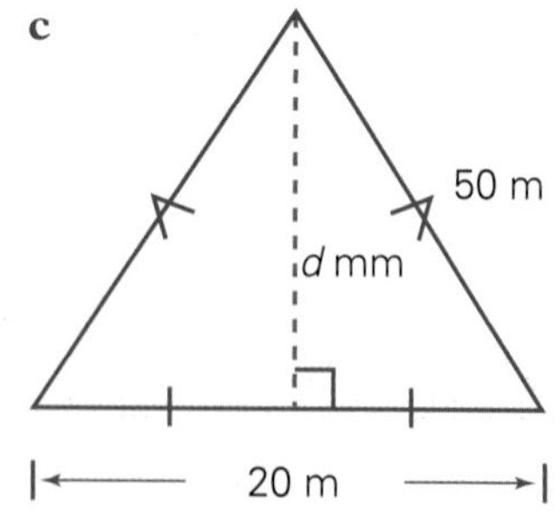

6.3 **4** *ABCD* is a rectangle. Find *AD*.

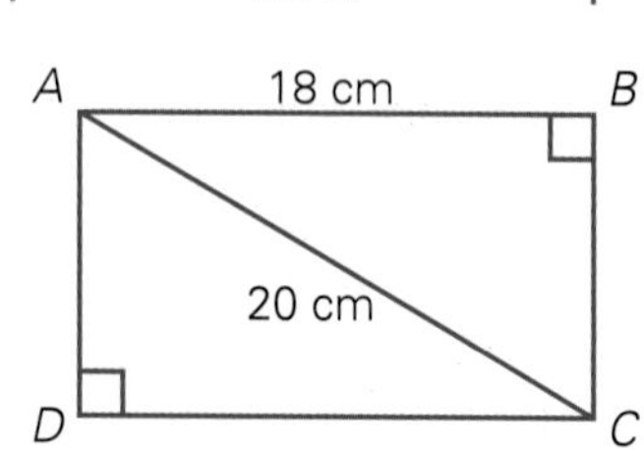

5 Test to see if a triangle of sides 7·4 m, 6·2 m and 3·1 m is right-angled. 6.5

6 If the diagonal of a square measures 6·5 cm, how long is each side? 6.6

7 20 m, 5 m, 2 m, ?

The cross-section of a symmetrical drainage ditch is shown at left. How wide is the base? (Correct to 1 decimal place.) 6.6

8 Find the value of x and y in: 6.6

a

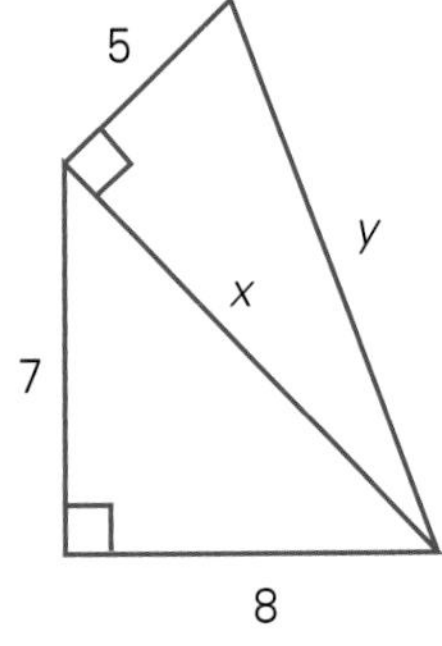

b

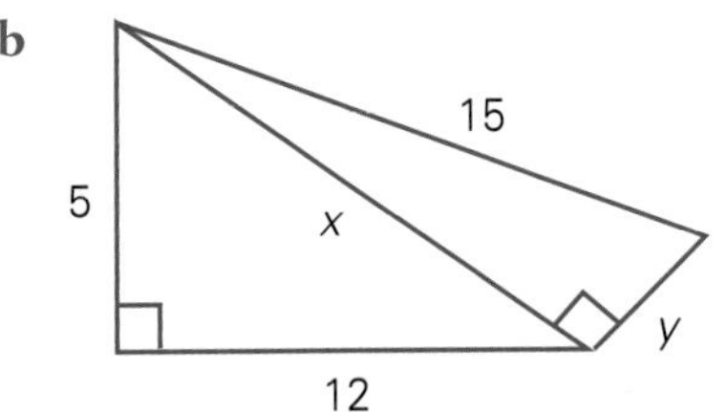

9 John and Rowan leave school from the front gate at the same time. John walks on a bearing of 300° T for 10 km to get home. Rowan walks on a bearing of 030° T to reach his house 4·5 km away. How far apart are the two houses? 6.6

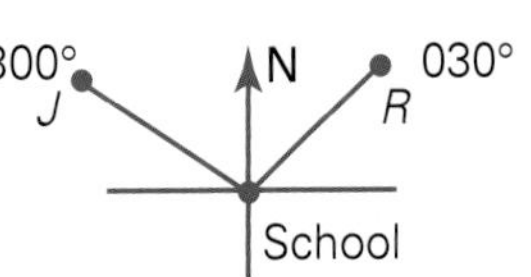

10 Find the length of the diagonal AG in: 6.6

a

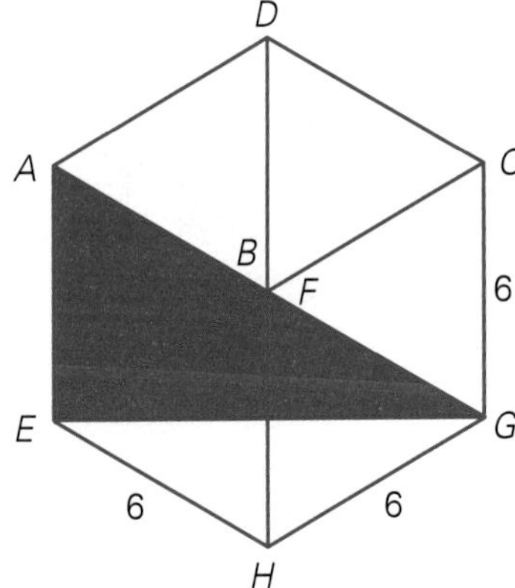

b

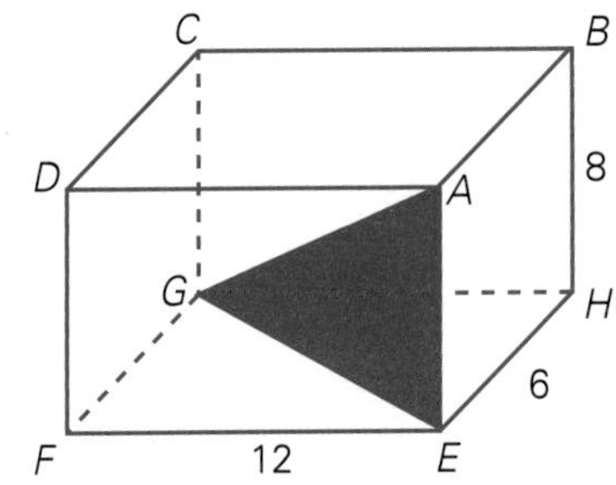

11 Find the exact value of x in: 6.6

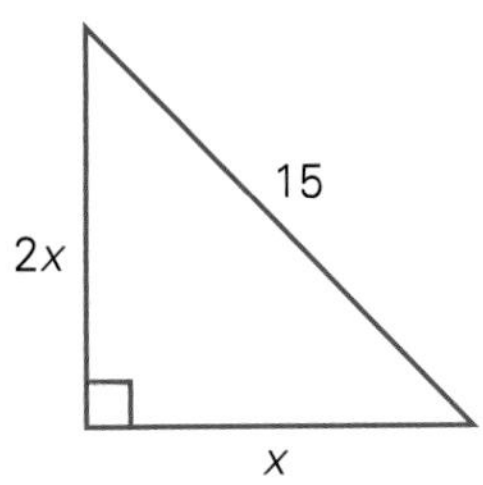

Keeping Mathematically Fit

Part A: Non-calculator

1 Evaluate $25 \times 497 \times 40$.

2 If $36 \times 36 = 1296$, write down the answer to $3{\cdot}6 \times 0{\cdot}36$.

3 Given 16, $\square$, 24 and 52 have an average of 31, find the missing number.

4 If $5x - 1 = 9$, find the value of $3x^2$.

5 Between which two consecutive integers does $\sqrt{93}$ lie?

6 Which is larger: $^-9$ or $^-12$?

7 Write the next two terms in this sequence: 1, 2, 3, 5, ______, ______.

8 If 5% of a number is 7, what is the number?

9 Increase 568 by 10%.

10 Find the reciprocal of $2\frac{2}{3}$.

Part B: Calculator

1 a What was the favourite colour of most students surveyed?
  b How many students were surveyed?
  c What fraction of the students choose blue as their favourite colour?

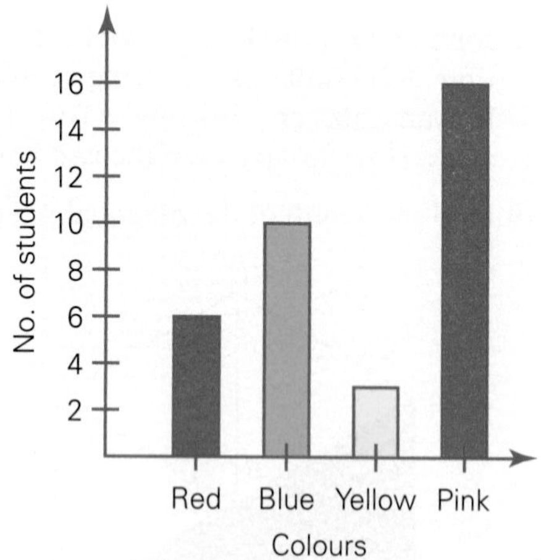

2 Evaluate $\dfrac{9{\cdot}6^3 + 2{\cdot}01^2}{4{\cdot}7}$ correct to 3 decimal places.

3 Add $3\frac{1}{4}$ to the product of $\dfrac{4}{5}$ and $2\frac{1}{2}$.

4 Solve $3{\cdot}2x - 1{\cdot}4 = 0{\cdot}6$.

5 A car valued at \$19 900 was offered for sale for \$18 500. What was the percentage discount?

6 Find 18% of \$3.56.

7 Solve for x:

a $7x = 420$ b $5 - x = {}^-6$ c $3(2x + 1) = 21$

8 Zayd was asked to calculate the cost of petrol to the nearest 5 cents. He wrote down \$27. What could the petrol have cost before rounding?

Cumulative Review 2

Part A: Multiple-choice questions

Write the letter that corresponds to the correct answer in each of the following.

1 $\sqrt[3]{\dfrac{8{\cdot}04}{1{\cdot}3 \times 4{\cdot}8}}$ correct to 3 decimal places is:

A 29·686 **B** 3·093 **C** 1·135 **D** 1·088

2 $2^{10} \div 2^2$ is equivalent to:

A 1^5 **B** 2^{12} **C** 2^8 **D** 1^8

3 When $x = {}^-1$, $3x^2 - x$ equals:

A $^-4$ **B** 4 **C** 2 **D** $^-2$

4 $(2x) - ({}^-x)$ is the same as:

A x **B** $2x^2$ **C** 2 **D** $3x$

5 0·3% is equivalent to:

A $\frac{3}{100}$ **B** $\frac{1}{3}$ **C** $\frac{3}{10}$ **D** $\frac{3}{1000}$

6 6% of an amount is \$300. The amount is:

A \$18 **B** \$5000 **C** \$180 **D** \$500

7 The value of b which makes $2b + 7$ take the value 19 is:

A 6 **B** $2\frac{1}{2}$ **C** 13 **D** $16\frac{1}{2}$

8 If $7x + 3 = 51 + 4x$, then x equals:

A 16 **B** 18 **C** $4\frac{4}{11}$ **D** none of these

9 $x \leq 2$ is a solution to:

A $x + 1 < 3$ **B** $2x - 3 \leq 1$ **C** $\frac{x}{2} - 1 \geq 0$ **D** $x - 1 \geq 1$

10 $x = 4$ is the solution of:

A $\frac{x+1}{3} = 1$ **B** $3x - 2 = x + 6$ **C** $3(x - 1) = 12$ **D** $5 - x = 9$

11 $y = 2x + 7$. If $x = {}^-3$ then $y =$

A 30 **B** $^-1$ **C** 1 **D** 13

12 The correct solution to $4a - 6 = 2a + 10$ is:

A $a = 16$ **B** $a = 8$ **C** $a = 4$ **D** $a = 2$

13 Which set represents numbers in a Pythagorean triad?

A 6, 7, 8 **B** 9, 41, 40 **C** 3·1, 4·1, 5·1 **D** 12, 15, 16

14 The length of the interval AB is:

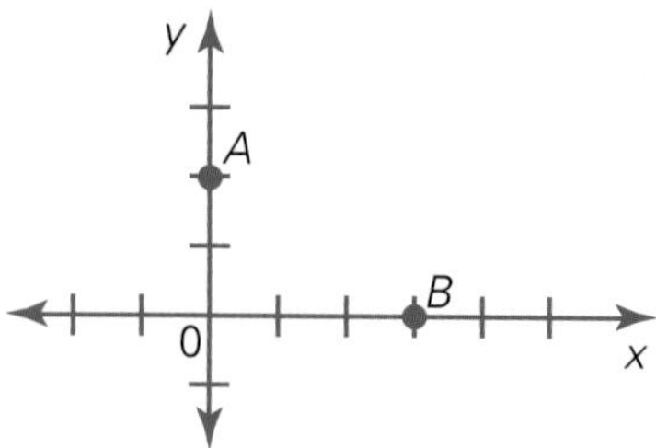

A 3 units B 2 units C 13 units D 3·6 units

15 Which of these is a Pythagorean triad?

A 6, 8, 12 B 8, 15, 16 C 7, 40, 41 D 10, 24, 26

Part B: Short-answer questions

Show full working for each of the following.

1 Find the value of the following, correct to 2 decimal places:

a $\dfrac{10}{4{\cdot}7 \times 3{\cdot}6}$ b $\sqrt[3]{\dfrac{8{\cdot}9}{4{\cdot}5}}$ c $\dfrac{\sqrt{14{\cdot}6 + 3 \times 5^2}}{2}$

2 Simplify:

a $2^9 \times 2^3$ b $4^7 \div 4^3$ c $(2^5)^3$

d $5ab + 7a - ab$ e $3(a + 4) + 2a + 1$ f $x(x - 3) + 5(x + 1)$

g $\dfrac{12a^2}{36ab}$ h $\dfrac{x+1}{3} - \dfrac{x}{3}$ i $\dfrac{x}{10} \times \dfrac{5}{2x}$

3 Find:

a 7% of \$496 b $8\frac{1}{2}$% of 72

4 Decrease \$947.50 by 6%.

5 \$2400 is invested at $5\frac{3}{4}$% p.a. for $3\frac{1}{2}$ years. Calculate the simple interest.

6 The following table shows the points won by each house at Mathstown High over a 5-year period in the inter-house swimming carnival. From the table, answer the following questions:

Year	Cook	Flinders	Hume
2000	231	182	87
2001	201	203	96
2002	257	130	113
2003	223	151	126
2004	158	169	173

a Which house won in 2001?
b In which year did Hume win?
c How many points did Flinders score in 2003?
d Which house won the carnival two years in a row? Which years?
e What was Flinders' worst year?
f How many more points than Hume did Cook score in 2000?

g How many more points did Hume score in 2002 than in 2000?
h What was the total number of points awarded at each carnival?

7

a What fraction of your body is muscle?
b Disregarding the 'Other' slice, list the sectors in increasing order of size.
c What percentage of your body is brain?
d Grant weighs 60 kg. What do his bones weigh? (Assume his body matches the proportions shown.)
e Christine's brain weighs 2 kg. What is her total body mass?

Adolescent body mass

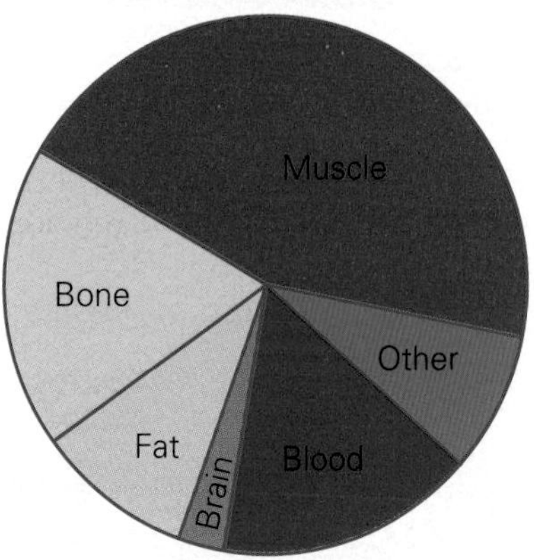

8 Find the value of each pronumeral, correct to 1 decimal place where necessary:

a

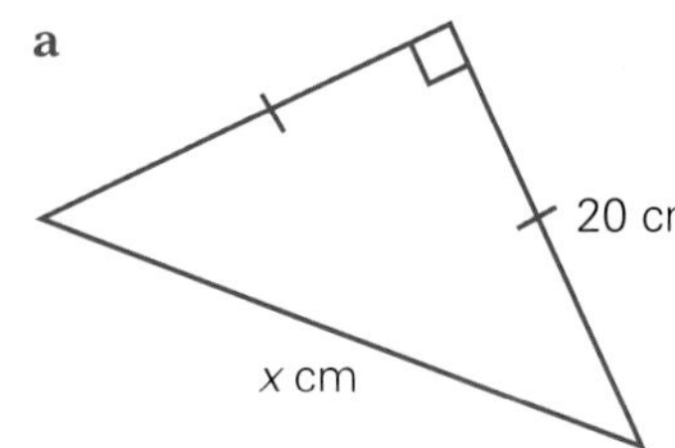

b

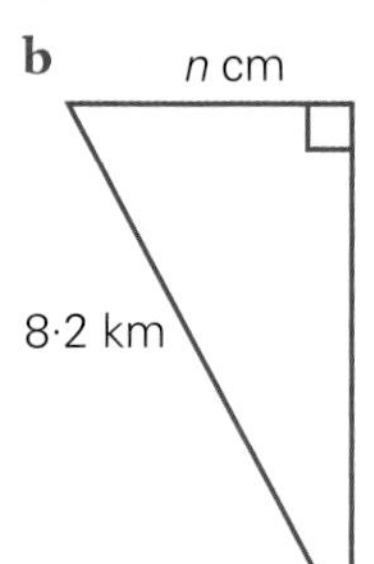

c

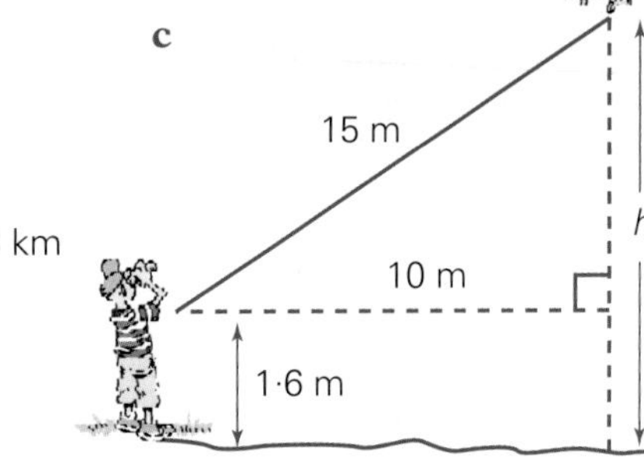

9 Solve for n:

a $7 + n = 18$
b $\frac{n}{5} = {}^{-}2$
c $\frac{2}{3}n = 4$
d $3n - 4 = 8$
e $3n + 4 + 2n = 29$
f $23 - 2n = 1$
g $6n + 3 = n + 8$
h $4n - 1 = 5 - 2n$
i $9n + 1 = 7(n + 3)$

10 When 7 is added to the product of 8 and a certain number, the result is 79. What is the number?

11 Write an inequality to represent each of the following:

a

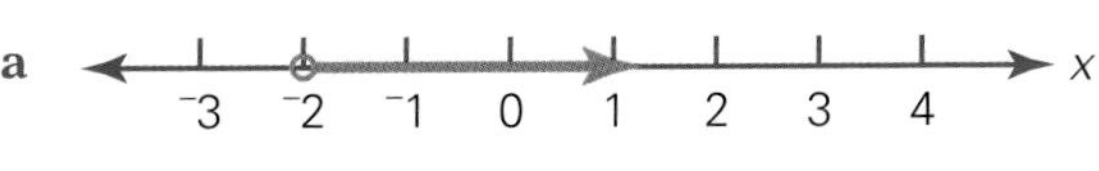

b x

⁻4 ⁻3 ⁻2 ⁻1 0 1 2 3 4 5 6

12 Solve:

a $4x + 12 \geq 20$
b $\frac{x}{3} - 1 < 9$

1 What you need to know and revise

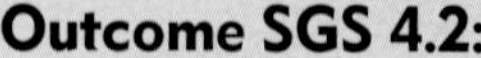

Outcome SGS 4.2:
Identifies and names angles formed by the intersection of straight lines, including those related to transversals on sets of parallel lines, and makes use of the relationships between them

Outcome SGS 4.3:
Classifies, constructs and determines the properties of triangles and quadrilaterals
- names angles
- adjacent angles, vertically opposite angles, straight angles and revolutions
- complementary and supplementary angles

2 What you will learn in this chapter

Outcome SGS 4.2:
Identifies and names angles formed by the intersection of straight lines, including those related to transversals on sets of parallel lines, and makes use of the relationships between them

Outcome SGS 4.3:
Classifies, constructs and determines the properties of triangles and quadrilaterals
- identifies a pair of parallel (//) lines and a transversal
- uses the common conventions to indicate parallel lines on diagrams
- identifies, names and measures the alternate, corresponding and co-interior angle pairs for two parallel lines cut by a transversal
- recognises the equal and supplementary angles formed when a parallel line is cut by a transversal
- uses angle properties to identify parallel lines
- uses angle relationships to find unknown angles in diagrams

Working Mathematically outcomes WMS 4.1–4.5
Students will be asked to ***question, apply strategies, communicate, reason*** and ***reflect*** in the sections of this chapter.

Geometry

Key mathematical terms you will encounter

acute
adjacent
alternate
complementary
co-interior
construction
corresponding
equilateral
exterior
interior
intersection
isosceles
obtuse
opposite
parallel
polygon
proof
quadrilateral
reasoning
reflex
revolution
right
scalene
straight
supplementary
transversal
triangle
vertex
vertically opposite

MathsCheck
Geometry

1 Complete the table of angles and terms studied last year.

a $\angle$ ______ or $\angle$ ______	**b** ______ angles lie between 0° and 90°.	**c** A ______ angle = 90°.	**d** An ______ angle lies between 90° and 180°.
e A ______ angle measures 180°.	**f** A ______ angle lies between 180° and 360°.	**g** A ______ = 360°.	**h** ______ are equal.
i $x°$ and $y°$ are ______ angles; they add to 90°.	**j** $x°$ and $y°$ are ______ angles; they add to 180°.	**k** $x + y + z =$ ______	**l** $\angle ABD$ and $\angle DBC$ are ______ angles.
m ΔPQR is a ______ triangle.	**n** ΔABC is an ______ triangle.	**o** ΔXYZ is an ______ triangle.	**p** ΔDEF is a ______ triangle.

2 Classify these triangles according to both their angle types and side lengths:

a

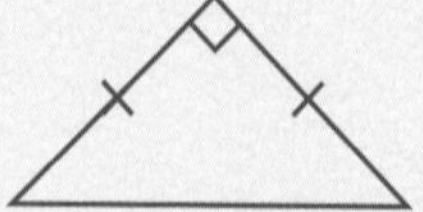

b

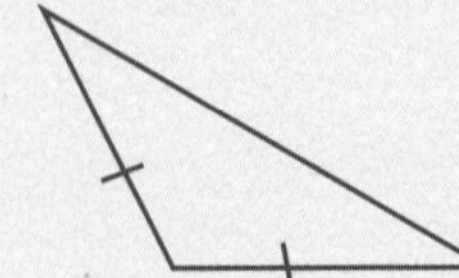

3 In the diagram:

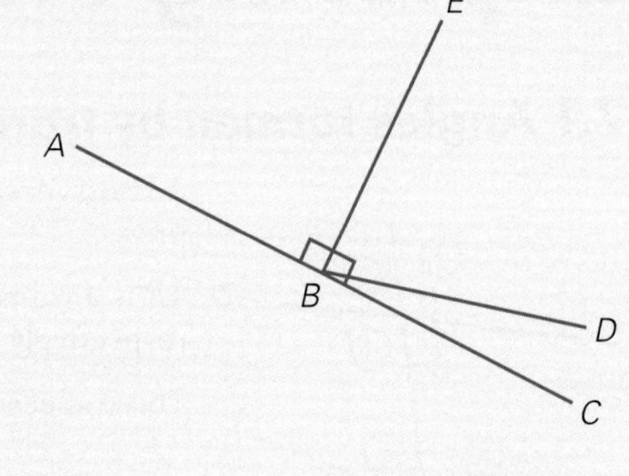

a name:

i an acute angle
ii a right angle
iii a straight angle
iv an obtuse angle
v a pair of adjacent angles
vi two complementary angles
vii two supplementary angles

b measure with a protractor:

i $\angle ABD$ **ii** $\angle CBD$

4 Find the value of each pronumeral in the diagrams below and choose an appropriate reason from the list supplied:

a

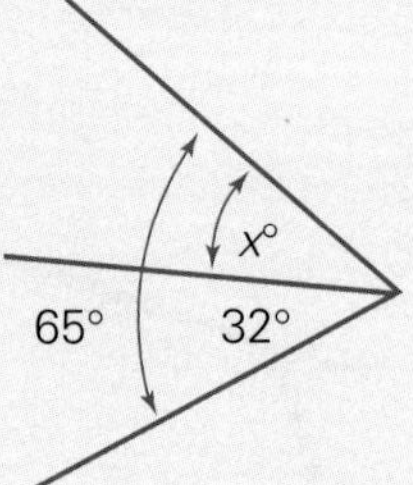

b

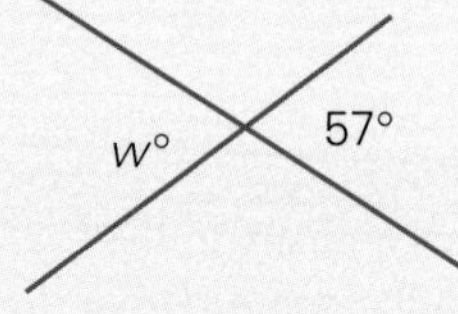

c

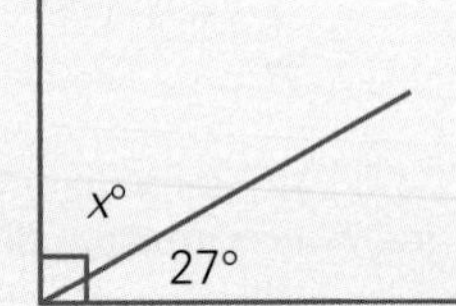

d

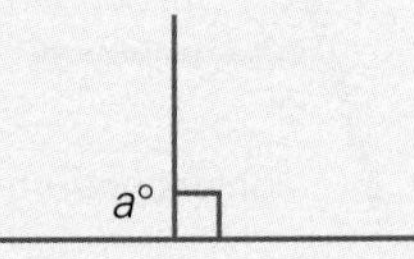

e

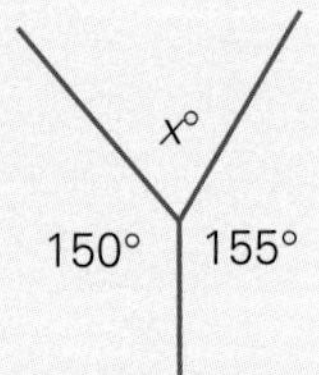

f

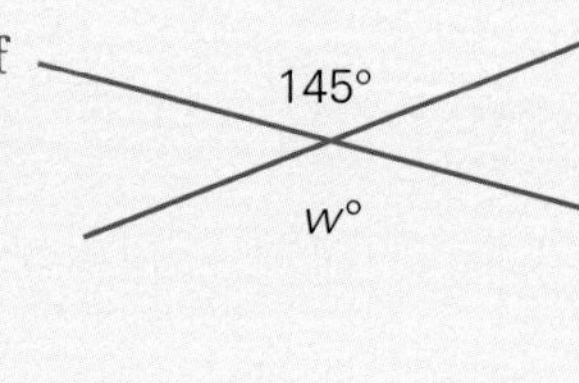

g

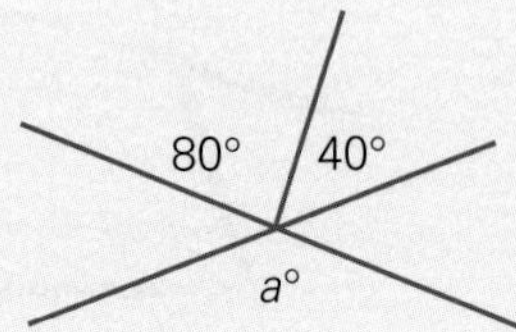

h

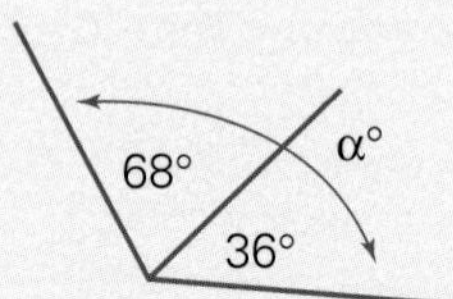

i

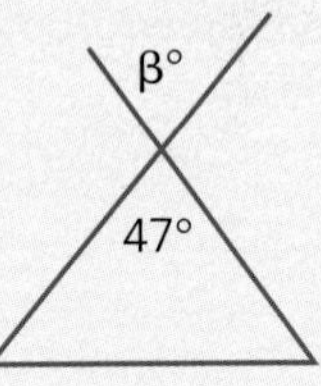

j

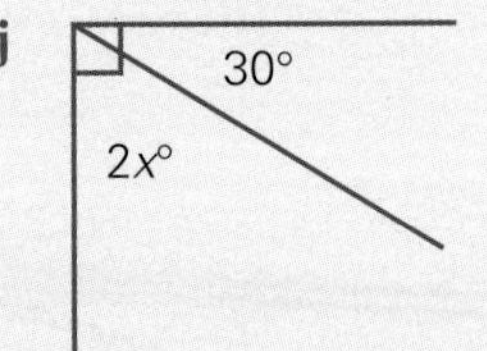

k

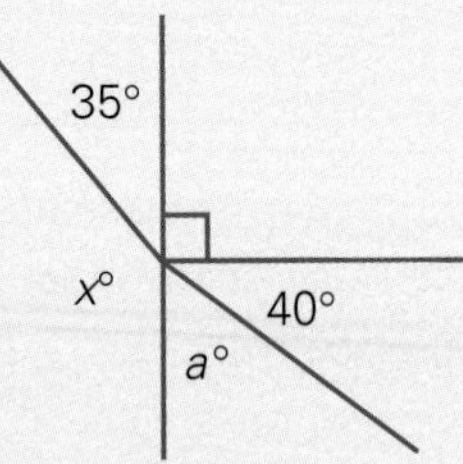

l

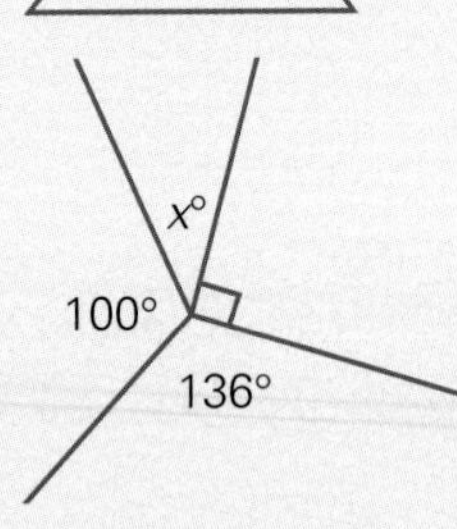

Reasons

A vertically opposite angles **B** angles on a straight line **C** revolution equals 360°

D sum of adjacent angles **E** adjacent complementary angles

TEACHER

Exploring New Ideas

7.1 Angles formed by transversals

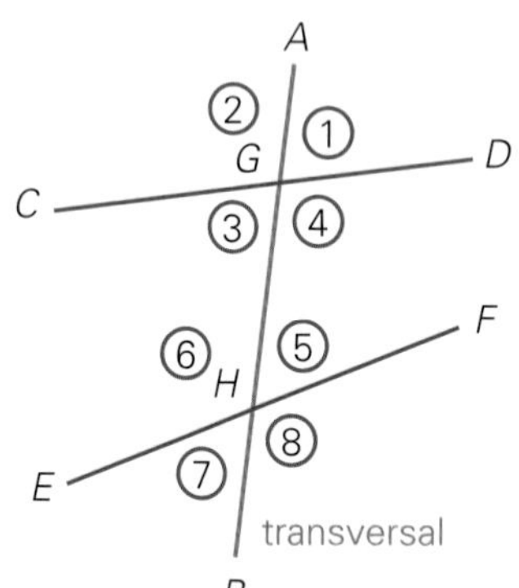

A **transversal** is a line which cuts across two or more other lines.

AB is a transversal, inter-secting *CD* and *EF* at *G* and *H* respectively. Can you name the eight angles formed?

Three special patterns of angles are formed.

a and ***b*** are **corresponding angles** as they are in the corresponding *positions*; both lie on the *same side of the transversal* and both lie on the *same side of the other two lines* (underneath, in this case). They form an F-shape.

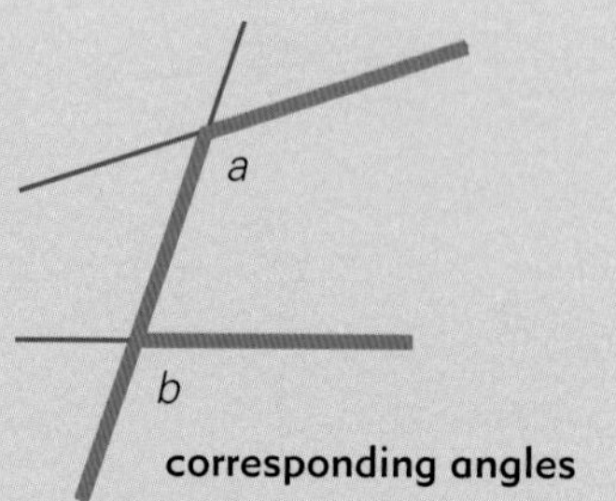

c and ***d*** are **alternate angles** as they are on *alternate sides of the transversal* and *between the other two lines.* They form a Z-shape.

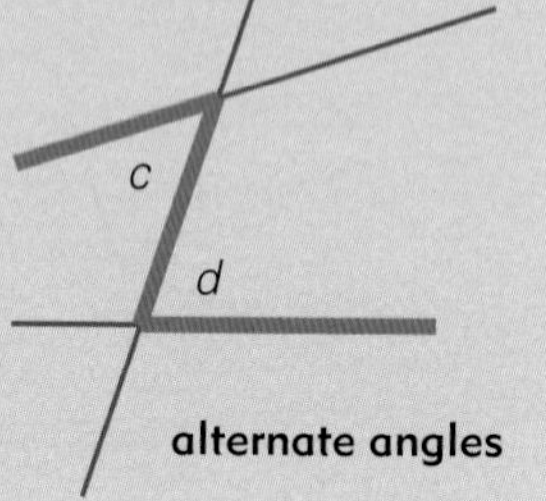

e and ***f*** are **co-interior angles**; both lie *inside* the other two lines and on the *same side of the transversal.* They form a [shape.

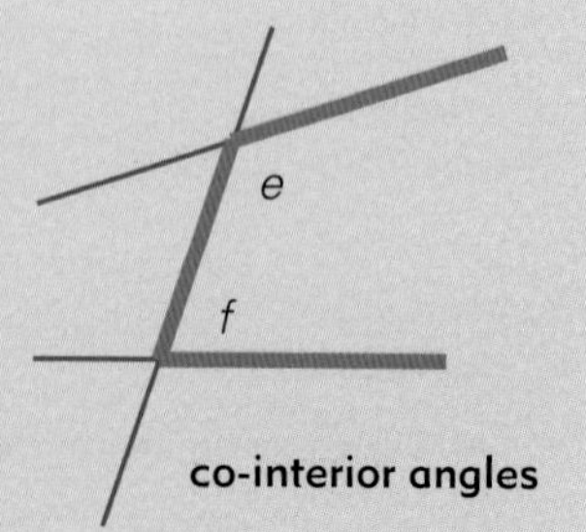

Four distinct pairs of *corresponding* angles are formed. Can you follow the transformation of the F-shape?

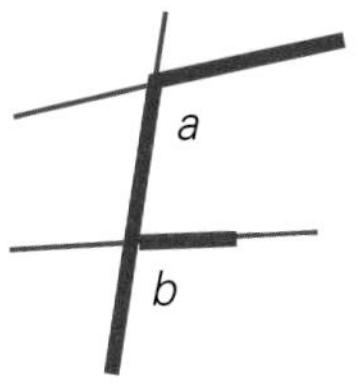

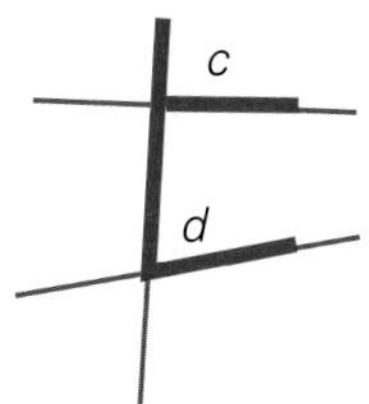

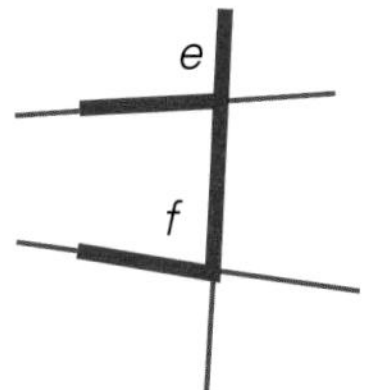

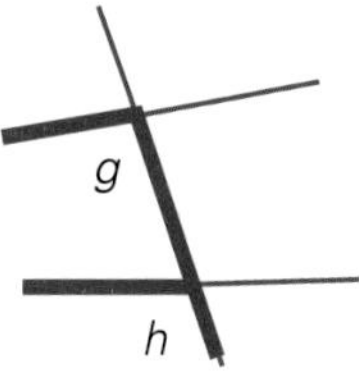

Two distinct pairs of *alternate angles* show up. Two distinct pairs of *co-interior angles* exist.

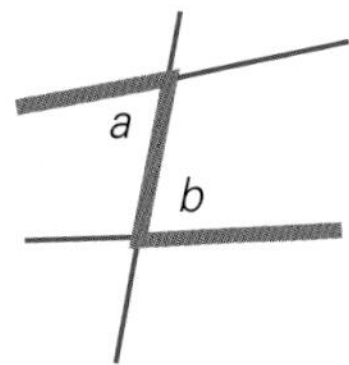

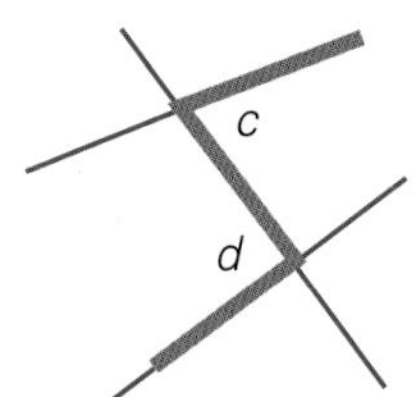

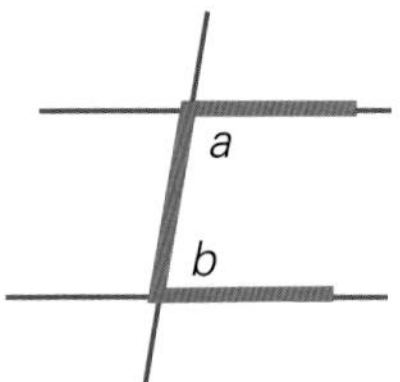

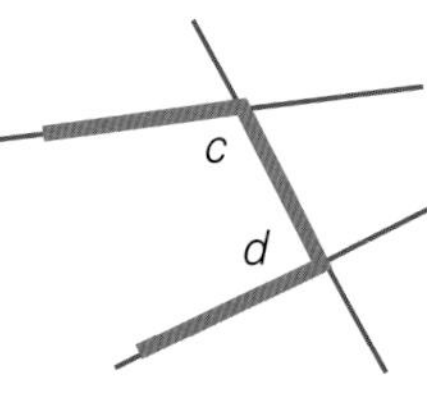

EXAMPLE 1

Name the angle alternate to $\angle PBD$.

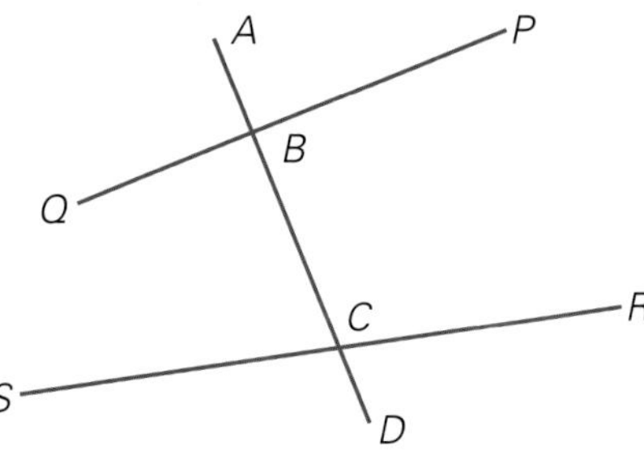

Solution

Follow the Z-shape and locate the alternate angle.
$\angle SCA$ is alternate to $\angle PBD$.

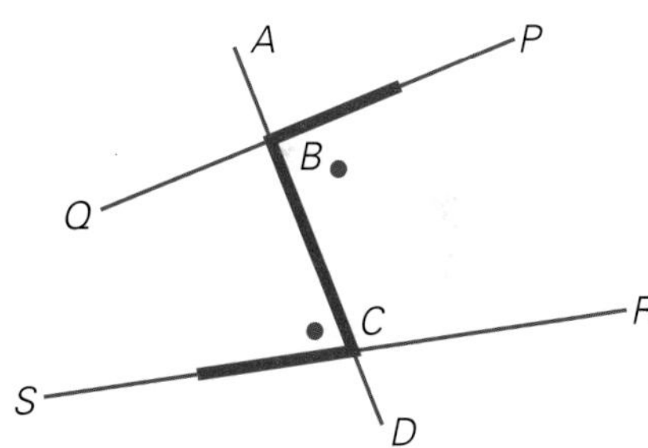

EXERCISE 7A

1 Classify the following pairs of angles as either corresponding, alternate or co-interior:

a *c* and *g* **b** *b* and *e*
c *c* and *e* **d** *d* and *f*

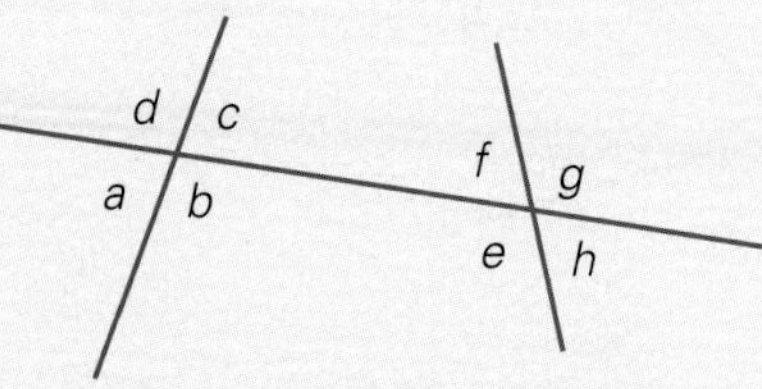

2 What relationship exists between **i** *a* and *c*? **ii** *e* and *f*?

3 In which diagram are m and p *alternate*?

A

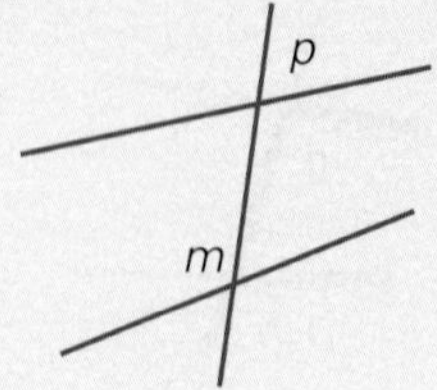

B

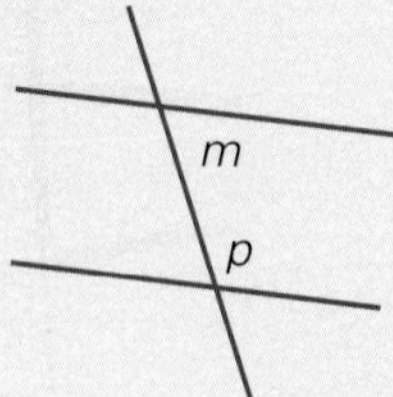

C

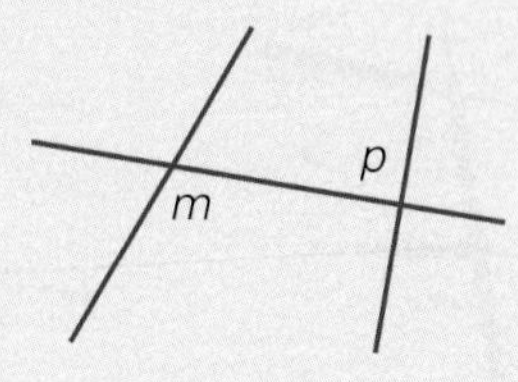

4 In which diagram are a and b *corresponding* angles?

A

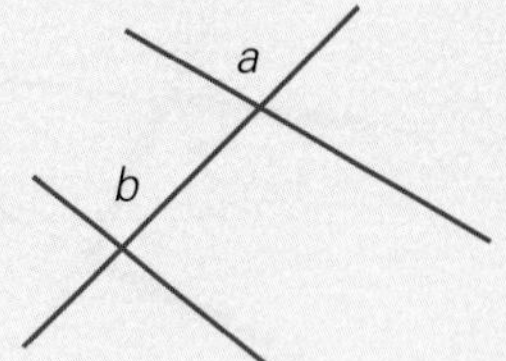

B

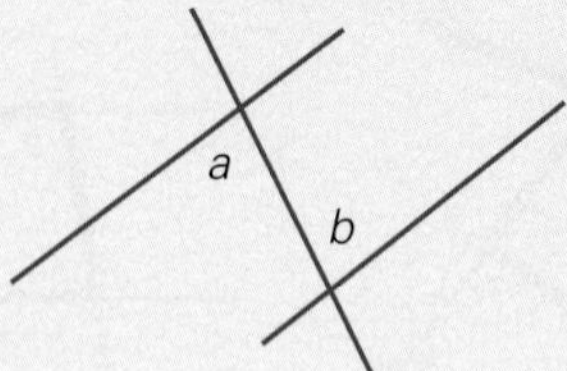

C

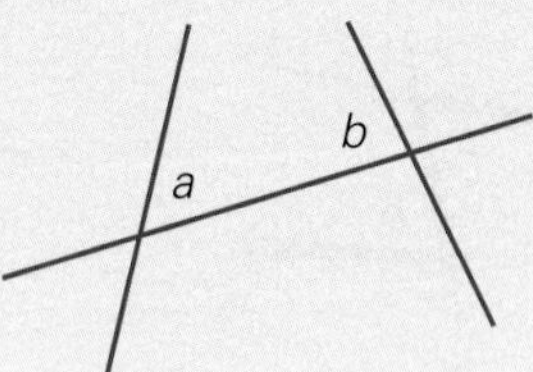

5 Name the angle co-interior with angle c in each case:

a

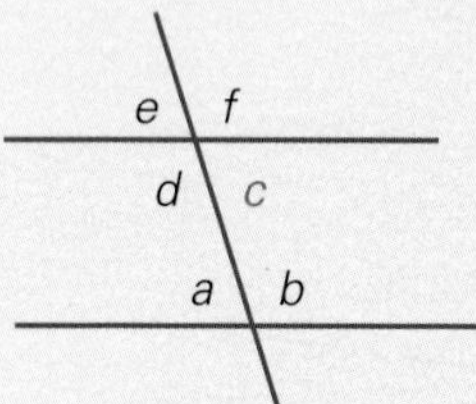

b

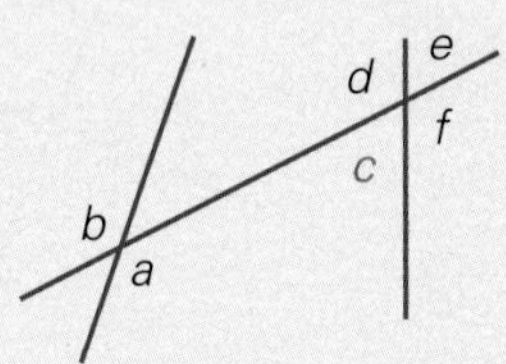

c

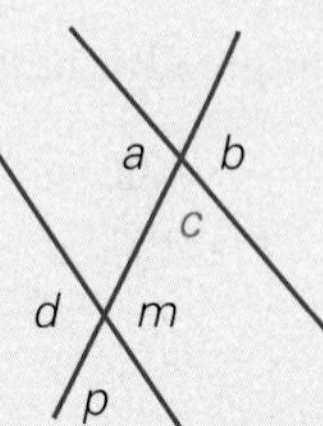

Level 2

6 Use the diagram at right to name:

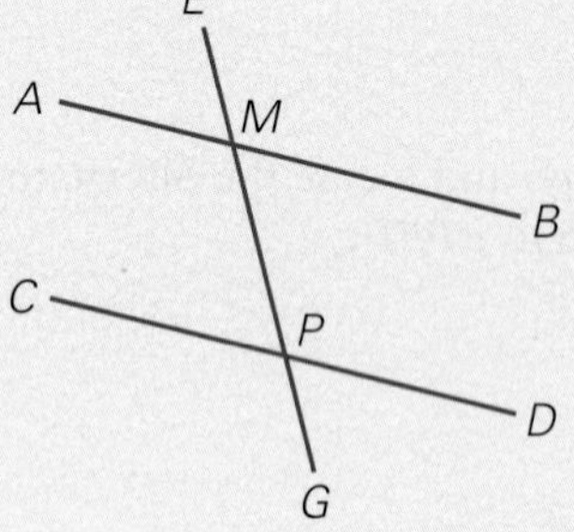

a the transversal
b the two lines it intersects
c the points of intersection
d the angle corresponding to $\angle EMB$
e the angle alternate to $\angle AMP$
f the angle co-interior with $\angle DPM$
g the angle corresponding to $\angle EMA$

7 Name all the pairs of alternate angles in:

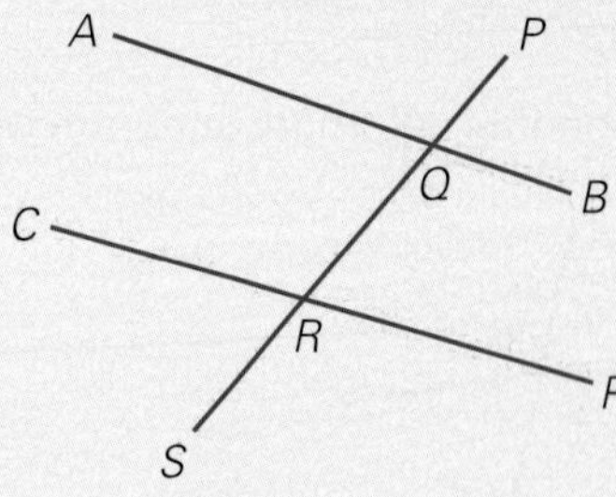

8 Name all the pairs of corresponding angles in:

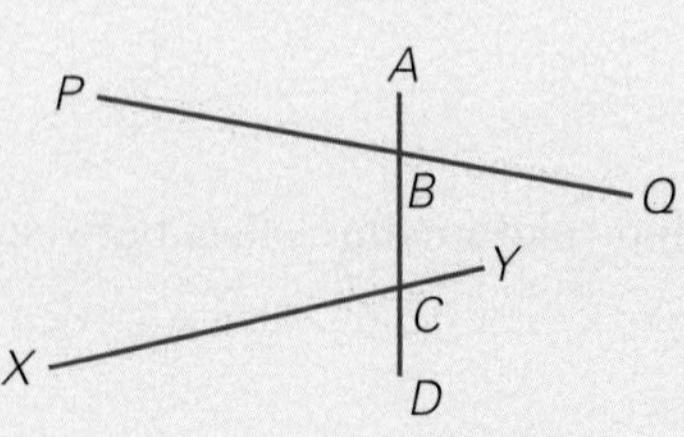

Investigation Angles on parallel lines

A Corresponding angles on parallel lines

You will need: set square, ruler, protractor

1 Hold a ruler firmly at an angle across a piece of paper and draw a line (dotted) along its edge. Place your set square against that edge of the ruler and draw a line along its hypotenuse as shown. Slide the set square down the ruler's edge to a new position (coloured) and draw a second line.

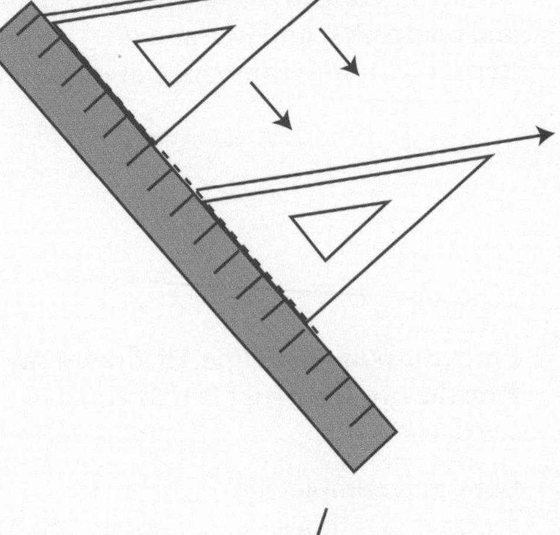

Take the instruments away. What shape has been created? Since the set square–ruler angle remained fixed, corresponding angles were made equal. What sort of lines were produced?

2 Select two widely spaced lines in your exercise book; draw in a pair of parallel lines and any *transversal*. Use a protractor to measure and compare the size of each pair of corresponding angles you can find.

Copy and complete the table.
Copy and complete the statement:
______ lines make corresponding angles ______.

Angle	Size	Corresp. ∠	Size
a		b	
c			
e			
g			

3 At the point A on line BA, draw a ray. At P on the ray, construct $\angle TPR$ equal to $\angle PAB$ as shown. Test to see if $PR \parallel AB$.
Copy and complete:

When ______ angles are equal, the lines cut by the ______ are ______.

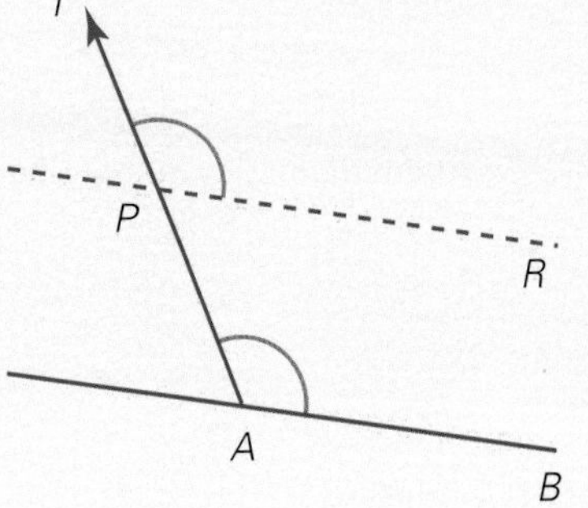

B Alternate angles on parallel lines

You will need: ruler, protractor, compasses

1 Select any two well-separated parallel lines in your exercise book and draw a transversal to cut them. Label the angles as shown. Measure and compare a and b, c and d.
Repeat with the transversal at different angles.

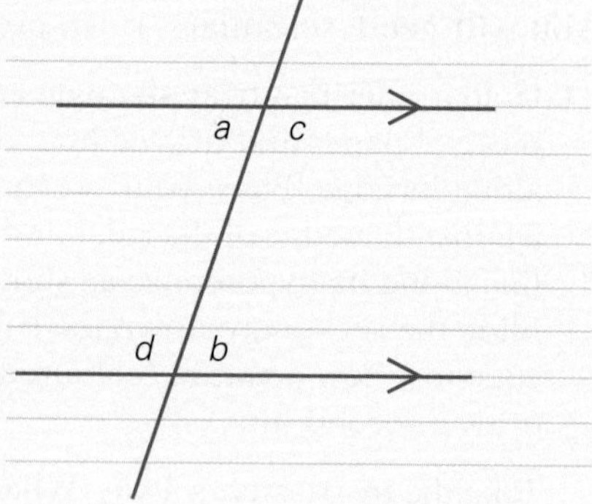

2 From the point A on line AB, draw a ray. At the point P on the ray, construct $\angle APM$ equal to $\angle BAP$. Test to see if $MP \parallel AB$.

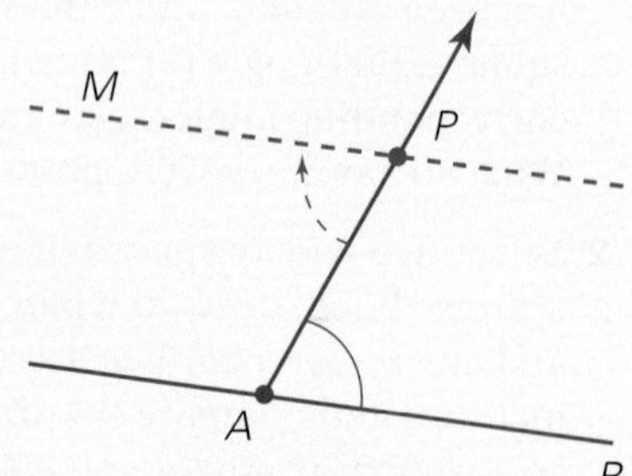

Copy and complete:

a On ______ lines, ______ angles are ______.

b If ______ angles are ______, the lines cut by the ______ are ______.

C Co-interior angles on parallel lines

1 Compare a and d, c and b from step 1 above. Write a *statement* of your *findings*.

2 Repeat step 2 above, this time constructing a *supplementary* $\angle APX$ at P, on the same side of the transversal. See if $AB \parallel PX$.

7.2 Angles formed by parallel lines and transversals

Euclid (c. 330–275 BC), a Greek mathematician known for his work on geometry, saw the need to prove the equality of angles associated with parallel lines. The angles associated with parallel lines are looked at in this section.

When parallel lines are crossed by a transversal alternate, corresponding and co-interior angles are formed.

Alternate angles on parallel lines are equal.

α

α

There are two pairs of alternate angles formed when parallel lines are cut by a transversal:

If the alternate angles are equal, then the lines that form them must be parallel.

Corresponding angles on parallel lines are equal.

There are four pairs of corresponding angles formed when parallel lines are cut by a transversal:

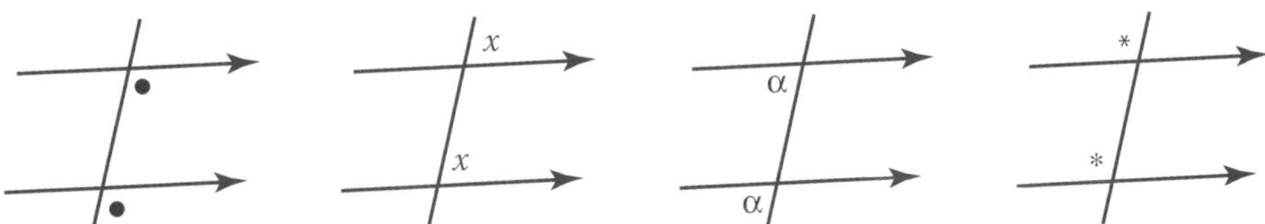

If the corresponding angles are equal, then the lines that form them must be parallel.

Co-interior angles are supplementary.

$x + y = 180$

There are two pairs of co-interior angles formed when parallel lines are cut by a transversal:

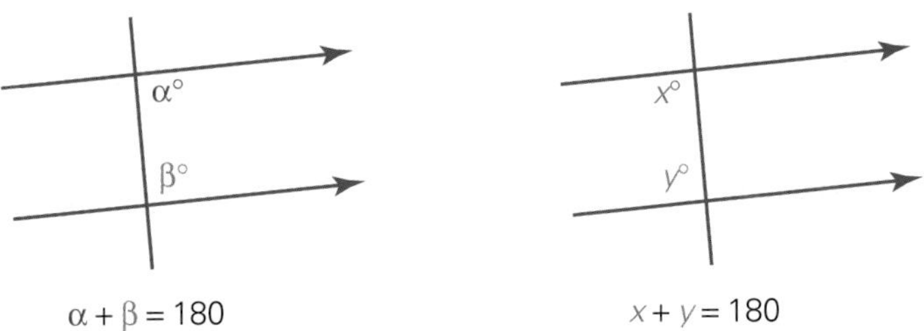

$\alpha + \beta = 180$ $\qquad$ $x + y = 180$

If the co-interior angles add to 180°, then the lines that form them must be parallel.

EXAMPLE 1

Find the value of a and b. Give a reason for each.

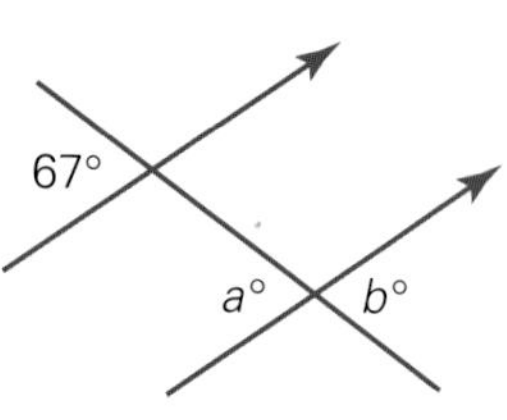

Solution

$a = 67$ (corresponding angles on || lines)
$b = a$ (vertically opposite angles)
$\therefore b = 67$

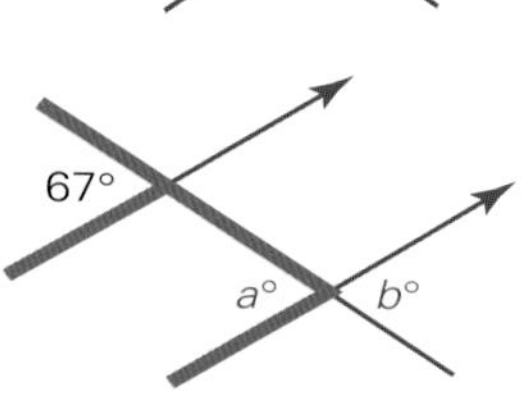

EXAMPLE 2

Find the value of x in each of the diagrams below:

a

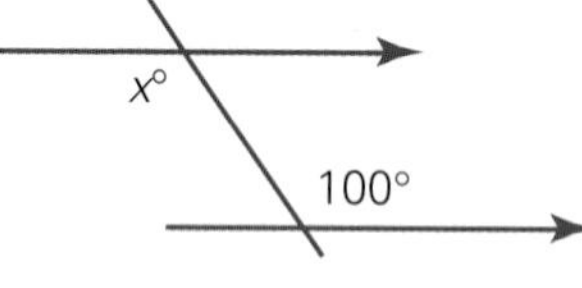

b

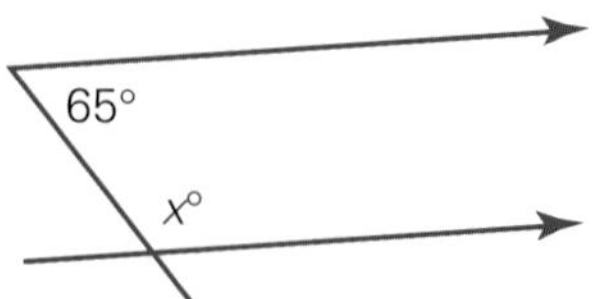

Solution

a

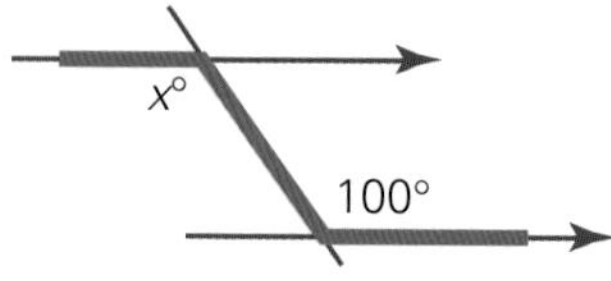

$x = 100$ (alternate angles on || lines)

b

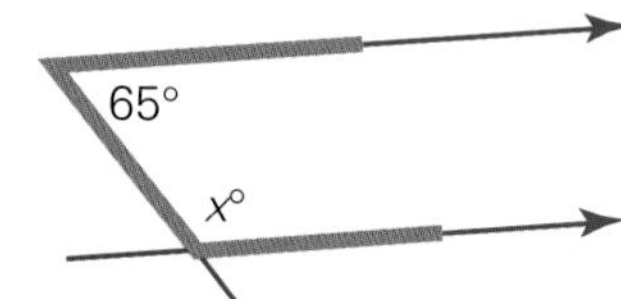

$x + 65 = 180$ (co-interior angles on || lines)
$\therefore x = 115$

EXAMPLE 3

Is PQ parallel to ST in:

a

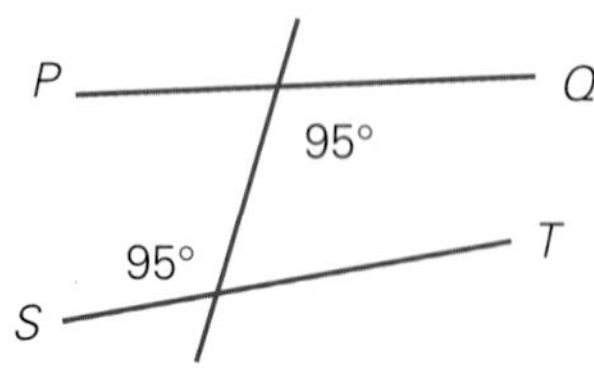

b

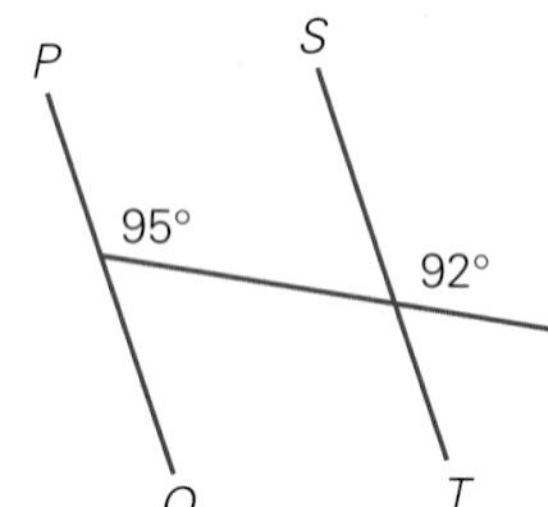

Solution

a Yes, $PQ \parallel ST$ as the alternate angles are equal.

b No, the lines are not parallel as the corresponding angles are not equal.

EXERCISE 7B

1 Find the value of n in each figure. Give a reason.

a

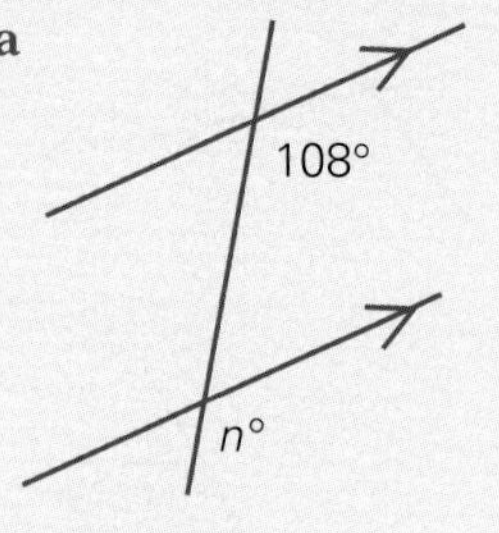

b

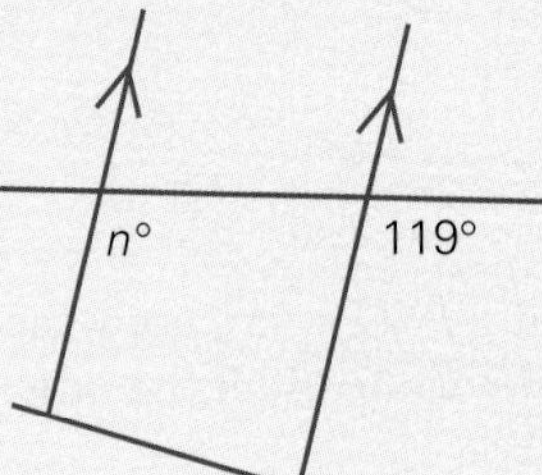

c

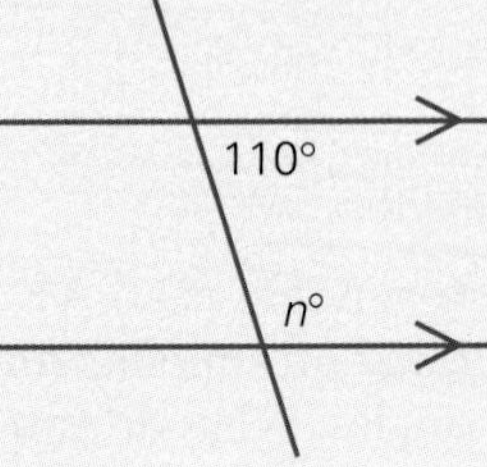

d

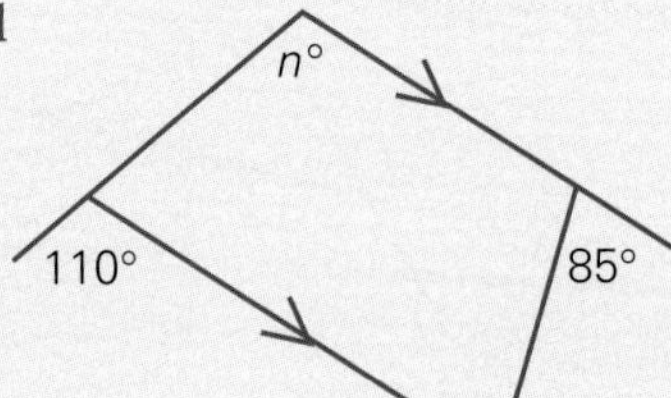

e

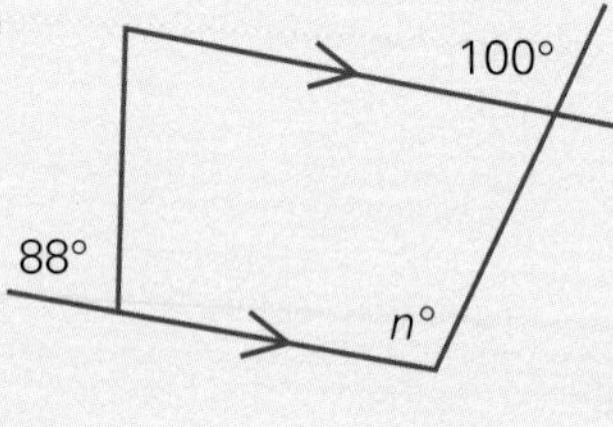

f

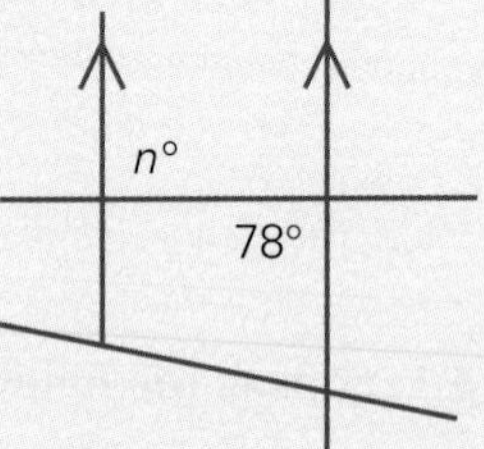

g

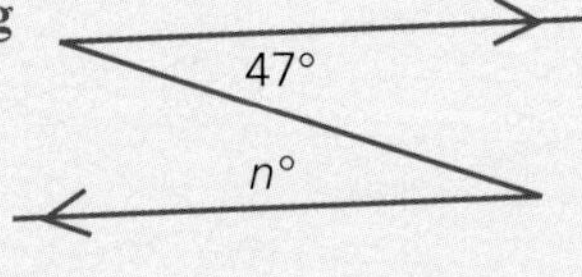

h

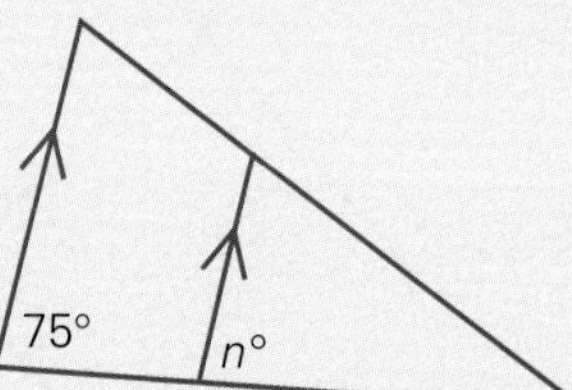

i

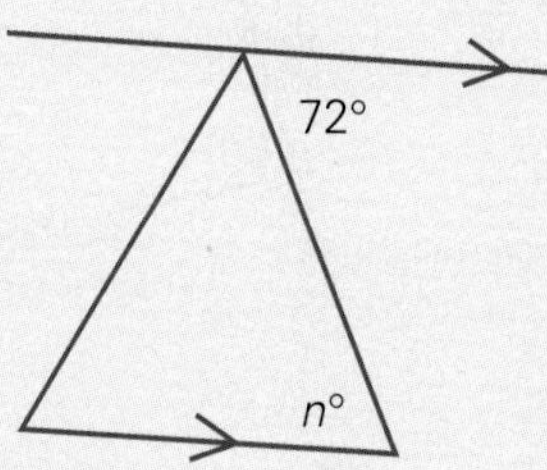

j

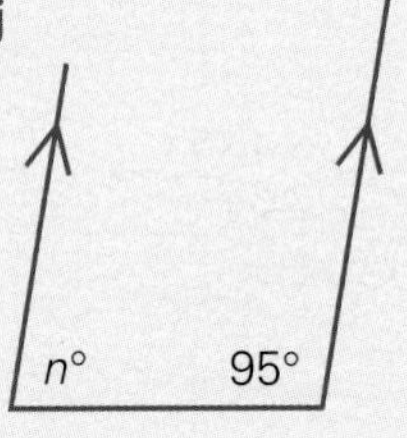

k

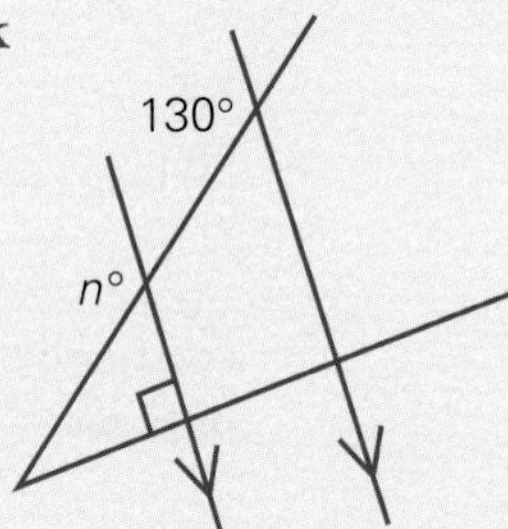

l

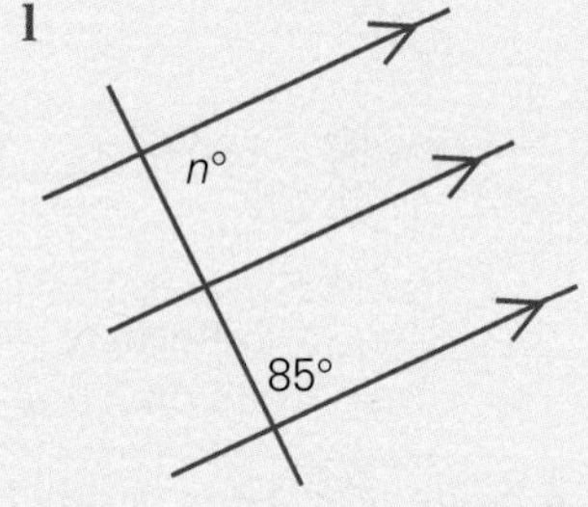

2 Copy the diagram and mark in all four pairs of corresponding angles using the symbols •, ∡, ∅, ∝.

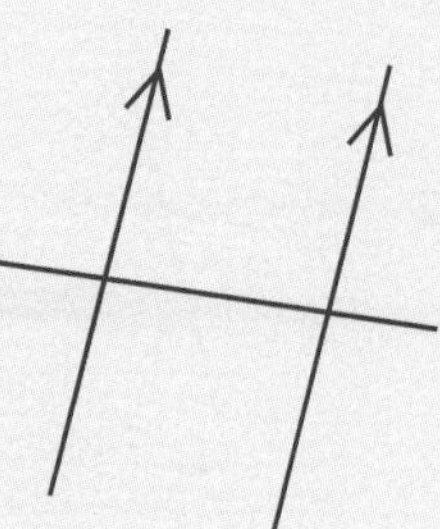

3 Copy the diagram and mark in the two pairs of alternate angles using the symbols α and β.

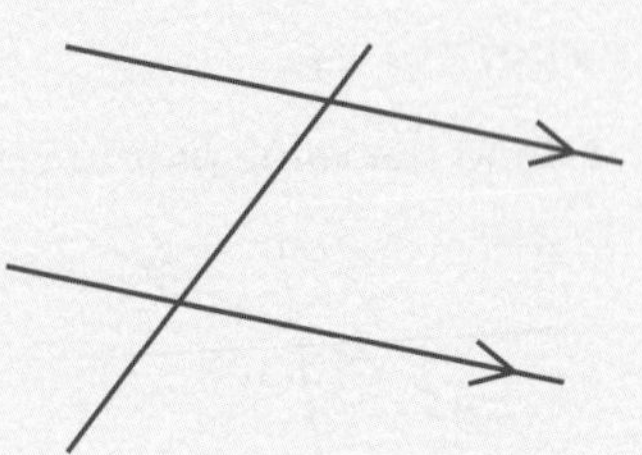

4 Given that $PQ \parallel XY$ and that $\angle RSQ = 55°$, copy the diagram supplied and find the value of all the other angles. What do you notice?

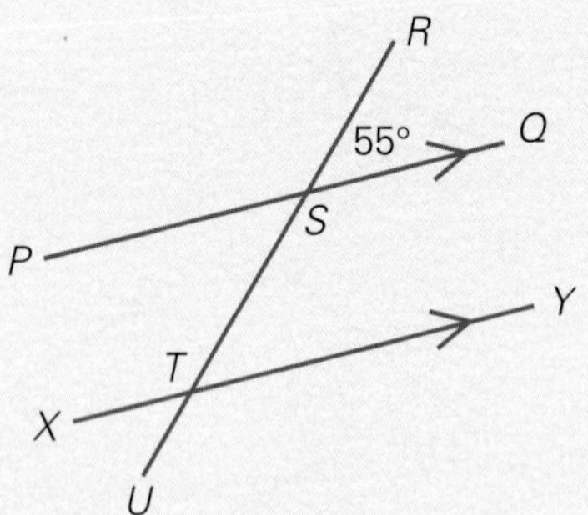

5 Complete the following:

a

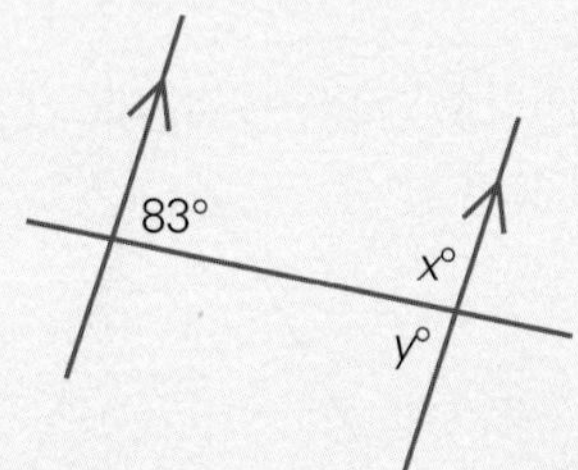

$x + 83 = 180$ (______)
$x =$ ______
$\therefore y =$ ______
(alternate $\angle$s on $\parallel$ lines)

b

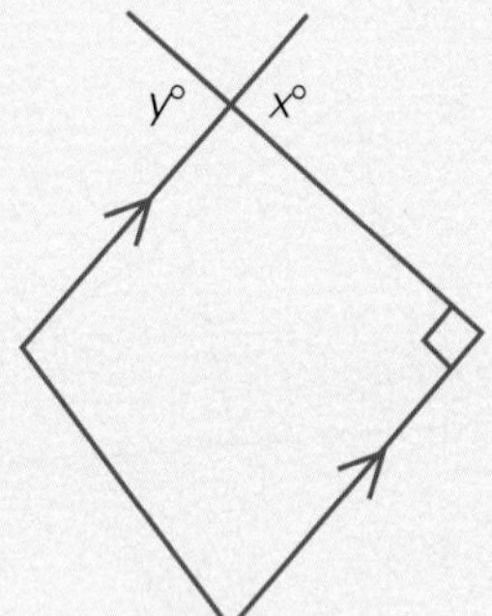

$x = 90$ (______)
$x = y$
$\therefore y =$ ______
(vertically opp. $\angle$s)

c

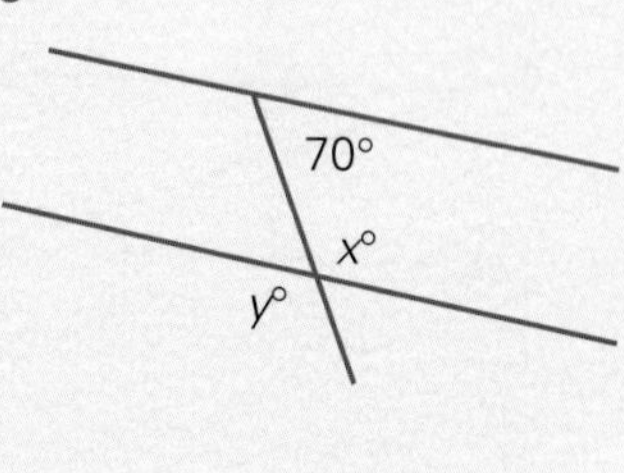

$x +$ ______ $= 180$ (______)
$x = y$ (______)
$\therefore y =$ ______

6 State why AB is parallel to PQ:

a

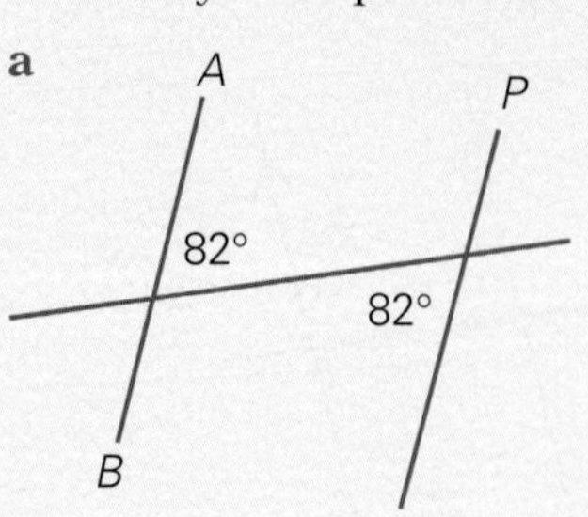

b

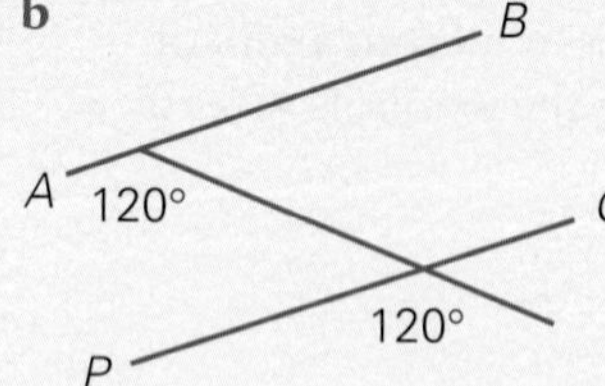

c

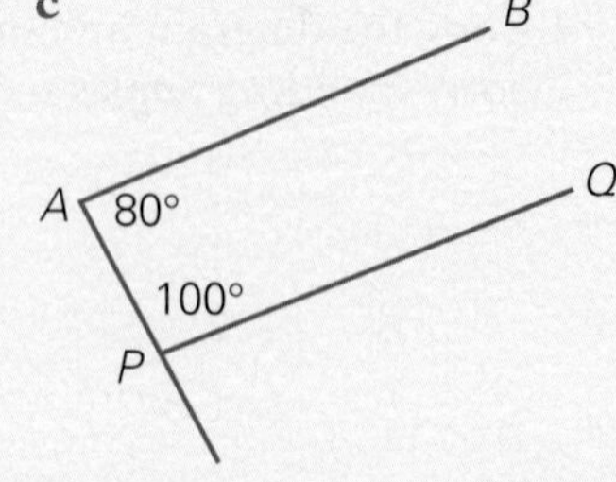

Level 2

7 Find the value of the pronumerals in each of the diagrams below. Giving **reasons**.

a

b

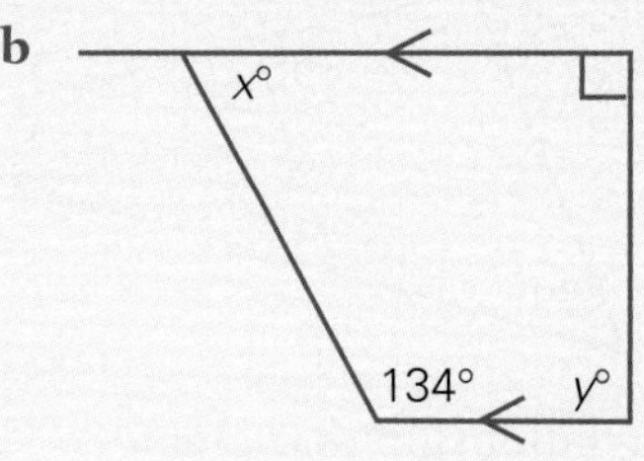

c

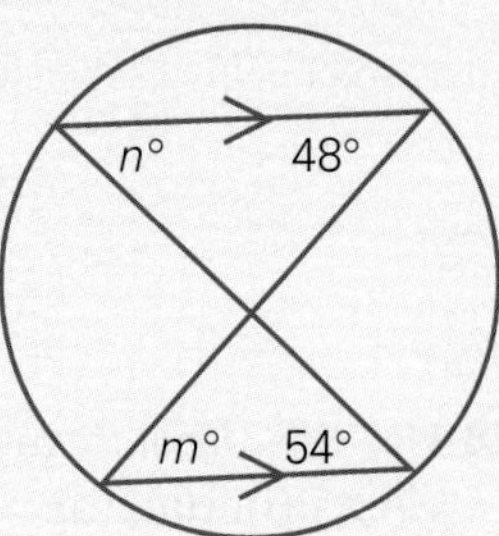

d

e

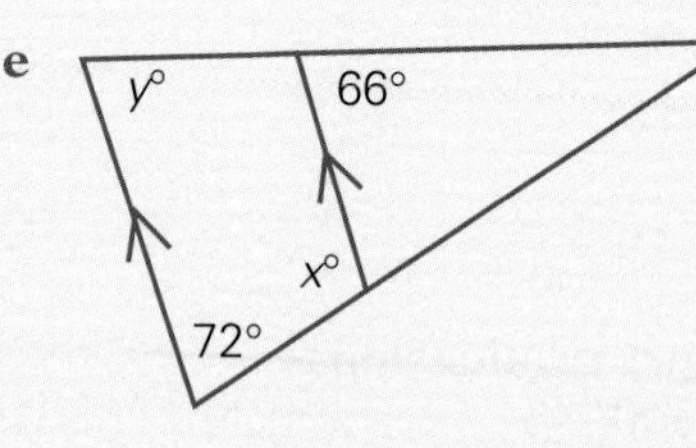

f

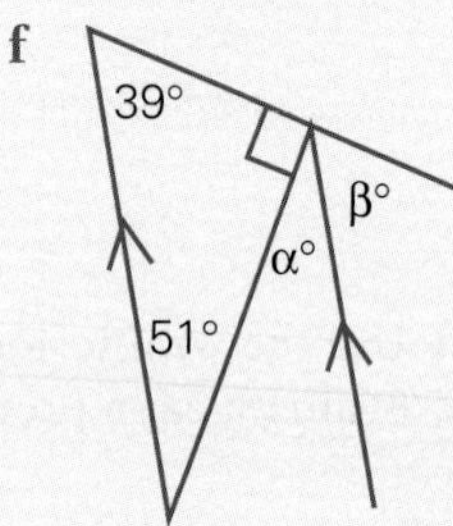

g

h

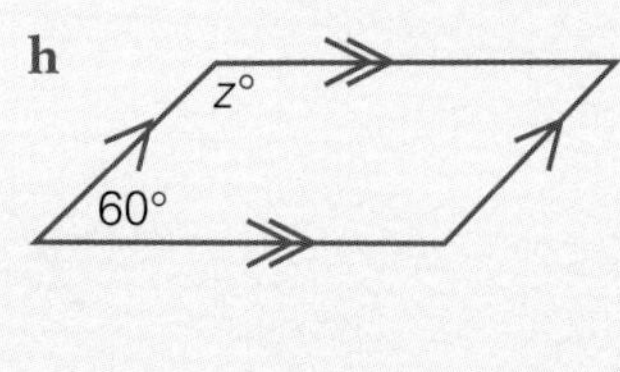

i

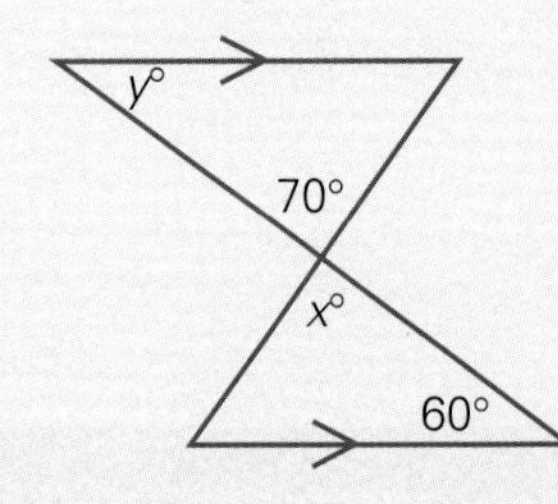

j

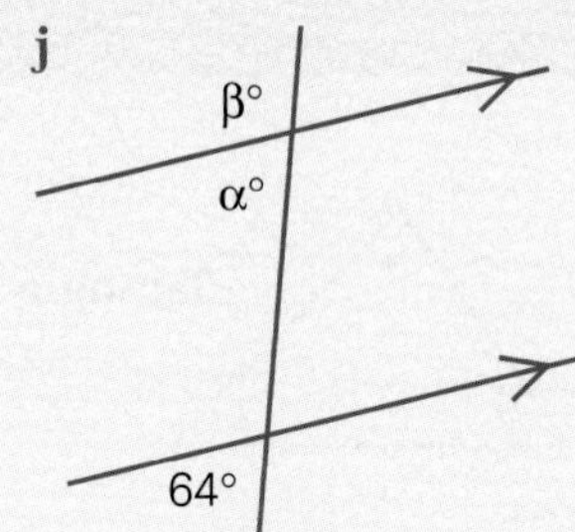

k

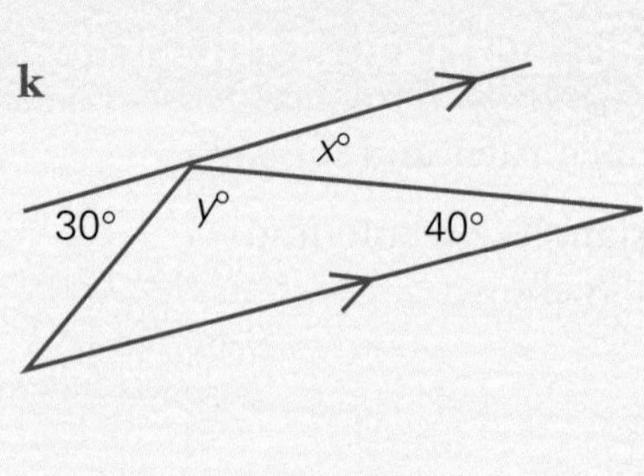

l

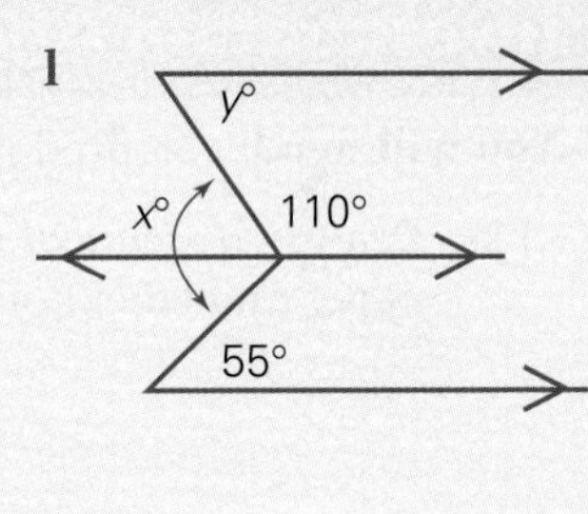

8 State whether $AB \parallel CD$ in each of the following diagrams, giving a reason to support your decision.

a

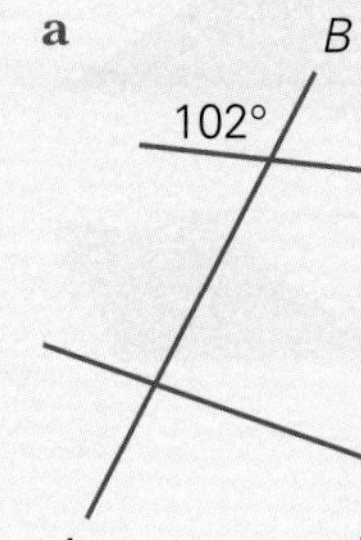

b

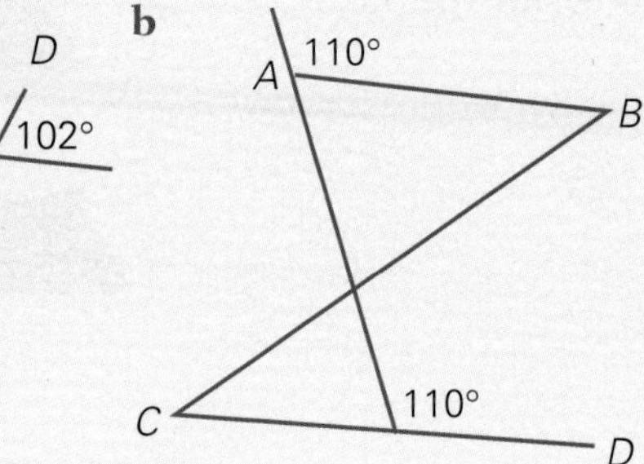

c

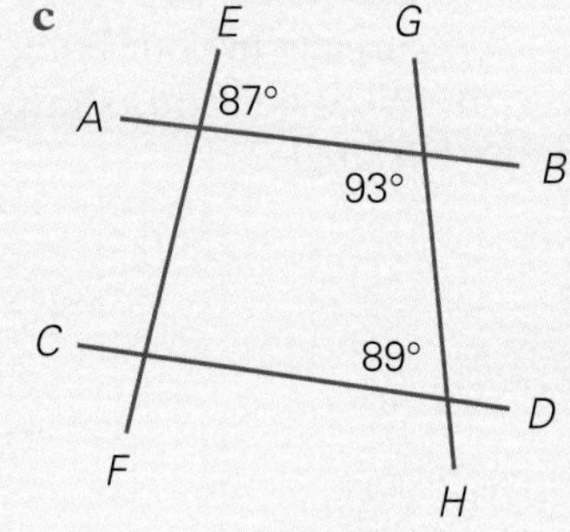

9 Use the corresponding angles on parallel lines relationship to write the value of each pronumeral:

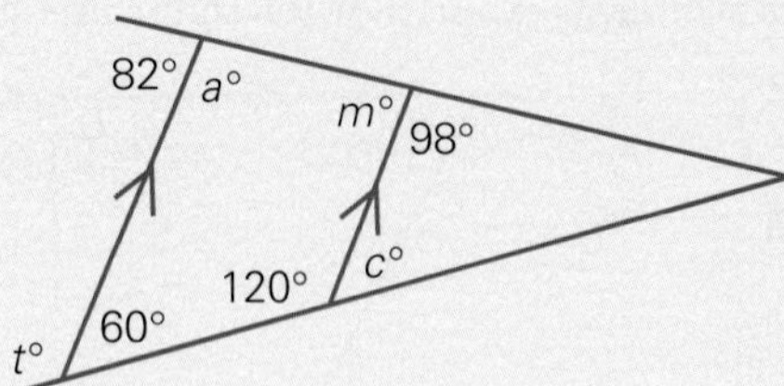

10 Use the *alternate* angles relationship to evaluate each pronumeral:

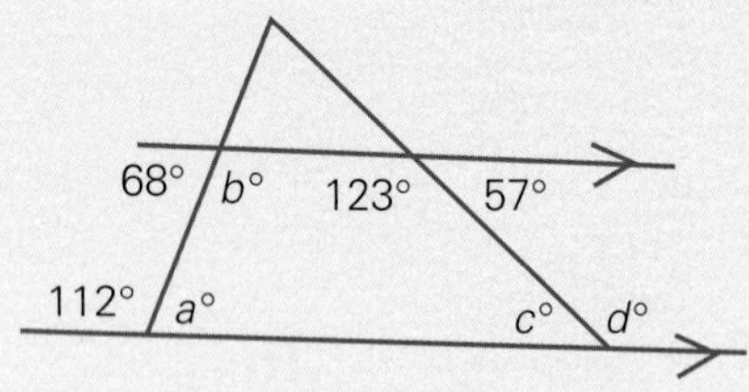

11 Use the co-interior angles relationship to evaluate each pronumeral:

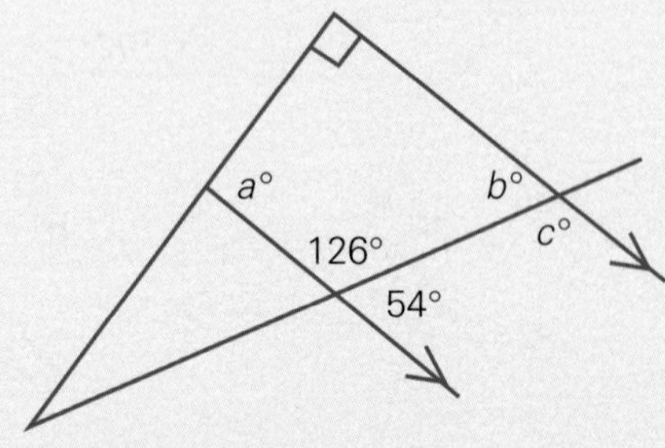

Investigation Angle sum of a triangle

You will need: coloured paper, ruler and scissors

1 a Draw *two identical* triangles on coloured paper and label both as shown.

b Paste one complete triangle into your workbook.

c Cut the other triangle as shown and rearrange, then paste it beside the one already in your book.

d What does this show about $a + b + c$ for *any* triangle?

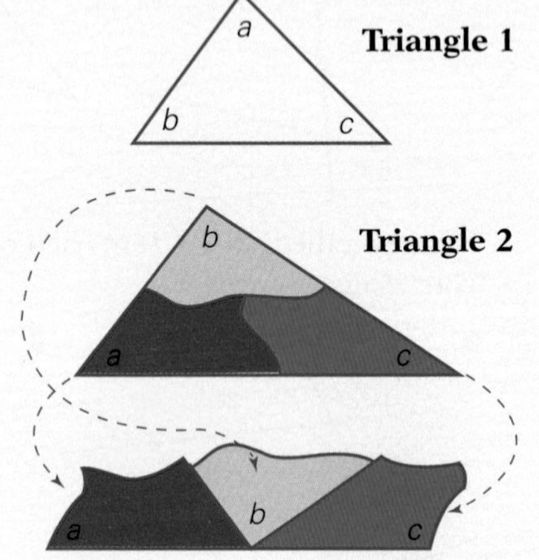

2 a Draw any acute triangle like the one shown.
b Measure angles a, b and c carefully with a protractor and find their sum.
c Repeat with an obtuse triangle and a right triangle.
d Compare your answer with those of other classmates.

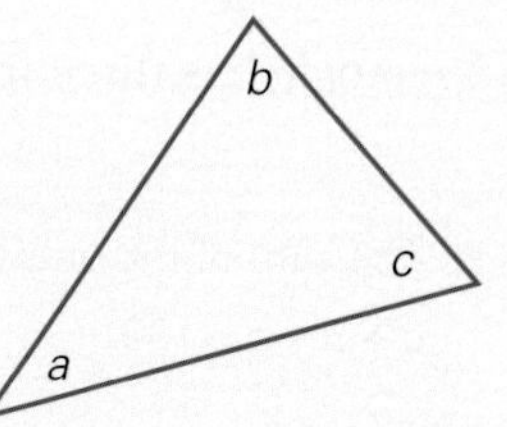

3

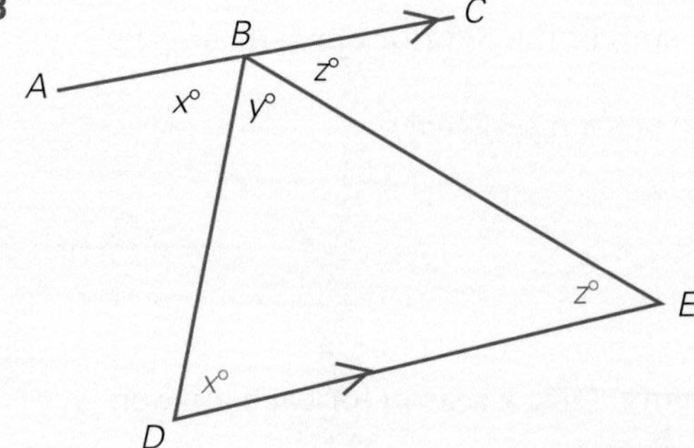

Complete the following proof:

$x + y + z =$ ______ (angles on a straight line)
$\angle BDE = x$ (______)
$\angle BED = z$ (______)
$\angle DBE = y$ (given)
$\therefore$ in ΔDBE
$\angle BDE + \angle BED + \angle DBE$
$= x + y + z$
$= 180$

4 Complete: The angle sum of any triangle is ______.

Investigation The exterior angle relationship

1 Draw any triangle and label the angles as shown. Cut out angles a and b and reposition them over the *exterior* angle x as shown. Record your findings.

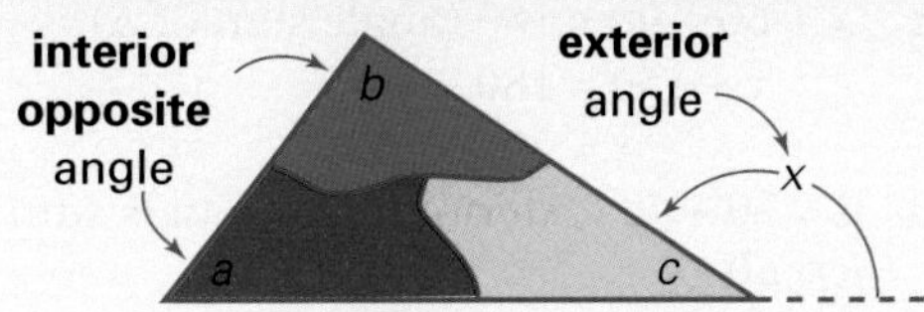

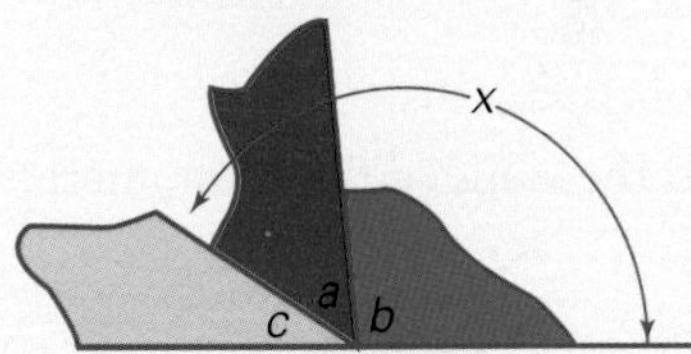

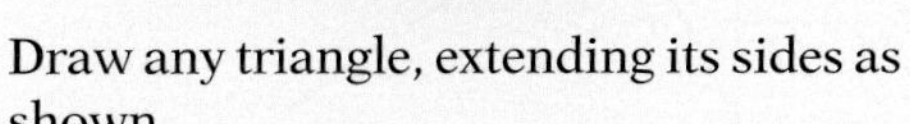

2

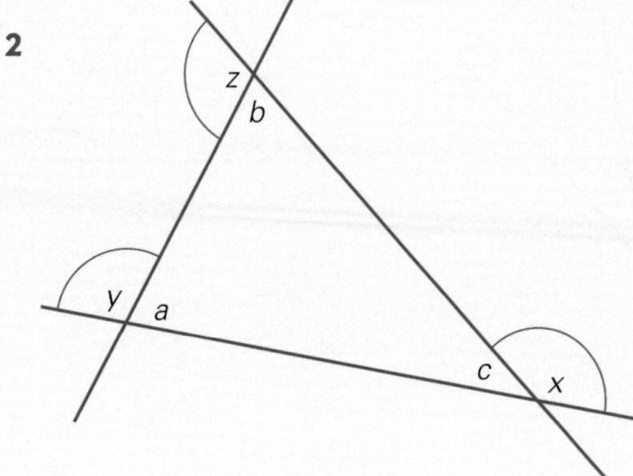

Draw any triangle, extending its sides as shown.

a Measure x, then a and b.
b Measure y, then b and c.
c Measure z, then a and c.

What do you notice?

7.3 Angle relationship in triangles

A triangle has three interior angles; the sum of these three angles always equals 180°.

The **sum** of the angles of a triangle is 180°.

$a + b + c = 180$

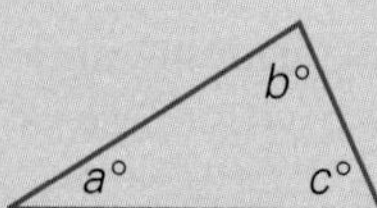

The exterior angle of any triangle equals the sum of the interior opposite angles.

Any **exterior** angle of a triangle equals the sum of the two **interior opposite** angles.

$x = a + b$

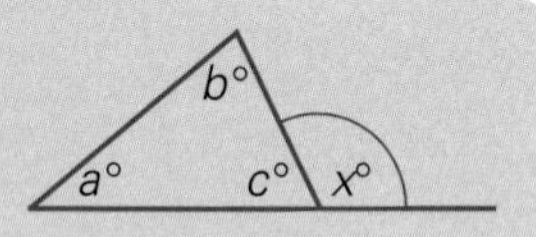

EXAMPLE

Find the value of the pronumerals in each figure. Give a reason for each answer.

a

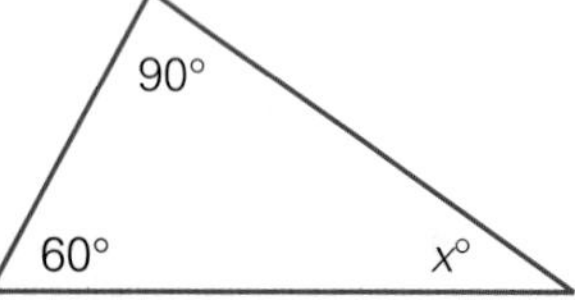

b

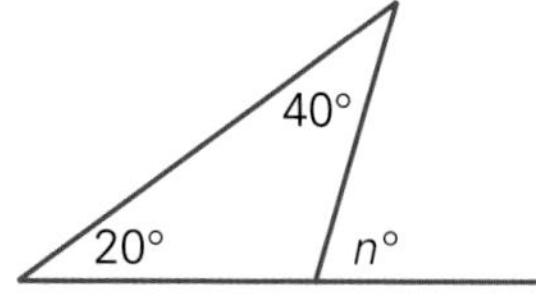

Solution

a $x + 60 + 90 = 180$ (angle sum of Δ)

$x + 150 = 180$

$x = 30$

b $n = 20 + 40$ (exterior angle equals sum of two opposite interior angles)

$n = 60$

EXERCISE 7C

1 Find the value of the pronumeral in each figure. Give a reason for each answer.

a

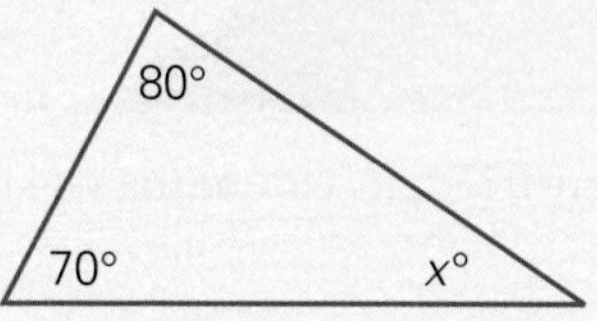

b

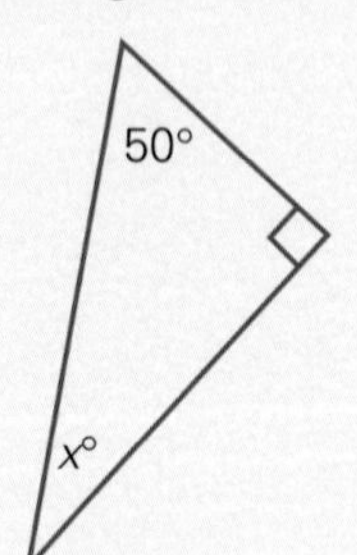

c

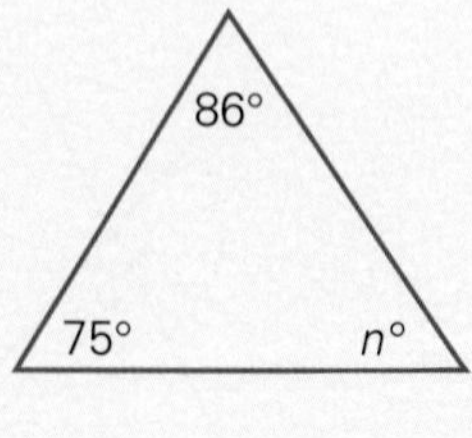

d

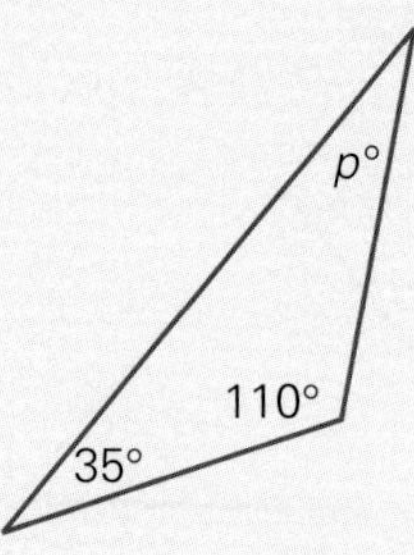

e

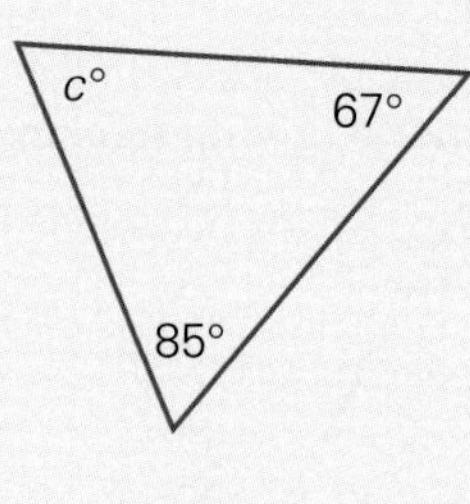

f

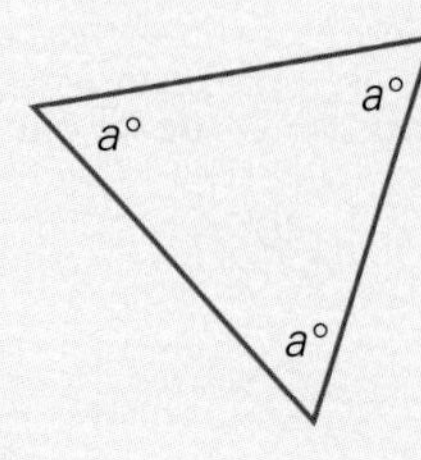

g

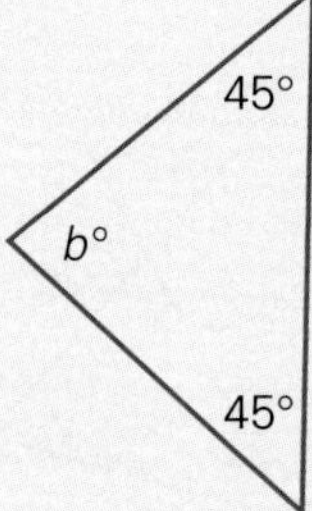

h

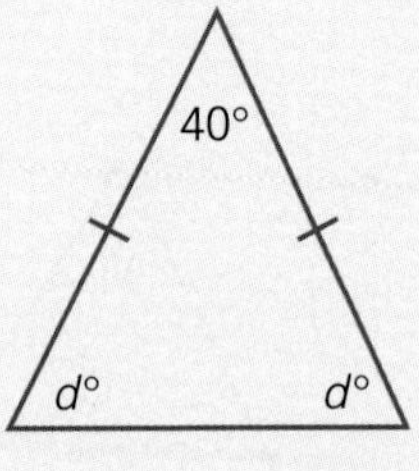

i

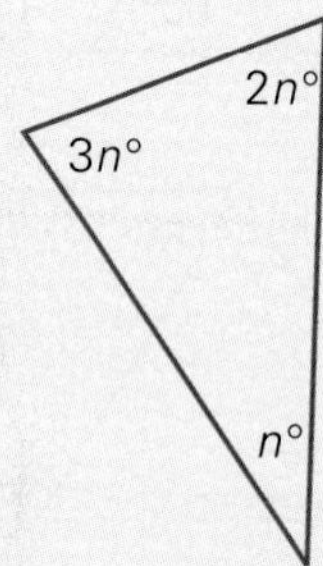

2 Use the exterior angle relationship of a triangle to find the value of each pronumeral:

a

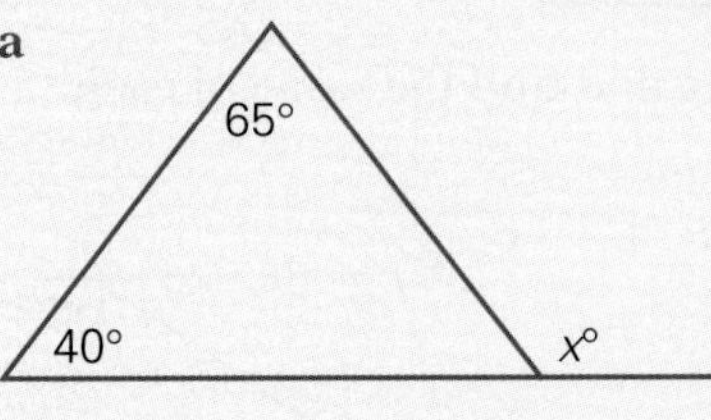

b

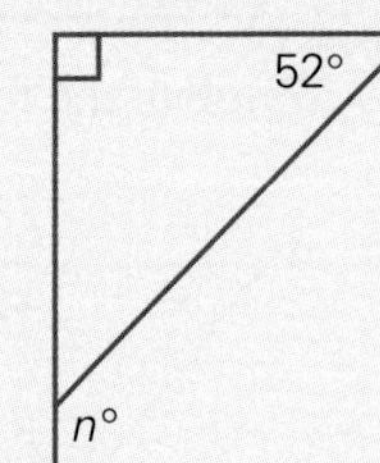

c

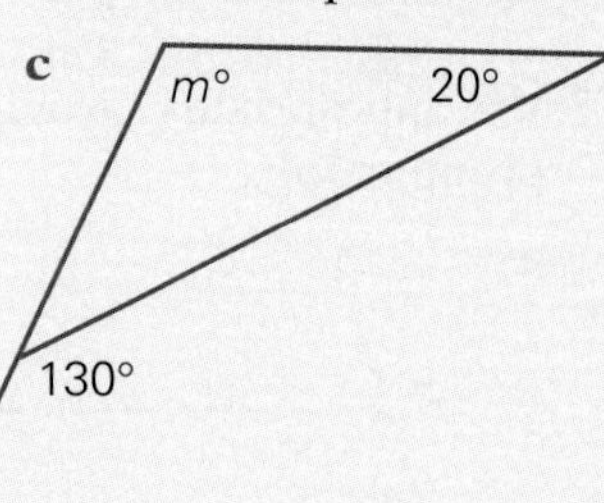

d

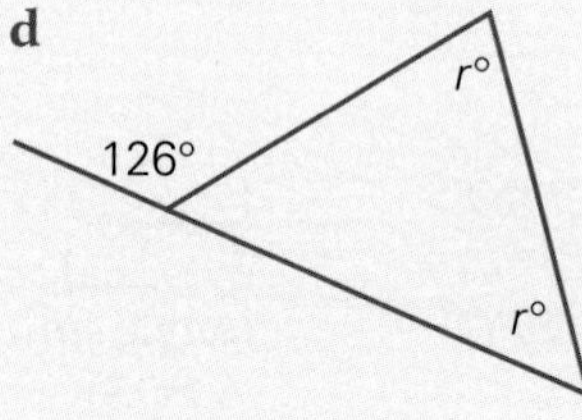

e

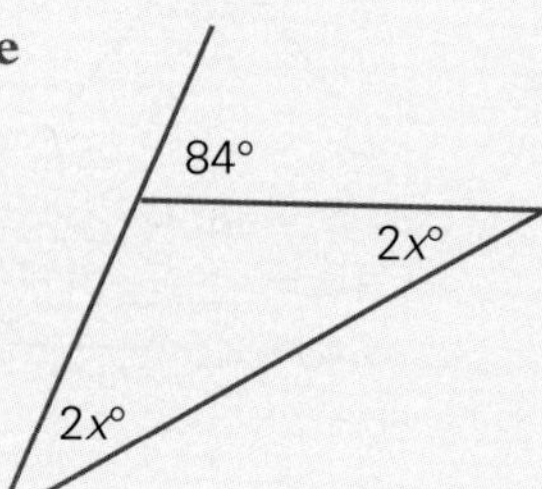

f

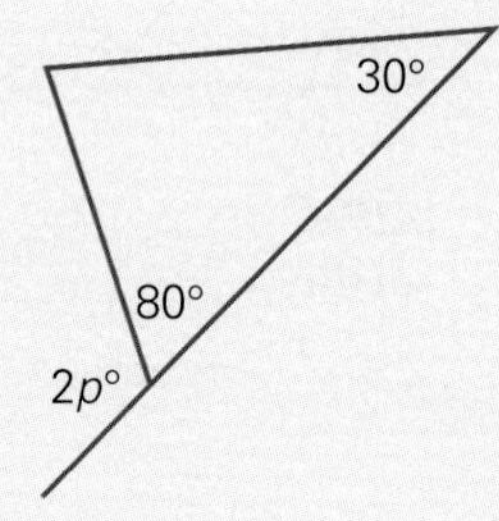

3 Copy and complete:

a

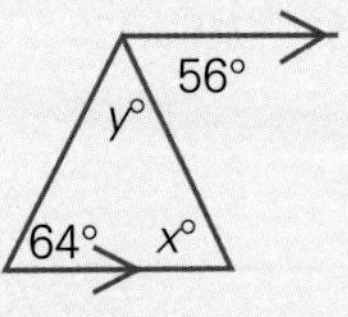

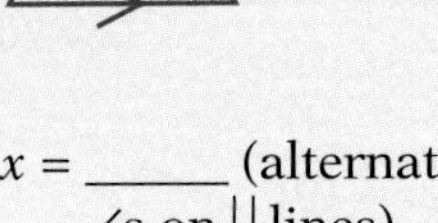

$x =$ ______ (alternate ∠s on || lines)

$y = 60$ (______)

b

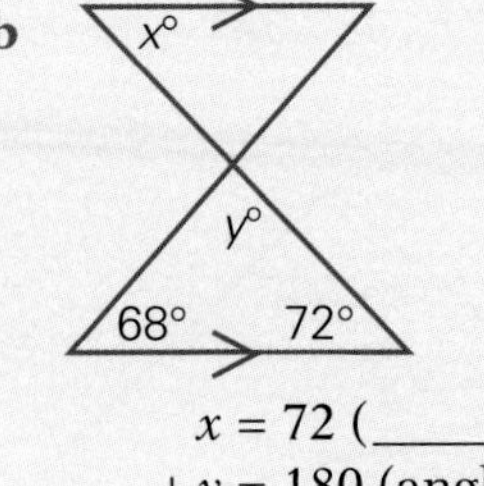

$x = 72$ (______)

______ $+ y = 180$ (angle sum of Δ)

$\therefore y =$ ______

c

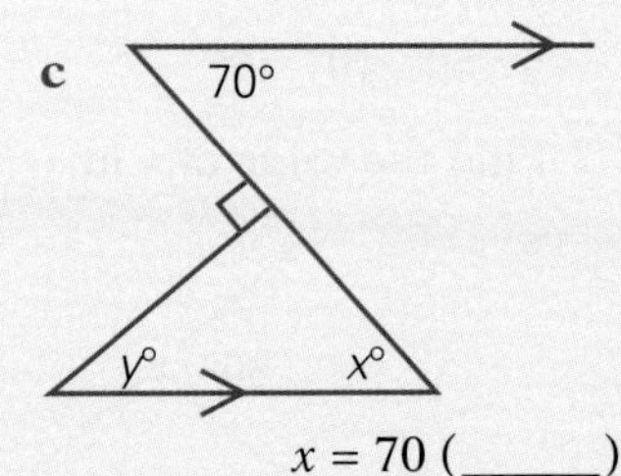

$x = 70$ (______)

$90 + 70 +$ ______ $= 180$ (______)

$\therefore y =$ ______

Level 2

4 Find the value of each pronumeral, giving reasons.

a 40° x° y° 80°

b y° 70° x° 130°

c z° y° x° 130° 150°

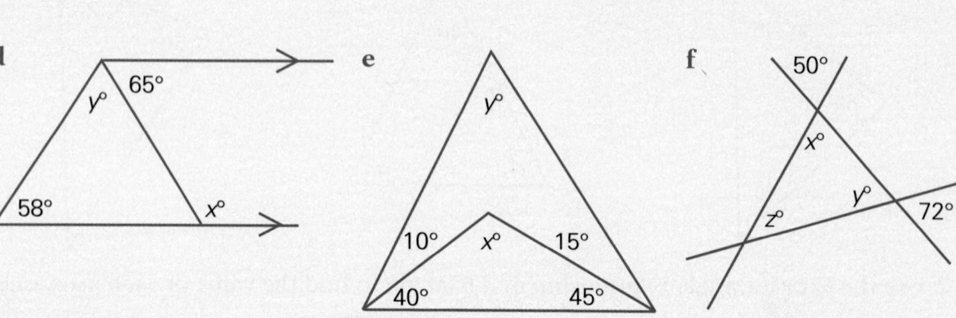

5 Use appropriate reasoning to complete the following and find the value of each pronumeral:

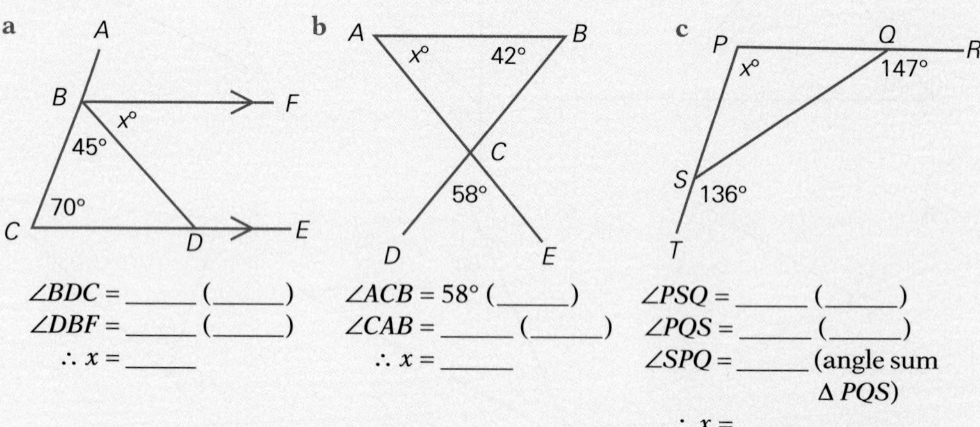

a
$\angle BDC =$ ______ (______)
$\angle DBF =$ ______ (______)
$\therefore x =$ ______

b
$\angle ACB = 58°$ (______)
$\angle CAB =$ ______ (______)
$\therefore x =$ ______

c
$\angle PSQ =$ ______ (______)
$\angle PQS =$ ______ (______)
$\angle SPQ =$ ______ (angle sum ΔPQS)
$\therefore x =$ ______

6 Can triangles be constructed with these angles? Give a reason for your answer.

a 60°, 10°, 20° **b** 71°, 45°, 74° **c** 35°, 110°, 35°

7 Find the value of x in:

a 80° $(x + 30)$° x°

b $(x + 50)$° $(x + 40)$° x°

c $(x + 20)$° $(2x + 10)$°

8 Find the value of the pronumerals in alphabetical order.

a

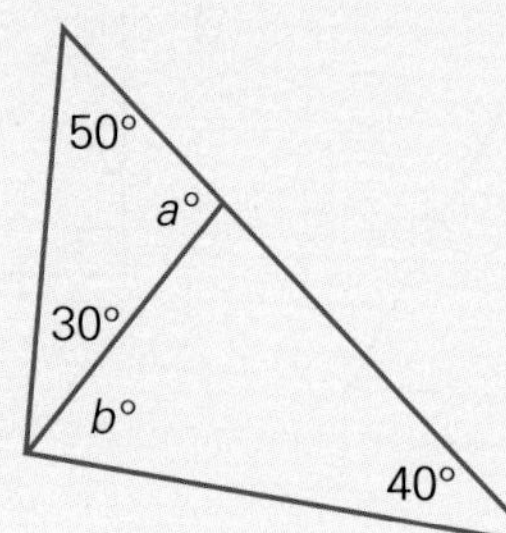

b

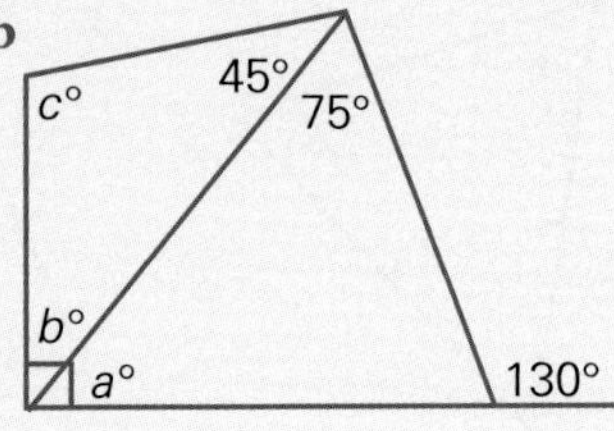

c

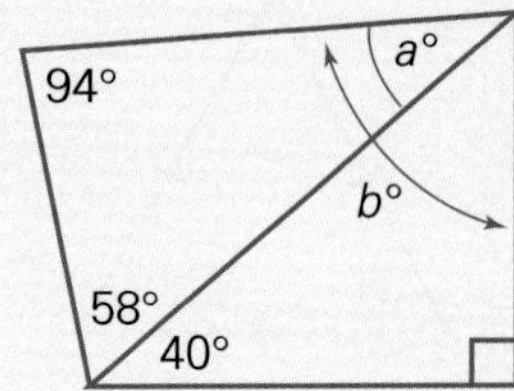

7.4 Isosceles and equilateral triangles

As you are aware, triangles can be classified according to the number of equal sides.

An *isosceles* triangle has *two equal* sides.

If two sides of a triangle are equal, then the angles opposite those sides are equal.

If two angles of a triangle are equal, then the sides opposite those angles are equal.

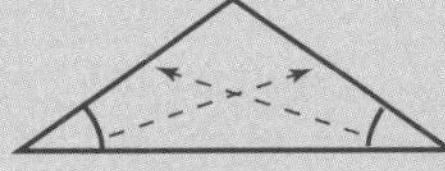

An *equilateral* triangle has *all sides equal.*

∴ an *equilateral* triangle has *all three angles equal,* each 60°.

If a triangle has all sides equal, it is equilateral.

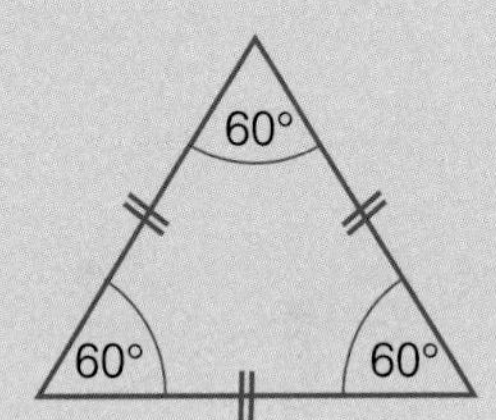

It sometimes helps to remember the axes of symmetry of isosceles and equilateral triangles.

An isosceles triangle:

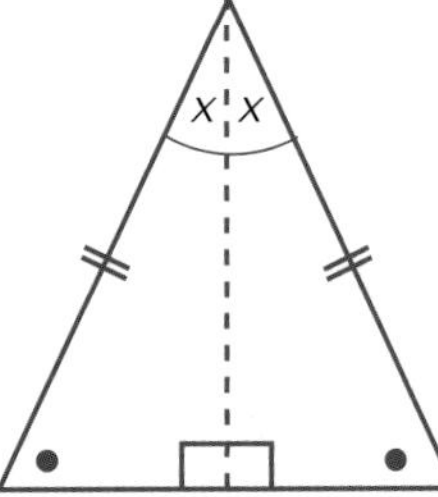

One axis of symmetry divides the isosceles Δ into two identical right triangles.

An equilateral triangle:

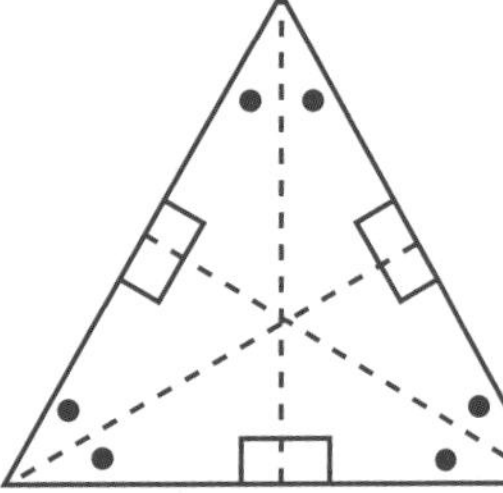

Three axes of symmetry each divide an equilateral triangle into two identical right triangles.

EXAMPLE

Find the value of x in:

a

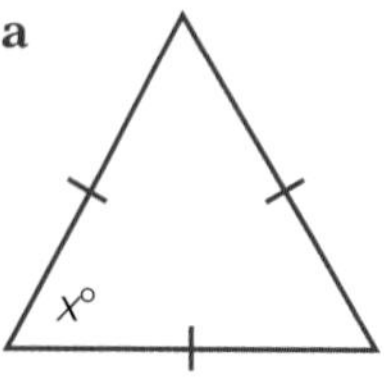

b

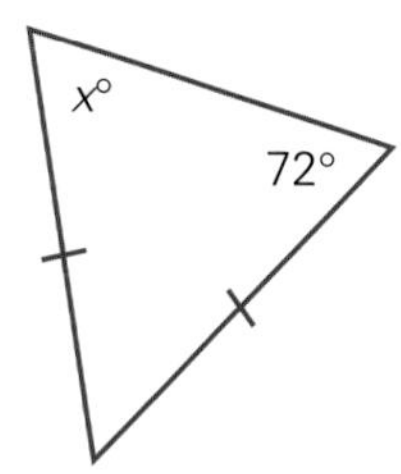

c

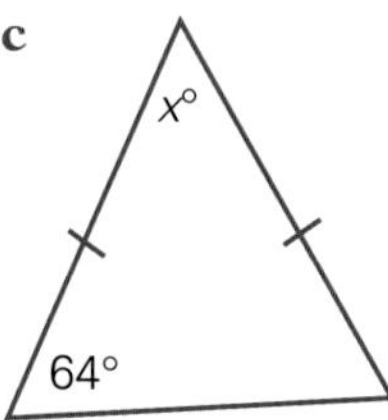

d

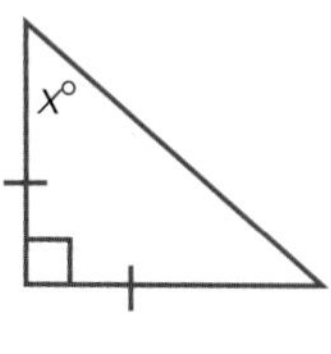

Solution

a $x = 60$
(angles in an equilateral Δ)

b $x = 72$
(angles opposite equal sides in an isosceles Δ)

c $x + 64 + 64 = 180$
(angle sum of isosceles Δ)
$x + 128 = 180$
$x = 52$

d $90 + x + x = 180$
(angle sum of isosceles Δ)
$2x = 90$
$x = 45$

EXERCISE 7D

1 Find the value of each pronumeral, giving reasons (other relationships could also be needed).

a

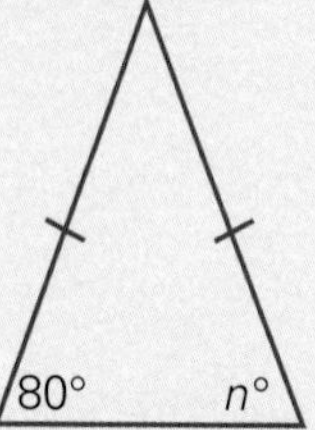

b

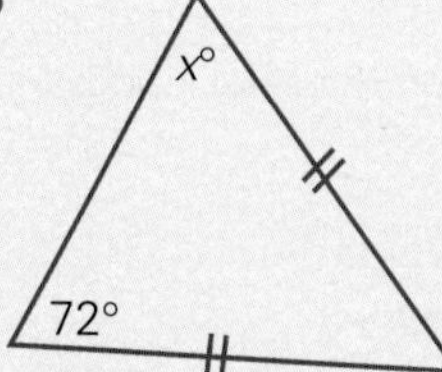

c

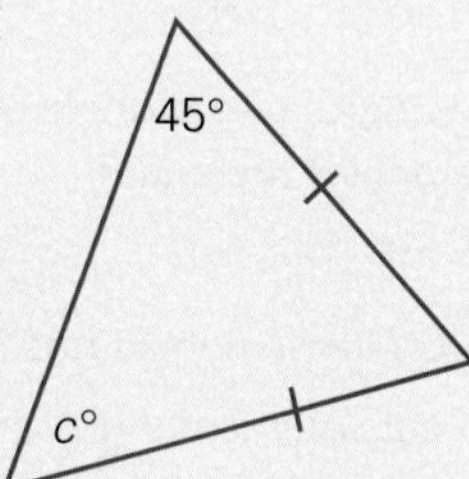

d

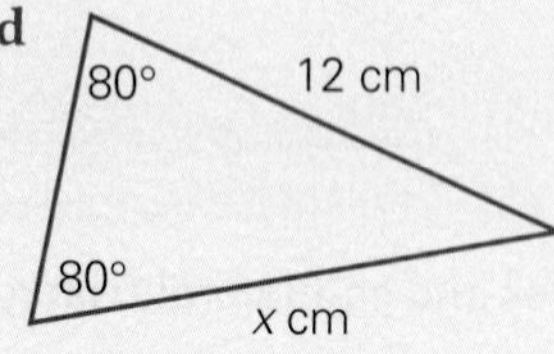

e

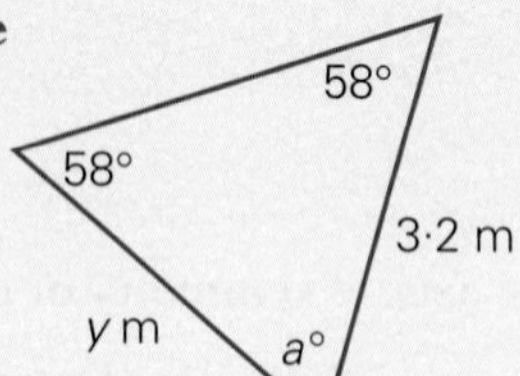

f

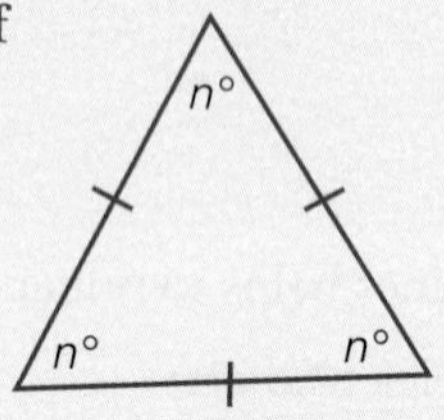

g

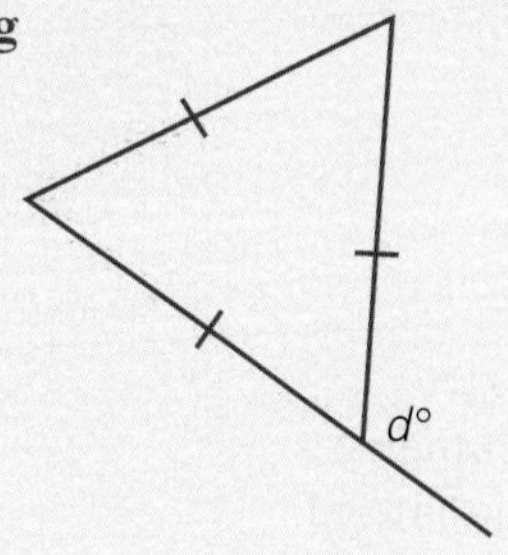

h

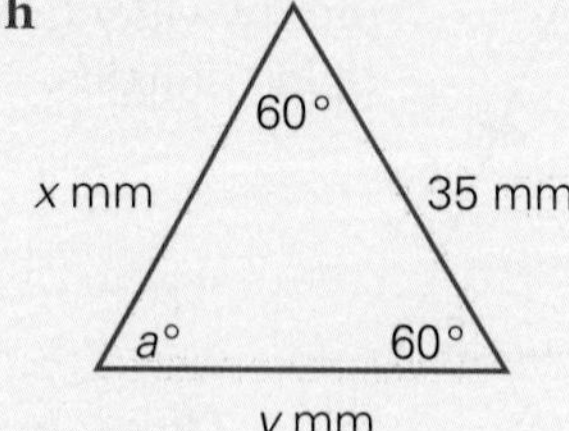

i

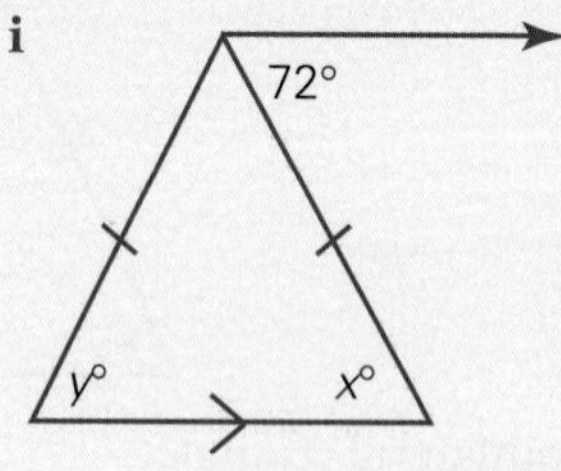

Level 2

2 Find the value of each pronumeral in:

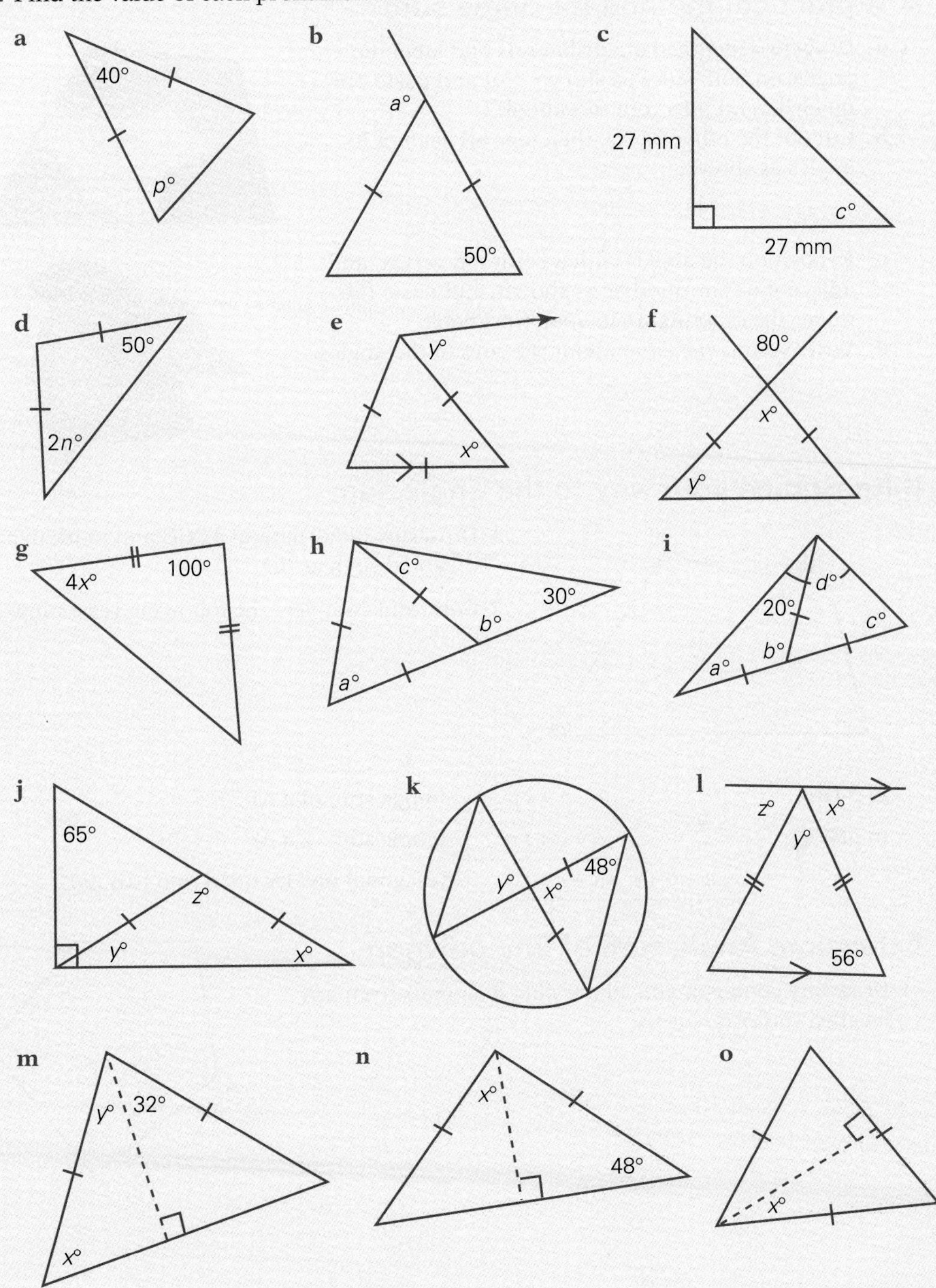

Investigation Angle sum of a quadrilateral

A A practical method for angle sum

1 a Draw two identical quadrilaterals and label the angles on both sides as shown. Cut and paste one quadrilateral into your workbook.

b Cut out the other figure, then tear off each of its angles as shown.

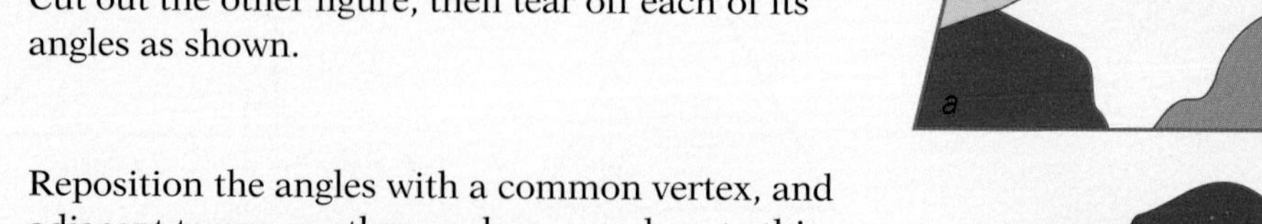

c Reposition the angles with a common vertex, and adjacent to one another as shown, and paste this under the other figure in your workbook.

d Write your *conclusion* about the sum of the angles.

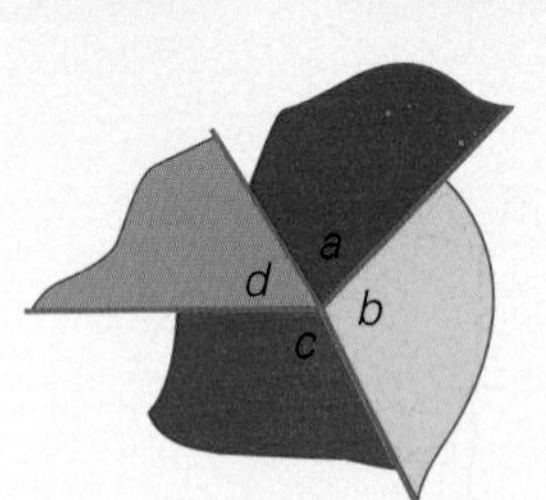

B Reasoning your way to the angle sum

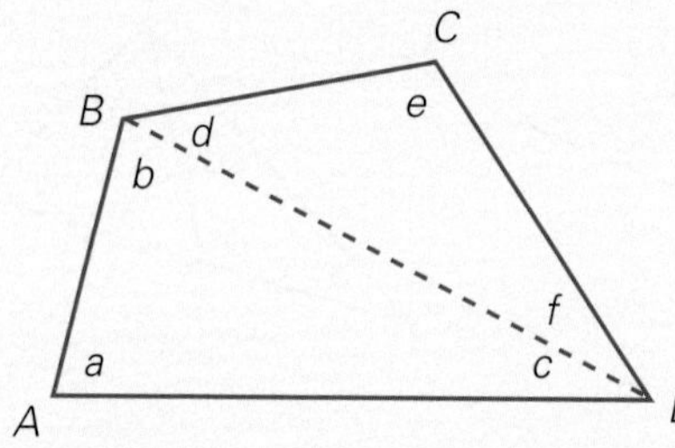

1 Draw any quadrilateral *ABCD* and mark in a diagonal such as *BD*.

2 Copy and complete the following reasoning:

In ΔABD, $a + b + c =$ ___ (angle sum of a Δ)

In ΔBCD, $d + e + f =$ ___ (angle sum of a Δ)

$\therefore a + b + c + d + e + f =$ ___ (diagonal divides quad into two Δs)

Extension: Angle sum of any polygon

1 Draw any pentagon and all possible diagonals from any labelled vertex or corner.

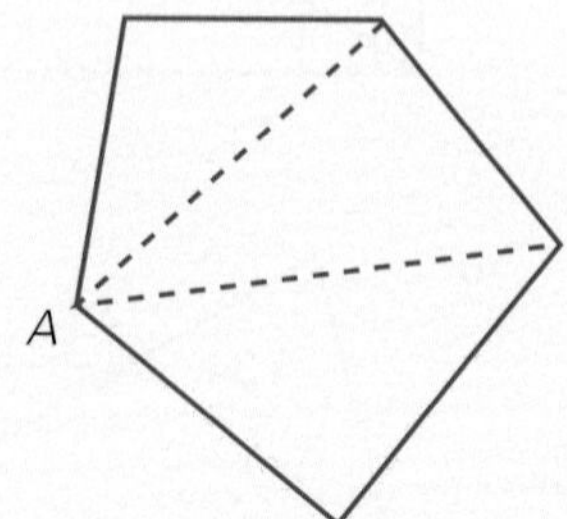

2 By repeating step 1 for other polygons, complete this table:

Polygon	No. of sides	No. of diagonals	No. of triangles	Angle sum
Quadrilateral	4	1	2	$2 \times 180° = 360°$
Pentagon	5	2		
Hexagon				
Octagon				
n-agon				

3 Summarise your findings in a sentence.

WEB RESEARCH

INTERACTIVE GEOMETRY

INTERACTIVE GEOMETRY

7.5 Angle sum of a quadrilateral

All four-sided polygons are called quadrilaterals. All quadrilaterals have four interior angles.

The sum of all the angles of any quadrilateral is 360°.

$a + b + c + d = 360$

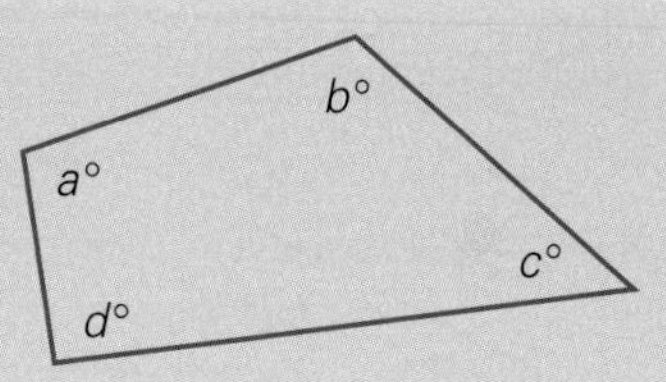

EXAMPLE

Find the value of x, giving reasons, in:

a

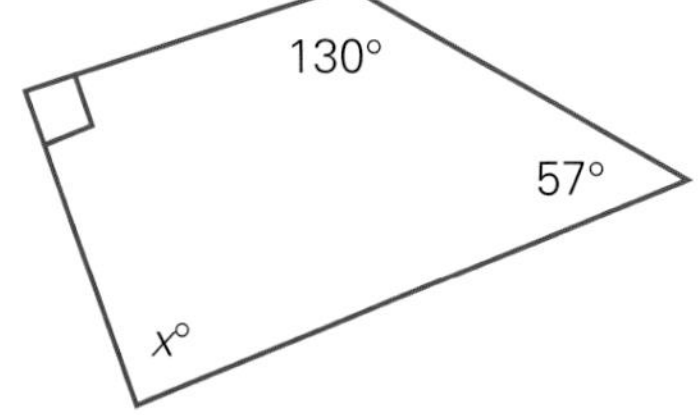

b

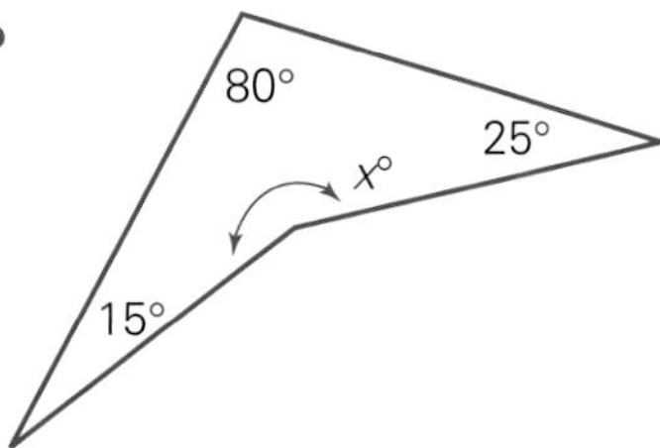

Solution

a $x + 90 + 130 + 57 = 360$ (angle sum of quadrilateral)

$\therefore x = 83$

b $x + 80 + 15 + 25 = 360$ (angle sum of quadrilateral)

$x = 240$

EXERCISE 7E

1 Find the value of each pronumeral, giving a reason (angle sum of a quadrilateral, or other/s):

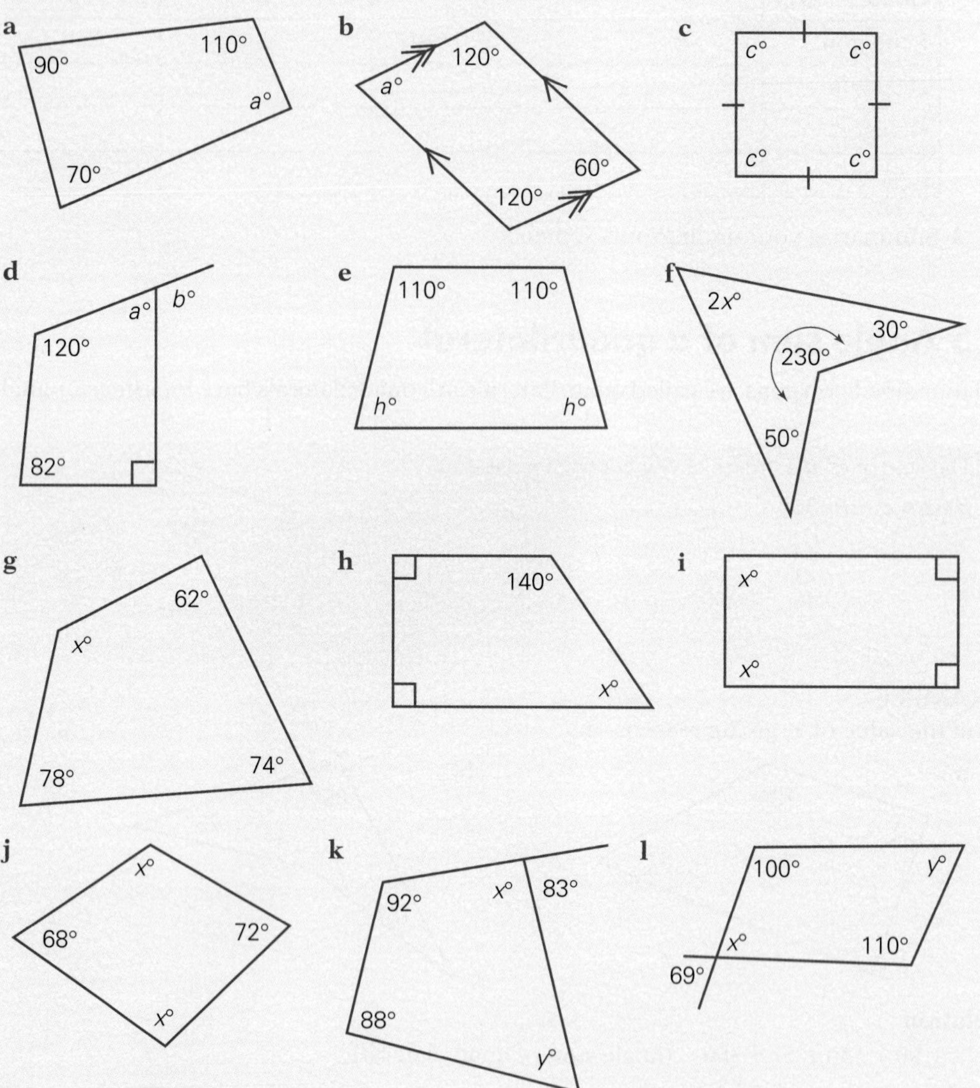

Level 2

2 Find the value of each pronumeral in:

a

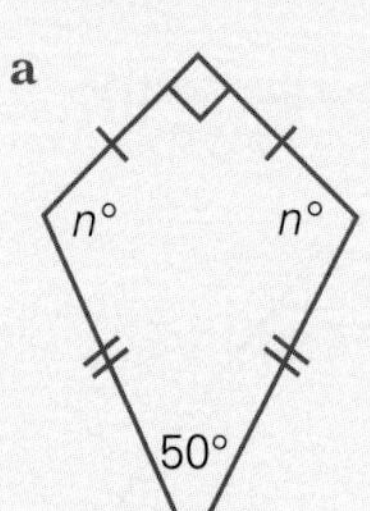

b

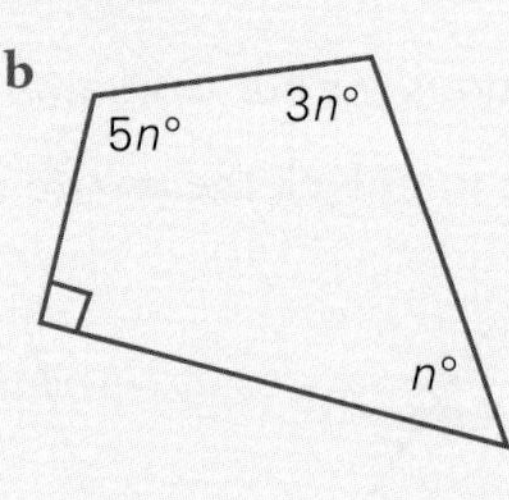

c

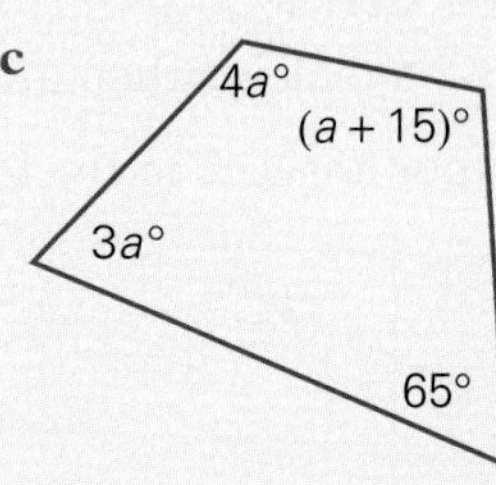

d

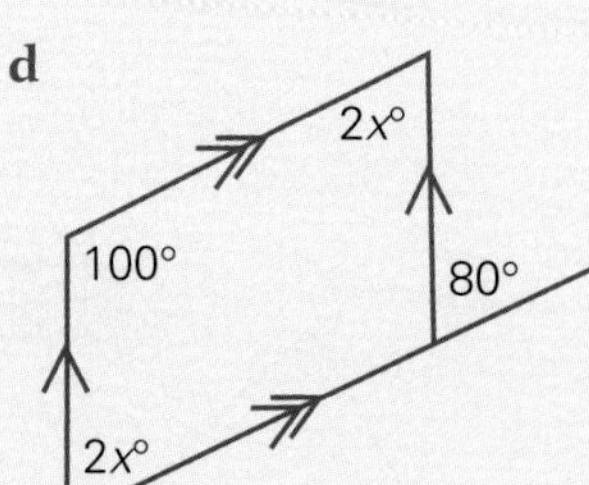

e

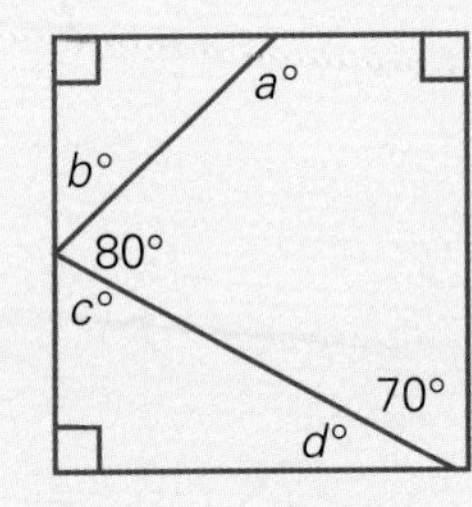

f

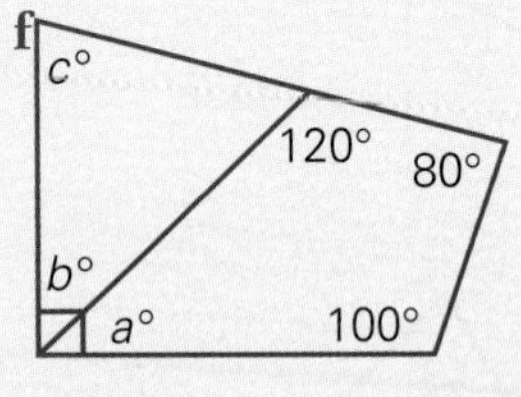

3 Find the value of each pronumeral, giving reasons:

a

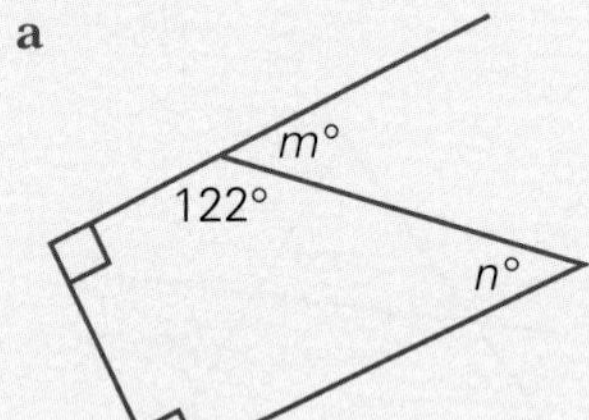

b

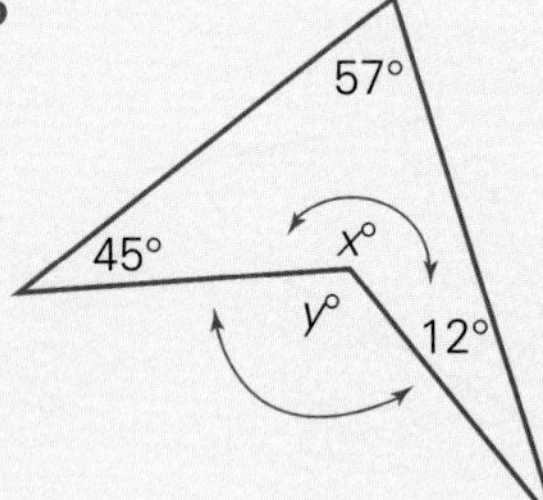

c

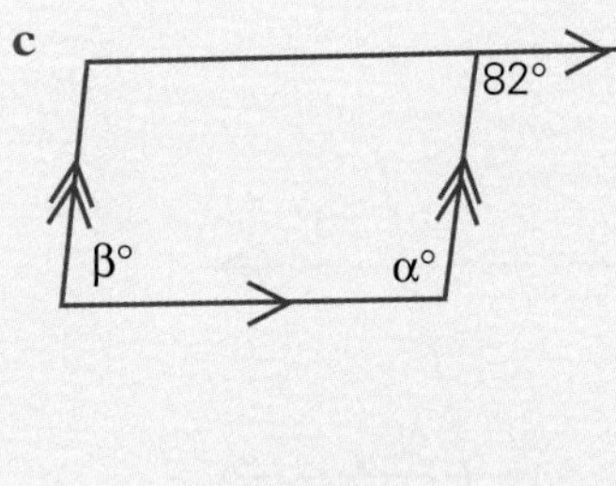

d

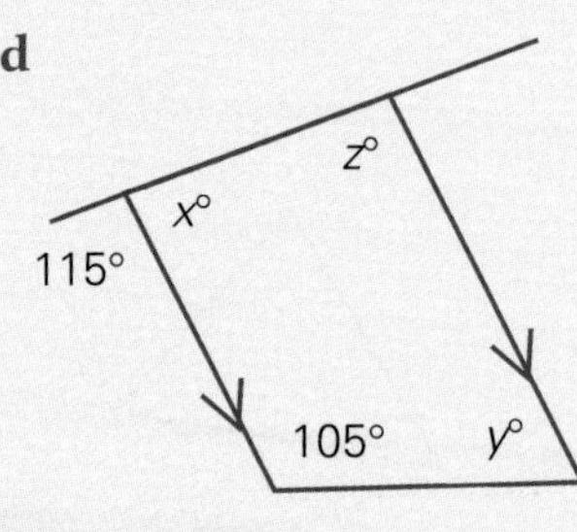

e

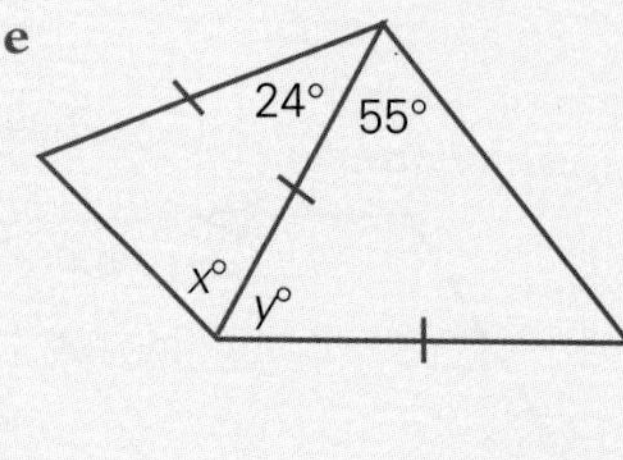

f

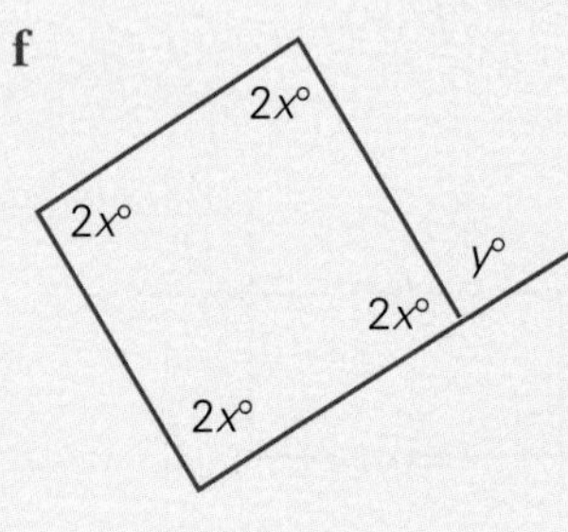

g

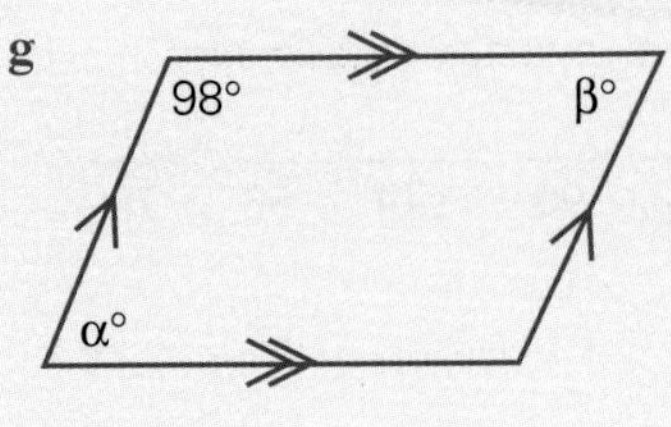

h

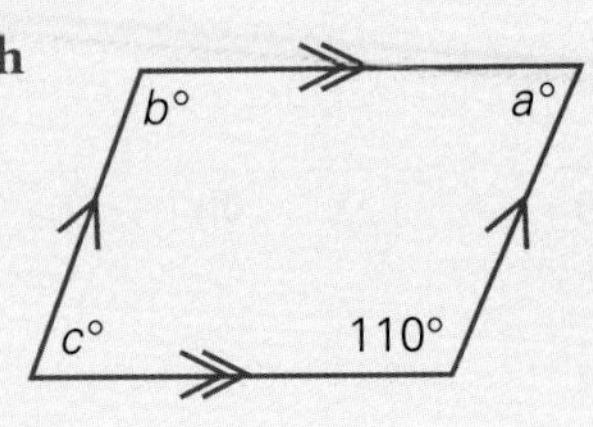

What does this tell us about the angles of a parallelogram?

Investigation Puzzle

Who am I?

I was a female mathematician famous for her work and publications on geometry.

Use your answers to the following to unlock the puzzle code.

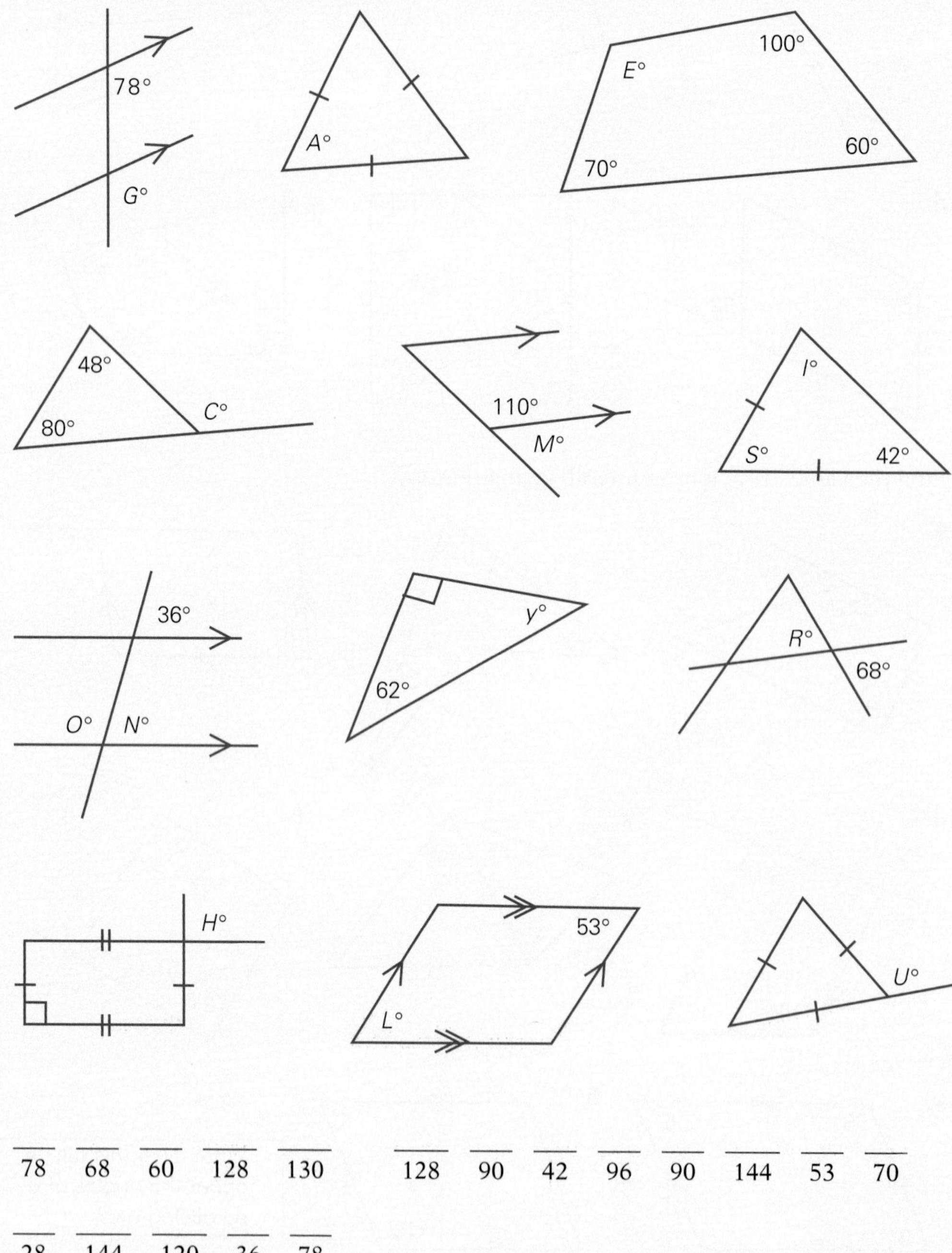

___ ___ ___ ___ ___   ___ ___ ___ ___ ___ ___ ___ ___

78 68 60 128 130   128 90 42 96 90 144 53 70

___ ___ ___ ___ ___

28 144 120 36 78

7.6 Getting it all together

You now have at least 12 geometric relationships that fit specific conditions. Which of these to choose in a given situation depends on the clues or given information. Then linking the relevant steps logically will produce a chain of evidence leading to a conviction!

Thales (640 BC) was a Greek mathematician interested in geometry. He founded a school of mathematics and he even taught Pythagoras.

EXAMPLE

Find the value of the pronumeral in each diagram, giving a reason for each step:

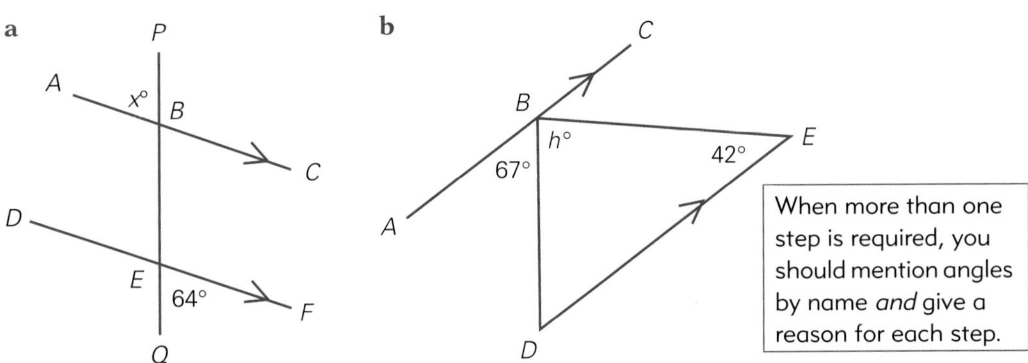

When more than one step is required, you should mention angles by name *and* give a reason for each step.

There may be other pathways to the same end result. Try finding them.

Solution

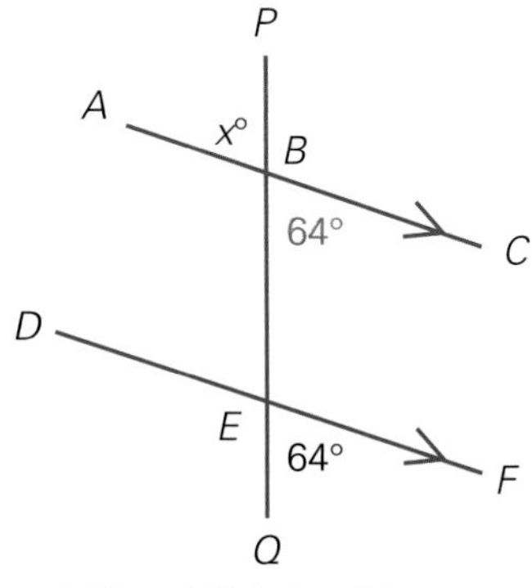

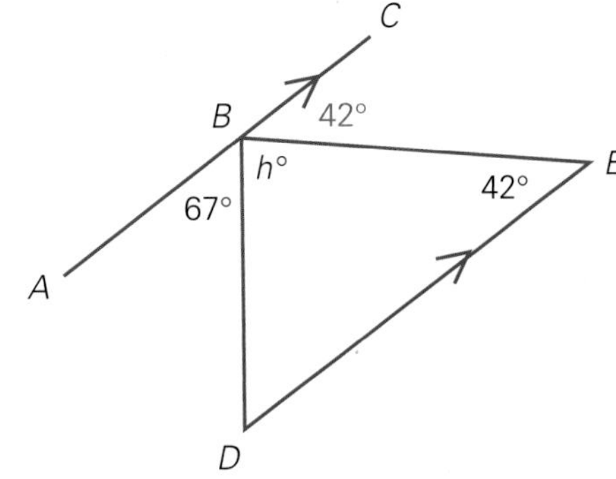

a $\angle QEF = 64°$ (given)

$\angle QBC = 64°$ (corresponding angles, $AC \parallel DF$)

$\angle APB = 64°$ (vertically opposite angles)

$\therefore x = 64$

b $\angle DEB = 42°$ (given)

$\angle EBC = 42°$ (alternate angles $BC \parallel DE$)

$67 + h + 42 = 180$ (angles on a straight line)

$\therefore h = 71$

EXERCISE 7F

1 Complete the following scaffolds:

a

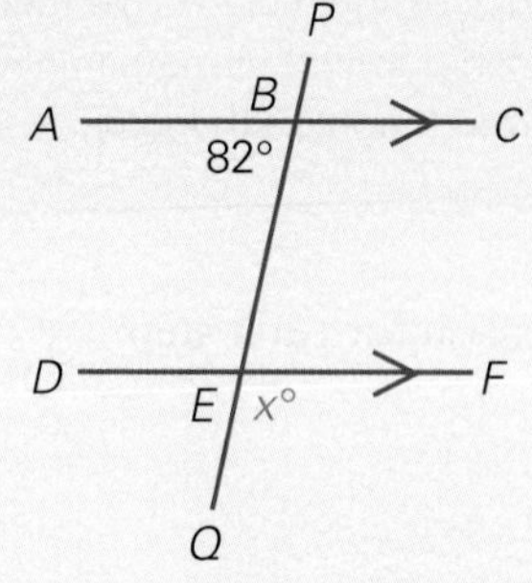

$\angle ABE =$ ______ (given)
$\angle PEF = 82°$ (______)
$x +$ ______ $= 180$ (angles on a straight line)
$\therefore x =$ ______

b

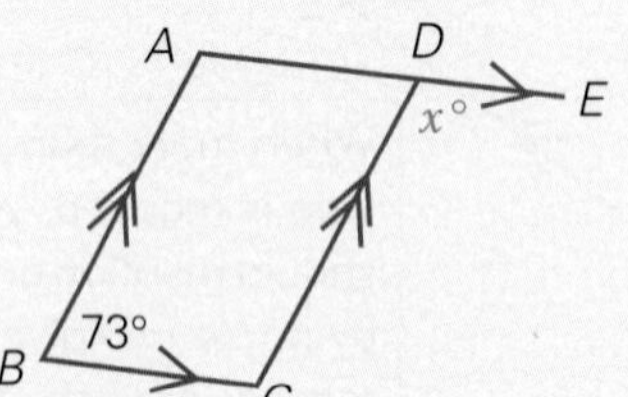

$\angle ABC = 73°$ (______)
$\angle BCD =$ ______ (co-interior angles are supplementary, $AB \parallel CD$)
$\angle CDE = 107°$ (______ on $AE \parallel BC$)
$\therefore x =$ ______

c

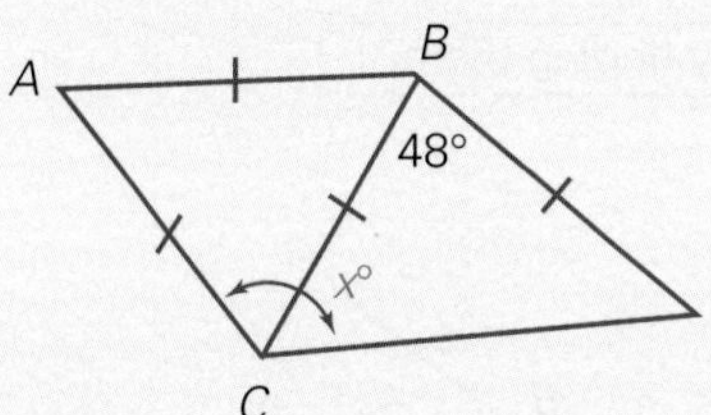

$\angle CBD =$ ______ (given)
$\angle BCD = \angle$______ (angles opposite equal sides in an isosceles Δ)
$\angle BCD + \angle$______ $+ 48° = 180°$ (angle sum of ΔBCD)
$\therefore \angle BCD =$ ______

ΔABC is ______ as all sides are equal.
$\therefore \angle ACB =$ ______
$\angle ACD = x$
$x =$ ______ $+$ ______ (sum of adjacent angles)
$=$ ______

Level 2

2 Find the value of each pronumeral, setting out your work and reasoning carefully:

a

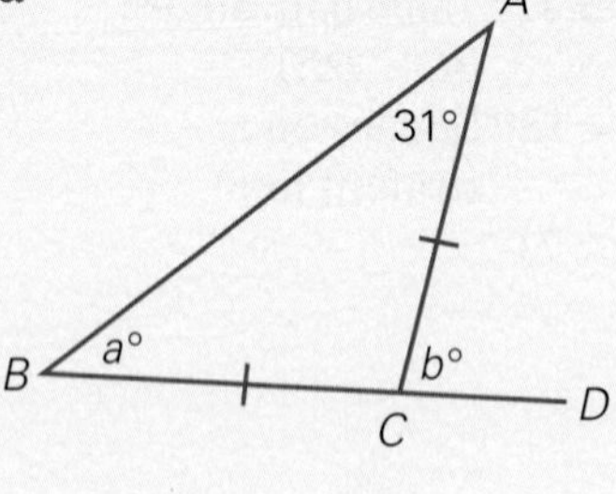

b

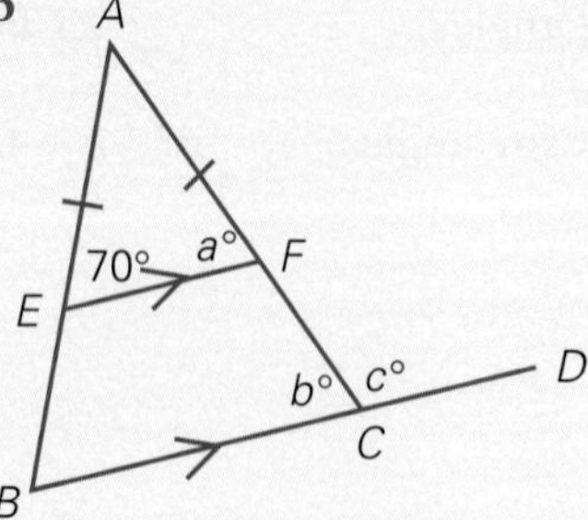

c

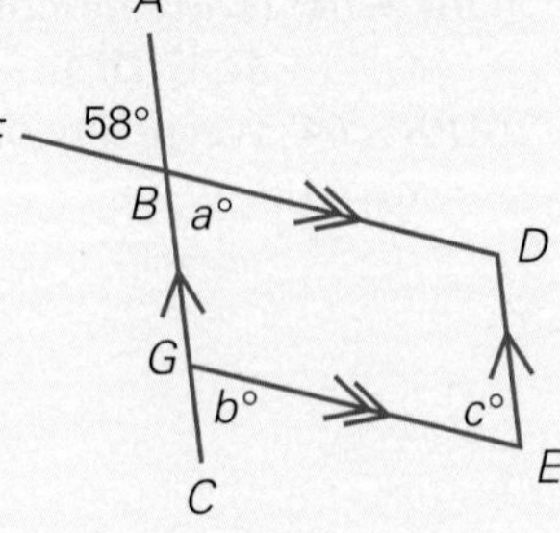

d **e** **f**

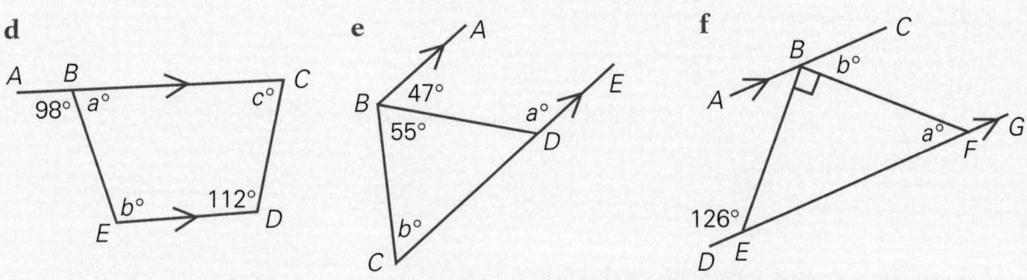

3 Find the value of each pronumeral, setting your work out carefully:

a **b** **c**

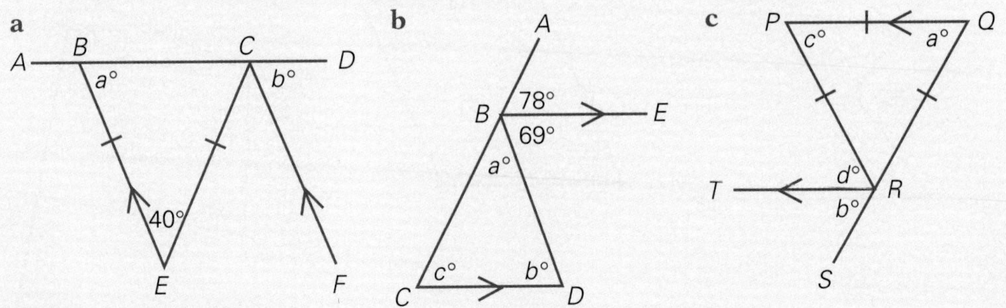

4 Find the value of each pronumeral, giving a reason for each step:

a A, B, C, D; $x°$, 140°

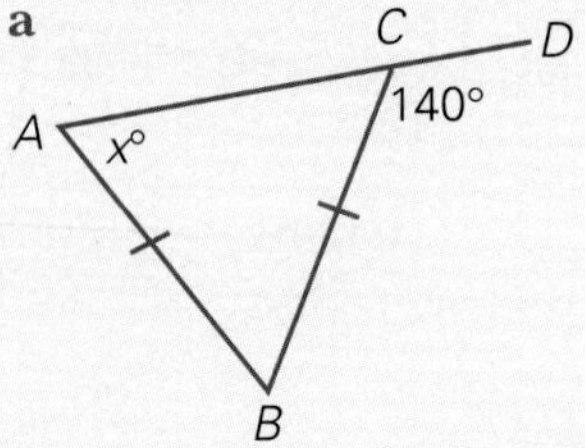

b A, B, C, D, E; $x°$, 86°

c P, Q, R, S, T; $x°$, 45°, 55°

d A, B, C, D, E; $x°$, 110°, 120°, 65°

e A, B, C, D, E; $x°$, 115°, 42°

f A, B, 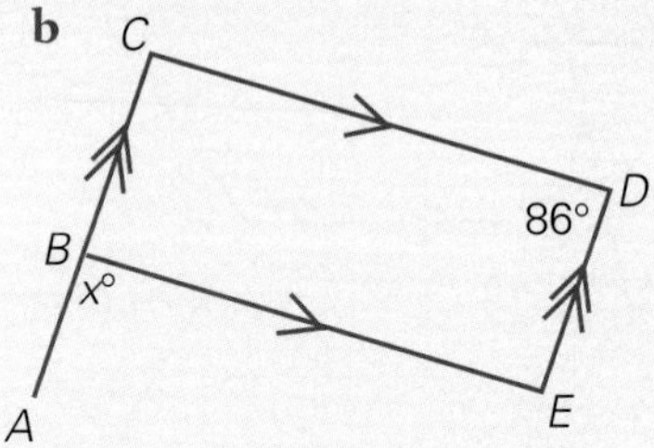 C, 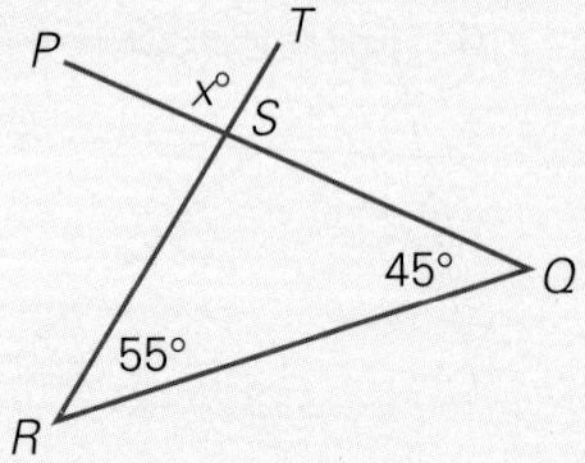 D, 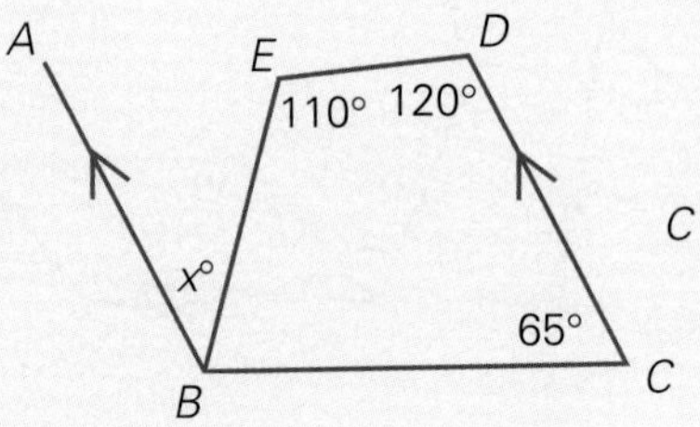 E, 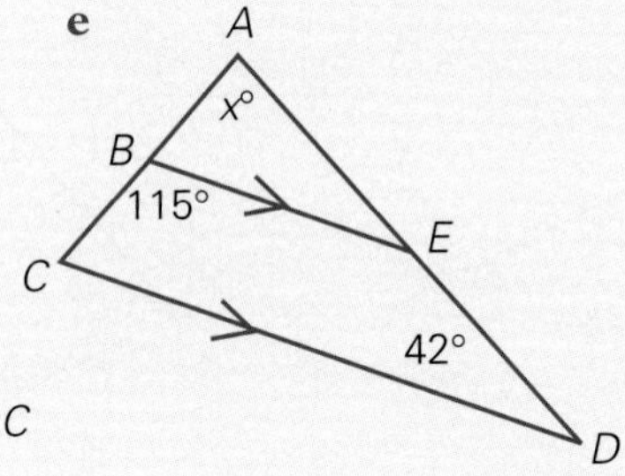 F, G; $x°$, 107°, 53°

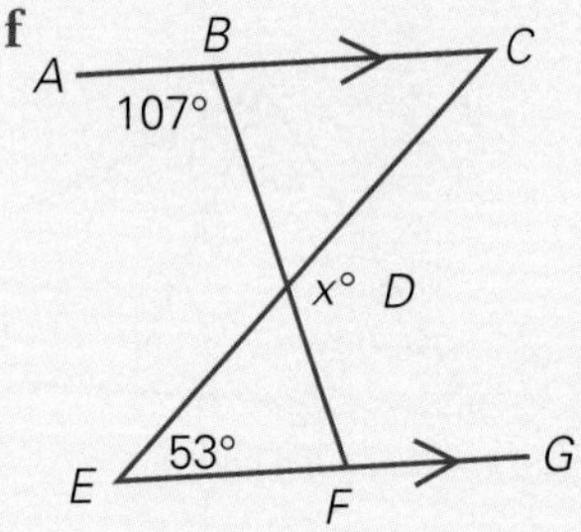

5 a $\alpha°$, $\beta°$, $\gamma°$

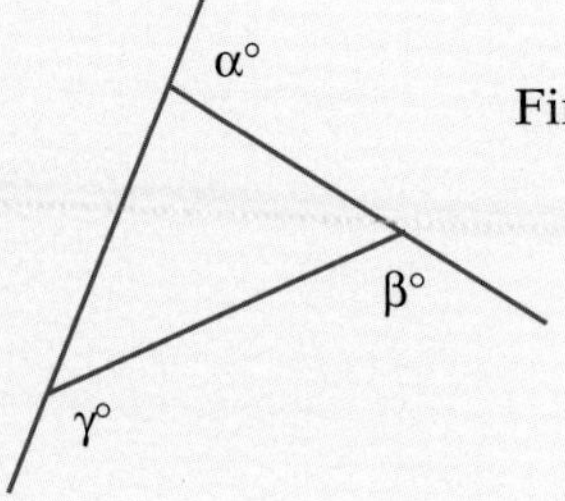

Find the value of $\alpha + \beta + \gamma$.

b

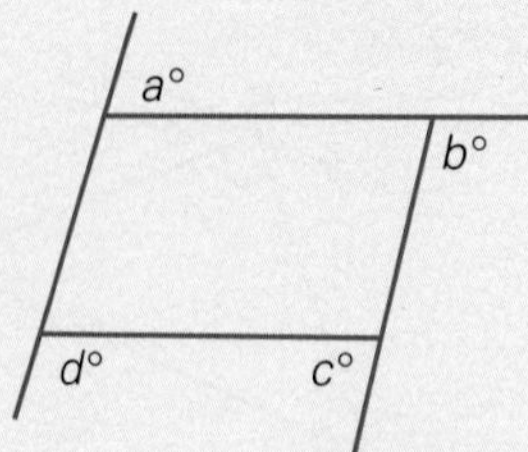

Find the value of $a + b + c + d$.

6 Find the value of x in each of the following:

a

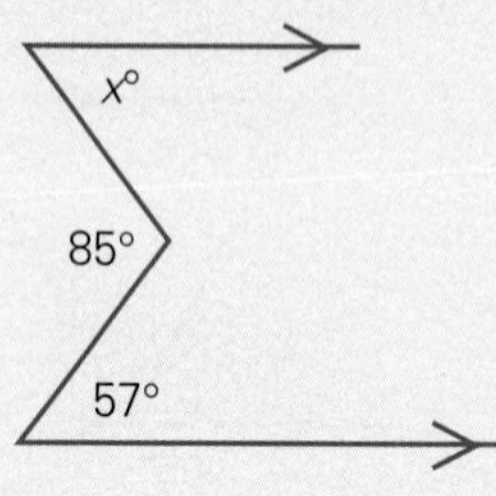

b

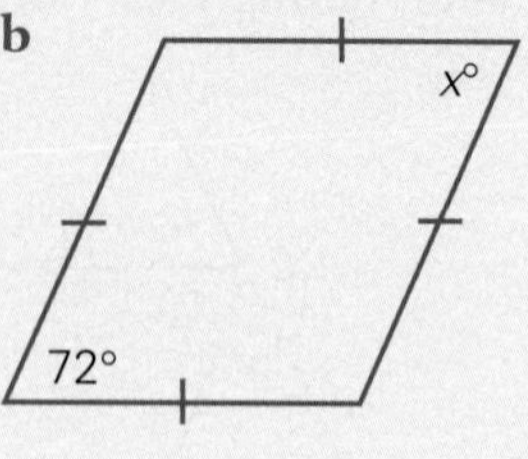

c

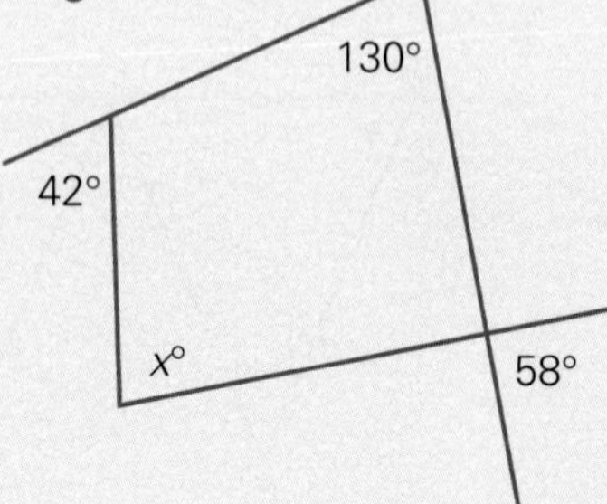

d

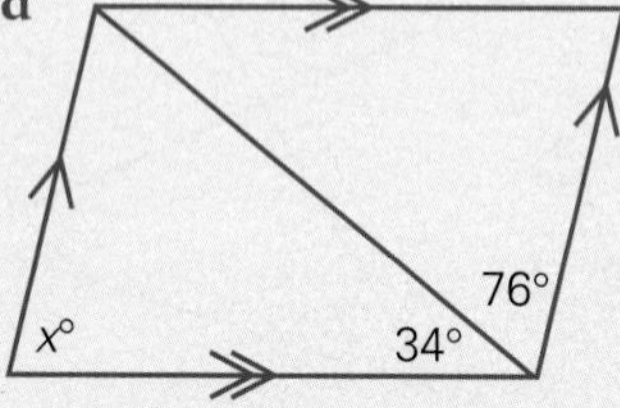

e

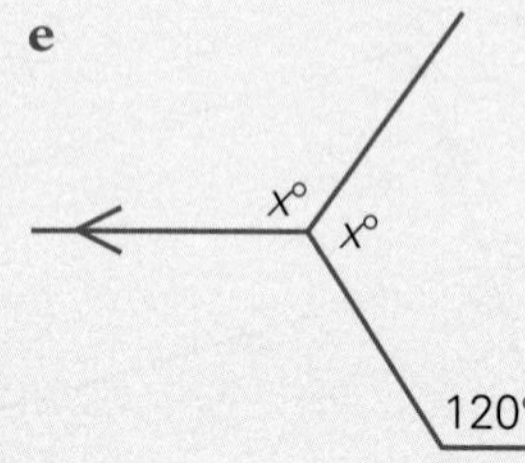

f

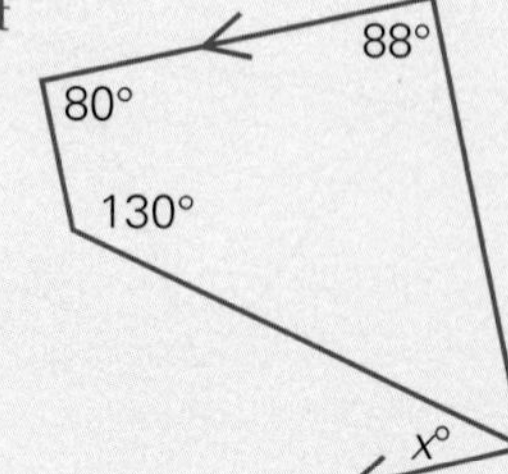

g

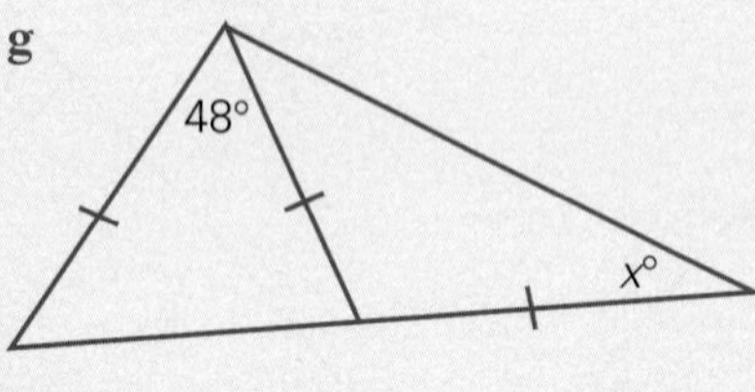

h

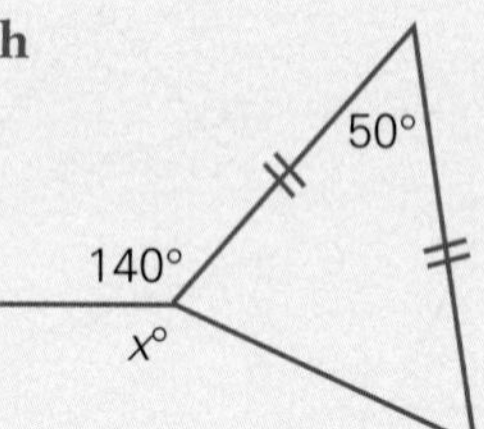

i

x°

140°

7.7 Reviewing constructions

90° AND 45° ANGLES

1 To construct a right angle (90°)

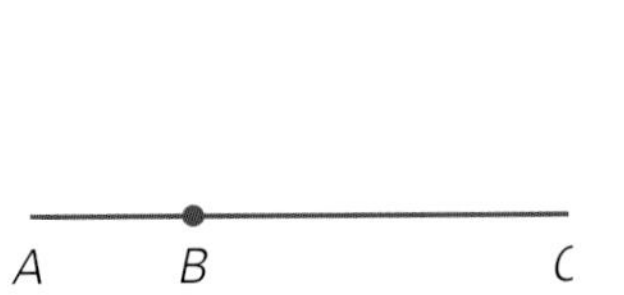

Draw a straight angle, ∠*ABC*.

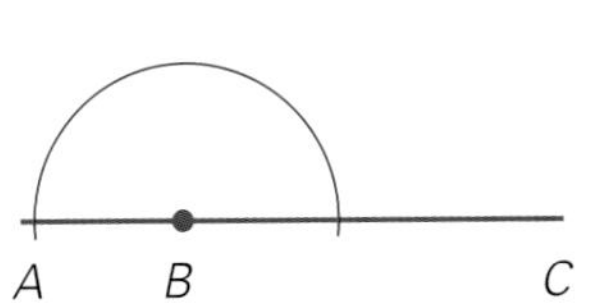

With centre *B*, draw an arc to cut both arms of ∠*ABC*.

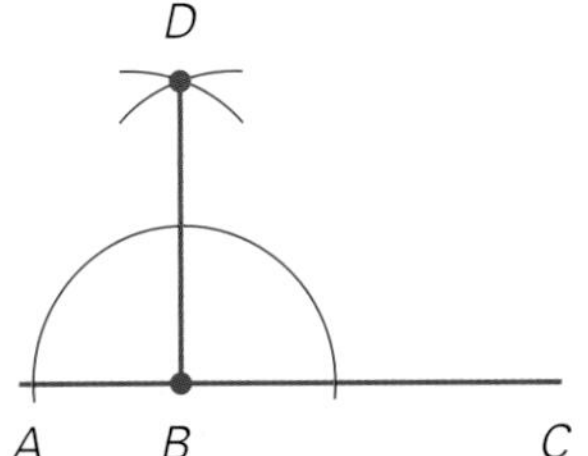

With centre at each point of arc intersection, draw two further arcs to cut at *D*. Join *DB* to form two 90° angles, ∠*ABD* and ∠*CBD*.

Discussion point: Why must this construction produce right angles?
Hint: Isosceles triangle.

2 To construct a 45° angle

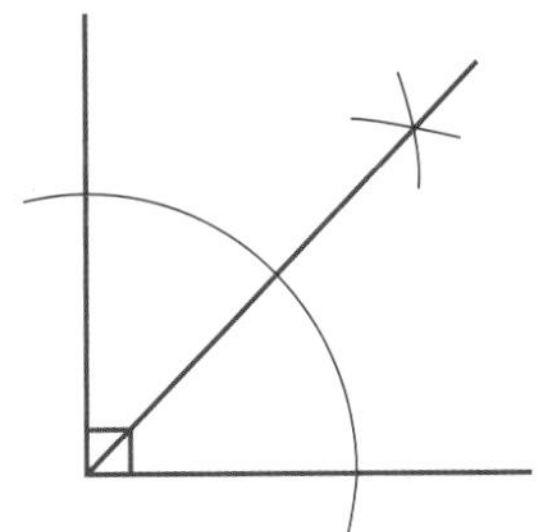

Construct a 90° angle as before, and *bisect* it.

3 To construct a 135° angle

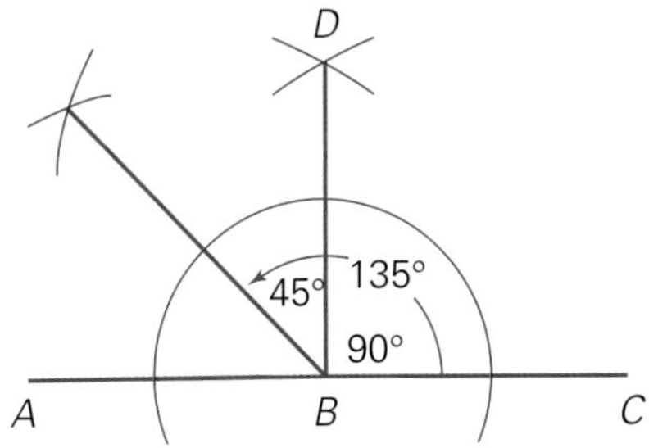

Combine a 90° and 45° construction.

60°, 30° AND 120° ANGLES

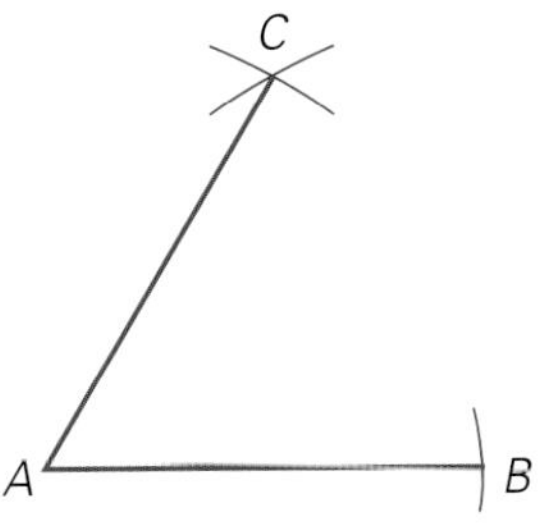

Draw one arm *AB*.
With centre *A* and radius *AB*, draw an arc.
Repeat with centre *B*. *C* is their point of intersection.
Join *AC*. ∠*CAB* = 60° Why?

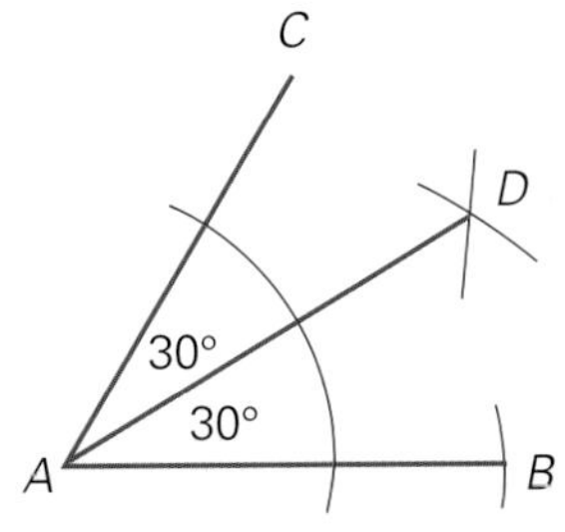

Bisect a 60° angle.

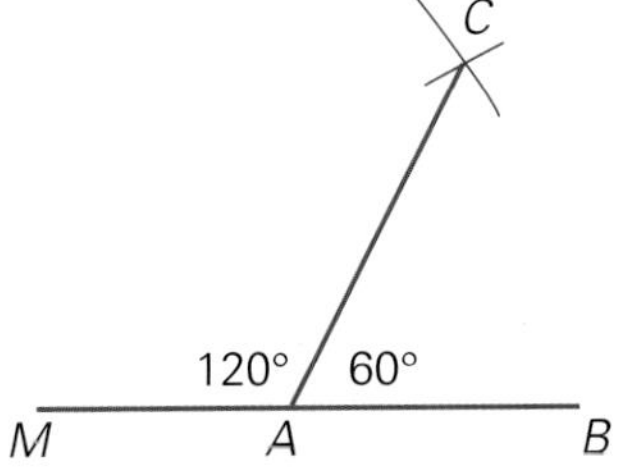

Draw a straight ∠*MAB*.
Draw a 60° ∠*CAB*.
Then ∠*MAC* is 120°, (adjacent angles on a straight line).

EXERCISE 7G

1 Construct an angle to match each of the following in size and position:

a 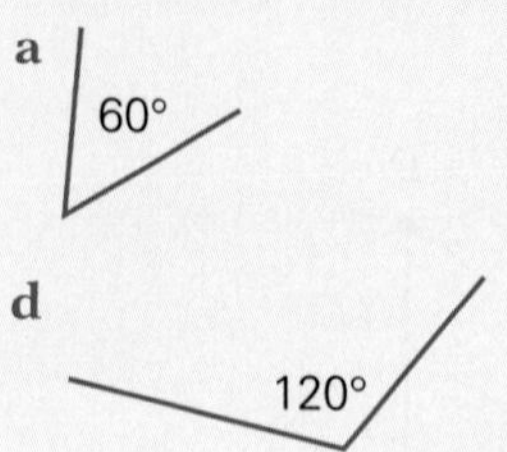

b 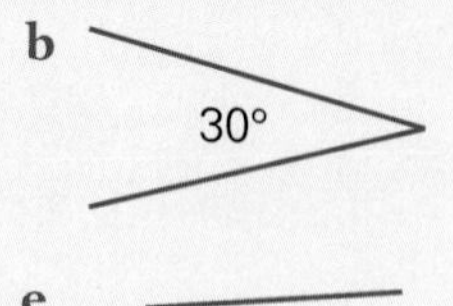

c 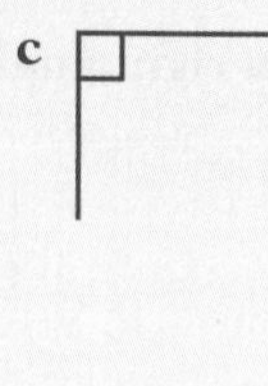

e

Level 2

2 Construct the following:

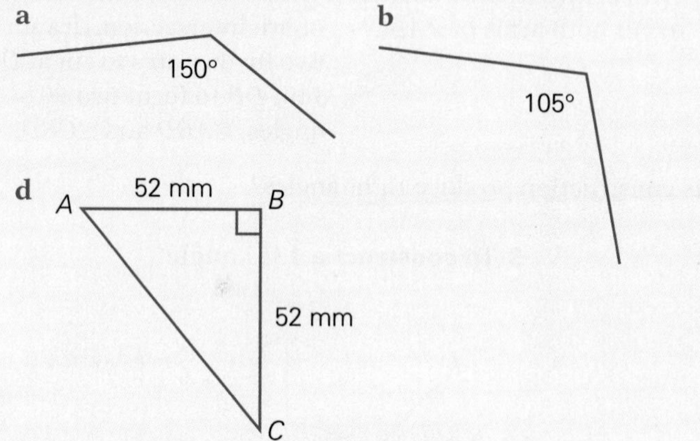

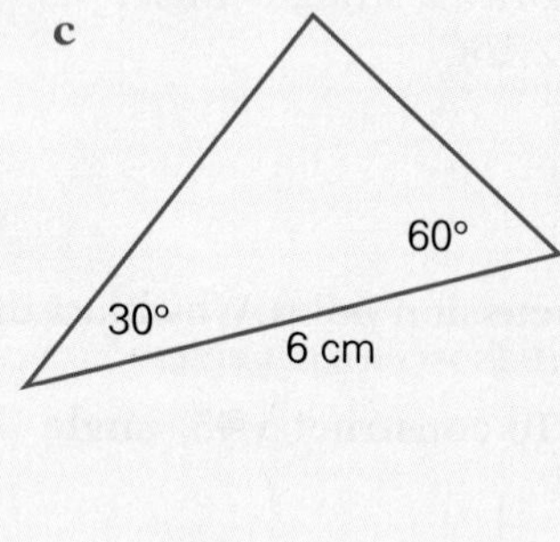

3 In ΔABC constructed in question **2d**, measure AC to the nearest millimetre. What must be the size of $\angle CAB$? $\angle ACB$? Give reasons for your answers.

4 Construct the trapezium $MPQR$ as shown. Measure MP, PQ and $\angle MPQ$.

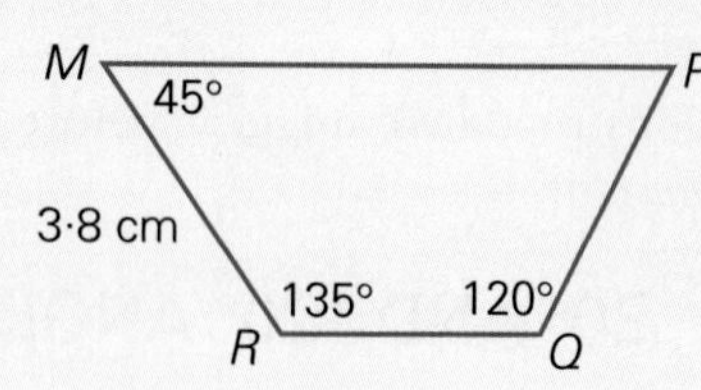

5 a Use only compasses and ruler to construct a figure identical to this one.
b Measure BC and CD.
c What must be the size of $\angle CDA$? Give a reason for your answer.

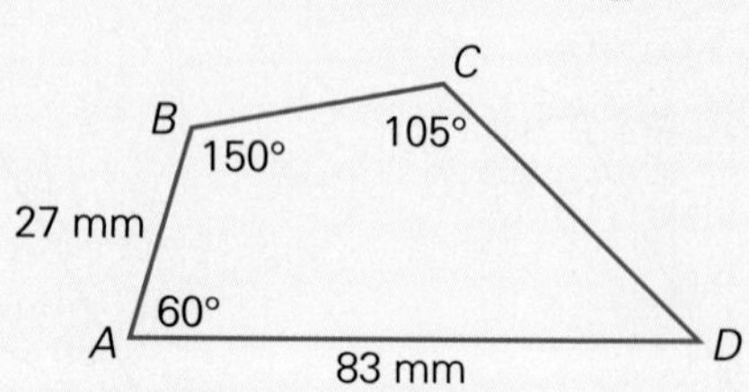

6 Make an acute construction of each of the following triangles:

a 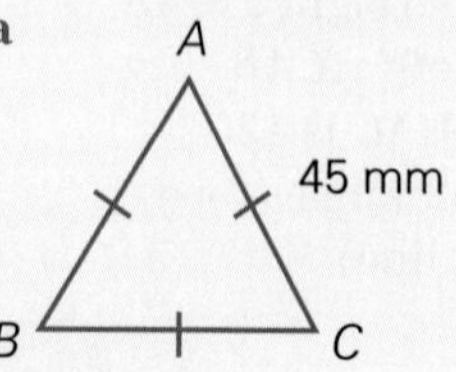

b 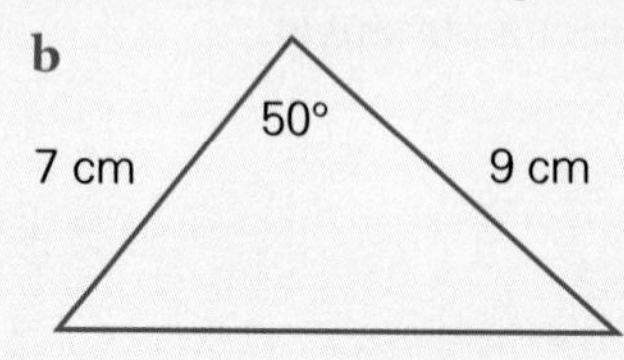

c 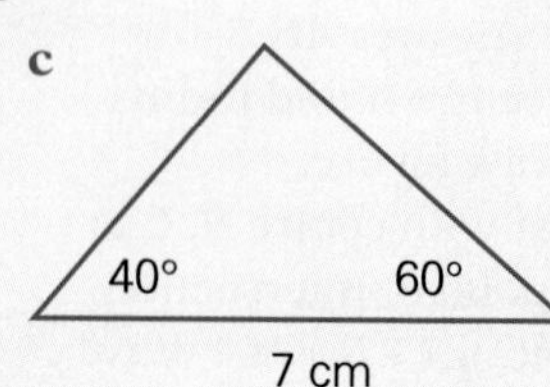

Chapter Review

Language Links

acute	exterior	proof	supplementary
adjacent	interior	quadrilateral	transversal
alternate	intersection	reasoning	triangle
complementary	isosceles	reflex	vertex
co-interior	obtuse	revolution	vertically opposite
construction	opposite	right	
corresponding	parallel	scalene	
equilateral	polygon	straight	

1 Draw a diagram to match the following terms:

a vertically opposite angles
b alternate angles
c corresponding angles
d co-interior angles
e isosceles triangle
f exterior angle
g adjacent angles
h transversal

2 Write down everything you know about each of the following diagrams:

a
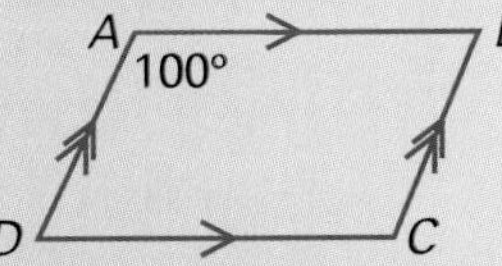

b
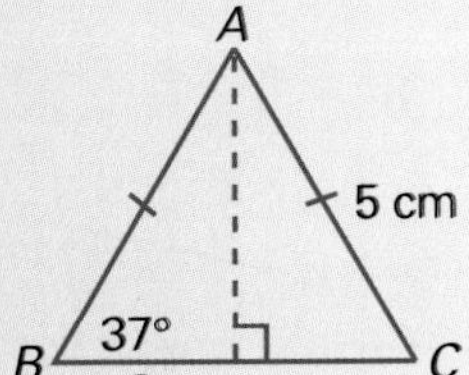

c
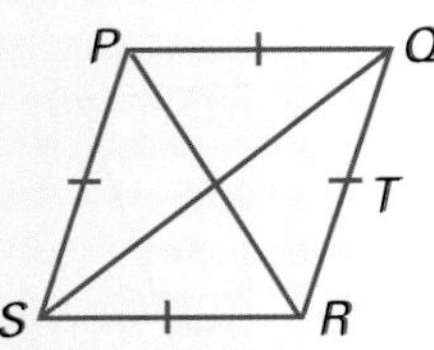

Chapter Review Exercises

1 Find the value of each pronumeral, giving a reason for each step:

a
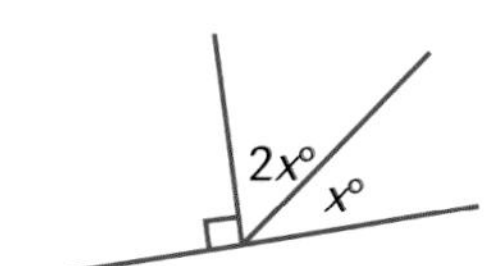

b
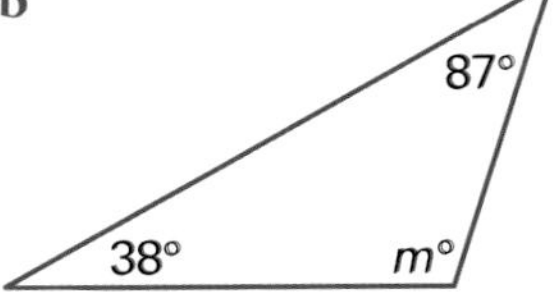

c
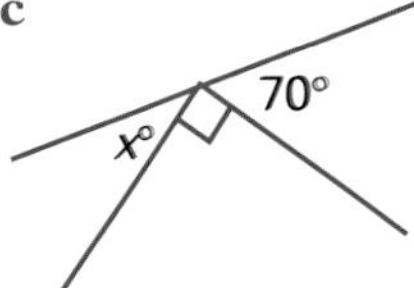

d
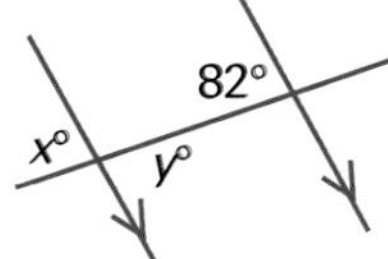

e
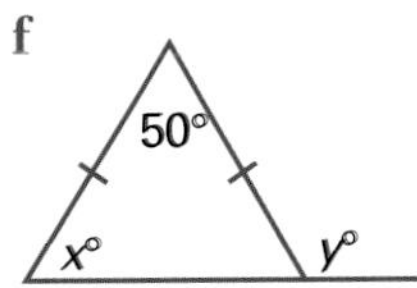

f

50° x° y°

7.3
7.4

2 Find the value of each pronumeral, giving a reason for each answer:

a

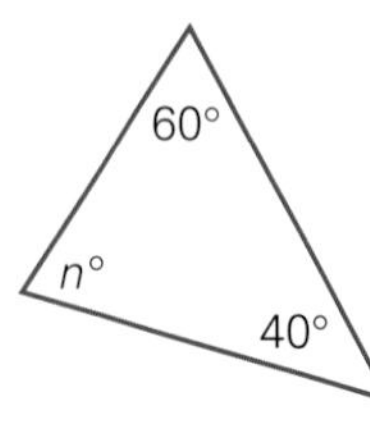

b

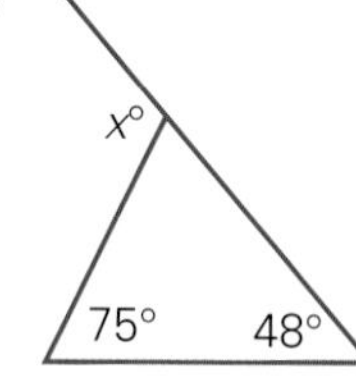

c

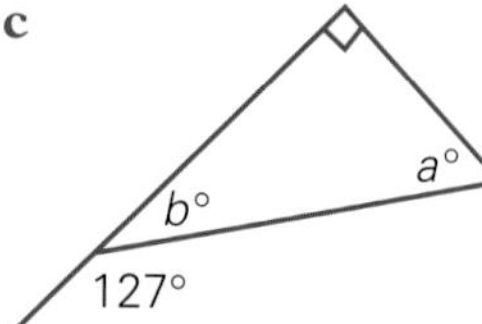

d

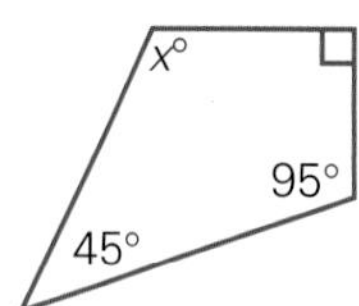

e

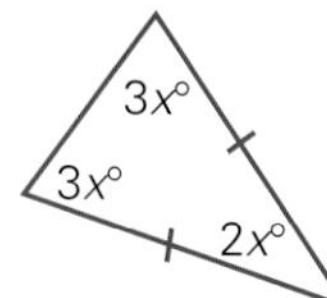

f

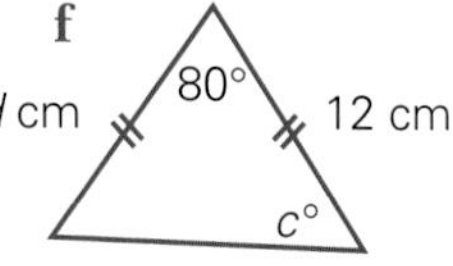

7.7

3 **i** Construct each of the following triangles using only ruler and compasses.
ii Classify each.

a

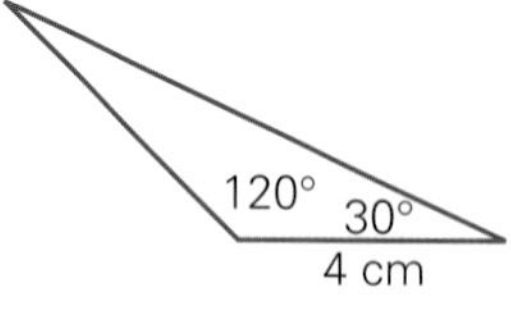

b

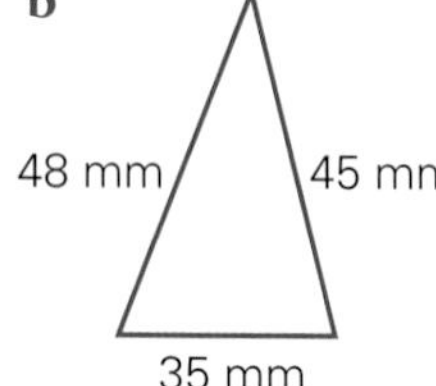

c

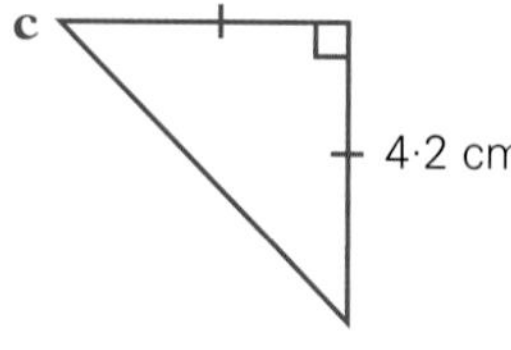

7.2

4 Is $AD \parallel CB$? Give a reason.

a

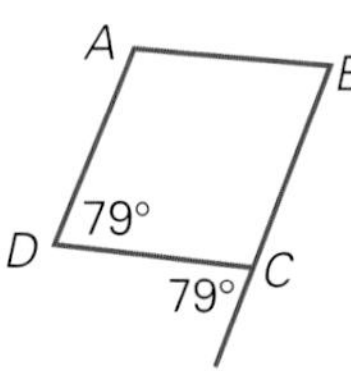

b

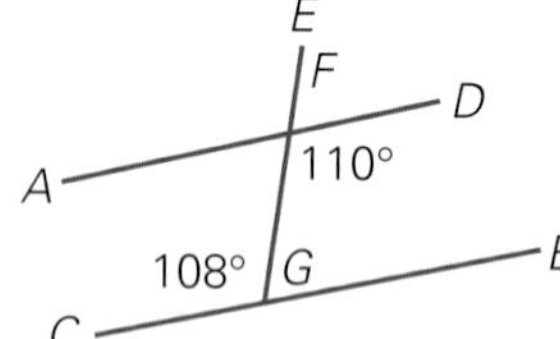

7.5
7.6

5 Find the value of a:

a

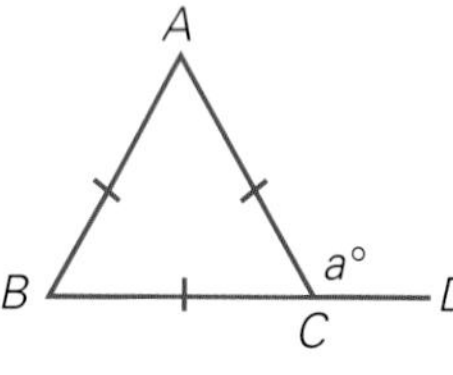

b

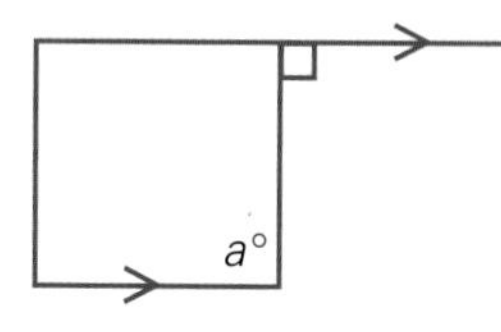

c

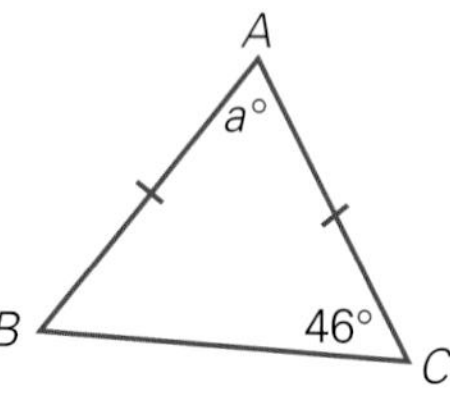

d

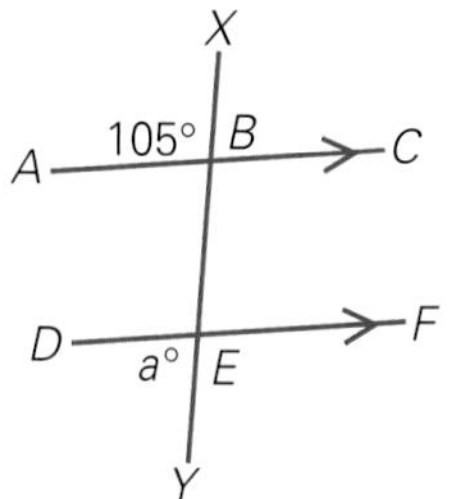

e

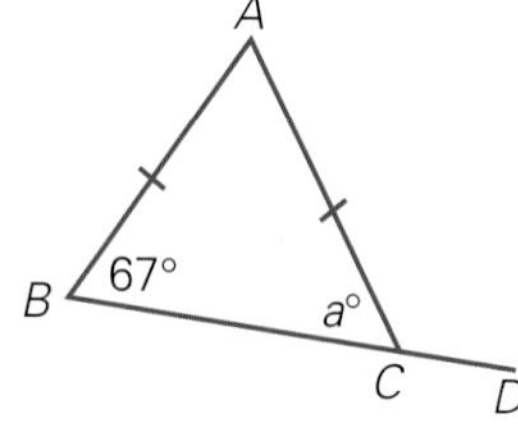

f

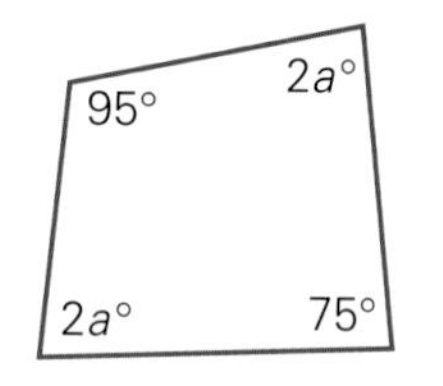

TEACHER

Keeping Mathematically Fit

Part A: Non-calculator

1 If $2\frac{1}{2}\%$ of a number is 4, what is the number?

2 Find $\sqrt[3]{64 \times 27}$.

3 Find the value of $6x - 4$, given $x = {}^-2$.

4 Between which two consecutive integers does $\sqrt{60}$ lie?

5 Find $1\frac{1}{3} \div 4$.

6 Estimate the value of $\frac{19 \cdot 3 + 9 \cdot 8}{0 \cdot 3}$. Give your answer as a whole number.

7 Insert +, −, ×, ÷ to make 15 ☐ 6 = 100 ◯ 10.

8 Solve $3a = 2a + 10$.

9 Complete: A bearing of 270° is the same as due ______.

10 Find the missing term in the pattern: 2, 6, 12, ☐, 30.

Part B: Calculator

1 Copy and complete:

Cost price	\$24	\$15	\$80	\$72		\$1.50
Selling price	\$32	\$20			\$100	\$2.10
% profit			10%	$12\frac{1}{2}\%$	10%	

2 Find the value of x in:

a
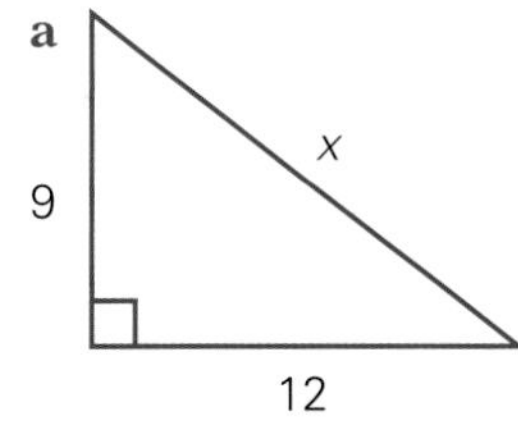

b
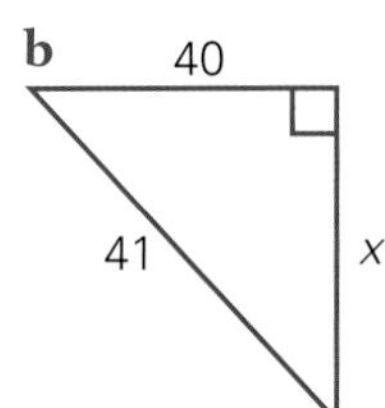

c
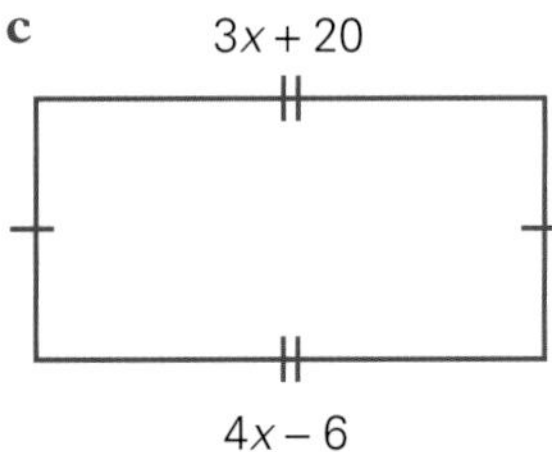

3 Find the simple interest on \$550 at 7% p.a. for 3 years.

4 Evaluate $\sqrt{6 \cdot 8} - 1 \cdot 02^2$ correct to 1 decimal place.

5 Find the average of $\frac{2}{3}$, $\frac{4}{5}$ and $1\frac{1}{2}$.

6 Factorise:

a $16a - 24$ b $16a + 2b$ c ${}^-16a - 16$

7 The area of a square is 204 mm². Find the perimeter of the square.

8 A number is squared and the result is 49. What is the value of the number?

TEACHER

1 What you need to know and revise

Outcome MS 3.2:
Selects and uses the appropriate unit to calculate area, including the area of squares, rectangles and triangles

Outcome MS 4.1:
Uses formulae and Pythagoras' theorem in calculating perimeter and area of circles and figures composed of rectangles

2 What you will learn in this chapter

Outcome MS 4.1:
Uses formulae and Pythagoras' theorem in calculating perimeter and area of circles and figures composed of rectangles

Outcome MS 4.2:
Calculates surface area of rectangular and triangular prisms and volume of right prisms and cylinders

- describes the limits of accuracy of measuring instruments ($\pm \frac{1}{2}$ unit of measurement)
- reviews areas of squares, rectangles, triangles, parallelograms and simple composite figures
- converts between metric units of area
- identifies the surface area and edge lengths of rectangular and triangular prisms
- finds the surface area of rectangular and triangular prisms, e.g. by using a net
- calculates the surface area of rectangular and triangular prisms
- converts between units of volume
- uses kilolitre as a unit in measuring large volumes
- identifies and uses the formula for finding the volume of right prisms
- calculates the volumes of prisms with cross-sections that are simple composite figures

Working Mathematically outcomes WMS 4.1–4.5
Students will be asked to ***question***, ***apply strategies***, ***communicate***, ***reason*** and ***reflect*** in the sections of this chapter.

Measurement

Key mathematical terms you will encounter

accuracy
altitude
area
breadth
capacity
composite
conversions
cross-section
cubic
dimensions
error
hectare
height
kilolitre
length
limits
measurement
perpendicular
prism
surface area
volume
width

MathsCheck

Measurement

1 Change to metres (m):

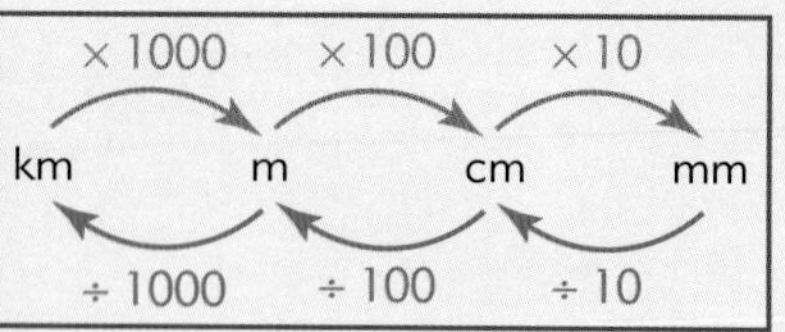

a 5 km　　**b** 1800 mm　　**c** 200 cm
d 2·8 km　　**e** 80 cm　　**f** 750 mm
g 3·06 km　　**h** 0·9045 km

2 Convert to kilograms (kg):

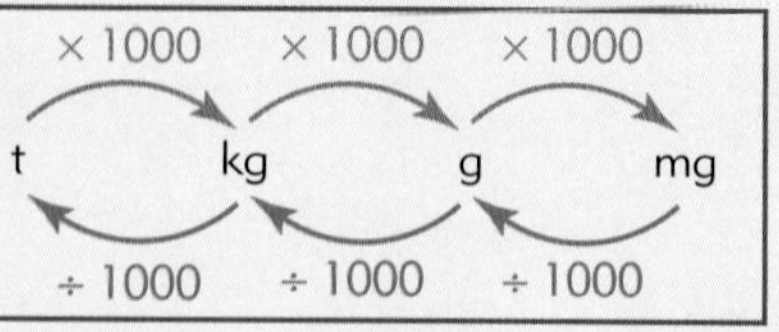

a 2 t　　**b** 3·7 t　　**c** 1200 g
d 454 g　　**e** 80 g　　**f** 1·03 t
g 0·005 t　　**h** 17 500 mg

3 Express in litres (L):

× 1000　× 1000　× 1000
ML　kL　L　mL
÷ 1000　÷ 1000　÷ 1000

a 6 kL　　**b** 1·54 kL
c 1250 mL　　**d** 2·4 ML

4 Complete the following conversions:

a 140 cm = □ m　　**b** 4·5 cm = □ mm　　**c** 800 m = □ km
d 10 g = □ kg　　**e** 6 kg = □ t　　**f** 20 mg = □ g
g 0·05 kL = □ L　　**h** 50 000 L = □ kL　　**i** 20 mL = □ L

5 Convert to the same unit, then decide which quantity is the larger:

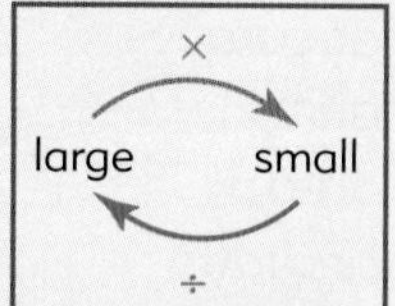

a 2·4 m or 290 cm　　**b** 780 L or 1·2 kL
c 1200 mg or 1·1 g　　**d** 2·5 cm or 26 mm
e 0·8 km or 820 m　　**f** 0·05 t or 500 kg

6 $1 coins are placed in a pile. If each is 2·3 mm thick, how high (in cm) is a pile of:

a 10 coins?　　**b** 40 coins?　　**c** 200 coins?

7 Find the distance round the boundary of each figure:

a

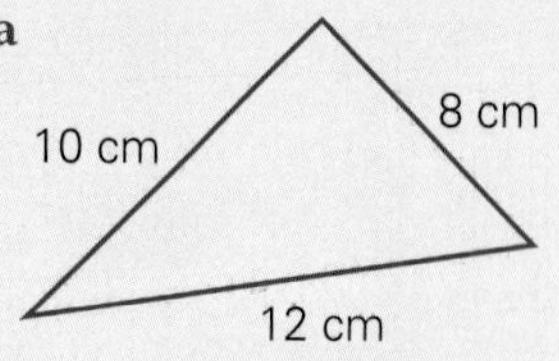

b

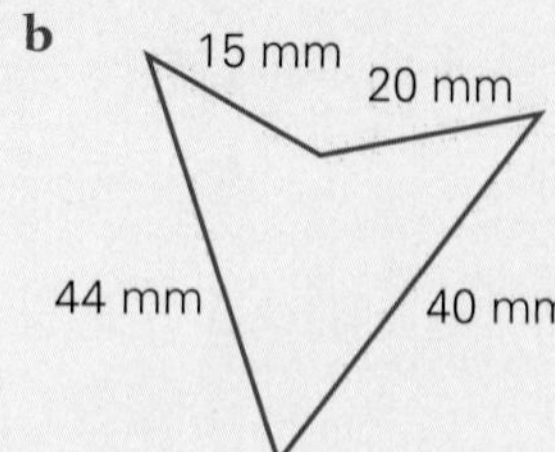

c

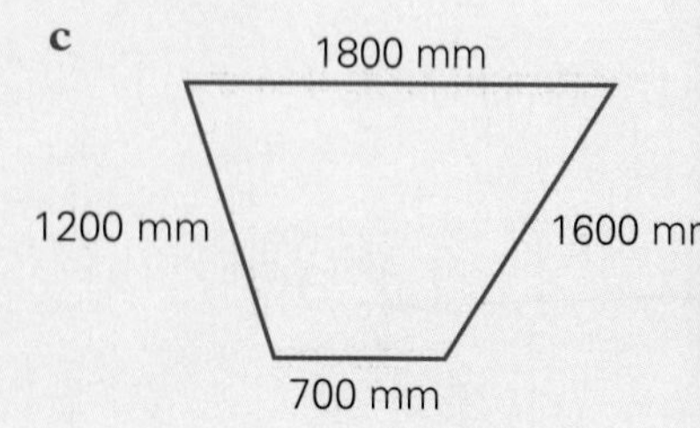

d

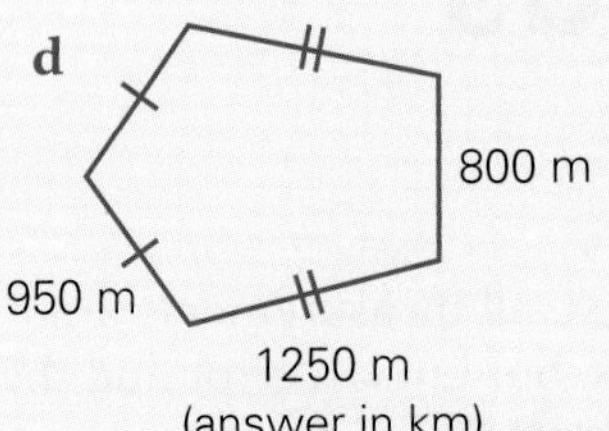

e

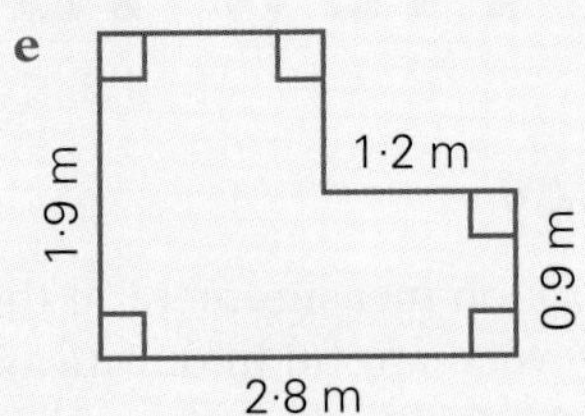

f

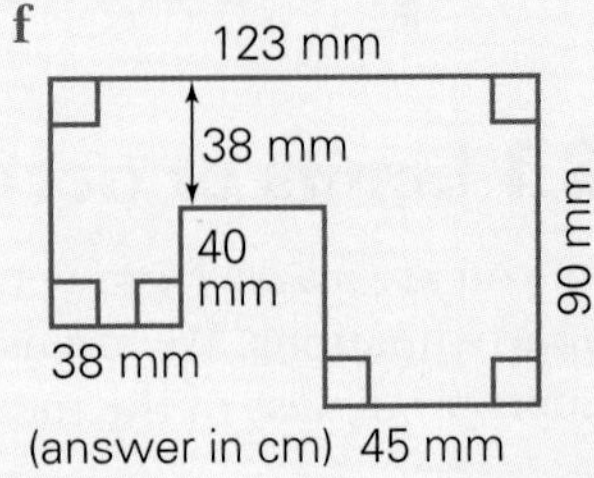

8 Calculate each perimeter using the appropriate formula:

a

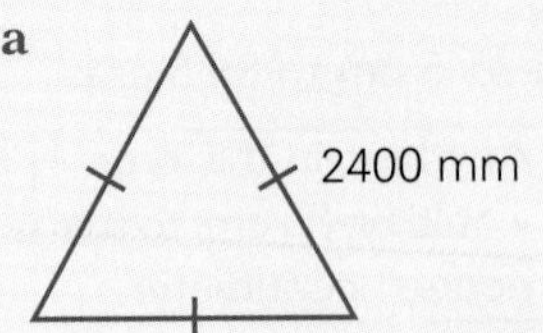

b

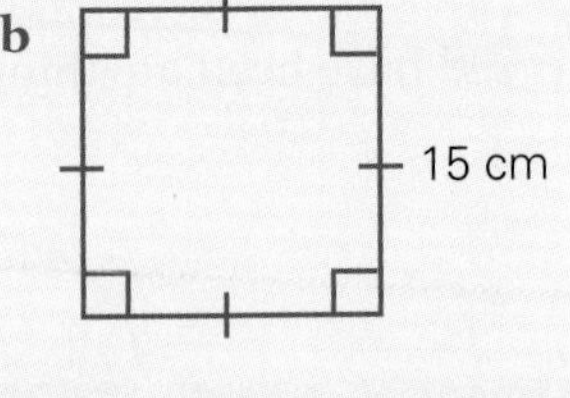

c

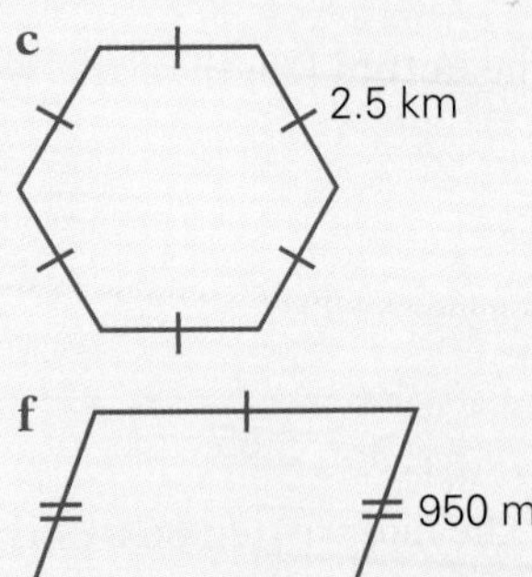

d

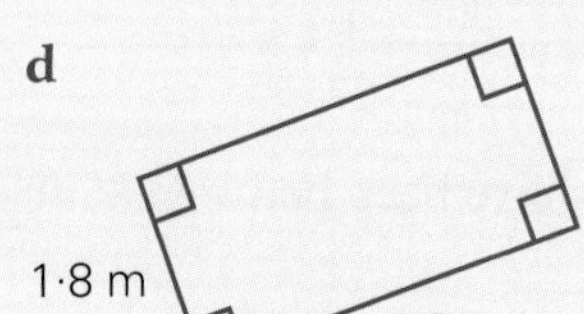

e

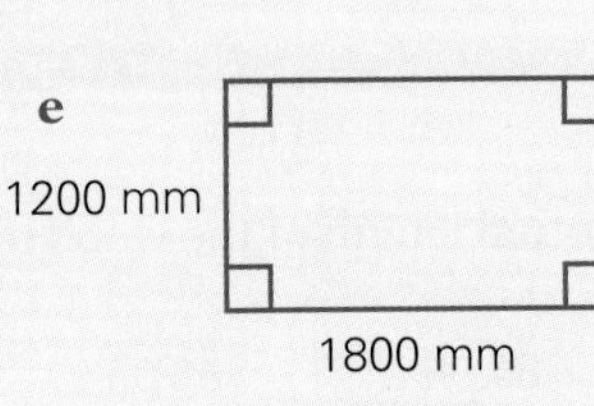

f

9 The perimeter of a square field is 460 metres. How long is each side?

10 A rectangular field has a perimeter of 672 m and a width of 136 m. How long is it?

11

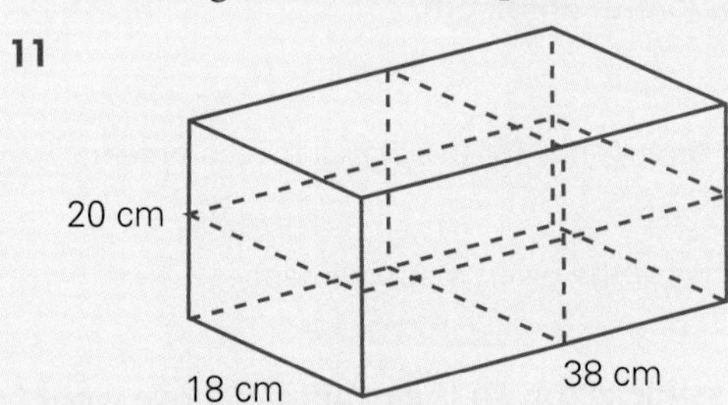

a What length of adhesive tape is needed to secure this box along the colour-dotted lines?

b How many such boxes can be secured from a 30-metre roll of tape?

c How much tape is left?

12 For each of the solids below:

i what is the name given to the shaded face?

ii what is the name of the solid figure shown?

iii draw the net of each of the solids below.

a

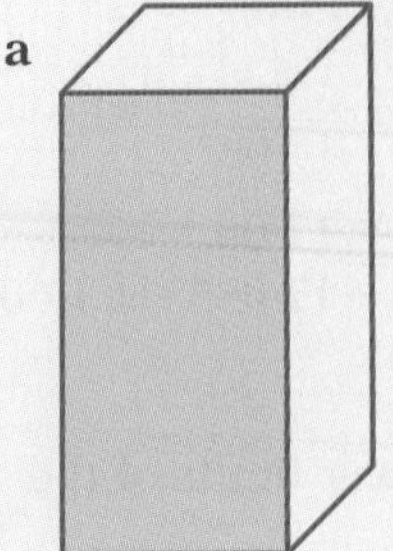

b

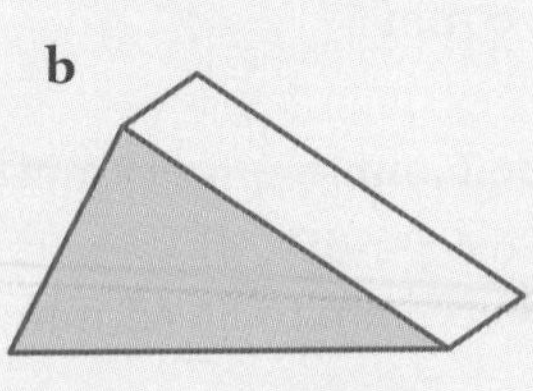

c

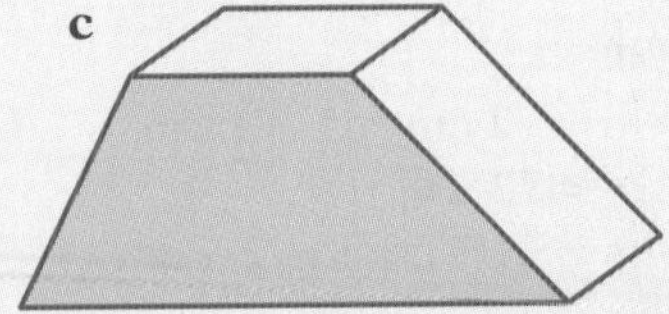

Exploring New Ideas

8.1 Limits of accuracy

As you are aware from your work on measurement in the past, all measurements are approximations. We are usually very careful in our measures, but human error and the appropriateness of the measuring device chosen are always factors.

Let us consider a piece of wood Christine measured for her Design and Technology project. She wrote down the length as 86 cm.

The **actual** length of the wood could have been anything from 85·5 cm to 86·5 cm.

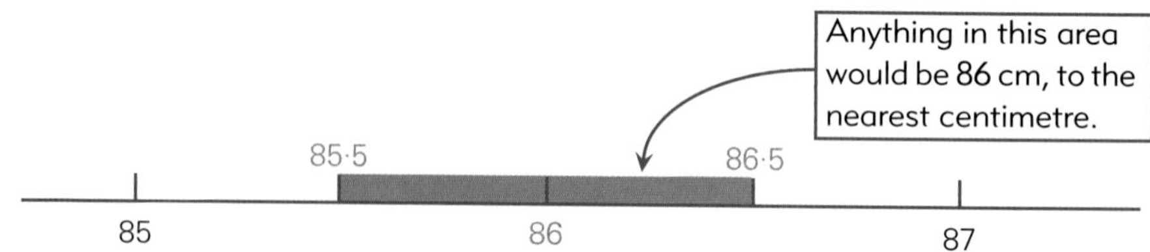

If she had written 86·0 cm, the actual length of the wood must have been between 85·95 and 86·05 cm.

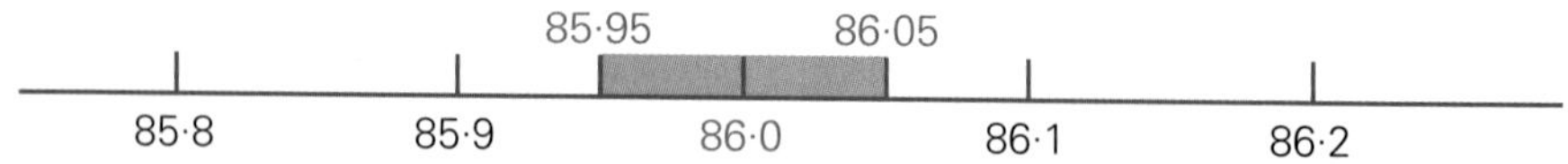

We say the **limits of accuracy** for a length given as 86 cm are 85·5 cm; 86·5 cm. Similarly the limits of accuracy for a length of 86·0 are 85·95 cm; 86·05 cm.

> The limits of accuracy tell you the lower and upper boundaries for the actual measurement.
> i.e. $\pm \frac{1}{2}$ **the smallest unit of measure**
> They are stated to *one* place of decimals *beyond* that of the given measurement.

Accuracy is sometimes stated in terms of the possible error, so the example above would appear as **86 cm ± 0·5 cm**. This means the actual measurement could be up to 0·5 cm larger or smaller than 86 cm.

EXAMPLE 1

Give the limits of accuracy for the following measurements:

a 72 cm **b** 86·6 mm **c** 17·06°

Solution

a $72 \text{ cm} \pm \frac{1}{2} \times 1 \text{ cm}$
$= 72 \pm 0{\cdot}5$
Limits = 71·5 cm to 72·5 cm

b $86{\cdot}6 \text{ mm} \pm \frac{1}{2} \times 0{\cdot}1 \text{ mm}$
$= 86{\cdot}6 \pm 0{\cdot}05$
= 86·55 mm to 86·65 mm

c $17{\cdot}06° \pm \frac{1}{2} \times 0{\cdot}01°$
$= 17{\cdot}06 \pm 0{\cdot}005$
= 17·055° to 17·065°

EXAMPLE 2

The sides of a square were measured as 6 cm. Find the upper and lower limits of the perimeter.

Solution

Side length limits $= 6 \text{ cm} \pm \frac{1}{2} \times 1 \text{ cm}$

$= 6 \pm 0{\cdot}5$

$= 5{\cdot}5 \text{ cm to } 6{\cdot}5 \text{ cm}$

$\therefore$ limits of accuracy for the perimeter are:

$4 \times 5{\cdot}5 \text{ cm to } 4 \times 6{\cdot}5 \text{ cm}$

$= 22 \text{ cm to } 26 \text{ cm}$

EXERCISE 8A

1 Give the upper and lower of limits of accuracy for the following measurements:

a 5 m b 8 cm c 78 m d 1 cm

2 Give the limits of accuracy for the following measurements:

a 23 cm b 3·9 kg c 19·5 m d 8·7 m

3 Give the limits of accuracy for:

a 18·56 cm b 19·0 km c 0·1 km d 4·20 t

Level 2

4 Give the limits of accuracy for a length of 150 cm that is given to the nearest:

a centimetre b millimetre

5 Write the following as a measurement, given the lower and upper limits are:

a 29·5 m; 30·5 m b 140 g; 150 g c 8·5 m; 9·5 m

6 The length of a piece of copper pipe is given as 25 cm.

a What is the minimum length of 16 pieces of pipe joined end to end?
b What would be the minimum length if the length of the pipe was given as 25·0 cm?

7 The side of a square is recorded as 9·2 cm.

a What is the minimum length the side of the square could actually be?
b What is the maximum length the side of the square could actually be?
c Find the lower and upper limits for the square's perimeter.

8 A rectangle has its length measured as 10 cm and its breadth as 8 cm.

a Write down the limits of accuracy for the rectangle's length.
b Write down the limits of accuracy for the rectangle's breadth.
c Find the lower limit for the rectangle's perimeter.

9 Janis measured the mass of an object to be 5·8 kg to 1 decimal place. Peter measured the same object to be 5·85 kg to 2 decimal places. If they are both correct, what are the limits of accuracy for the mass of the object?

10 Write down a sentence explaining the need to accurately measure items in our everyday lives. Give three examples of items needed to be measured correct to the nearest:

a kilometre b metre c millimetre

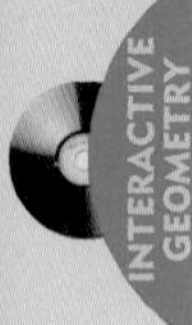

8.2 Review of area

As we know from last year, the area of a figure is the number of square units covering the surface of the shape.

The units used in area are:

1 cm^2 = 100 mm^2 1 m^2 = 10 000 cm^2 1 ha = 10 000 m^2 1 km^2 = 100 ha

It is possible to covert between units of area:

larger to smaller: **multiply**

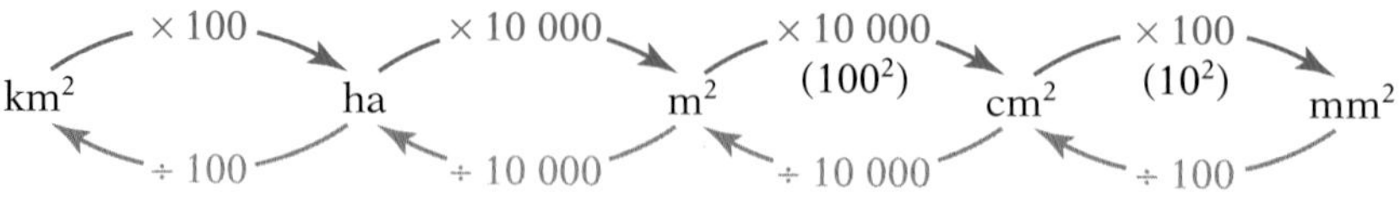

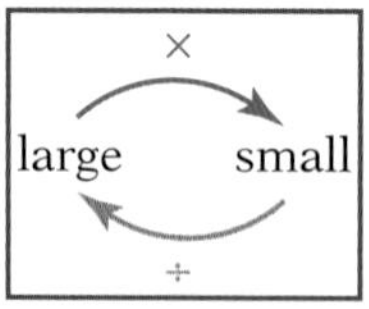

smaller to larger: **divide**

FORMULAE

<table>
<tr><td>Rectangles
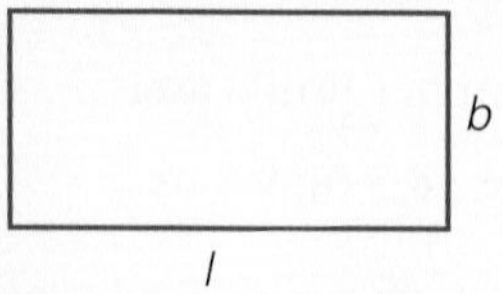</td><td>Area equals length times breadth:
$A = lb$</td></tr>
<tr><td>Squares
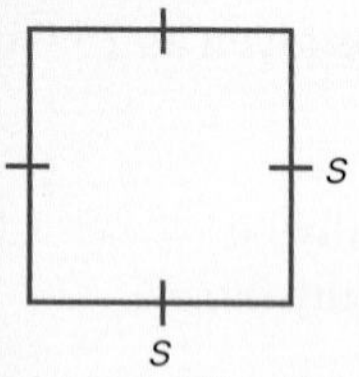</td><td>Area equals length times breadth or side times side (side squared):
$A = s^2$</td></tr>
<tr><td>Triangles (half rectangles)
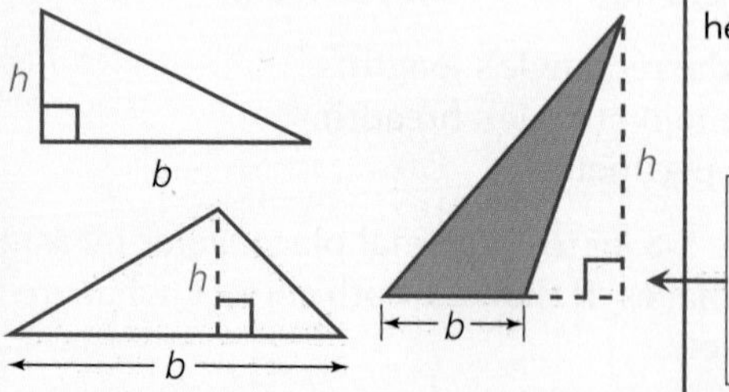</td><td>Area equals half of base times perpendicular height (altitude):
$A = \frac{1}{2}bh$ or $\frac{bh}{2}$
Note three possible positions of h, the altitude line. It is always at right angles to the base.</td></tr>
<tr><td>Parallelogram
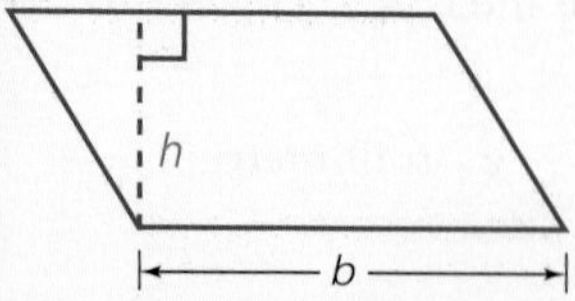</td><td>Area equals the product of the base and the perpendicular height:
$A = hb$</td></tr>
</table>

EXAMPLE 1

Convert the following to square metres:

a 3·5 ha **b** 12 000 cm^2

Solution

a 3·5 ha = 3·5 × 10 000 m^2
= 35 000 m^2

× 10 000
ha → m^2

b 12 000 cm^2 = 12 000 ÷ 10 000 m^2
= 1·2 m^2

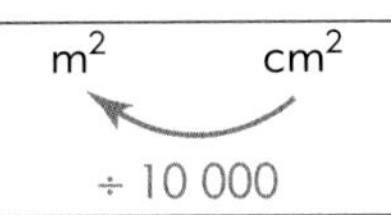

EXAMPLE 2

Find the area of:

a

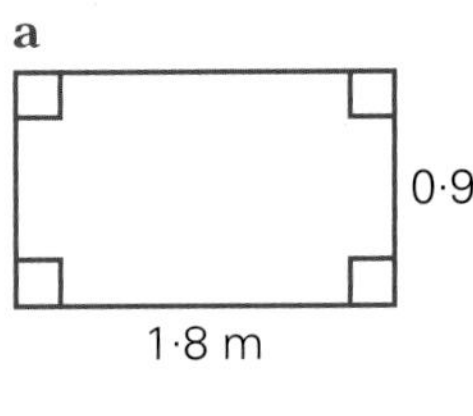

b

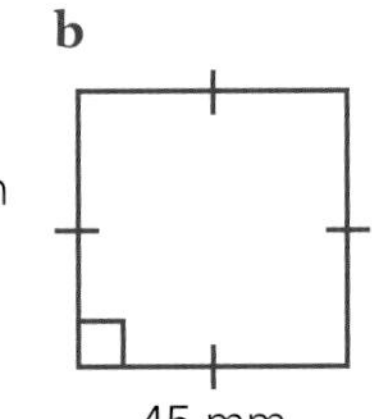

c

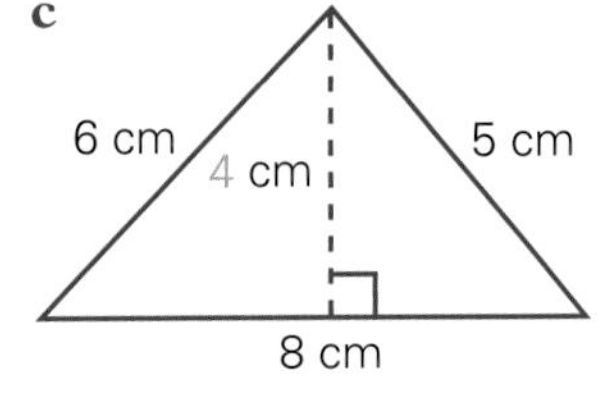

d

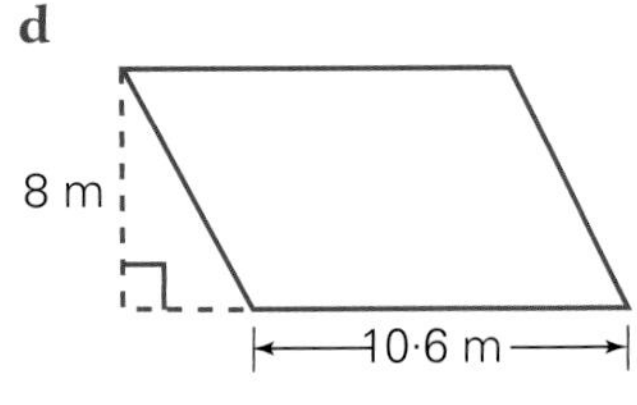

Solution

a $A = lb$
$= 1{\cdot}8 \times 0{\cdot}9$
$= 1{\cdot}62$
Area is 1·62 m^2

b $A = s^2$
$= 45^2$
$= 2025$
Area is 2025 mm^2

c $A = \frac{bh}{2}$
$= \frac{8 \times 4}{2}$
$= 16$
Area is 16 cm^2

Altitude is always perpendicular to the **base**.

d $A = hb$
$= 8 \times 10{\cdot}6$
$= 84{\cdot}8$
Area is 84·8 m^2

EXERCISE 8B

Did you realise that a hectare is approximately two football fields next to each other?

1 Convert each of the following to hectares:

a 50 000 m^2 **b** 5000 m^2 **c** 8000 m^2

2 Convert the following to square metres:

a 3 ha **b** 1·7 ha **c** 30 000 cm^2 **d** 6000 cm^2

3 Copy and complete the following conversions, using the flow diagram to help:

a 50 ha = ☐ km^2 **b** 400 mm^2 = ☐ cm^2 **c** 4070 m^2 = ☐ ha

4 Calculate the area of each figure:

a

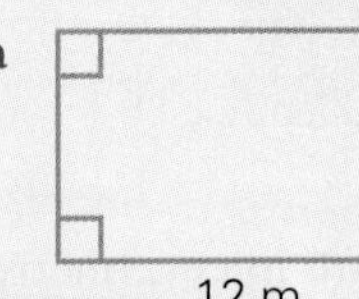

b

30 mm

c

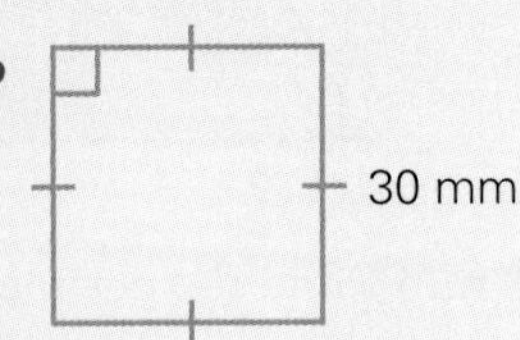

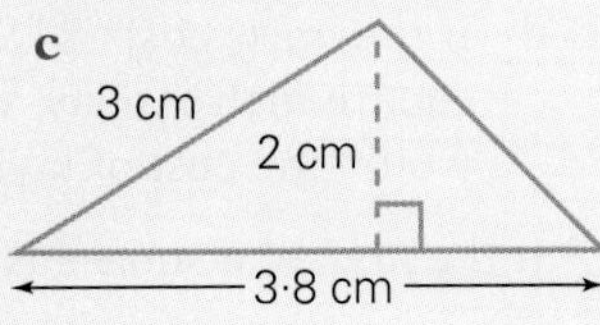

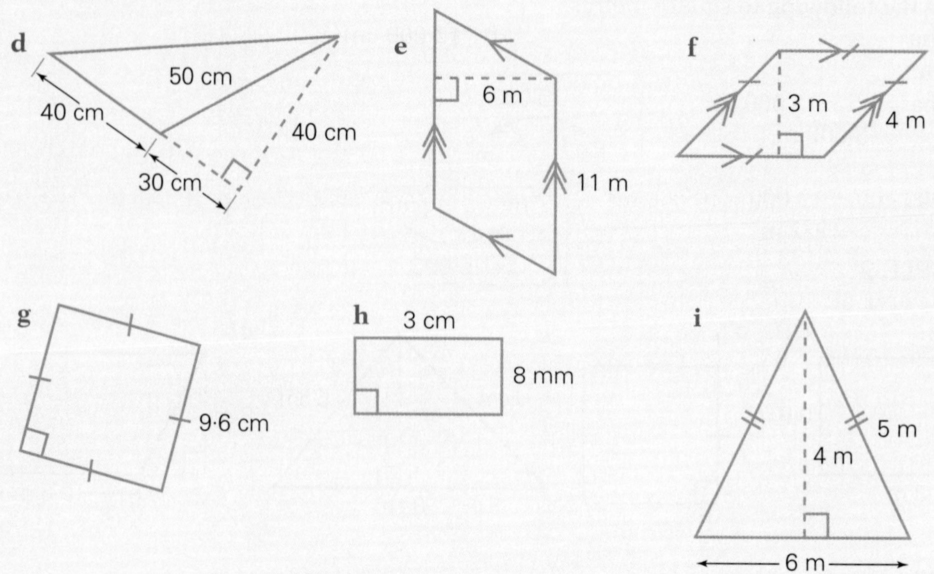

Level 2

5 Calculate the area occupied by the parking lot shown in the diagram below.

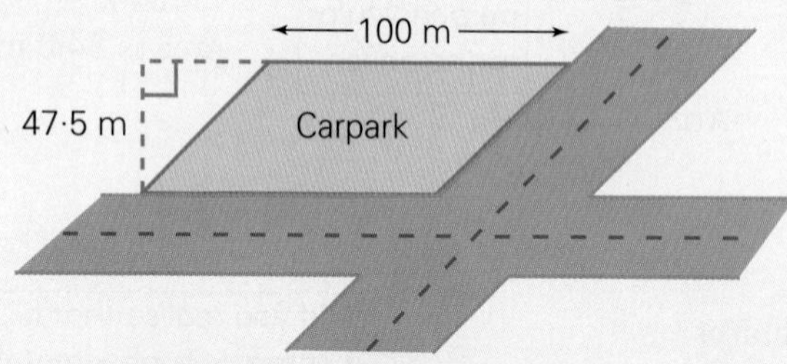

6 A developer wishes to build a house on the block of land shown below and to lay grass on the remaining area.

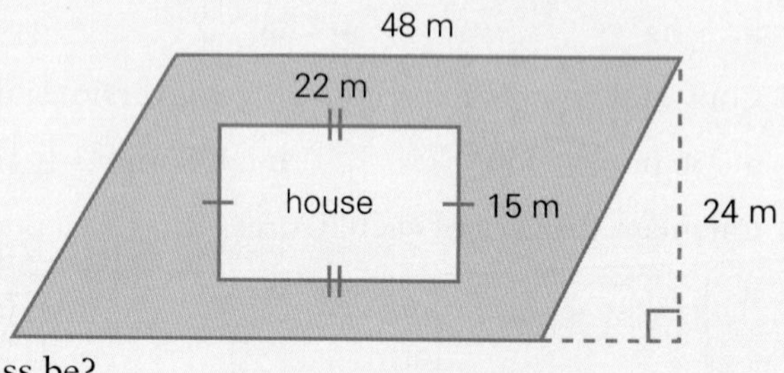

a Find the area of the block of land.
b Find the area the house would occupy.
c What percentage of the land is left for the yard?
d If the grass costs \$4.95 per square metre to lay, what would the cost of laying the grass be?

7 If the area of a square is 3600 mm^2, find the perimeter of the square in centimetres.

8 A rectangle has a perimeter of 44 cm. Give the dimensions of three rectangles that satisfy this description and find the area of each one.

9 Convert 1000 000 000 mm^2 into square metres.

10 a How many hectares are in one million square centimetres?
b Convert 1 hectare to square centimetres.

11 A rectangular piece of land has an area of 1 hectare.

a Give the dimensions of three possible rectangular pieces of land with such an area and calculate the perimeter of each one.

b Find the area of a square with the same perimeter as the rectangle. Calculate its area.

12 A parallelogram has an area of 38 cm^2. If the base is twice its height, find the height of the parallelogram.

13 A rectangle has a perimeter of 46 cm. If the width of the rectangle is 3 cm more than its length, find the area of the rectangle.

14 The dimensions of a rectangle are given as 5 m by 3 m.

a What are the lower boundaries of the length and breadth of the rectangle?

b What are the upper boundaries of the length and breadth of the rectangle?

c Between which two values does the true area lie?

15 Find the shaded area of:

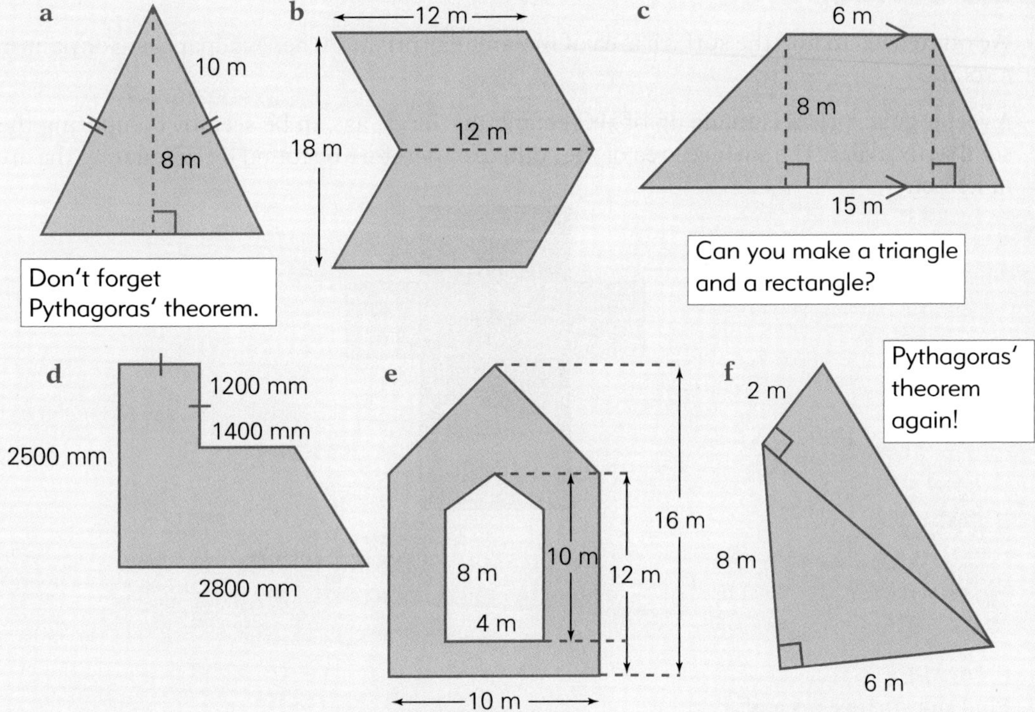

16 Find the area of the shaded face for each of the following solids:

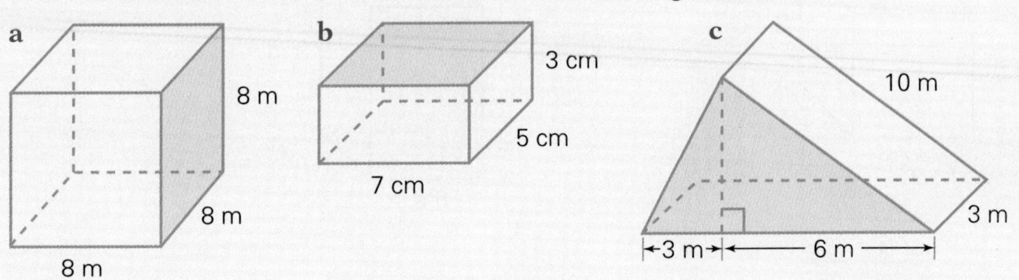

SPREADSHEET ACTIVITY

Try this

Given the right-angled triangle has a hypotenuse of 10 cm and one side is twice the length of the other, calculate the triangle's area, correct to the nearest square millimetre.

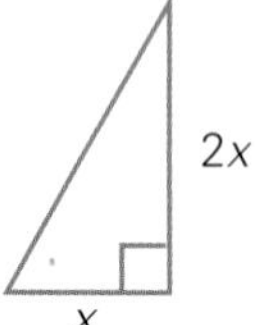

Investigation Money carpet

Imagine you are to cover the entire floor of your classroom in 20 cent pieces. How many would you need to cover the entire floor in one layer as shown?

Explain your reasoning.

8.3 Surface area

We often need to find the surface area of rectangular prisms, when wallpapering or painting a room.

A rectangular prism is made up of six rectangular faces, as can be seen by comparing the solid with its net. The surface area of the solid can therefore be found by calculating the area of its net.

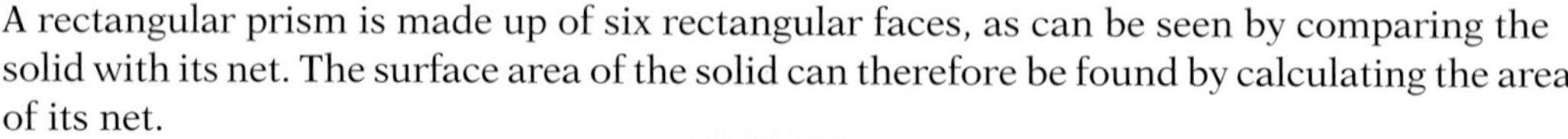

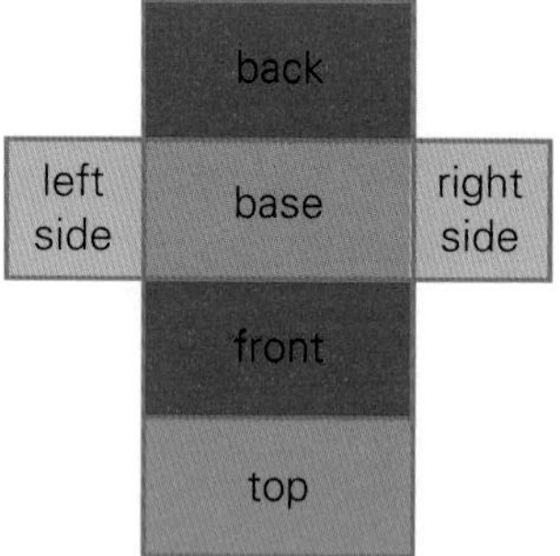

Surface area = the total area of all faces of the solid

EXAMPLE 1

Find the surface area of:

a

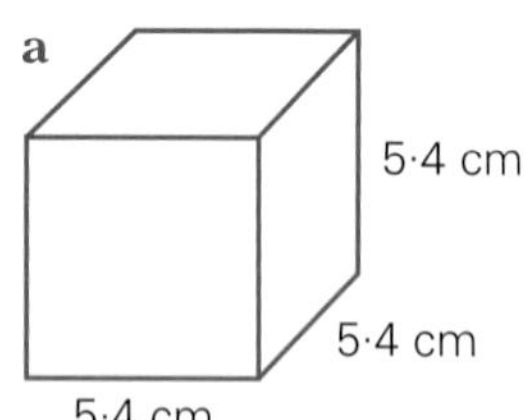

b

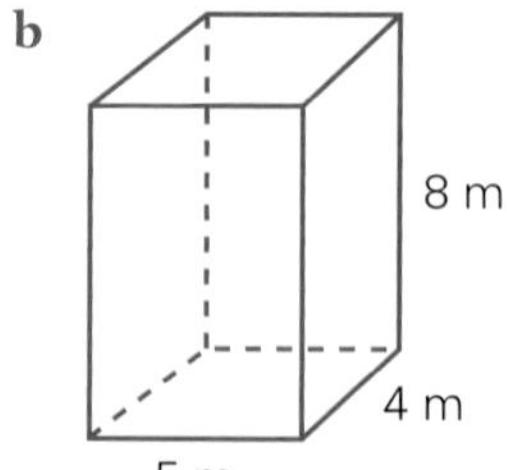

Solution

a Total surface area of a cube
= 6 × area of a square face
= $6 \times 5{\cdot}4^2$
= 174·96
∴ surface area is 174·96 cm^2

Rather than doing six individual faces, look for equal areas.

b Total surface area of rectangular prism
= (2 × area of front) + (2 × area of top) + (2 × area of side)
= (2 × 5 × 8) + (2 × 5 × 4) + (2 × 4 × 8)
= 184
∴ surface area is 184 m^2

EXAMPLE 2

Find the surface area of:

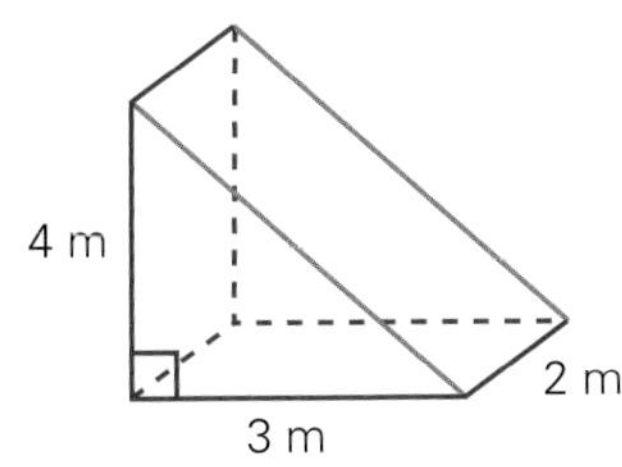

Solution

- Calculate the **hypotenuse** first.

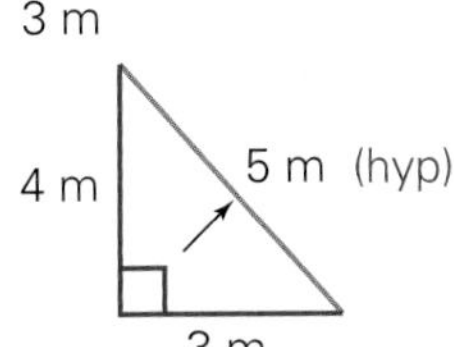

- Surface area of triangular prism
= (2 × area of front) + (area of base) + (area of right side) + (area of left side)
= $(2 \times \frac{1}{2} \times 4 \times 3) + (3 \times 2) + (5 \times 2) + (4 \times 2)$
= 12 + 6 + 10 + 8
= 36
∴ surface area is 36 m^2

EXERCISE 8C

1 Find the area of the shaded face in each of the following solids:

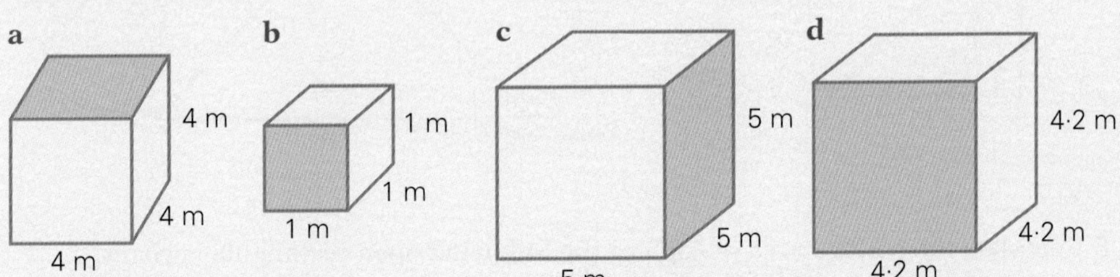

2 Calculate the surface area of each of the following prisms:

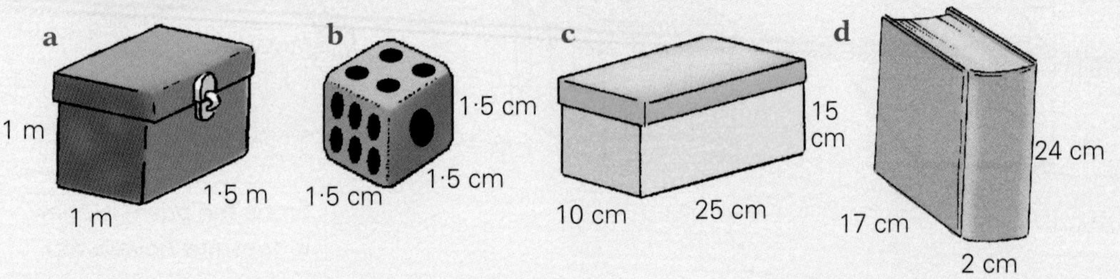

3 Calculate the surface area of each of the following closed rectangular prisms:

a

4 m
8 m
40 m

b

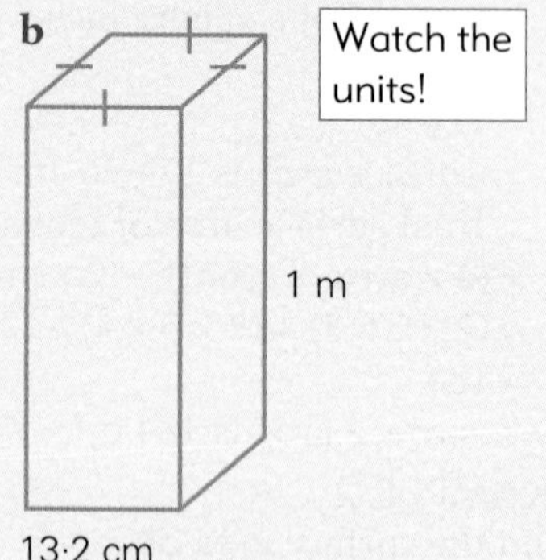

4 Find the surface area of the prisms below, given their net:

a

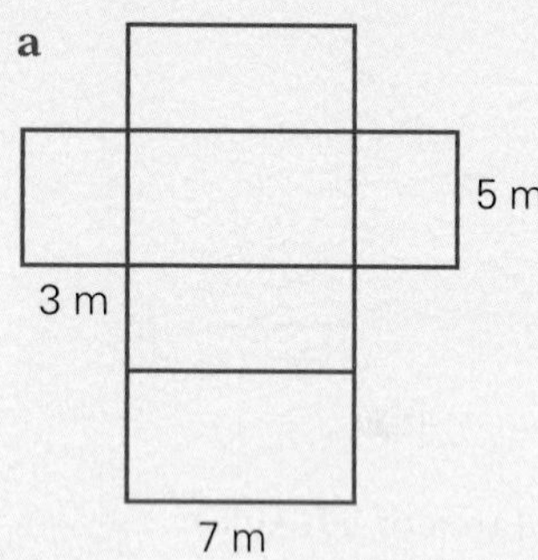

b

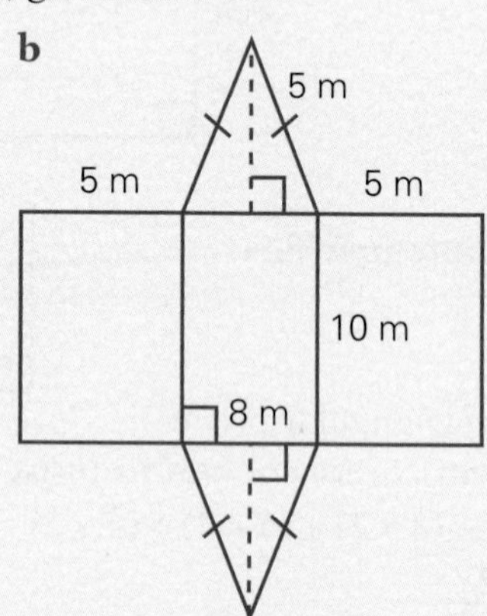

5 Find the surface area of the following triangular prisms:

a

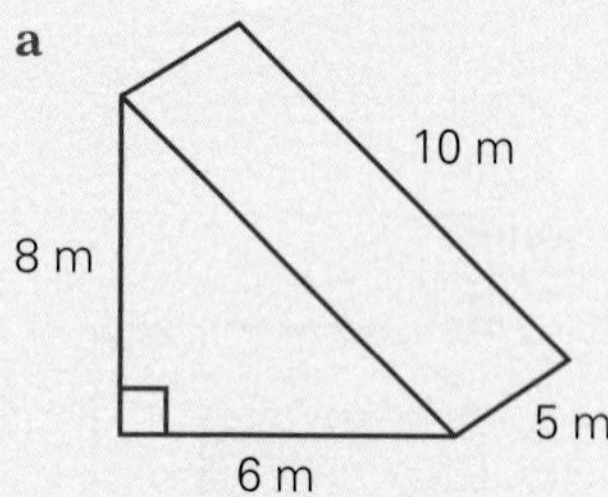

b

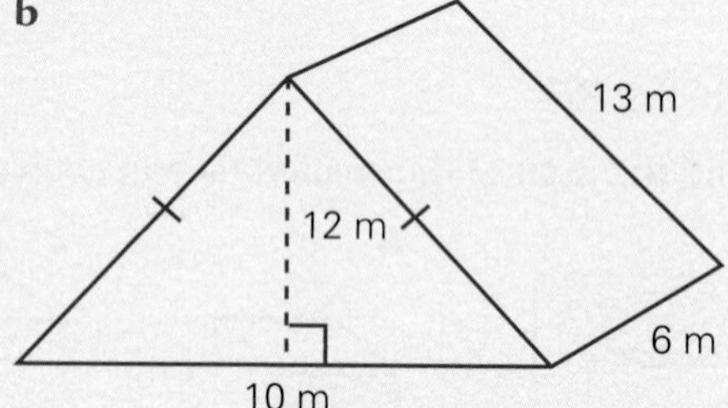

Level 2

6 Calculate the surface area of each of the following open rectangular prisms:

a

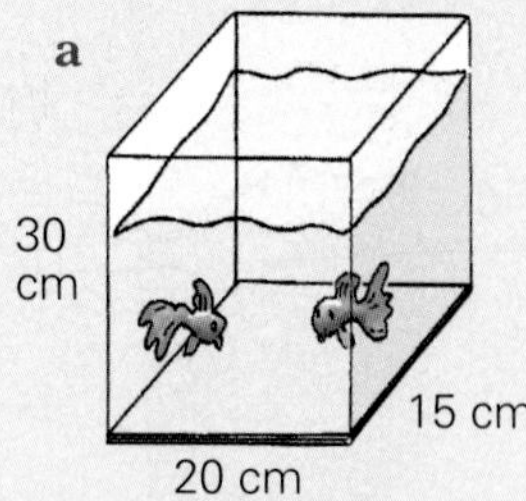

b

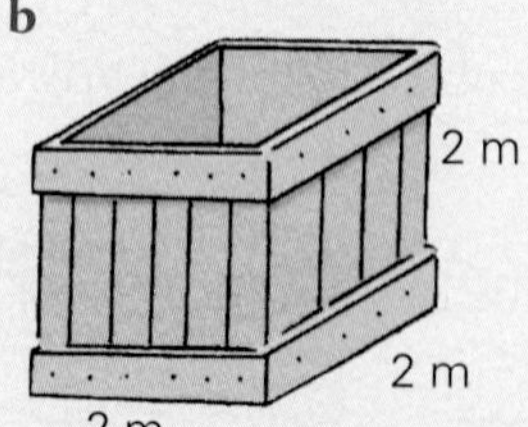

c

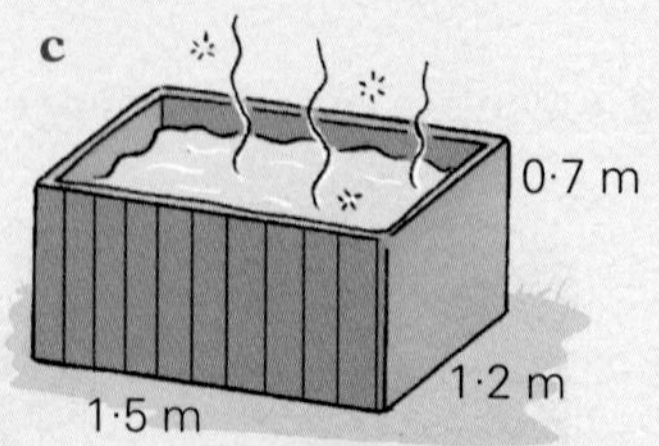

Since the prism is open, it does not have a top.

7 Find the surface area of:

a

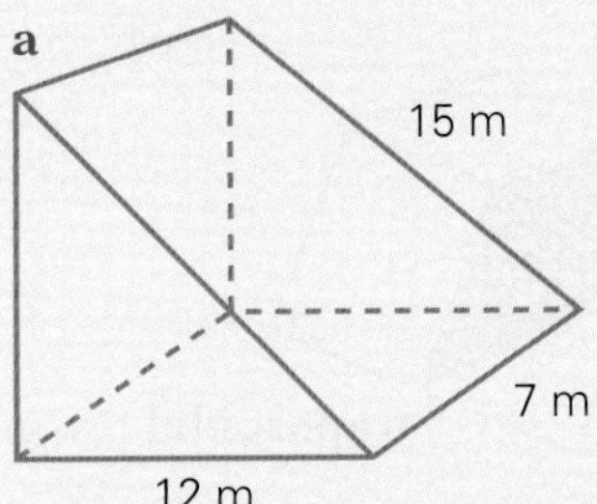

b

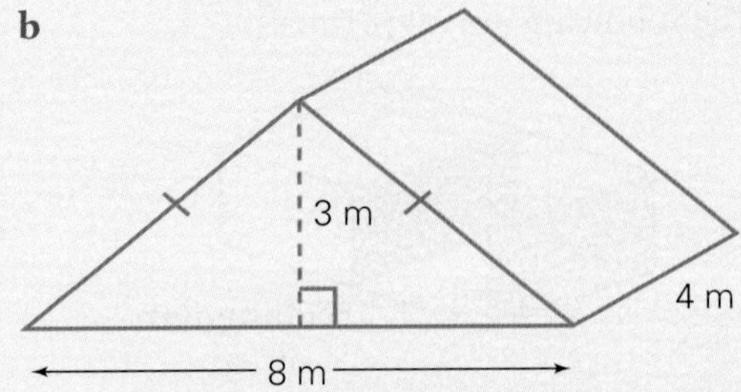

8 Find the total surface area of each of the following shapes:

a a rectangular pill box with dimensions of 30 mm × 2 cm × 10 mm

b a rectangular shipping container with dimensions of 1·8 m × 2·1 m × 2 m

c a package to be wrapped and sent overseas with dimensions of 0·3 m × 0·4 m × 56 cm

9 Thomas wants to make an open storage container for all his swap cards. If the container is to have a base of 30 cm × 45 cm and a height of 10 cm, what area of cardboard does Thomas need?

10 Allison wants to paint her son's bedroom walls, door and roof. The floor of the room is 2·5 m × 1·8 m and its height is 3 m. There is one window, 1 m^2, not being painted.

a Find the surface area to be painted.

b If 1 litre of paint covers 3 m^2, how many litres are needed to paint the bedroom?

8.4 Volume

Every three-dimensional figure has volume. The volume of a solid is the amount of space it occupies. To calculate a solid's volume we count the number of **cubic units** inside the shape. You may remember from last year that a solid that can be filled has both a volume and a capacity.

The capacity of a container is the amount of liquid it can hold, measured in litres, etc.

Container volume	Liquid capacity
1 cm^3	1 mL
1000 cm^3	1 L
1 000 000 cm^3 (1 m^3)	1 kL

holds **1 KL** of liquid

one cubic metre

100 cm

100 cm

100 cm

$$1 \text{ m}^3 = 1\,000\,000 \text{ cm}^3$$
$$= 1\,000\,000 \text{ mL}$$
$$= 1000 \text{ L}$$
$$= 1 \text{ kL}$$

You will recall that a prism is a three-dimensional shape with a uniform cross-section. That is, its end faces are polygons, parallel and the same shape and size.

The following are all prisms:

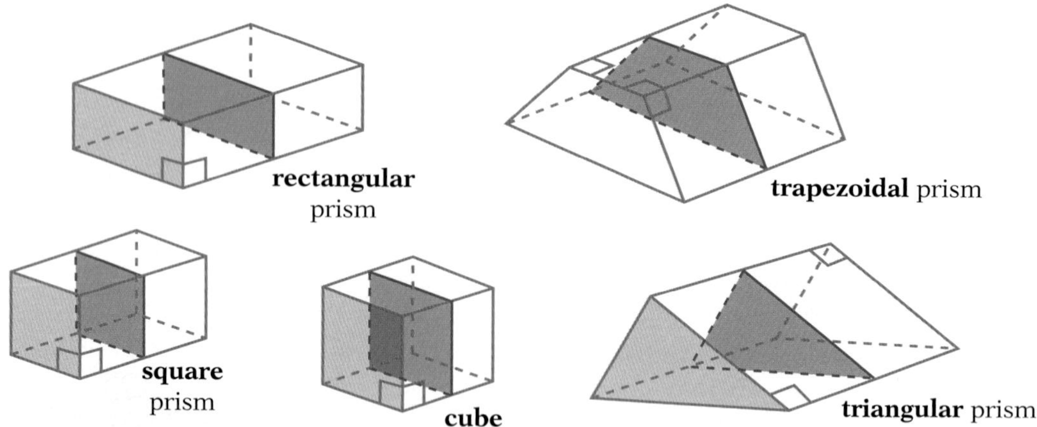

The prisms above are all right prisms. Right prisms have sides that are at right angles to the base or cross-section.

VOLUME OF RIGHT PRISMS

All right prisms may be thought of as made up of **layers of cubes**.

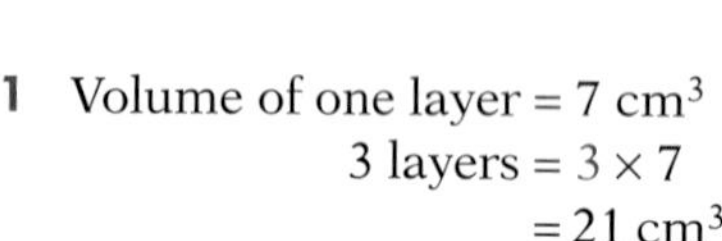

1 Volume of one layer = 7 cm^3
3 layers = 3×7
= 21 cm^3

2 1 layer = 10 cm^3
4 layers = 4×10
= 40 cm^3

3 1 layer = 4×5 = 20 cm^3
3 layers = 3×20
= 60 cm^3

In rectangular prism **3**, the top layer contains 4×5 or 20 cubes. Also, the area of the cross-section is 4×5 or 20 cm^2.

> Volume = area of cross-section $\times$ height

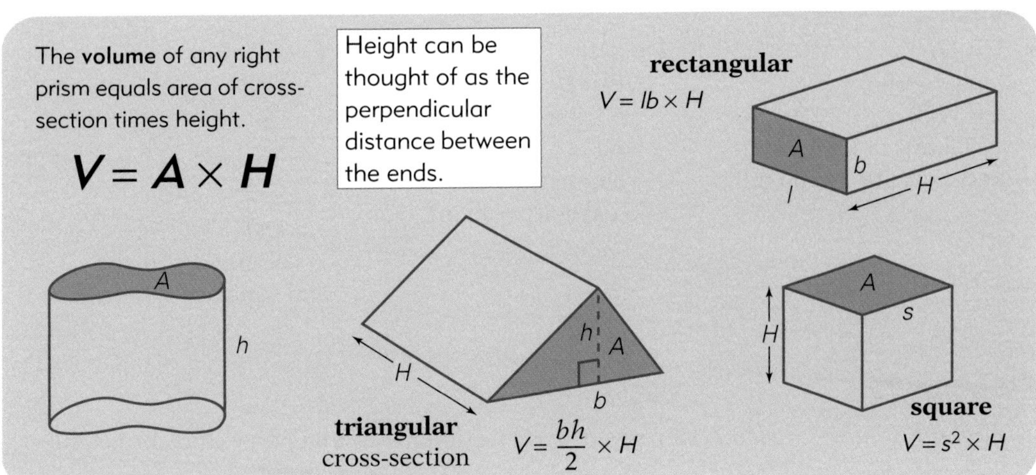

To convert units of volume, we consider:

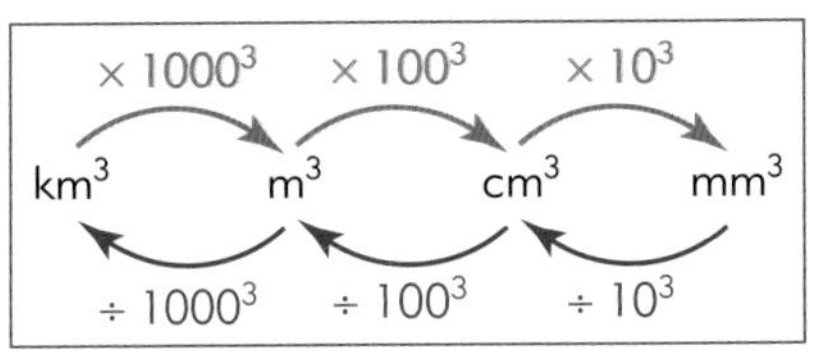

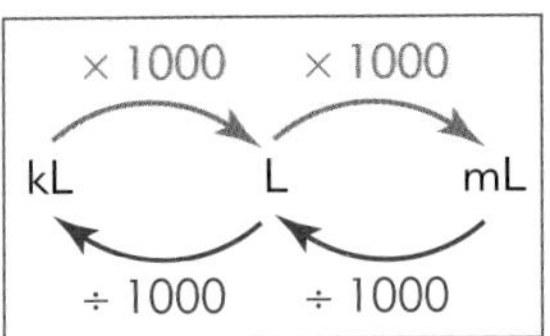

EXAMPLE 1

Complete the following conversions:

a $2{\cdot}7\ \text{m}^3 =$ ________ cm^3 **b** $7000\ \text{mm}^3 =$ ________ cm^3 **c** $7\ \text{L} =$ ________ cm

Solution

a $2{\cdot}7\ \text{m}^3$
$= 2{\cdot}7 \times 100^3\ \text{cm}^3$
$= 2{\cdot}7 \times 1\ 000\ 000\ \text{cm}^3$
$= 2\ 700\ 000\ \text{cm}^3$

(× 100³: m³ → cm³)

b $7000\ \text{mm}^3$
$= 7000 \div 10^3\ \text{cm}^3$
$= 7000 \div 1000\ \text{cm}^3$
$= 7\ \text{cm}^3$

(÷ 10³: mm³ → cm³)

c 7 L
= 7000 mL
$= 7000\ \text{cm}^3$

(× 1000: L → mL)

1 mL = 1 cm^3

EXAMPLE 2

Find the volume and capacity, in millimetres, of the following prisms:

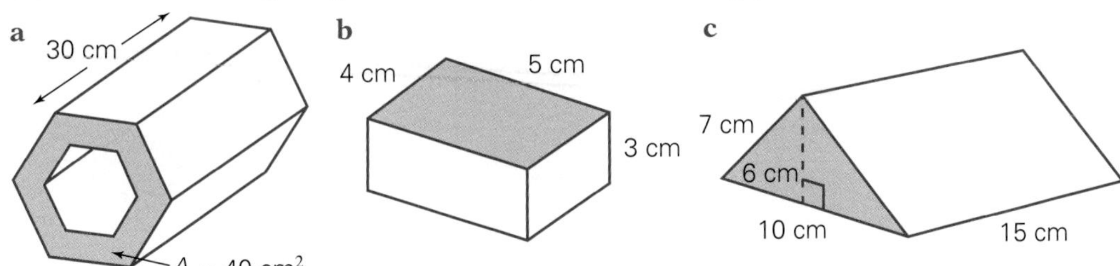

Solution

a $V = AH$
$= 40 \times 30$
$= 1200$
$\therefore$ volume is 1200 cm^3
and capacity = 1200 mL

b $V = AH$
$= lb \times H$
$= 5 \times 4 \times 3$
$= 60$
$\therefore$ volume is 60 cm^3
and capacity = 60 mL

c $V = AH$
$= \frac{bh}{2} \times H$
$= \frac{10 \times 6}{2} \times 15$
$= 450$
$\therefore$ volume is 450 cm^3
and capacity = 450 mL

EXERCISE 8D

1 Find the volume of each prism, given cross-sectional area and perpendicular height:

a

$A = 20$ cm^2

4 cm

b

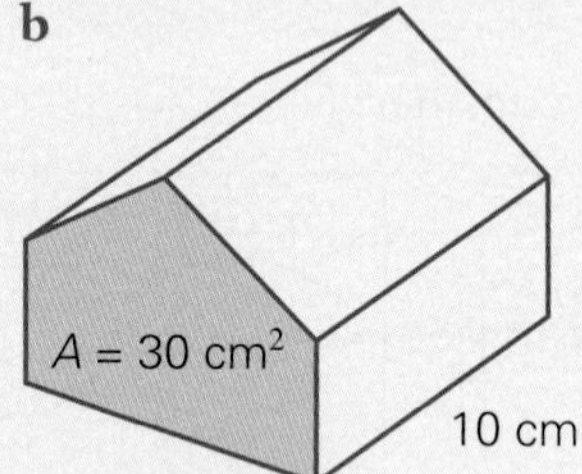

c

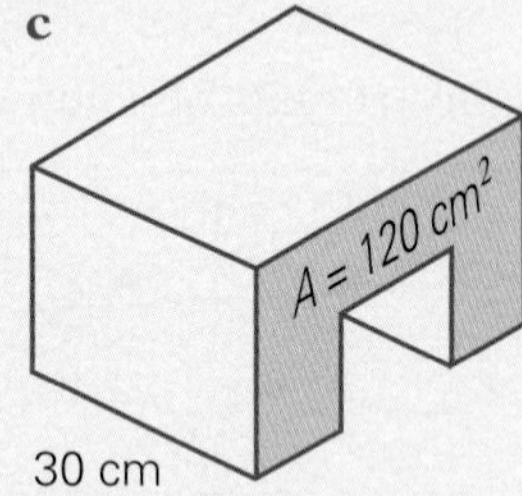

d

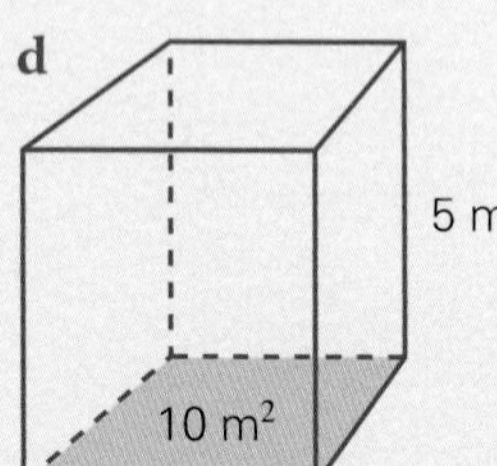

e

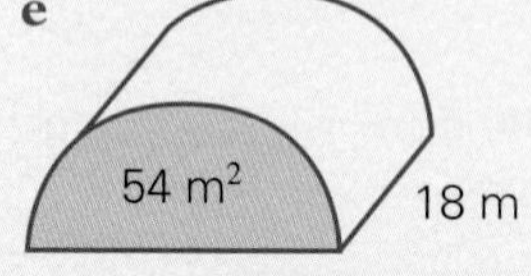

f

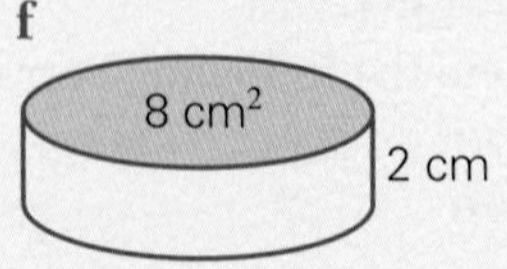

g

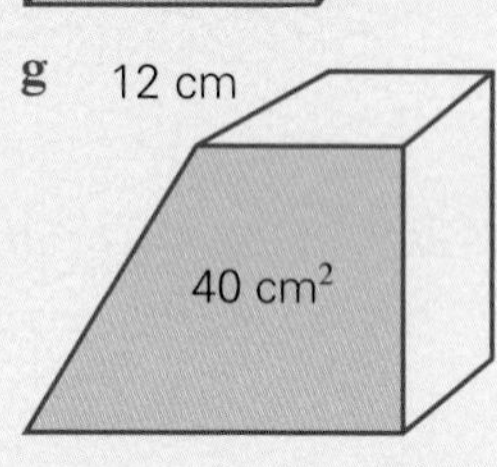

h

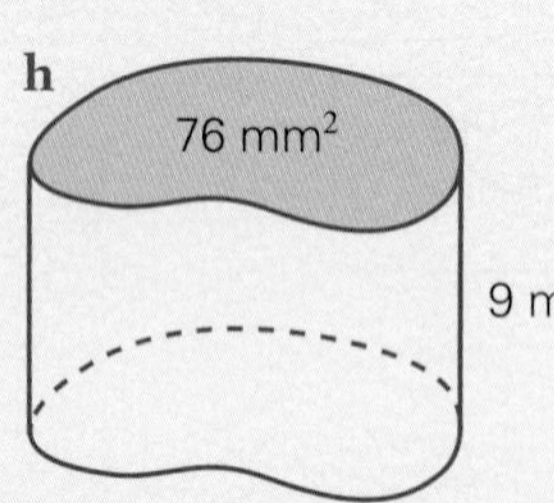

i

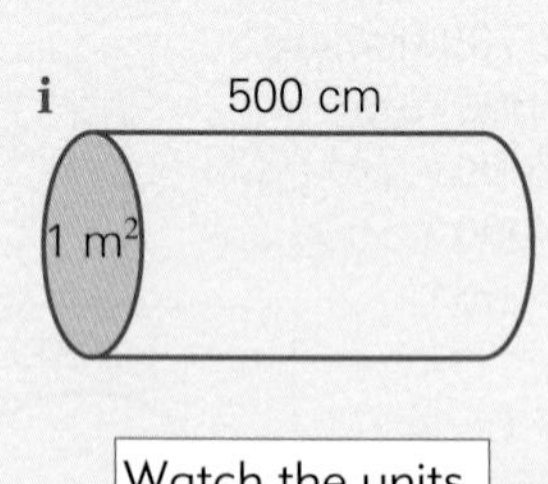

Watch the units.

2 The bases of the following prisms are shaded. Calculate the volume of each of the prisms shown below.

a

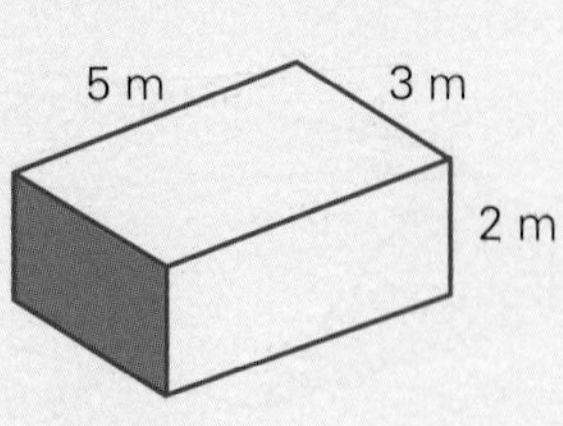

b

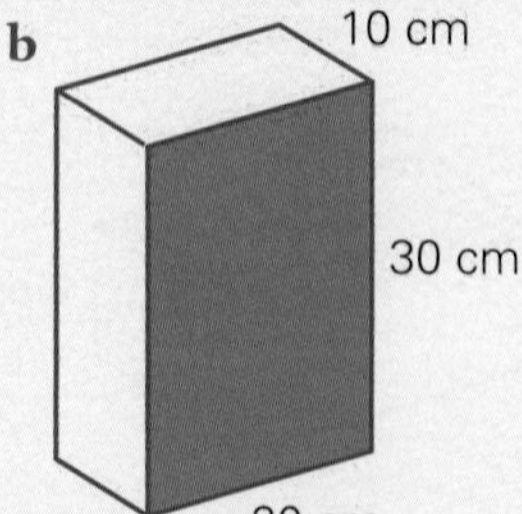

c

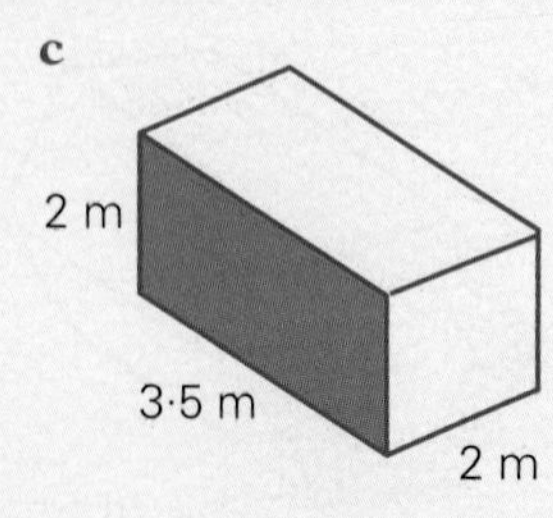

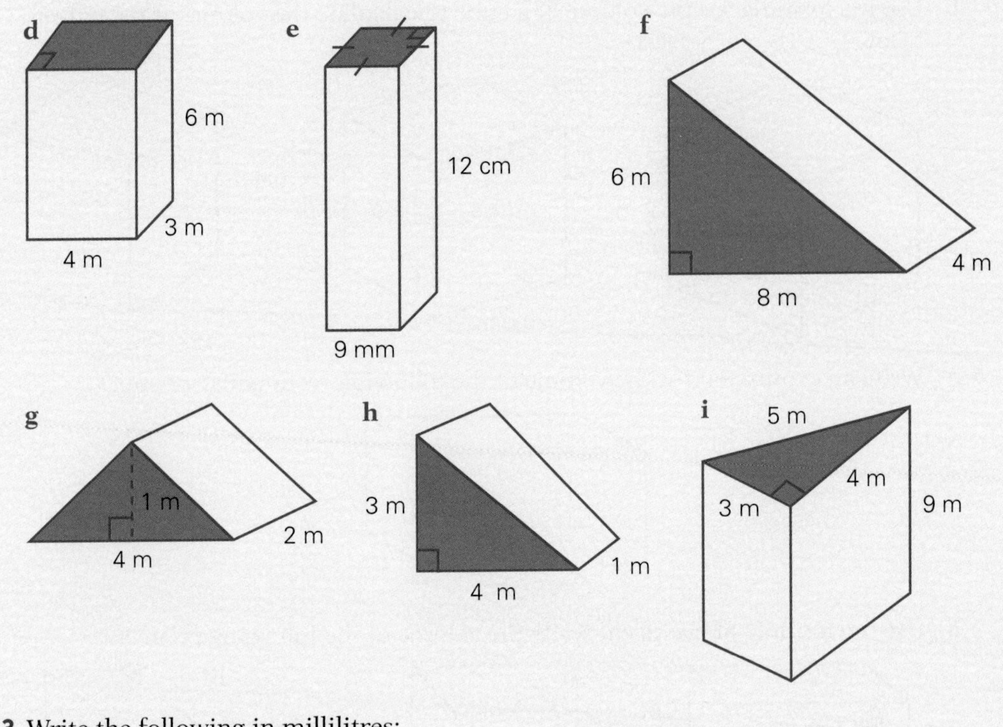

3 Write the following in millilitres:

a 6 L **b** 15 cm^3 **c** 8000 cm^3 **d** $\frac{3}{4}$ L

4 Find the capacity of the following prisms, in litres:

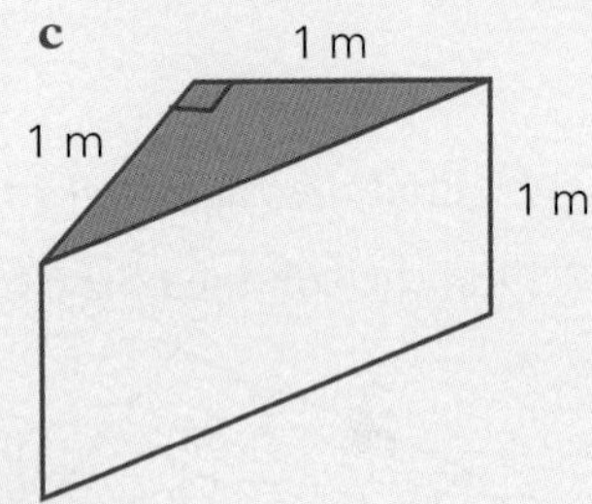

Level 2

5 a Given the sides of a cube are all equal, write a formula for the volume of:

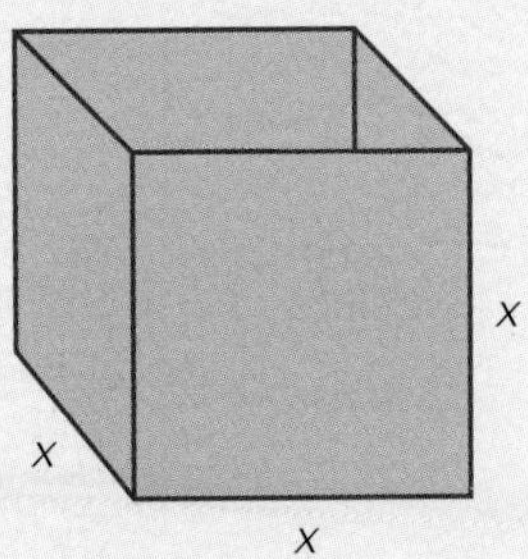

b Use the formula for the volume of a cube to calculate the volume of the cubes below:

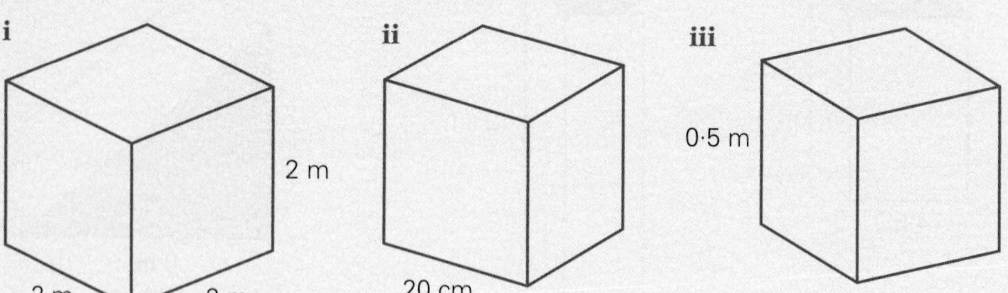

6 a Write an expression for the volume of the following rectangular prism:

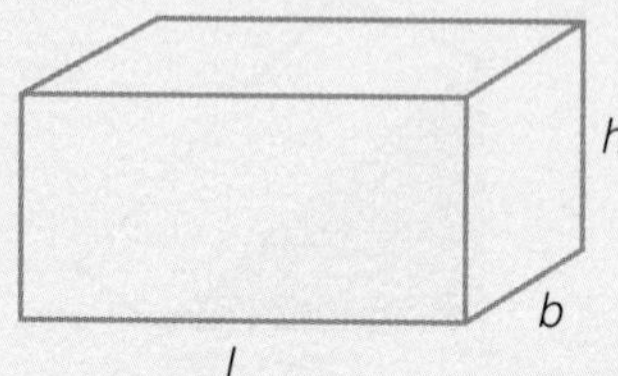

b Use the formula above to calculate the volume of the following prisms:

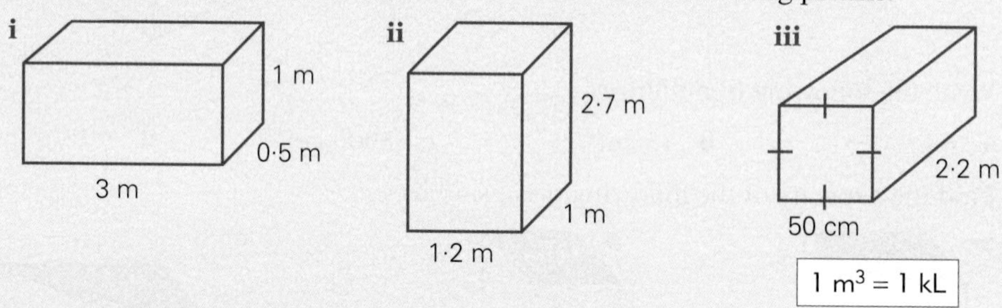

c Find the capacity of the prisms above, in kilolitres.

7 Find the volume of the following prisms:

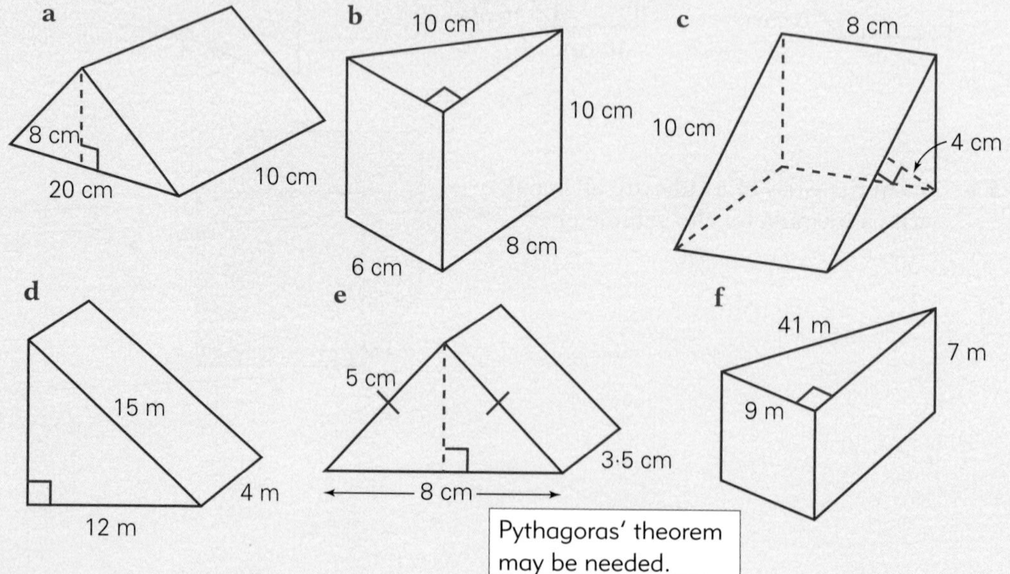

8 First find the area of each composite cross-section, then find the volume of the prism:

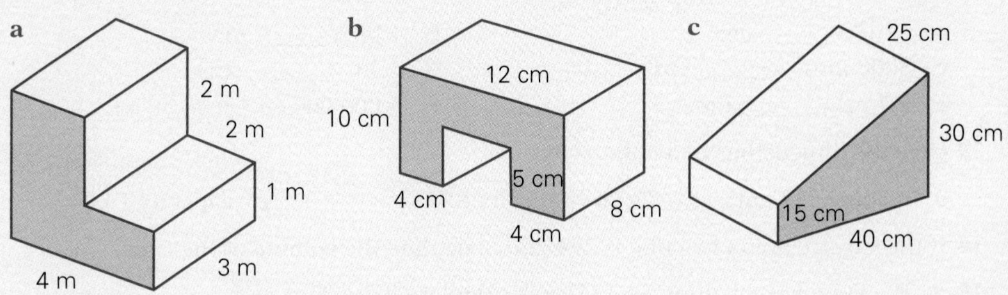

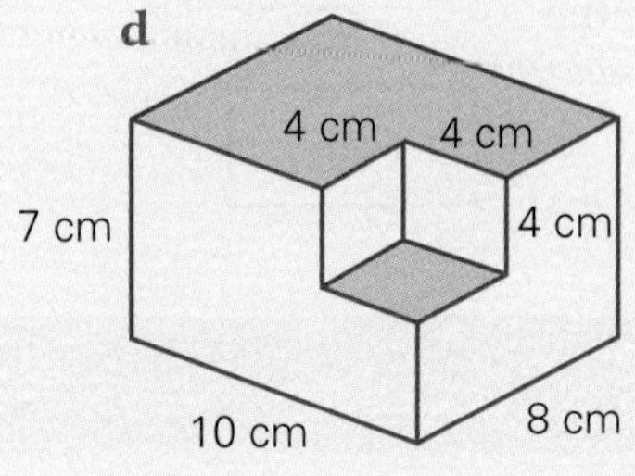

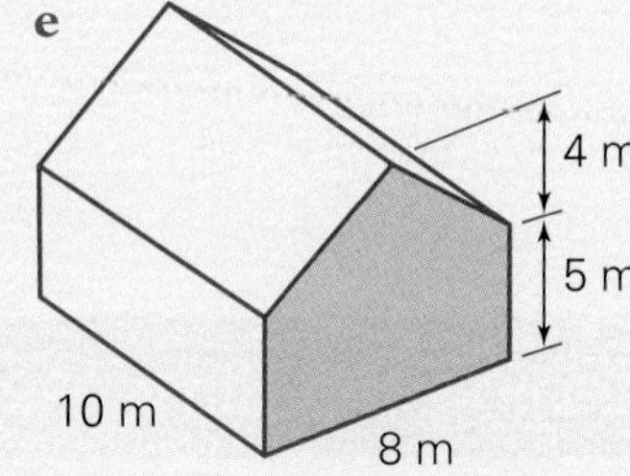

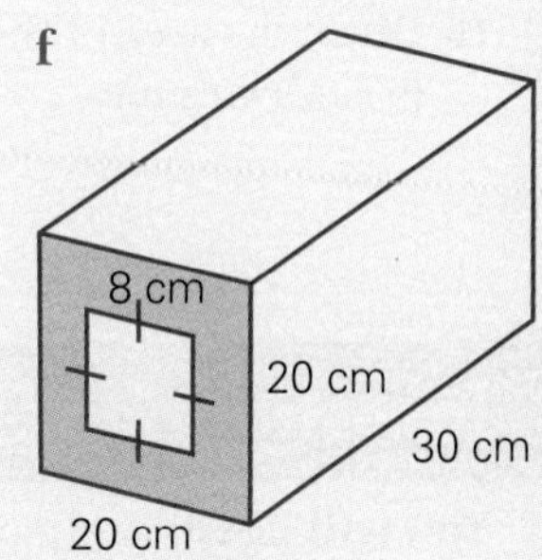

9 This dump truck transports oil, sand or gravel. Find:

a the volume of its tray
b the volume that can be carried if heaping up the load adds 10% to its capacity
c the number of heaped truckloads needed to deliver 90 m^3 of gravel

10 A square prism of volume 360 cm^3 has a height of 10 cm.

a What is the area of the cross-section?
b What is the side of the square base?

11 Find the capacity in litres of each container.

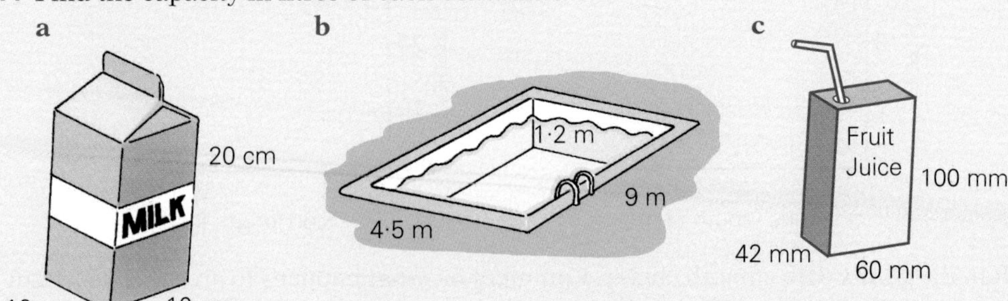

12 Copy and complete:

a $2\text{ m}^3 = \_\_\_\_\_\_ \text{ cm}^3$

b $1\text{ kL} = \_\_\_\_\_\_ \text{ m}^3$

c $8000\text{ mm}^3 = \_\_\_\_\_\_ \text{ cm}^3$

d $\frac{1}{5}\text{ kL} = \_\_\_\_\_\_ \text{ cm}^3$

e $0{\cdot}7\text{ m}^3 = \_\_\_\_\_\_ \text{ mm}^3$

f $5\,000\,000\text{ cm}^3 = \_\_\_\_\_\_ \text{ kL}$

13 Give the dimensions of a cube with:

a volume 512 cm^3 b capacity 8 L c capacity 1 kL

14 If the surface area of a cube is 294 cm^2, calculate the volume of the cube.

15 a If a cube has a volume of 1331 m^3, calculate its surface area.

b Would this cube fit into your classroom?

SPREADSHEET ACTIVITY

16 Given the area of the prism's faces are as shown, calculate the prism's volume.

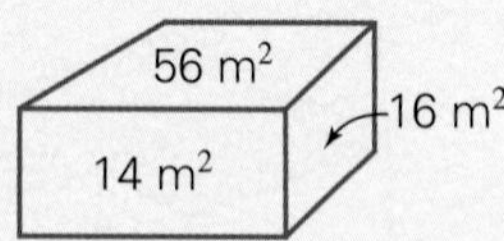

Investigation Changing the dimensions of a cube

You will need: centicubes or unifix blocks, pencil, graph paper

1 In small groups, construct cubes of sides 1, 2, 3, 4, 5, … cm.

2 Copy and complete the chart:

Side length (cm)	Area of face (cm^2)	Volume (cm^3)	Relationship length/area	Relationship length/volume
1	1	1	1/1	1/1
2	4	8	2/4	2/8
etc.				

3 Use the data from the chart to copy and complete these graphs.

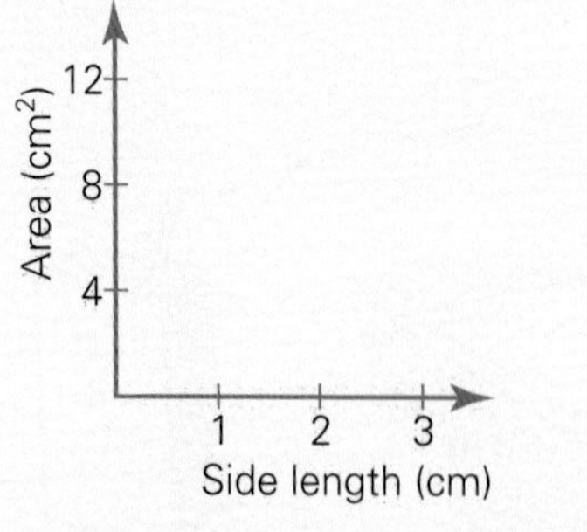

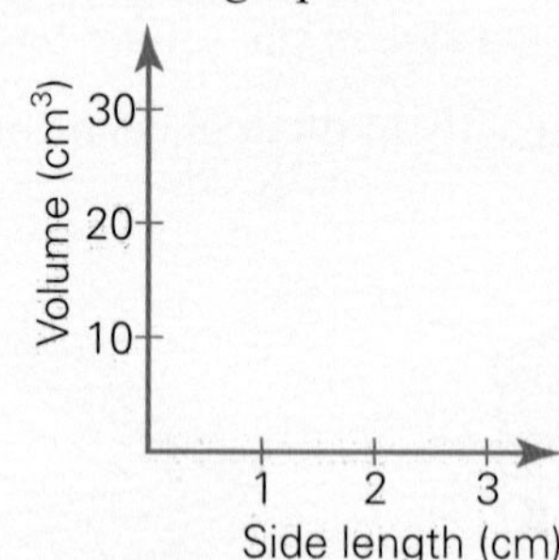

Join the points with smooth curves. Comment on what happens to area as side length increases. Repeat for volume.

Write a sentence explaining the overall relationship between **a** length and area **b** length and volume.

Chapter Review

Language Links

accuracy
altitude
area
breadth
capacity
composite
conversions
cross-section
cubic
dimensions
error
hectare
height
kilolitre
length
limits
measurement
perpendicular
prism
surface area
volume
width

1 List the names of the three dimensions of a rectangular prism.

2 What is the name given to a rectangular prism with the three dimensions equal?

3 **Describe** the difference between the capacity and the volume of a solid.

4 Write another name for the altitude of a shape.

5 What is true about all cross-sections of right prisms?

6 Write a sentence explaining why all measurements are approximations.

7 Write a **procedure** for finding the following:

a the limits of accuracy of a measurement
b the surface area of any prism
c the volume of any prism

8 Write the meaning, in words, of the following formulas:

a $A = \frac{1}{2}hb$ b $A = 6s^2$ c $V = s^3$ d $V = lbh$

9 Explain the need for the unit kilolitre. What is it used for?

Chapter Review Exercises

1 Convert measurements as shown: MC 8.2

a 3600 mm = ______ m
b 1·75 L = ______ mL
c 2·04 kg = ______ g
d 510 m = ______ km
e 4250 kg = ______ t
f 208 cm = ______ m
g 2·3 m^3 = ______ L
h 15 000 m^2 = ______ ha
i 0·02 L = ______ cm^3
j 1 m^2 = ______ cm^2
k 4 cm^3 = ______ mm^3
l 7000 mm^2 = ______ cm^2

2 Find **i** the perimeter and **ii** the area of each figure: 8.2

a
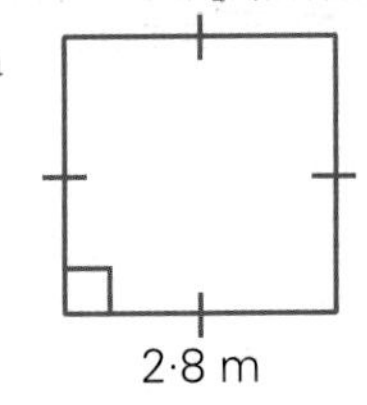

b
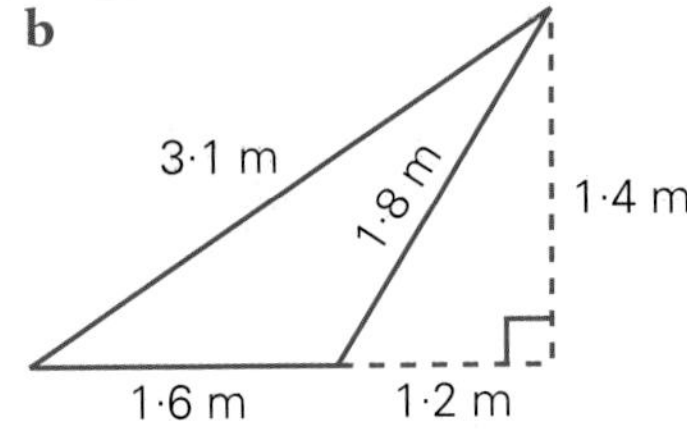

c
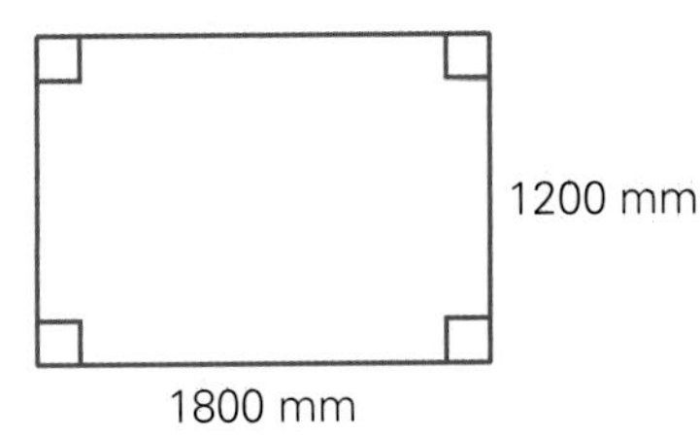

d

2 m
90 cm
2 m
160 cm

e

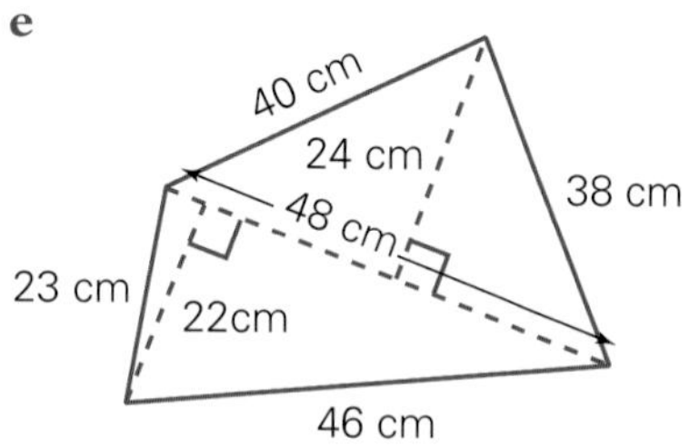

f

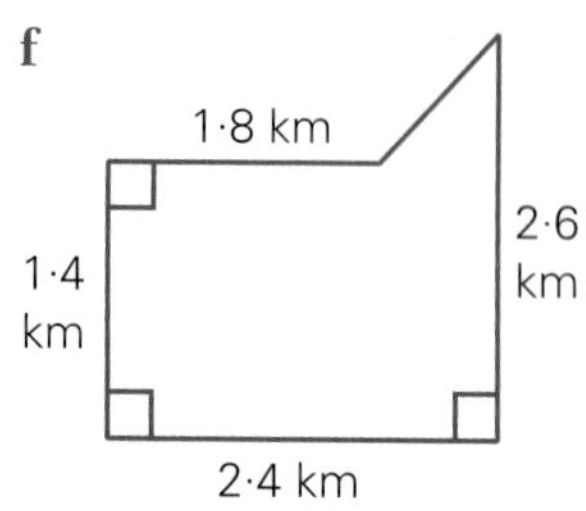

8.2 **3** Calculate the shaded area for:

a

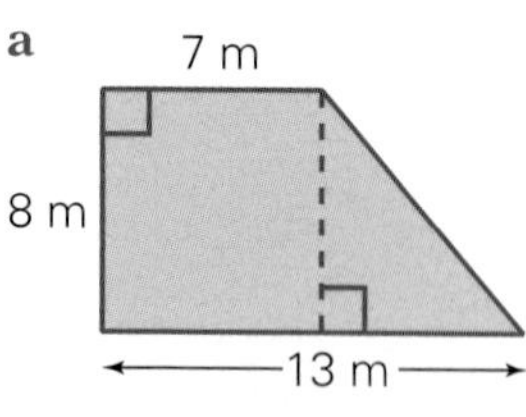

b

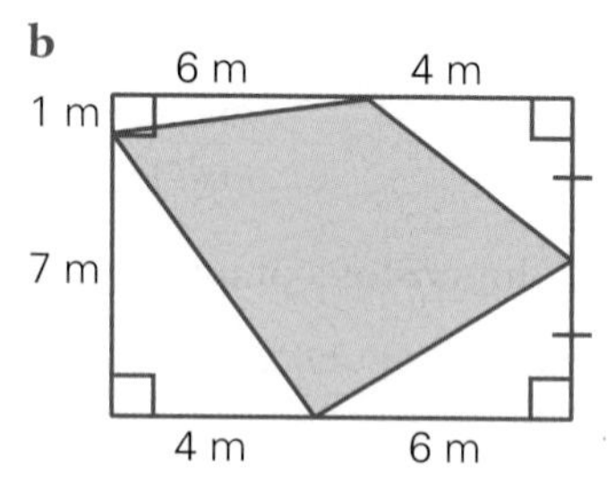

c

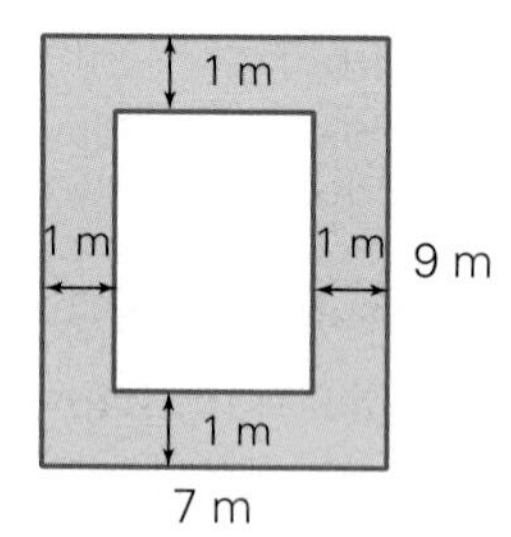

8.1 **4** Give the limits of accuracy for the following measurements:

a 7 m   **b** 7·4 m   **c** 7·44 m

8.3 **5** Find the surface area of the following prisms:

a

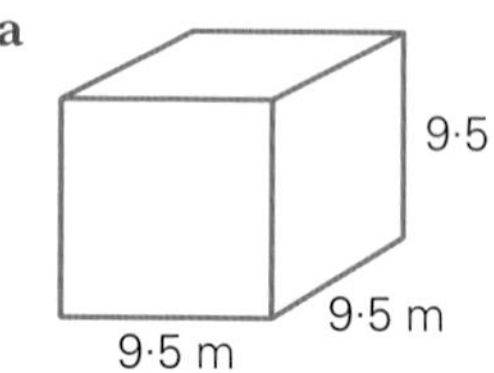

b

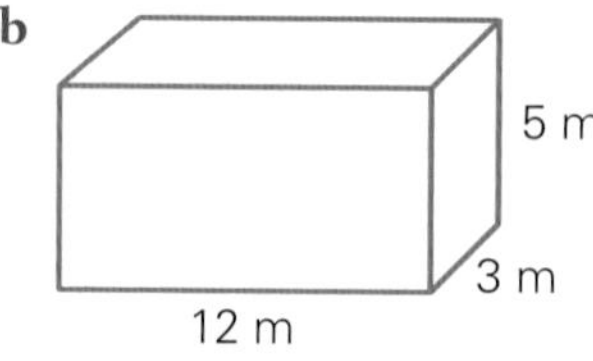

c

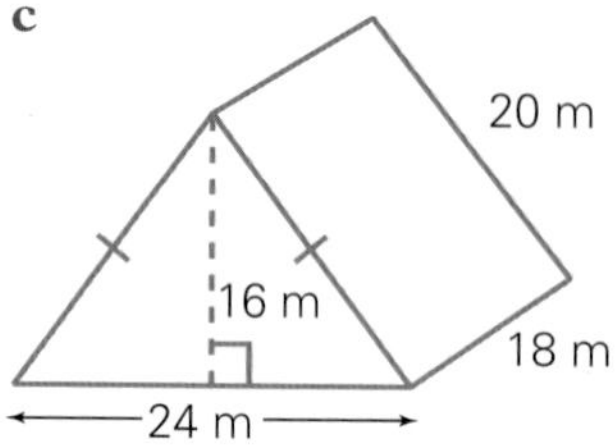

8.4 **6** Calculate the volume of each prism:

a

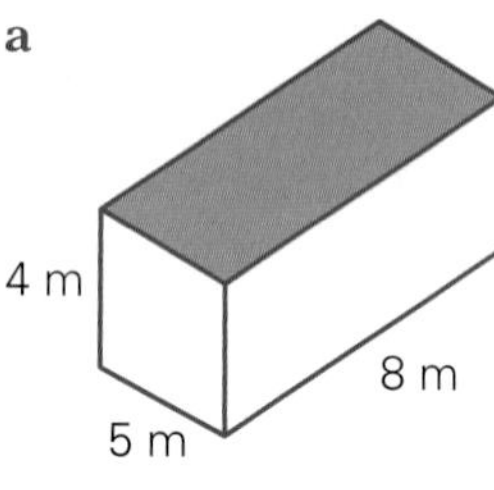

b

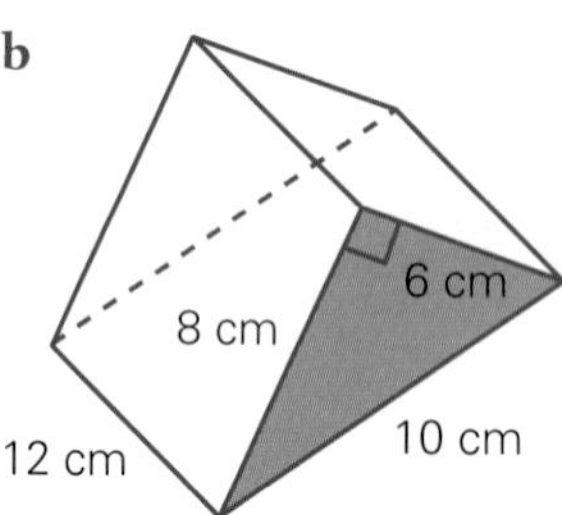

c

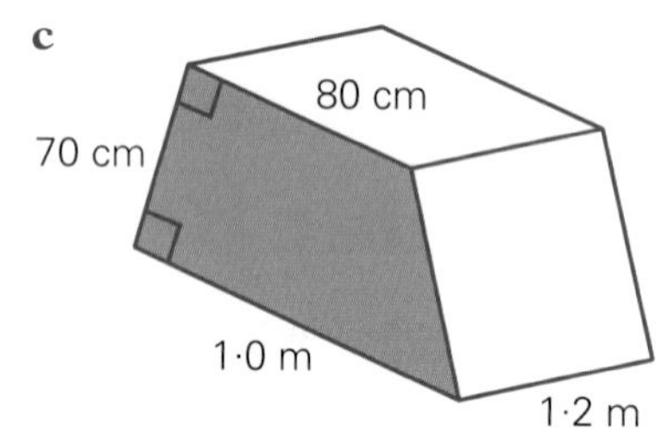

8.4 **7** A 30 m by 20 m rectangular lawn is top dressed with 7 m^3 of soil. What is the depth of soil if it is spread evenly over the lawn, to the nearest centimetre?

8.4 **8** An open wooden box is constructed as shown. Calculate the box's:

a volume, in cubic metres
b capacity in kilolitres
c surface area of the outside sides and floor

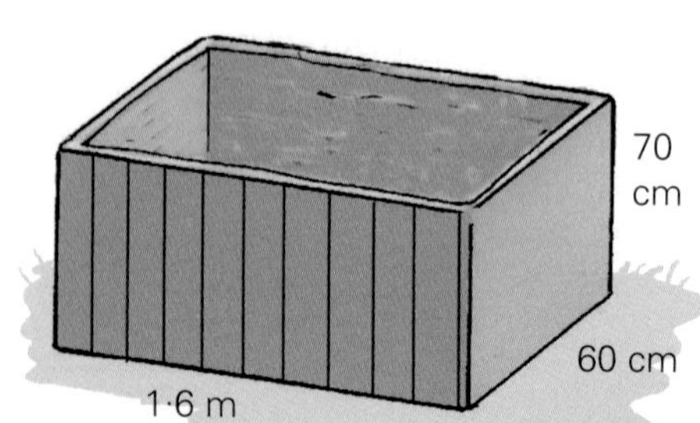

Keeping Mathematically Fit

Part A: Non-calculator

1 Evaluate $0{\cdot}7 \times 0{\cdot}4$.

2 If Claire is 10 years old now, how old will she be in x years?

3 Find $17 - 9{\cdot}605$.

4 Convert 9·8 km to metres.

5 If 10 books cost $\$x$, what is the cost of y books?

6 Simplify $2\frac{1}{4} \times \frac{4}{5}$.

7 Find 200% of \$1.50.

8 If $x = {}^{-}4$, find the value of $3x^3$.

9 Find the value of $9 \times 8 - 42 \div 7$.

10 If $0{\cdot}63 \times 0{\cdot}25 = 0{\cdot}1575$, find the value of $6{\cdot}3 \times 250$.

Part B: Calculator

1 Find the value of x in:

a

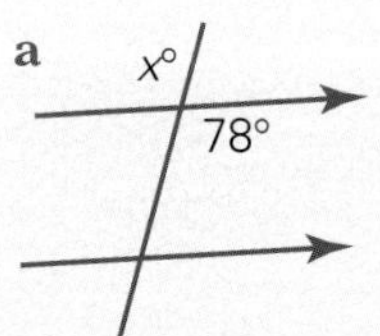

b

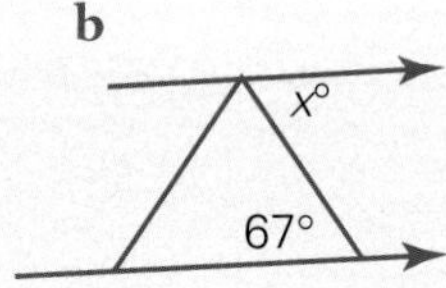

c

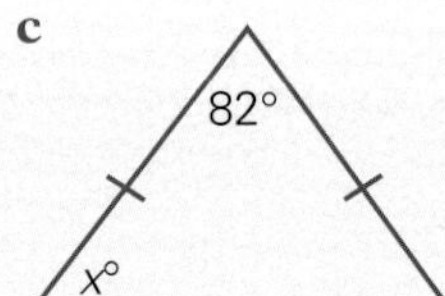

d

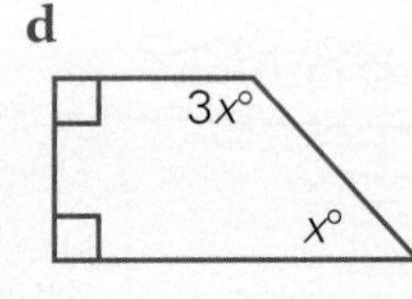

2 Calculate the area of:

a

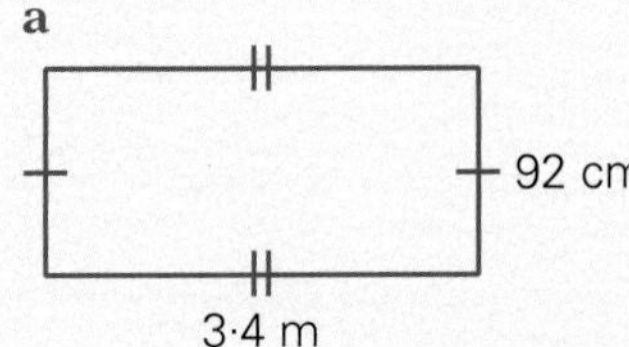

b

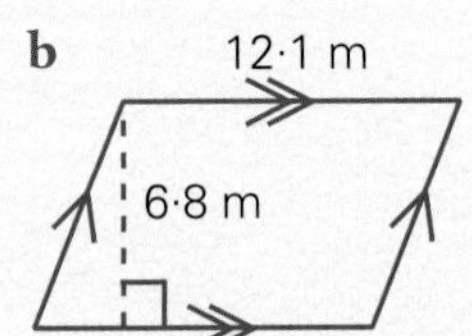

c

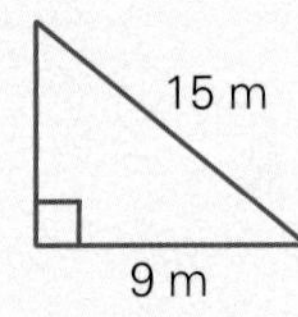

3 Find the value of $3\sqrt{\dfrac{4{\cdot}06}{3{\cdot}1 - 2{\cdot}4^2}}$ correct to 2 decimal places.

4 Margaret is three times as old as her daughter Sarah. In 10 years' time the sum of their ages will be 68. How old was Margaret when Sarah was born?

5 Expand and simplify:

a $5(2x - 1) + 5$ **b** $\frac{1}{2}(2x + 6) + x + 1$ **c** $3(2x + 4) - 2(x + 1)$

6 Evaluate $xy^2 + y$, where $x = 2$ and $y = {}^{-}3$.

7 Solve $4 + 3x \geq 13$.

8 The perimeter of a square was measured as 48·8 cm. What could the length of each side have been?

TEACHER

1 What you need to know and revise

Outcome NS 3.4:
Compares, orders and calculates with decimals, simple fractions and simple percentages

Outcome NS 4.3:
Operates with fractions, decimals, percentages, ratio and rates

2 What you will learn in this chapter

Outcome NS 4.3:
Operates with fractions, decimals, percentages, ratio and rates

- uses ratio to compare quantities of the same type
- writes ratios in various forms
- simplifies ratios
- applies the unitary method to solve ratio problems
- divides into a given ratio
- interprets and calculates ratios that involve more than two numbers
- calculates speeds given distance and time
- calculates rates from given information

Working Mathematically outcomes WMS 4.1–4.5
Students will be asked to ***question***, ***apply strategies***, ***communicate***, ***reason*** and ***reflect*** in the sections of this chapter.

Ratio and rates

Key mathematical terms you will encounter

average	equivalent	quantity	slope
comparison	gradient	rate	speed
distance	ordered	ratio	

MathsCheck
Ratio and rates

1 Express each fraction in simplest form:

a $\frac{2}{4}$ b $\frac{6}{9}$ c $\frac{8}{12}$ d $\frac{15}{20}$ e $\frac{45}{60}$

2 Write in the missing number to form a pair of equivalent fractions:

a $\frac{1}{2} = \frac{\square}{6}$ b $\frac{2}{5} = \frac{4}{\square}$ c $\frac{5}{15} = \frac{\square}{3}$ d $\frac{18}{24} = \frac{3}{\square}$

3 Express the first quantity as a fraction of the second:

a 50 cm, 1 m b 15 min, 1 h c 750 g, 1·25 kg

4 Find:

a $\frac{2}{3}$ of \$30 b $\frac{3}{5}$ of 100 kg c $\frac{3}{4}$ of 80 m

5 The cost of 5 kg of apples is \$19.95. Find the cost of:

a 1 kg b 3 kg c $7\frac{1}{2}$ kg

6

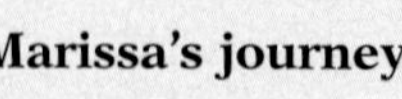

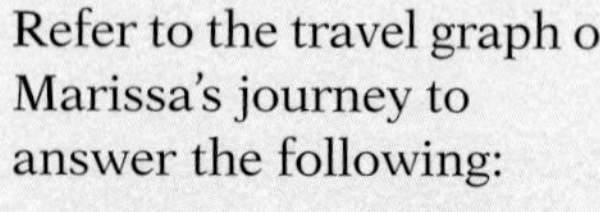
Refer to the travel graph of Marissa's journey to answer the following:

a At what time did Marissa start her journey?
b At what time did Marissa first stop and for how long?
c What do the horizontal sections of the graph represent?
d How far did Marissa travel between her first stop and her second stop?
e How far did Marissa travel during her journey, as shown on the graph?
f Did Marissa return to her starting point? How can you tell by looking at the graph?

Exploring New Ideas

9.1 Introducing ratios

SPREADSHEET ACTIVITY

In mathematics, two things can be compared by division and when this occurs it is called a ratio.

> Ratios are used to compare quantities of the same units in a definite order.
>
> A ratio is written as $a : b$ or $\frac{a}{b}$.

- A ratio does not have units because they are the same and cancel out.
- As a ratio is a fraction, we can simplify ratios in the same way we simplify fractions; that is, divide by the HCF of both/all quantities.

In a recipe for chocolate cake there are 60 g of butter, 250 g of flour and 15 g of cocoa. We can use ratios to **compare** the number of grams of flour with the number of grams of butter.

The ratio of flour to butter is 250 : 60 or $\frac{250}{60}$. This ratio can be **simplified**.

$$\begin{aligned} &250 : 60 \\ &\div 10 \quad \div 10 \\ &= 25 : 6 \end{aligned}$$

Similarly, the ratio of cocoa to flour = 15 : 250

$$\begin{aligned} &\div 5 \quad \div 5 \\ &= 3 : 50 \end{aligned}$$

We can combine the above to form one ratio.

Quantity of butter to the quantity of flour to the quantity of cocoa

$$\begin{aligned} &= 60 : 250 : 15 \\ &\div 5 \quad \div 5 \quad \div 5 \\ &= 15 : 50 : 3 \end{aligned}$$

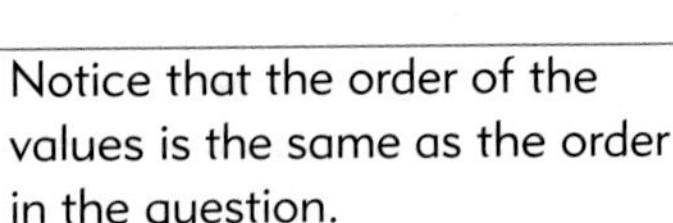
Notice that the order of the values is the same as the order in the question.

EXAMPLE 1

A class consists of 13 girls and 15 boys. Write the ratio of:

a girls to boys **b** boys to girls **c** boys to the class total

Solution

a Girls to boys = 13 : 15 **b** Boys to girls = 15 : 13 **c** Boys to class total = 15 : 28

EXAMPLE 2

Simplify the ratio 9 : 12 in simplest terms.

Solution

$$\begin{aligned} 9 : 12 &= 3 \times 3 : 3 \times 4 \\ &\div 3 \quad \div 3 \\ &= 3 : 4 \text{ or } \frac{3}{4} \end{aligned}$$

Divide by the HCF.

EXAMPLE 3

Simplify the following ratios:

a

to

b 2·5 to 4·5   **c** $3\frac{1}{2} : 1$ **d** 45 min to $1\frac{1}{4}$ h

Solution

a

30c to \$1 — Notice the order.

30c to 100c — Write with common units.

$= 30 : 100$ — Write as a ratio.

$\div 10 \quad \div 10$ — Divide by HCF.

$= 3 : 10$

b

2·5 to 4·5 — Multiply by 10 to remove all decimals.

$= 2{\cdot}5 : 4{\cdot}5$

$\times 10 \quad \times 10$

$= 25 : 45$ — Divide by HCF.

$\div 5 \quad \div 5$

$= 5 : 9$

c

$3\frac{1}{2} : 1$

$\times 2 \quad \times 2$ — Multiply by LCM of denominators.

$= 7 : 2$

d

45 min to $1\frac{1}{2}$ h — Write with common units.

45 min to 75 min

$= 45 : 75$ — Write as a ratio.

$\div 15 \quad \div 15$ — Simplify.

$= 3 : 5$

EXERCISE 9A

1 The following symbol uses squares of purple and green. Write the ratio of:

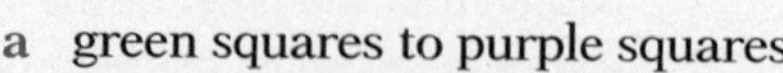

- **a** green squares to purple squares
- **b** green squares to total squares
- **c** total squares to purple squares

2

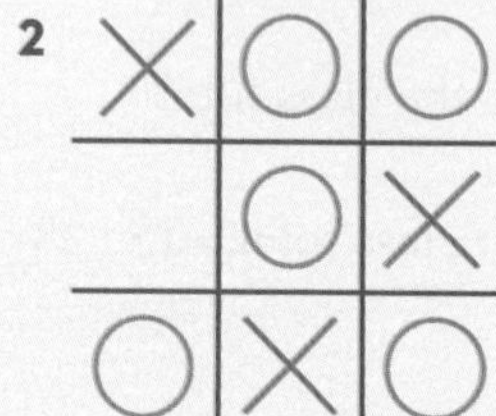

- **a** Write the ratio of noughts to crosses.
- **b** What is the ratio of empty squares to crosses?

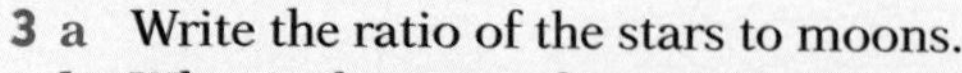

3
- **a** Write the ratio of the stars to moons.
- **b** What is the ratio of moons to stars?
- **c** Find the ratio of stars to all heavenly bodies.

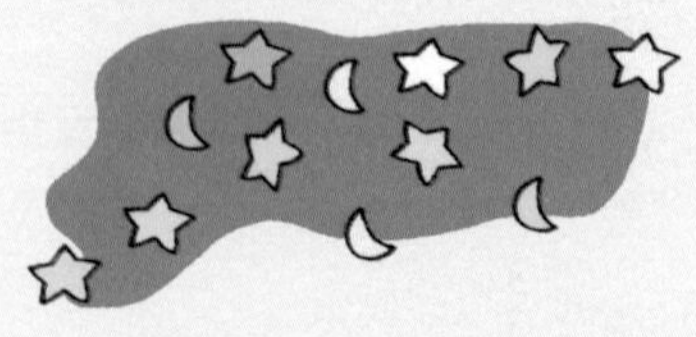

4 MATHEMATICS

- **a** Write the ratio of vowels to consonants in the word above.
- **b** What is the ratio of Ms to vowels?

5 Copy and complete:

a $\frac{5}{10} = \frac{\square}{2}$ b $\frac{16}{24} = \frac{4}{\square}$ c $\frac{3}{\square} = \frac{15}{20}$

d $5 : 8 = 10 : \square$ e $3 : 7 = \square : 21$ f $2 : 3 : 4 = 8 : \square : \triangle$

g $9 : \square = 3 : 10$ h $\square : 1 = 12 : 4$ i $\frac{7}{4} : \frac{3}{4} = \square : \triangle$

6 Express each ratio in simplest terms:

a $\frac{10}{20}$ b $8 : 12$ c $21 : 12$

d $16 : 30$ e $49 : 14$ f $35 : 42$

g $3000 : 500$ h $45 : 105$ i $39 : 52$

j $75 : 100$ k $25 : 100$ l $75 : 25$

m $76 : 40$ n $9 : 15 : 18$ o $12 : 20 : 24$

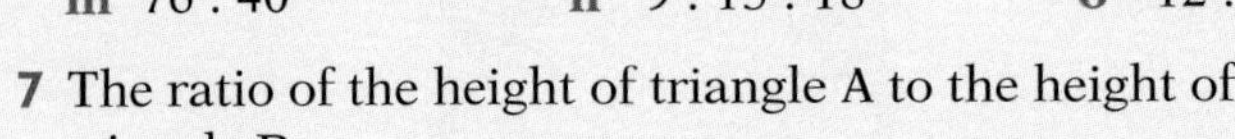

7 The ratio of the height of triangle A to the height of triangle B.

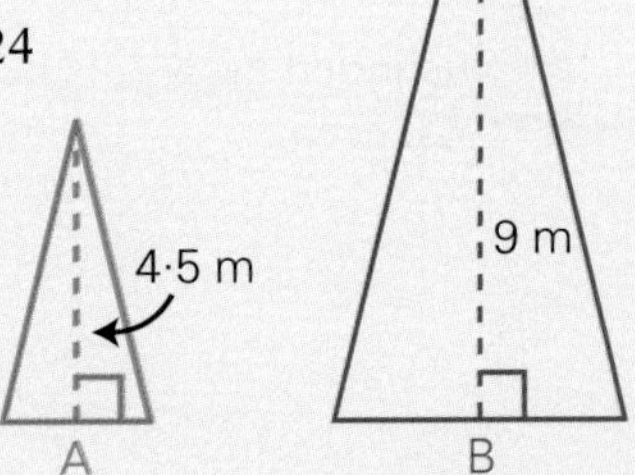

8 Write the ratio suggested by each statement:

a There are 10 Indonesians to every Australian.
b Brett spends \$3 out of every \$8 he earns on his car.
c For each \$5 Lisa earned, her parents put \$10 in her account.
d 1 cm on the drawing represents 1 m on the real house.

Remember to use the same units.

Level 2

9 Simplify:

a \$5 to \$7.50   b 28c to \$2   c $\frac{3}{4}$ kg to 2000 g

d 70 m to $\frac{1}{2}$ km   e 15 days to 1 year   f 720 g to $\frac{1}{5}$ kg

10 Simplify the following ratios:

a $\frac{1}{2} : 3$ b $\frac{2}{3} : 4$ c $\frac{3}{5} : 3$

d $\frac{8}{9} : \frac{2}{3}$ e $4\frac{1}{2} : 1\frac{4}{5}$ f $1\frac{4}{5} : 2\frac{1}{3} : \frac{3}{10}$

11 Simplify:

a $0{\cdot}5 : 3$ b $1{\cdot}2 : 1{\cdot}5$ c $12 : 0{\cdot}6$

d $0{\cdot}8 : 4$ e $1{\cdot}02 : 0{\cdot}2$ f $4{\cdot}2 : 6 : 1{\cdot}4$

12 Find the ratio of the area of figure A to the area of figure B:

a

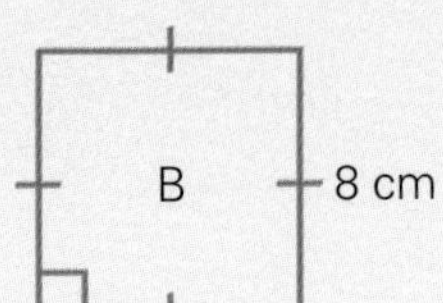

B 8 cm

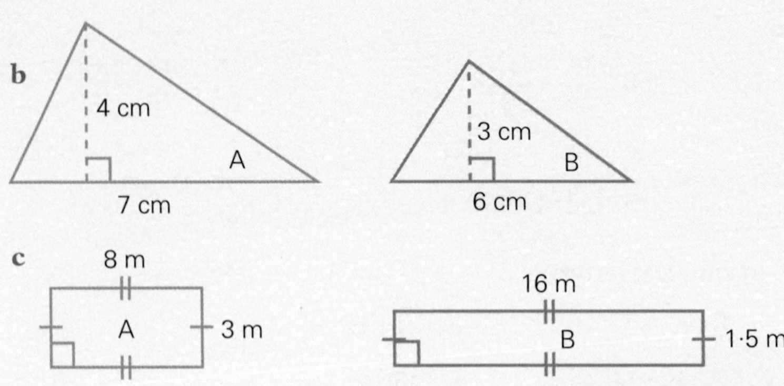

13 Write the ratio of the shaded area to the:

a unshaded area
b total area

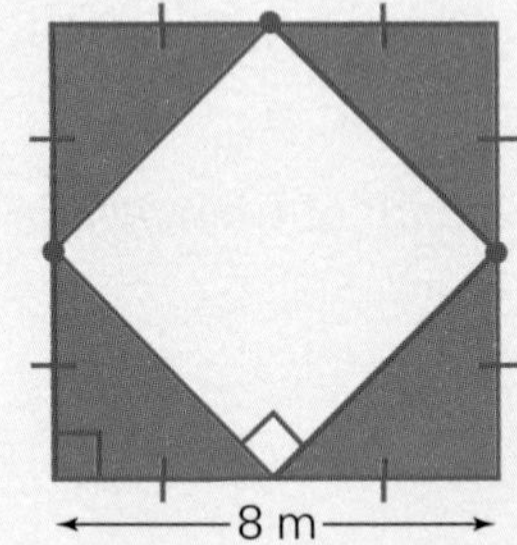

14 Souths' WIN : LOSS ratio was 4 : 3, while Easts' was 5 : 4. Which team has the better record?

Investigation The golden ratio

INTERACTIVE GEOMETRY

You will need: rulers, compass

Some buildings, both ancient and modern, show rectangular shapes that are more pleasing to the eye than others. The Parthenon in Athens is an example. Picture frames also show this 'balance' between length and width.

1 Measure the sides *AB* and *BC* of each rectangle to the nearest millimetre. In each case, calculate the ratio *AB* : *BC* as a unit ratio of the form *x* : 1.

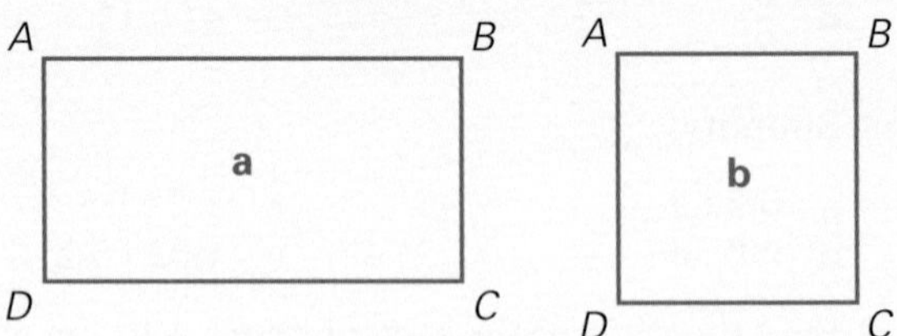

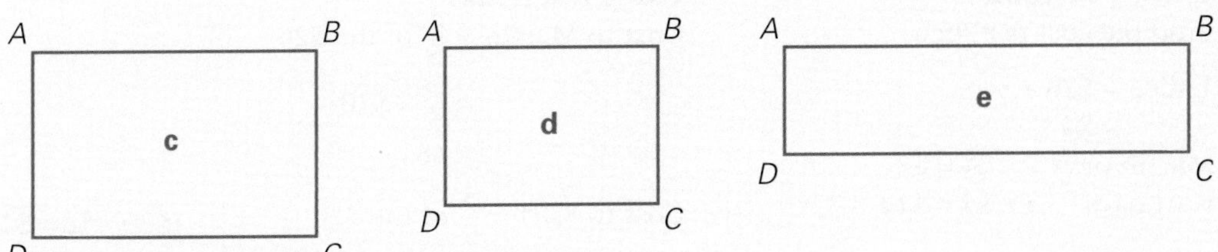

Most find rectangle **e** the most 'balanced' shape. Its $L : W$ ratio was named the **golden ratio** by the Greeks.

2 Construct a **golden rectangle** with the length : width ratio of 1·62 : 1.

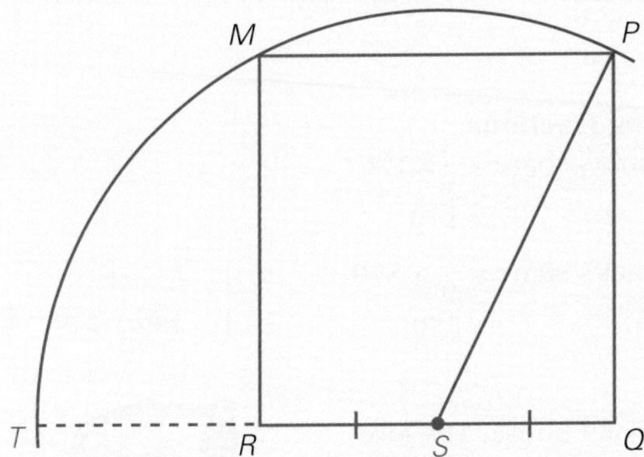

a Start with *any* square *MPQR*. Bisect *RQ* at *S*. Join *SP*.

b Extend *QR* through *R*. With centre *S* and radius *SP*, draw an arc to cut *QR* extended at *T*.

c Complete the rectangle with length *QT* and width *QP*. This is a golden rectangle.

d Find its $L : W$ ratio.

3 **Research**. Many natural spirals such as those of animal shells, plant features and proportions of the human body display the golden ratio. Find out some specific examples, sketch and report.

9.2 Dividing a quantity into a given ratio

A cake costing \$20 was cut into five equal-sized pieces. Of this cake, Martin ate two pieces and Karl had the other three slices. How much should each of the boys contribute to the cost of the cake if it is to be shared in the ratio of the amounts both had eaten?

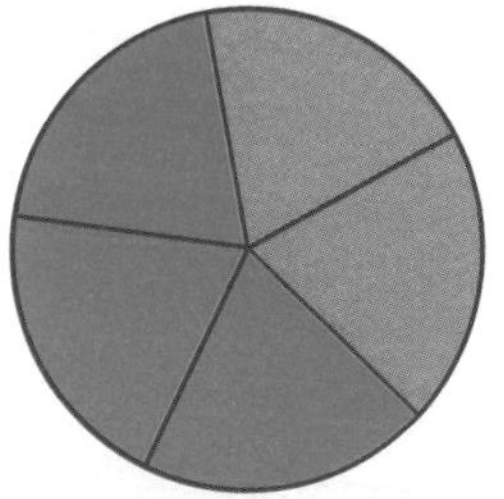

To divide an entire unit, in this case the \$20, in the ratio of 2 : 3, we need 2 + 3 or 5 parts.

Martin ate two pieces out of five pieces, i.e. $\frac{2}{5}$ of the cake.

Unitary method

Find the cost per slice.

$1 \text{ slice} = \$20 \div 5$
$= \$4$

Martin pays $2 \times \$4 = \$8$
Karl pays $3 \times \$4 = \$12$

Using fractions

Cost to Martin $= \frac{2}{5}$ of the \$20
$= \frac{2}{5} \times \$20$
$= \$8$

Cost to Karl $= \frac{3}{5}$ of the \$20
$= \frac{3}{5} \times \$20$
$= \$12$

(\$20 – \$8 = \$12, the remainder of the cost!)

EXAMPLE 1

Emma and Jack were given \$90 by their parents to share in the ratio 4 : 5. How much did each receive?

Solution

Unitary method

$4 + 5 = 9 \text{ shares}$
$1 \text{ share} = \$90 \div 9$
$= \$10$
$\therefore$ Emma's 4 shares $= 4 \times \$10 = \$40$
Jack's 5 shares $= 5 \times \$10 = \$50$

Using fractions

Emma's share $= \frac{4}{9} \times \$90$
$= \$40$
Jack's share $= \frac{5}{9} \times \$90$
$= \$50$

Check:
\$40 + \$50 = \$90

EXAMPLE 2

Margaret, Sarah and Tony paint the inside of a house. The job pays \$3600. If they divide the money in the ratio of hours worked, find Sarah's share given Margaret worked 5 h, Sarah 8 h and Tony 7 h.

Solution

5 + 8 + 7 = 20 hours

Unitary method

$\$3600 \div 20$ hours
$1 \text{ hour} = \$180$
$\therefore$ Sarah's share
$= 8 \times \$180$
$= \$1440$

Using fractions

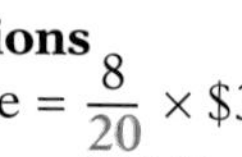

Sarah's share $= \frac{8}{20} \times \$3600$
$= \$1440$

EXERCISE 9B

1 Divide \$10 in the ratio 3 to 7.

2 Share \$100 in the ratio 1 : 9.

3 Divide:

a 20 in the ratio 2 : 3
b 180 in the ratio 7 : 2
c \$84 in the ratio 3 : 4
d 450 m in the ratio 7 : 8
e 400 in the ratio of 3 : 5
f 96 in the ratio of 7 : 1

4 Copy and then divide each line in the given ratio:

a ____ (2 : 3)
b ______________________________ (4 : 3)

5 Divide:

a 360 kg in the ratio of 7 : 5
b 4 days in the ratio of 3 : 1
c 1 t in the ratio of 3 : 2
d 0·7 m in the ratio of 3 : 3
e $7.20 in the ratio 1 : 3 : 4

6 A drink recipe lists 1 part fruit juice to 5 parts water. How much fruit juice is needed to make up 2·4 L of the drink?

7 Fuel for a two-stroke motor mower contains petrol and oil mixed in the ratio 19 : 1. In a 4-litre can of fuel, how many litres of oil are there?

8 In a concert audience there were five teenagers for every three adults. How many teenagers were in the audience of 24 800?

Level 2

9 Thomas wins Lotto and decides to share half his winnings between two charities in the ratio of 3 : 4. If Thomas won $2·5 million, how much did each charity receive?

10 A concrete mix is made by combining sand, gravel and cement in the ratio of 5 : 3 : 2. If a new driveway requires 80 kg of concrete mix to be made, how much gravel is required?

11 A mine produces 500 tonnes of ore each week. The ratio of copper to zinc in the ore is 3 : 2. Determine the amount of copper produced in a single week.

12 A 10-metre length of wire is used to fence a rectangular field with dimensions in the ratio of 3 : 1. Calculate the area of the rectangle.

13

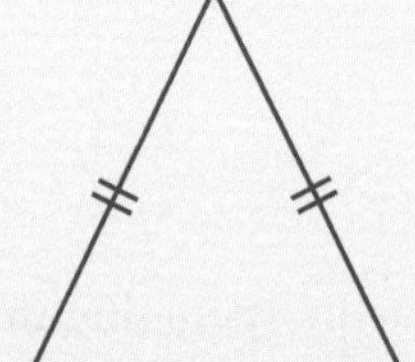

An isosceles triangle has angles in the ratio 2 : 3 : 3. Find the size of the two equal angles.

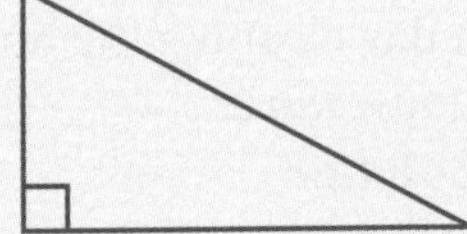

14 The sides of a right-angled triangle are in the ratio 3 : 5 : 4. If the perimeter of the triangle is 70 m, how long is the hypotenuse?

15 Michael, Dagmar and Sirie buy a $12.00 Lotto special entry by contributing $5, $4 and $3 respectively. How much more than Dagmar should Michael receive from winnings of $588 000?

16 Alcohol is mixed with fruit juice in the ratio 1 : 14, then that solution is mixed with water in the ratio 1 : 4. How much alcohol would there be in 250 mL of the final mixture?

17 In a class of 28 students, the ratio of boys to girls is 4 : 3. If two boys leave the class and are replaced by two new girls, what does the ratio of boys to girls become?

18 White paint is mixed with blue in the ratio of 2 : 7. This is then mixed with green paint in the ratio of 3 : 5. What is the ratio of:

a white paint to green paint?
b blue paint to green paint?

19 Orange juice is mixed with pineapple juice in the ratio of 3 : 2. This pine–orange mix is then added to lemonade in the ratio of 3 : 5. What is the ratio of orange juice to lemonade in the final mixture?

20 The ratio of A to B is 5 : 7 while the ratio of B to C is 4 : 9. Find the ratio of A to C.

9.3 Applications of ratios

It is possible to find one quantity in a ratio given the ratio and the other quantity. The solution to these ratio problems can be done in a variety of ways.

EXAMPLE 1

The ratio of cordial to water in a drink mix is 2 : 9. How much water should be mixed with 3 litres of cordial?

Solution

Unitary method

C : W
2 : 9
3 L : ____
∴ 2 shares = 3 L
1 share = 1·5 L
∴ water = 9 shares
= 9 × 1·5 L
= $13\frac{1}{2}$ L

An algebraic solution

$\frac{C}{W} = \frac{2}{9} = \frac{3}{x}$ Introduce pronumeral.

or $\frac{9}{2} = \frac{x}{3}$ Invert so that the pronumeral is on top.

×3 ×3

$\frac{27}{2} = x$ Solve.

∴ water is $13\frac{1}{2}$ L

EXAMPLE 2

If Tom and Martha shared the housework in the ratio of 3 : 2 and Martha spent 40 minutes a day cleaning the house, how much of the day did Tom spend at his share of the housework?

Solution

Unitary method

Tom : Martha
3 : 2
x : 40
∴ 2 parts = 40
÷2 ÷2
1 part = 20
∴ Tom spends 3 × 20 = 60 minutes on his share of the housework.

An algebraic solution

$\frac{T}{M} = \frac{3}{2} = \frac{x}{40}$

$\therefore \frac{3}{2} = \frac{x}{40}$

×40 ×40

$\frac{120}{2} = x$

$x = 60$

∴ Tom spends 60 minutes on housework.

EXAMPLE 3

Concrete is made by mixing sand, gravel and cement in the ratio 5 : 3 : 2. If 1 t of sand is available, how much gravel is needed?

Solution

1 t = 1000 kg

Unitary method

Sand : gravel : cement

5 : 3 : 2

1000 kg : x : y

$\therefore$ 5 parts = 1000 kg

$\div 5 \quad \div 5$

1 part = 200 kg

Gravel requires 3 parts.

$\therefore$ gravel = 3 × 200 kg

600 kg of gravel is required for 1 t of sand.

An algebraic solution

$$\frac{5}{3} = \frac{1000}{x}$$

$$\therefore \frac{3}{5} = \frac{x}{1000}$$

$\times 1000 \quad \times 1000$

$$\frac{3000}{5} = x$$

$$\therefore x = 600$$

600 kg of gravel is needed.

EXERCISE 9C

1 Daniel and Kimberly share money in the ratio of 1 : 2. If Daniel's share is $150, what is the value of Kimberly's share?

2 Mary and Julie share the profits of their printing company in the ratio of 2 : 1. If Mary receives $4000, how much of the profit does Julie receive?

3 The ratio of lemonade to dry ginger ale in a summer punch is 3 : 1. If 1 litre of dry ginger ale is used, how many litres of lemonade are required?

4 Cordial and water are mixed in the ratio of 1 : 5. How much water is required for 25 mL of cordial?

5 A video cassette has two programs recorded. The ratio of the length of each program is 2 : 5. If the shorter of the two programs lasts for 30 minutes, find:

a the length of the longer program, in minutes

b the total length of time (in minutes) the video cassette has been recorded on.

6 Sand and cement for mortar between bricks are mixed in the ratio 4 : 1. How much sand should be mixed with five shovels of cement?

7 Two lengths of pipe are in the ratio of 4 : 5. If the longer pipe is 45 cm, what is the length of the shorter pipe?

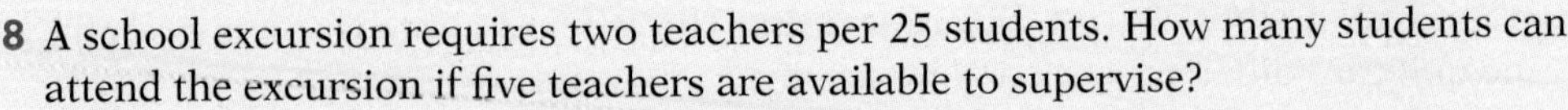

8 A school excursion requires two teachers per 25 students. How many students can attend the excursion if five teachers are available to supervise?

9 A business finds it must keep the ratio of costs to profit at 7 : 5. For costs of $560, what should the profits be?

10 A model yacht is built to a scale of 1 : 50.

  a If the length of the model's hull is 350 mm, what is the length of the real yacht's hull?

  b If the real yacht's mast is 13·5 m tall, what is the height of the model's mast?

11 The ratio of length to width of a rectangle is 5 : 3. If the length of the rectangle is 45 mm, what is the perimeter of the rectangle?

12 The prices of unleaded petrol to leaded petrol are in the ratio 4 : 5. If unleaded petrol is 85.2 cents per litre, what is the cost of the leaded petrol?

13 A length of timber is cut into two pieces in the ratio of 3 : 4. The shorter piece is 450 mm. What was the original length of timber?

Level 2

14 The ratio of advertisements to television show is 2 : 7. There are 6 minutes 40 seconds of ads. If a person starts watching at 4:30 pm, what time will they have finished watching the show, including the ads?

15 The three sides of a triangle are in the ratio of 1 : 2 : 3. The longest side of the triangle is 21·75 cm. Calculate the perimeter of the triangle.

16 The angles of a triangle are in the ratio of 2 : 3 : 1. If the triangle has a right angle, what is the size of the smallest angle?

17 Chemicals X, Y and Z are combined in the ratio of 1 : 4 : 2. If 60 mL of Y are used, what are the measures of the other two chemicals?

18 The heights of three friends Zara, Kevin and Peta are in the ratio of 53 : 55 : 47. If Kevin is 175 cm tall, what is the average height of the three friends?

19 Zara, Kevin and Peta share a lottery prize in the ratio of 7 : 5 : 11. If Peta's share of the prize was $45 000, how much was the lottery win?

20 Zara, Kevin and Peta invest $4500, $6000 and $6600 respectively in their business and plan to share the profits in the same ratio. Zara earns $500 in the first month in profits. Calculate the total profits made by the business in the first month.

9.4 Rates

Rates compare quantities with **different units**. The order of a rate is still important. In a rate the units are different and must be specified.

> A rate is written as:
>
> first quantity **per one** of the second quantity

Some common rates include speed and pay rates.

For example, \$12/h means:
\$12 **per** hour
each
every
for one

EXAMPLE 1

George earns \$120 in 15 hours. Find George's hourly rate of pay.

Solution

Rate = \$120 in 15 hours
\$120/15 hours
÷15 ÷15
= \$8/1 hour
= \$8/h

EXAMPLE 2

Water flows at the rate of 150 L/min through a hose.

a How much water would flow in 2 hours?

b How long would it take to fill a 60 kL pool?

Solution

a 150 L/1 min
× 60 × 60
= 9000 L/60 min
= 9000 L/1 h
× 2 × 2
= 18 000 L/2 hours
∴ 18 000 L will flow through the hose in 2 hours

b 60 kL = 60 000 L (1 kL = 1000 L)
∴ time taken = 60 000 L ÷ 150 L/min
= 400 min
= 6 h 40 min

To convert minutes to hours, ÷ by 60 and use the ˚ ' '' or DMS button on your calculator.

EXAMPLE 3

Express 50 g/min in kg/h.

Solution

50g/min
× 60 × 60
= 3000 g/60 min
= 3000 g/1 h
÷ 1000 ÷ 1000
= 3 kg/h

EXERCISE 9D

1 Express each rate in simplest form:

a \$40 in 4 h	**b** \$3.90 for 3 cans	**c** 86 people in 2 buses
d \$2.50 for 5 kg	**e** 120 km for 4 h	**f** 56 runs from 14 overs
g 480 km on 60 L	**h** 30 L in $\frac{1}{2}$ min	**i** 200 emus in 8 ha
j \$60 in 2 h	**k** 400 km on 32 L	**l** 182 points in 10 games
m \$480 for 5 m	**n** 5 g in 2 min	**o** \$1200 in $\frac{1}{2}$ a day

2 Use your calculator to find the *unit costs* in the units indicated in brackets, for the following (round to the nearest cent where necessary):

a 3 kg onions for $2.85 (cents/kg)
b 4 m of fabric for $31.80 ($/metre)
c 200 g chocolate for $2.50 (cents/grams)
d 50 L of petrol for $39.45 (cents/litre)
e 800 m^2 of land for $125 000 ($/square metre)

3 Recipes are also based on ratios. For each recipe adjust the quantities as stated.

a **Savoury tomato cups**
(serves 4)
4 medium tomatoes
garlic salt + pepper
$\frac{1}{2}$ cup mushrooms, finely chopped
1 cup grated cheese
4 poached eggs

What quantities for:
i eight people?
ii two people?
iii one person?
iv six people?

b **Garlic prawns**
(serves 2)
500 g green prawns
$\frac{1}{3}$ cup peanut oil
2 cloves garlic, crushed
2 tablespoons shallots, chopped
2 tablespoons dry sherry
salt and pepper to taste

What quantities for:
i four people?
ii one person?
iii five people?

4 Copy and complete each of the following:

a 10 g/min = ______ g in 2 min
b 80c/min = ______ $/h
c $\frac{3}{4}$ km/min = ______ m/min
d $8 in 20 min = ______ $/h
e 1·3 km in 10 min = ______ m/min
f $700 in 5 days = ______ $/day
g 7 g in $\frac{1}{2}$ min = ______ g/min
h 1800 g/h = ______ kg/h

5 Martha earns $17.50 per hour working at a local bookstore.

a How much does Martha earn in an 8-hour shift?
b How many hours does Martha work in a week if she earns $420?
c If Martha works 30 hours a week, 42 weeks a year, calculate her yearly income.

6 A shearer sheared 80 sheep in 2 hours.

a Express this as a rate in sheep per hour.
b If the shearer continues at the same rate for 5 hours, how many sheep have been shorn?
c How many sheep can be shorn in an 8-hour day?
d How long does it take to shear 300 sheep?

7 A mobile phone company charges 10 cents per 30 seconds. Find the cost of a phone call lasting:

a 1 minute **b** 7 minutes **c** 32 minutes

Level 2

8 The electricity supplier charges 8.9 c/unit of electricity. What is the charge for 1240 units?

9 An overseas phone company charges $1.85/min for an STD call. What is the cost of a call lasting:

a 4 min? **b** $\frac{1}{2}$ h? **c** 45 s?

10 A car uses 50 litres of fuel to cover 420 kilometres.

a What is the rate of fuel consumption in km/L?
b How far could the car travel on 20 L of fuel?
c Express the fuel consumption in L/100 km.
d How much fuel would it use in travelling 3500 km?

11 An animal's heart beats at the rate of 48 beats/min.

a How many times will it beat in: **i** an hour? **ii** 15 s?
b How long does it take to beat: **i** 96 times? **ii** 8 times?

12 If Giorgio can save $120/month:

a How much will he save in $2\frac{1}{2}$ years?
b How long will it take for him to save $2000?

13 Fertiliser is spread at the rate of 3 kg/100 m^2.

Remember:
1 ha = 10 000 m^2

a How many tonnes are needed for 10 hectares?
b What area can be fertilised by **i** 30 kg? **ii** 1·8 kg?

14 A tap drips water at the rate of 30 mL/min.

a How much water is wasted in 1 week?
b How long does it take to waste a kilolitre of water?

15 Australia's birth rate is 16 babies for every 1000 people in the country. If India's 850 million people produced 34 million babies in the same year, compare the two countries' birth rates.

16 Robert's council rates bill is $1350 p.a. for land valued at $120 000, while Anne pays $1560 on land worth $140 000 to another council. Which council charges the higher rate? (Use $/$1000 of valuation to decide.)

17 Copy and complete:

a 45 kg/s = ______ kg/min **b** $240/day = ______ $/h
c 60 g/min = ______ g/h **d** 15c second = ______ $/h

18 Fuel consumption of cars is usually quoted as the number of litres of fuel required to travel 100 km. Amanda's car's manual quotes its fuel consumption as 100 km/8 L.

a Express this as a rate in km/L.
b How many litres of petrol does Amanda use in a week where she travels 400 km?
c At the end of the week, Amanda fills her car at a petrol station charging 92 cents/L. How much does it cost Amanda to fill her car?
d If Amanda only has $25 in her wallet, how many litres can she afford at the petrol station in part **c**?

Try this

Pump A takes 10 minutes to fill the water tank, while pump B can fill the same water tank in 15 minutes. If both pumps are operating together, how long does it take to fill the water tank?

9.5 Distance, speed and time

As mentioned in the previous section, one of the most common rates is speed. Speed compares the distance travelled to the time taken.

Write: 45 km/h.

Say: 45 kilometres **per** hour
an
each
every
for one

Calculations involving distance, speed and time can be related as follows:

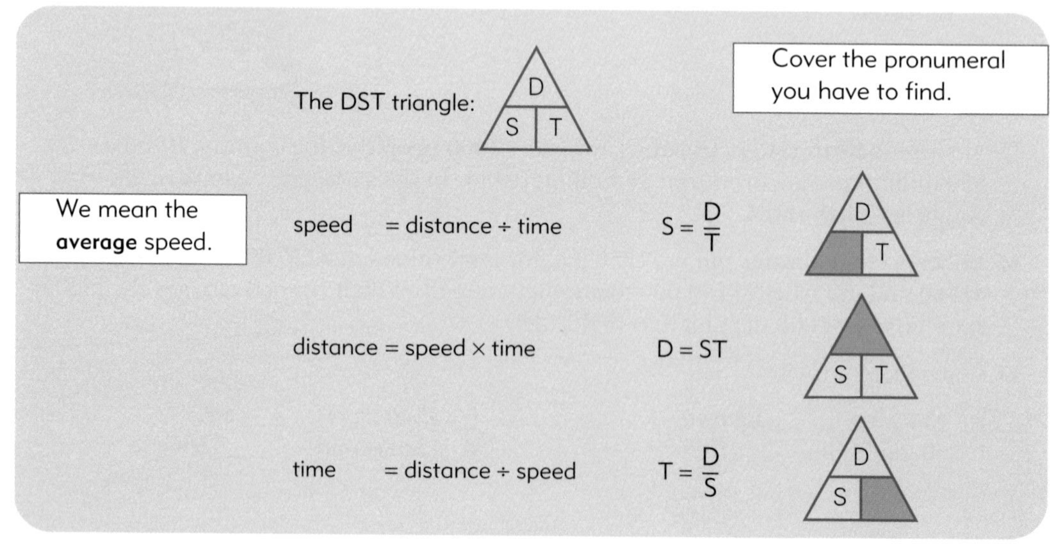

EXAMPLE 1

a Find the average speed of a vehicle which travels 180 km in 3 hours.
b How far would it travel in 4 hours at that speed?
c How long would it take to travel 100 km at that speed?

Solution

a $S = \dfrac{D}{T}$

$= 180\text{ km} \div 3\text{ h}$
$= 60\text{ km/h}$

b $D = ST$

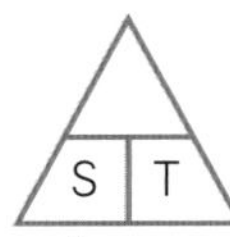

$= 60\text{ km/h} \times 4\text{ h}$
$= 240\text{ km}$

c $T = \dfrac{D}{S}$

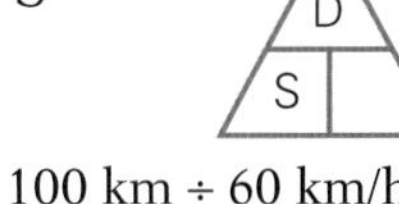

$= 100\text{ km} \div 60\text{ km/h}$
$= 1{\cdot}\dot{6}$ ______ h
$= 1\text{ h } 40\text{ min}$

Remember the 'time' key.

EXAMPLE 2

Express 100 km/h in metres/second.

Solution

100 km/1 h
  × 100 × 100
= 100 000 m/60 min
  ÷ 60 ÷ 60
$= 1666\frac{2}{3}$ m/1 min
$= 1666\frac{2}{3}$ m/60 s
  ÷ 60 ÷ 60
$= 27\frac{7}{9}$ m/s

EXERCISE 9E

1 Express each of the following as rate in km/h:

a 50 km in 1 hour **b** 140 km in 2 hours **c** 250 km in 4 hours
d 150 km in 3 hours **e** 90 km in 3 hours **f** 750 km in 5 hours

2 Find the average speed (km/h) of a vehicle which travels:

a 150 km in 2 hours **b** 450 km in $4\frac{1}{2}$ hours **c** 1500 km in 2·4 hours
d 6 km in $1\frac{1}{2}$ hours **e** 30 km in $\dfrac{1}{2}$ hour **f** 1200 km in 10 hours
g 40 km in $\dfrac{1}{2}$ hour **h** 15 km in 20 minutes **i** 90 km in $1\frac{1}{2}$ hours

3 Find the distance travelled by a car whose average speed is 65 km/h if the journey lasts:

a 2 hours **b** 5 hours **c** $3\frac{1}{2}$ hours **d** 8 hours **e** $\dfrac{1}{2}$ hour

4 How long will it take a vehicle to travel:

a 270 km at a speed of 90 km/h?
b 80 km at a speed of 100 km/h?
c 250 m at a speed of $12\frac{1}{2}$ m/min?

TEACHER

5 The following travel graph shows the journey of three different people.

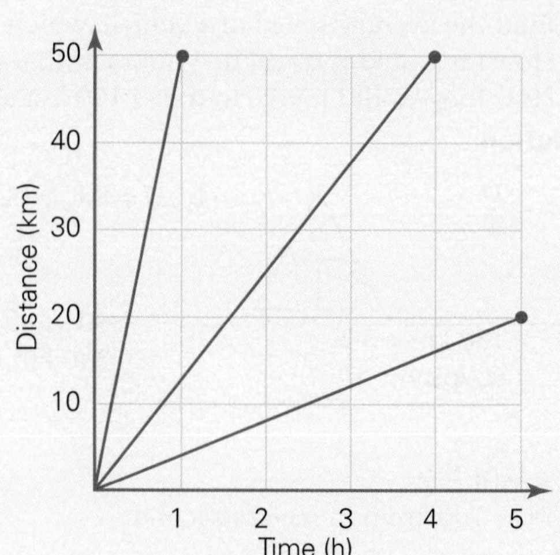

a Match the following three statements to the three journeys shown.

A The *walker* moves at $\frac{20}{5}$ km/h or 4 km/h.

B The *cyclist* has a speed of $\frac{50}{4}$ km/h or $12\frac{1}{2}$ km/h.

C The *car* travels at $\frac{50}{1}$ km/h or 50 km/h.

b Which transportation method is the fastest? Describe the feature of the graph that indicates this.

c Find the ratio (as a fraction) of the vertical height to the horizontal height of each journey shown in the graph. What can you conclude about the ratios found (known as the slope or gradient of the graph) and the speeds of each person.

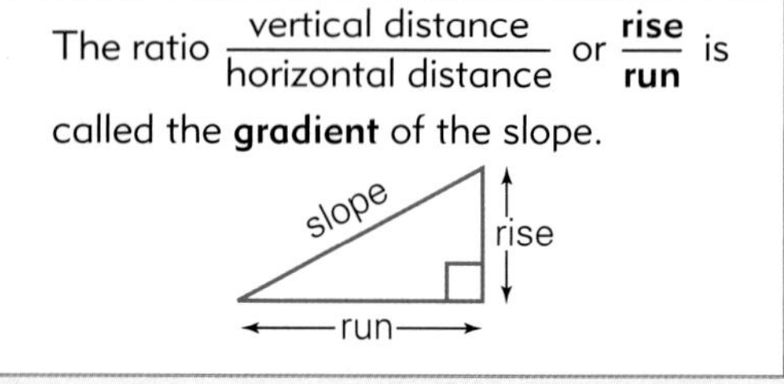

6

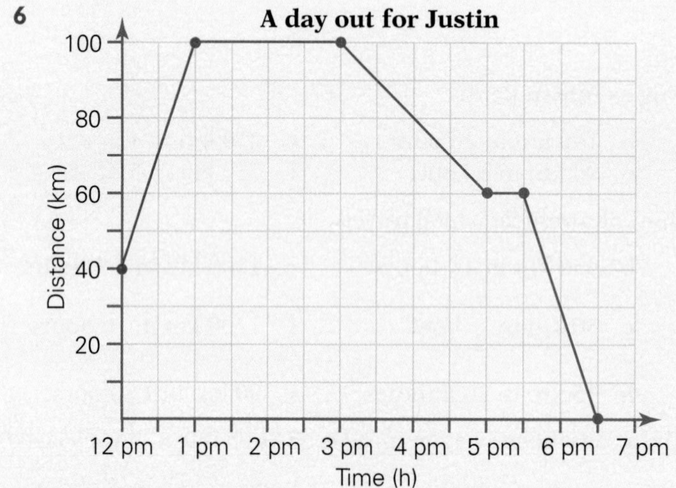

a What is the scale used on the horizontal time axis?

b What is the scale used on the vertical distance axis?

c At what time did Justin first stop and for how long?

d How far did Justin travel during the trip?

e What was Justin's average speed on the return journey?

Level 2

7 Find the average speed of a vehicle which travels:

a 650 km in 3 h 15 min **b** 500 m in 2·5 s **c** 38·5 m in 4 min 45 s (1 dec. pl.)

8 How long will it take a vehicle to travel:

a 2 km at a speed of 150 m/min? **b** 0·8 km at 28·6 m/s (nearest second)?

c 100 m at a speed of 10·2 m/s?

9 How far will a bicycle travelling at 15 km/h go in:

a 3 h? **b** $\frac{1}{2}$ h? **c** 10 min? **d** 1 s?

10 Light travels at around 300 000 km/s.

a Express this speed in km/h.
b How far does light travel in 2 minutes?
c How long does light from the sun take to travel the 149 million kilometres to Earth?
d If telephone conversations travel at the speed of light, how long does a word spoken in Newcastle, NSW, take to reach Perth in Western Australia, 4000 km away?

SPREADSHEET ACTIVITY

11 The speed of sound at sea level (called Mach 1) is approximately 340 m/s.

a Change this to km/h.
b How far does a shout travel in half a minute?
c How long (to 1 decimal place) would it take to hear the sound of an approaching aircraft when it was 5 km away?
d If the aircraft was flying directly towards you at 1300 km/h, how long does it take to cover the 5 km? Compare answers **c** and **d** and write your conclusion.

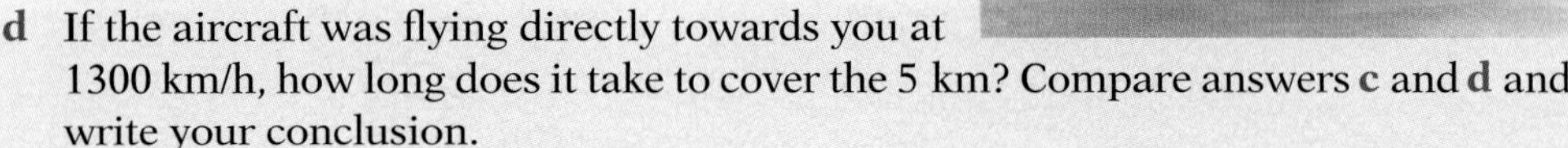

12

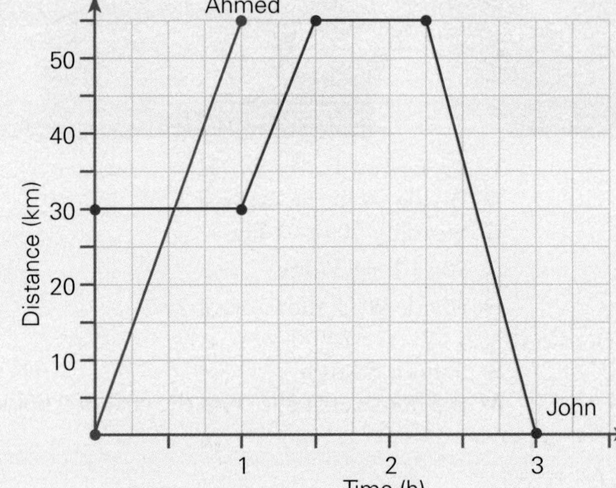

a What was Ahmed's speed?
b At what time did John leave and when did he arrive at the destination?
c What was John's speed on the return journey? $\left(S = \frac{D}{T}\right)$
d For how many minutes did John stop?

13 Make the rate name changes shown:

a 5 km/h = ______ m/h **b** 600 m/h = ______ m/min
c 1200 km/h = ______ km/min **d** 18 m/s = ______ km/h

14 Matt Shervington ran the 100 m sprint in 10·03 s in the 1998 Commonwealth Games.

a Express his speed in metres per second.
b If Matt was able to maintain this speed for an hour, how far would he be able to run? (Answer in kilometres.)

15 In August 2001 Ian Thorpe swam his 400 m freestyle event in 3:47·49 (3 minutes and 47·92 seconds).

- **a** Express Ian's speed in m/s.
- **b** If Ian was able to keep this pace over 1500 m, how long would it take him to swim the 1500 m race?
- **c** Express Ian's speed in kilometres per hour.

16 The world record for the men's 100 m is 9·86 s, held by Carl Lewis of the USA. For the 1500 m, it is 3:28·82, held by N. Morceli of Algeria.

- **a** Who runs at the faster rate?
- **b** If Lewis could maintain his pace over 1500 m, how long would he take to run that distance?

Investigation It's a puzzling time for ratios

Complete the questions to discover what bridge is shown in the photograph.

C Simplify 16 : 20
O 4 : 7 = O : 28
L Divide 20 in the ratio of 1 : 4
H Simplify $\frac{1}{2}$: 3
T Divide 54 in the ratio of 5 : 1
B 60 km/min = B km/h
E Simplify 50c : $6
R Divide 350 in the ratio of 3 : 4
A Simplify 0·2 : 3
I 2 : 3 = 16 : 1
D Divide 99 in the ratio of 2 : 9
G Simplify 50 m : 1 km
N 90c/1 h = $$N$/day
O Divide 90 in the ratio of 2 : 7
U 1 : 7 = U : 700
S 5 m/s = S km/h
M If 5 books cost $95, find the cost of 9 books.

45 : 9   1 : 6   1 : 12

171   1 : 15   45 : 9   1 : 6   1 : 12   171   1 : 15   45 : 9   24   4 : 5   1 : 15   4 : 16

3600   150 : 200   24   18 : 81   1 : 20   1 : 12   24   21·60

20 : 70   100   1 : 12   1 : 12   21·60   180

4 : 5   16   4 : 16   4 : 16   1 : 12   1 : 20   1 : 12

4 : 5   1 : 15   171   3600   150 : 200   24   18 : 81   1 : 20   1 : 12

Chapter Review

Language Links

average	equivalent	quantity	slope
comparison	gradient	rate	speed
distance	ordered	ratio	

1 Explain the difference between a rate and a ratio.

2 Copy and complete the following sentences:

- **a** The ______ the line on a travel graph, the faster the journey.
- **b** On a travel graph, the horizontal line indicates ______ distance travelled over time.
- **c** Another word for the slope of a line is its ______ .
- **d** When simplifying a ratio, it is important to ensure that the units are the ______ .
- **e** 2 : 5 and 4 : 10 are called ______ ratios.

3 Explain how to use the DST triangle. Use a diagram to assist your explanation.

4 For the travel graph below, write a story explaining its key features. Include the speeds of each section and the times the journey starts and stops.

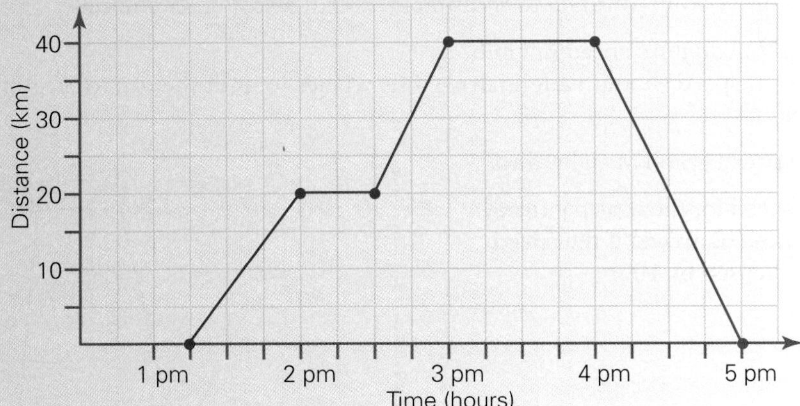

Chapter Review Exercises

1 a Find the ratio of the shaded to the unshaded parts. 9.1
b Find the ratio of shaded parts to the total number of parts.

2 A Scout troop has three adult leaders, 14 boys and 12 girls as members. Find the ratio of: 9.1

a boys to girls **b** girls to adults **c** leaders to troop members

3 Simplify each ratio: 9.1

a 12 : 27 **b** 10 : 2·5 **c** $1\frac{1}{2} : 4\frac{1}{2}$

d 6 cm to 20 mm **e** 75c to $2.25 **f** $3\frac{1}{2}$ years to 18 months

9.1 **4** Copy and complete these equivalent ratios:

a $\frac{7}{2} = \frac{\square}{6}$ **b** $\frac{48}{21} = \frac{16}{\square}$ **c** $6{\cdot}3 : 5 = \square : 1$

d $3 : 5 : 8 = \square : 20 : \triangle$ **e** $\frac{\square}{15} = \frac{7}{50}$ **f** $1 : \square : 4 = 3 : 15 : \triangle$

9.1 **5** Team A scored 12 goals in five matches, while team B managed 15 goals from seven games. Which team has the better goals ratio?

9.2 **6** Divide:

a 40 in the ratio of 2 : 3 **b** 300 in the ratio of 1 : 5 **c** 6 m in the ratio of 4 : 1

9.2 **7** Share $660 in the ratio 2 : 3 : 5.

9.2 **8** Kim, Mark and Jed invest $5, $4 and $11 respectively to purchase a lottery ticket. If they share their prize of $400 in the same ratio, how much does Jed receive?

9.3 **9** If three video cassettes cost $8.94, what will five cost?

9.3 **10** Metholated spirits is diluted with water in the ratio of 3 : 20. If 1 litre of water is used, how many millilitres of metholated spirits are used?

9.4 **11** Simplify these rates:

a 150 L in 6 hours **b** 36·4 g in 20 cm^3

9.5 **12** Vaughn drove from Miranda to Strathfield, 28 km away, in 35 minutes.

a Express Vaughn's speed in km/h.
b At what speed should Vaughn drive if he wishes to limit the trip to Strathfield to half an hour?

9.5 **13** For the travel graph at right, find:

a the speed for the return journey
b the time of arrival if the speed is increased by 10%

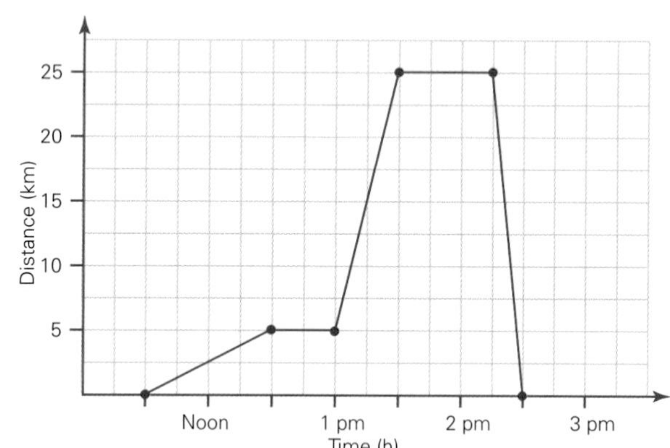

9.5 **14** Express each rate in the new unit in brackets:

a 30 km/h (km/min) **b** 20 t/d (kg/h) **c** 50 m/min (m/h)

d 50 m/min (km/h) **e** $\frac{3}{4}$ km/min (km/h) **f** 30 km/h (m/min)

Keeping Mathematically Fit

Part A: Non-calculator

1 A number is increased by 9 and the result is the same as 15 less the number. What is the number?

2 Find the value of $\sqrt{\dfrac{5^6}{5^4}}$.

3 If $\square \times 3{\cdot}694 = 36\,940$, find $\square$.

4 If the average of x, y and z is 10, what is the average of $2x$, $2y$ and $2z$?

5 Mark walks on a bearing of NW for 5 km. If John walks in the opposite direction to Mark, on what bearing does John walk?

6 A TV is bought for \$360 and sold 1 year later for \$324. Calculate the loss as a percentage of the cost price.

7 If $3a - 1 = 8$, find the value of ^{-}a.

8 Write a decimal that lies between $\frac{1}{2}$ and $\frac{1}{3}$.

9 80 cm² / 90 cm² / 72 cm²

The area of three sides of a rectangular prism is shown. Calculate the volume of the prism.

10 Convert $5\frac{1}{4}$ m² to cm².

Part B: Calculator

Questions 1–4 refer to the graph supplied.

1 At what time did the journey start?

2 Between what times did the travellers stop?

3 How many kilometres did they travel during the trip?

4 What was their speed on the return journey?

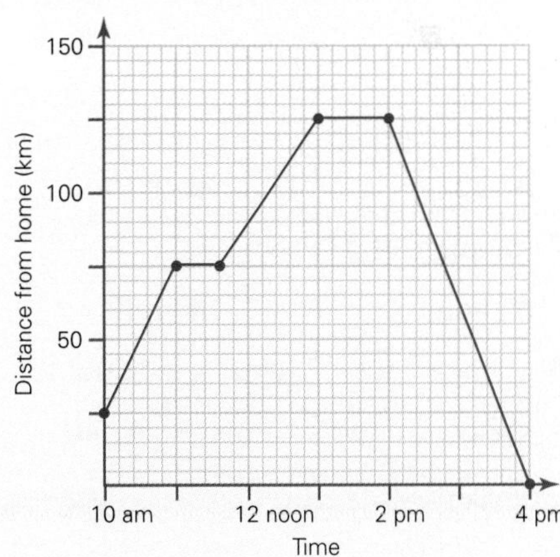

5 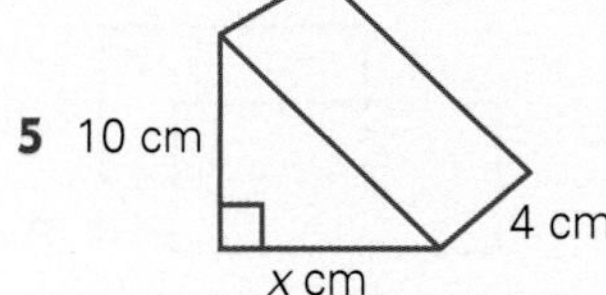

Find the value of x in the diagram, given the volume of the prism is 120 cm.

6 Increase \$964.90 by 10% and then decrease the result by 5%.

7 Evaluate $4(a - b)$, where $a = {}^{-}3$ and $b = 2$.

8 Give the dimensions of a prism if the prism's volume is 1000 cm³.

TEACHER

Cumulative Review 3

Part A: Multiple-choice questions

Write the letter that corresponds to the correct answer in each of the following.

1 $\sqrt{3^2 + 4^2}$ is:

A 7 **B** 5 **C** 25 **D** 3

2 $5a(2b - 7)$ is the same as:

A $7ab - 7$ **B** $7ab - 2a$ **C** $10ab - 35a$ **D** $10ab - 35$

3 If $a = {}^-3$ and $b = 5$, then the value of a^2b is:

A 225 **B** 45 **C** $^-45$ **D** $^-225$

4 $8\frac{3}{4}\%$ of \$7000 is:

A \$6125 **B** \$612.50 **C** \$61.25 **D** \$6.13

5 Which graph represents the solution to $x + 5 > 8$?

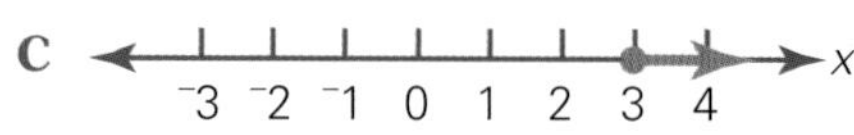

A: ⁻3 ⁻2 ⁻1 0 1 2 3 4 x

B: ⁻3 ⁻2 ⁻1 0 1 2 3 4 x

C: ⁻3 ⁻2 ⁻1 0 1 2 3 4 x

D: ⁻3 ⁻2 ⁻1 0 1 2 3 4 x

6 The exact value of x in ΔABC is:

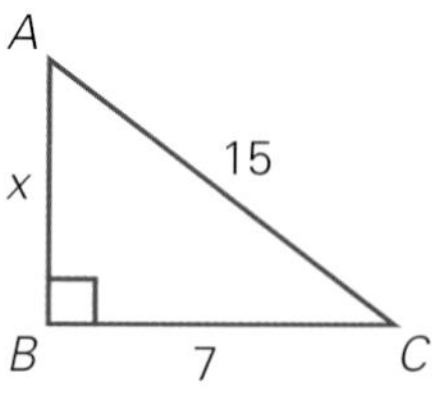

A 22 **B** 8 **C** $\sqrt{176}$ **D** $\sqrt{274}$

7 An angle of 135° is:

A acute **B** reflex **C** straight **D** obtuse

8 140° 50° 90° x°

x is equal to:

A 40 **B** 90 **C** 80 **D** none of these

9 The surface area of the open rectangular prism is:

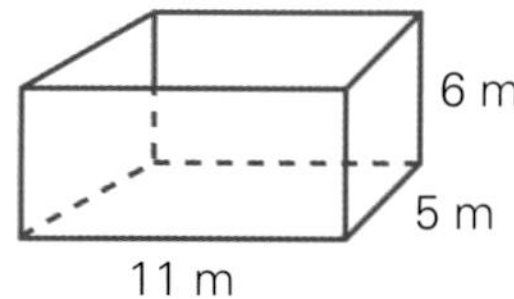

A 247 m² **B** 302 m² **C** 330 m² **D** 151 m²

10 A cube has a volume of 8 cm³. The surface area of this cube is:

A 2 cm² **B** 4 cm² **C** 24 cm² **D** 384 cm²

11 What is the ratio of 5 hours to 15 minutes in simplest form?

A 1 : 3 **B** 20 : 1 **C** 1: 20 **D** 3 : 1

12 The number of kilogmetres in x metres is:

A $100x$ **B** $\frac{x}{100}$ **C** $1000x$ **D** $\frac{x}{1000}$

13 In 5 hours I walk d kilometres. My speed is:

A $5d$ km/h **B** $\frac{d}{5}$ km/h **C** $\frac{5}{d}$ km/h **D** $(5 + d)$ km/h

14 The ratio of 1200 mm to 2 m is:

A 1200 : 2 **B** 600 : 1 **C** 3 : 5 **D** 6 : 1

15 \$100 shared in the ratio 3 : 2 is:

A \$30, 20 **B** \$70, 30 **C** \$60, \$40 **D** \$40, \$60

Part B: Short-answer questions

Show full working out for each of the following.

1 Find the value of:

a $\frac{9{\cdot}6^2}{0{\cdot}2}$ **b** $(4{\cdot}7 - 2{\cdot}3 \times 4{\cdot}5)^2$ **c** $\sqrt{7\,290\,000}$

2 Simplify:

a $5^6 \times 5^4$ **b** $5^6 \div 5^4$ **c** $(2^4 \times 3^2)^5$ **d** $6°$

e $6p + 9q + p$ **f** $7ab \times {}^-3a$ **g** $\frac{18a}{20ab}$ **h** $10(3a + b) - a$

3 Factorise:

a $15ab - 3a$ **b** ${}^-xy - 9x$ **c** $a^2b + 5a$

4 Find the value of x, correct to 1 decimal place:

a

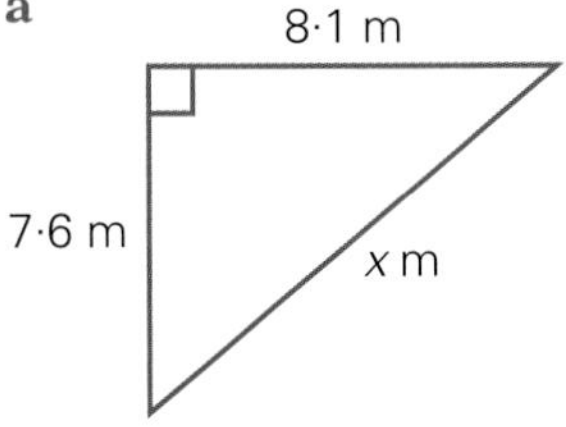

b

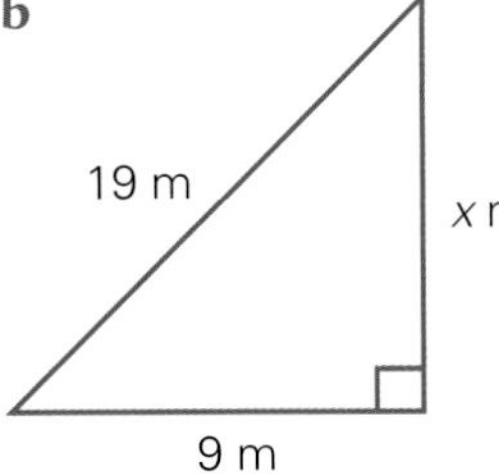

c

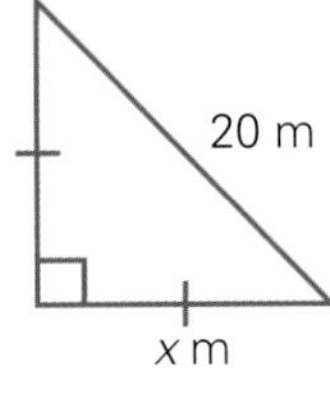

5

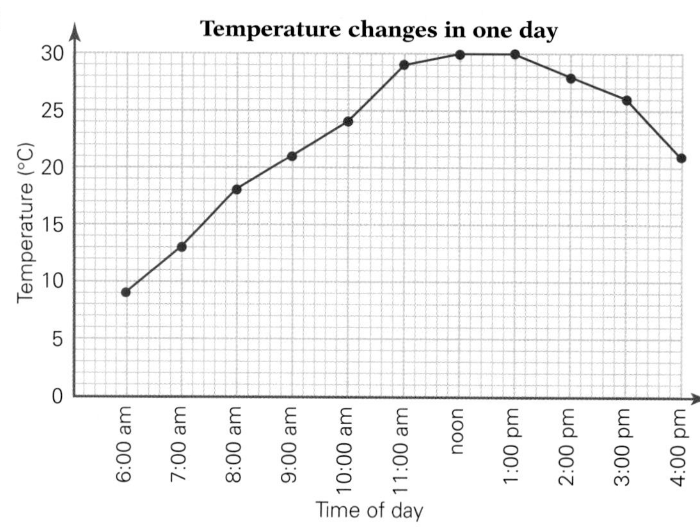

a What was the temperature at **i** 7:00 am **ii** 11:00 am?
b At what time was the temperature 18°C?
c Are in-between values meaningful? Can we estimate the temperature at 3:30 pm?
d What was the fall in temperature between 1:00 pm and 4:00 pm?
e What was the minimum temperature?

6 Solve:

a $7w + 9 = 5w - 3$

b $\frac{2 - 3a}{7} = 3$

c $4(x - 1) = 32$

d $3p + 7 \leq 20$

7 In the diagram at right, name an angle which is:

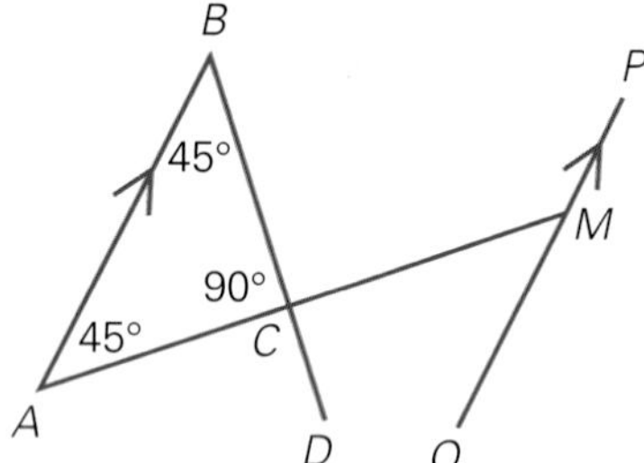

a equal to $\angle ABC$
b adjacent to $\angle CMQ$
c supplementary to $\angle ACB$
d vertically opposite $\angle MCB$
e complementary to $\angle CAB$
f alternate to $\angle BAC$
g co-interior with $\angle BAC$

8 Find the value of the pronumeral in each case:

a

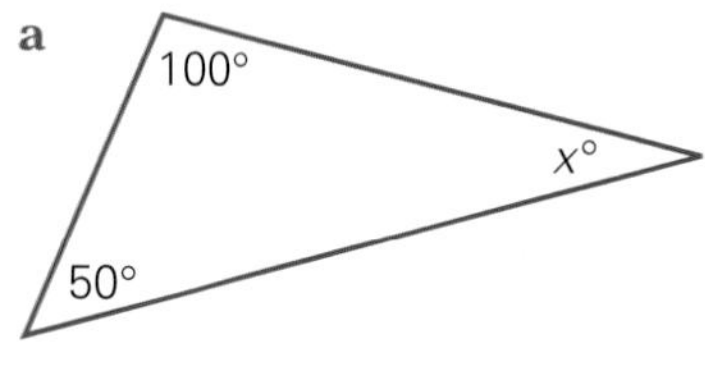

b

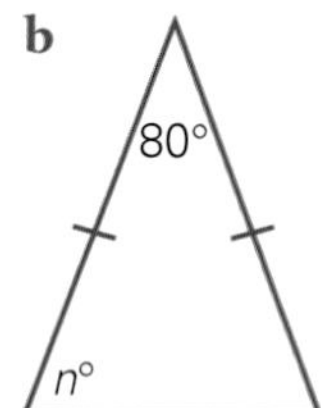

c

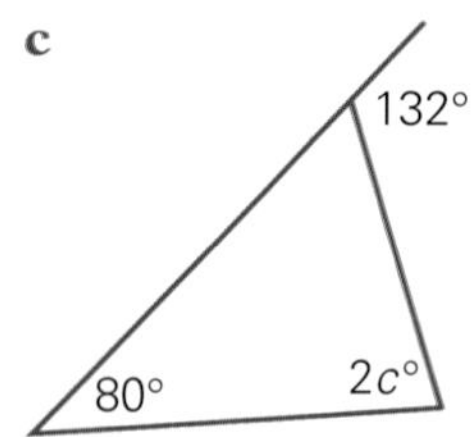

9 Construct each of the following triangles using just ruler and compasses. Classify each according to angles and sides.

a

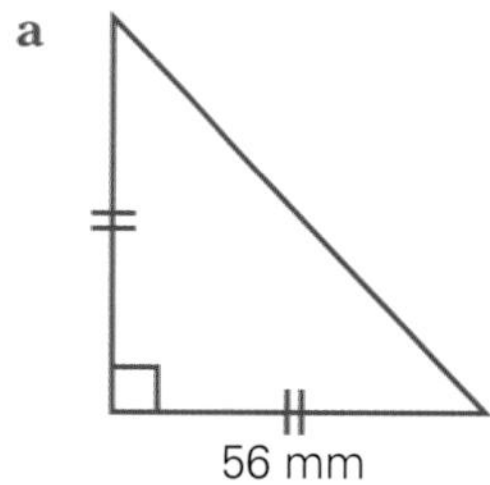

b

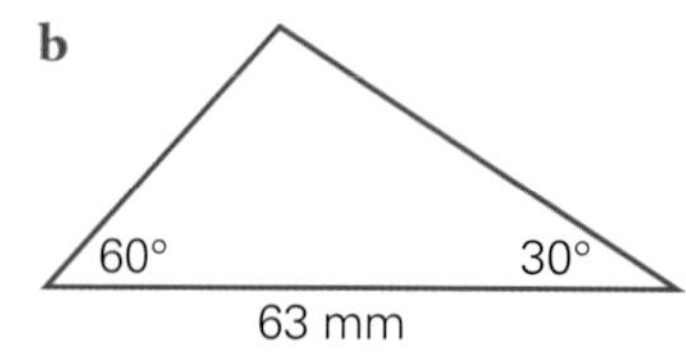

10 Find **i** the volume **ii** the capacity in litres of each container:

a

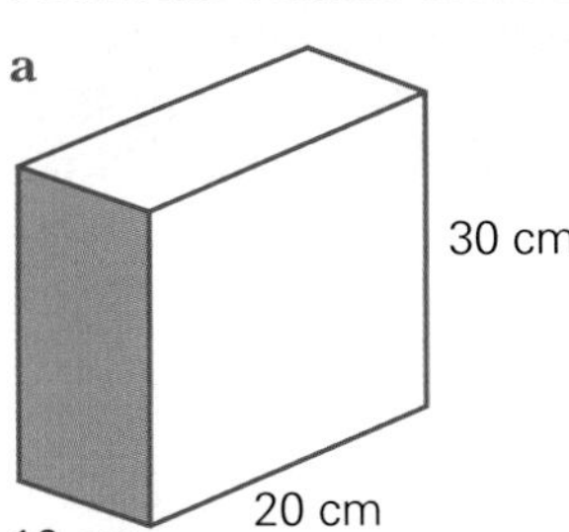

b

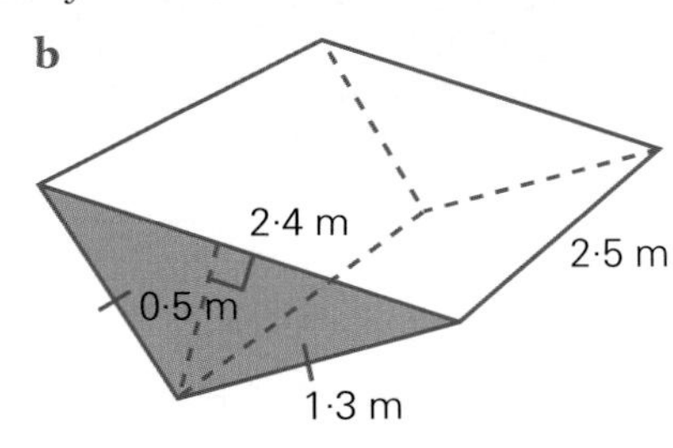

c

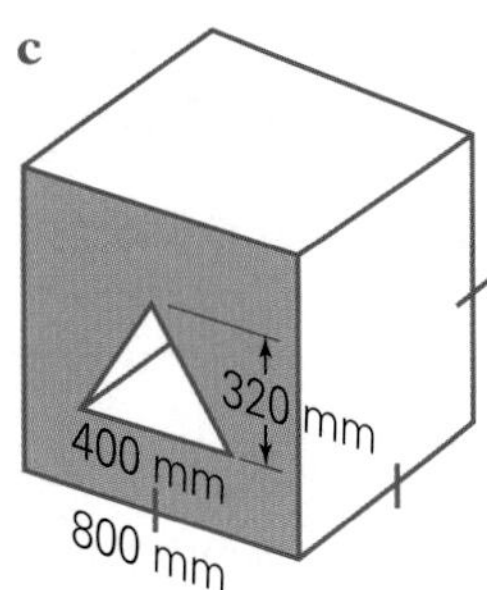

11 Find the surface area of:

a

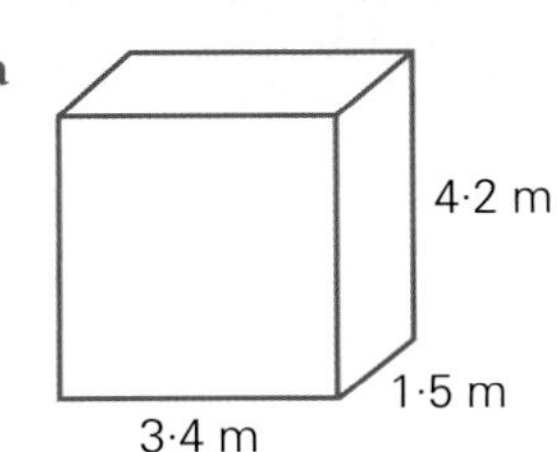

b

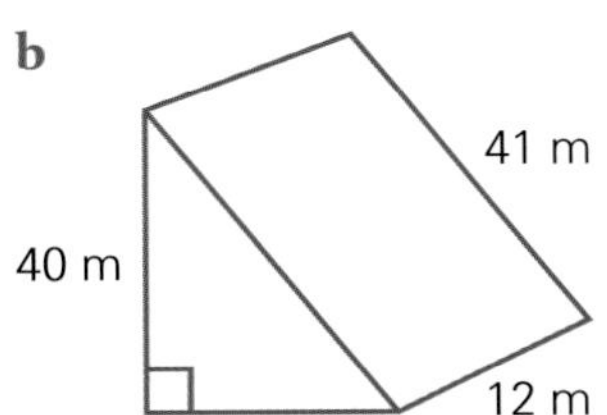

12 500 mL of cordial, 2 L of fruit juice and 6 L of lemonade were mixed in a fruit punch. Write the ratio of:

a fruit juice to lemonade
b cordial to fruit juice
c cordial to lemonade
d lemonade to total mixture

13 Simplify these ratios:

a 10 : 6
b 3·5 : 7·5
c $2\frac{1}{4} : 1\frac{1}{2}$

14 Find the missing number in each pair of equivalent ratios:

a 3 : 4 = 15 : □
b 7 : 3 = □ : 21
c 160 : □ = 5 : 2

15 If 7 : 8 = x : 2, find x.

16 Divide $840 in the ratio 2 : 3.

17 Share 720 kg in the ratio 2 : 3 : 4.

18 The ratio of scouts : tents is 4 : 1. If there are 15 tents, how many scouts will fill them?

19 A cake mixture has 3 cups of flour and $\frac{1}{2}$ a cup of sugar as part of the recipe for a dozen small cakes. How much of each ingredient is needed for 18 cakes?

20 Simplify these rates:

a 250 g for 50 cents
b 60 m in 8 seconds
c $8 for 25 pencils

1 What you need to know and revise

Outcome DS 4.1:
Constructs, reads and interprets graphs, tables, charts and statistical information

2 What you will learn in this chapter

Outcome DS 4.1:
Constructs, reads and interprets graphs, tables, charts and statistical information

Outcome DS 4.2:
Collects statistical data using either a census or a sample and analyses data using measures of location and range

- recognises data as quantitative (discrete or continuous) or categorical
- uses a tally to organise data into a frequency distribution table
- draws frequency histograms and polygons
- draws and uses dot plots
- draws and uses stem-and-leaf plots
- uses the terms 'cluster' and 'outlier' when describing data
- formulates and refines key questions to generate data
- recognises the difference between a census and a sample
- finds the measures of location and range
- collects data
- makes predictions from a sample
- uses spreadsheets to tabulate and graph data
- analyses categorical data

Working Mathematically outcomes WMS 4.1–4.5
Students will be asked to ***question***, ***apply strategies***, ***communicate***, ***reason*** and ***reflect*** in the sections of this chapter.

Statistics

Key mathematical terms you will encounter

analysis
biased
categorical
census
clusters
continuous
correlation slope
cumulative
data
discrete
distribution
dot plot
frequency
grouped
histogram
information
leaf
mean
median
mode
outliers
polygon
population
predictions
quantitative
random
range
representative
sample
statistics
stem
summary
survey
tally

MathsCheck

Statistics

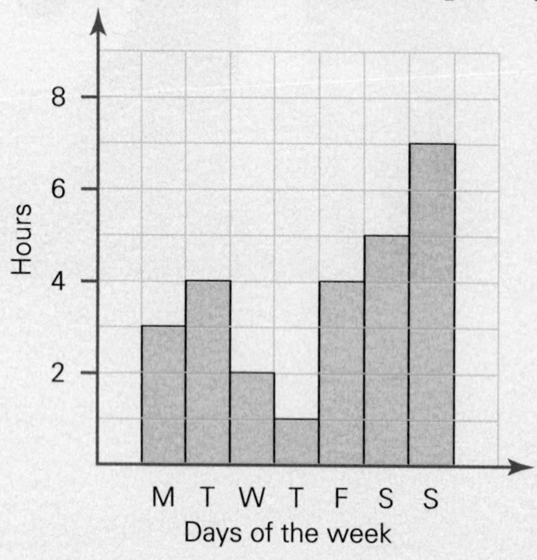

1 What sort of graph is shown?

2 Is the title clear in its meaning?

3 Are the scales accurate?

4 On which day was the most TV watched?

5 How many hours of TV are watched by Johanna over the week?

6 What percentage of her Saturday is spent watching TV?

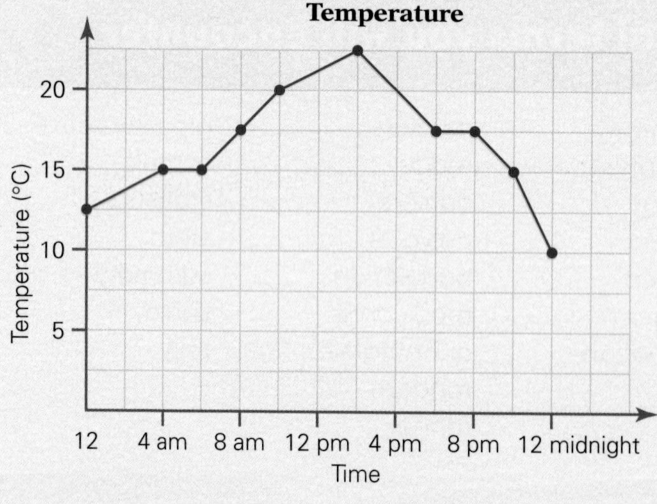

7 Is the data shown at left discrete or continuous? How can you tell?

8 What was the lowest temperature for the day?

9 What was the highest temperature for the day?

10 At what time did the temperature first reach 20 degrees?

Score	Frequency
1	2
2	5
3	7
4	16
5	3
6	2

11 The data shown in this frequency distribution table relates to the number of beds in each of the houses surveyed. Display this data in a column graph.

Exploring New Ideas

10.1 Data and its collection

Statistics involves the collection and display of information.

Governments, businesses, town planners and other organisations require information so that they can decide where facilities such as schools, hospitals and roads are needed. Companies employ market researchers to find out what people are buying, and how best to present and advertise their products. Opinion polls are conducted to find out what people think on many issues.

Data is information. Data could be class marks, the number of cars per family, incomes, heights ... anything that results from counting or measuring.

Data can be:

- **Categorical**: As the name suggests, this is when the data can be classified in groups, for example colours (red, blue, orange), gender (male or female), quality (poor, average, good or excellent).
- **Quantitative**: This is when the data is numerical. It can be either **discrete** or **continuous**.
  - **Discrete data** consists of individual (distinct) values or scores. This means that there are no in-between values, for example the number of students in a class, the number of TV sets per household.
  - **Continuous data** is when the full range of values are possible, for example heights and temperatures.

DATA COLLECTION

Data is collected by taking a **survey**. A survey is a set of questions designed to find out information.

When the *whole population* (the entire group you are interested in) is surveyed it is called a **census**.

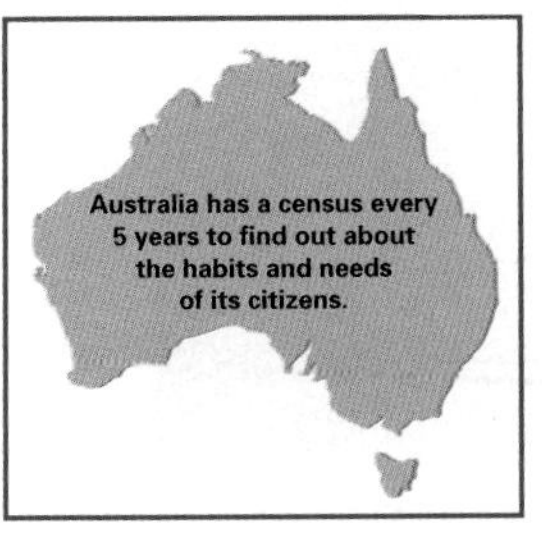

If only a section of the target population is surveyed it is called a **sample**.

A sample should be:

- **random**: this means decided by chance
- **representative**: this means large enough to give results that can be considered to *represent* the entire population

If a sample is neither of these two things, it is said to give a **biased** result.

EXAMPLE 1

Classify the following data as categorical, quantitative discrete or quantitative continuous:

a a question asking a person to indicate Male or Female

b the number of students enrolled in public schools in NSW

Solution

a Gender is categorical data.

b The number of students in NSW public schools (measured as individual students) is quantitative discrete data.

EXERCISE 10A

1 Classify the following data as categorical, quantitative discrete or quantitative continuous:

a the number of cars per household

b the weights of a group of elephants

c the time of arrival for planes at Sydney's Kingsford Smith airport

d the temperature of the ocean during the day

e the number of members present for a sports club training day

f the colours of cars in the school's staff car park

2 A suitable survey question will gather information on the topic only, and gain responses that can be grouped under common headings and then analysed to find trends.

Marcus needs to find out about the shopping habits of people in his local area to discover if there is a need for a large shopping centre to be built.

Decide on the **suitability** of each of the following questions for use by Marcus in his survey.

a How many cars do you own?

b What language do you speak at home?

c How much do you spend each week on food?

d What high school did you attend?

e What is your favourite type of music?

f Do you like mathematics?

g How often do you travel to the shops?

h Do you own a credit card?

i What colour is your hair?

j Do you receive junk mail?

3 You want to find out the favourite recreation activity of students at your school. Which of these would make the most unbiased sample?

A

B all Year 8s in the school

C those in the computer rooms in the lunch break

D those buying lunch at the canteen

E every sixth student on the school roll

4 Who might be left out of each type of survey given?

A a telephone poll

B a postal questionnaire

C a survey of central city mall shoppers on a weekday

D door-to-door interviews during the week

E supermarket shoppers filling in questionnaires at the checkout

F television poll

G TV magazine poll

5 Discuss and decide on whether a random sample or a census is the most appropriate way to collect data on the following:

a your mathematics class

b the travelling habits of the Haton family each morning

c the modes of transport to school by the students of a school

d the world's drinking habits

e the height of 10 friends

6 Suppose the aim is to find out the favourite sport of Year 8 students at your school.

a Rank these surveys in order of accuracy:

A asking five students from each Year 8 class

B asking every Year 8 student

C interviewing 10 of your friends

D interviewing everyone in your class

b Which survey would give an acceptable result?

c Write a suitable question that could be used in a survey to gather data on favourite sports of Year 8 students.

Refine your question and carry out your survey of your classmates.

Level 2

7 **Discuss** and decide whether the samples below would give biased results, and if so how they could be changed so that the results would be truly representative:

a surveying a group of 20 people as they leave a heavy metal rock concert on their favourite music and using the results to represent the musical opinions of Sydney

b surveying five people on their shoe size and using the results to predict the shoe size of the average Year 8 student

c surveying 1000 people from the phone book on their voting habits and using them to predict the election result

8 **Identify** a possible problem with each of these survey questions:

a Do you think greedy, exploiting bosses should be made to give all loyal, hard-working employees an immediate pay rise? [Yes/No]

b If your answer to our phone call is '2KX is my favourite station' we pay you $100!

c Do you prefer this product in its beautiful, stimulating packaging or in a plain wrapper?

d What is your reaction to this new government law? [Hate it/Love it]

e Do you believe the Wik legislation on indigenous land rights will seriously disadvantage rural communities with respect to security of tenure? [Yes/No]

9 Give an example of the following types of data (different to the examples already met in this chapter):

a discrete **b** continuous **c** categorical

10 **Discuss** why two surveys on the same topic might obtain different results.

Investigation

A Ratings

- List three areas in which ratings are used.
- Why are they seen as important?
- How are the ratings decided upon?
- How is the data collected? Use newspaper or magazine examples.
- What are the weaknesses in any of the procedures?
- Is there any potential for bias?
- Include examples of ratings in the media.

Top 5 Books

1. The Life of Pythagoras
2. The Real Story
3. Tall Tales for Short People
4. Much Ado!
5. Spectrum 8

B Design your own survey

- Choose a suitable area to research. Try a survey of your class's favourite type of chocolate.
- Write down five suitable survey questions.
- Discuss and refine to one survey question.
- Survey 20 people and record their responses.
- Display this date colourfully in a suitable graph.
- What **conclusions** can you make?

10.2 Displaying data

It is usual, once you have collected the data, to display it in a way that makes it easy to read.

FREQUENCY HISTOGRAMS AND FREQUENCY POLYGONS

A frequency histogram is a special type of column graph used in statistics. The frequency polygon is a line graph (usually used when the data is continuous) and is often displayed on the same set of axes as the frequency histogram.

EXAMPLE

Display the following data in:

a a frequency distribution table
b a frequency histogram
c a frequency polygon

lowest score

5	6	7	9	8	3	5
5	7	7	7	(2)	5	6
8	7	5	6	4	3	7
7	9	4	7	6	2	(10)

highest score

Solution

a The **frequency distribution table**—we saw these in chapter 4.

Score	Tally	Frequency
2	II	2
3	II	2
4	II	2
5	~~IIII~~	5
6	IIII	4
7	~~IIII~~ III	8
8	II	2
9	II	2
10	I	1

A 'gate post' ~~IIII~~ stands for 5.

b The **frequency histogram**

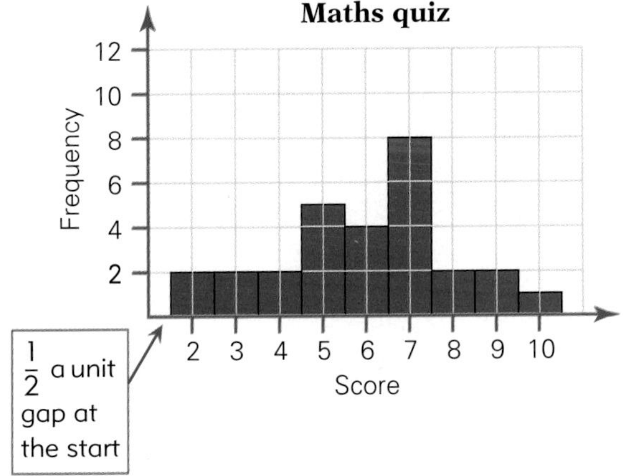

Columns are the same width.
The score is in the middle of the column.
No gaps between the columns.

c The **frequency polygon**

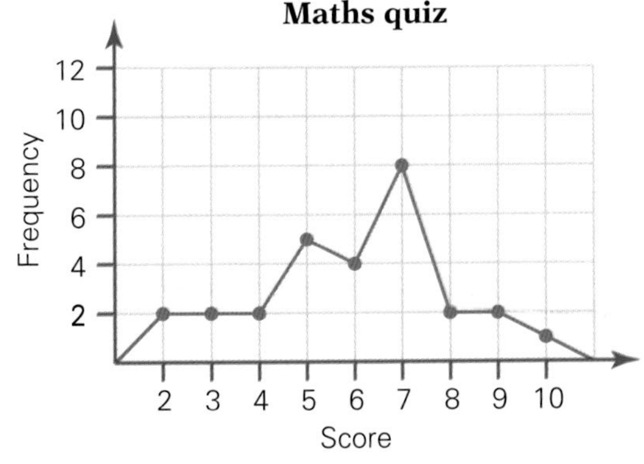

Each dot is directly above the score.

Note: The frequency histogram and the frequency polygon are sometimes displayed on the same set of axes.

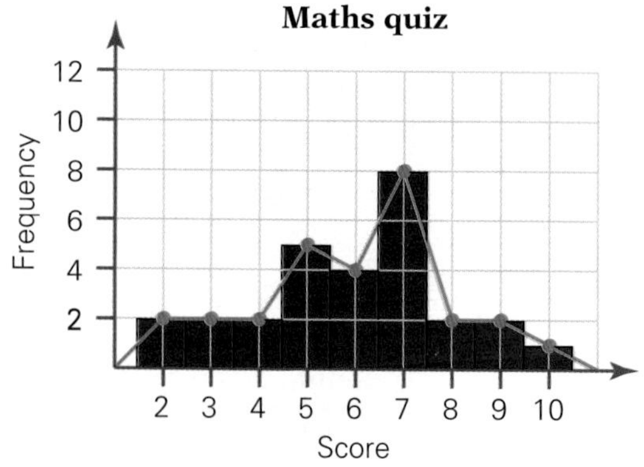

EXERCISE 10B

1 The following questions refer to the frequency polygon given below.

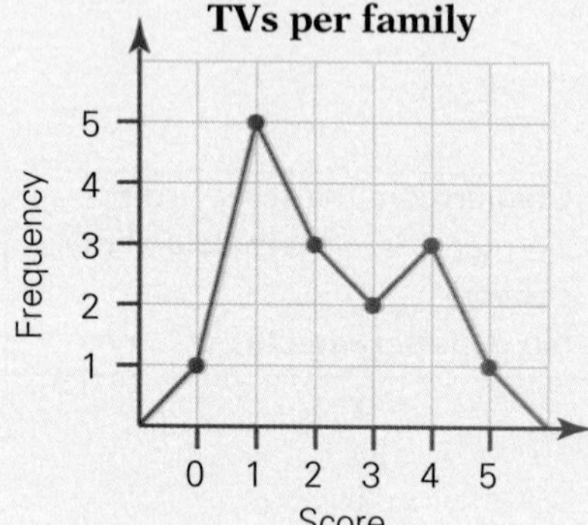

a What is the data about?
b What is the most common score?
c What is the least common score?
d How many scores are there?
e How many families have more than three TVs?
f How many families have fewer than three TVs?
g Is the data displayed in the graph quantitative or categorical?
h Display the data in a frequency distribution table.

Score	Frequency
0	
1	
2	
3	
4	
5	

2 The following questions refer to the frequency histogram.

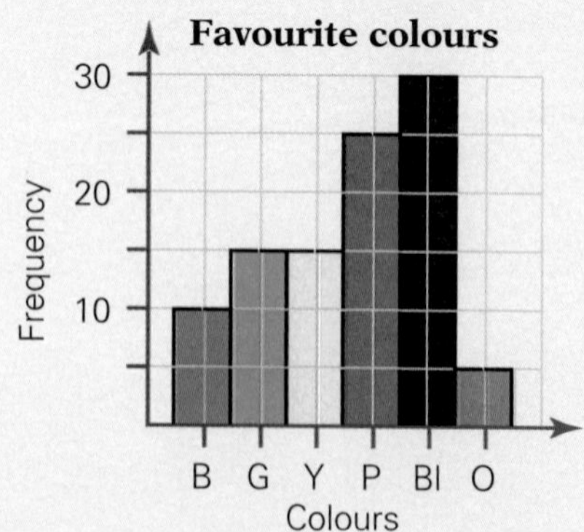

a What is the most popular colour?
b How many people chose green?
c How many people were surveyed?
d What percentage of people liked blue?
e Is the data displayed in the graph quantitative or categorical?
f Draw up a frequency distribution table for the data.

3 Draw a frequency histogram to show the following information:

a

Score (x)	Freq. (f)
0	5
1	6
2	2
3	1
4	2

b

Score (x)	Freq. (f)
130	2
131	3
132	3
133	8
134	3
135	5

4 Draw a frequency polygon for the data in question **3**.

5 The following data shows the number of children in each of the families surveyed.

4	5	3	6	7	2	1	1	1	5
3	4	4	3	2	2	2	2	0	0
2	2	4	3	6	5	4	5	1	2
1	1	2	2	4	7	8	3	7	0
2	2	2	5	3	1	1	0	1	3

a Sort this data in a frequency distribution table.
b What fraction of the families surveyed had fewer than three children?
c What percentage of the families surveyed had exactly three children.
d Present the data as a frequency histogram with a polygon overlay.
e Find the average number of children in the families surveyed.

Level 2

6 Two dice were tossed and the higher of the two numbers on the uppermost face was recorded.

5	6	1	2	3	6	4	6
3	2	5	6	1	3	6	3
5	3	4	4	4	3	2	1
5	4	6	6	6	2	5	4
2	5	6	5	6	5	5	3

a Organise the above data into a frequency distribution table.
b Display the data as a frequency polygon.
c What was the most common number recorded?
d What was the least common number recorded? Is this the number you would expect? **Explain**.
e How many times were the dice tossed?
f What fraction of the tosses accounted for a score of 3 or 4?
g Obtain your own set of data by repeating the process. **Compare** your results to the ones above. Are they similar?

TEACHER

TEACHER

7 *If only Adam had had his microphone that day, people across the school could have heard his speech on animal rights and the environment.*

a Using the above statement, complete a frequency distribution table on the vowels (a, e, i, o, u) used.
b Complete a frequency histogram on the above data.
c What vowel occurs most?
d What vowel occurs least?
e How many times does the letter 'i' occur in this statement?
f What is the total number of vowels in this statement?
g What percentage of vowels are 'e'?

DOT PLOTS

The dot plot is another easy way to display data. Each dot represents a score and its exact position on a horizontal axis.

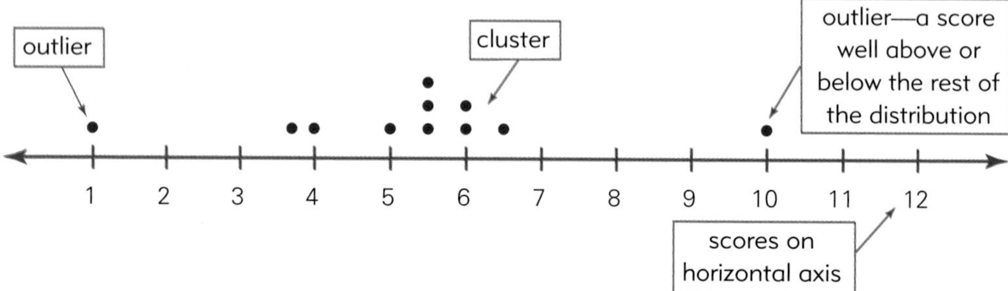

Dot plots clearly show any scores which are **outliers** as well as any **clusters** (groups of scores).

To construct a dot plot:

- Find the lowest and highest scores to construct a scaled number line.
- Space out the dots evenly above the appropriate score.

EXAMPLE

The following are the scores out of 20 for a Year 8 history quiz:

5	12	16	16	15	16	19
20	17	16	10	10	15	16
18	20	10	17	18	13	16
14	20	17	16	16	15	16

Use a dot plot to find:

a the lowest score
b how many students scored full marks
c how many students scored $\frac{16}{20}$
d the value of any score that could be considered an outlier

Solution

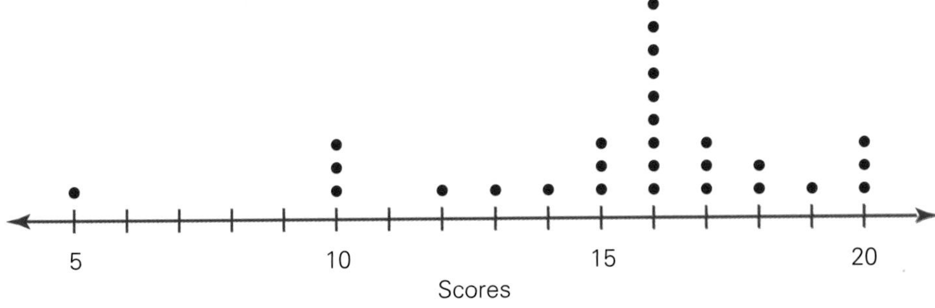

a The lowest mark was 5 out of 20.
b Three people scored full marks.
c The score of $\frac{16}{20}$ was achieved by nine students.
d The score of 5 is an outlier.

EXERCISE 10C

1 The following dot plot represents students' correct responses in a 30-question trivia quiz.

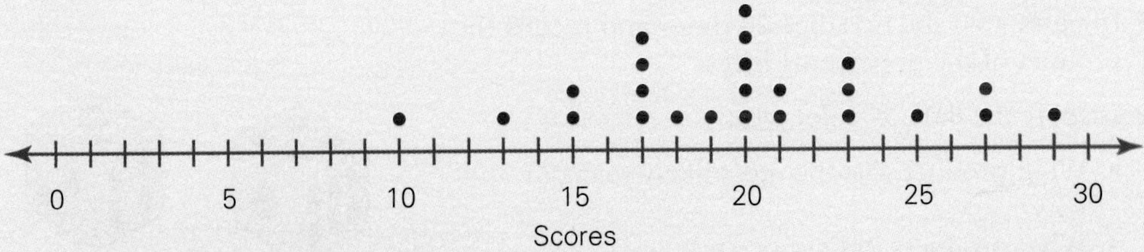

a How many students participated in the quiz?
b What was the lowest score?
c What was the most common score?
d How many students scored over 20?

2 The following dot plot indicates the number of TV sets in the households of the students surveyed.

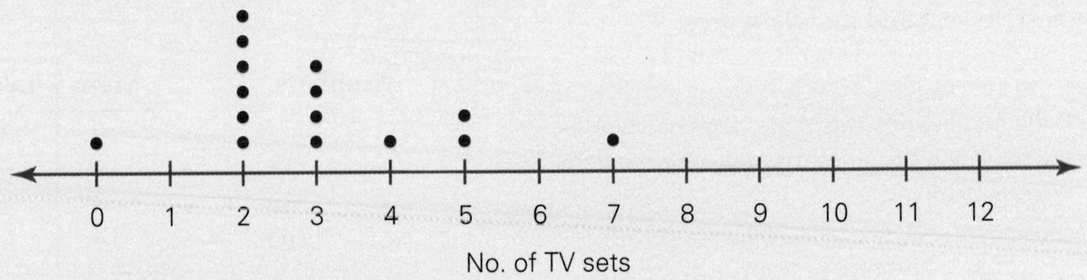

a How many students were surveyed?
b Were there any scores that could be classified as outliers? If so, what are they?
c What fraction of the students surveyed had more than three TV sets in their homes?
d What was the highest number of TV sets in any of the households surveyed?
e What was the most common number of TV sets in any of the households?

f Use these results to predict the number of TV sets in any household in Australia.

Level 2

3 Jeremy travels from Sutherland to Camden each weekday. Listed below is the time in minutes taken for Jeremy to drive to work each day during the month of April.

47	45	59	68	45	53	54	62	58	58	58
55	47	53	62	58	58	57	51	49	58	53

a Construct a dot plot for the above data using a suitable scale (45–70).
b Are there any clusters of scores in the data above? Justify your conclusion.
c What is the shortest time taken for Jeremy to make the trip to Camden?
d What factors could account for the variation in travelling times for Jeremy?
e If Jeremy was due at Camden at 9:30 am, at what time should he begin his journey on any given day? Justify your answer.
f Discuss the occurrences of any clusters and outliers.

Investigation Rolling dice

You will need: a pair of dice for each group

1 In pairs, toss the two dice 20 times and record the product of the uppermost faces.

2 Display the data as a dot plot.

- What product was the most likely to occur and why?
- What products did not occur?

3 Repeat and look at the sum of the uppermost faces.

STEM-AND-LEAF PLOTS

A stem-and-leaf plot is a table that can be used to organise data. It allows us to place the data into groups. It gives a good idea as to the shape of the data and can be used to identify the highest, lowest and middle scores.

> If the number has two digits, such as 16 or 34, the tens digit becomes the **stem** and the units digit the **leaf**.

Numbers		**Stem**	**Leaf**
3	⟶	0	3
16	⟶	1	6
34	⟶	3	4
143	⟶	14	3

3 | 4 means 34

EXAMPLE

The ages of the people attending Jennifer's 30th birthday are shown below. Display the data in a stem-and-leaf plot.

2	6	38	43	30	31	35
22	32	47	45	29	27	26
29	30	30	34	44	15	19

Solution An ordered stem-and-leaf plot has the leaves arranged in ascending order.

Stem	Leaf
0	2 6
1	5 9
2	2 6 7 9 9
3	0 0 0 1 2 4 5 8
4	3 4 5 7

4 | 7 means 47 years old

EXERCISE 10D

1 The stem-and-leaf plot shows the number of Year 8 students ordering their lunch on any given day at the school canteen during the month of March.

Stem	Leaf
0	6 8
1	0 4 7 9 9
2	5 8 9
3	1 3 5 5 5 9
4	0 7 8 9
5	2

5 | 2 means 52 Year 8 students

a How many school days were there in March?
b How many lunch orders were placed by Year 8 students in March?
c What was the most number of school lunches ordered by Year 8 students on any given day in March?
d On how many days were more than 30 lunch orders placed by Year 8 students at the school surveyed?

2 Rewrite the stem-and-leaf plots below as **ordered** stem-and-leaf plots.

a

Stem	Leaf
0	3 7 5 3 2
1	5 3 7 5 6 1
2	6 3 2 8 5 8
3	0 9 3 5 3 5

b

Stem	Leaf
2	6 7 4 2 0 8
3	9 6 6 8 0 1 4
4	8 9 9 2 1 6 7 7 0 2

3 For each of the stem-and-leaf plots shown, write out a list of data in ascending order.

a

Stem	Leaf
1	0 0 5
2	1 3 5 8
3	3 7 8 9 9
4	
5	7 7 9

5 | 9 means 59

b

Stem	Leaf
10	2 3 6
11	5 8 9 9
12	0
13	3 5 7
14	0

14 | 0 means 14·0

c

Stem	Leaf
245	1 3 8 9
246	0 4 6
247	5 5
248	0 4 7

248 | 7 means 248·7

4 Organise the data below into stem-and-leaf plots.

a Age of women when they married (years)

18	24	35	33	20	23	23
38	24	28	29	30	42	19

b The number of years the women above were married (years)

25	30	47	1	12	6	34
40	5	50	22	14	2	17

c Average speeds of motorbikes in a race (km/h)

125	100	126	130	145	156	129	135	162	102
120	150	142	105	111	155	138	124	125	160

Level 2

5 The following are ordered stem-and-leaf plots. What could the missing digit be?

a

Stem	Leaf
0	4 7 9
1	3 □ 4 6
2	1 7 2

b

Stem	Leaf
5	1 7
6	7 □ 9
7	3 9
8	1

c

Stem	Leaf
41	0 1 3 4 8
42	0 2 2 □ 2

6 Two groups of students have their pulse rates (beats per minute) recorded. The results are shown in the back-to-back stem-and-leaf plot below.

Group B	Stem	Group A
7 7 6 5 5	6	0 2 4 9
8 6 4 2 2 0	7	2 4 6 7 8 8 9
6 4 2 2 1 1 1	8	0 1 3 6 8 8
	9	1

(Annotations: 65 with an arrow from stem 6 to leaf 5 in Group B; 91 with an arrow from stem 9 to leaf 1 in Group A.)

Remember to read the stem first.

a How many students were in **i** group A **ii** group B?
b List the pulse rates for the students in group B.
c In which group was the student with the:
  i highest pulse rate?
  ii lowest pulse rate?
d How many students in group A had a pulse rate over 85 beats per minute?
e If a 'normal' pulse rate for the students tested is considered approximately 60 to 75 beats per minute, which group represents the 'healthier' group. **Justify** your answer.

10.3 Analysing data

Once data is collected, it is usual for it to be analysed to see if there are any special tendencies, for example:

- Are the scores close together?
- What is the most common score?
- What is the average score?
- What do the scores lie between?
- What score lies in the middle?

The **mean**, **mode** and **median** are all measurements that describe the *centre of the distribution*. They are the measures of location.

The **range** describes the spread of the scores.

FINDING THE MEAN FOR SMALL SETS OF SCORES

$$\text{mean} = \frac{\text{total of all scores}}{\text{number of scores}}$$

or $\bar{x} = \dfrac{\Sigma x}{n}$

$\bar{x}$ mean

x scores

n number of scores

f frequency

Σ sum of

The **average** marks in your exams are calculated using this method.

EXAMPLE

Find the mean of 11, 16, 13, 15, 14, 19, 13.

Solution

$$\text{mean} = \frac{\text{total of all scores}}{\text{number of scores}}$$

$$\bar{x} = \frac{\Sigma x}{n}$$

Total:
(11 + 16 + 13 + 15 + 14 + 19 + 13) = 101

$$= \frac{101}{7}$$

There are 7 scores.

$= 14{\cdot}4$ (1 dec. pl.)

Note: The mean is often not one of the original scores.

The mean can be calculated on your calculator.

- Put your calculator into statistics mode (SD).
- Make sure the memory is clear (scl).
- Enter each score pressing M^+ or DATA after each one.
- To find the mean use the $\bar{x}$ function key.

Always clear the statistics memory before starting on a new set of data.

EXERCISE 10E

1 Find the mean for each of the following sets of scores.

Check these on your calculator.

- **a** 5, 6, 7, 8, 9, 10
- **b** 56, 92, 86, 101, 95
- **c** 0·6, 1·3, 0·9, 1·4
- **d** 47·6, 47·8, 47·0, 47·9, 47·5
- **e** 6, 9, 15, 16, 15, 20

2 Ahmed scored the following number of goals per game during a soccer season. What was Ahmed's mean number of goals per game?

0	1	2	0	1	1	1	0	2
1	3	0	0	0	3	1	3	1

3 The rainfall in Maxville (in mm) during the month of February was recorded each day as follows:

2	12	2	0	0	4	7
8	2	4	6	0	0	0
2	8	7	12	16	2	1
4	6	8	0	0	4	6

Calculate the mean rainfall for Maxville for the month of February.

Level 2

4 Find an expression for the mean of x, $x + 1$, y, $y - 3$, 4

5 If the mean of a set of nine scores is 32, what is the total of the scores?

6 In a cricket season, a batter's average run rate was 34. How many runs did the batter score in a season of 12 matches?

7 The mean height of a group of five students is 143·75 cm. Another student whose height is 170 cm joins the group.

- **a** What is the total height of the five students?
- **b** What is the new total when the sixth student is added?
- **c** What is the average height of the six students?

8 The average of 10 scores is 13. What is the new average when a score of 5 is included?

9 The scores 5, 6, 7, 9, 8 and x have a mean of 12. What is the value of x?

10 The scores 100, 245, 234, x and $2x$ have a mean of 247. What is the value of x?

11 Cheyne wishes to end the year with a mean examination mark of at least 90%. In the four tests so far this year, Cheyne has averaged 88.5%. What is the lowest mark that Cheyne can receive in the final exam if he wishes to reach his goal?

12 a What number would you add to these to get a mean of 20?

17 24 19 36 ☐

b What number would you add to these to get a mean of 70?

90 50 75 130 ☐

c Which four of these numbers together have a mean of 6·5?

7·2 5 8·3 6·5 12

FINDING THE MODE, MEDIAN AND RANGE FOR SMALL SETS OF DATA

The mode

The **mode** is the **most popular score**:

- the score with the **highest frequency**
- the score that occurred the **most**

M.O. for Mode
M.O. for Most

The median

The **median** is the **middle** score when all the scores are *arranged* in either *ascending* or *descending* order:

- the middle score for an odd number of scores
- the average of the two middle scores for an even number of scores

The median strip is also in the middle of the road.

The same number of scores will occur on either side of the median.

The range

The **range** describes the **spread** of scores.

range = highest score − lowest score

EXAMPLE 1

Find the modal score for the following:

5	8	5	6	1	5
9	8	5	6	5	1

Solution

The mode is 5.

The score of 5 occurred more than any other score.

EXAMPLE 2

Find the median in each of the following sets of data.

a 16, 11, 15, 14, 13, 19, 13

b 28, 20, 12, 25, 16, 5, 10, 10, 27

Solution

a Arrange in ascending order.

11 13 13 14 15 16 19

↑

The median is 14.

b Arrange in ascending order.

↓

5, 10, 10, 12, 16, 20, 25, 25, 27, 28

$\text{median} = \frac{16+20}{2}$

$= 18$

Average the two middle scores.

The median is 18.

EXAMPLE 3

Find the range of 5, 6, 1, 12, 7, 3.

Solution

Range = 12 – 1
= 11

The range is 11.

> **range = highest score – lowest score**
> highest score = 12
> lowest score = 1

EXERCISE 10F

1 Find the mode, median and range for each of the following sets of data:

a 2, 4, 5, 2, 3, 4, 2, 2
b 56, 57, 58, 89, 32
c 0, 0, 3, 0, 2, 1, 4, 5, 3, 0, 4
d 99, 110, 34, 56, 100
e 7, 8, 7, 8, 7, 4, 6, 8, 6, 9, 10, 7, 8, 7
f 0·8, 0·7, 2·1, 3·2, 0·51, 2·4, 1·9, 0·1

2 Calculate the mean for the sets of data given in question **1** using your calculator.

3 The following are the prices of the different types of drinks sold at the school canteen:
$1.20 $1.40 $1.20 $1.20 $1.80 $1.20

a What is the range of prices for drinks at the canteen?
b What is the modal price?
c What is the median price?
d How many different types of drinks are sold at the canteen?

4 Find the mean, mode, median and range for:

a 7, 12, 0, 56, 19, 192, 29, 0
b 101, 200, 100, 200, 97

5 **a** Find the mean for the following data set:
3, 5, 8, 5, 8, 7, 1, 3
b A score of 5 is added to the set above. What effect does this have on the mean?
c A score of 7 is added to the original set of data above. What effect does this have on the mean?
d Find the median and mode for the original set of data. What effect does the addition of a new score of 5 have on the median and mode?

6 Find the mean, mode, median and range for each of the following sets of data:

a 7, 5, 8, 7, 9, 11, 1
b 14, 10, 16, 14, 18, 22, 2
c 8, 6, 9, 8, 10, 12, 2
d 6, 4, 7, 6, 8, 10, 0

Level 2

7 When five scores are analysed they are found to have a median of 3, a mean of 3, and the score of 2 has a frequency of 2. What are the five scores?

8 Who has the better results?
Leigh: 90, 87, 63, 90, 75
Justin: 55, 90, 80, 90, 90

What measure best describes these results and **why**?

9 A real estate agency has eight properties on its books for one suburb in Sydney. The prices listed are:
$497 000, $286 000, $314 000, $400 000,
$390 000, $1 500 000, $395 000, $410 000

a Find the mean price for the properties listed.
b Find the median price for the properties listed.
c Which best represents the centre of the data, the mean or the median, and **why**?
d Find out which statistic is commonly used to summarise house prices in the media.

10 Two soccer teams both play eight games each. Their results are summarised in the table below.

Team A	0	1	1	4	2	0	1	2
Team B	0	1	1	3	1	2	2	2

a Calculate the mean and range for the results of both teams.
b Which team is more consistent? **Explain** your answer.

11 a Which three of these numbers together have a median of 20 and a range of 20?
30 50 10 40 20
b Which four of these numbers together have a mean of 15 and a median of 15?
20 10 7 18 14 16
c Which four of these numbers together have a median of 5 and a mode of 4?
2 4 6 9 4 3

12 Find a set of five scores whose mean, mode, median and range all equal 5.

10.4 Further analysing data

As we have seen, data can be summarised using the summary statistics mean, mode, median and range. In this section we will find the summary statistics from frequency distribution tables, stem-and-leaf plots and dot plots.

EXAMPLE 1

Find the mean, mode, median and range for the data displayed in the dot plot.

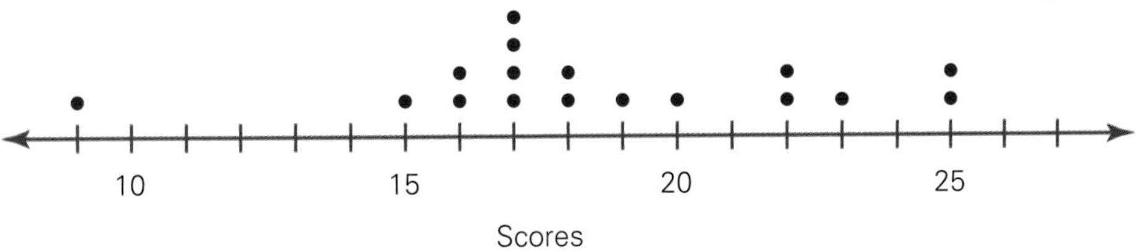

Solution

Mean $\bar{x} = \frac{\Sigma x}{n}$

$= \frac{316}{17}$

$= 18{\cdot}6$ (1 dec. pl.)

Mode = 17 (highest column)

Median

The median of 17 scores is the 9th score.

Median is 18.

Range = 25 − 9
= 16

EXAMPLE 2

Find the mean, mode, median and range for the data displayed in the stem-and-leaf plot.

Stem	Leaf
0	0 7 9
1	1 1 2 3 5 8
2	1 3 3 5
3	4 8

Solution

Mean $\bar{x} = \frac{\Sigma x}{n}$

$= \frac{260}{15}$

$= 17\frac{1}{3}$

Modes are 11 and 23.

The data is bimodal.

Median

The median of 15 scores is the 8th score.

Median is 15.

Range $= 38 - 0$

$= 38$

EXAMPLE 3

Find the mean, mode, median and range for the data displayed in the frequency distribution table.

Score (x)	Frequency (f)
1	5
2	8
3	6
4	5
5	1
Total	25

Solution

Mode is 2.

Range $= 5 - 1$

$= 4$

Mean

The mean may be calculated by extending the frequency table as shown.

x	f	fx
1	5	$1 \times 5 = 5$
2	8	$2 \times 8 = 16$
3	6	$3 \times 6 = 18$
4	5	$4 \times 5 = 20$
5	1	$5 \times 1 = 5$
	$\Sigma f = 25$	$\Sigma fx = 64$

Freq. × score

A quick way of adding all the scores together

$\bar{x} = \frac{\Sigma fx}{\Sigma f}$

$= \frac{64}{25}$

$= 2{\cdot}56$

Median

The median of 25 scores is the 13th score.

x	Frequency	
1	5	
2	8	$5 + 8 = 13$
3	6	$13 + 6 = 19$
4	5	
5	1	

Count down the freq. column until you reach the 13th score.

The 13th score has a freq. of 8; that is, the **score** of 2.

The median is 2.

EXERCISE 10G

1 The ages of the students on the Mathsville High student council are shown in the following frequency distribution table.

Use the table to calculate the mean age of the students on the council.

Score (x)	Frequency (f)	Freq. × score (fx)
12	3	$12 \times 3 = 36$
13	5	$13 \times 5 = 65$
14	4	$14 \times 4 = 56$
15	7	$15 \times 7 = 105$
16	2	$16 \times 2 = 32$
17	3	$17 \times 3 = 51$
Total	24	345

Add on the fx column.

Three scores of 12 account for 36 towards the total of all scores.

2 Complete each of the following tables and calculate the mean:

a

x	f	fx
0	3	
2	8	
4	12	
6	5	
8	1	

b

x	f	fx
132	1	
133	3	
134	7	
135	2	
136	3	
137	2	

c

x	f	fx
7	8	
8	15	
9	24	
10	9	

d

x	f	fx
15	100	
16	97	
17	56	
18	172	
19	93	

3 State the mode for each of the frequency distribution tables given in question **2**.

4 Give the range for each of the frequency distribution tables given in question **2**.

5 Find the median of each of the following sets of data shown below:

a

x	f	
0	4	
1	7	4 + 7 = 11
2	8	11 + 8 = 19
3	4	19 + 4 = 23

Median is the 12th score.

b

x	f	
10	3	
11	2	3 + 2 = 5
12	5	5 + 5 = 10
13	4	10 + 4 = 14
14	4	14 + 4 = 18

Median is the average of the 9th and 10th scores.

6 Find the mean of each of the following sets of data shown in stem-and-leaf plots.

a

Stem	Leaf
0	4 4 7 9
1	1 3 4 7 8 9
2	0 1 5

b

Stem	Leaf
10	1 2 4
11	4 4 7 8
12	0 5 7 9
13	1 6

c

Stem	Leaf
7	1 1 4 7 8 9
9	0 7 7 8 8 9 9

7 Find the mean, mode, median and range for the ages of guests at an engagement party displayed on the dot plot below.

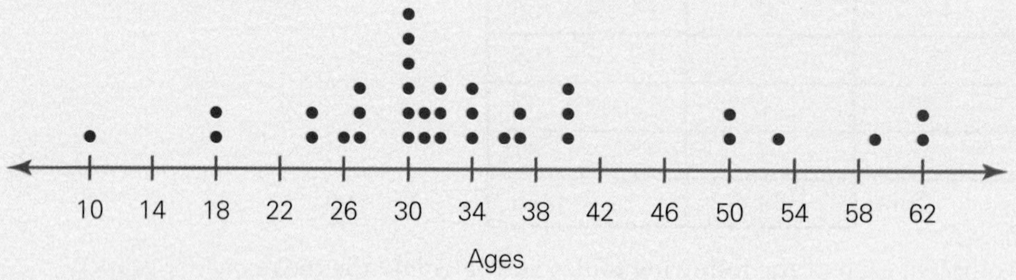

Level 2

8 Fifty bags of lollies were inspected to find the number of lollies each contained. The results were:

36	42	38	38	39	40	45	38	42	44
37	38	38	39	42	43	42	37	36	37
40	43	41	40	39	39	38	42	45	38
38	37	38	43	44	38	36	38	40	39
38	38	41	37	36	43	41	38	38	43

a Display the data in a frequency distribution table.
b What was the modal number of lollies?
c What was the range of lollies in the 50 bags?
d What was the mean number of lollies?
e What is the median number of lollies?
f **i** If you were to purchase a bag of mixed lollies, how many lollies could you expect it to contain?
ii Which measure gives you the best indication of this and **why**?

9 Find the median and the mode of each of the following data sets shown in the stem-and-leaf plots.

a

Stem	Leaf
0	4 7 6
1	1 1 5 7
2	4 9

b

Stem	Leaf
12	1 1 1
13	2 3 7 9
14	4 7 7
15	0 2

c

Stem	Leaf
0	1 4 7
1	0 2 2 4 8 8
2	1 3 9
3	0 4

10 Find the median and the mode of each of the following sets of data shown in the frequency distribution table.

a

x	f
0	1
1	7
2	3
3	4
4	2

b

x	f
6	4
6·5	5
7	9
7·5	7
8	6
8·5	4

c

x	f
12	8
13	12
14	24
15	6
16	10

11

Score	Frequency
4	4
5	6
x	6

Find the value of x if the given set of scores has a mean of $5\frac{1}{2}$.

12

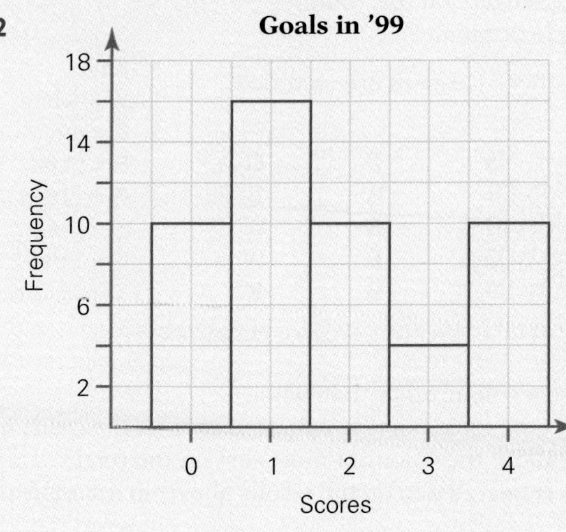

In the frequency histogram at left:

a What is the range of goals scored?

b Find the mode.

c Display in a frequency distribution table.

d Calculate the mean score.

e Find the median score.

13 Passengers on a European tour were surveyed and the following statistics were gathered.

Place of origin	
Score	**Frequency**
NSW	16
VIC	9
QLD	7
WA	4
SA	1
NT	2
NZ	6

Age of passengers	
Score	**Frequency**
20	4
21	2
22	6
23	10
24	3
25	4
26	3
27	2
28	4
29	3
30	3
31	1

a If the company wished to advertise to increase the number of tours operating each season, where should they concentrate their advertising? Explain your answer.

b What percentage of the passengers were under 23 years of age?

c Construct a frequency histogram to show the places of origin.

d Calculate the mean age of the passengers on this tour.

e From which place did most people originate?

14 The colours of cars found in the teachers' car park are recorded as follows:

W	R	W	W	B	G
S	W	R	R	W	B
G	S	O	G	R	W
B	B	O	G	B	W
S	W	W	S	B	R

Key:
W = white
R = red
B = blue
G = green
S = silver
O = other

a Construct a frequency distribution table for the categorical data above.

b What is the modal colour?

c What percentage of the cars were a colour other than white?

d If these results are used to predict the most popular colour of cars on the roads in NSW, what would you conclude about the colour of most cars on the road?

e How many white cars would you expect, based on the results above, in a sample of 2 million cars?

f Carry out your own survey. Does it **support** your conclusions from above?

15 Calculate the mean for each of the following:

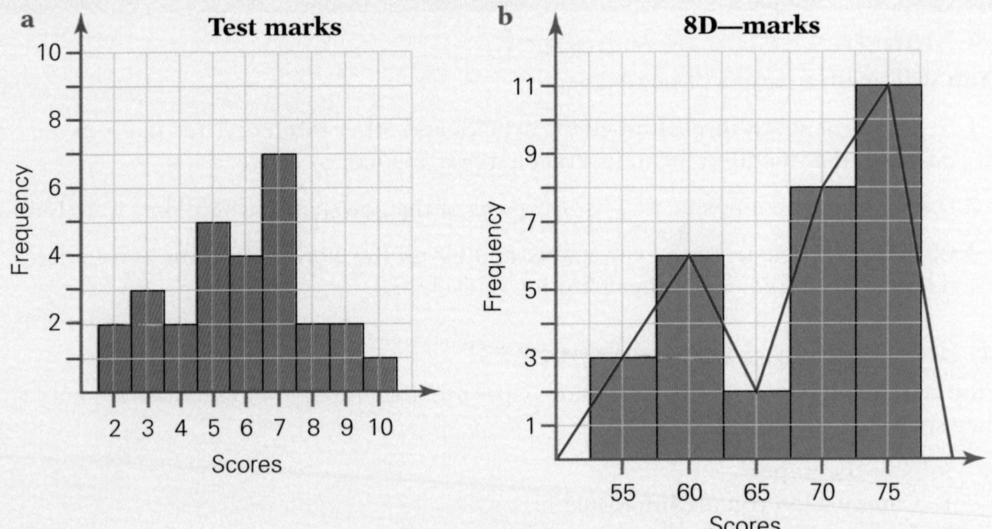

16 Two basketball teams from two different high schools compete in a 15-game competition. Their results are shown in the table below.

Game	1	2	3	4	5	6	7	8	9	10	11	12	13	14	15
Team A	24	36	40	35	44	50	37	29	36	52	29	30	34	40	40
Team B	30	33	42	44	34	38	45	41	43	36	38	44	30	30	28

a Copy and complete the following back-to-back stem-and-leaf plot.

Team B	Stem	Team A
8	2	
8 8 6 4 3 0 0 0	3	
5 4 4 3 2 1	4	
	5	
	6	

b Find the **i** mean **ii** mode **iii** median **iv** range for team A's results.

c Find the **i** mean **ii** mode **iii** median **iv** range for team B's results.

d **Reflect** on the similarities and differences in the means, modes, medians and range and **comment** on the abilities and consistencies of both teams.

17 a Use the **random number function** on your calculator to find 20 integers lying between 0 and 10.

b Display your results in a frequency distribution table and find the four summary statistics.

c Repeat and **compare** your two data sets.

Investigation

A Statistics in the newspaper I

You will need: a variety of newspapers

1 In groups of about three, find all the articles and other references (as many as possible) that mention mean (average), mode, median or range.

2 Display these on a poster, and list the parts of the newspaper where they were found.

3 Choose an article from the newspaper and list all the words you come across that relate to this topic. Write the definition of each one.

B Statistics and the newspaper II

You will need: copies of the real estate pages and the employment sections from a newspaper

a **Jobs in the paper**
  1 Choose a job you are interested in.
  2 Find as many references to this job as possible in the paper.
  3 Record the income for each of the jobs listed.
  4 Calculate:
    - the mean income for this job
    - the range of the incomes for this job

b **Areas to live**
  1 Research the prices of property in your local area and calculate:
    - the mean price
    - the range of prices
  2 List any reasons you can think of that explain the variations in property prices.

C Random sample survey—sport

1 Survey 30 people on their:

  a age b favourite sport

2 Display these results in the ways discussed in this chapter. Include at least two different types of graphs.

3 What **conclusions** can be made after analysing your results?

10.5 Predictions and scattergraphs

Having surveyed a random sample large enough to truly represent a whole population, we collect and display the data. We then analyse it by finding mode, median, mean and range to get a good picture of how the data clusters and spreads.

Finally we **estimate** what that means for the whole population at this time, and **predict** future results.

Scattergraphs can also be used to see relationships and make predictions.

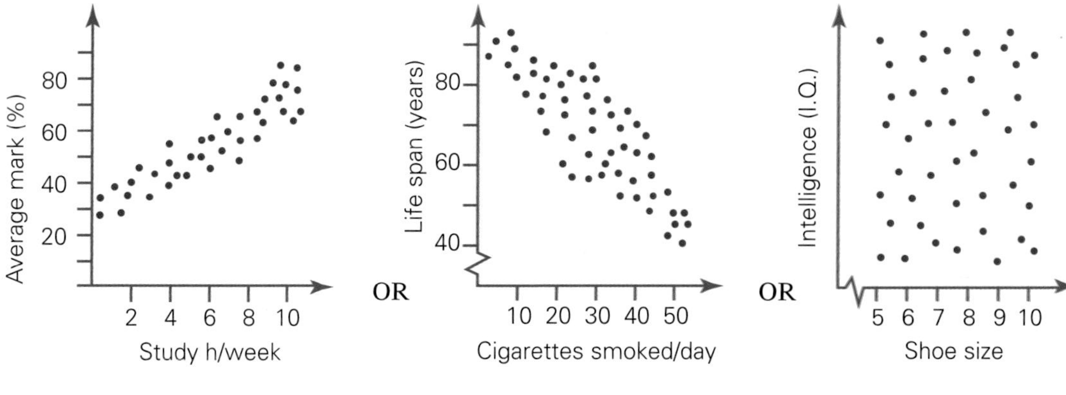

A Positive correlation **B Negative correlation** **C No correlation**

Scattergraphs graph two variables to see if there is a relationship. The three situations shown above illustrate common results.

A shows *the more you study, the better your results*: **positive correlation** graphs as a **rising slope**, with points clustering to suggest a straight line.
B shows *the more cigarettes you smoke, the shorter your lifespan*: **negative correlation** graphs as a **falling slope**, with points clustering to suggest a straight line.
C shows **no relationship** between the variables; no clustering of points suggests no alignment, either rising or falling.

Where a clear correlation shows up on a scattergraph, the graph can be used to predict probable outcomes.

The dots are from data already collected about a Year 8 class.

Erik puts in 7 hours per week of quality study time. If he is of reasonable intelligence, predict what his average mark should be.

By ruling up from the 7-hour mark on the horizontal axis to the **line of best fit** (a straight line drawn through the middle of the dots to show their trend), then across to the vertical axis a reading of about 72% is obtained.

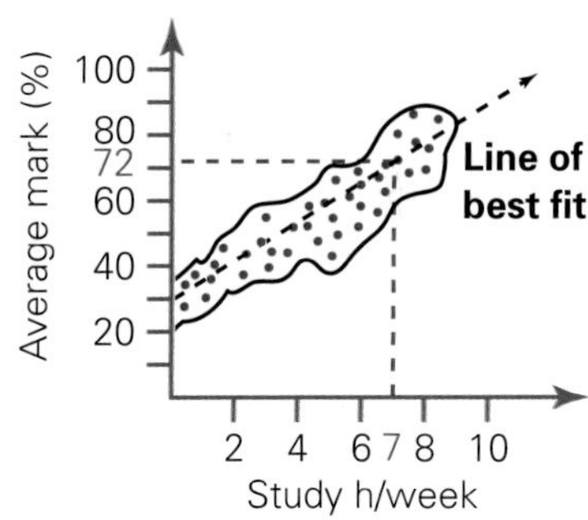

EXAMPLE

A nursery assistant selects 20 trays at random from a batch of 1000, each containing 10 seeds. After 1 week, an average of nine seeds per tray have sprouted. Predict how many seeds will sprout over the entire batch.

Solution

Sample average $= \dfrac{9}{10}$

Prediction: $\dfrac{9}{10}$ of 1000×10

$= 9000$ seeds will sprout

EXERCISE 10H

1 The average number of rainy days in September in a certain town has been 12 over the last 20 years.

a How many rainy days would you predict for next September?
b On what percentage of the September days do you expect rain to fall?

2 A biscuit factory selects a random sample of 50 biscuits from the production line for close inspection. It is found that three have no chocolate chips in them and should be rejected.

a Express the sample rejection rates as **i** a fraction and **ii** a percentage.
b At this rate, estimate how many of the day's production of 20 000 will be rejected.

3 Fifteen out of the 25 in 8A went to the Easter Show. Of the 200 students in the whole of Year 8 at the school, how many would you estimate went to the show?

4 In a survey of 5000 people, 47% said they would vote for the Laboral Party in the next election. If there are 8 000 000 voters in the whole population, how many can be expected to vote Laboral?

5 The results for a class in Maths and English are recorded on the scattergraph below. Amy was absent for the English exam but scored 60 out of 80 in Maths.
Use the line of best fit to estimate Amy's result in English.

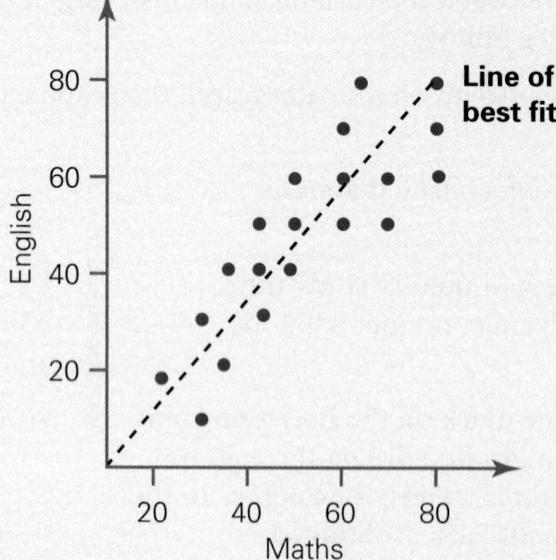

Level 2

6 There are approximately 19 million people living in Australia. Predict the number of people who:

a eat Wheatmix if it is advertised that 9 out of 10 eat Wheatmix
b are not retired if 26% of Australians have retired from the workforce

7 *The younger the driver, the more accidents he/she has*, say the motor vehicle insurance companies.

a Do you agree with their claim? Do you think there is a correlation between the two variables, age and number of accidents?

b Which of these scattergraphs fits the insurance companies' assertion?

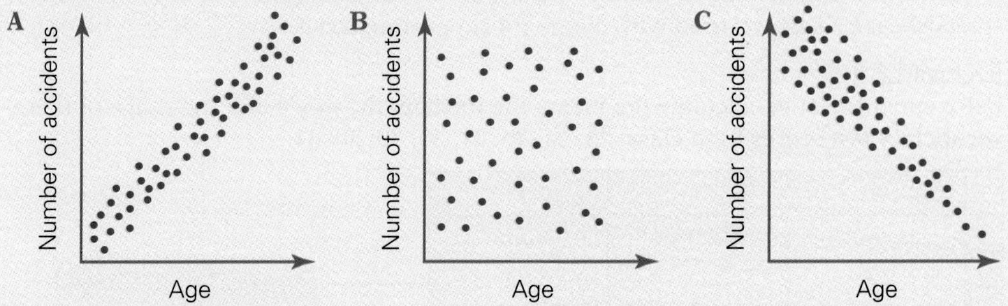

c If you try to insure your car at age 18, what do the companies *predict* about you? How is this reflected in the premiums they charge?

8 Here are the Maths and Science test results for a Year 8 class:

Name	Maths %	Science %
John A.	30	25
Anna B.	68	59
Huan C.	76	62
Esme D.	43	51
Fred E.	52	50
Gino F.	74	69
Julio G.	44	57
Agnes H.	83	79
Irina I.	94	96
Jim K.	75	77
Karl L.	64	68
Mimi M.	80	75
Alfonse N.	87	90
Olga P.	36	24
Deng Q.	94	75
Sean R.	48	54
Rita S.	63	68
Jack T.	55	61
Peter V.	67	74
Gina W.	86	82

a Draw a scattergraph of the data, with Maths % on the horizontal axis, Science % on the vertical.

b If the plotted points seem to show a correlation trend, use eye judgement to draw in a line of best fit.

c Would it be true to say for this class: 'the better you are at Maths, the better you are at Science'?

d Valmai scored 70% in the Maths test, but was away for her Science exam. **Predict** from the line of best fit what her result might have been.

10.6 Spreadsheets and statistics

A spreadsheet can be used to analyse and display sets of data. (These instructions are for the spreadsheet EXCEL, as used with chapter 4 support material.)

EXAMPLE 1

Use a spreadsheet to calculate the mean, the median, the mode and the range of these vocabulary test scores for a class: 25, 25, 26, 27, 30, 30, 30, 31.

	A	B	C	
1	Scores	Analysis	Value	
2	25	mean	28	=AVERAGE(A2:A9)
3	25	min	25	=MIN(A2:A9)
4	26	max	31	=MAX(A2:A9)
5	27	range	6	=C4 – C3
6	30	median	28·5	=MEDIAN(A2:A9)
7	30	mode	30	=MODE(A2:A9)
8	30			
9	31			

Solution

The mean of the scores is 28, the median of the scores is 28·5, the mode of the scores is 30, and the range of the scores is 6.

EXAMPLE 2

Use a spreadsheet to draw a column graph using the data in the given frequency table.

	A	B
1	Score	Frequency
2	1	2
3	2	4
4	3	8
5	4	4
6	5	2

In order to construct a graph, highlight the cells A1:B6, and then select CHART from the INSERT menu. On Chart Wizard 1, select COLUMN. On Chart Wizard 2, select the SERIES tab at the top of the Chart Wizard window. Under the sample graph, in the series window ensure the heading of the left-most data column is highlighted, then press REMOVE. Then click the mouse in the Category (X) label box near the bottom of the Chart Wizard window and type in =(the same label as showing in the values box) ! (first cell:last cell of VALUES in left-most data column—for our example, A1:A6). On Chart Wizard 3, press the TITLE tab, and label the axes appropriately. Press other tabs to complete the graph as required.

Solution

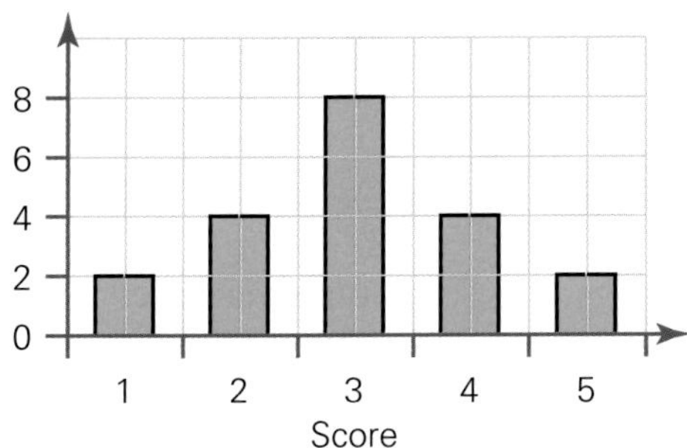

EXERCISE 10I

1 Use a spreadsheet to find the mean, the median, and the range for each of the following sets of data:

a 2, 3, 3, 4, 4, 6, 7, 7, 8, 9

b 3, 3, 4, 4, 6, 6, 7, 8, 9, 10

c 12, 12, 13, 13, 17, 20, 20, 21

d 68, 56, 77, 95, 59, 52, 86, 82, 84, 70

e 425, 439, 476, 451, 401, 465

f 1·7, 3·1, 1·9, 2·4, 3·4, 1·5, 3·9, 4·3, 4·8

Level 2

2 Enter the data shown in the frequency tables below into a spreadsheet and draw a column graph. Manually calculate the mean, the median and the range for each set of data.

a

Score	Frequency
1	1
2	6
3	3
4	4
5	2

b

Score	Frequency
21	3
22	5
23	8
24	9
25	0
26	3
27	4

c

Score	Frequency
10	10
20	15
30	19
40	18
50	14
60	11

TEACHER

Chapter Review

Language Links

analysis
biased
categorical
census
clusters
continuous
correlation slope
cumulative
data
discrete
distribution
dot plot
frequency
grouped
histogram
information
leaf
mean
median
mode
outliers
polygon
population
predictions
quantitative
random
range
representative
sample
statistics
stem
summary
survey
tally

1 Group as many of the above words as possible under these headings: *Collecting data*, *Displaying data*, *Analysing data*.

2 **Explain** the differences between discrete and continuous data.

3 List the features of a frequency histogram.

4 Write down the difference between a sample and a census.

5 Choose the correct term from the list above to help you complete the FIND THE WORD puzzle.

- a column graph used in statistics
- a set of questions designed to gather information
- when only a section of the population is asked
- the most common score
- the highest score minus the lowest score
- by chance
- when using a lot of information we often do this
- the number of times an event occurs
- the marking of each score with a dash
- asking the entire population
- when the scores can be in the full range of values
- a name given to the information collected
- the middle score
- another word for average
- when the scores are individual items or numbers

E	U	I	G	V	I	U	E	E	Z	Q	N	T	N	Q	W	A
O	B	D	D	Z	A	T	D	D	Y	A	R	T	B	T	V	D
O	K	T	V	G	E	O	E	E	I	B	Y	J	S	A	M	B
L	G	J	G	R	M	P	V	D	W	L	W	D	T	O	C	U
M	L	C	C	M	U	R	E	C	L	E	K	A	D	W	O	T
O	E	S	U	O	U	M	R	A	E	S	D	N	G	O	W	C
Q	I	A	R	S	R	C	T	A	N	N	A	S	K	L	R	C
D	T	G	N	H	X	N	L	C	N	R	S	M	K	H	E	Z
C	O	N	T	I	N	U	O	U	S	G	A	U	P	V	Q	X
F	R	E	Q	U	E	N	C	Y	W	S	E	C	S	L	Y	S
H	I	S	T	O	G	R	A	M	O	M	O	V	W	P	E	J

Chapter Review Exercises

1 Display each of the following sets of data in a frequency distribution table: 10.2

a ages

15	14	12	13	10
15	10	11	12	13
13	11	13	15	13
10	9	14	14	13

b correct answers out of 5

0	5	1	3	2
4	1	1	3	5
0	4	4	2	1
3	4	1	0	5

2 Display each of the sets of data in question **1** in a frequency histogram and polygon. 10.2

3 Calculate the mean for each set of data in question **1**.

4 The stem-and-leaf plot shows the number of phone calls made by a business each week. 10.2

Stem	Leaf
0	5 9 9
1	4 7 7 9
2	1 4 5 8 9
3	4 5
4	8

4 | 8 means 48 phone calls

a For how many weeks was the data recorded?
b What was the smallest number of phone calls made in 1 week?
c What was the largest number of phone calls made in 1 week?
d How many weeks were there more than 20 phone calls made?
e What was the total number of phone calls made over the period?

5 The number of cans of soft drink sold from a vending machine each day during February was recorded. 10.2

17	24	23	19	24	36	29
16	24	28	20	35	36	36
25	26	30	37	36	24	17
18	20	40	9	33	40	25

Draw a dot plot to show the data and comment on any outliers and clusters.

6 Find the mean, mode, median and range for: 10.3

a 0, 1, 3, 1, 1, 3, 2, 0, 3, 4
b 132, 265, 192, 56, 317, 192
c 5, 6, 7, 8, 9, 10, 11
d 0·75, 0·68, 0·77, 0·75, 0·93

7 10.4 10.1

x	f	fx
9	7	
10	15	
11	6	
12	5	
13	10	
14	2	
15	6	

a Copy and complete this frequency distribution table.
b Find the mean.
c Find the mode.
d Find the range.
e Find the median.
f Explain why the data is discrete.

10.3 **8** Display the following data in a dot plot or a frequency distribution table and find the mean, mode and range:

4, 6, 7, 6, 7, 8, 7, 5, 5, 6, 7, 8, 9, 12, 4, 1, 6, 6, 6, 7, 4, 5, 7, 8, 6, 6

10.4 **9** Find the mean and median for the sets of data displayed in the following stem-and-leaf plots:

a

Stem	Leaf
5	1 2 4
6	0 1 1 3 7
7	5 9 9

b

Stem	Leaf
12	0 4 8
13	1 3 3 4 9
14	8 9 9
15	5

10.3 **10** Eight local coffee shops were surveyed on the price of a cappucino. The prices were recorded as:

$1.75 $2.20 $2 $2.20 $1.80 $2 $2.20 $1.90

a What is the mean price for a cappucino?
b What is the range of the prices?
c What is the modal price?
d What is the median price?

10.4 **11**

Score	Frequency
11	5
12	4
13	1
14	6
15	4

a How many scores are shown in the frequency distribution table?

b Find the:
i mean **ii** mode **iii** median **iv** range
for the scores displayed.

c A score of 13 is added to this sample. Recalculate the:
i mean **ii** mode **iii** median **iv** range

10.4 **12** Find the data if:

- there are seven scores (whole numbers)
- the mode is 1 and its frequency is 3
- the mean is 2
- the range is 3
- there is one 3

TEACHER

Keeping Mathematically Fit

Part A: Non-calculator

1 Which graph best illustrates the solution of $2x - 1 < 9$?

A

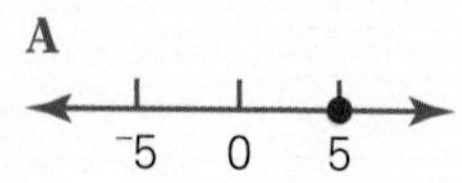

B

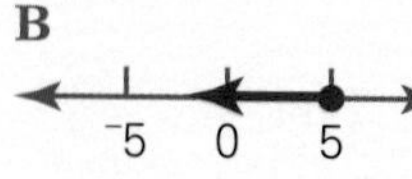

C

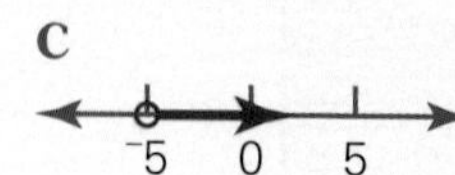

D

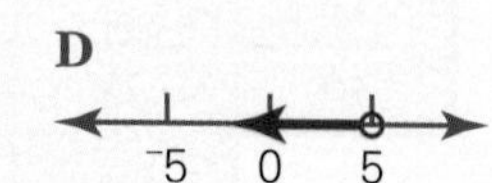

2 Give a whole number estimate for $\dfrac{3{\cdot}94 \times 9{\cdot}7}{0{\cdot}5}$.

3 Find the value of a and b if $\dfrac{3^5 \times 2^4 \times 3^2}{2} = 3^a \times 2^b$.

4 Divide 96 in the ratio of 2 : 1.

5 A tiler uses black and white tiles in his design as shown:

a How many black tiles are needed for four white tiles?
b How many black tiles are needed for n white tiles?

6 What is the supplement of 72°?

7 A train leaves Gymea Station at 10:17 am and arrives at Bondi Junction station 48 minutes later. At what time did the train arrive at Bondi?

8 If $\dfrac{5x}{2} = 15$, find the value of $\dfrac{2x}{5}$.

9 Simplify $4x + 3(2 - 3x)$.

10 Convert $\dfrac{4}{5}$ ha to m^2.

Part B: Calculator

1 A wholesaler buys a lounge for \$1546 and sells it to the public for \$2400. Calculate the percentage profit.

2 Solve $\dfrac{3 + x}{4} \geq 6$.

3 Find the volume of:

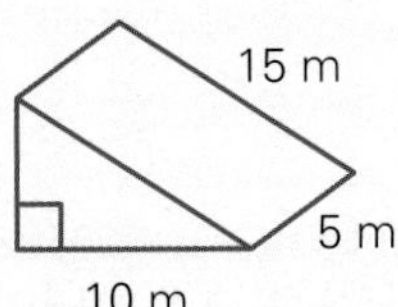

4 Evaluate $9.4 - (3{\cdot}06 - 2.1^2) \div 4{\cdot}8$, correct to 3 decimal places.

5 The angles in a triangle are in the ratio of 2 : 3 : 5. Find the size of the smallest angle.

6 A car leaves Sydney at 9:25 am and drives the 1062 km to Melbourne, arriving at 9:13 pm. Find the average speed for the trip.

7 Find the value of x in:

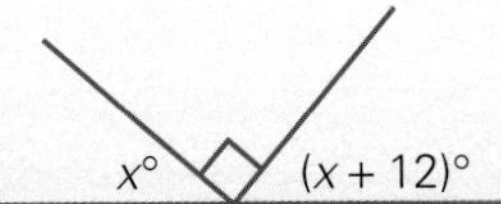

8 A set of four scores have a mode of 3, a range of 3, and a mean of 4.

a Find a set of four scores that satisfies these conditions.
b How many different sets can you find?

TEACHER

1 What you need to know and revise

Outcome PAS 4.2:
Creates, records, analyses and generalises number patterns using words and algebraic symbols in a variety of ways

Outcome PAS 4.5:
Graphs and interprets linear relationships on the number plane
- finds rules for geometric and number patterns
- interprets the number plane and plotting, reading and naming points
- locates the origin

2 What you will learn in this chapter

Outcome PAS 4.2:
Creates, records, analyses and generalises number patterns using words and algebraic symbols in a variety of ways

Outcome PAS 4.5:
Graphs and interprets linear relationships on the number plane
- represents relationships from number and geometric patterns on a graph
- graphs points on the number plane from a table of values
- extends the line joining a set of points
- reads values from the graph of a linear relationship
- derives the rule for a set of points that has been graphed on the number plane
- forms a table of values and then graphs on a number plane with an appropriate scale
- graphs more than one line on the same set of axes and compares the graphs to determine similarities and differences
- graphs two intersecting lines and reads off their point of intersection

Outcome PAS 5.1.2:
Determines the mid-point, length, and gradient of an interval joining two points on the number plane and simple non-linear relationships from equations
- graphs vertical and horizontal lines
- graphs a variety of linear relationships on the number plane by constructing tables of values
- graphs simple non-linear relationships

Working Mathematically outcomes WMS 4.1–4.5
Students will be asked to ***question***, ***apply strategies***, ***communicate***, ***reason*** and ***reflect*** in the sections of this chapter.

Linear relations and the number plane

Key mathematical terms you will encounter

axes	equation	intersection	reference
collinear	gradient	linear	relationship
constant term	graph	location	satisfies
continuous	grid	ordered pairs	slope
coordinates	horizontal	origin	straight
decreasing	increasing	plot	values
discrete	intercept	quadrant	vertical

MathsCheck

Linear relations and the number plane

1 For the first quadrant number plane supplied, answer the following questions.

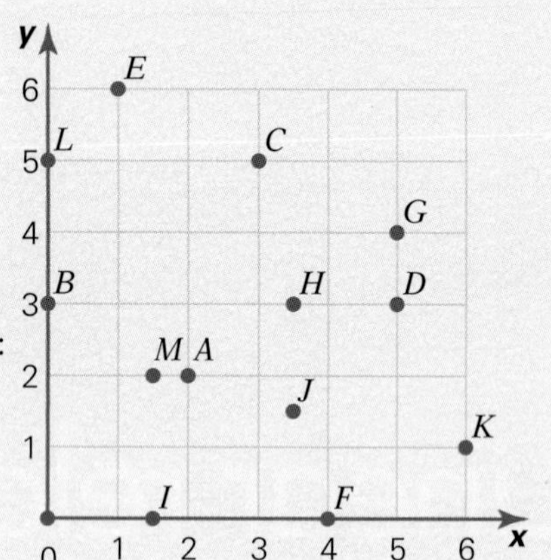

- **a** Write down the name given to the horizontal axis.
- **b** Give the coordinates of the following points:
  i *A* **ii** *B* **iii** *D* **iv** *F*
  v *I* **vi** *C* **vii** *G* **viii** *J*
- **c** Name the point with the following coordinates:
  i (1, 6) **ii** (6, 1) **iii** $(1\frac{1}{2}, 0)$
  iv $(3\frac{1}{2}, 3)$ **v** (0, 0) **vi** (0, 5)
- **d** Name all points with a y-coordinate of 3.
- **e** Name all points with an x-coordinate of 0.

2 Grids can also be used to locate places on the Earth's surface. The location of a place is given as an ordered pair: the **latitude** (angle north or south of the equator) followed by the **longitude** (angle east or west of the prime meridian).

Equator

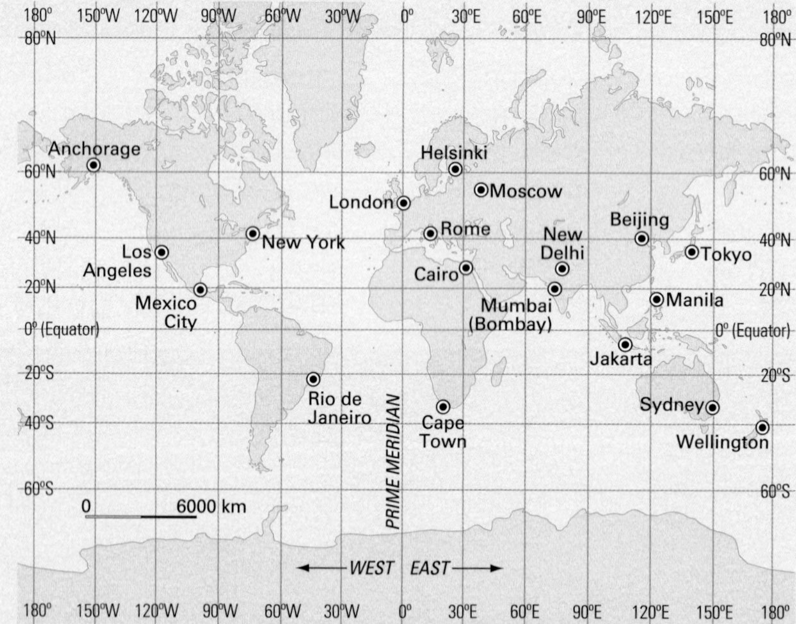

Sydney is at 34°S, 151°E

What major cities have the following coordinates?

a 43°N, 75°W b 52°N, 0° c 28°N, 32°E
d 20°S, 42°W e 35°N, 140°E f 18°N, 103°W

3 Give the coordinates for the following cities:

a Jakarta b Bombay c Los Angeles
d Rome e Beijing f Moscow
g Manila h Cape Town i Anchorage

The **horizontal** reference axis is the ***x*-axis**; the vertical is the ***y*-axis**. **Coordinates** follow the order (x, y).

The axes divide the plane into four **quadrants** or regions.

The point of intersection of the axes (0, 0) is called the **origin**.

Refer to the number plane supplied.

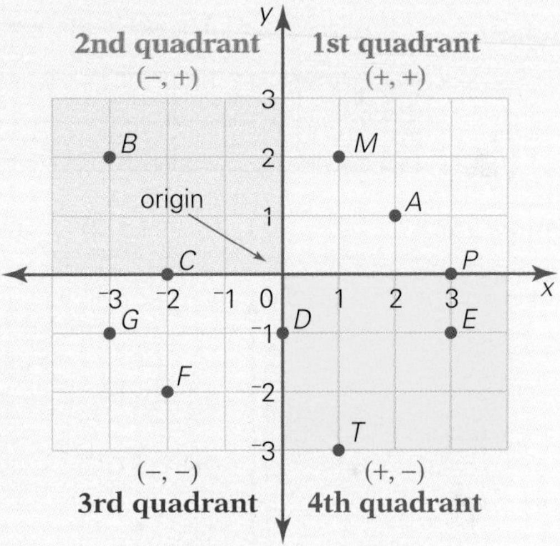

4 Give the coordinates of:

a B b C c D d E e F

5 Name the point whose coordinates are:

a $(1, {}^{-}3)$ b $(1, 2)$
c $({}^{-}3, {}^{-}1)$ d $(3, 0)$

6 Which axis must each point lie on? (n, c are positive integers.)

a $(n, 0)$ b $(0, c)$

7 In which quadrant does each point lie?

a $({}^{-}2, {}^{-}1)$ b $(4, 3)$ c $({}^{-}3, 3)$

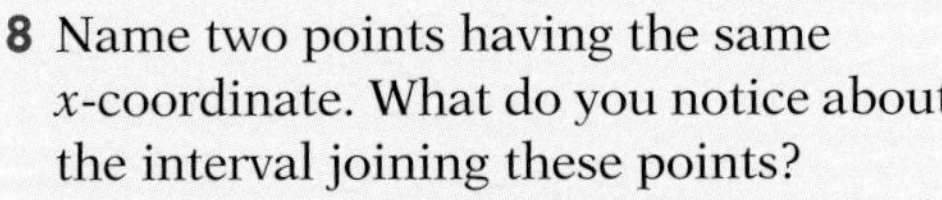

8 Name two points having the same x-coordinate. What do you notice about the interval joining these points?

9 Name two points with the same y-coordinate. What is true of the interval joining these points?

10 If $GBMS$ is a rectangle, write the coordinates of S.

11 If $OAEQ$ is a parallelogram, write the ordered pair for Q.

12 How long is interval BM? BG?

13 Copy and complete each of the following tables:

a

x	0	1	2	3	4
$3x$					

b

n	3	4	5	6	7
$10 - n$					

c

x	0	1	2	3	4
$4x + 5$					

d

n	$^{-}1$	0	1	2
$3 - 2n$				

TEACHER

Exploring New Ideas

11.1 Number plane graphs

The number plane can be used to represent relationships between different quantities and variables.

We have already completed tables of values for given equations last year. These form the ordered pairs that are then plotted onto a number plane.

WEB RESEARCH

WEB RESEARCH

EXAMPLE 1

From the table of values, write down the coordinates of the points and plot them on a first quadrant number plane.

x	1	2	3	4
y	3	2	1	0

Solution

x	1	2	3	4
y	3	2	1	0

Points are: (1, 3) (2, 2) (3, 1) (4, 0)

Remember: a point is written (x, y).

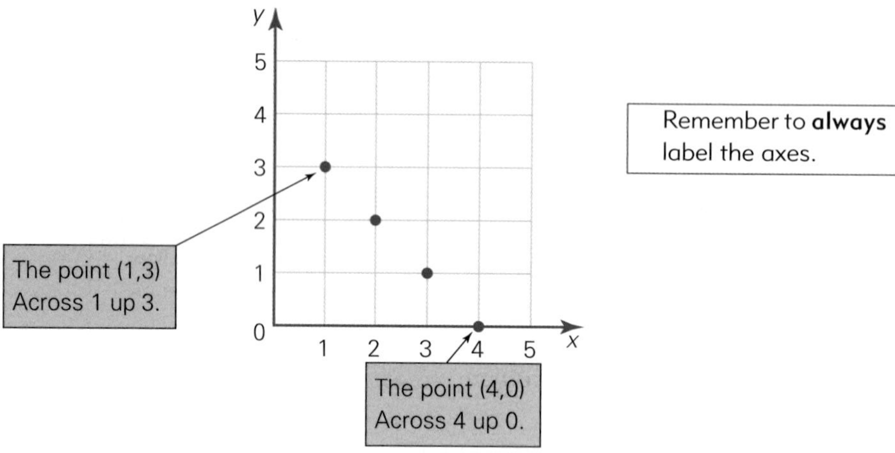

Remember to **always** label the axes.

EXAMPLE 2

The Pacific Plunge water ride uses cars that each hold six people. Park management uses computers to record the number of cars (x) passing through the system and thus the number of people (y).

It can be described by the rule $y = 6x$.

a Complete the table.

No. of full cars (x)	0	1	2	3	4	5
No. of people (y)	0	6	12	18		

b Write down the ordered pairs from the table.
c Graph the formula from the Pacific Plunge water ride on a number plane for values of x from 0 to 5.

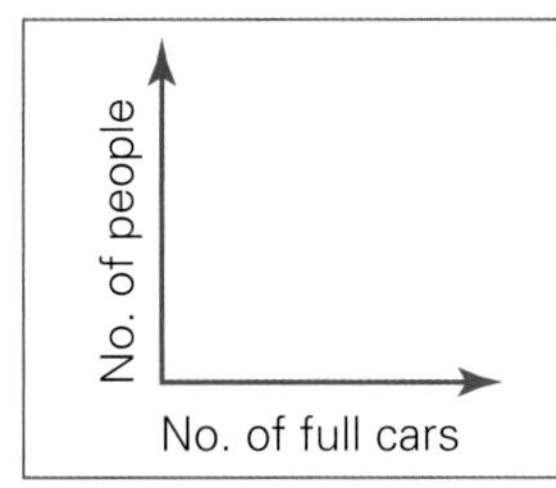

Solution

a $y = 6x$

x	0	1	2	3	4	5
y	0	6	12	18	24	30

$p = 6 \times 5$

$p = 6 \times 4$

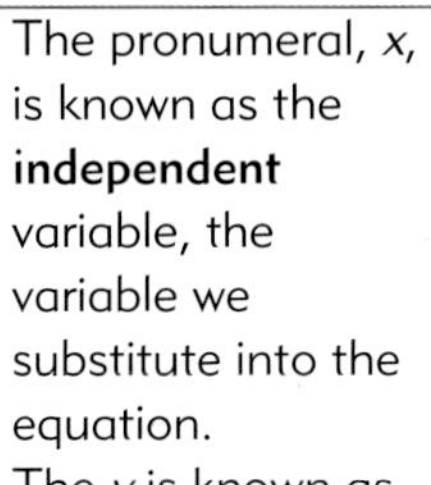

The pronumeral, x, is known as the **independent** variable, the variable we substitute into the equation.
The y is known as the **dependent** variable as its value depends on what value we substitute into the equation.

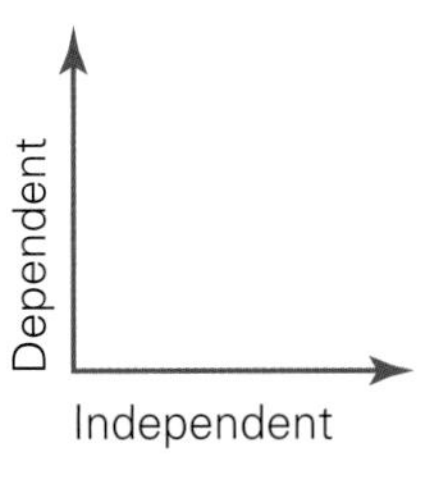

b (0, 0) (1, 6) (2, 12) (3, 18) (4, 24) (5, 30)

c

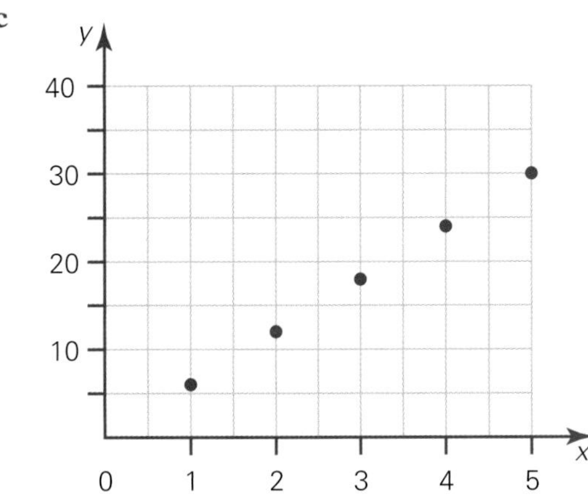

Draw up a set of axes, scaled to suit the range of values in the table.

Plot the ordered pairs as clear, bold points.

Note:

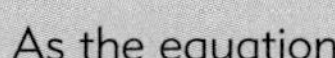

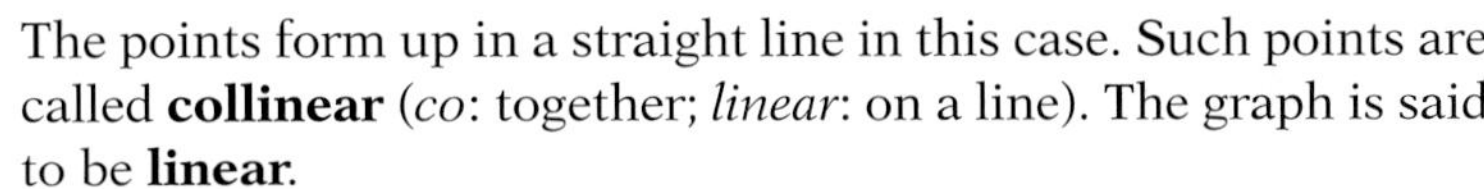

The points form up in a straight line in this case. Such points are called **collinear** (*co*: together; *linear*: on a line). The graph is said to be **linear**.

As the equation

$$y = 6x$$

in the example above requires the values for x to be counting numbers or positive integers (we are counting the number of passenger cars), no *in-between* values are possible, giving a graph of separated points called a **discrete** graph.

In this case we do not draw a line through the points.

y
discrete points
x

EXAMPLE 2

For the equation $y = 2x$:

a Complete the table of values.

x	$^-2$	$^-1$	0	1	2
y					

b Write down the ordered pairs from the table.
c Graph the equation $y = 2x$ on a number plane.
d By extending your graph, find the value of x that has a y-value of 9.

Solution

a By substituting each of the x-values into the equation $y = 2 \times x$ the table is completed.

x	$^{-}2$	$^{-}1$	0	1	2
y	$^{-}4$	$^{-}2$	0	2	4

$y = 2 \times {}^{-}2$
$= {}^{-}4$

$y = 2 \times 1$
$= 2$

b The ordered pairs are $({}^{-}2, {}^{-}4)$, $({}^{-}1, {}^{-}2)$, $(0, 0)$, $(1, 2)$, $(2, 4)$.

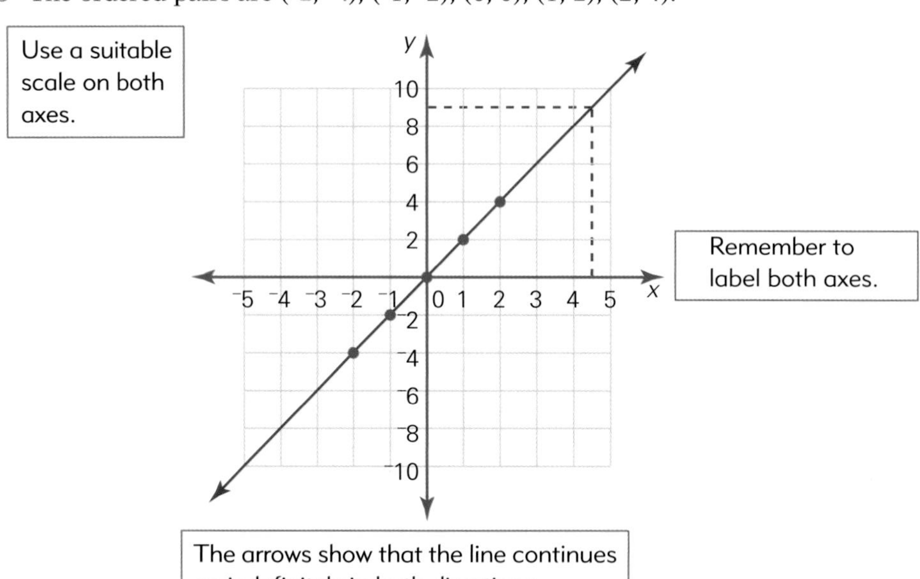

c From the graph we can read off the x-value.
$\therefore$ when $y = 9$, $x = 4\frac{1}{2}$

As the pronumerals x and y can be any values, not just whole numbers, the pattern of points formed is really **continuous**. This is shown by drawing a *continuous line* through the points.

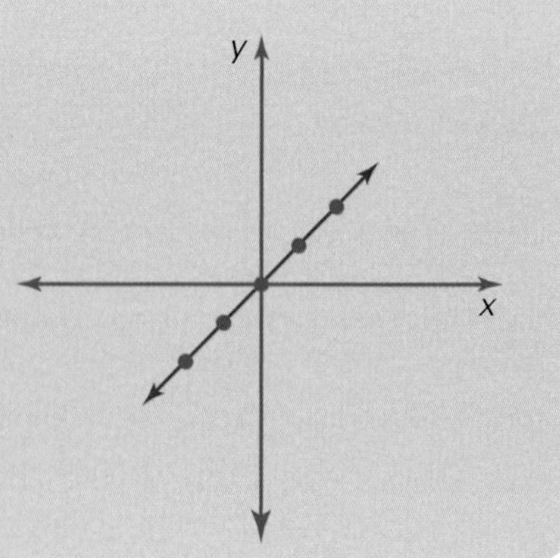

EXAMPLE 3

Graph the areas of squares of sides to 5 metres.

Solution

$A = s^2$

s	1	2	3	4	5
A	1	4	9	16	25

Note: Sides of squares *may have measurements in between* the whole numbers plotted, so the pattern of points is really **continuous**.

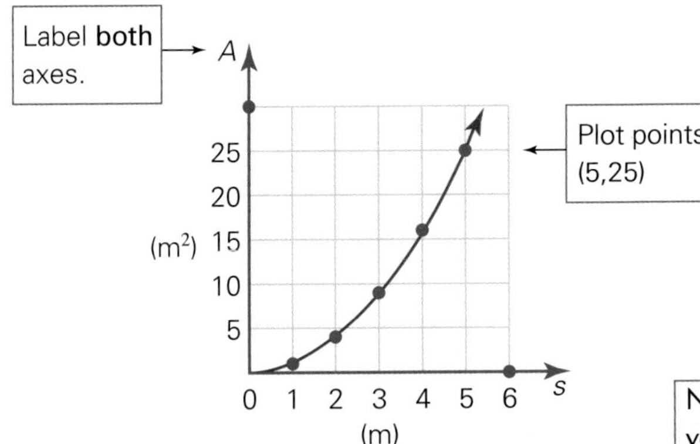

Not all relationships between variables are straight lines.

EXERCISE 11A

1 The graph shown here gives the perimeter of a square (P cm) with side lengths x cm. Copy the graph into your workbook.

a From the graph write down the coordinates of the points A, B, C, D.

b From the coordinates above, complete the table shown:

S	0	1	2	3	4
P					

c What does the point C tell us about the square?

d By extending the graph, find the perimeter of the square when the side length is:

i 8 cm **ii** 10 cm

e **Why** is the graph continuous?

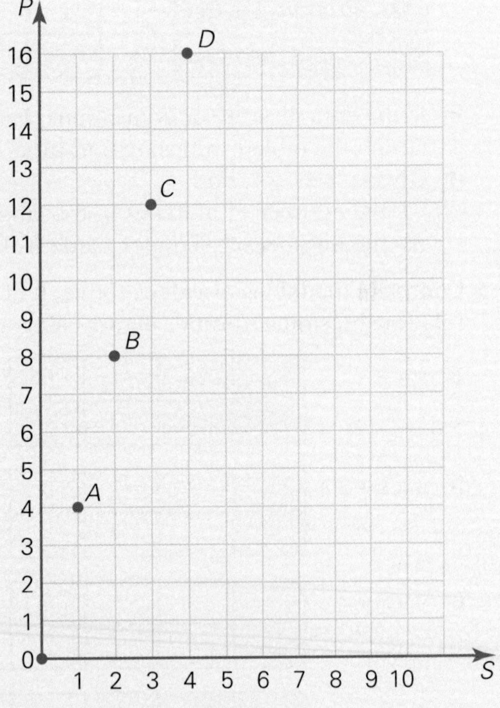

2 Yuan and Sophia are friends who were born 3 years apart. Their ages are related by the rule $y = x + 3$.

a Complete the table that shows the relationship between their ages.

Yuan's age (x)	0	2	4	8	11
Sophia's age (y)	3				

b Graph the data on a number plane. Use only the first quadrant.
c Does it make sense to join the points? **Explain**.
d Use the graph to find:
 i Sophia's age when Yuan is $3\frac{1}{2}$ years
 ii Yuan's age when Sophia is 6 years 9 months

3 The surface area of a cube can be found by finding the area of one face and multiplying this result by six; that is, $A = 6x^2$.

a Copy and complete the table of values showing the relationship between the side length of one face and the surface area.

Side length (x)	0	1	2	3	4
Surface area (A)					

b Graph the data on a number plane.
c **Explain** why the data is continuous.

SPREADSHEET ACTIVITY

4 a Use the diagram of the geometric shapes to copy and complete the table of values below. You may remember this pattern from last year.

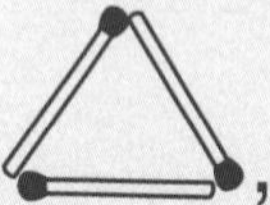, 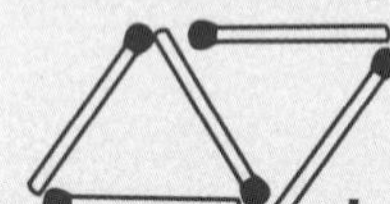, 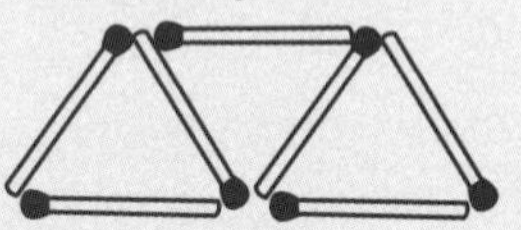, . . .

No. of triangles (x)	1	2	3	4	5
No. of matches (y)	3				

b Is the data discrete or continuous. **Explain**.
c Graph the tabled values on a number plane.
d Given $y = 2x + 1$, find:
 i the number of matches in 8 triangles
 ii the number of triangles made from 43 matches

5 Complete the tables of values for each of the rules below and then graph the relationships on a number plane, using a suitable scale.

a $y = x$

x	$^-3$	$^-1$	0	1	3
y	$^-3$				

b $y = x + 2$

x	$^-3$	$^-1$	0	1	3
y	$^-1$				

c $y = x - 2$

x	$^-3$	$^-1$	0	1	3
y	$^-5$				

d $y = 3x$

x	$^-3$	$^-1$	0	1	3
y	$^-9$				

e $y = 2x + 3$

x	$^-2$	$^-1$	0	1	2
y					

Level 2

6 a Here is a set of (x, y) coordinates in which the y-value is four more than the x-value each time: $(^-2, 2)$, $(^-1, 3)$, $(0, ?)$, $(1, ?)$, $(2, 6)$. Find the missing values.

b Use the points above to copy and complete the table below:

x	$^-2$	$^-1$	0	1	2
y					6

c Complete the rule $y = x +$ ______

d List the coordinates of two other points that lie on the same line.

7 a Make a list of five ordered pairs of numbers in which the y-coordinate is three less than the x-coordinate.

b Plot your points on a set of axes.

c Complete the rule $y = x -$ ______

8 Sara is paid \$8 an hour.

a Complete the table of values below.

x (hours)	1	2	3	4	5
y (pay)					

b Plot the graph of y against x.

c Use your graph to find:

i Sara's wage if she works 6 hours

ii how long she needs to work to earn \$44

9 Complete a suitable table of values, i.e. $(^-2, ^-1, 0, 1, 2)$, for each of the following relationships and sketch on a number plane:

a $y = {^-x}$ **b** $y = x - 1$ **c** $y = x^2$ **d** $y = \frac{x}{2}$

10 a Which of the equations in question **9** form linear relationships (straight lines)?

b Describe the differences in the equations that form straight lines and the ones that do not.

c Without graphing the following, predict which of them are linear relations.

i $y = 3 - 5x$ **ii** $x + y = 8$ **iii** $y = {^-x^2}$ **iv** $y = 4x - x^2$

v $y = 3 - 2x$ **vi** $y = x^3$ **vii** $y = 4x$ **viii** $y = 2x + 1$

Investigation Optimisation problem

Geoffrey wishes to build a rectangular garden for his organic vegies using only 40 metres of fencing.

1 Draw three rectangles that meet Geoffrey's conditions; that is, have a perimeter of 40 metres.

2 Find the area of each of your rectangles above. Which has the largest area? Can you draw another rectangle with an even larger area?

TEACHER

3 Consider the rectangle below:

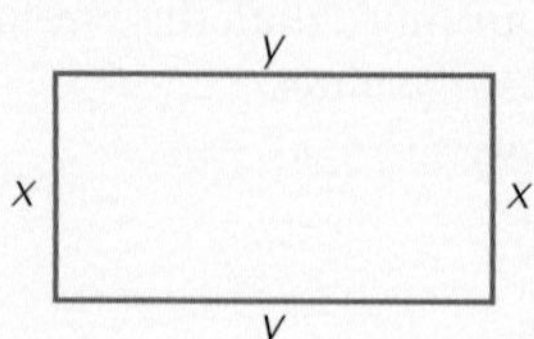

Write down the value of $2x + 2y$ ____________________

i.e. $x + y =$ ________________

The area of this rectangle is $A = xy$

$= x \times (20 - x)$

Copy and complete the table of values:

x (metres)	0	5	10	15	20	25	30	40
Area (m²)								

Display these values on a number plane.

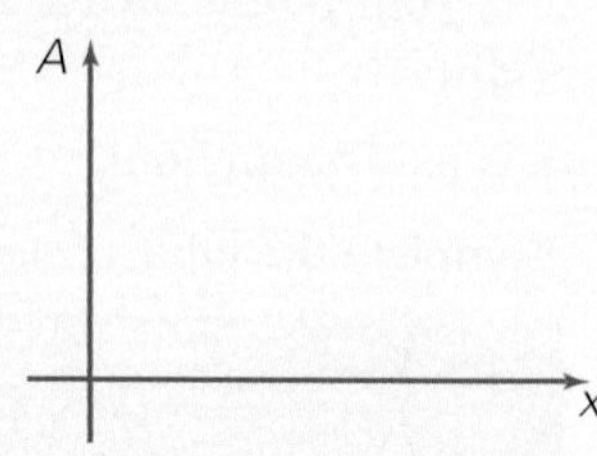

Join the points freehand to show the graph of $A = x(20 - x)$.

By reading off the value of x which produces the maximum area, find the dimensions of the rectangle Geoffrey needs to use if he hopes to produce a garden with the largest area.

Comment on what **conclusions** can be drawn by Geoffrey about which rectangles have the largest area.

11.2 Finding rules for linear relations

A relation is a set of ordered pairs connected by a rule or equation. A **linear** relation forms a straight line when plotted.

EXAMPLE 1

Do the points A (1, 4) and B (⁻3, ⁻5) lie on the line whose equation is $y = 2x + 1$?

Solution

Substitute $x = 1$ in $y = 2x + 1$.

$y = 2 \times 1 + 1$

$y = 3$

$y \neq 4$

∴ the point A (1, 4) does *not* lie on the line.

Substitute $x = {}^-3$ in $y = 2x + 1$

$y = 2 \times ({}^-3) + 1$

$y = {}^-5$

∴ the point B (⁻3, ⁻5) *does* lie on the line.

A point that lies on a line satisfies its equation.

EXAMPLE 2

Find a rule connecting x and y.

x	0	1	2	4
y	1	3	5	9

Solution

The difference between the consecutive y values is 2, and therefore the rule is of the form:

$y = 2x +$ ______

Substitute $x = 1$ and $y = 3$ into this equation: $3 = 2 \times 1 + \_\_\_\_\_\_$

$3 = 2 + \_\_\_\_\_\_$

$\_\_\_\_\_\_ = 1$

$\therefore$ the rule is $y = 2x + 1$

> This rule can be used to find the coordinates of any point on the line.

EXERCISE 11B

1 Which of the two points A or B lie on the line whose equation is given?

a $y = x$ $\quad A\,(3, 3), B\,(4, {}^-4)$

b $y = x - 4$ $\quad A\,(0, {}^-4), B\,(3, {}^-1)$

c $y = x + 1$ $\quad A\,({}^-5, {}^-6), B\,(2, 3)$

d $y = 2x$ $\quad A\,(3, 6), B\,(2, 8)$

e $y = {}^-x$ $\quad A\,({}^-1, {}^-1), B\,({}^-3, 3)$

f $y = 2x + 1$ $\quad A\,(2, 5), B\,({}^-3, {}^-7)$

g $y = 2x + 5$ $\quad A\,(0, 5), B\,(5, 10)$

h $y = 3 - 4x$ $\quad A\,(1, {}^-1), B\,(0, {}^-3)$

2 Find a rule connecting x and y each time.

a

x	0	1	2	3
y	3	4	5	6

b

x	0	1	2	3
y	$^-3$	$^-2$	$^-1$	0

c

x	0	1	2	3
y	0	3	6	9

d

x	0	1	2	3
y	5	6	7	8

Level 2

3 For each of the following tables of ordered pairs, write a rule (formula) to show the algebraic pattern:

SPREADSHEET ACTIVITY

a

x	3	2	1	$^-2$
y	$^-3$	$^-2$	$^-1$	2

b

x	10	9	8	7
y	6	5	4	3

c

x	$^-2$	0	2	3
y	$^-5$	1	7	10

d

x	3	2	$^-1$	$^-4$
y	9	4	1	16

4 Complete a table of values for each graph and find the rule (equation) for the line:

a

b

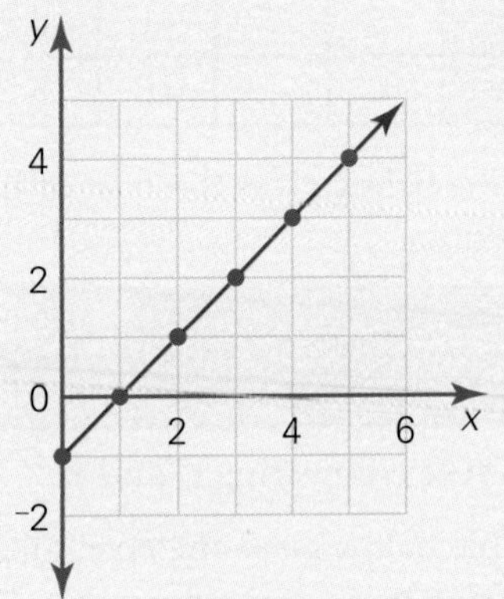

c

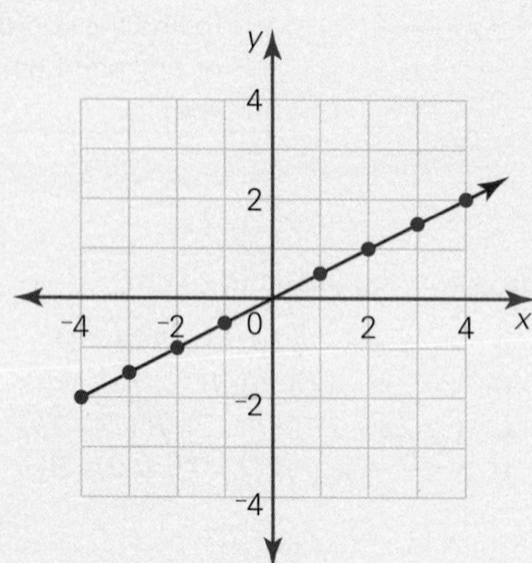

d

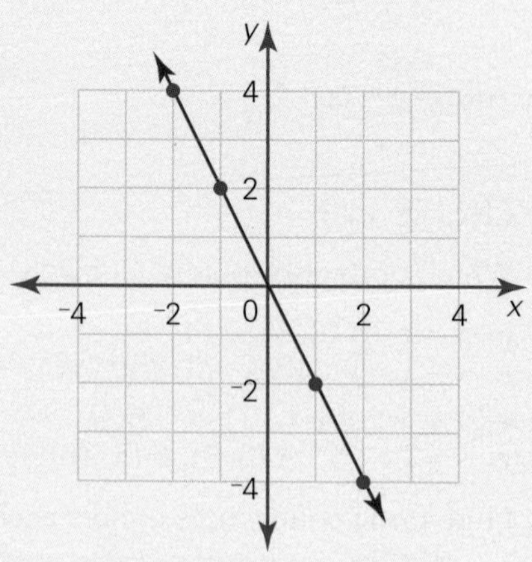

5 Give a set of three points that lie on each of the following lines:

a $y = 5x$ **b** $y = 10 - x$ **c** $y = {}^{-}3x$

6 Find the rule connecting x and y for each set of points:

a (1, 3), (2, 6), (3, 9) **b** (⁻1, ⁻3), (⁻2, ⁻6), (⁻3, ⁻9)
c (1, 4), (2, 3), (3, 2), (4, 1) **d** (9, 10), (10, 11), (11, 12), (12, 13)

7 Find the rule connecting x and y and use it to complete the table each time:

a

x	⁻3	⁻2	⁻1	0	1	2	3
y				7	8	9	10

b

x	⁻3	⁻2	⁻1	0	1	2	3
y	13	12		10			

c

x	⁻3	⁻2	⁻1	0	1	2	3
y		⁻11		⁻3	1		

d

x	⁻3	⁻2	⁻1	0	1	2	3
y			10	8	6		

8 The table below gives the cost of hiring a boat for a navigation excursion for x students.

Number of students (x)	10	12	14	16	18
Cost of boat (C)	55	65	75	85	95

Find a rule connecting C and x.

9 The table below gives the cost, $\$C$, of buying n litres of petrol.

Number of litres (n)	10	20	30	40
Cost of petrol (C)	9·50	19	28·50	38

Find the rule connecting C and n.

10 The table below gives the cost, $\$C$, of hiring a taxi for n km.

n (km)	10	20	30	40	50
C (\$)	17·50	32·50	47·50	62·50	77·50

Find the rule connecting C and n.

11.3 Graphing linear relations using a rule

The graph of any relation between two variables can be plotted if we have an equation, by constructing a table of values.

The **highest power** of x and y in the equation of a **linear relation** is **one**.

The equations $y = 3x$, $2x + y - 9 = 0$, $y = {}^{-}2x - 8$ are examples of **linear relations**.

$y = x^2$ and $y = 2 + 3x^3$ are **not** linear.

To graph a linear relation:

1. Draw up a table of three ordered pairs that fit the rule. Choose integer values for x unless told otherwise, as they are easier to work with. You may be given a range of values for x.
2. Plot the points on a number plane scaled to accommodate the values from the table.
3. Assume 'in-between' values are possible, unless told otherwise (i.e. **continuous** data). Join the points. Place arrows at the ends to show the line extends in both directions.

INCREASING LINES

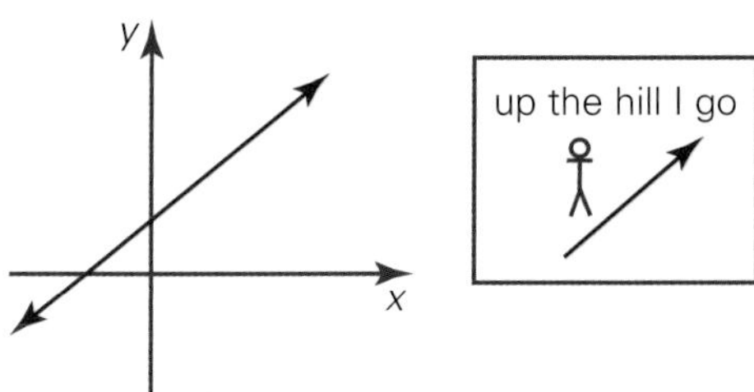

As the values of x increase, so do the values of y.
i.e. from left to right the line '**rises**'.

DECREASING LINES

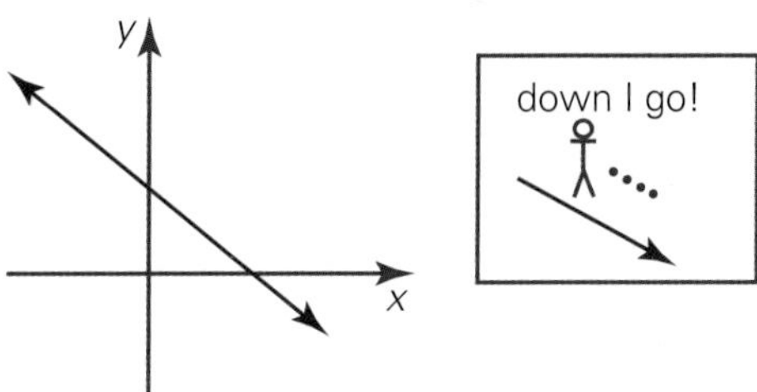

As the values of x increase, the values of y decrease.
i.e. from left to right the line '**falls**'.

EXAMPLE
Graph the line with the equation $y = 2x + 1$.

Solution

x	$^-3$	0	3
y	$^-5$	1	7

Step 1

Step 2: **Plot the points** using the coordinates from the table.

Step 3: Rule the line through the points to show all points on the line. **Show arrows**.

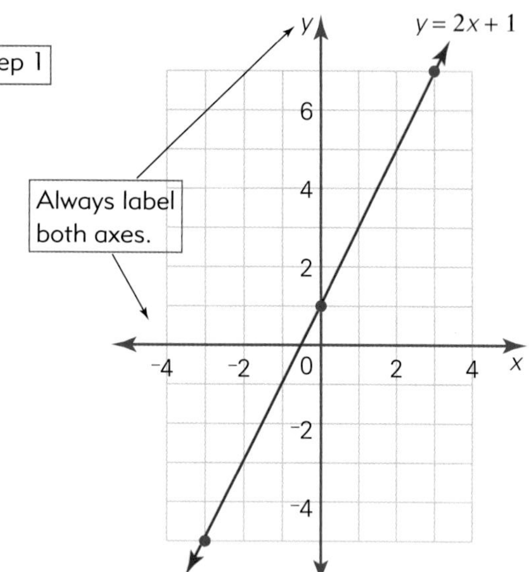

EXERCISE 11C

1 Which of the following equations are linear relations?

a $y = 3x - 8$ **b** $y = 2 - 3x$ **c** $y = 2 - 3x^2$ **d** $y = x^2 - 2x + 1$

e $y = {}^-2x$ **f** $y - x^2 = 9$ **g** $y - 4x + 1 = 0$ **h** $x = 7y$

2 Complete the given tables, then graph each line:

a $y = x + 3$

x	$^-3$	0	3
y	0	3	

b $y = 2 - x$

x	3	0	$^-1$
y			3

c $y = 4x - 1$

x	$^-1$	0	1
y	$^-5$		

3 By constructing a table of values such as:

x	$^-2$	$^-1$	0	1	2
y					

for each of the following equations, display them on the same number plane.

The number in front of the x is known as the coefficient of x.

a $y = 2x$ **b** $y = 3x$ **c** $y = 4x$ **d** $y = \frac{1}{2}x$ **e** $y = x$

4 a Through which point do all the straight lines in question **3** pass?

b What effect does the number in front of the x have on the slope of the line? A line's slope is often referred to as its gradient.

5 By constructing tables of values, similar to those in question **3**, display the following equations on the same number plane:

a $y = {}^-x$ **b** $y = {}^-2x$ **c** $y = {}^-3x$ **d** $y = \frac{^-1}{2}x$

6 a Through which point do all the lines in question **5** pass?

b What effect does the number in front of the x have on the slope (gradient) of the line?

c **Compare** the equations in question **3** to those in question **5**. What effect does the negative sign have on the direction of the line?

d **Describe** the gradient/slope of the lines in question **3** as increasing or decreasing.

e **Describe** the gradient/slope of the lines in question **5** as increasing or decreasing.

7 Graph these equations on the same number plane, by first completing a table of values:

a $y = 2x$ b $y = 2x - 3$ c $y = 2x + 1$ d $y = 2x - 1$

8 By comparing the graphs in question **7**, answer the following questions:

a Are the above lines parallel?

b What do the equations have in common?

c By referring to the equations above what is the significance of the constant term?

The constant term has no pronumeral.

9 Do your conclusions from question **8** hold true for the following equations? Check by sketching each on the same number plane.

a $y = x - 5$ b $y = x + 1$ c $y = x - 2$ d $y = x + 3$

Level 2

10 Using a separate number plane for each, graph each linear relationship (use $x = {}^{-}2, 0, 2$):

a $y = 2x - 2$ b $y = 3 - 2x$ c $y = \frac{1}{2}x + 1$

11 For each of the following equations, complete a table of values, plot the graph and state the point where the line crosses the y-axis:

The point where a graph crosses the y-axis is called its y-intercept.

a $y = 2x + 5$ b $y = 1 - 2x$ c $y = \frac{x}{2} - 2$

12 In each of the following pairs, decide which has the largest gradient (steepest graph).

a **A**: $y = 2x - 4$ **B**: $y = 3x - 2$

b **A**: $y = 3x + 1$ **B**: $y = x + 1$

c **A**: $y = 5x$ **B**: $y = x + 5$

d **A**: $y = x$ **B**: $y = 2x$

e **A**: $y = 8 + 7x$ **B**: $y = 8 + x$

13 Which of the following equations represent decreasing lines? Check by graphing each one.

a $y = 3x + 8$ b $y = {}^{-}x + 2$ c $y = 4 - 2x$

d $y = 4x$ e $y = 1 - x$

14 Which of the following represent increasing lines? Check by graphing each equation.

a $y = 1 + 3x$ b $y = 1 - 3x$ c $y = \frac{x}{2}$ d $y = {}^{-}4x$ e $y = {}^{-}4x$

15 Rewrite each of the following equations in the form of $y =$ ______ before drawing them on the same number plane:

a $2x - y = 4$ b $x + y - 4 = 0$ c $x - 2y + 2 = 0$

16 Using your deductions from the questions above, **predict** how the graph of the equation $y = x + 3$ would appear. Write a **description** of it. Do not draw the graph.

11.4 Horizontal and vertical lines

Horizontal lines

The graph of a line that is parallel to the x-axis has an equation of the form $y = a$, where a is a constant number.

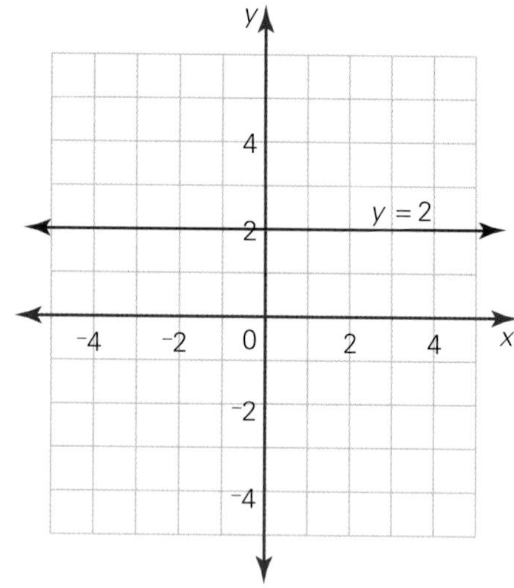

For example, the line $y = 2$ contains the points $(^-1, 2)$, $(0, 2)$, $(1, 2)$, $(2, 2)$, $(3, 2)$ etc.

That is, the x-value changes while the y-value is always 2.

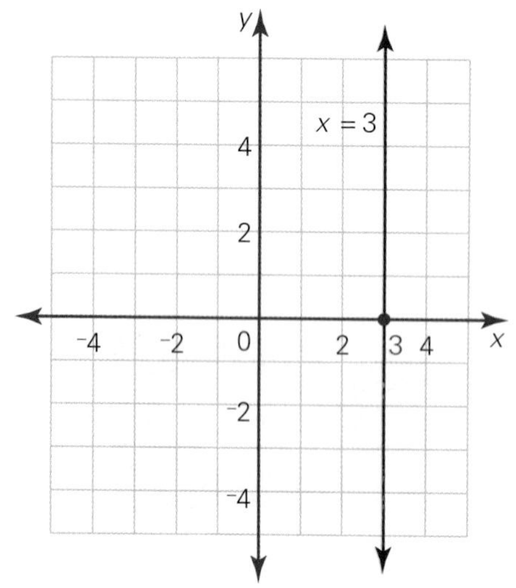

Vertical lines

The graph of a line that is parallel to the y-axis has an equation in the form of $x = b$, where b is a constant number.

For example, the line $x = 3$ contains the points $(3, ^-1)$, $(3, 0)$, $(3, 1)$, $(3, 2)$ etc. That is, the x-value is always 3 and the y-value can change.

EXAMPLE

Sketch the graphs of the following equations:

a $x = 1$ **b** $y = ^-2$

Solution

a

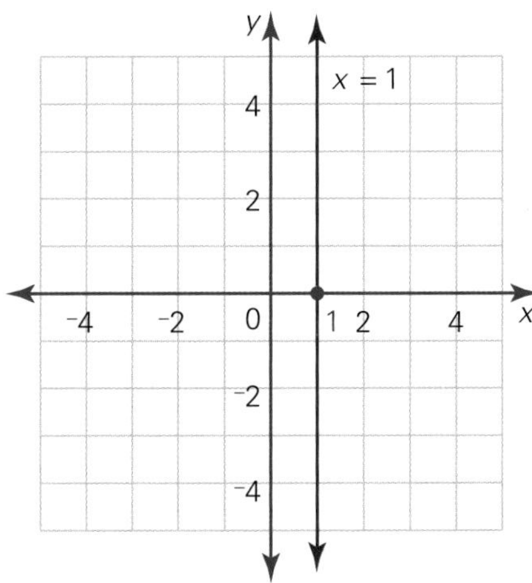

b

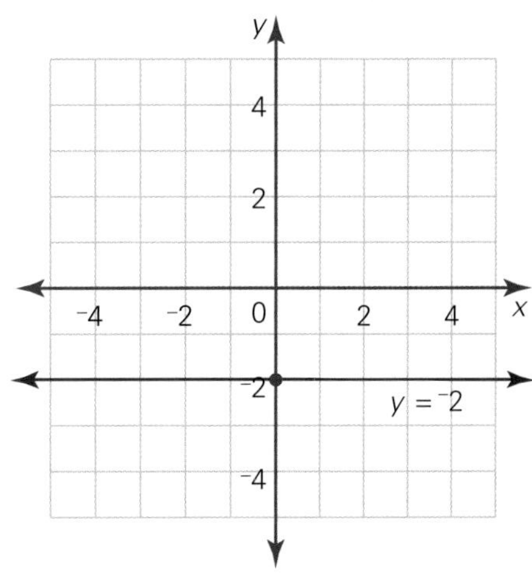

EXERCISE 11D

1 Which of the following points lie on the line $y = 1$?

(2, 1), (⁻1, ⁻1), (1, ⁻1), (3, 1), (⁻4, 1), (12, 1), (⁻6, 1), (1, ⁻6)

2 Which of the following points lie on the line $x = {}^-3$?

(3, 1), (⁻3, 1) (3, 3), (⁻3, 4), (⁻3, 10), (1, ⁻3), (16, ⁻3), (⁻3, 3)

3 Match each of the lines shown on the number line with one of the equations listed.

A: $x = {}^-1$	**B**: $x = 1$
C: $y = {}^-3$	**D**: $y = 4$
E: $y = {}^-4$	**F**: $x = {}^-4$

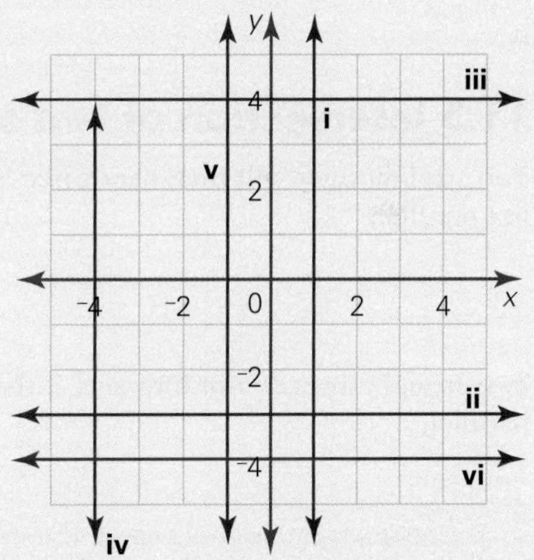

4 Sketch the graph of each of the following equations on the same number plane:

a $x = 6$ **b** $x = {}^-3$ **c** $x = 3$ **d** $x = {}^-5$ **e** $x = 0$

5 Sketch the graph of each of the following equations on the same number plane:

a $y = 6$ **b** $y = {}^-1$ **c** $y = 1$ **d** $y = {}^-5$ **e** $y = 0$

Level 2

6 What is the equation for each of the following lines?

a

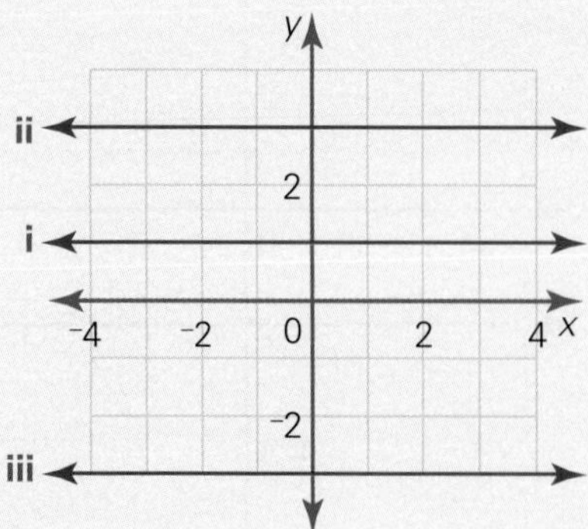

b

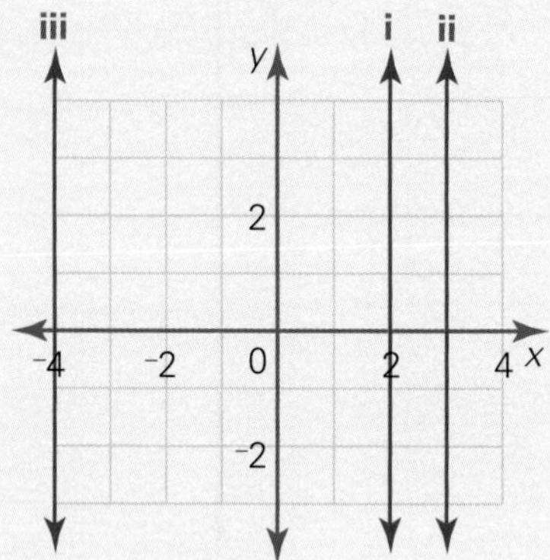

7 a Give the equations of two lines that are parallel to the x-axis.
b Give the equation of two lines that are parallel to the y-axis.

8 Find the equation of the line that is:

a parallel to the x-axis and passes through the point (1, 6)
b parallel to the x-axis and passes through the point $(^-1, 1)$
c parallel to the y-axis and passes through the point (3, 2)
d parallel to the y-axis and passes through the point $(^-4, 3)$
e parallel to the x-axis and passes through the origin

11.5 Intersection of two straight lines

Two straight lines will **intersect once** if they are not parallel.

Two straight lines do not intersect if they are parallel.

The point where two lines intersect, known as the **point of intersection**, can be found by sketching their graphs on one set of axes, and then reading off the coordinates.

In business, the point of intersection represents the ‘break-even’ point, where the company spends as much as it obtains. Graphs are used in a variety of subjects and contexts. It is important to be able to find the point where two graphs intersect.

EXAMPLE
Sketch the lines $y = 2x$ and $y = x - 3$ on the same set of axes and find their point of intersection.

Solution

$y = 2x$

x	$^-1$	0	1
y	$^-2$	0	2

$y = x - 2$

x	$^-1$	0	1
y	$^-3$	$^-2$	$^-1$

$\therefore$ the lines $y = 2x$ and $y = x - 3$ intersect at the point $(^-2, ^-4)$.

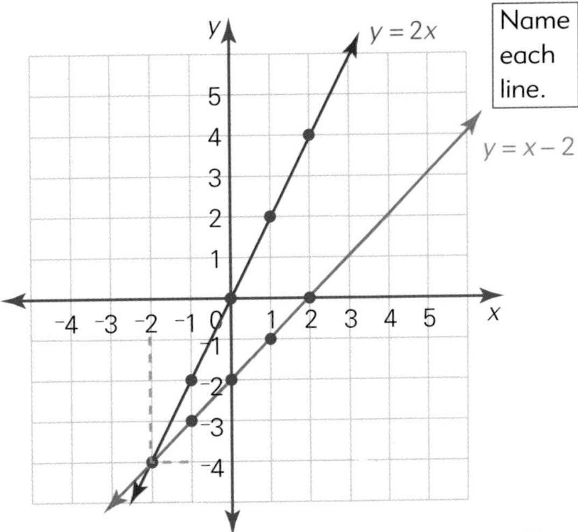

You can check by substitution: when $x = ^-2$

$y = 2x$ | $y = x - 2$
$y = 2x - 2$ | $y = ^-2 - 2$
$= ^-4$ | $= ^-4$

The point satisfies **both** equations.

EXERCISE 11E

1 For each of the following pairs of equations, read off their points of intersection from the given graphs. Check your coordinates by substitution into the equations.

a

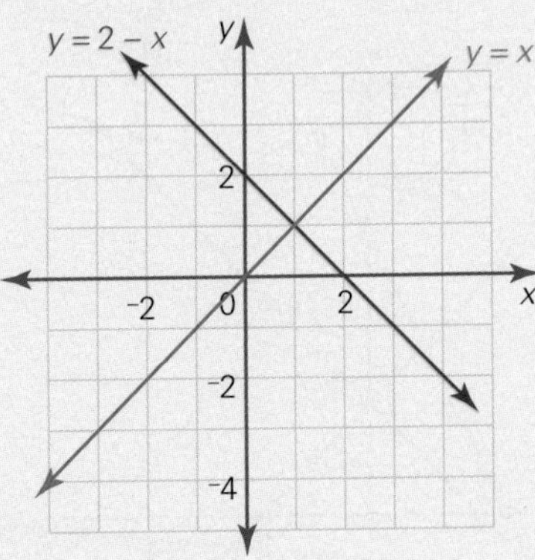

b

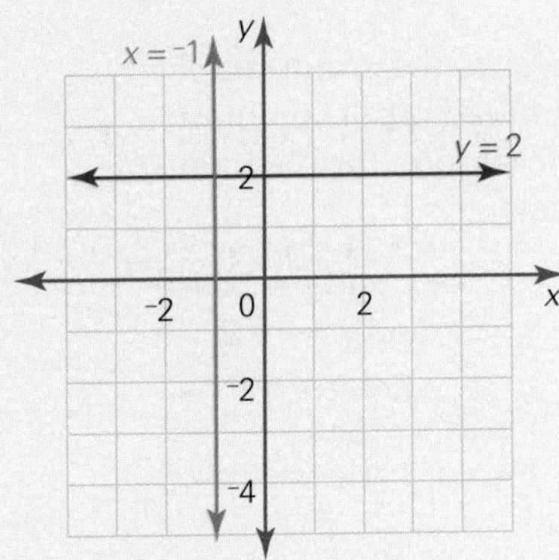

c

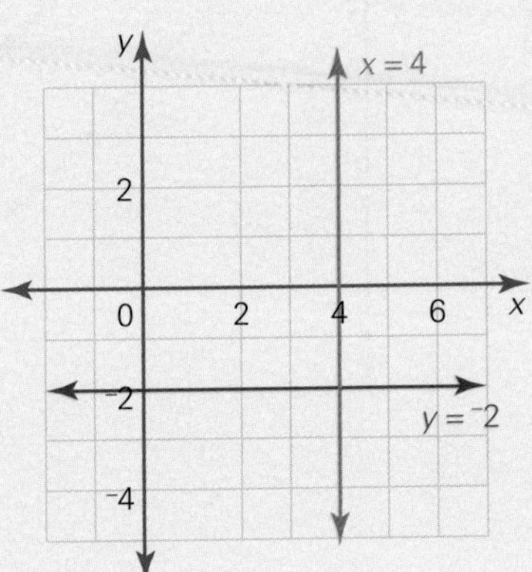

d

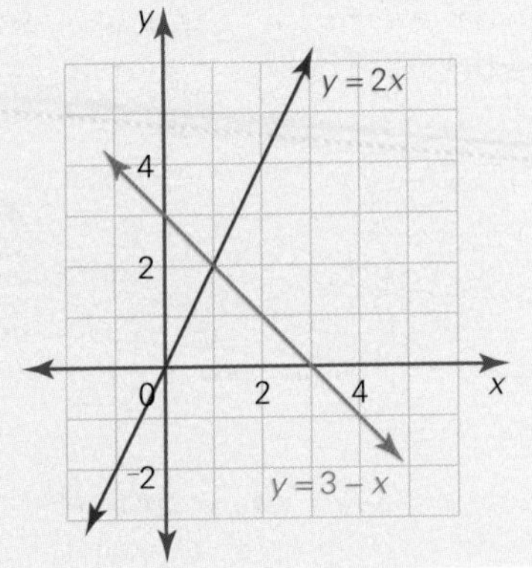

e

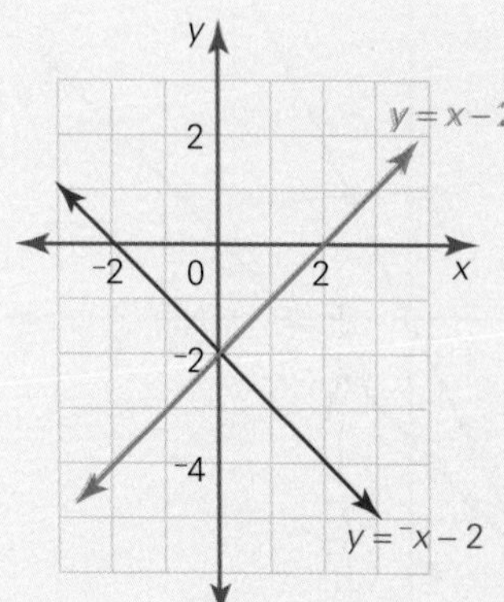

f

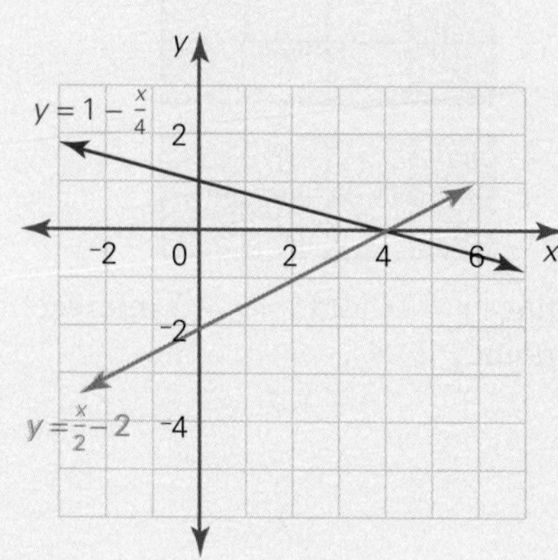

2 The number plane to the right shows the graph of $y = x$.

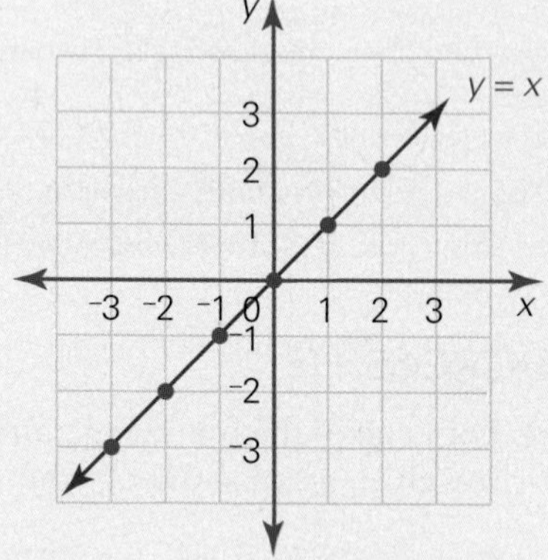

a Copy the graph into your workbook and on the same set of axes sketch the line given by the equation $y = 2 - x$ and state the point of intersection.

b On the same set of axes, graph the line $y = 3x + 2$. State the point of intersection of the graphs:

i $y = x$ and $y = 3x + 2$

ii $y = 2 - x$ and $y = 3x + 2$

3 Refer to the graph supplied and read off the point of intersection for the following lines:

a $y = 5 - x$ and $y = 2x + 5$

b $y = 2x$ and $y = x + 2$

c $y = 2 - x$ and $y = x + 2$

d $y = 2x + 5$ and $y = 2 - x$

e $y = x + 2$ and $y = 2x + 5$

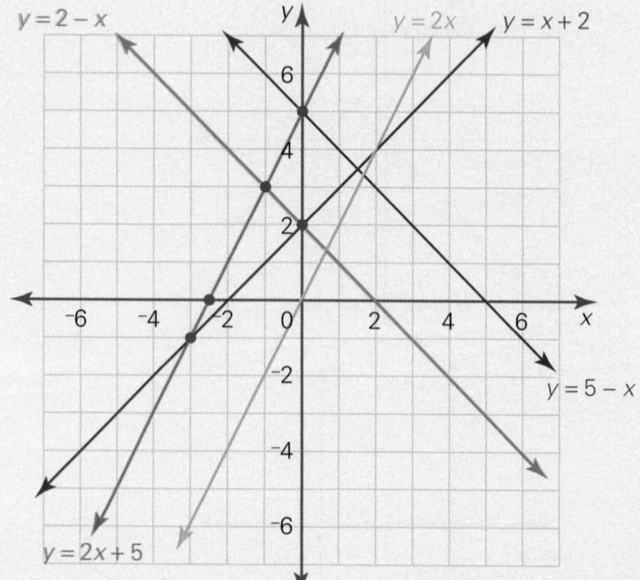

Level 2

4 For each of the following equations, sketch on a suitable number plane and state their point of intersection:

a $y = x - 2$ and $y = 6 - x$
b $y = 3x$ and $y = 2x$
c $y = x + 2$ and $y = {}^{-}2x - 4$
d $y = {}^{-}x$ and $y = x + 2$
e $y = 2x + 3$ and $y = x - 1$
f $y = 2x + 4$ and $y = {}^{-}6x$

5 Answer the following as either true or false:

a (1, 2) is the point of intersection of the lines $y = 2x$ and $y = x + 1$
b (⁻2, 0) is the point of intersection of the lines $y = x - 2$ and $y = x + 2$
c (1, 1) is the point of intersection of the lines $y = 3x$ and $y = x + 3$
d (0, 0) is the point of intersection of the lines $y = x$ and $y = {}^{-}x$
e (⁻2, 2) is the point of intersection of the lines $y = {}^{-}x$ and $y = x + 4$

6 The cost, $\$C$, of running a tutorial firm is given by the rule $C = n + 250$, and its revenue, $\$R$, is given by the rule $R = 26n$, where n is the number of students.

Sketch the graphs of C versus n and R versus n on the same set of axes and determine the point where the cost is equal to the revenue.

7 The cost, $\$C$, of running a coffee shop is given by the rule $C = 400 + 2n$, where n is the number of customers on any given day. On average, a customer spends \$6.

a Write a rule for the coffee shop's daily revenue.
b By sketching graphs of the coffee shop's daily expenses and daily revenue on the same set of axes, state the number of customers required for the coffee shop to break even.
c If the coffee shop has 400 customers in one day, find the day's profit.

TEACHER

SPREADSHEET ACTIVITY

Investigation Linear relations

1 Use a graphics calculator to sketch each of the following sets of linear relations.

Set A

$y = 3x - 5$ $\quad y = 3x - 2$ $\quad y = 3x$ $\quad y = 3x + 7$

Set B

$y = 2x - 3$ $\quad y = x - 3$ $\quad y = {}^{-}x - 3$ $\quad y = \frac{{}^{-}x}{2} - 3$

Investigate the **similarities** and **differences** between the graphs of the linear relations in:

a set A
b set B

2 Write a **summary** of your findings. Select any two linear relations from set B. How many points of intersection can you find between the two linear relations chosen from set B and all those in set A?

Use the graphics calculator to determine these points of intersection.

Chapter Review

Language Links

axes
collinear
constant term
continuous
coordinates
decreasing
discrete
equation
gradient
graph
grid
horizontal
increasing
intercept
intersection
linear
location
ordered pairs
origin
plot
quadrant
reference
relationship
satisfies
slope
straight
values
vertical

1 What is the difference between discrete and continuous graphs?

2 What do we mean by a *grid reference*?

3 Explain the difference between discrete and continuous data.

4 Draw a line, and give the equation of an increasing line. What feature in the equation tells us the line will be increasing?

5 Use the words from the list above to complete the following close passage.

The number plane is made up of two __________. The x-axis is known as the __________ axis, where the y-axis is __________. The point where the two axes meet is called the __________.

The equation $y = 2x + 1$ can be graphed onto the number plane by first completing a table of __________. The points are then __________ to form a __________ line. The equation represents an example of a __________ relation. The data in this case is __________, as all possible x-values can be used. The graph of $y = 2x + 1$ is __________ as it has a positive __________ . If the __________ of a line is negative then it is said to be __________ .

6 List three applications or subjects where the ability to graph relationships between variables is important.

7 Write a sentence explaining how you can tell, without sketching, that an equation represents a linear relationship.

8 **Compare and contrast.**

a Write down the similarities of the following linear relations:
$y = 2x + 1,\ y = x + 1,\ y = 1 - 2x$

b How do they differ?

c What about the following? How are they similar?
$y = 1 - x,\ y = 4 - x,\ y = {}^{-}x$

9 Write down everything you know about the linear equation $y = 4x - 8$.

Chapter Review Exercises

1 Give the coordinates of: MC

a A b B

2 Which point has the coordinates: MC

a $(^-2, ^-4)$ b $(0, 3)$

3 In which quadrant does Y lie? MC

4 Name a point in quadrant 4. MC

5 Calculate the distance from: MC

a T to M b S to P c V to W

6 Choose the linear equations from: 11.3

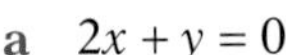

a $2x + y = 0$ b $y = x^2 - 2$ c $y = 5 - x$ d $y = \frac{x}{5}$

7 Complete the table of values, and then graph the relation on a number plane: 11.1

$y = x^2 - 1$

x	$^-3$	$^-2$	$^-1$	0	1	2	3
y							

8 Complete each table of ordered pairs and graph the **discrete** points: 11.3

a $y = x + 3$

x	$^-2$	$^-1$	0	1	2
y					

b $y = 3x - 2$

x	$^-3$	$^-1$	0	1	3
y					

9 a Graph the continuous linear relationship $y = 2 - 3x$. Use the values of x from $^-2$ to 2. 11.3
b Is the equation $y = 2 - 3x$ increasing or decreasing?
c Where does the line cross the y-axis?

10 a Copy and complete each of the following tables: 11.3

i $y = 2x$

x	$^-1$	0	1
y			

ii $y = {^-2x}$

x	$^-1$	0	1
y			

iii $y = 8 - 2x$

x	$^-1$	0	1
y			

b Sketch the lines $y = 2x$, $y = {^-2x}$ and $y = 8 - 2x$ on the same set of axes.
c Which two lines pass through the origin?
d Write down the coordinates of the points where the line $y = 8 - 2x$ crosses the y-axis.
e Which two lines are parallel?
f Write down the coordinates of the point of intersection of the line $y = 2x$ and $y = 8 - 2x$.
g Write down the equation of another line that also passes through the point of intersection of $y = 2x$ and $y = 8 - 2x$.

11 Sketch on separate number planes the graphs of: 11.4

a $x = 5$ b $y = {^-3}$ c $x = 0$ d $x = {^-1\frac{1}{2}}$ e $y = 2$

11.5 **12** Find the point of intersection of the following pairs of lines by sketching each pair on a separate number plane:

a $x = 4$ and $y = 2x$
b $y = x - 3$ and $y = {}^{-}2x$
c $y = 2x + 4$ and $y = x + 2$
d $y = 3x$ and $y = {}^{-}x$
e $x + y = 4$ and $x - y = 0$

11.2 **13** Find the equation of the line for:

a

x	$^{-}2$	$^{-}1$	0	1	2
y	$^{-}6$	$^{-}3$	0	3	6

b

x	$^{-}2$	$^{-}1$	0	1	2
y	9	7	5	3	1

11.4 **14** By completing a table of values, or otherwise, find the equation for each line:

a

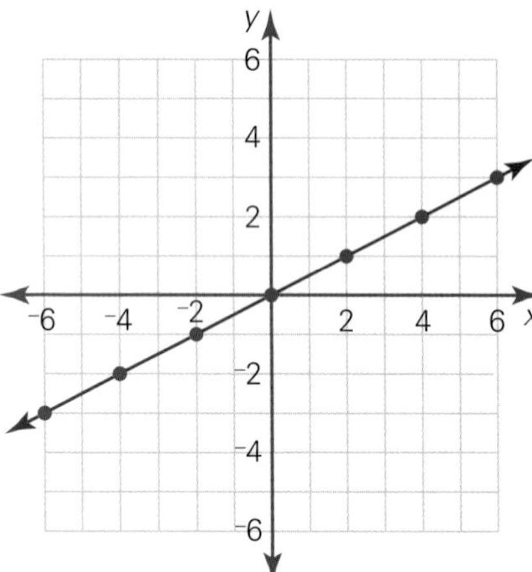

b

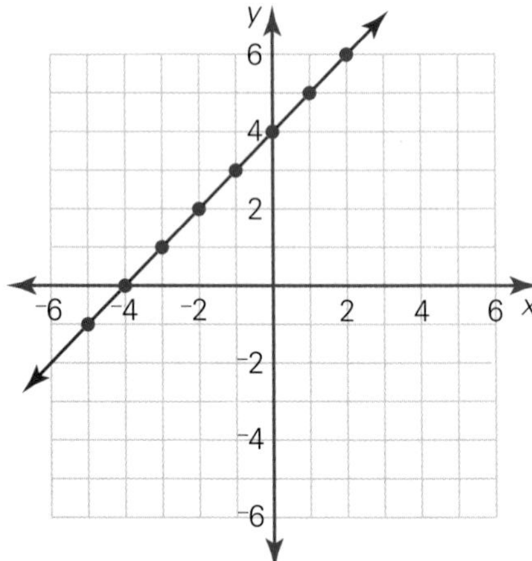

c

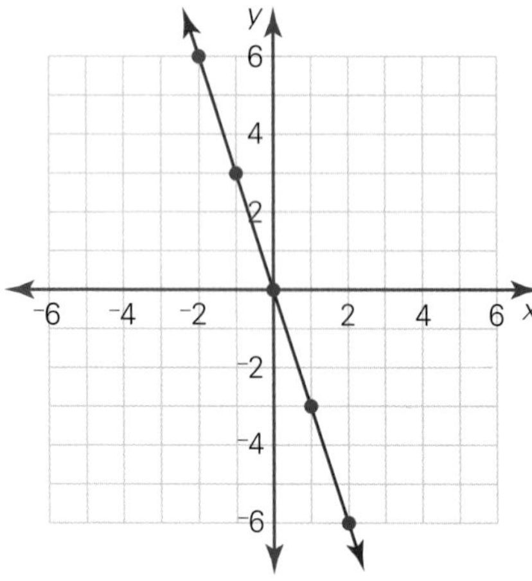

d

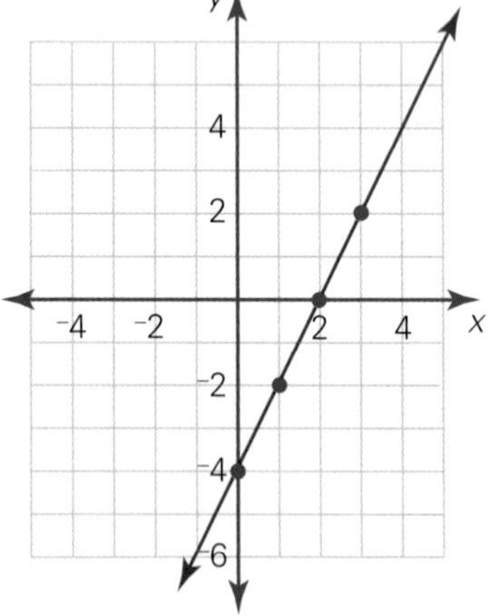

Keeping Mathematically Fit

Part A: Non-calculator

1 If p is an odd number, what is the next odd number greater than p?

2 Express $\frac{9}{20}$ as a decimal.

3 Find the value of $9{\cdot}8 - 1{\cdot}92$.

4 Find $33\frac{1}{3}\%$ of 1 hour.

5 How many degrees between south-west and due east?

6 Divide 845 in the ratio of 2 : 3.

7 Simplify $8x + 4(3 - 2x)$.

8 If $94{\cdot}2 \times 36{\cdot}5 = 3438{\cdot}3$, find the value of $\frac{34\,383}{365}$.

9 Twice a number less 10 is the same as two more than the number. What is the number?

10 A car depreciates at a rate of 10% p.a. Find the value of a $24 000 car in 1 year's time.

Part B: Calculator

1 Find the value of $\frac{1}{4 \times \sqrt{0{\cdot}19}}$ correct to 2 decimal places.

2 For the scores 1, 6, 9, 1, 3, 4, 12 find the:

a mean b mode c median d range

3 Subtract $2x + 4$ from $3x - 9$.

4 An open rectangular prism is 80 cm long, 30 cm wide and 40 cm high. Find the outside surface area of this prism.

5 A car travels 476 km on 32 litres of petrol. How many litres does the car need to travel 1000 km?

6 Expand and simplify:

a $3(x + 4) + 5x$ b $3(a + 2b) - a + b$ c $4(x + y + z) - 3(x - y)$

7 Find the area and perimeter of:

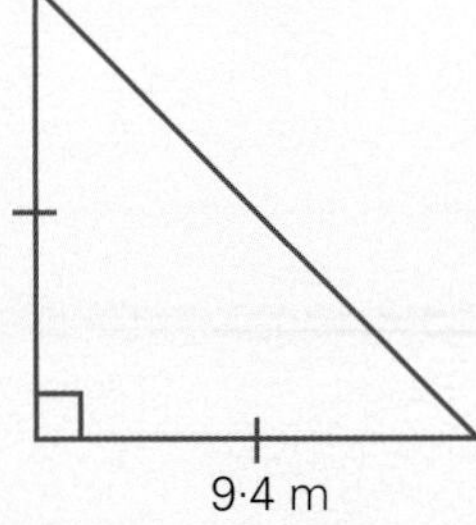

8 Two lines intersect at the origin. One line is $x - y = 0$. The other line is $y = mx$. What values can m take? What conclusions can you make about the equations of straight lines in the form of $y = mx$?

TEACHER

1 What you need to know and revise

Outcome MS 4.1:
Uses formulae and Pythagoras' theorem in calculating perimeter and area of circles and figures composed of rectangles

Outcome MS 4.2:
Calculates surface area of rectangular and triangular prisms and volume of right prisms and cylinders

- perimeter and area of rectangles, squares, triangles and parallelograms
- volumes of prisms
- units of length, area and volume

2 What you will learn in this chapter

Outcome MS 4.1:
Uses formulae and Pythagoras' theorem in calculating perimeter and area of circles and figures composed of rectangles

Outcome MS 4.2:
Calculates surface area of rectangular and triangular prisms and volume of right prisms and cylinders

- demonstrates by practical means that the ratio of the circumference of a circle to its diameter is a constant
- defines the number pi/π
- develops and uses the formula for the circumference and area of a circle
- develops and uses the formula for the volume of a cylinder

Working Mathematically outcomes WMS 4.1–4.5
Students will be asked to ***question***, ***apply strategies***, ***communicate***, ***reason*** and ***reflect*** in the sections of this chapter.

Circles and cylinders

Key mathematical terms you will encounter

arc
area
chord
circle
circular
circumference
concentric
cross-section
cylindrical
cylinder
diameter
estimate
height
perimeter
pi (π)
quadrant
radius
sector
segment
semicircle
tangent
volume

MathsCheck
Circles and cylinders

1 Convert each of the following to metres:

a 400 cm b 1200 cm c 3000 mm d 45 cm e $\frac{1}{2}$ km

2 Convert each of the following to millimetres:

a 5 cm b 12 cm c 2 m d $\frac{1}{2}$ m e 34 cm

3 Convert:

a $2\ m^2 =$ ______ cm^2 b $3\ m^2 =$ ______ cm^2 c $4000\ cm^2 =$ ______ m^2
d $3\ 000\ 000\ mm^2 =$ ______ m^2

4 Find the perimeter and area of each of the following:

a
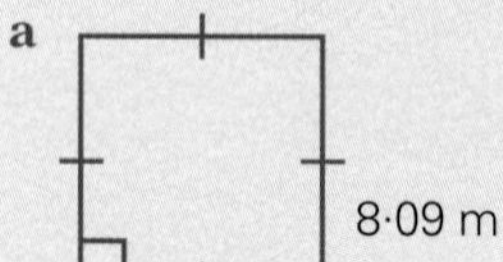

b
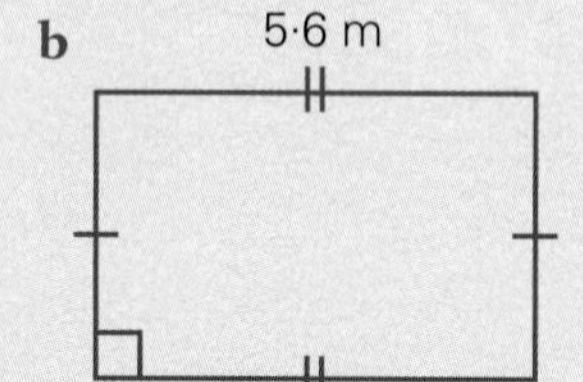

c
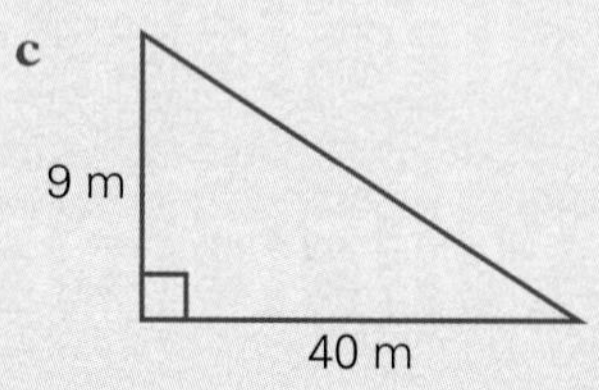

5 Find the volume of each of the following prisms:

a
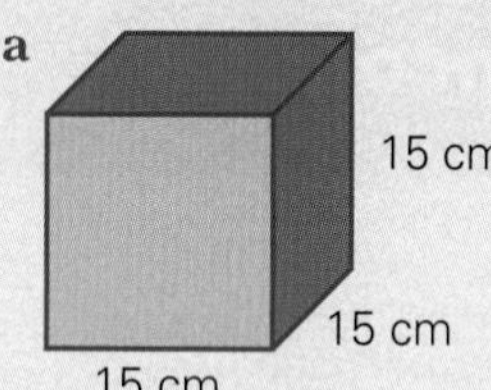

b
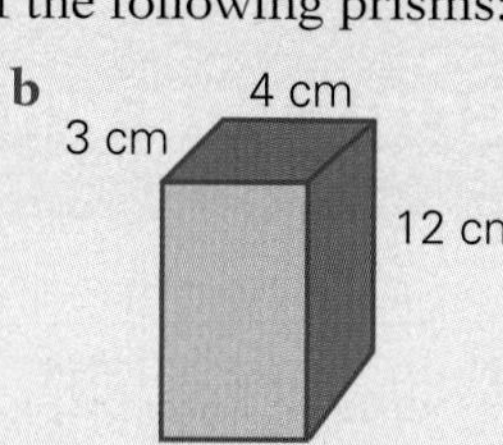

c
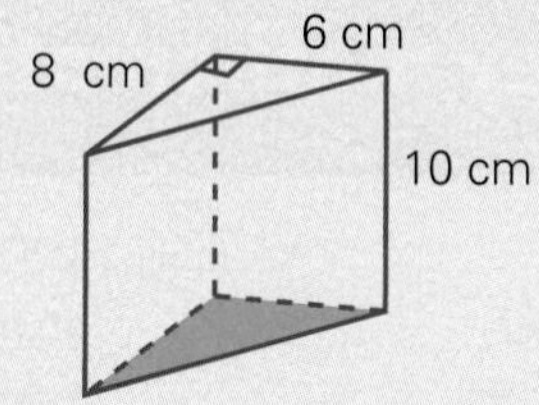

6 Convert the following to litres:

a 5000 mL b $5000\ cm^3$
c $\frac{1}{2}$ kL d $\frac{1}{2}\ m^3$
e 750 mL f $1800\ cm^3$
g $1\frac{1}{2}$ kL h $1\frac{1}{2}\ m^3$

Remember:
1000 mL = 1 L
$1\ cm^3 = 1$ mL
1000 L = 1 kL
$1\ m^3 = 1$ kL

7 Convert:

a $1\ cm^3 =$ ______ mm^3
b $\frac{1}{2}\ cm^3 =$ ______ mm^3
c $70\ 000\ mm^3 =$ ______ cm^3
d $1\ m^3 =$ ______ cm^3
e $\frac{1}{4}\ m^3 =$ ______ cm^3
f $5\ 000\ 000\ cm^3 =$ ______ m^3

$\times 1000^3$ $\times 100^3$ $\times 10^3$
$1\ km^3$ → m^3 → cm^3 → mm^3
$\div 1000^3$ $\div 100^3$ $\div 10^3$

8 Find the capacity, in millilitres, of the solids given in question **5** above.

Exploring New Ideas

12.1 Parts of a circle

Circles are used all over the world. Road signs, flags and even the Olympic rings are circles.

A circle is the set of all points that are the same distance (equidistant) from a fixed point, the centre O.

The following table shows commonly used parts of a circle.

Name	Diagram	Definition
Circumference	circumference	The boundary of a circle is known as its **circumference**.
Radius	O	A line from the centre of the circle to the circumference is a **radius** (plural: radii).
Chord		A **chord** is an interval joining any two points on the circumference of the circle.
Diameter	O	A **diameter** is a special chord that passes through the centre of the circle.

Name	Diagram	Definition
Semicircle	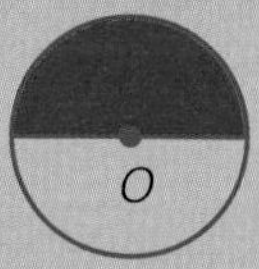	Half a circle. The diameter divides the circle into two **semicircles**.
Sector	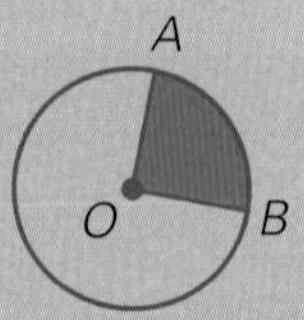	A **sector** is a slice of a circle bounded by two radii.
Arc	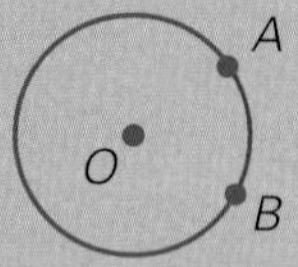	An **arc** is a section of the circle's circumference.
Segment	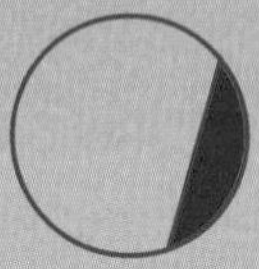	A **segment** is a region bounded by a chord and an arc.
Angle at centre	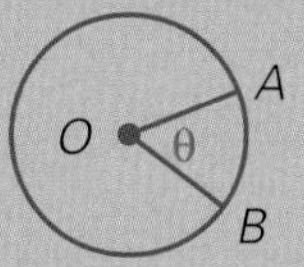	The angle at the centre, between two radii, is known as the **central angle**.
Tangent	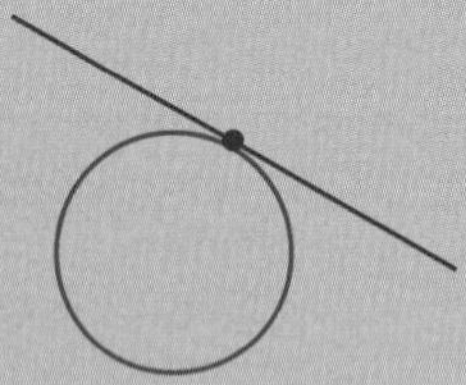	A line that intersects the circle in only one point is known as a **tangent** to the circle. It can be thought of as touching at just one point.
Concentric circles	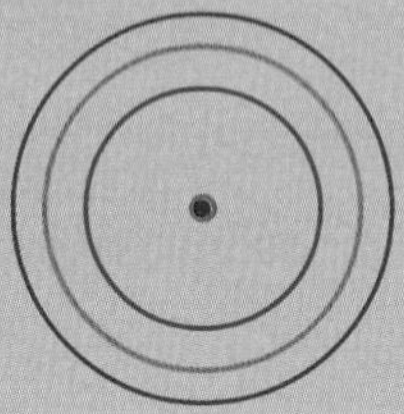	Two or more circles drawn with the same centre, but with different radii, are said to be **concentric circles**.

EXAMPLE 1

Name the feature indicated in each of the following diagrams.

a

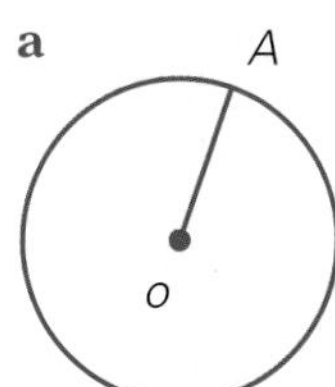

b

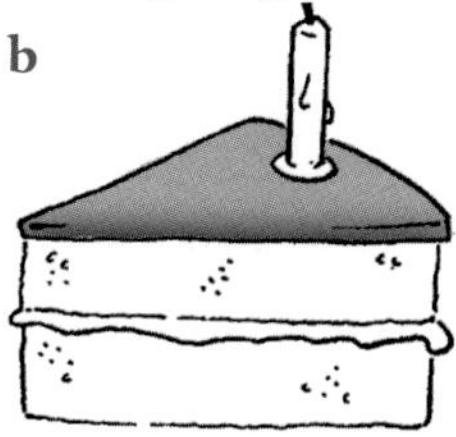

Solution

a Radius

b Sector

EXAMPLE 2

What fraction of the circle is shaded in:

a

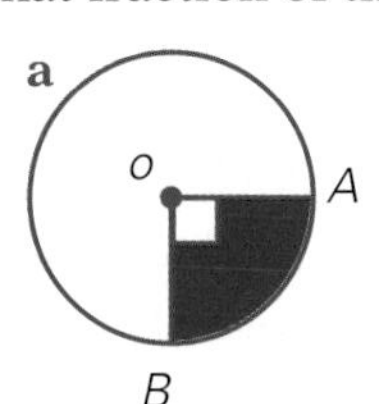

b

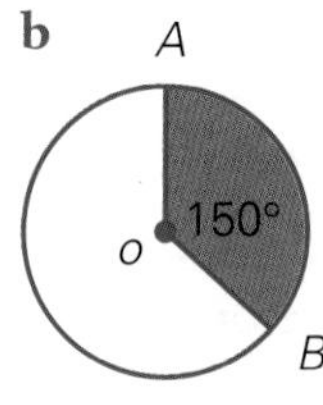

Sector fraction $= \frac{\text{central angle}}{360°}$

- Don't forget to simplify.

Solution

a Fraction shaded $= \frac{90°}{360°}$

$= \frac{1}{4}$

b Fraction shaded $= \frac{150°}{360°}$

$= \frac{5}{12}$

EXERCISE 12A

1 Write the name of the circle part/s featured in each object:

a

b

c

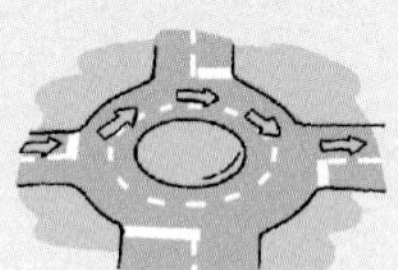

d

2 Name the features described below:

a interval AB

b interval CD

c region ACD

d interval EF

e region EGF

f region ADB

g $\angle ACD$

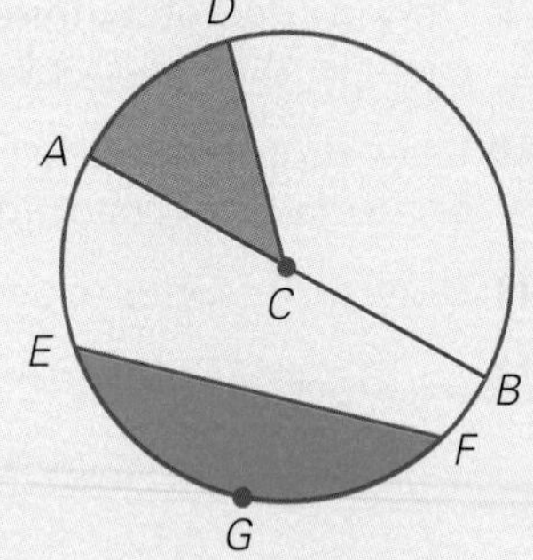

3 Draw a circle of radius 5 cm and show, using different colours, a chord, a sector, a segment, a diameter, a tangent and a radius.

4 How many times the length of a radius is the length of a diameter?

5 What parts form the boundary of:

a a sector? **b** a segment?

6 a Use a compass to draw a circle, centre C, radius 35 mm. Mark any point B on the circle. Draw in the following:

i diameter AB **ii** radius CP
iii arc APX (use colour) **iv** tangent TP

b Shade in the following:

i a sector **ii** a semicircle

Level 2

7 If A and B are any two points on a circle, what two parts of a circle show paths from A to B?

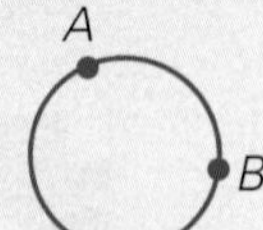

8 What fraction of the whole circle is each sector below?

a

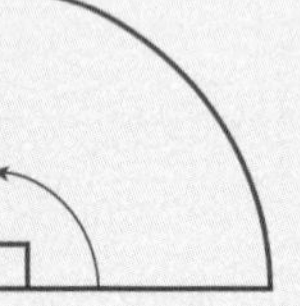

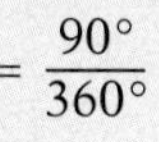

sector fraction

$= \dfrac{90°}{360°}$

$= ?$

b

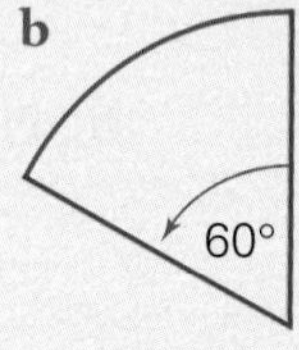

c

d

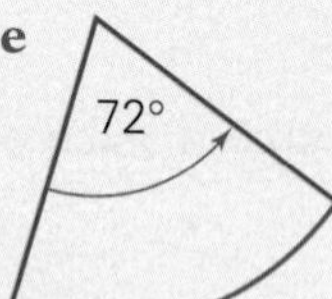

e

f

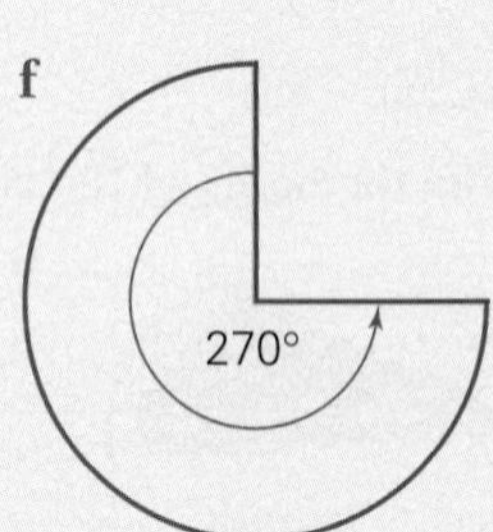

9 Draw a circle of radius 4 cm. By drawing two radii and a chord, construct an isosceles triangle. Why *must* it be isosceles?

10 Draw three *concentric* circles of radius 3 cm, 4 cm, 5 cm respectively. List three occurrences of concentric circles in real life.

11 Draw a 70° sector in a circle of radius 35 mm.

12

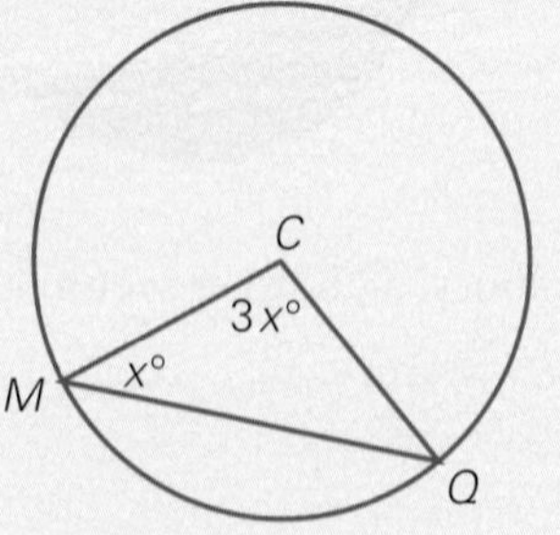

a What type of triangle is ΔCMQ? Why?

b Find x.

Investigation Finding the circumference of a circle

A different kind of pi

You will need: a tailor's tape measure, a marking pen, graph paper, several circular shapes, such as drink can, cake tin, saucer, disposable cup, 20c coin

1 In small group discussion, decide on the most accurate ways of finding the perimeter and diameter of the circular shapes. List all your practical ideas.

2 Compare your methods with these:

a **Perimeter** i

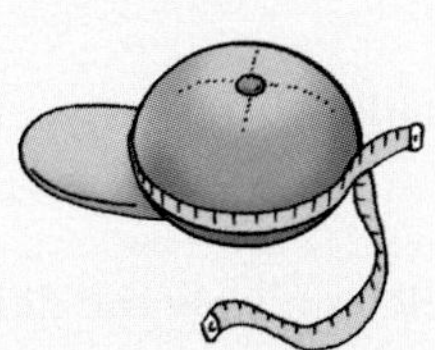

ii

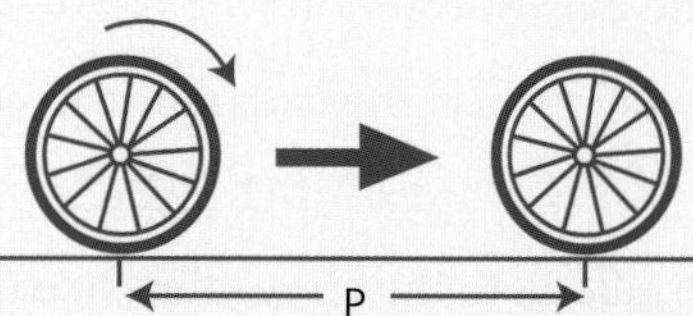

b **Diameter** i

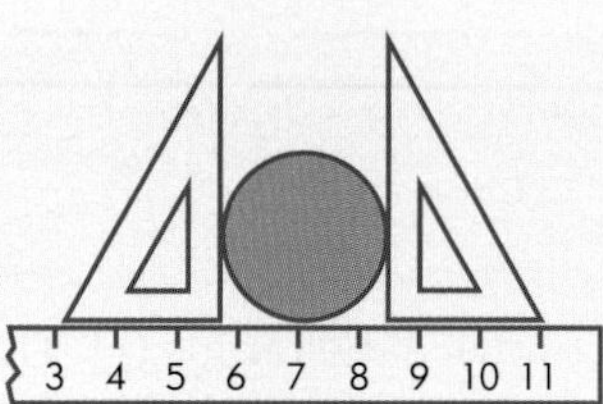

ii

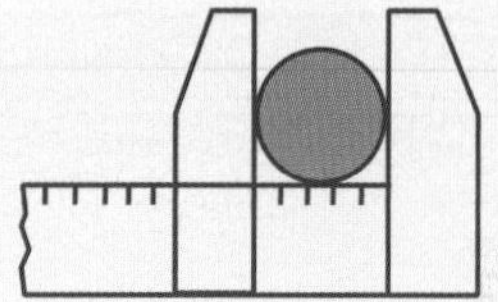

3 As accurately as possible, copy and complete this chart for at least five objects (some samples are shown).

	Object	Perimeter (circumference)	Diameter	$\frac{\text{Circumference}}{\text{Diameter}}$
a	Mower wheel	47 cm	15 cm	$\frac{47}{15}=$ ___
b	4 L paint tin	56 cm	18 cm	$\frac{56}{18}=$ ___ $\frac{56}{18}=$ ___
c				

Express this quotient correct to 4 decimal places.

4 Graph the circumference against the diameter for each object listed in the chart above.

i What do you notice about the shape of the graph?

ii Use the graph to predict the circumference of an object, given its diameter, and vice versa.

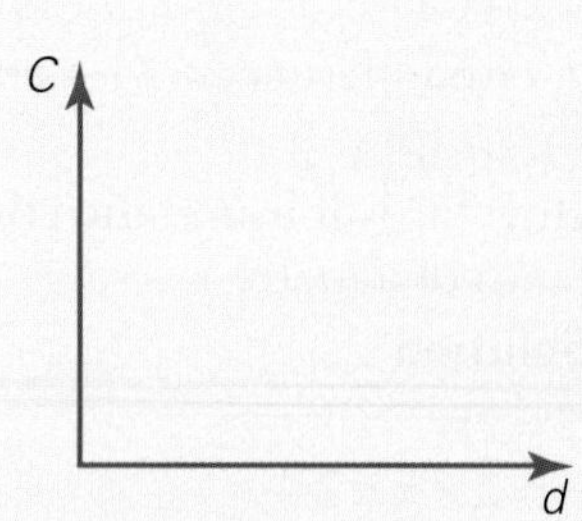

5 Develop a formula for finding the circumference of *any* circle.

12.2 Circumference of a circle

For many centuries, it has been known that the circumference of a circle is approximately three times the diameter.

We use the Greek letter π (pi) for the value of $\dfrac{\text{circumference}}{\text{diameter}}$.

The value of circumference ÷ diameter is the same for any circle, so that $C \div D = \pi$.

Pi (π) is one of those peculiar numbers that cannot be expressed exactly as a fraction or a decimal. Its decimal digits go on forever without repeating or showing any pattern. Such numbers are called **irrational**.

The closest fraction to π is $\frac{22}{7} = 3{\cdot}142\ 857\ 143$

Today we use the calculator in our working with pi.

$$\pi = 3{\cdot}141\ 592\ 654\ \ldots$$

To calculate the circumference (C) of any circle, multiply the diameter (D) by the value of π.

> Circumference = π × diameter
>
> $C = \pi \times D$

As we are aware, the diameter is equal to twice the radius (r). This fact allows us to rewrite the formula above in terms of the radius.

> Circumference = 2 × π × radius
>
> $C = 2\pi r$

Remember to round your answers to a sensible number of decimal places.

EXAMPLE 1

Find the circumference of the circle with diameter 10 cm ($\pi \approx 3{\cdot}14$).

Solution

$C = \pi D$
$\quad \approx 31{\cdot}4 \times 10$
$\quad \approx 31{\cdot}4$
$\therefore$ circumference $\approx 31{\cdot}4$ cm.

EXAMPLE 2

How far is it round the edge of a circular pond of radius 2400 mm? (Answer to the nearest tenth of a metre.)

Solution

$C = 2\pi r$
$\quad = 2 \times \pi \times 2{\cdot}4$
$\quad \doteqdot 15{\cdot}079\ 644\ 74$
$\quad \doteqdot 15{\cdot}1$ (to 1 decimal place)
$\therefore$ it is 15·1 m around the edge of the pool.

2400 mm = 2·4 m

Use a calculator and round as required:

EXAMPLE 3

A wheel has a revolution measuring 24·5 cm. What is the radius of this wheel, to the nearest centimetre?

Solution

$C = 2\pi r$ | A revolution = the circumference.

$24{\cdot}5 = 2 \times \pi \times r$

$\div 2 \quad \div 2$

$12{\cdot}25 = \pi \times r$ | Solve the equation.

$\div \pi \quad \div \pi$

$r = 3{\cdot}8992 \ldots$

$\therefore$ radius is 4 cm (nearest cm).

EXAMPLE 4

Find the perimeter of the sector shown:

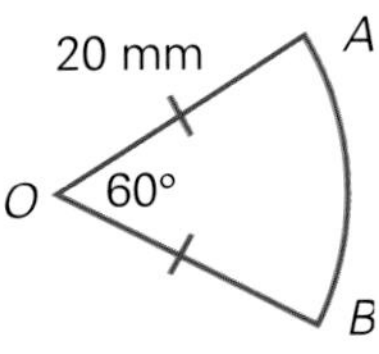

Solution

Perimeter = arc length AB + radius OA + radius OB

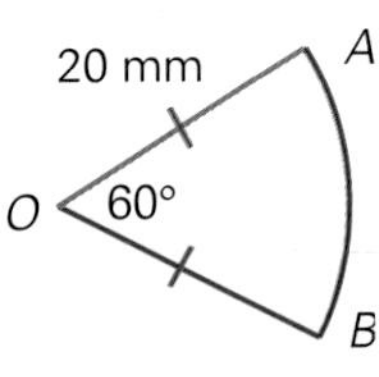

$$\text{arc } AB = \frac{60}{360} \times 2 \times \pi \times r$$

$$= \frac{1}{6} \times 2 \times \pi \times 20$$

$$\doteqdot 20{\cdot}943\ 95$$

$$\therefore P \doteqdot 20{\cdot}943\ 95 + 20 + 20$$

$$\doteqdot 60{\cdot}9 \text{ (1 decimal place)}$$

$\therefore$ perimeter is 60·9 mm (1 decimal place)

EXERCISE 12B

$\doteqdot$ is equivalent to $\approx$

1 *Estimate* the circumference of each circle, using $\pi \doteqdot 3$:

a diameter 10 cm **b** diameter 20 mm **c** diameter 38 m

d radius 31 cm **e** radius $19\frac{1}{2}$ m **f** radius 1·25 km

2 Use $\pi \approx 3{\cdot}14$ to find each circumference, correct to 1 decimal place:

a diameter 20 cm **b** diameter 80 mm **c** radius 15 cm **d** radius 50 m

3 Use $\pi \approx 3\frac{1}{7}$ to calculate each circumference:

a diameter 14 m **b** radius $10\frac{1}{2}$ mm

4 Use a calculator to find each circumference, correct to 2 decimal places:

a

82 mm

O

b

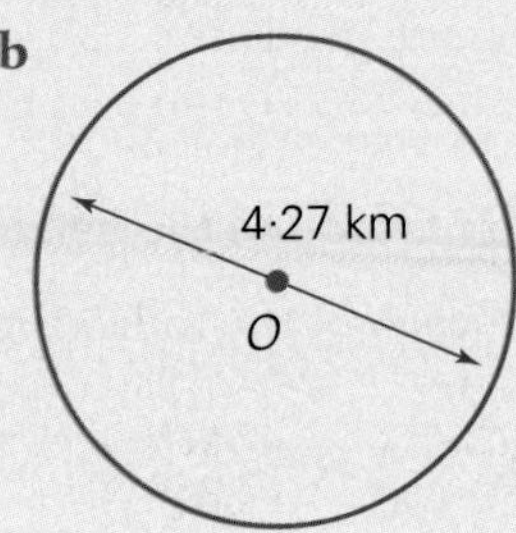

c

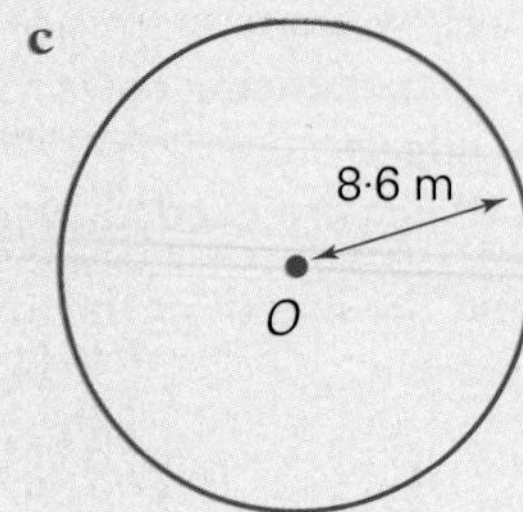

d

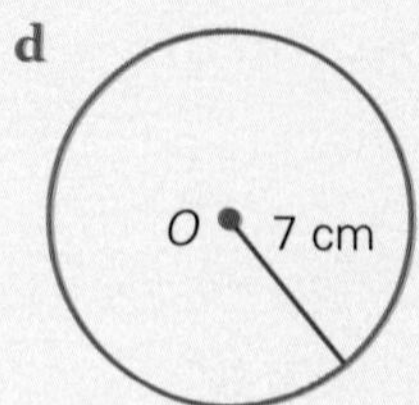

e

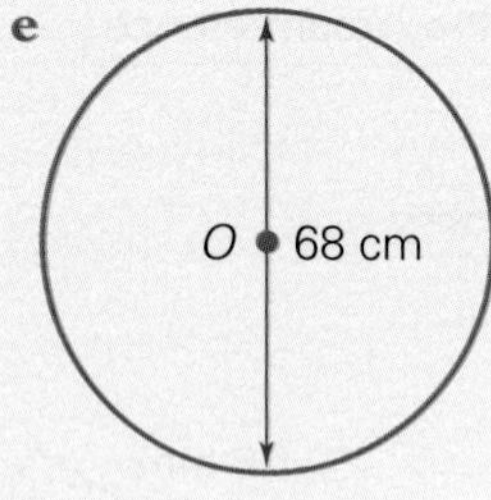

f

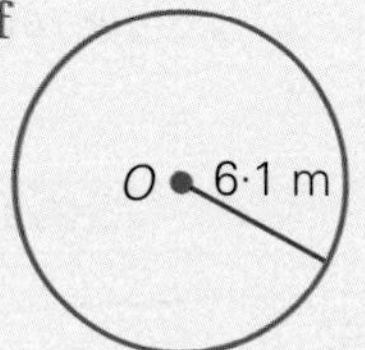

g

O
3 cm

h

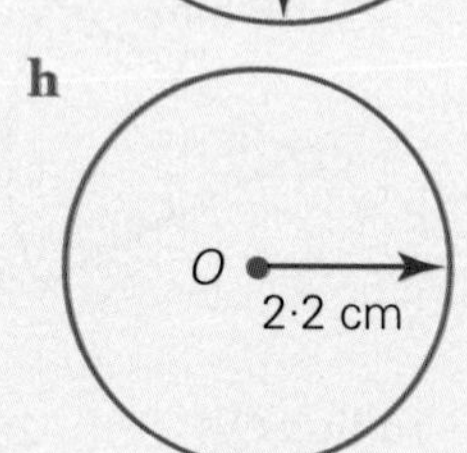

i

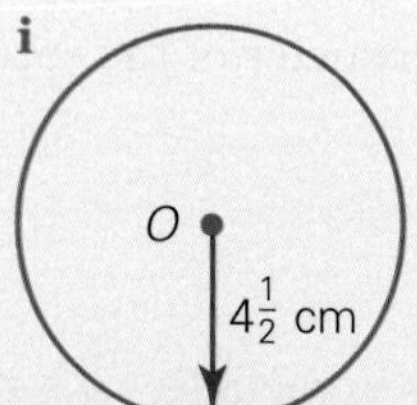

5 The radius of our planet Earth is approximately 6400 km. How far is it around the equator, to the nearest hundred kilometres?

6 How much wire is needed to form a circle of diameter 14·5 mm? (Answer to 2 decimal places.)

7 Holly's bike wheel has a diameter of 58 cm.

- **a** How far (to the nearest cm) does the wheel travel in one full revolution?
- **b** What distance in metres (to the nearest metre) is covered in 1000 turns of the wheel?
- **c** How many revolutions are made in travelling $1\frac{1}{2}$ km?

8 The radius of a circle was measured as 8·5 cm.

- **a** Between which two values does the true radius lie?
- **b** Using these two values, calculate the upper and lower limits for the circle's circumference.

Level 2

9 A circular pool of diameter 15 m needs to be fenced 1 metre from the pool's edge.

- **a** What is the radius of the circle whose circumference is that of the fence?
- **b** How many metres of fencing are needed?
- **c** If the fencing costs \$16 a metre, what is the cost of fencing 1 metre from the pool's edge?
- **d** Posts are needed every 2 metres around the fence. How many posts are needed?

10 Find the length of the arc *AB* in each of the following sectors (answer to 1 decimal place):

a

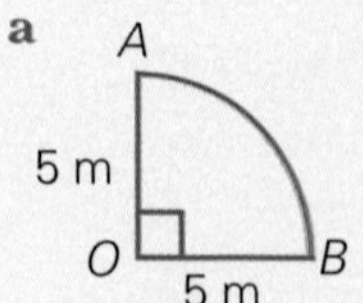

b

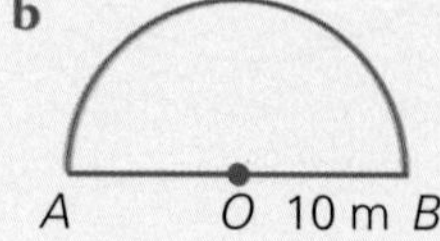

c

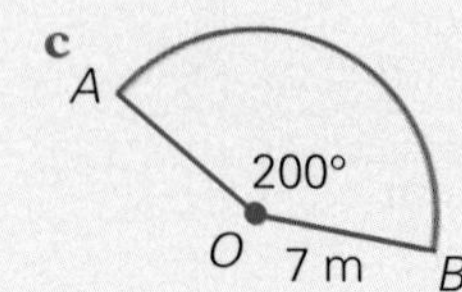

11 Find the perimeter of each of the sectors below (answer to the nearest metre):

a

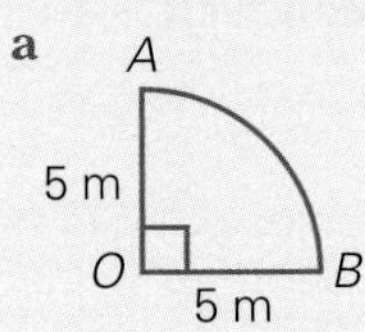

b

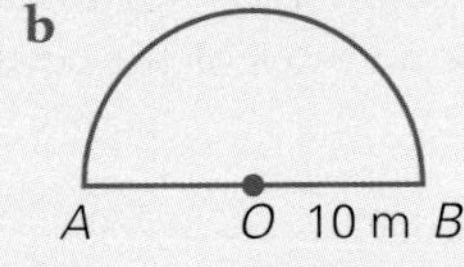

c

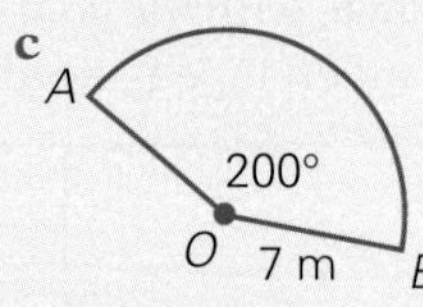

d

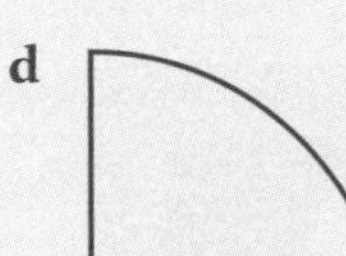

e

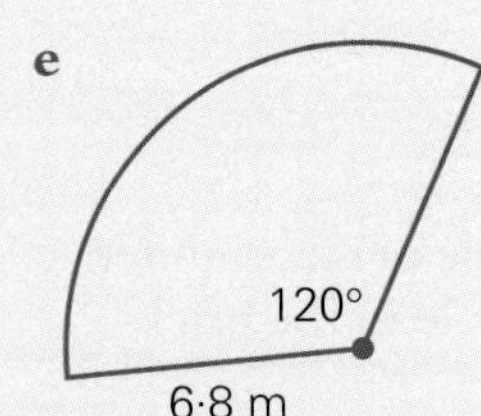

f

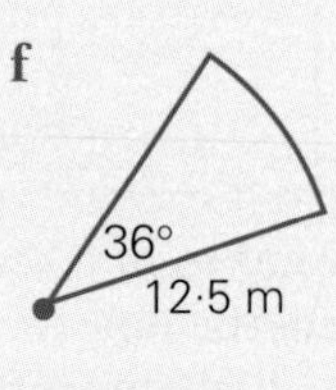

12 Find the perimeter of the figures below (answer to 1 decimal place):

a

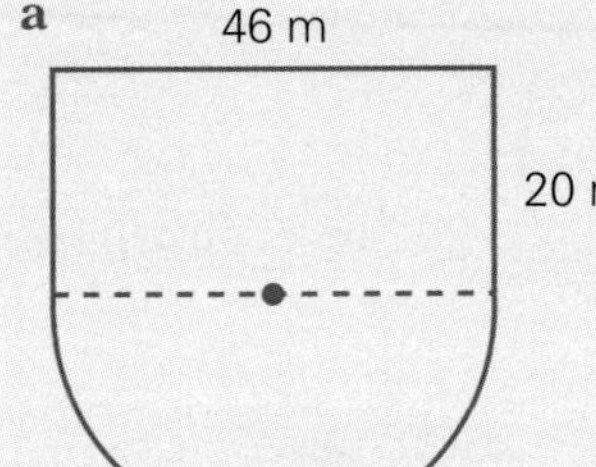

b

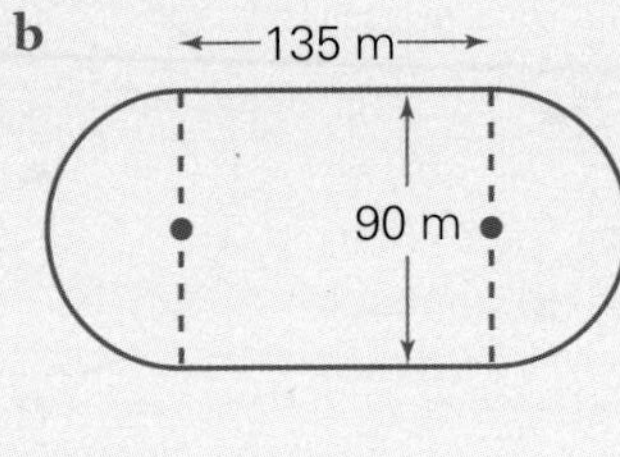

c

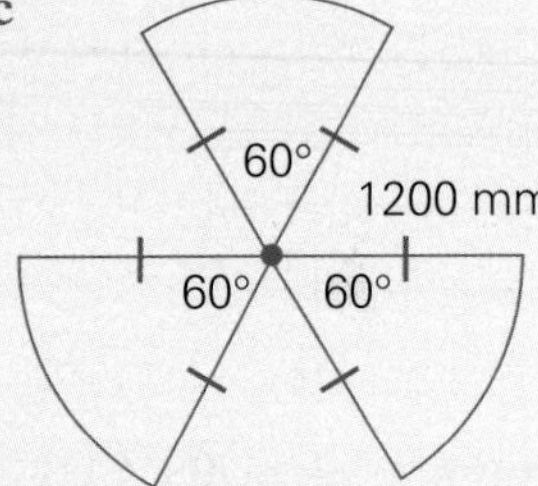

d

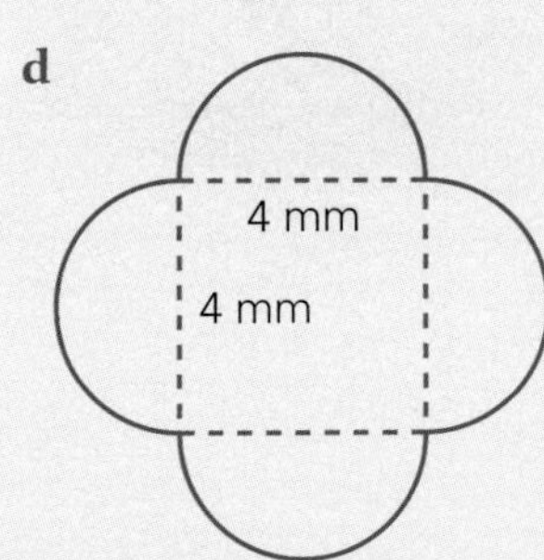

e

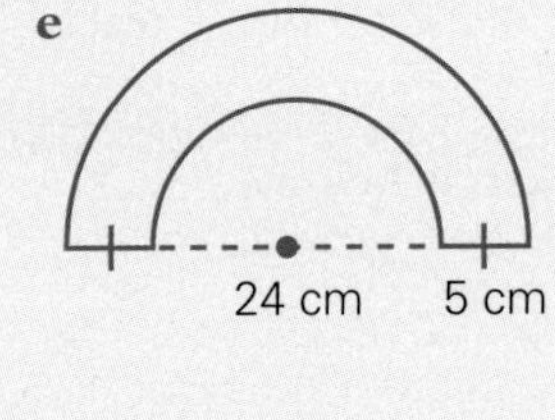

f

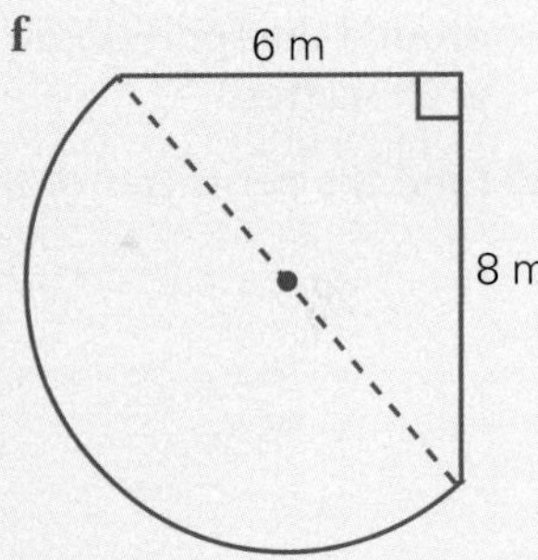

13 The circumference of a circle is given as 34·7 cm.

a Find the length of the radius.

b Discuss the accuracy of the circumference and decide on the number of decimal places that you believe the radius should have.

14 a What is the diameter of a circle of circumference 628 m? (Answer to 1 decimal place.)

b What is the radius of a semicircle whose arc length is 157 cm? (Use $\pi = 3{\cdot}14$)

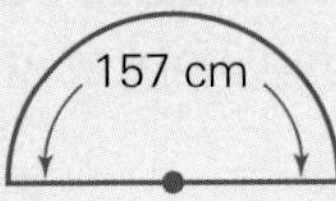

15 The minute hand of a City Hall clock is 3·4 m long. How far does its outer tip travel in 1 week, to the nearest hundred metres?

16 A netball court is 30·5 m long and half as wide. It is divided into three equal regions, as shown, with the goal circles in the end thirds. The radius of each semicircle is 4·9 m. How much tape is needed to mark all lines of the court?

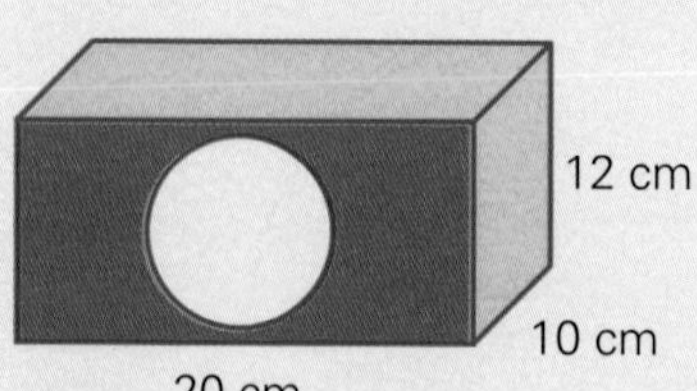

17 A builder has to fit a sewerage pipe of circumference 15 cm. Will the pipe fit into the hole in the brick as shown? Explain your answer.

18 Find the ratio of the circumference of circle A to that of circle B.

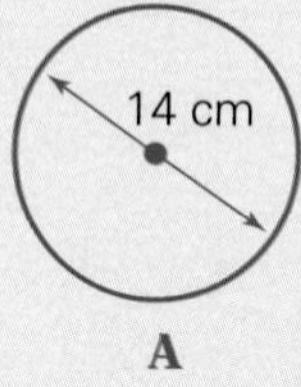

19 If Mariah walks 3 kilometres in 1 hour, how long will it take her to complete one lap of the circular track around the sports centre if the diameter of the track is 25 metres?

20 Find the perimeter of the semicircle below.

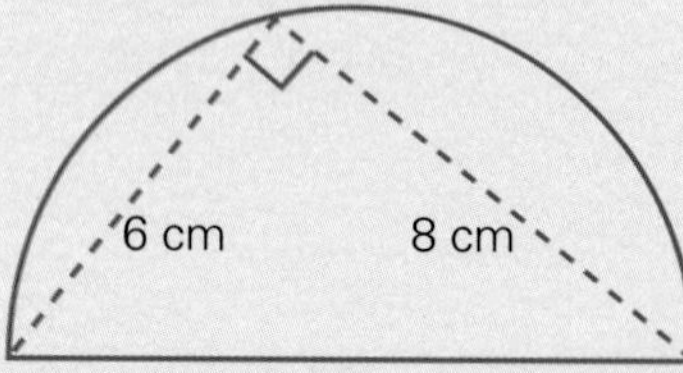

21 A running track has five lanes each 2 metres wide. Find the difference in the distance travelled by an athlete running along the very inside of the track and an athlete running along the outside boundary of the track.

Investigation Changes in the diameter

Investigate what happens to the circumference of any circle when its diameter is:

a doubled
b halved
c tripled

To help, you may like to start by looking at a circle of diameter 1 m.

Investigation The area of a circle

In search of a rule for area

You will need: compass, 1 cm grid paper, protractor, plain paper, scissors

1 Counting squares

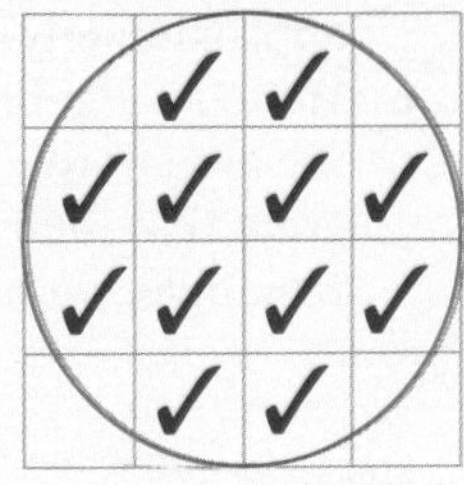

Using the standard procedure of counting only squares where more than half lies within the shape, find an *approximate* area of a circle of radius 2 cm.

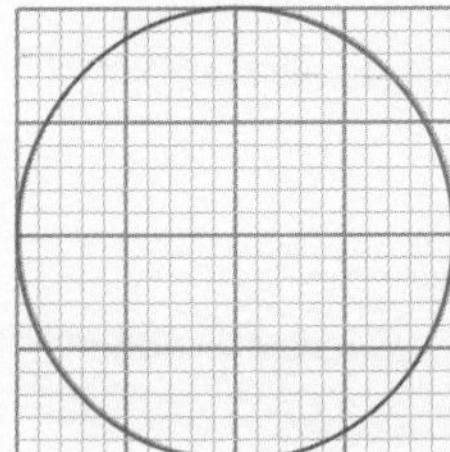

For greater accuracy, we could use smaller squares. Using the same counting procedure, find a better approximation for the area of the same circle.

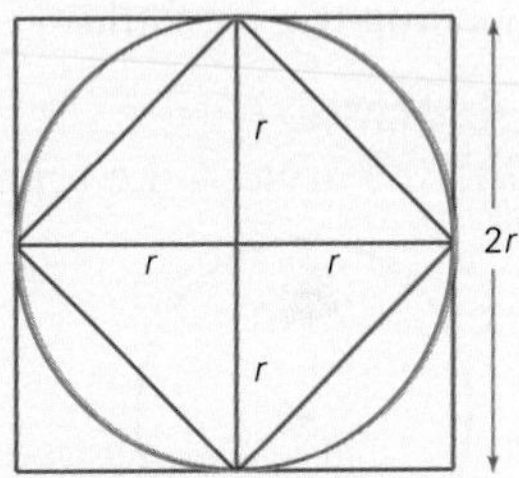

2 Archimedes' method (250 BC)

The area of the circle of radius r must lie between the area of the outer square (side $2r$) and that of the inner square (made up of four identical triangles). Find algebraic expressions for the areas of both squares.

Split the difference and write an approximate area of the circle algebraically.

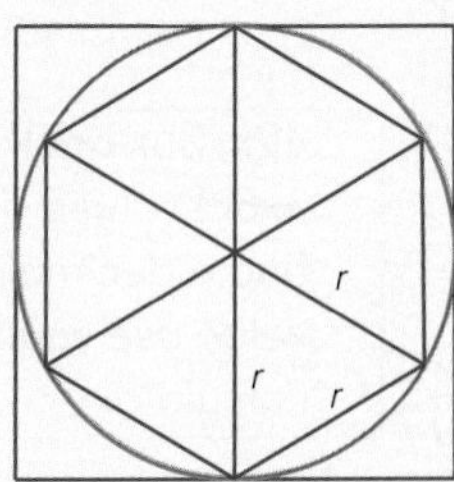

Can we improve the accuracy of the inner figure, to more closely approximate a circular shape? Let's try a hexagon. Inscribe a hexagon as shown. (Challenge: Find its area algebraically!) Again, the circle's area must be *between* that of the square, $4r^2$, and that of the inner hexagon.

Archimedes constructed polygons progressively with more and more sides, eventually arriving at area = $3\frac{10}{71} \times r$. Not bad!

3 Slicing for pi

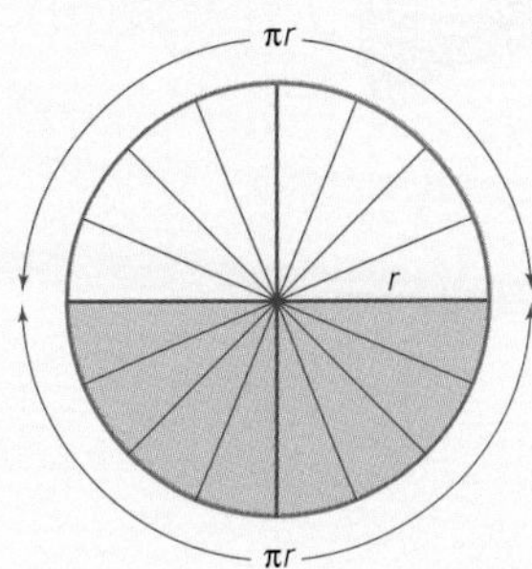

a Draw a circle of radius 6 cm and divide it into 16 identical sectors, shading it as shown.

b Cut out the circle, then cut it into the shaded and unshaded semicircles.

c Beginning at the centre, cut along each radius, stopping just before reaching the circumference each time. Open out the sectors carefully, like a row of shark's teeth.

d Carefully fit the two portions together as shown, to form a shape that is very close to a rectangle. What is the length of the rectangle? What is its breadth, approximately?

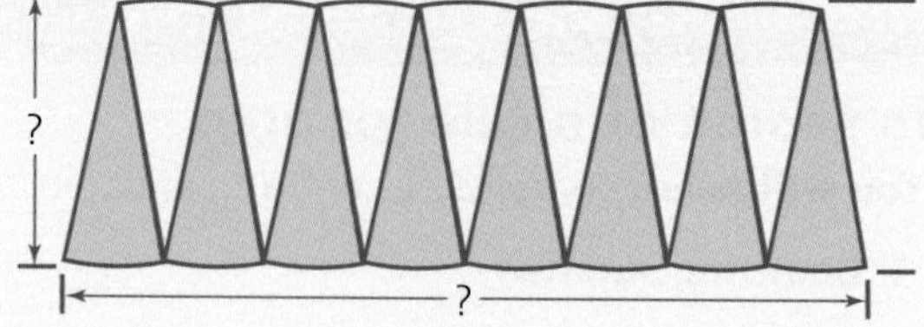

e Write an algebraic expression for the area of the rectangle of sectors. Since we could make the number of slices so great that their combined shape would get closer and closer to that of a true rectangle, we now have a *formula* for area of a circle.

12.3 Area of a circle

The area of a circle can be calculated by using the formula:

Area = πr^2 where r = radius

EXAMPLE 1

Find the area of the following circles:

INTERACTIVE GEOMETRY

a

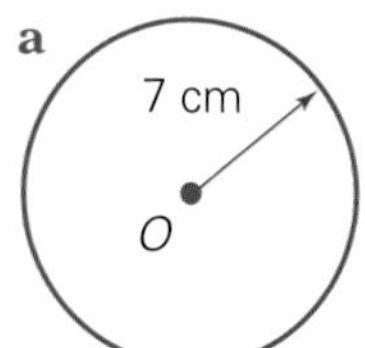

b

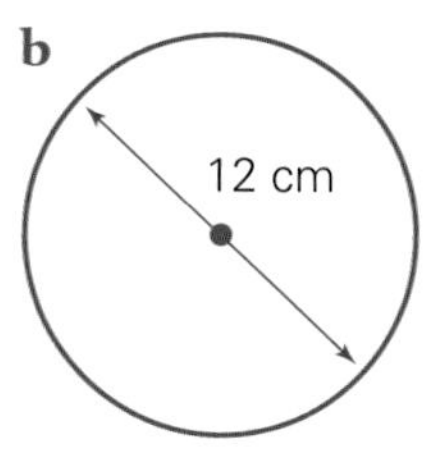

Solution

a $r = 7$

$A = \pi r^2$

$= \pi \times 7^2$

$= 49\pi$

Area = $153{\cdot}9$ cm^2 (1 dec. pl.)

b diameter = 12, $\therefore r = 6$

$A = \pi r^2$

$= \pi \times 6^2$

$= 36\pi$

Area = $113{\cdot}1$ cm^2 (1 dec. pl.)

> 49π, 36π are the **exact** values. To find a decimal value use your calculator.

EXAMPLE 2

Find the area of the sector with central angle 45°.

12·5 cm
45°

Solution

Sector = $\frac{45}{360}$ of the circle ← Fraction of the circle

$\therefore$ Area = $\frac{45}{360} \times \pi r^2$

$= \frac{45}{360} \times \pi \times 12{\cdot}5^2$

$= 61{\cdot}3 \ldots$ cm^2

Area is 61 cm^2 (nearest cm).

EXERCISE 12C

1 Estimate the area of each circle, using $\pi \approx 3$:

a radius 10 m **b** radius 20 m **c** radius 3·9 m
d diameter 8 m **e** diameter 4·3 m **f** diameter 64 m

2 Use $\pi \approx 3{\cdot}14$ to find area, correct to 1 decimal place where necessary:

a radius 20 cm **b** radius 30 cm **c** diameter 4 m

3 Use $\pi \approx 3\frac{1}{7}$ to find area:

a radius 7 cm **b** radius $3\frac{1}{2}$ m **c** diameter $3\frac{1}{2}$ cm

4 Use a calculator to find each area, correct to 2 decimal places:

a

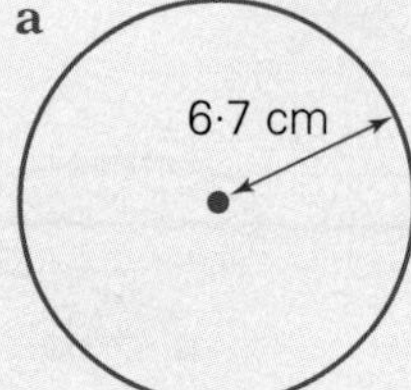

b

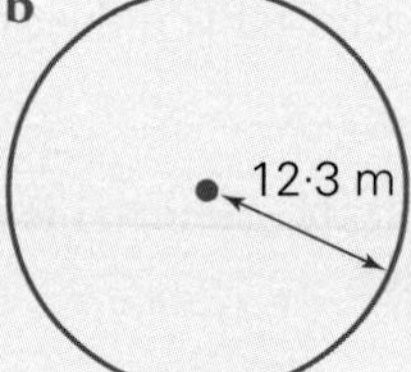

c

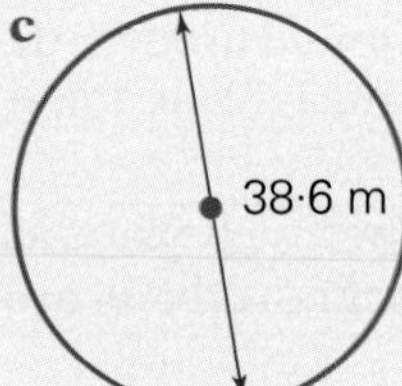

d

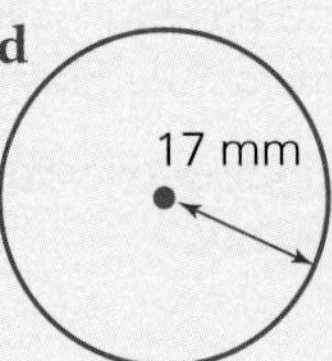

e

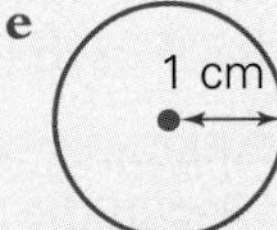

f

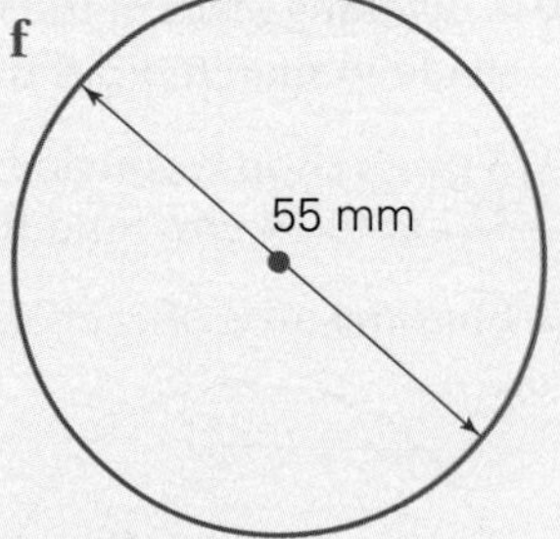

5 Find the area of the following circles, answers to be left in *exact* form.

a

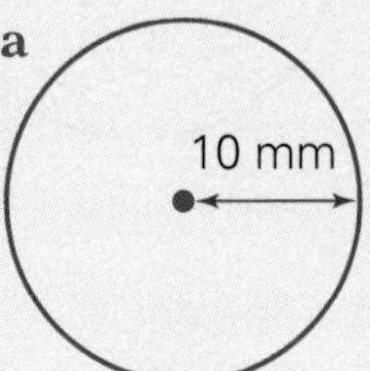

b

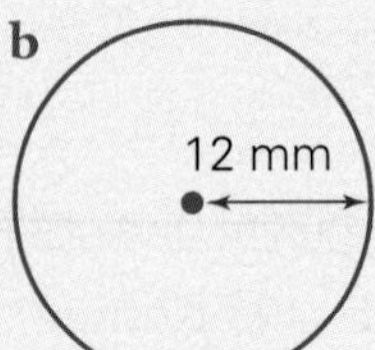

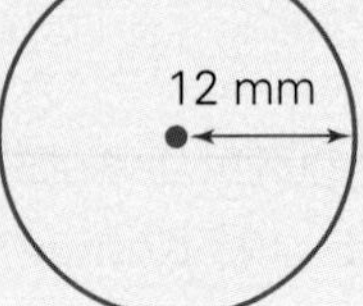

c

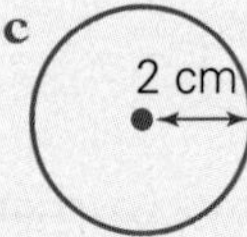

d

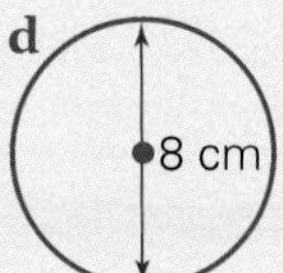

e

14 cm

f

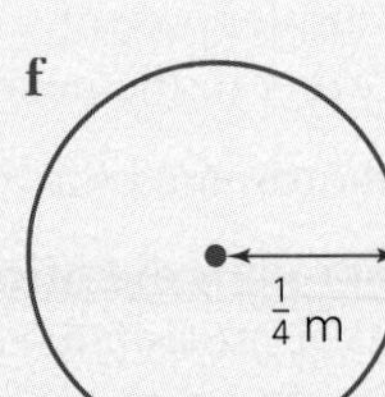

6 A lawn sprinkler waters a circle 7·5 m in diameter. What area in square metres does it cover?

7 A park ranger can see for 20 km in all directions from her fire observation tower. What area is covered by her field of view, to the nearest square kilometre?

Level 2

8 A circular landscaped area of diameter 32 metres is covered with turf at \$2.95 per square metre, then top-dressed with soil at a quoted price of 45 cents per square metre. Find the total cost of the job.

9 **a** How much material (to the nearest centimetre) is required to cover the top of a circular table if the diameter of the table is 120 cm?

b How much material is required if the material is to allow for a 10 cm overhang around the table?

10 A drinking glass in the shape of a cylinder with a diameter of 6 cm is to have a coaster made to match. Find the area of the coaster if its diameter is the same as the glass.

11 A pizza shop is considering increasing the diameter of its family sized pizza tray from 32 cm to 34 cm. Find the percentage increase in the area of the pizza tray.

12 Find the area of each sector, to the nearest square metre:

a

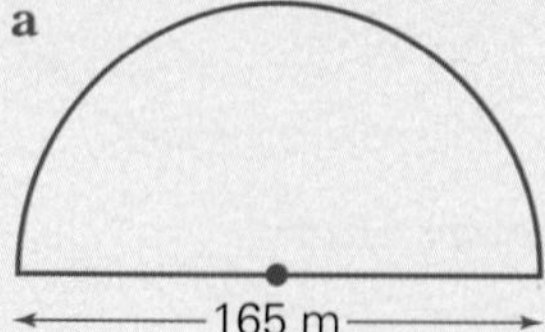

b

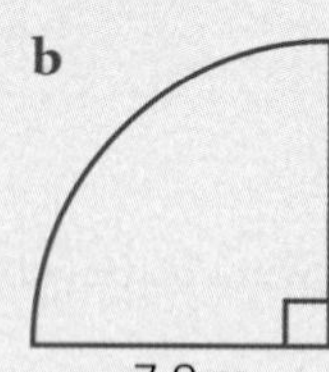

c

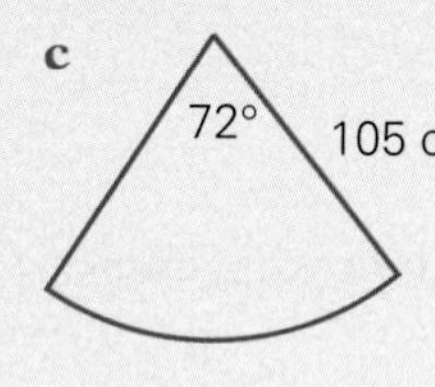

d

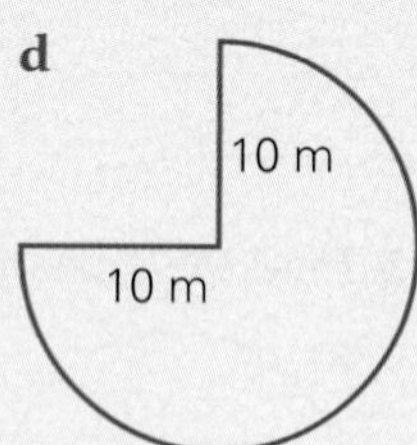

13 Find the radius (correct to 1 decimal place) of a circle with:

a area = 50 cm^2 **b** area = 123·67 mm^2 **c** area = 36π cm^2

14 The circumference of a circular playing field is 315 m. Find the area of the playing field, correct to the nearest square metre.

15 Michael estimated the area of the circle he had drawn to be 3×4^2.

a What radius did Michael's circle have?
b What approximate value of pi did Michael use?
c Without using your calculator, write down another approximate value of the circle's area.
d Estimate the value of the circle's circumference.

Try these

1 The area of the unshaded portion of the circle is 40 cm^2 and the angle AOB is 100°. Calculate the area of the shaded sector of the circle.

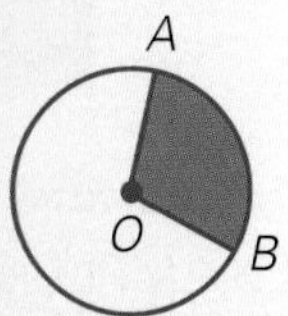

2 Since diameter is easier to measure accurately than radius, engineers prefer an area formula based on *diameter*. Can you produce the adjusted engineer's formula from πr^2?

Investigation Accuracy of measure

You will need: ruler, tape measure, metre ruler, copy of the circle

1 Divide the class into four groups:

- Group A—ruler
- Group B—tape measure
- Group C—metre ruler
- Group D—estimation

2 Each group is to measure the radius of the circle drawn by the teacher and use this value to calculate its circumference and area.

3 Each member of the group is to record their own answers and **discuss** any discrepancies and **reasons** for these differences.

4 Complete a table in your workbook with the results from all the groups. Which group had the highest and lowest values? Who do you believe had the most accurate results? Discuss.

12.4 Composite areas with circles

Many composite two-dimensional shapes are made up of circles or portions of circles. In this section we will look at calculating the area of such shapes.

EXAMPLE 1
Find the area of this composite shape.

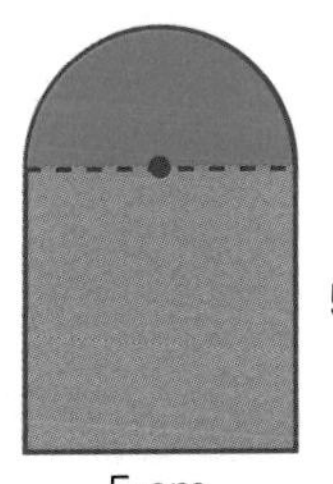

Solution

Area 1

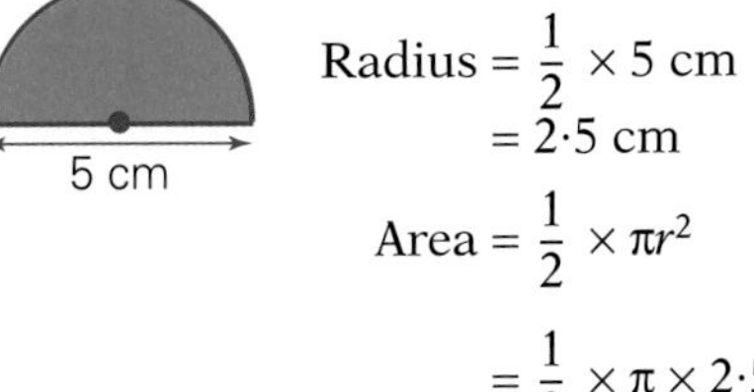

$$\text{Radius} = \frac{1}{2} \times 5 \text{ cm}$$
$$= 2{\cdot}5 \text{ cm}$$
$$\text{Area} = \frac{1}{2} \times \pi r^2$$
$$= \frac{1}{2} \times \pi \times 2{\cdot}5^2$$

Area 2

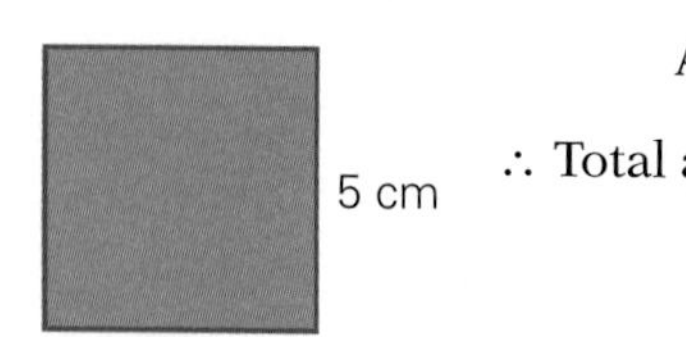

$$\text{Area} = 5 \times 5$$

$$\therefore \text{Total area} = \frac{1}{2} \times \pi \times 2{\cdot}5^2 + 5 \times 5$$
$$= 34{\cdot}817 \ldots$$

$\therefore$ Total area is $34{\cdot}8\ \text{cm}^2$ (1 decimal place).

EXAMPLE 2

Find the area of the composite figure at right:

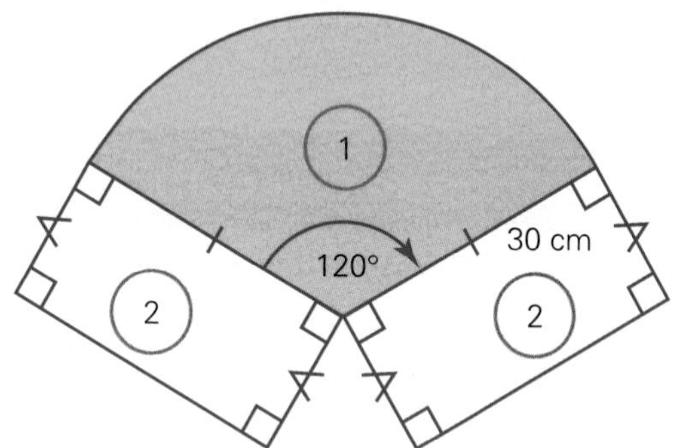

Solution

$$A_{①} = \frac{120}{360} \times \pi r^2$$
$$= 120 \div 360 \times \pi \times 30^2$$
$$= 942{\cdot}47 \ldots \text{cm}^2$$
$$2 \times A_{②} = 2 \times lb$$
$$= 2 \times 30 \times 15$$
$$= 900\ \text{cm}^2$$
$$\therefore \text{total area} = 942 + 900\ \text{cm}^2$$
$$= 1842\ \text{cm}^2 \text{ (nearest square cm)}$$

EXERCISE 12D

1

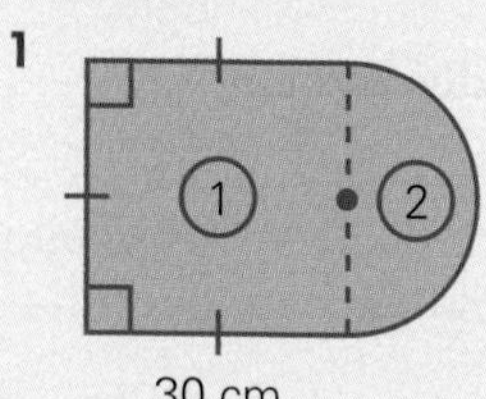

a What is the length of the diameter?
b Calculate the area of the square ①.
c Find the area of the semicircle ②.
d What is the total area of the shape ① + ②?

2 a Find the area of the triangle.
b Find the area of the semicircle.
c By adding the areas in **a** and **b** above, find the area of the composite figure supplied.

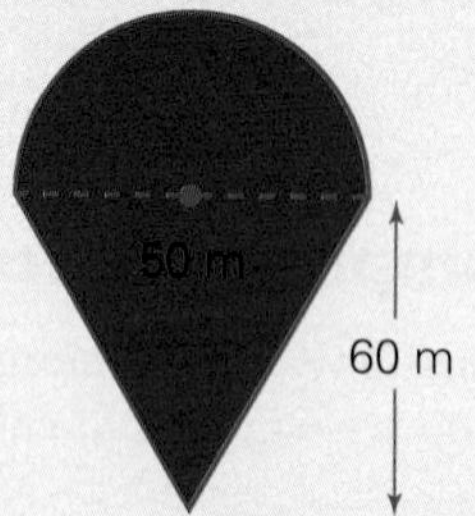

3

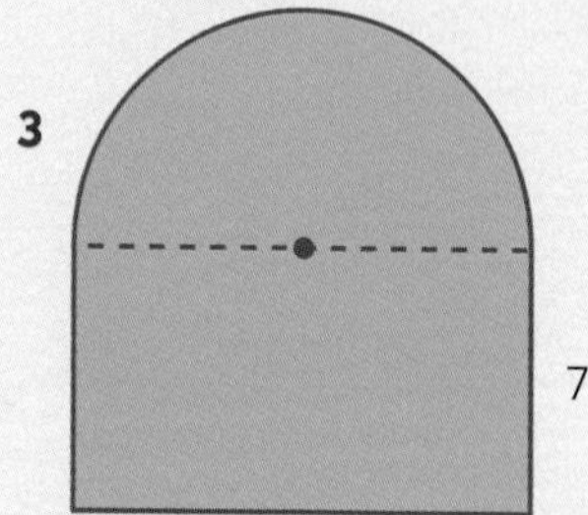

Find the area of the figure at left by adding the area of the rectangle to the area of the semicircle.

4 Find the shaded area by subtracting the area of the circle from the area of the square.

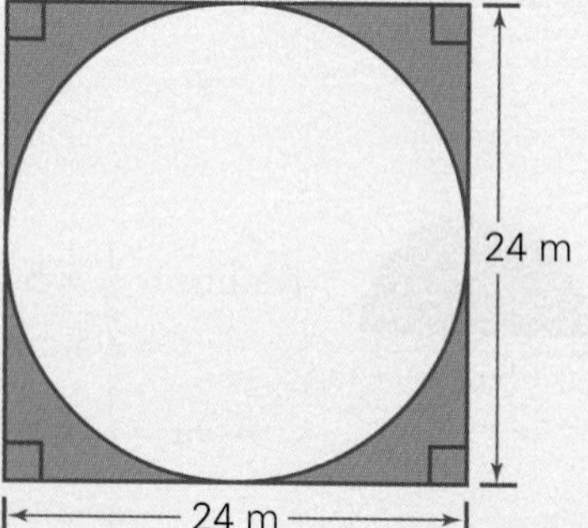

5 The shaded area in the figure below is found by subtracting the area of the smaller circle from that of the larger circle. Calculate the shaded area.

Annulus is the Latin word for ring. In Maths an annulus is the region between two concentric circles.

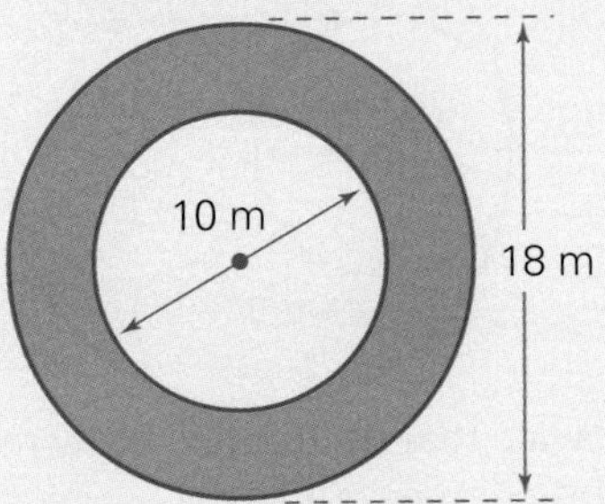

This shape is known as an annulus. Can you come up with a formula for the area of any annulus?

Level 2

6 Find the area of the shaded region of each of the figures given below (answers to 1 decimal place).

a

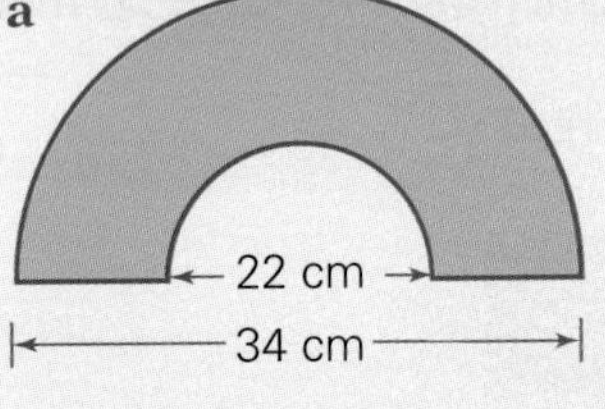

b

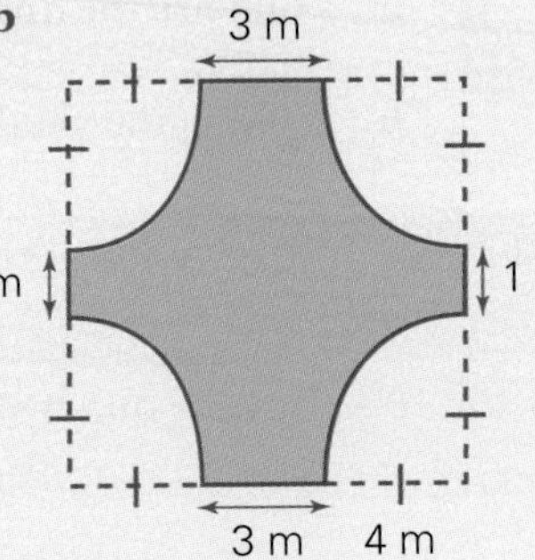

c

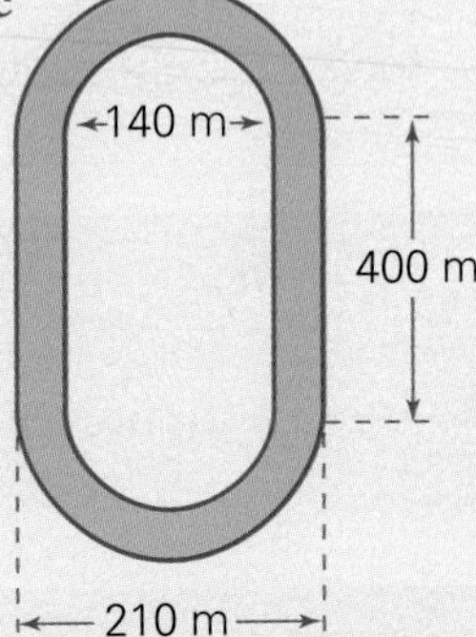

7 By using Pythagoras' theorem, where appropriate, calculate the shaded area of each of the figures below (answer to 2 decimal places).

a

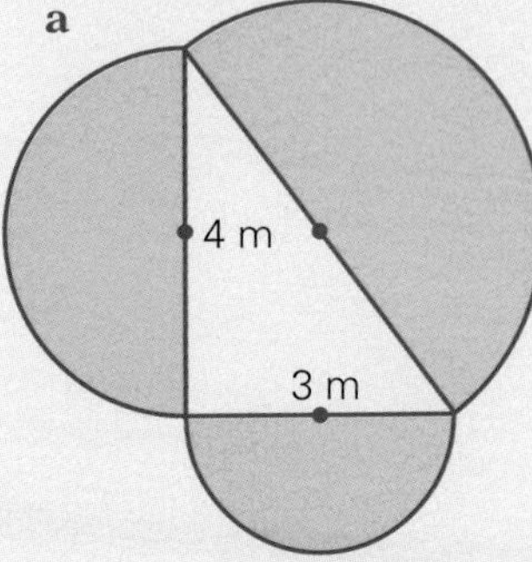

b

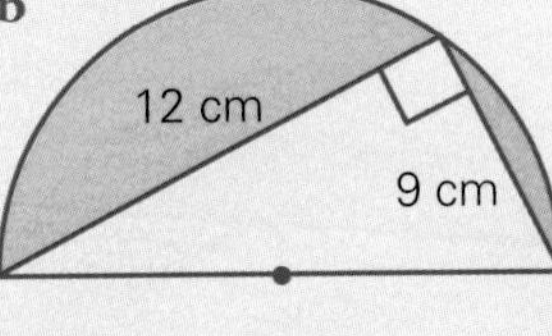

c

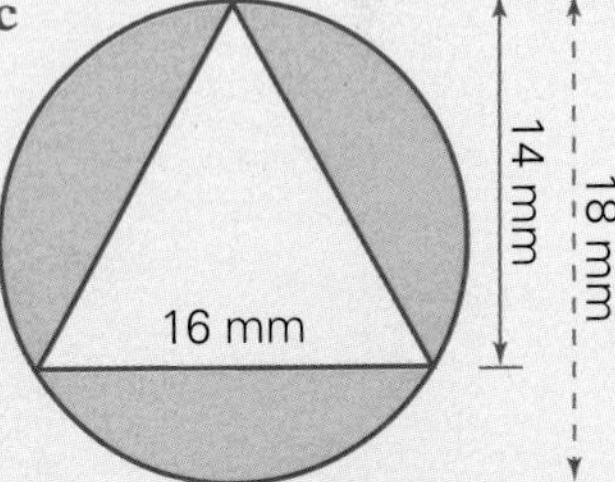

8 Find an expression, in terms of π, for the shaded area.

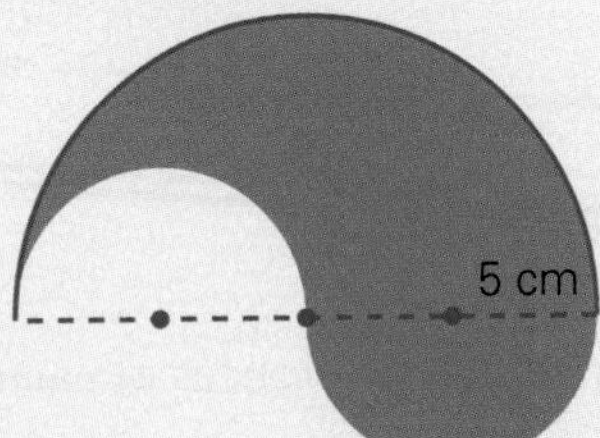

9 Calculate the shaded area:

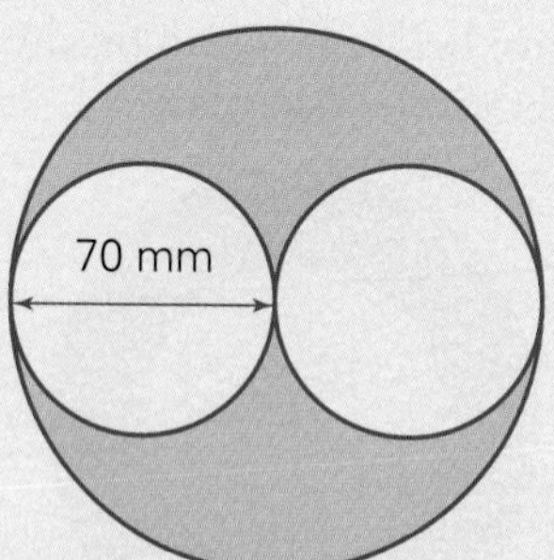

TEACHER

10 The radius of one circle is twice that of another. How many times the area of the smaller circle is the area of the larger circle?

11

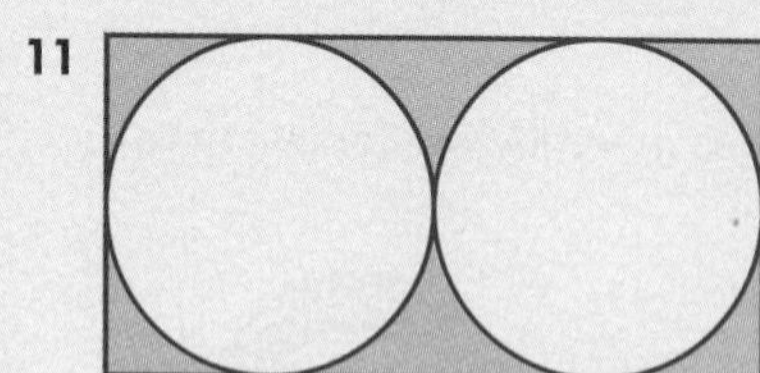

INTERACTIVE GEOMETRY

Two identical circles are to be cut out of a sheet metal rectangle as shown. Their combined areas are 5652 cm^2, found using $\pi \approx 3{\cdot}14$.

a What are the dimensions of the rectangular metal sheet?

b What is the shaded area?

Try this

To increase the area of a circle by 10%, by how much should the radius be increased?

Investigation Magic circles

Over 300 years ago, a Japanese mathematician called Seki Kowa discovered this magic circle.

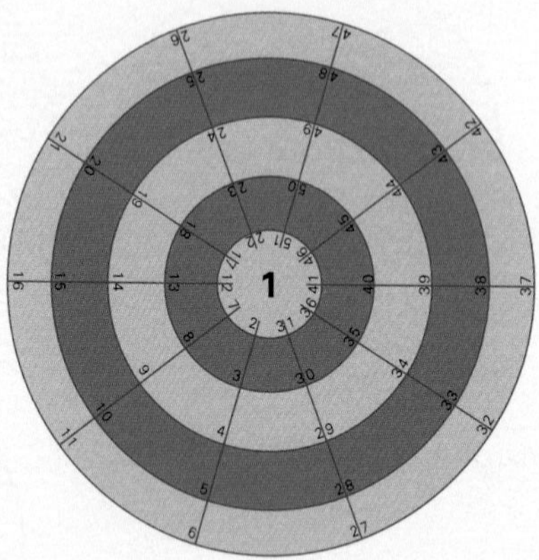

The circumferences of each of the concentric circles total the same number.

The numbers, excluding 1, along each diameter also total the same number.

1 What is the magic number for each diameter?

2 What is the total of each circumference?

3 Complete the following magic circle using the numbers from 1 to 21. Some of the numbers have been placed for you already.

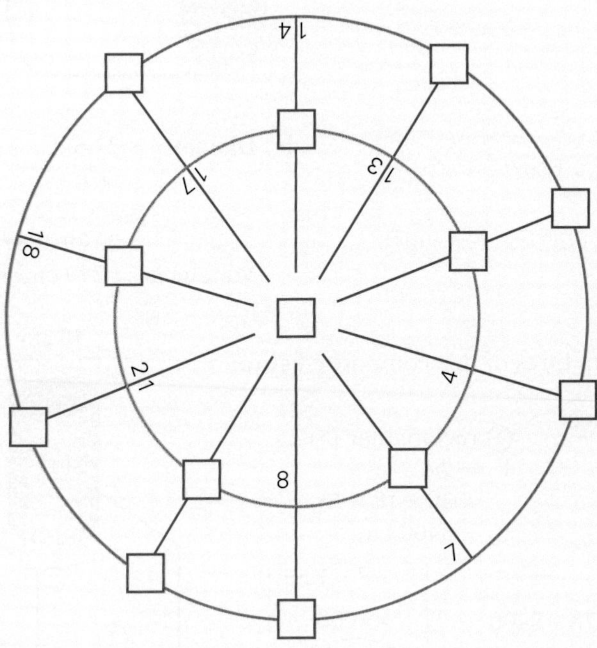

12.5 Volume of a cylinder

We saw in chapter 8 that the volume of a prism can be found by finding the area of its uniform cross-section and multiplying this by the perpendicular height of the prism.

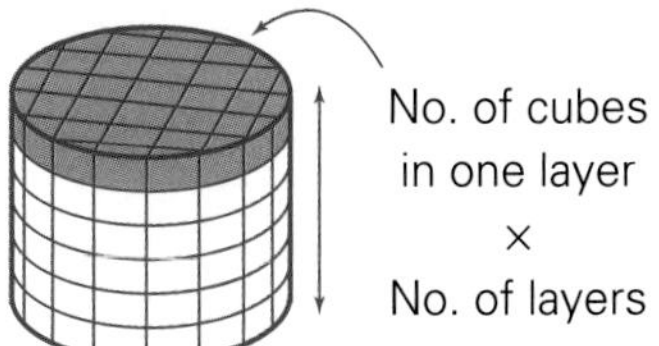

A **cylinder** has a circular cross-section staying the same size and shape from one end to the other, like a *prism*. However, it has curved sides. A *right cylinder* has sides perpendicular to the base (cross-section). So its volume can be found in the same way as that of a prism.

The volume of a right cylinder equals area of base (cross-section) times height (distance from end to end).

$V = A \times h$

$$V = \pi r^3$$

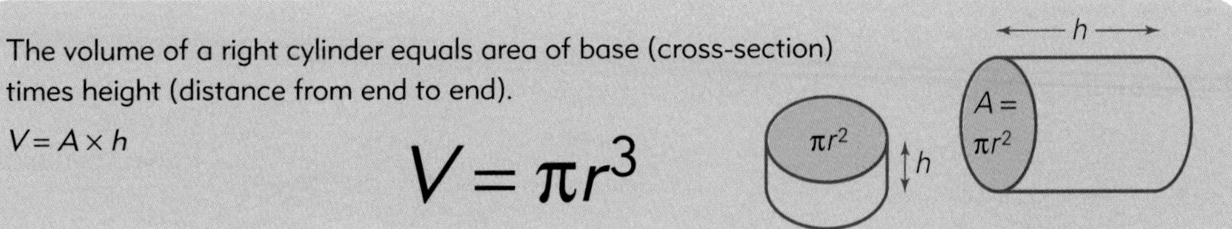

EXAMPLE 1

Find the volume of each cylindrical solid (answer to the nearest cubic centimetre).

a

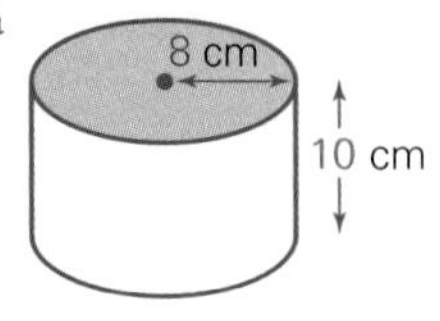

b

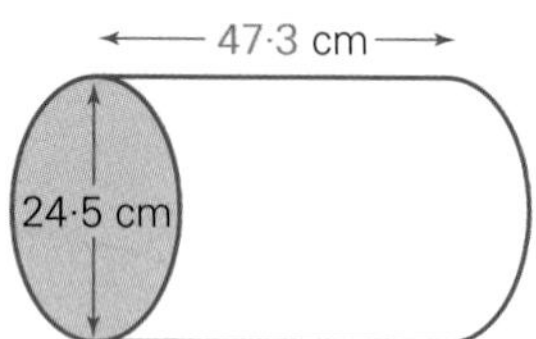

Solution

a $V = \pi r^2 h$

$V = \pi \times 8^2 \times 10$

$V \doteqdot 2010 \cdot 61 \ldots$

Volume is 2011 cm^3 (nearest cm^3).

b Diameter = 24·5 $\therefore r = 12 \cdot 25$

$V = \pi r^2 h$

$= \pi \times 12 \cdot 25^2 \times 47 \cdot 3$

$\doteqdot 22\,298 \cdot 887 \ldots$

Volume is 22 299 cm^3 (nearest cm^3).

EXAMPLE 2

Find the capacity in litres of this composite figure:

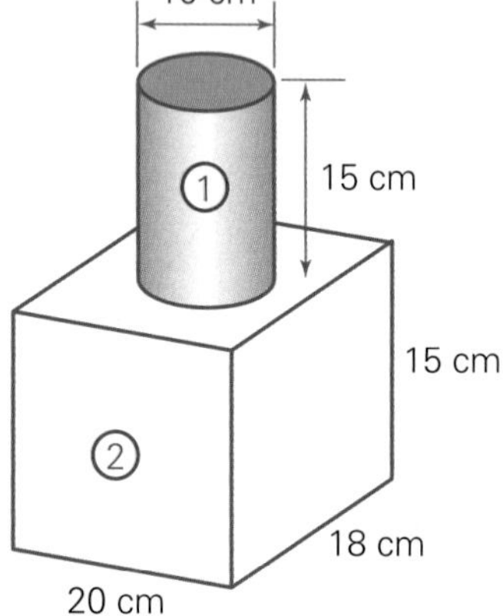

Solution

① cylinder:

$V_1 = \pi r^2 h$

$= \pi \times 5^2 \times 15$

$= 1178 \cdot 09 \ldots$

$V_1 \doteqdot 1178$ cm^3

② rectangular prism

$V_2 = lbh$

$= 20 \times 18 \times 15$

$V_2 = 5400$ cm^3

Total volume = 1178 + 5400

= 6578 cm^3

$\therefore$ capacity = 6578 mL

= 6·578 L

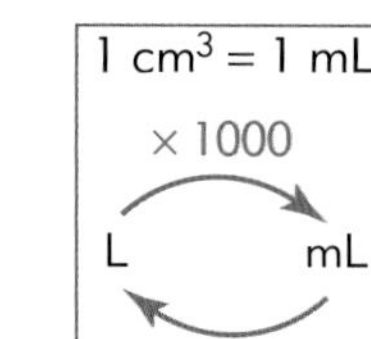

EXERCISE 12E

1 Copy and complete the following table (answer to 2 decimal places):

Radius of cylinder	Height of cylinder	Volume of cylinder
4 cm	8 cm	________ cm^3
1 mm	5 mm	________ mm^3
0·4 m	1·2 m	________ m^3
34 cm	102 cm	________ cm^3
0·1 m	1 m	________ m^3

2 Calculate the volume of each cylinder to the nearest cm^3:

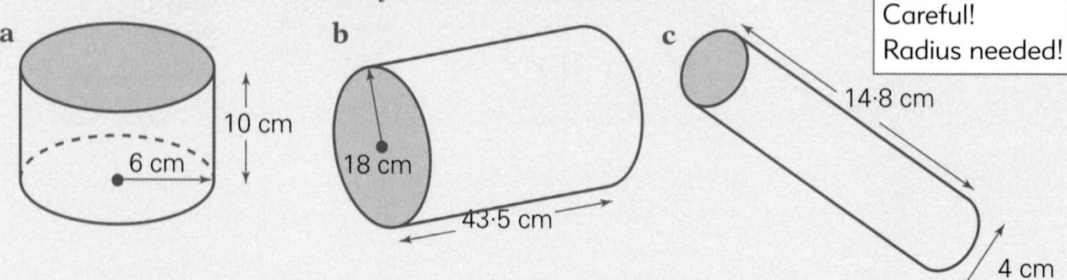

d

38 cm

10 cm

e

20 cm

2 cm

f

20 cm

40 cm

3 Find each volume, correct to 2 decimal places. All measurements are in centimetres.

a

$D = 7, h = 5$

b

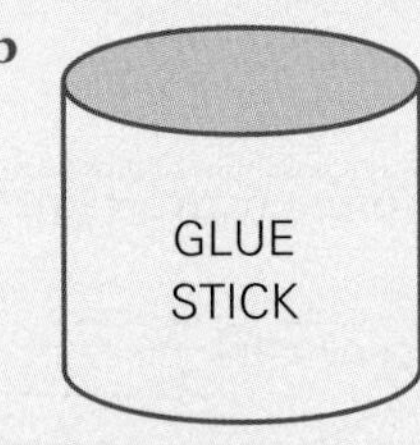

$D = 2{\cdot}8, h = 12$

c

$D = 5, h = 18{\cdot}6$

4 Which of these cylinders has the greater volume? Show working to support your answer.

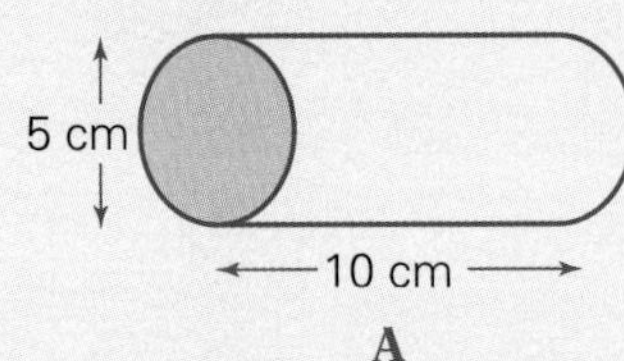

A

10 cm

5 cm

B

5 Calculate the capacity, to the nearest mL, of the following cylinders:

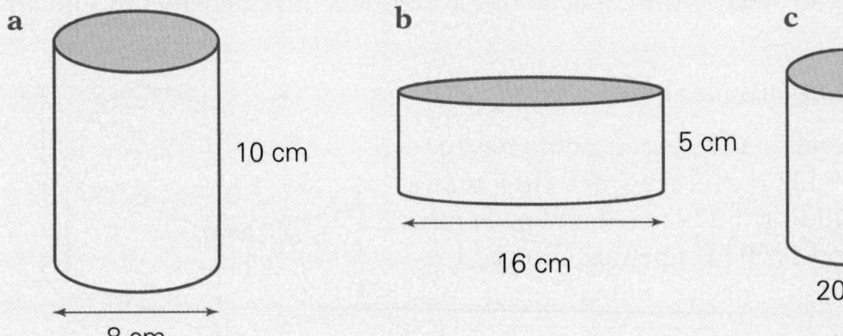

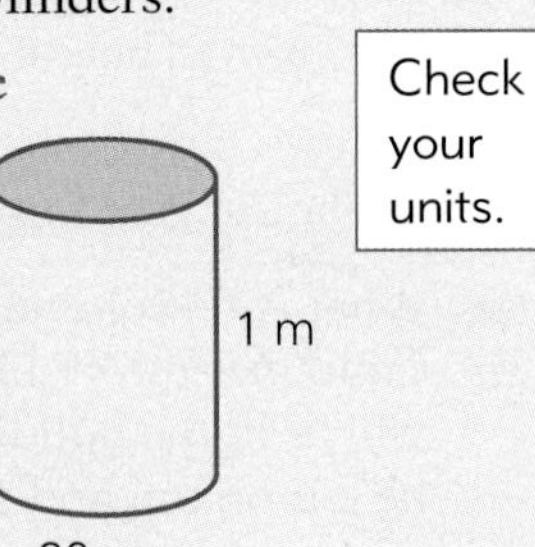

Check your units.

Level 2

6 This box just holds six cans. Each can is 12 cm high and has a radius of 5 cm.

a What is the volume of each can?

b What is the volume of the box?

c What volume of the box remains unused when the cans are packed as shown?

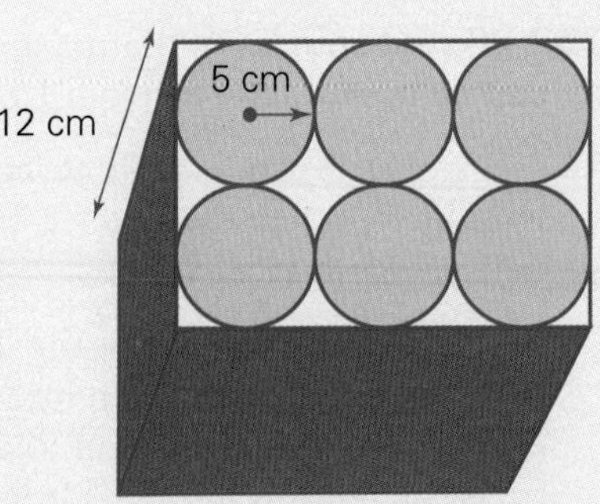

7 The volume of a cylinder is given as 240π cm^3. If the radius of the base is 2 cm, what is the height of the cylinder?

8 Given the net of the cylinder below, calculate:

a the area of each circular end

b the volume of the cylinder, once formed

Leave your answers in exact form.

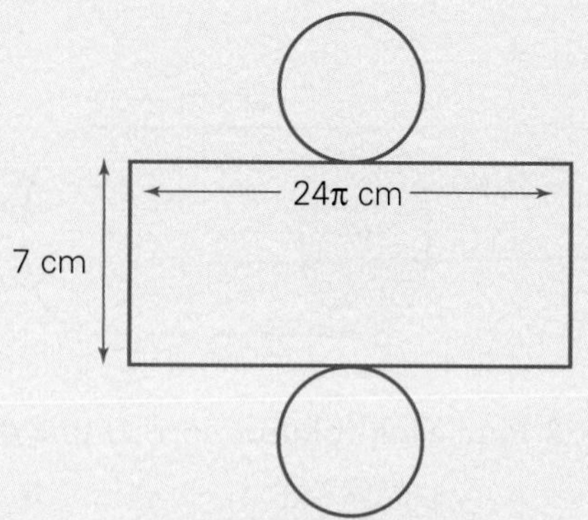

9 What is the volume of each solid, to the nearest cm^3?

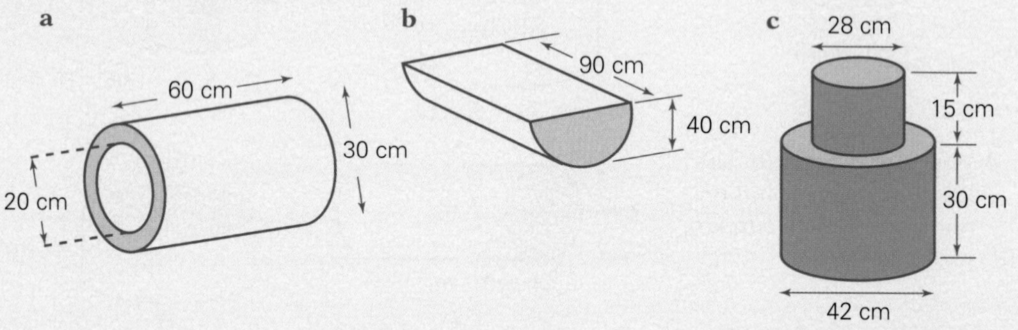

10 Each of the six cylinders in a popular family car engine measures 91 mm in diameter and 100 mm in height. What is the capacity of the engine in litres, correct to 1 decimal place?

11 A warehouse is of dimensions shown.

a What is its volume, to the nearest cubic metre?

b Containers stored in it average 2 m^3. How many will fit in the building if 229 m^3 of air-space must be left on top and 1000 m^3 must be allowed for access aisles?

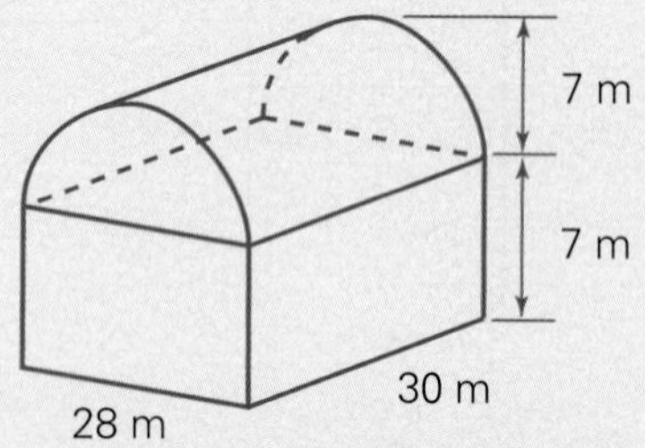

12 A 20-cent coin has a diameter of 2·8 cm and a thickness of 2 mm. How many of them can be manufactured from 1 m^3 of metal?

13 A petrol tanker has a cylindrical tank of length 6·4 m and diameter 2·1 m, and delivers to service stations.

a What is the volume of the tank?

b Express its capacity in litres.

c If a litre of fuel sells for 99.7c, what is the retail value of its full load?

d If each car takes an average of 40 L to fill, how many cars can be refuelled from the tanker's load?

14 A farmer wants a cylindrical concrete water tank to have a capacity of 60 000 L, with height of 3 m for convenient access. What should be its diameter, to the nearest centimetre?

15 Terry and Jerry have been given two measuring devices in their Science class. Terry fills his container to the rim before pouring the contents into Jerry's container. If the dimensions of Terry's container are as shown, how high up the sides of Jerry's container will the liquid reach?

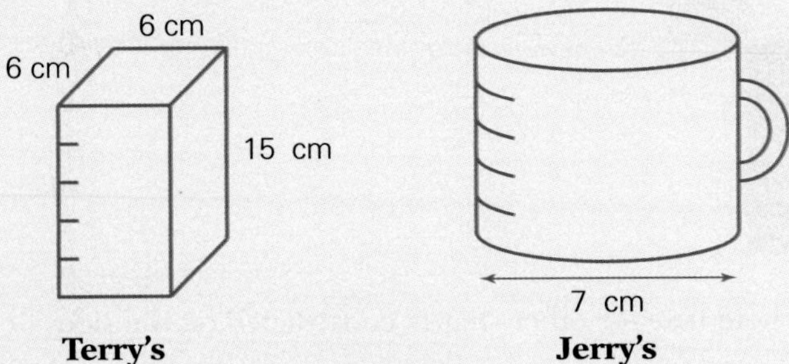

16 A cylindrical beaker of radius 4 cm is filled to a level of 500 mL.

a How high up the cylinder's wall does the liquid reach?

b If a pair of scissors is placed in the beaker and the water rises 3 cm, what is the volume of the scissors?

This method of finding the volume of a solid by displacement is sometimes used in science.

TEACHER

17 Develop a formula for volume of a cylinder based on diameter rather than radius.

SPREADSHEET ACTIVITY

Try this

A packaging problem

Soft drink comes in 375 mL cans of diameter 65 mm and height 130 mm. The cans are packaged in cartons of 24.

1 In how many different ways can the 24 cans be arranged in a rectangular carton?

2 How much space is unoccupied in each type of carton?

3 Express the waste space as a percentage of the volume of each carton.

4 Summarise your findings in a report on the best packaging solutions. Show all diagrams and calculations.

Investigation Extending Pythagoras

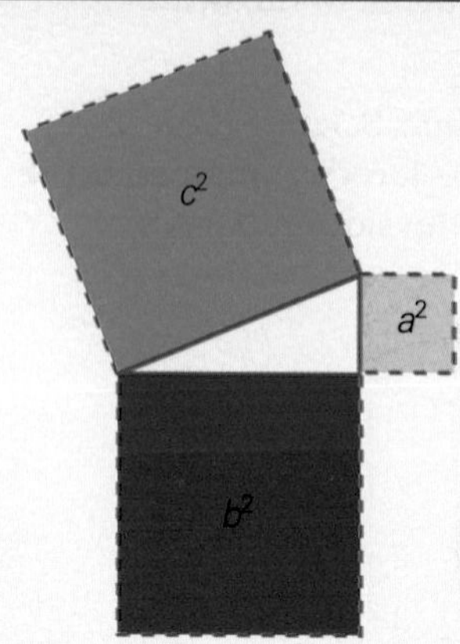

Pythagoras stated that *the* **square** *on the hypotenuse is equal to the sum of the* **squares** *on the other two sides*.

In the diagram $c^2 = a^2 + b^2$.

The pattern may hold true for other shapes constructed on the sides of right triangles.

Investigate

Hint: Start with semicircles.
Then try equilateral triangles.

(Scale drawing or grid paper will help.)

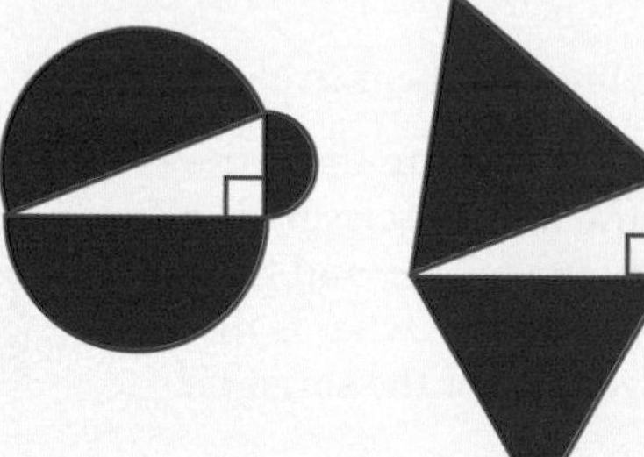

Chapter Review

Language Links

arc
area
chord
circle
circular
circumference
concentric
cross-section
cylindrical
cylinder
diameter
estimate
height
perimeter
pi (π)
quadrant
radius
sector
segment
semicircle
tangent
volume

1 What is the difference between a *sector* and a *segment*?
2 Complete the sentence: *Semicircle* is to *circle* as *radius* is to ______________
3 Use these words all in one sentence: *circular, cylinder, cross-section, volume, height.*
4 Explain the difference between these formulas: $2\pi r$, πr^2.
5 Find the name given to each diagram in the find-a-word supplied.

a
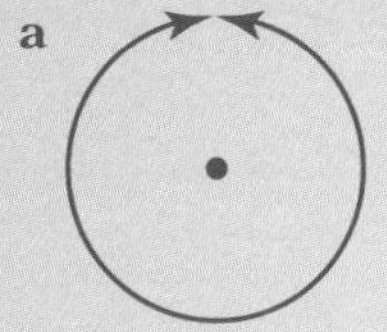
b
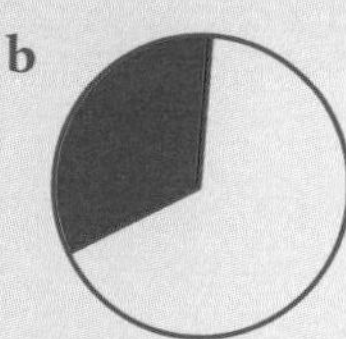
c
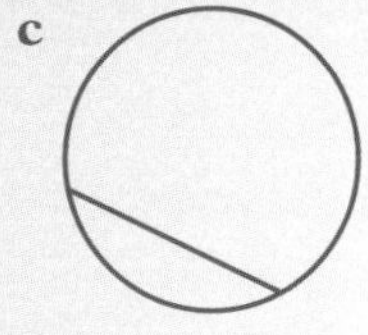
d
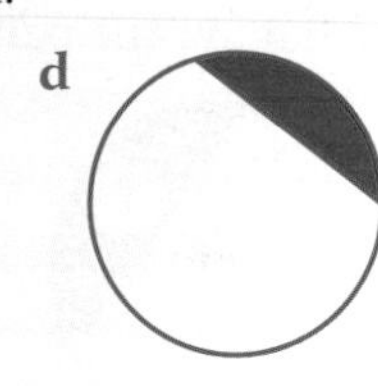
e
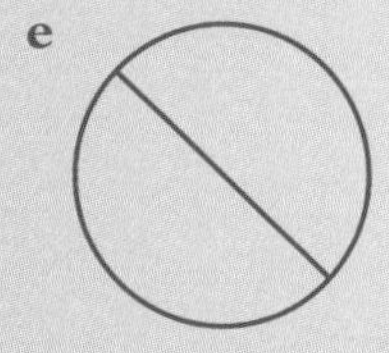
f

g
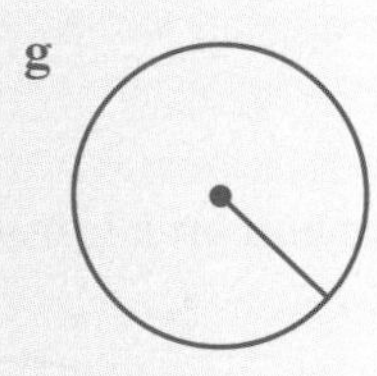
h
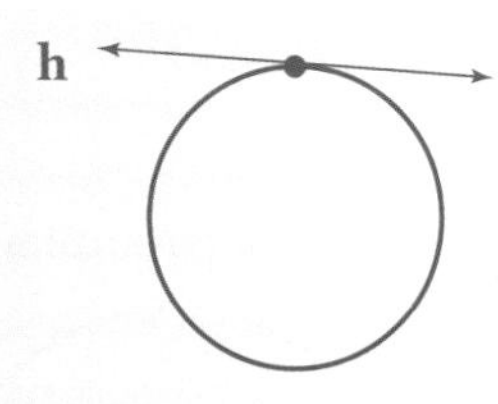
i
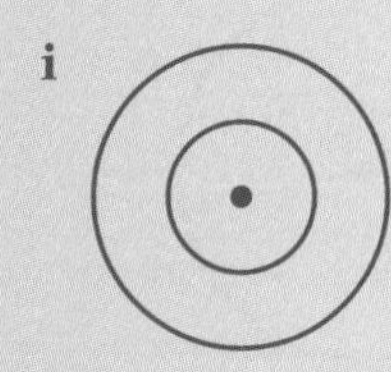
j

k
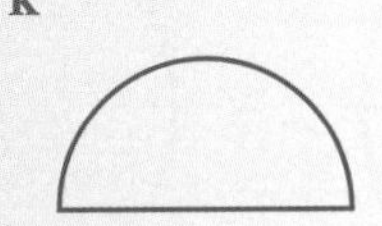
l
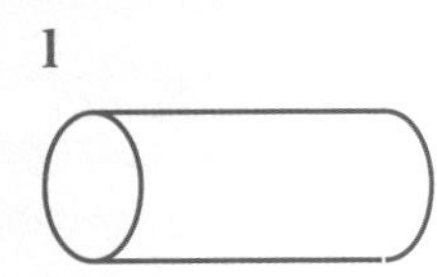

FIND A WORD

S	A	Q	M	R	V	K	C	C	L	A	E	V	S	B	V	W	E	C	N	Y	W
J	H	Z	R	E	T	R	G	A	W	M	L	K	C	J	S	X	R	O	T	F	T
I	Y	I	O	T	A	X	V	C	I	R	C	U	M	F	E	R	E	N	C	E	I
S	E	G	M	E	N	T	F	U	E	Y	R	D	S	N	I	Q	D	C	I	P	A
Q	V	G	M	M	G	Q	V	I	O	O	I	R	U	E	Q	C	N	E	P	I	S
V	X	R	Z	A	E	D	G	I	T	F	C	O	I	W	A	C	I	N	S	E	A
V	M	O	E	I	N	Y	A	C	A	K	I	H	D	C	V	J	L	T	K	G	J
U	T	X	V	D	T	H	E	E	Y	B	M	C	A	H	A	Q	Y	R	K	F	K
S	S	P	E	B	P	S	J	V	Z	S	E	O	R	C	U	Y	C	I	P	S	C
R	M	G	B	G	S	K	J	X	O	N	S	X	E	S	B	M	A	C	J	T	K

Chapter Review Exercises

12.2 **1** Give three approximate values of pi.

12.1 **2**

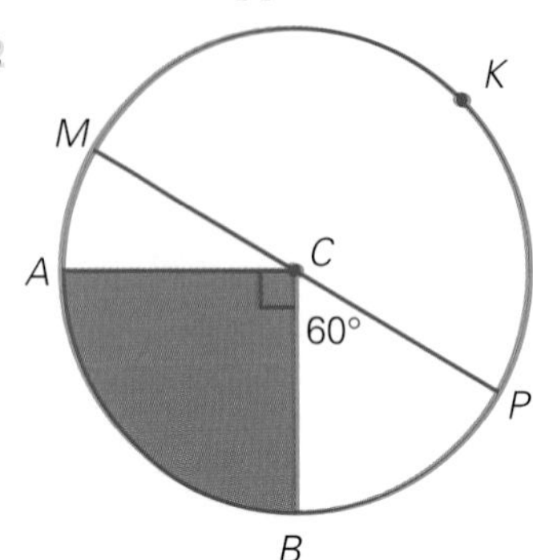

a Give the correct geometric name for:

i interval *MP** **ii** region *ABC*
iii arc *MKP* **iv** interval *CB*

e What fraction of the circumference of the circle is cut off by the shaded region *ACB*?

f What fraction of the area of the circle is covered by the region *MAC*?

12.1 **3** Two circles are touching at a point. What is the distance between the points *A* and *B*, if *A*, *T* and *B* are collinear?

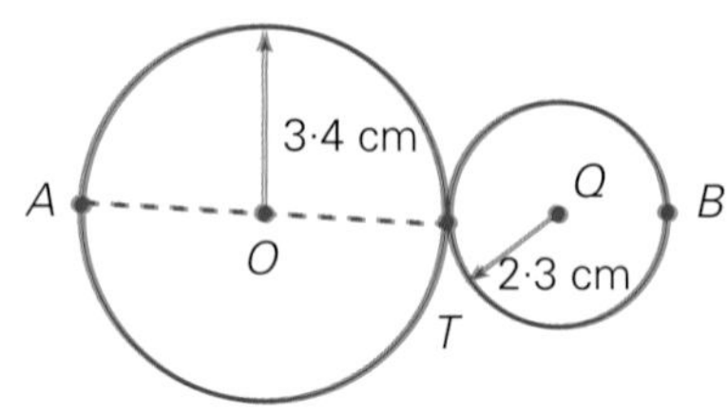

4 What are the parts of the circle labelled 1 and 2?

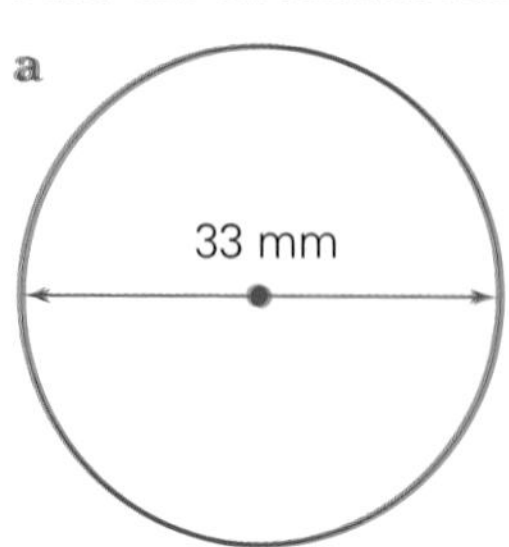

12.2 12.3 **5** Find the circumference and area (answer to 1 decimal place) of:

a

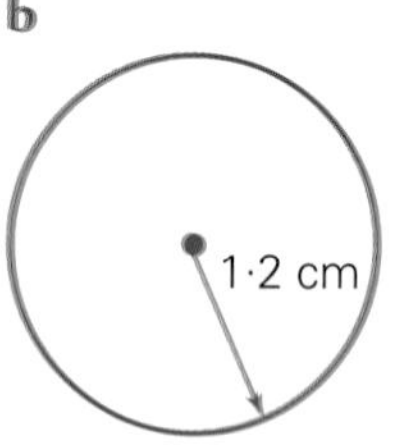

b

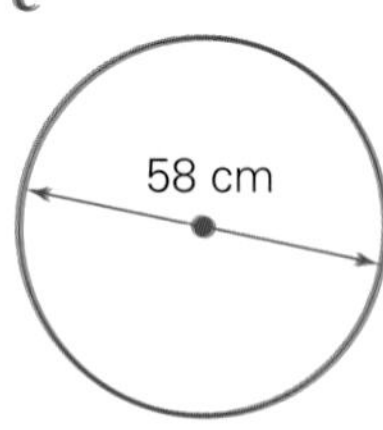

c

58 cm

12.2 **6** Find the perimeter and area (answer to 1 decimal place) of:

a

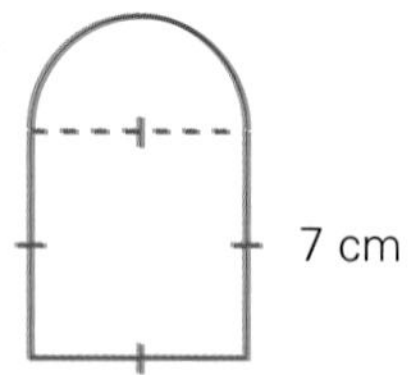

b

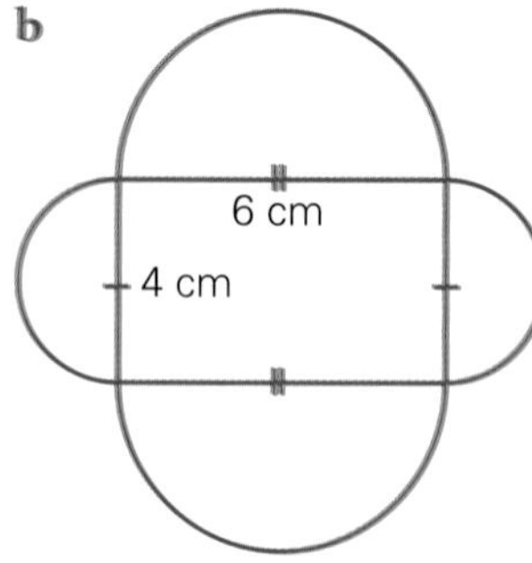

c

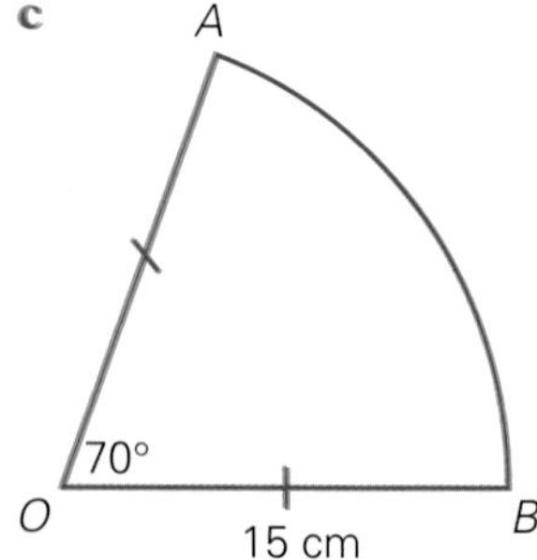

7 Find the area of the shaded region, correct to 1 decimal place: 12.4

a 40 cm **b** 30 m, 20 m **c** 20 cm, 20 cm **d** 20 cm

8 Find the volume of these cylinders. (Leave your answers in terms of pi.) 12.5

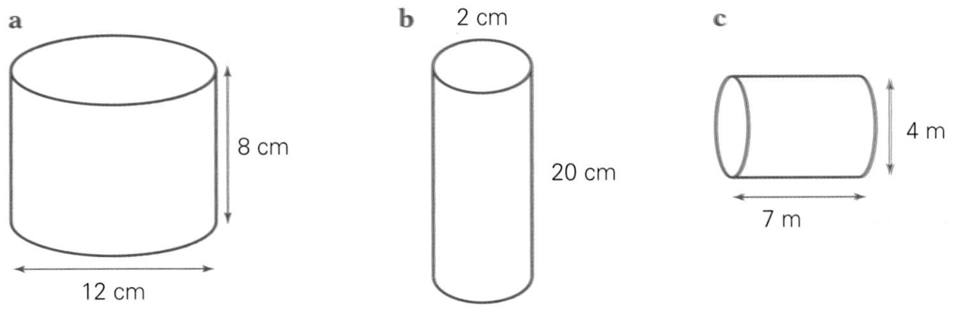

9 Find the volume to the nearest cubic centimetre of: 12.5

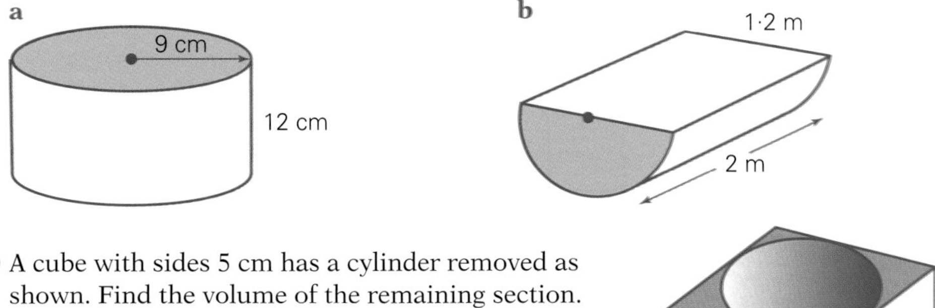

10 A cube with sides 5 cm has a cylinder removed as shown. Find the volume of the remaining section. 12.5

5 cm, 5 cm, 5 cm

11 A cylinder with a diameter of 7 cm has a capacity of 500 mL. What is the height of the cylinder? 12.5

12 Six cubic metres of topsoil spread over a circular lawn of diameter 24 metres will cover it to what depth in centimetres, correct to 1 decimal place? 12.5

Keeping Mathematically Fit

Part A: Non-calculator

1 How many metres in m kilometres?

2 Express Todd's mark of 19 out of 25 as a percentage.

3 Find x given $3 : 2x = x : 6$, and x is a positive integer.

4 Write an expression for the sum of three consecutive even numbers given the smallest of the numbers is x.

5 If a car travels 90 km/h, how far can it travel in 40 minutes?

6 Give the coordinates of the point half way between A ($^{-}3$, 7) and B (5, 7).

7 The mean of x, y and z is 4. What is the mean of $2x$, $2y$ and $2z$?

8 Which is larger, and by how much: $\frac{2}{3}$ or $\frac{4}{7}$?

9 Decrease \$850 by 5%.

10 If $x = \frac{^{-}1}{2}$, find the value of $4x^2$.

Part B: Calculator

1 Find the surface area of:

a

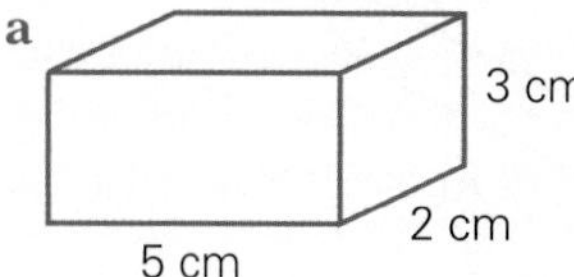

b

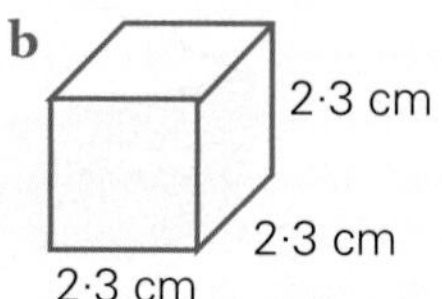

c

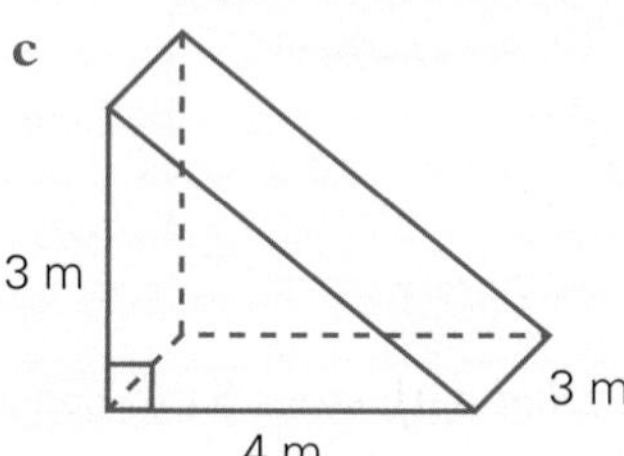

2 Given $V = u + at$, find V when $u = 20{\cdot}4$, $a = 9{\cdot}6$ and $t = \frac{1}{2}$.

3 Find $3\frac{1}{2} \div \frac{2}{5} \times 1\frac{1}{2}$.

4 Solve: **a** $3(x + 1) > 18$ **b** $4x - 7 = x + 2$ **c** $\frac{x - 4}{6} = {}^{-}2$.

5 If $a + b = 17{\cdot}6$, find the value of $5a + 5b$.

6 Graph $y = 2x - 2$ on a number plane.

TEACHER

7 A tap loses 5 mL of water every 2 minutes. How many litres of water does it lose in 1 week?

8 Name all quadrilaterals whose diagonals bisect.

Cumulative Review 4

Part A: Multiple-choice questions

Write the letter that corresponds to the correct answer in each of the following.

1 The simplest expression for $(2 + x) - (x - 3)$ is:

A $2x - 1$ **B** $^{-}1$ **C** $2x + 5$ **D** 5

2 In which of the following is $(x + 2)$ *not* a factor?

A $5x + 10$ **B** $x^2 + 2x$ **C** $xy + 2xy$ **D** $^{-}x - 2$

3 If $p \times q - 3 = m$, then:

A $p = m + \frac{3}{q}$ **B** $p = m + 3 \div q$ **C** $p = \frac{m + 3}{q}$ **D** $p = m - 3 \times q$

4

$x°$

56°

The value of x is:

A 56 **B** 34 **C** 158 **D** 124

5

$x°$

75° 125°

The value of x is:

A 75 **B** 55 **C** 50 **D** 25

6 The number of centimetres in x metres is:

A $100x$ **B** 100 **C** $\frac{x}{100}$ **D** $\frac{1}{100}$

7 The area of triangle PQR is:

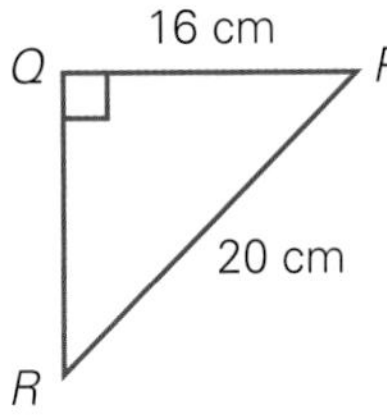

A 192 cm^2 **B** 320 cm^2 **C** 160 cm^2 **D** 96 cm^2

8

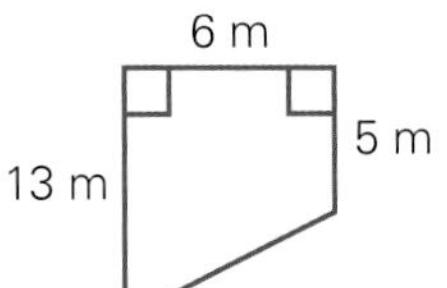

The area of this figure is:

A 34 m^2 **B** 54 m^2 **C** 108 m^2 **D** 78 m^2

9 The volume of:

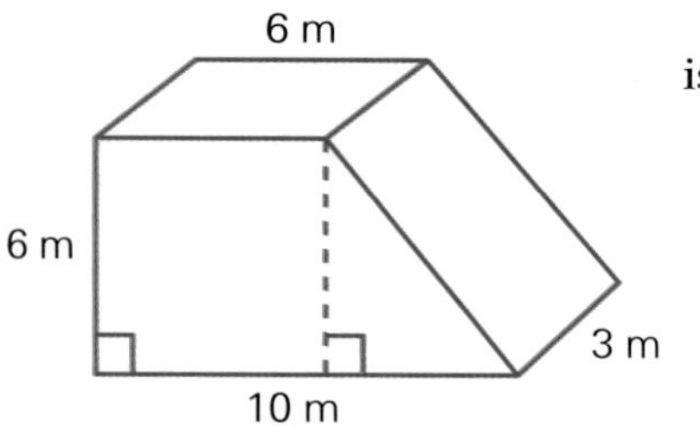

is:

A 288 m^3 **B** 144 m^3 **C** 1080 m^3 **D** 540 m^3

10 Which of the following is in the ratio of 2 : 5?

A \$4 to 20 cents **B** 4 m to 1000 cm **C** \$1.50 to 60 cents **D** $\frac{2}{3} : \frac{5}{6}$

11 A train travelling at 140 km/h leaves Sutherland station at 11:12 am. It travels, without stopping, the 56 km to Central station. The train arrives at Central station at:

A 11:37 am **B** 11:37 pm **C** 11:53 am **D** 11:53 pm

12 For 3, 2, 1, 5, 7, 3, 1, 3 the mode is:

A 3 **B** 1 **C** 4 **D** 6

13 Which of the following has a mean of 2 and a mode of 3?

A 1, 1, 2, 2, 3 **B** 2, 2, 2, 4, 5 **C** 1, 2, 3, 4, 5 **D** 1, 3, 3, 3, 0

14

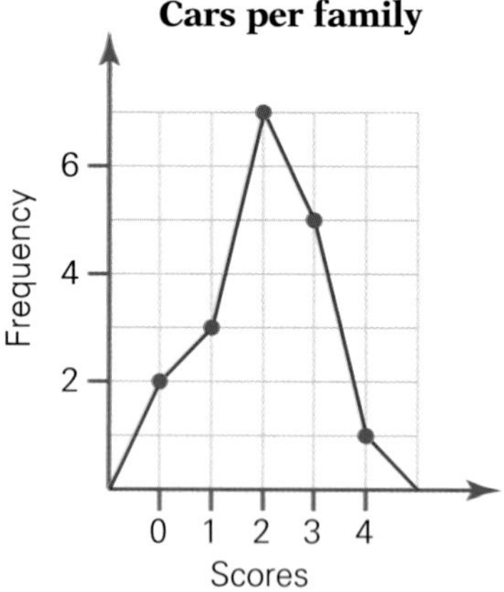

The frequency polygon shows the results of a survey on the number of cars per family. The mean number of cars per family is:

A 36 **B** 7 **C** 2 **D** 18

15 The distance between A and B is:

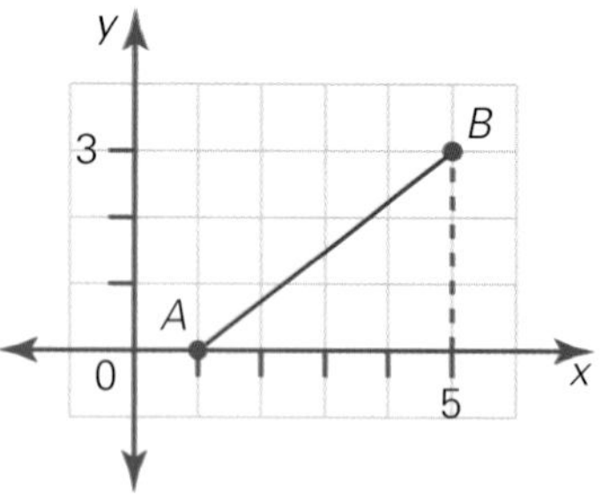

A 4 units
B 3 units
C 5 units
D 7 units

16 The graph of $x = 3$ is:

A

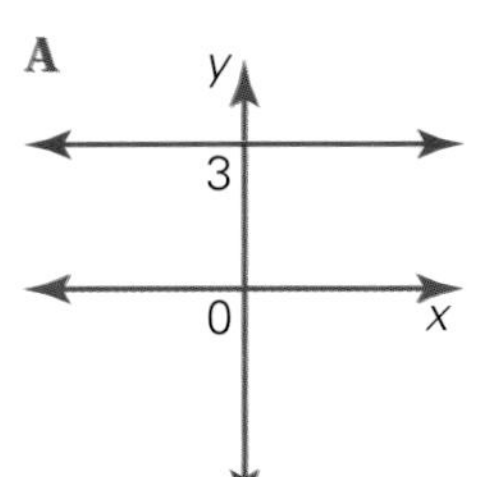

B **C** **D**

17 Which point is *not* on the line $2x + y = 6$?

A (3, 0) **B** (1, 4) **C** ($^{-}2$, 10) **D** ($^{-}3$, 3)

18 Which of these statements about the point ($^{-}2$, 4) is correct?

A It lies in the third quadrant.
B It has a y-coordinate of $^{-}2$.
C It has an x-coordinate of $^{-}2$.
D It is in the same location as (4, $^{-}2$).

19 This shaded region of a circle is called:

A semicircle **B** sector **C** segment **D** arc

20 The boundary length of any circle is found using the formula:

A $2\pi D$ **B** πr^2 **C** $2\pi r^2$ **D** πD

21 The shaded region's share of the area of the circle is:

A $\frac{1}{3}$ **B** $\frac{4}{5}$ **C** $\frac{3}{4}$ **D** $\frac{2}{3}$

22 The volume of the cylinder (pi = 3·14) is:

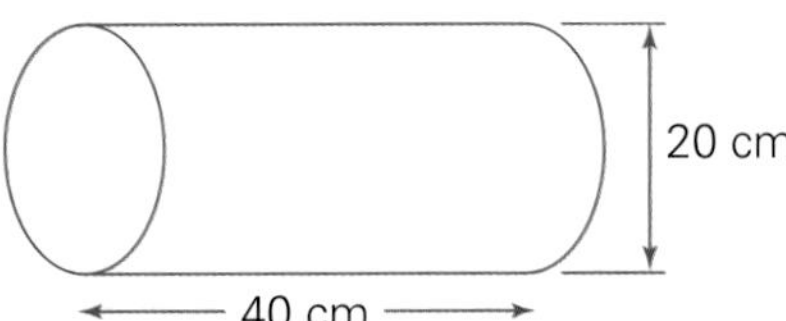

A 50 240 cm^3 **B** 2512 cm^3 **C** 12 560 cm^3 **D** 1256 cm^3

Part B: Short-answer questions

Show full working for each of the following.

1 Find the value of $\sqrt[3]{\frac{19{\cdot}406}{7{\cdot}08}}$ correct to 2 decimal places.

2 Simplify:

a $5a + 3b - a - 3b$ **b** $x(x - 3) + x(5 - x)$ **c** $3^4 \times 9$ **d** $\frac{3ab \times 4a}{6ab}$

3 Increase $9700 by 5%.

4 A class of 50 students has 15 students from non-English-speaking backgrounds. Express this as a percentage.

5 A salesman earns $285 p.w. plus 2% commission on the value of all sales. Find the value of sales made in a week where the income is $400.

6 Solve:

a $7 - 3x = 11 + x$

b $10(x + 7) = 5x + 50$

c $4 + 3x > 7$

d $\frac{x-4}{5} \leq {}^{-}1$

7 Find the value of x in:

a
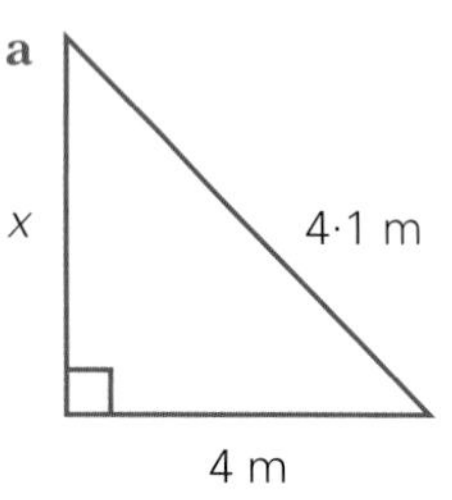

b
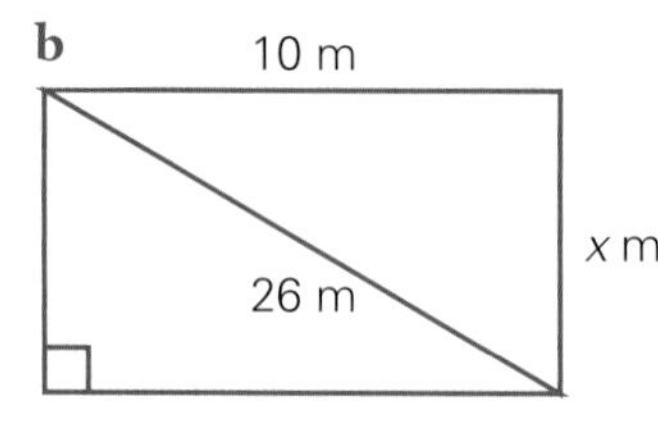

c
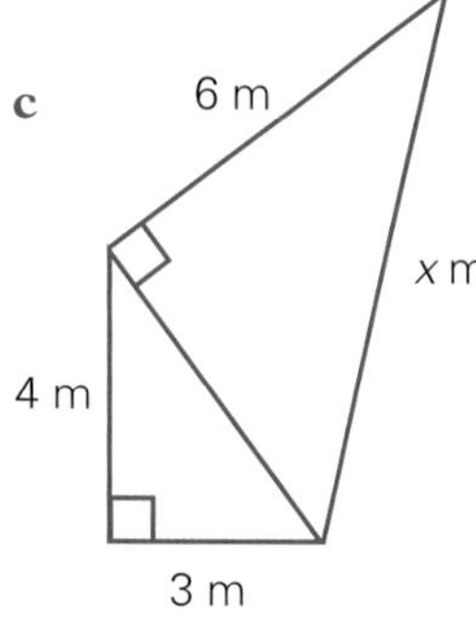

8 Find the value of each pronumeral, giving a reason:

a
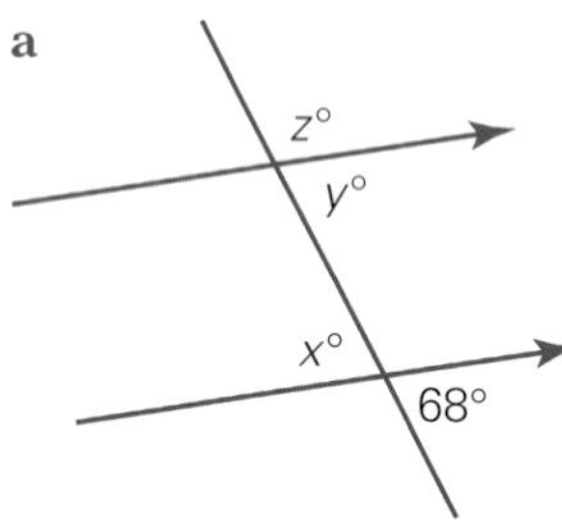

b
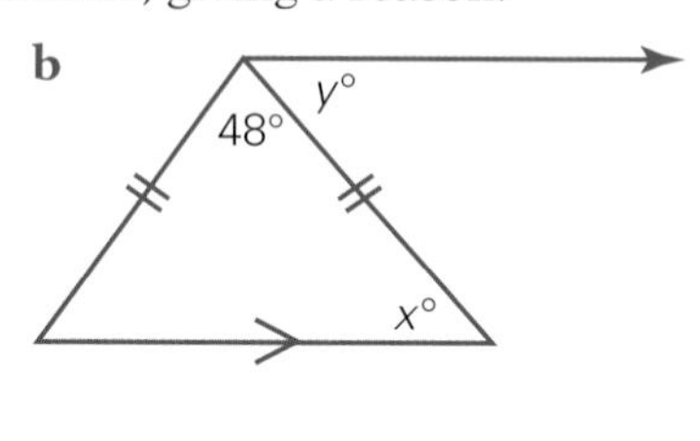

9 a Compare the percentages of program time devoted to drama by the ABC and commercial channels.

b What percentage of time does the ABC assign to sport?

c Which type of program do the commercial channels devote least time to? What percentage?

d How much greater is the percentage of time used by the ABC for children's programs than that used by commercial channels?

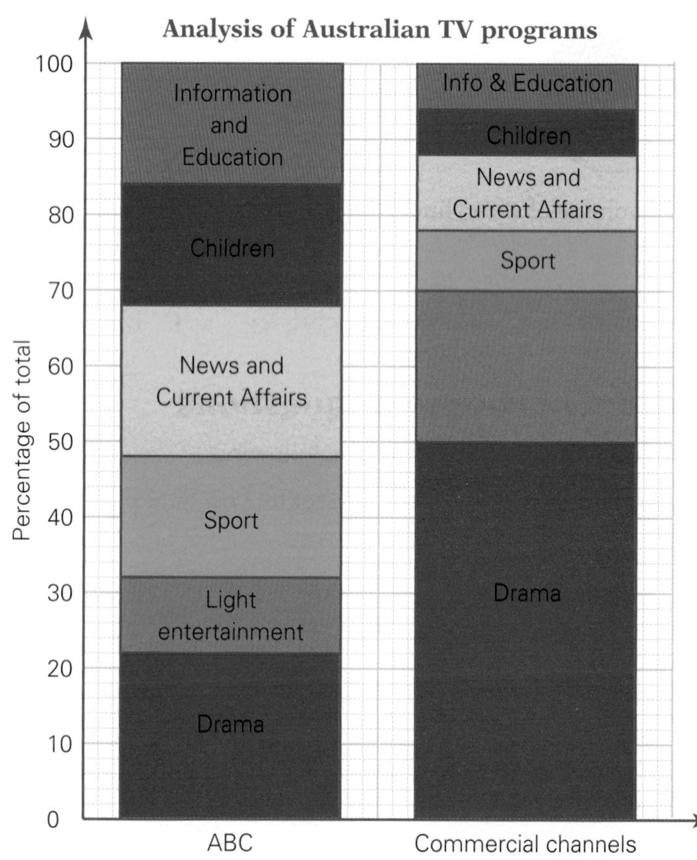

10 Find the mean, mode, median and range for the following sets of data:

a 6, 9, 5, 4, 6
8, 9, 10, 12, 4
0, 10, 15, 6, 9

b

x	f
20	40
21	55
22	15
23	10

c

Stem	Leaf
0	1 5 7
1	2 2 4 9 9
2	0 5 6 8
3	1 6

11 **a** Display the following scores in a frequency distribution table and a frequency polygon.
Children per family
0 3 2 3 1 0 2 3 2 1 1 0 4 0 2 2 1 1 1 2 5 1 2 2 4

b **i** How many families were surveyed? **ii** Find the range.
iii Find the mode. **iv** Find the mean.

12 For {47·2, 81, 93·6, 87·5, 47·2, 96·1} calculate the:
a mean **b** mode **c** median **d** range

13 Find the rule for:

a

x	1	2	3	4
y	2	4	6	8

b

x	$^{-}1$	0	1	2
y	$^{-}3$	$^{-}1$	1	3

14 The average of five students' ages is 12 years. Find the average if two more students aged 15 and 17 join the group.

15 Find the point of intersection of the straight lines $y = 2x + 3$ and $x + y - 6 = 0$.

16 Find: **i** circumference **ii** area (correct to 1 decimal place where necessary):

a

30 cm

b

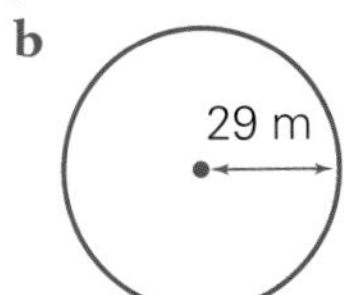

c

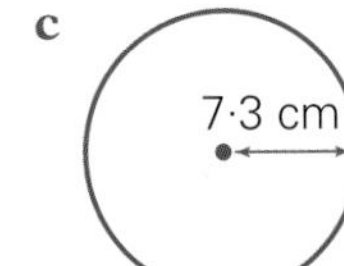

17 Find: **i** perimeter **ii** area to the nearest whole unit:

a

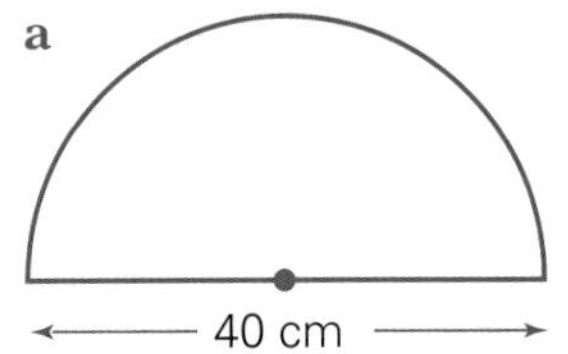

b

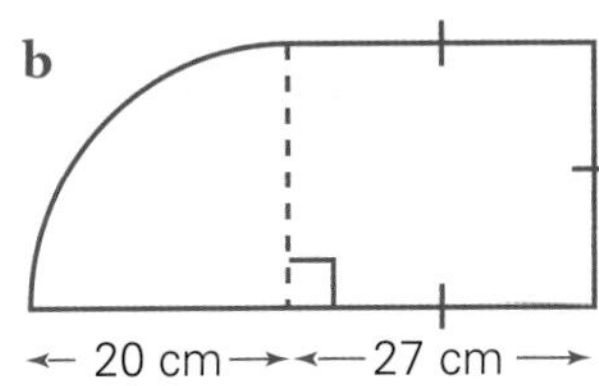

c

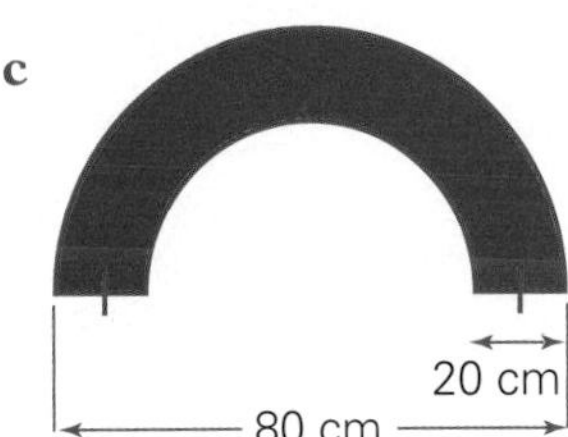

18 **i** Calculate the volume of each shape, to the nearest m^3:

a

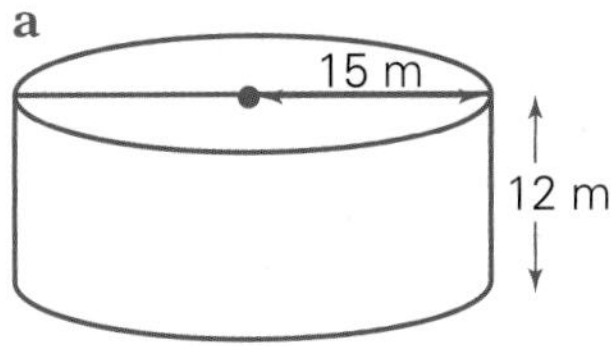

b

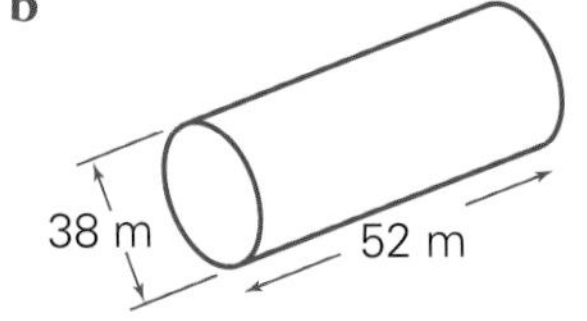

c

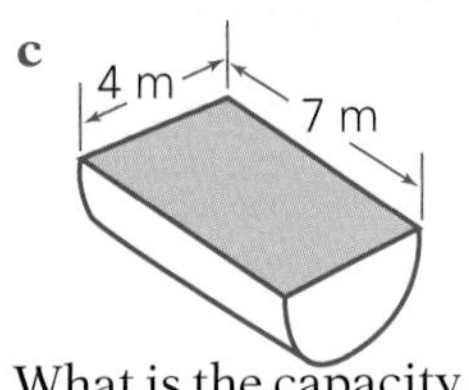

ii What is the capacity of this trough in litres?

1 What you need to know and revise

Outcome NS 4.3:
Operates with fractions, decimals, percentages, ratio and rates

- writes, simplifies, and uses fractions

2 What you will learn in this chapter

Outcome NS 3.5:
Orders the likelihood of simple events on a number line from 0 to 1

Outcome NS 4.4:
Solves probability problems involving simple events

- orders the likelihood of simple events on a number line from zero to one
- lists possible outcomes of a simple event
- uses the term 'sample space'
- assigns probabilities to simple events by reasoning about equally likely events
- expresses the probability of a particular event as a fraction between 0 and 1
- assigns a probability of 1 to events that are certain and a probability of 0 to events that are impossible
- recognises that the sum of the probabilities of all possible outcomes of a simple event is 1
- identifies and finds the probability of complementary events

Working Mathematically outcomes WMS 4.1–4.5
Students will be asked to ***question***, ***apply strategies***, ***communicate***, ***reason*** and ***reflect*** in the sections of this chapter.

Probability

Key mathematical terms you will encounter

almost certainly	equally likely	impossible	probability
always	even	likelihood	random
certain	event	likely	sample space
chance	favourable	never	scale
complementary	highly likely	outcomes	unlikely

MathsCheck
Probability

1 Simplify the following fractions:

a $\frac{12}{40}$ **b** $\frac{6}{8}$ **c** $\frac{12}{100}$ **d** $\frac{34}{50}$ **e** $\frac{76}{100}$

2 Find:

a $\frac{2}{3}$ of 60 **b** $\frac{4}{5}$ of 1000 **c** $\frac{1}{9}$ of 2700

d $\frac{1}{5}$ of 750 **e** $\frac{8}{9}$ of 2700 **f** $\frac{1}{5}$ of 1000

3 A drawer contains 5 red socks, 7 black socks, 8 blue socks and 10 white socks. What fraction of the socks in the drawer are:

a white? **b** blue? **c** red? **d** not red?
e blue or white? **f** black? **g** not black? **h** orange?

4 The ages of 20 children at an afternoon daycare centre are given below:

3	5	7	6	5
5	7	5	7	6
5	7	8	9	8
4	6	7	8	5

a Display their ages in a frequency distribution table.
b What is the age of the youngest child in afternoon care?
c What fraction of the children are aged:
i 5? **ii** 8? **iii** less than 5?
iv greater than 5? **v** 2? **vi** 5 or 6?

5 A group of school children were asked: 'How many children, including yourself, do you live at home with?' Their ages are summarised in the frequency table supplied.

Children	Frequency
1	5
2	15
3	10
4	4
5	6

a How many children were surveyed?
b How many children have no brothers or sisters living at home with them?
c What fraction of the group had no brothers or sisters living at home with them?

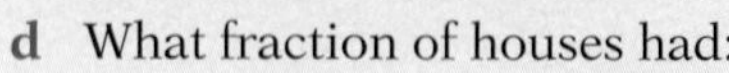

d What fraction of houses had:
i 4 children? **ii** 2 children? **iii** more than 2 children?

Exploring New Ideas

13.1 Introducing probability

Benjamin Franklin, a great American scientist and diplomat, said: *'But in this world nothing can be said to be certain but death and taxes!'*

Benjamin Franklin may well have been correct in his statement. However, there is a way of representing the **likelihood** or **chance** of an event occurring and that is to calculate its **probability**.

WEB RESEARCH

Probability is a measure of the likelihood of an event occurring.

The probability that an event occurs is usually expressed as a fraction, decimal or percentage, and is a number lying between 0 and 1.

0 represents the event **never** occurring (impossible) and 1 represents the event **always** occurring (certain).

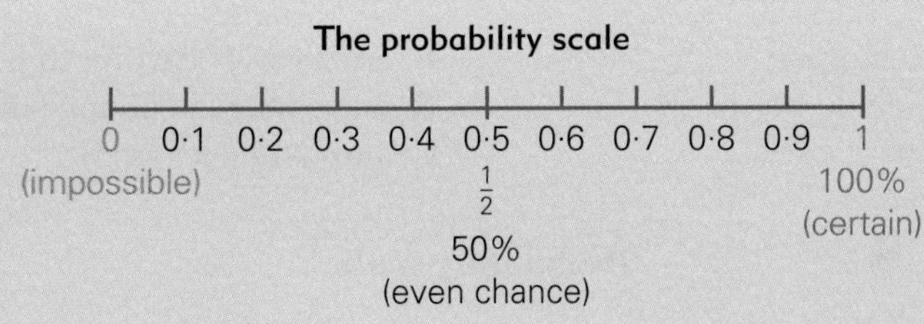

Many words are used to describe the probability of an event, for example impossible, unlikely, even chance, likely, highly likely and certain.

What chance words do you know?

EXAMPLE

Give the probability of:

a finding an elephant in your letterbox

b the day after Sunday being Monday

Solution

a P(elephant in the letter box) = 0

An elephant wouldn't fit in the letterbox.

b P(Monday after Sunday) = 1

The day after Sunday is always Monday.

EXERCISE 13A

1 Copy the probability scale and show the correct positioning of the words and phrases given. More than one word or phrase may have the same position.

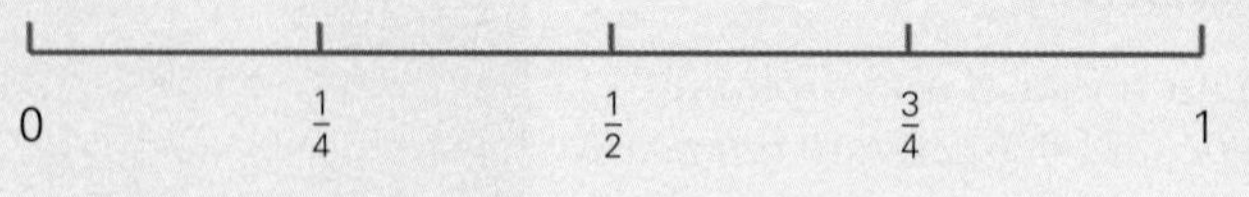

likely, certain, very likely, always, highly likely, no chance, almost always, 50/50, highly unlikely, unlikely, never

2 State whether the probability of each event listed below is 0 or 1:

vowels: A, E, I, O, U

- **a** selecting a vowel from the word WHY
- **b** winning lotto without buying a ticket
- **c** selecting a day ending in the letters DAY
- **d** school finishing at 10 pm at night
- **e** walking from Sydney to Newcastle in 1 hour
- **f** rolling a zero on a six-sided die
- **g** the sum of 1, 2, 3 and 4 being 10
- **h** selecting the letter P from the word NORMAL
- **i** Anzac Day being 25 April next year
- **j** selecting an orange marble from a box of 20 white marbles
- **k** your teacher winning the gold medal in the 100 m freestyle event at the next Olympics

3 Write three events, not mentioned so far, which:

- **a** are certain to happen
- **b** are almost certain to happen
- **c** are highly likely to happen
- **d** have an even chance of occurring
- **e** are unlikely to occur
- **f** almost never occur
- **g** are impossible

4 **Probability scale**

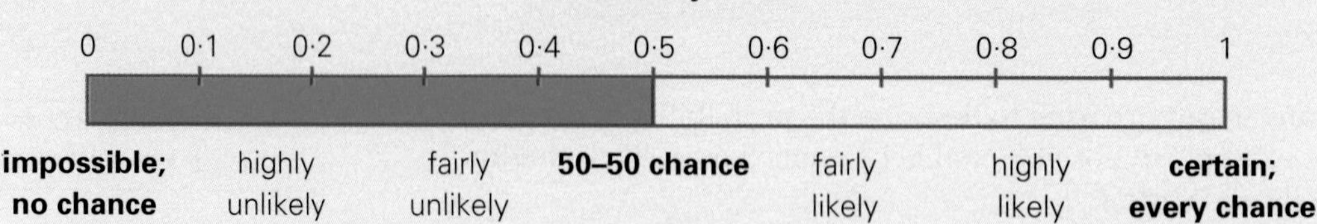

Use the probability scale above to describe the chance of each event happening:

- **a** The world will end tomorrow.
- **b** The baby will be a boy.
- **c** A summer's day will be hot.
- **d** You'll draw out an ace from a well-shuffled pack of cards.
- **e** You'll get maths homework.
- **f** You'll have a maths test sometime this term.

5 Use the words unlikely, even chance, likely or almost certainly to describe the following events:

- **a** a pair of snake eyes (double 1) appearing when two dice are rolled
- **b** getting a tail when tossing a coin
- **c** an even number appearing when a die is rolled
- **d** choosing a vowel from the word SAID
- **e** choosing a vowel from the word MAMMALS
- **f** winning Lotto having bought only one ticket
- **g** choosing the king of spades if one card is chosen from a deck of playing cards
- **h** choosing a red card if one card is chosen from a deck of playing cards

Level 2

6 Choose an estimate from the list below for the probabilities of the following:

0	0·01	0·3	0·5	0·7	0·99	1

- **a** it raining every day this year
- **b** choosing an S from the word IS
- **c** a boy being chosen if one student is picked from your class
- **d** an M being chosen from the word MUM
- **e** rolling an odd number when rolling a die

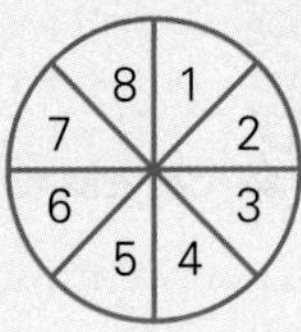

7 The spinner is spun once. Choose a term that best describes the probability the number it lands on:

a 1	**b** even	**c** odd
d 8	**e** prime	**f** less than 10
g even or prime	**h** not 8	**i** 3 or 6

impossible, unlikely, even chance, likely, very likely, certain

8 A spinner was spun and it landed on the colour green. List the spinners below in order from the most likely to least likely to land on the colour green.

A

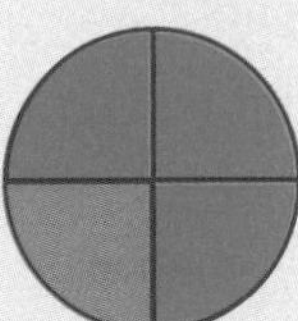

B

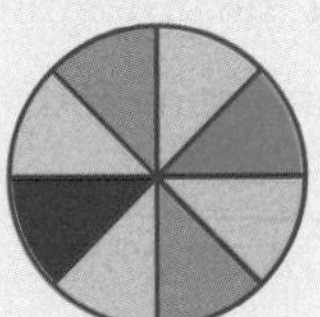

C

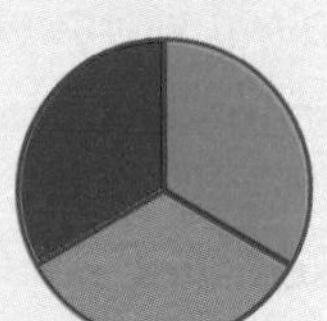

D

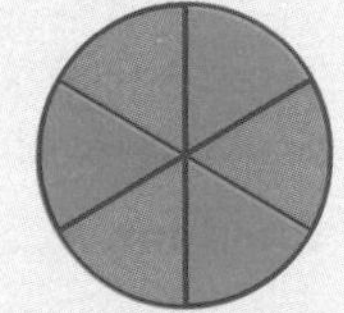

9 A spinner was spun and it landed on the number 1. List the spinners below in order from most likely to least likely to land on the number 1.

A

B

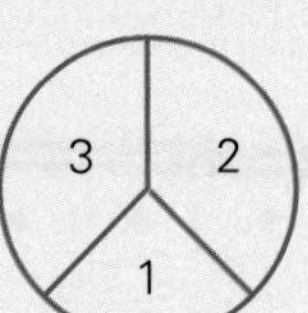

C

D

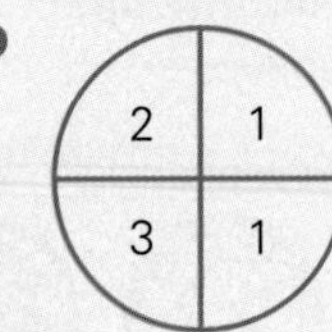

Investigation Practical probability

A Tossing a coin

You will need: a larger coin (20c or 50c), tally sheet, clear desktop

SPREADSHEET ACTIVITY

1 Work in pairs to toss a coin 20 times. Allow the coin to fall on the desktop. (Careful it doesn't roll off.) Record the results in a frequency table.

2 Add the entire class's results together.

3 What fraction of the total tosses came up heads? (Express it correct to 2 decimal places.) What fraction came up tails? What is the total of the two? Are the results what you **expected**? If a coin is tossed 20 times, how many heads would you expect?

B Rolling a dice

You will need: large, clearfaced dice, tally sheet, clear desktop

1 What possible **outcomes** are there when you roll a dice? Work in pairs to roll a dice 20 times on the clear desktop. Record the results in a list: 1 5 3 1 4 6 2 1 6 … Then **present** them in a frequency table.

2 Add the entire class's results together.

3 What fraction of the total tosses came up a 1? (Express it correct to 2 decimal places.) Repeat for each possible outcome. What is the total of all the fractions? Are the results what you expected? If you rolled a dice 18 times, how many 6s would you expect to get? **Compare** what you expect with what actually happened in your list in Step 1. Write your **observations**. Look for any unexpected patterns in the list.

13.2 Calculating probabilities

The following terms are used when calculating the probabilities of events:

- **Trial:** A trial is one performance of an experiment or activity where the outcome or result depends on chance.
- **Random:** As seen in statistics, random means determined purely by chance.
- **Equally likely events:** The outcomes of a random experiment are equally likely if each outcome has the same chance of occurring, for example tossing a coin.
- **Sample space:** The list of all possible outcomes in a situation is known as the sample space.

Tossing a coin
Sample space:
H or T

Rolling a die
Sample space:
1, 2, 3, 4, 5, 6

Suits in a deck of playing cards
Sample space:
hearts, clubs, spades or diamonds

Gender of a baby
Sample space:
girl, boy

The probability of an event occurring is the number of favourable outcomes possible divided by the total number of outcomes.

$$P(\text{event}) = \frac{\text{number of favourable outcomes}}{\text{total number of outcomes}}$$

$$P(e) = \frac{n(e)}{n(s)}$$

P(e) = probability of an event

n(e) = the number of favourable outcomes

n(s) = the number of outcomes in the sample space

EXAMPLE 1

A marble is chosen, randomly, from a bag containing 4 blue marbles, 3 green marbles and 1 red marble.

a List the outcomes in the sample space.

b What is the probability that the chosen marble is:

i red? **ii** blue? **iii** green?

Solution

a Sample space = B B B B G G G R

b **i** $P(R) = \frac{1}{8}$ — One red marble out of a sample space with 8 outcomes.

ii $P(B) = \frac{4}{8}$ — Four blue marbles out of a sample space with 8 outcomes.

$= \frac{1}{2}$

iii $P(G) = \frac{3}{8}$ — Three green marbles out of a sample space with 8 outcomes.

EXAMPLE 2

A die is rolled once.

a List all the outcomes in the sample space.

b What is the probability that the number on the uppermost face is:

i 6? **ii** 5? **iii** 5 or 6? **iv** even?

Solution

a Sample space = 1, 2, 3, 4, 5, 6

b **i** $P(6) = \frac{1}{6}$ — One 6 out of a sample space with six outcomes.

ii $P(5) = \frac{1}{6}$ — One 5 out of a sample space with six outcomes.

iii $P(5 \text{ or } 6) = \frac{1}{6} + \frac{1}{6}$ — In probability, 'or' means addition.

$= \frac{2}{6}$

$= \frac{1}{3}$

iv $P(\text{even}) = \frac{3}{6}$ — Three even numbers out of a sample space with six outcomes.

$= \frac{1}{2}$

EXERCISE 13B

1 The sample space for tossing two coins is given as: HH HT TH TT

a How many outcomes does the sample space contain?

b State the probability of the two coins showing:

i 2 heads **ii** 2 tails **iii** 1 tail and 1 head

iv a tail, then a head **v** at least 1 head **vi** no heads

c Find the value of P(HH) + P(HT) + P(TH) + P(TT).

2 The sample space for the rolling of a four-sided (tetrahedron) die is given as 1, 2, 3, 4.

1
4
2 3
net of a tetrahedron

a How many outcomes does the sample space contain?

b State the probability that the number the die lands on after one roll is:

i 1 **ii** 3 **iii** less than 4

iv greater than 4 **v** 4 **vi** even

vii odd **viii** prime

c What is another name for a tetrahedron?

3 A class raffle has tickets numbered 1 to 30. If one ticket is chosen at random, find the probability that the number on the chosen ticket is:

a 1 **b** 30 **c** 10 **d** even **e** greater than 20

f a single-digit number

4 If one letter is selected at random from the letters in the word WONDERFUL, what is the probability that the letter will be:

a W? **b** F? **c** a vowel? **d** a consonant? **e** I?

5 A bag contains 10 marbles of equal size and shape. If one marble is chosen at random what is the probability that the colour of the marble is:

a pink? **b** green? **c** not green?

d blue? **e** pink or blue?

6

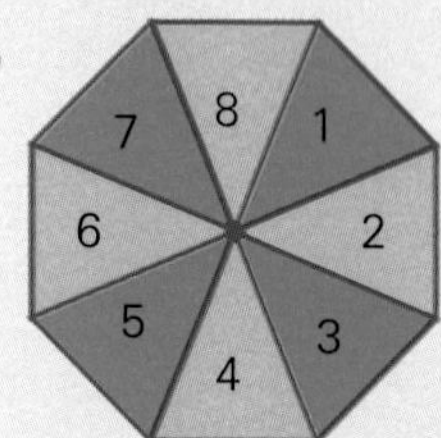

A spinner is spun once. Find the probability that the number that it lands on is:

a 3 **b** 8 **c** 1

d even **e** 2 or 3 **f** not 1

7 The numbers from 1 to 5 are written on separate cards. One card is chosen at random. What is the probability that the card chosen shows the number:

a 1? **b** 3? **c** 4 or 5?

d prime? **e** even? **f** odd?

g divisible by 3? **h** less than 10?

8 A bag contains 6 red balls, 4 black balls and 5 orange balls. One ball is chosen randomly from the bag. What is the probability that the ball chosen is:

a red? **b** black? **c** green? **d** red, black or orange?

Level 2

9 Probability is useful when dealing with gambling and it is therefore handy to be familiar with a standard deck of playing cards. In a standard deck of 52 playing cards, how many cards are:

a black? **b** red?
c in the suit of hearts? **d** in the suit of clubs?
e in the suit of spades? **f** in the suit of diamonds?
g tens? **h** queens?
i red sevens? **j** black tens?
k aces? **l** jacks?
m black threes? **n** the king of diamonds?

10 A card is chosen at random from a standard deck of playing cards. Find the probability that the card is:

a black **b** red **c** a heart
d a club **e** a spade **f** a diamond
g a ten **h** a queen **i** a red seven
j a black ten **k** an ace **l** a jack
m a black three **n** the king of diamonds **o** a king
p a king or queen **q** not a king **r** not a club
s a red ace **t** the queen of diamonds

11 The numbers 1, 2, 3 are arranged to form a three-digit number.

a List the possible three-digit numbers contained within the sample space.
b What is the probability that the number formed:
i is 321? **ii** ends in a 3? **iii** is even?
iv starts with a 2? **v** is greater than 300? **vi** is divisible by 3?
vii is odd? **viii** is less than 100?

12 A class of 30 students were asked to state their favourite flavour of ice-cream. The results are shown in the frequency table. If one student is chosen at random what is the probability that the student chosen likes:

Flavour	No. of students
Chocolate	5
Vanilla	8
Strawberry	10
Toffee	4
Other	3

a chocolate ice-cream?
b vanilla ice-cream?
c strawberry ice-cream?
d toffee ice-cream?

13 A survey of 100 households was taken to determine how many used certain shampoos and conditioners.

Brand	No. households
Pears	40
Funky	35
Teen	17
Tweed	5

a Based on these results, what is the probability of a household chosen at random:
i using the Pears brand?
ii using the Funky brand?
iii using the Teen brand?
iv not using one of the four brands listed?

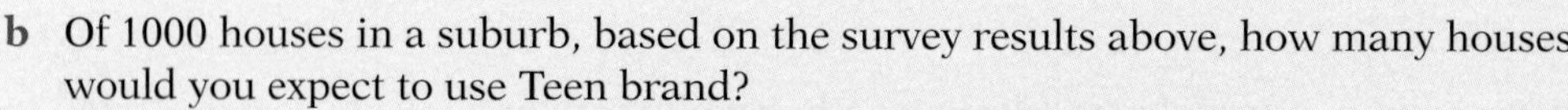

b Of 1000 houses in a suburb, based on the survey results above, how many houses would you expect to use Teen brand?

14 The letters of the word WAS are arranged to form a three-letter word.

SPREADSHEET ACTIVITY

a Write down all the possible three-letter words that can be formed.
b If one word is chosen at random, what is the probability that the word:
  i starts with S? **ii** doesn't start with S?
  iii ends in S? **iv** is the word SAW?

13.3 Complementary events

Everything in life seems to have an opposite; this is also the case in probability. The opposite of any event is its **complement**.

When a six-sided die is rolled, the probability of rolling a 4 is $\frac{1}{6}$.

The probability of a 1, 2, 3, 5 or 6 (i.e. *not* a 4) is $\frac{5}{6}$.

These probabilities add to 1.

$\therefore$ P(4) + P(*not* a 4) = 1

$$\frac{1}{6} + \frac{5}{6} = 1$$

These are **complementary** events as their probabilities add to 1 and so P(4) = 1 – P(*not* a 4) or P(*not* a 4) = 1 – P(4).

The complement of an event is the number of outcomes in the sample space that are *not* in the event. The complement of event E is denoted as $\overline{E}$.

The sum of an event and its complement is always 1.

$\therefore$ **P($\overline{E}$) = 1 – P(E)**

Complementary events can be thought of as opposites.

EXAMPLE

The probability of it raining on any given day is given as $\frac{1}{4}$. What is the probability that it does not rain on any given day?

Solution

$$\text{P(rain)} = \frac{1}{4}$$
$$\therefore \text{P(no rain)} = 1 - \frac{1}{4} = \frac{3}{4}$$

The complement of it raining is it *not* raining.
P(rain) + P(no rain) = 1

EXERCISE 13C

1 Write the event which is the complementary event to:

a obtaining a head when tossing a coin
b choosing a red card from a standard deck of playing cards
c obtaining an odd number when rolling a die
d winning a prize in a raffle
e testing positive in a tetanus test
f drawing a black card from a standard deck of cards

2 The probability of Carol winning a game of chess is $\frac{4}{9}$.

What is the probability that Carol will *not* win her next game of chess?

3 **a** The probability of an event, E, is given as $P(E) = \frac{1}{9}$. Find the value of $P(\overline{E})$.

b The probability of an event, E, is given as $P(E) = \frac{3}{5}$. Find the value of $P(\overline{E})$.

c The probability of an event, E, is given as $P(E) = \frac{3}{4}$. Find the value of $P(\overline{E})$.

d The probability of an event, E, is given as $P(E) = 0{\cdot}7$. Find the value of $P(\overline{E})$.

e The probability of an event, E, is given as $P(E) = 0{\cdot}44$. Find the value of $P(\overline{E})$.

4 A six-sided die is rolled. A is the event of obtaining a number less than 3 on the uppermost face of the die.

a List the outcomes in the event A and calculate its probability.

b List the outcomes in the event complementary to A and calculate its probability.

5 The probability that Cronulla wins the premiership is given as 0·84. What is the probability that Cronulla won't win the premiership?

6 The probability of scoring full marks in a maths quiz is given as $\frac{1}{100}$. What is the probability of not scoring full marks in the quiz?

7 The probability of a marksman hitting the bullseye is given as $\frac{3}{7}$. What is the probability that the marksman missed the bullseye?

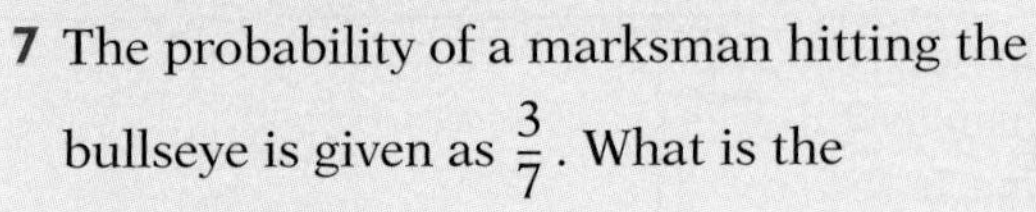

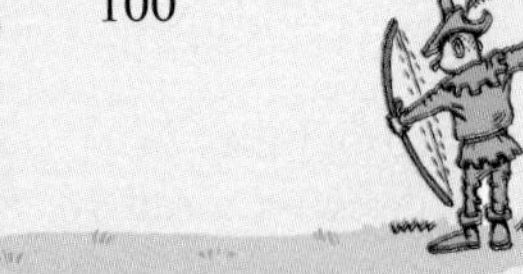

8 The probability of choosing a red ball from a bag of 10 balls is given as $P(R) = \frac{1}{5}$.

a How many red balls are in the bag?

b What is the probability of not choosing a red ball from the bag?

Level 2

9 A letter is selected at random from the letters in the word CHOCOLATE. Calculate the following probabilities:

a $P(C)$ **b** $P(\overline{C})$ **c** $P(O)$ **d** $P(\overline{O})$ **e** $P(\text{vowel})$

f $P(\overline{\text{vowel}})$ **g** $P(E)$ **h** $P(\overline{E})$ **i** $P(E) + P(\overline{E})$

10 A bag contains disks of three different colours, blue, pink and yellow. If $P(B) = \frac{3}{7}$ and $P(P) = \frac{2}{5}$, find:

a $P(\overline{B})$ **b** $P(\overline{P})$ **c** $P(Y)$

d $P(\overline{Y})$ **e** $P(B) + P(P) + P(Y)$ **f** $P(B) + P(\overline{B})$

11 One card is selected at random from a standard deck of 52 playing cards. If Q = [queen], K = [king], B = [black card], R = [red card], A = [ace], find:

a $P(K)$ b $P(Q)$ c $P(B)$

d $P(\overline{R})$ e $P(\overline{A})$ f $P(A) + P(\overline{A})$

TEACHER

Investigation Who am I?

I was born in Normandy, France in 1749 and was one of the first mathematicians to use probability to study ideas other than gambling.

Use your answers to unlock the puzzle code.

P The probability of choosing a club from a standard deck of playing cards.	**S** The probability of choosing a red 10 from a standard deck of playing cards.	**O** The probability of obtaining a 5 in a single throw of a six-sided die.
N The probability of choosing a vowel if one letter from the word MIRACLE is chosen.	**E** The probability of choosing a yellow ball from a bag with 4 green balls and 1 yellow ball.	**L** The probability of rolling a number less than 3 when a six-sided die is rolled.
R The probability of rain is given as 20%. What is the probability that it doesn't rain?	**C** If one letter is chosen at random from the word MUD, what is the probability that it is *not* a vowel?	**I** The chance of spinning an even number with one spin of the wheel shown. 
M The probability of not choosing a red 10 from a standard deck of playing cards.	**A** The probability of winning is given as 7 out of 10. What is the probability of not winning?	

$\frac{1}{4}$ $\frac{1}{2}$ $\frac{1}{5}$ $0{\cdot}8$ $0{\cdot}8$ $\frac{1}{5}$ $\frac{1}{26}$ $0{\cdot}5$ $\frac{25}{26}$ $\frac{1}{6}$ $\frac{3}{7}$

$\frac{1}{3}$ $0{\cdot}3$ $\frac{1}{4}$ $\frac{1}{3}$ $\frac{3}{10}$ $\frac{2}{3}$ $\frac{1}{5}$

Chapter Review

Language Links

almost certainly, always, certain, chance, complementary, equally likely, even, event, favourable, highly likely, impossible, likelihood, likely, never, outcomes, probability, random, sample space, scale, unlikely

1 Which of the words in the list above mean a probability of:

a 0? **b** 1? **c** $\frac{1}{2}$?

2 **Explain** what is meant by complementary events. Why do they always add to 1?

3 **Explain** the meaning of the phrase 'sample space'. How would you explain the process of finding a sample space to a classmate?

4 **Summarise** your probability words on the following scale:

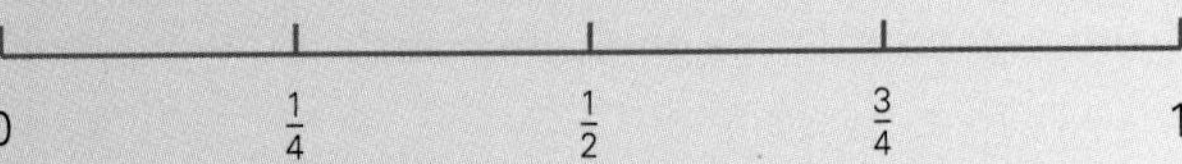

5 Define the following terms:

a random **b** sample space
c probability **d** equally likely

6 What is meant by an event having a probability of $\frac{1}{3}$?

7 Mark said that the probability of choosing a red ball from a bag was $\frac{2}{3}$. John thought that this meant that there were two red balls in a bag containing three balls. **Comment** on the validity of John's thinking.

8 Give an example of another event that has a probability of $\frac{2}{3}$.

9 **a** Using a page out of a newspaper—in pairs—highlight any probability words you find.
b Define each word.

Chapter Review Exercises

13.1 **1** Decide whether the probability of the following events is impossible, even chance or certain:

- **a** choosing a consonant from the word WHY
- **b** New Year's Day being 1 January next year
- **c** rolling a zero on a normal six-sided die
- **d** a horse winning a race it is not entered in
- **e** obtaining a number greater than 2 with one roll of a six-sided die

13.2 **2** What is the sample space of the following events?

a rolling a six-sided die **b** tossing a coin **c** choosing the gender of a baby

13.2 **3** Find the probability of:

- **a** obtaining a 4 with a single roll of a six-sided die
- **b** obtaining a head when a coin is tossed once
- **c** obtaining a number less than 2 when a six-sided die is rolled once
- **d** choosing the correct answer, by guessing, in a multiple choice question, given four choices
- **e** choosing a boy from a class of 12 girls and 16 boys

13.2 **4** One card is chosen at random from a standard deck of playing cards. Find the probability that the card is:

a black **b** a queen **c** a red 10 **d** an ace
e a black ace **f** a picture card (K, Q, J) **g** the 10 of clubs
h a heart **i** not a heart **j** a black 3 or the king of spades

13.2 **5** Margo has 4 red pens, 3 black pens and 5 blue pens in her pencil case. If Margo chose one pen at random to quickly sign her name to her test, what is the probability that the colour of the pen chosen by Margo was:

a red? **b** blue? **c** not black? **d** red or blue? **e** black or red?

13.2 **6** The ages of friends at a 21st party are given in the table. If one person is chosen at random to give a speech at the party, what is the probability that the person is:

a 21? **b** 25? **c** not 21?
d older than 21? **e** younger than 21?

Ages	No. of people
20	5
21	40
22	25
23	9
24	21
25	10

13.3 **7** The probability of a person catching a cold at some time during the year is given as 0·78. What is the probability that a person will remain cold-free for the year?

13.3 **8** The probability that Todd shooting a goal in Union is $\frac{3}{7}$. Find:

- **a** The probability that Todd will miss the goal.
- **b** If Todd made 21 attempts at goal in one training session, how many goals would you expect him to score?
- **c** What is the least number of goals Todd can score in the training session?

TEACHER

Keeping Mathematically Fit

Part A: Non-calculator

1 Divide 880 in the ratio 7 : 4.

2 What is the reciprocal of $3\frac{1}{3}$?

3 Find the average of 0·2, 0·01 and 0·003.

4 Find the value of $\dfrac{15 \times 40}{25}$.

5 Find $2\frac{1}{2}$% of 4800 mm.

6 Convert 1·6 m^2 to cm^2.

7 Choose the incorrect line in:
$$\frac{4x-5}{2} = 6$$
$$4x - 5 = 12$$
$$4x = 7$$
$$x = \frac{7}{4}$$
and rewrite the solution correctly.

8 Write two numbers that have a sum of 35 and a product of 250.

9 In the diagram AB is 4 units. **a** Write down the coordinates of B. **b** Find the area of the quadrilateral.

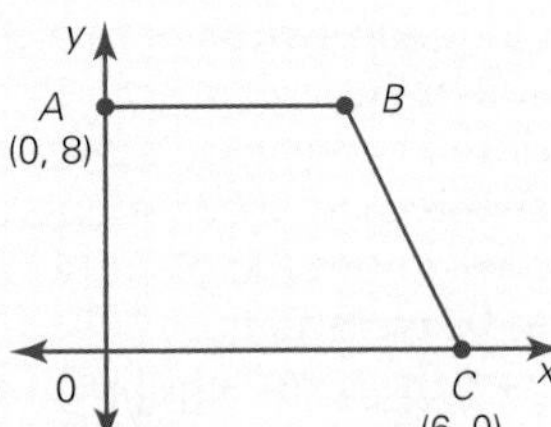

10 Simplify $\dfrac{6a^2}{2ab}$.

Part B: Calculator

1 6, 9, 5, 9, 7, 9, 1

Find the: **a** mean **b** mode **c** median

2 The volume of an open cube is 343 cm^3.
Calculate the surface area of the outside faces.

3 A car travels 150 km from Bilby to Zebra in 100 minutes.
What was the average speed the car was travelling?

4 Solve:

a $\dfrac{3-x}{4} = {}^{-}2$ **b** $4x + 9 = 2x - 11$ **c** $\dfrac{5}{x} = 6$

5 A profit of one million dollars was shared between three business partners in the ratio of 11 : 6 : 3. What was the largest profit received by any of the business partners?

6 A salesman earns $240 p.w. plus 3% commission on the value of all sales over $40 000. Find a salesman's income for a week where sales totalled $55 000.

7 Find the point of intersection of the lines $y = 2x - 2$ and $y = x$.

8 The product of two numbers is 512, and one number is half the other.

a What are the two numbers? **b** Is there more than one solution?

1 What you need to know and revise

Outcome SGS 4.2:
Identifies and names angles formed by the intersection of straight lines, including those related to transversals on sets of parallel lines, and makes use of the relationships between them

Outcome SGS 4.3:
Classifies, constructs and determines the properties of triangles and quadrilaterals

2 What you will learn in this chapter

Outcome SGS 4.4:
Identifies congruent and similar two-dimensional figures, stating relevant conditions

- identifies congruent figures
- matches sides and angles of two congruent polygons
- names the angles in matching order when using the symbol ≡ in a congruency statement
- draws congruent figures using geometrical instruments
- determines the condition for two circles to be congruent
- uses the term 'similar' and matches the sides and angles of similar (|||) figures
- determines the scale factor for a pair of similar figures—including circles
- calculates dimensions of similar figures using the enlargement or reduction factor
- chooses an appropriate scale in order to enlarge or reduce a diagram
- constructs similar figures
- draws similar figures using geometrical instruments

Working Mathematically outcomes WMS 4.1–4.5
Students will be asked to ***question***, ***apply strategies***, ***communicate***, ***reason*** and ***reflect*** in the sections of this chapter.

14 Congruency, similarity and scale drawings

Key mathematical terms you will encounter

congruent	enlargement	original	scale
corresponding	image	ratio	similar
drawing	matching	reduction	

MathsCheck

Congruency, similarity and scale drawings

1 What are the mathematical names of these shapes?

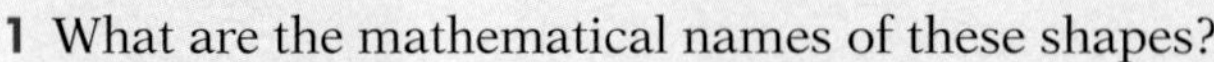

a

b

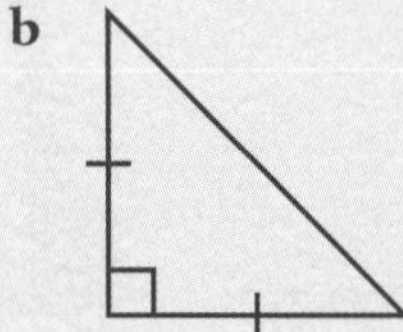

c

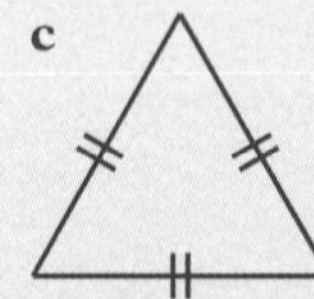

2 Find the value of the pronumerals in each of the following diagrams. Give a reason for each.

a

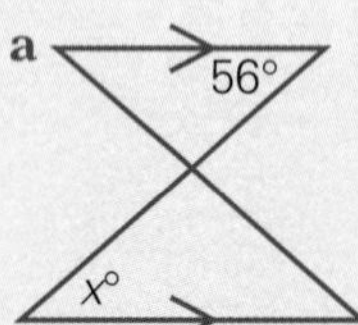

b

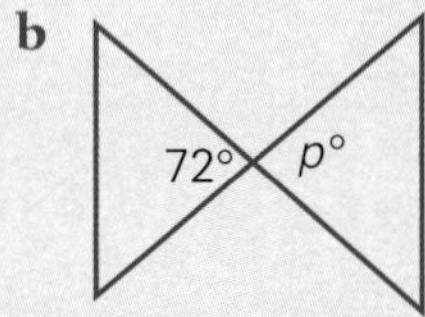

c

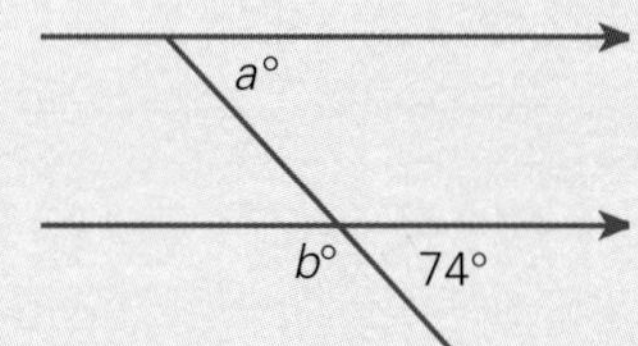

d

e

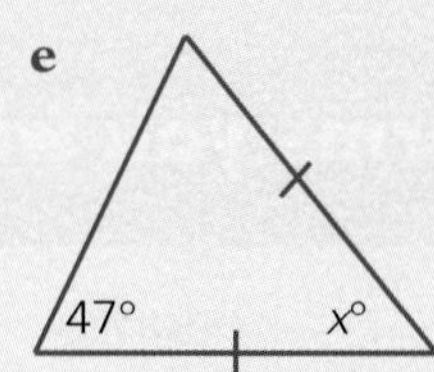

f

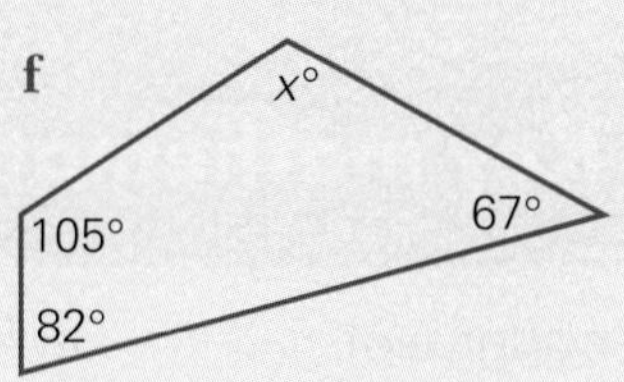

3 Simplify the following ratios:

a 3 cm to 1 m **b** 4 cm to 4 m **c** $\frac{1}{2}$ cm to 1 m

d 5 cm to $\frac{1}{2}$ m **e** 14 mm to 40 m **f** 20 cm to $\frac{1}{2}$ km

4 Write each of the following ratios in the form of $x : 1$.

a 2 : 500 **b** 3 : 3000 **c** 20 : 1500

d 15 : 3000 **e** 10 : 3000 **f** 2 : 5 000 000

5 Describe the following transformations:

a

b

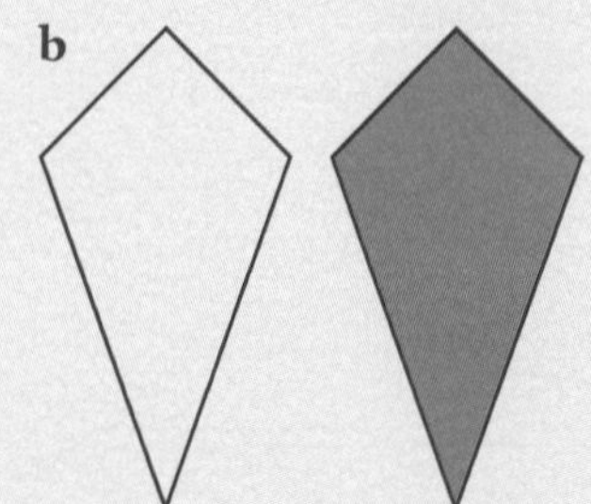

Exploring New Ideas

14.1 Constructing triangles

In the following section we look at constructing triangles and the minimum information needed to construct a triangle.

INTERACTIVE GEOMETRY

EXAMPLE 1

Construct a triangle, *given three sides*: 4 cm, 5 cm, 7 cm.

Solution

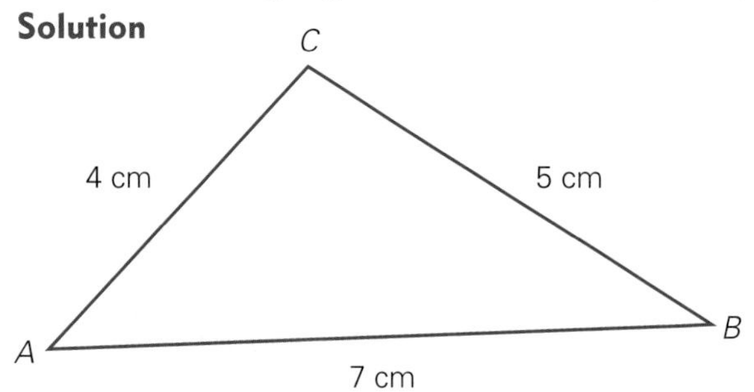

1. Draw the 7cm line as base *AB*. (Starting with the longest is easier.)
2. With radius 4 cm and centre *A*, draw an arc above *AB*.
3. With radius 5 cm and centre *B*, draw another arc to intersect the first at *C*.
4. Join *AC* and *BC* and mark in lengths.

EXAMPLE 2

Construct a triangle, *given two sides*, 6 cm and 4·5 cm, and the *angle between them* at 50°.

Solution

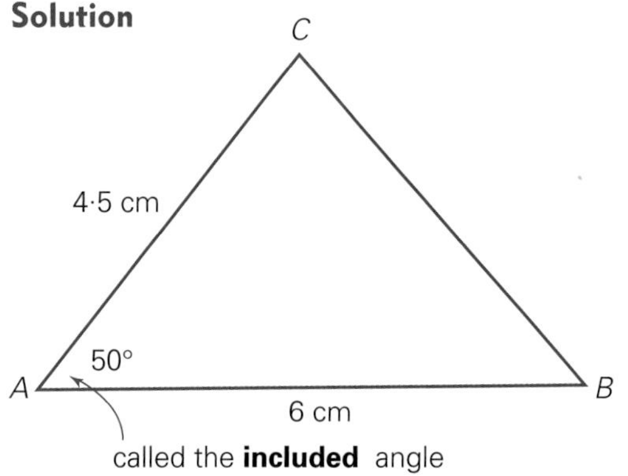

1. Draw the 6 cm line as base *AB*.
2. Use a protractor to construct a 50° angle at *A*. Mark off a point *C*, 4·5 cm along this second arm of the angle.
3. Join *CB*. Mark in all given measurements.

EXAMPLE 3

Construct a triangle, *given two angles* 40° and 55° and *one side* of 6 cm between the given angles.

Solution

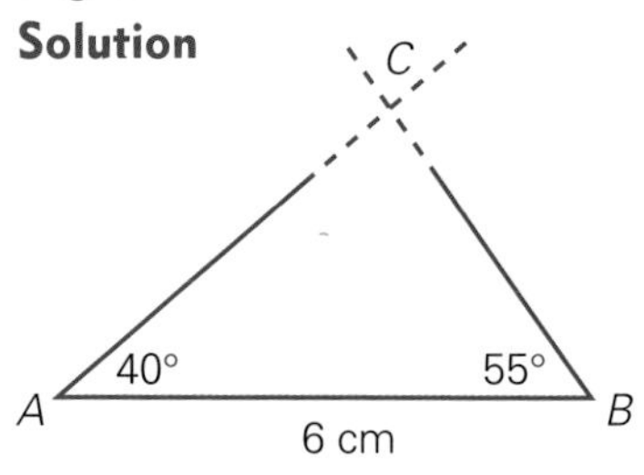

1. Draw the 6 cm line as base *AB*.
2. With a protractor, construct an angle of 40° at one end, and one of 55° at the other.
3. Extend the upper arms until they meet at *C*.

EXERCISE 14A

1 Use ruler and compasses to construct triangles with sides of:

a 4 cm, 5 cm, 5 cm **b** 4 cm, 4 cm, 4 cm **c** 30 mm, 45 mm, 55 mm

2 Build a triangle of dimensions 4 cm, 5 cm, 6 cm. Measure all angles with a protractor. Is the longest side opposite the largest angle?

3 Draw a triangle of sides 3·6 cm, 4·8 cm and 6 cm. Does the *largest angle opposite the longest side* pattern hold?

4 True or false?

a For *scalene* triangles, each angle is a different size.
b For *isosceles* triangles, all three angles are equal.
c For *equilateral* triangles, all sides and angles are equal.

5 Use ruler and protractor to construct triangles, given three features. Measure the other three parts not given.

a 3 cm, 4 cm, included angle 80°
b 6 cm, 6 cm, included angle 120°
c 40 mm, 52 mm, angle in between 90°

6 Use ruler and protractor to build triangles with the two given angles and the side in-between; measure the other three parts:

a 50 mm, 45°, 65° **b** 65 mm, 50°, 85° **c** 5·8 cm, 70°, 70°

Level 2

7 Is it possible to draw only one triangle given the following pieces of information?

a sides 3 cm, 4 cm and an angle of 60°
b sides 8 cm, 10 cm and 12 cm
c two angles of 60° and 50° respectively and one side 7 cm
d a right angle and two sides of 10 cm and 8 cm

8 Is it possible to draw a triangle with sides of 8 cm, 4·5 cm and 3 cm? **Explain.**

9 Rewrite the information given in question **7** for those triangles impossible to draw so that a unique triangle can be produced.

10 Construct each triangle and compare to a classmate's. They should now be identical triangles.

INTERACTIVE GEOMETRY

Investigation Sides of a triangle

Any two sides of a triangle together are greater than the third.

Is this true?

Investigate!

Try constructing or sketching each different type of triangle in a variety of sizes and then see what you find.

14.2 Congruent shapes

Shapes that are **identical** in every way are said to be **congruent**.

Congruent shapes make an exact match when they are superimposed; that is, placed on top of each other.

> **Thales** (640–546 BC) began the science of astronomy and the study of geometry and is regarded as the first philosopher of the Western world.
>
> As a merchant, his travels to Egypt taught him much about practical uses of geometry, but he developed many *general truths* based on their knowledge.
>
> He used *congruent triangles* to calculate a ship's distance from shore, and used *similar* triangles to calculate the height of the Egyptian pyramids.

The symbol to represent congruent shapes is ≡.

We write $AB \equiv CD$ and we say that AB **is congruent to** CD.

Superimpose for an exact match:

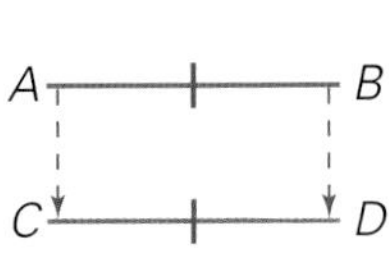

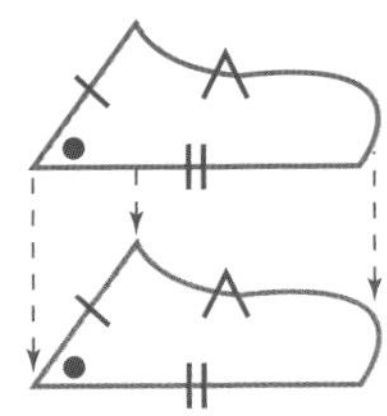

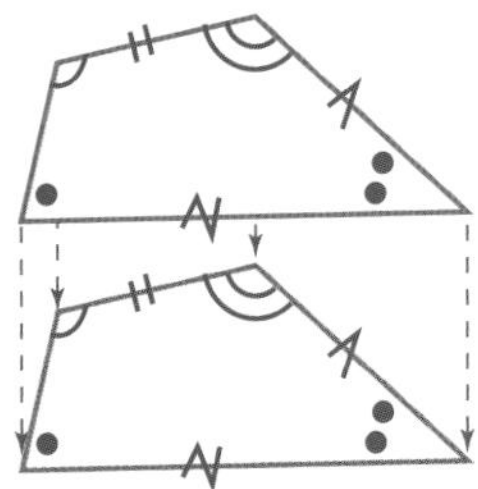

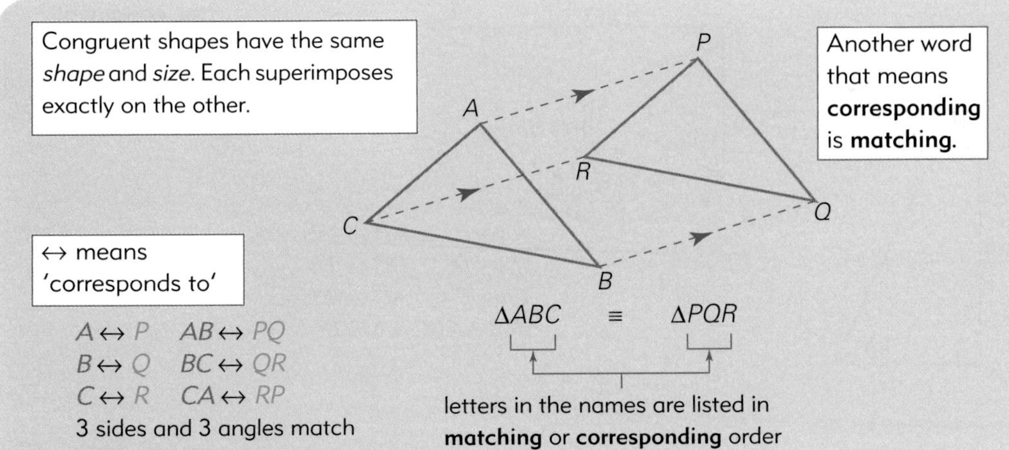

In congruent shapes, corresponding or matching angles are equal.

In congruent shapes, corresponding or matching sides are equal.

For congruent circles, their radii are equal.

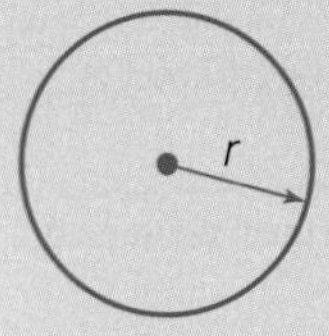

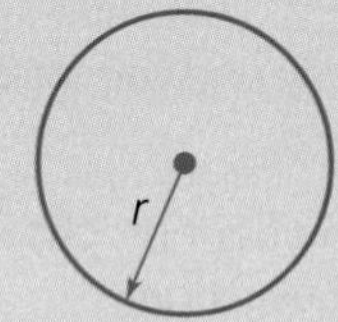

EXAMPLE

Show, by measuring the sides and angles, that the two shapes are congruent and write a statement showing the corresponding sides and angles.

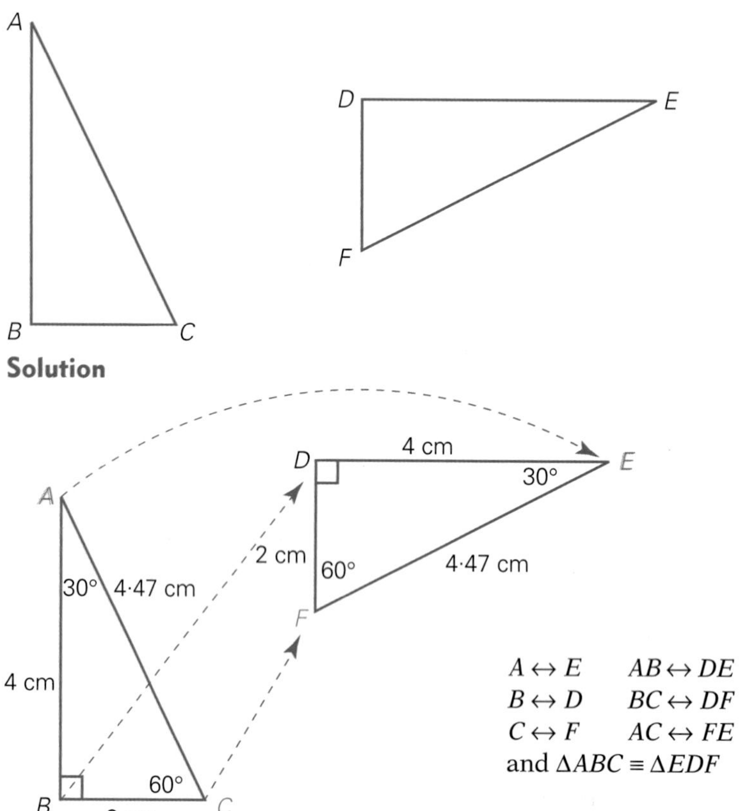

Solution

$A \leftrightarrow E \qquad AB \leftrightarrow DE$

$B \leftrightarrow D \qquad BC \leftrightarrow DF$

$C \leftrightarrow F \qquad AC \leftrightarrow FE$

and $\Delta ABC \equiv \Delta EDF$

> Note the areas of the congruent shapes are also equal.

EXERCISE 14B

1 Test to see which of the following pairs of shapes are congruent. Tracing may help.

a

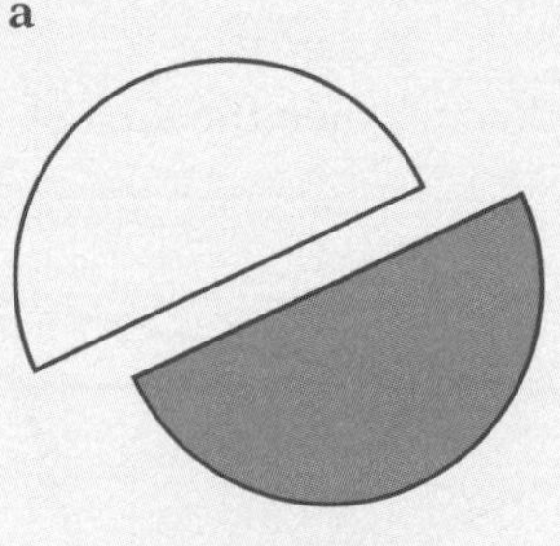

b

c

d

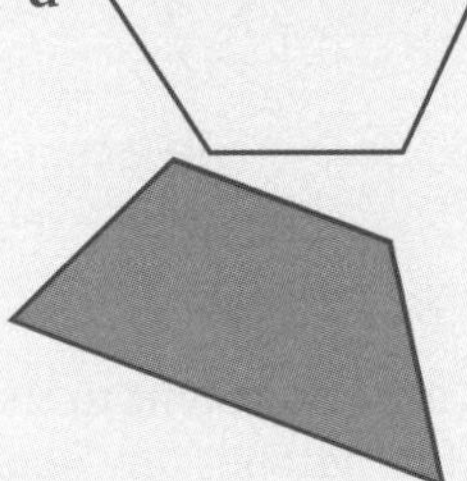

e

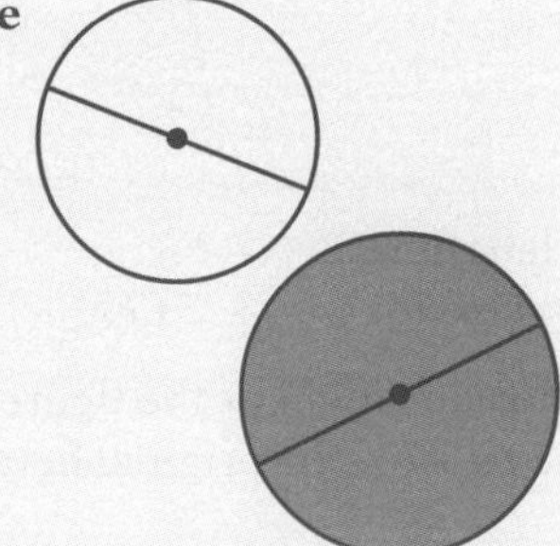

f

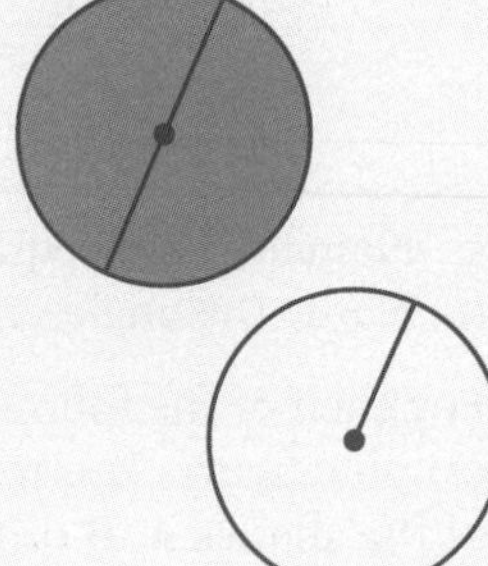

2 These two triangles are congruent. Sketch them and apply markings to show all congruent parts.

Note: One triangle is just a translation of the other.

3 a Trace these two shapes into your workbook.

b Measure each side and angle.

c Are the two shapes congruent?

d Name the two shapes in matching order.

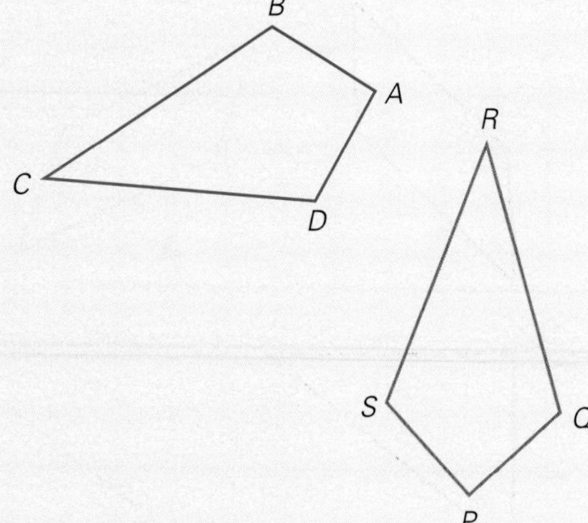

4 If $\Delta CMA \equiv \Delta PBT$, copy and complete the following:

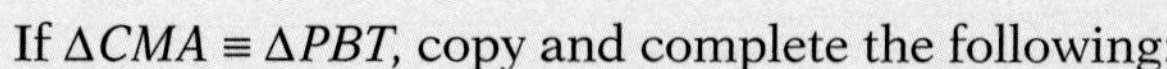

a $CM =$ ______ **b** $CA =$ ______

c $AM =$ ______ **d** $\angle CMA = \angle$ ______

e $\angle$ ______ $= \angle PTB$ **f** $\angle$ ______ $= \angle BPT$

g If the area of ΔACM is 2·25 cm², then the area of ΔTPB is ______.

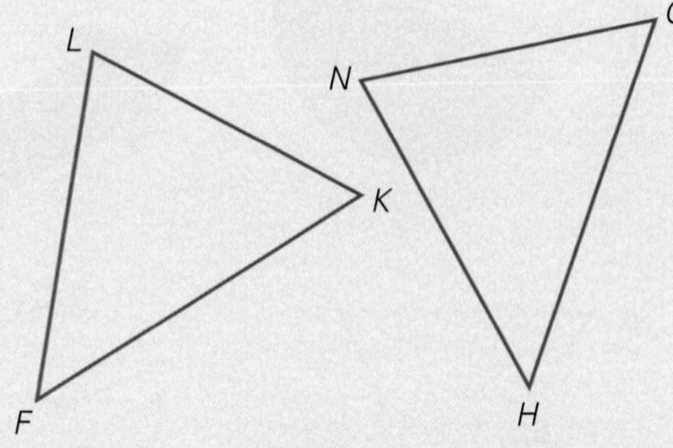

5 a Use measurement to show that these two triangles are congruent.

b Write a *statement of congruency*: $\Delta LKF \equiv$ ______

c Name the side equal in length to: **i** FK **ii** NH

d Name the angle equal in size to: **i** $\angle LFK$ **ii** $\angle GNH$

6 For each of the following, decide which of the figures **A**, **B** or **C** are congruent to the original figure. You may like to trace the original figure and see if it matches any or all of the shapes A, B or C.

a

A

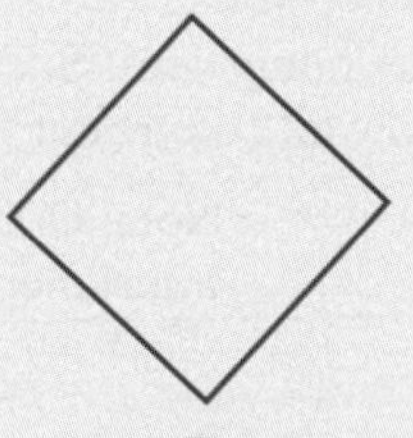

B

C

b

A

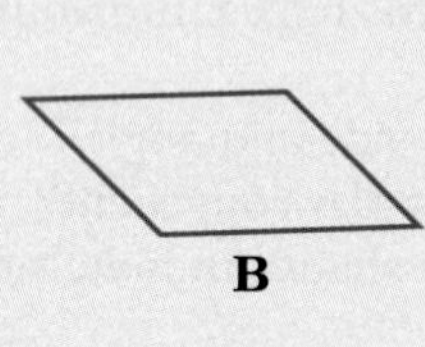

B

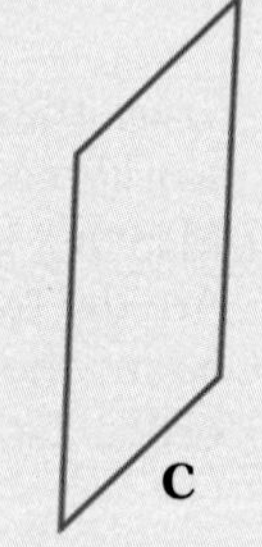

C

c

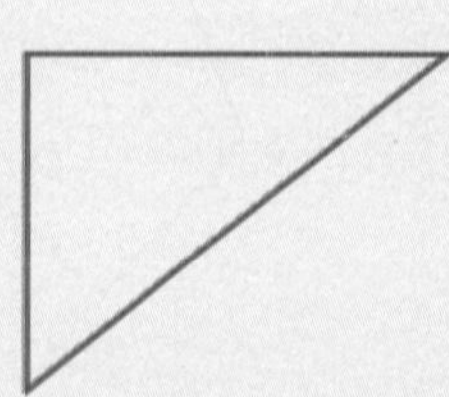

A

B

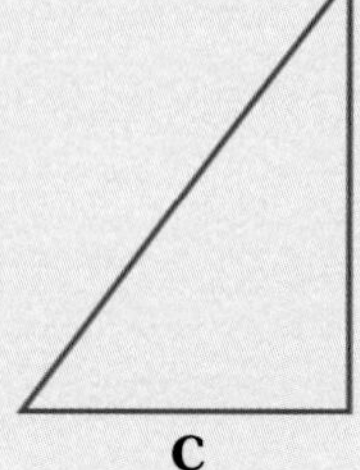

C

Level 2

7 If we know that $\Delta XZT \equiv \Delta MAE$, name a part to match the following:

a XT **b** MA **c** $\angle ZXT$ **d** $\angle MEA$

8 Construct $\Delta DMF \equiv \Delta QBT$. Cut out and superimpose to check congruency.

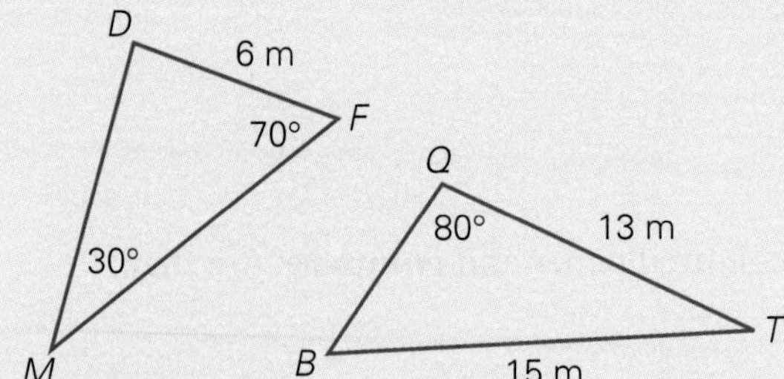

a The triangles at left are congruent. Copy and complete these statements:

i $\Delta MDF \equiv$ ______

ii $\angle D = \angle$ ______

iii $DF =$ ______

b Write the value of:

i $\angle T$ **ii** $\angle B$ **iii** BQ **iv** DM **v** MF

9 If a triangle SPN was constructed such that $\Delta SPN \equiv \Delta RVT$, what would be the value of:

a NS? **b** PN? **c** $\angle NSP$?
d $\angle NPS$? **e** PS? **f** $\angle N$?

R
20 cm
51°
V
92°
35 cm
28 cm
T

10 Copy each figure. Draw in one line from a vertex to create two congruent triangles. Write a statement of congruency for each pair of triangles.

a

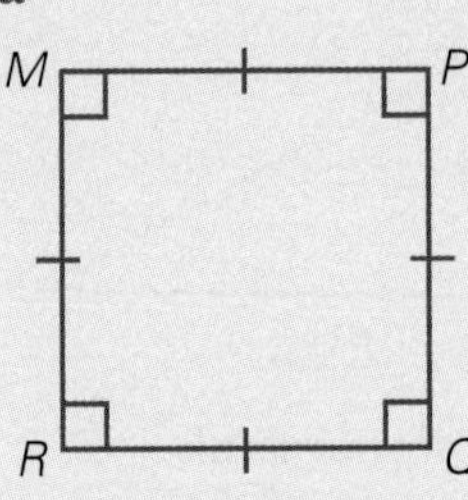

b

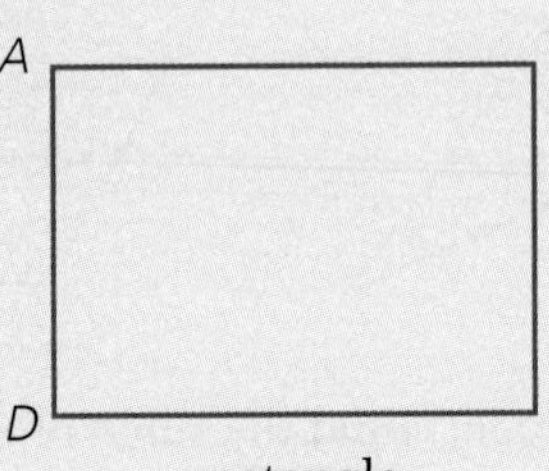

rectangle

c

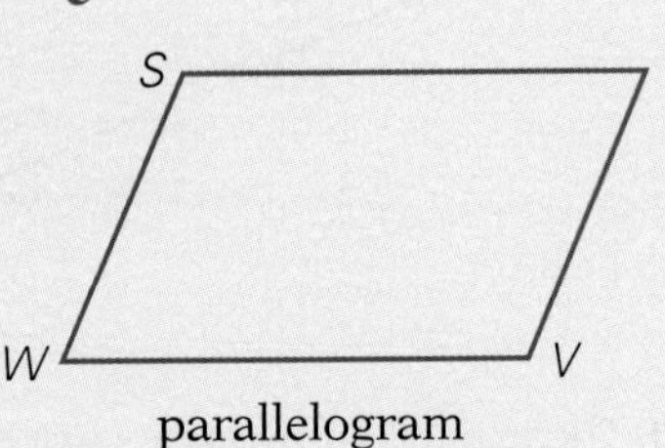

parallelogram

d

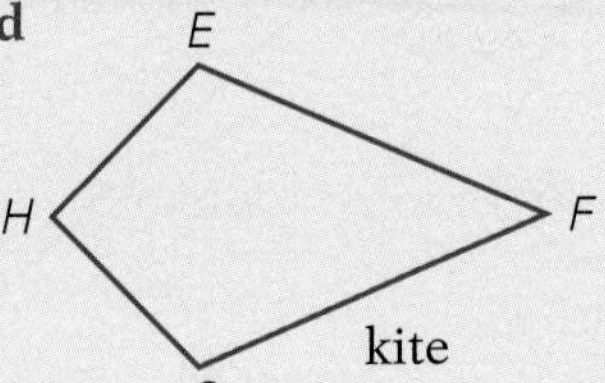

kite

e

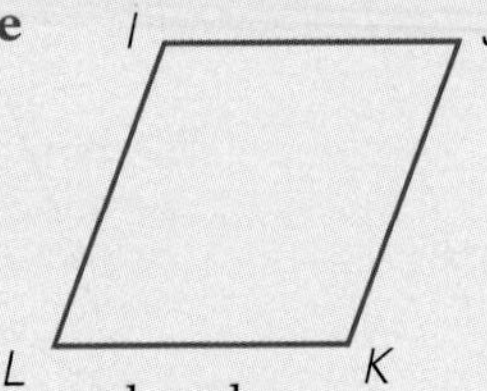

rhombus

f

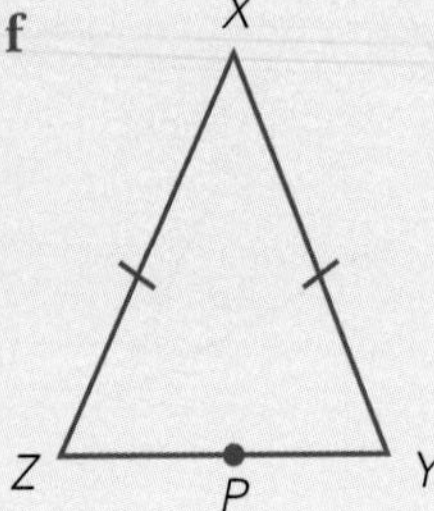

11 The diagram at left shows two sets of congruent triangles.

a Write two statements of congruency.

b Redraw triangles ABC and DCB in a non-overlapping position.

12 Construct the following shapes onto coloured paper and **compare**. Are they congruent?

a

6 cm

8 cm

6 cm

10 cm

8 cm

b

5 cm

80°

7 cm

60°

7 cm

80°

5 cm

60°

c

5 cm

40°

110°

30°

110°

5 cm

13 These pairs of triangles are congruent. Find the value of each pronumeral.

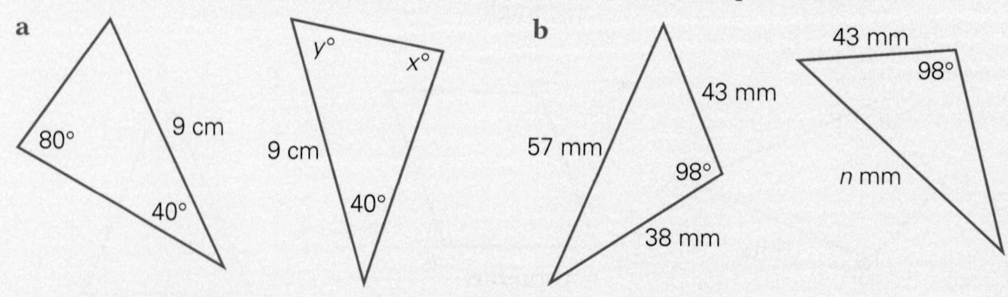

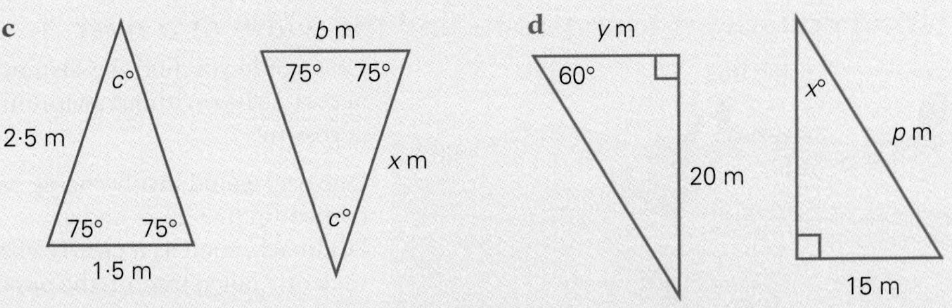

Investigation Using congruency

A Tessellating triangles

You will need: isometric and square grid paper, coloured pencils

Congruent triangles tessellate

1 Using *equilateral* triangles:
Extend the designs, then generate your own. Use colour effectively.

2 Using *isosceles* triangles:
Extend the designs shown, then design your own using whole or parts.

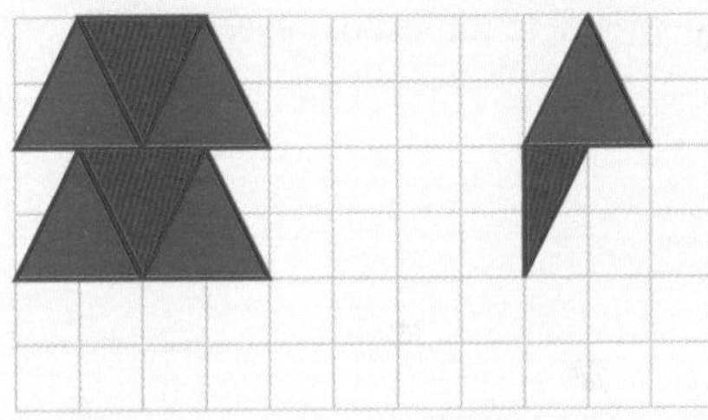

3 Using *scalene* triangles:
More of a challenge. Cut out a copy of the triangle and use it to help.

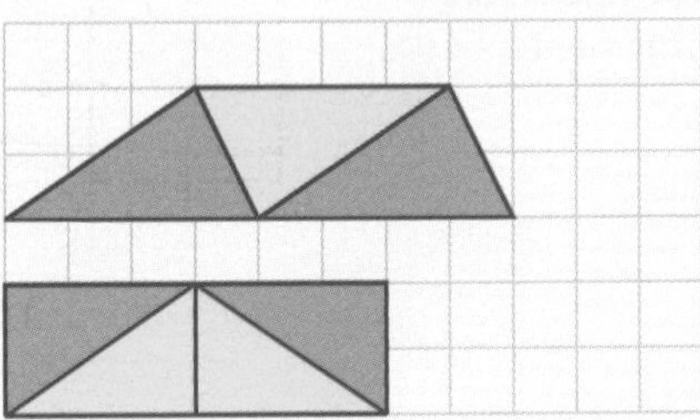

B Using congruent triangles to find the width of a river

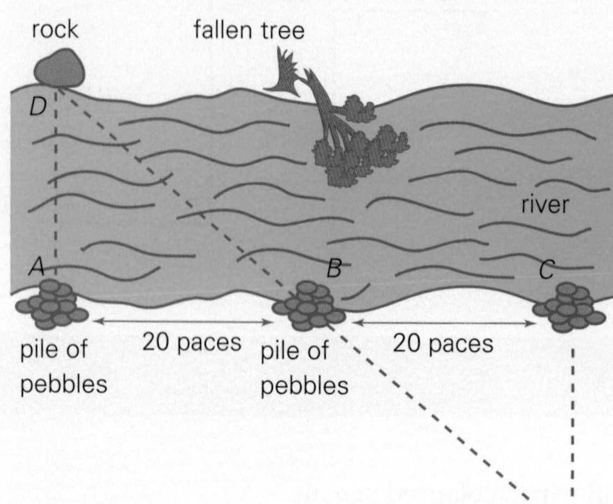

How could you find the distance across a river without swimming across it?

One way would involve using easily-picked-up materials and a landmark, such as a clearly visible rock or a fallen tree, on the opposite bank.

Making a pile of pebbles directly opposite a landmark (pile A), then a second pile 20 paces away at B and a third a matching distance away at C could serve as the basis of a pair of congruent triangles.

Extend this idea, complete a diagram and show how the distance AD could be calculated.

14.3 Similar figures

INTERACTIVE GEOMETRY

Similar figures have the *same shape* but not the same size.

The symbol to represent similarity is |||.

Similar figures are enlargements.

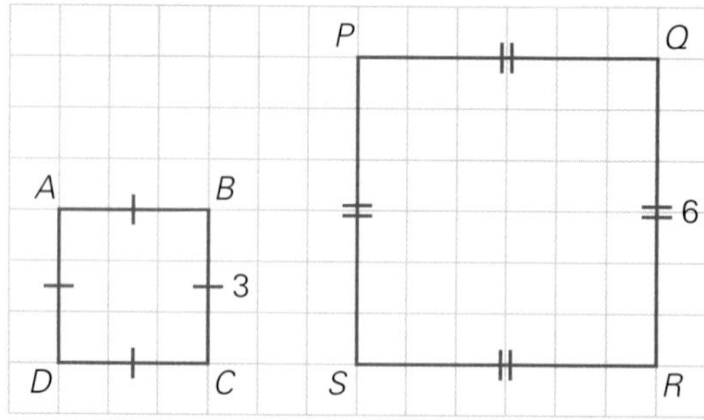

$ABCD$ ||| $PQRS$

Ratio of sides is 1 : 2
i.e. **scale factor** is 2.

Similar figures are reductions.

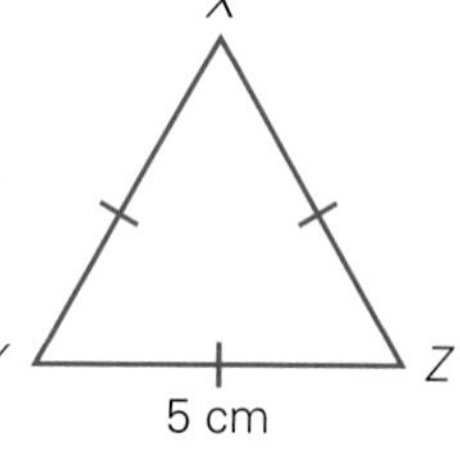

A
B
C
2·5 cm

ΔXYZ ||| ΔABC

Ratio of sides is 2 : 1 or 1 : $\frac{1}{2}$
i.e. **scale factor** is $\frac{1}{2}$.

Similar figures have **corresponding angles equal** and **ratios of corresponding sides equal**.

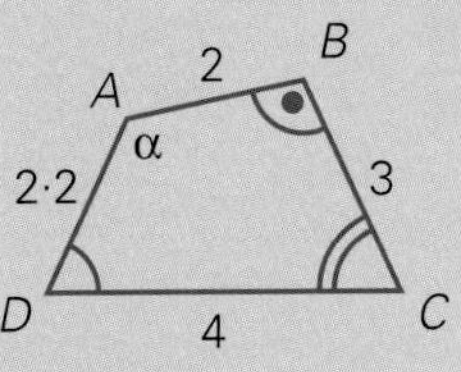

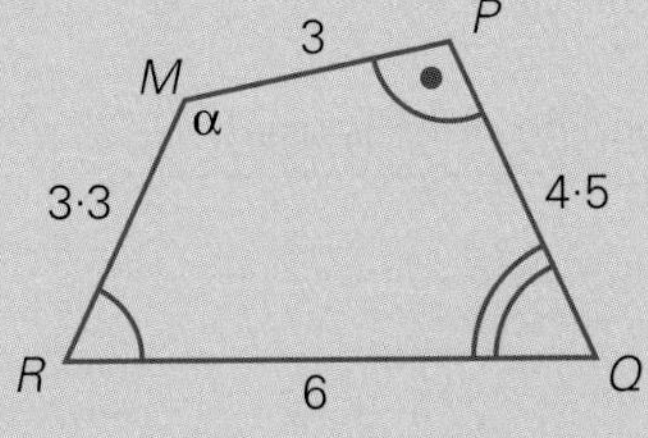

ABCD ||| *MPQR*
↑
is similar to

$\angle A = \angle M$
$\angle B = \angle P$
$\angle C = \angle Q$
$\angle D = \angle R$

$$\frac{AB}{MP} = \frac{BC}{PQ} = \frac{CD}{QR} = \frac{DA}{RM} = \frac{2}{3}$$

Congruent shapes are also similar.
The ratio of sides is 1 : 1.

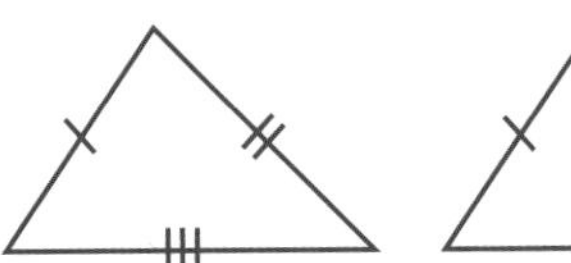

Meet Krys, the full-scale model of an 8·6 m savannah king saltwater crocodile, located in Normanton, Qld (scale 1 : 1).

EXAMPLE

Use a protractor and ruler to decide whether the following pairs of figures are similar. If they are, find the common ratio.

a

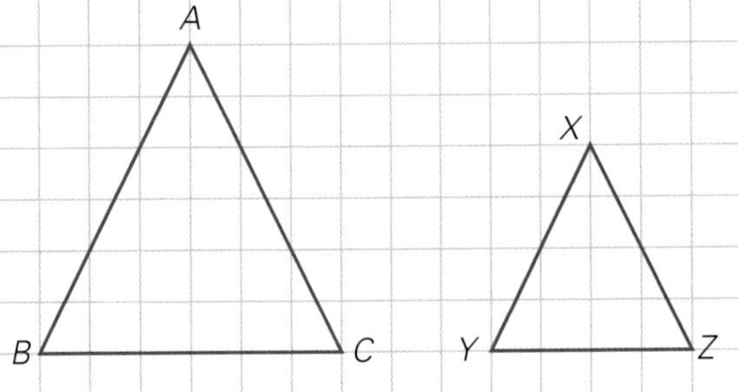

b

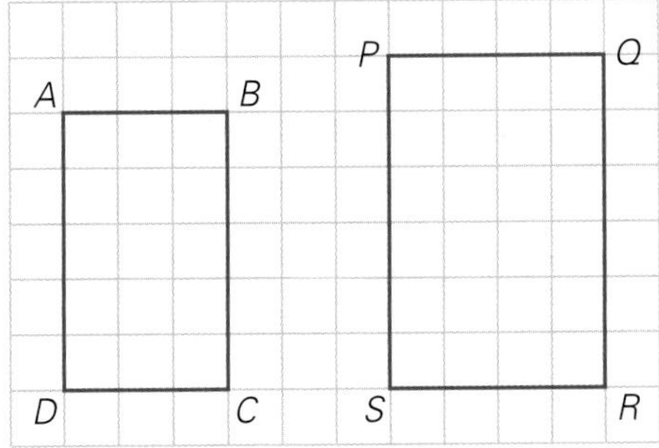

Solution

a $\Delta ABC \;|||\; \Delta XYZ$
ratio of sides: 3 : 2
reduction factor: $\frac{2}{3}$

$\angle A = \angle X$ $\angle B = \angle Y$ $\angle C = \angle Z$ corresponding order is used

b The shapes are *not* similar as the ratio of sides is not equal

i.e. $\frac{AB}{PQ} = \frac{3}{4} = 0{\cdot}75 \qquad \frac{AD}{PS} = \frac{5}{6} = 0{\cdot}8\dot{3}$

EXERCISE 14C

1 Use a protractor and ruler to decide whether the following pairs of figures are similar. If they are, find the common ratio.

a

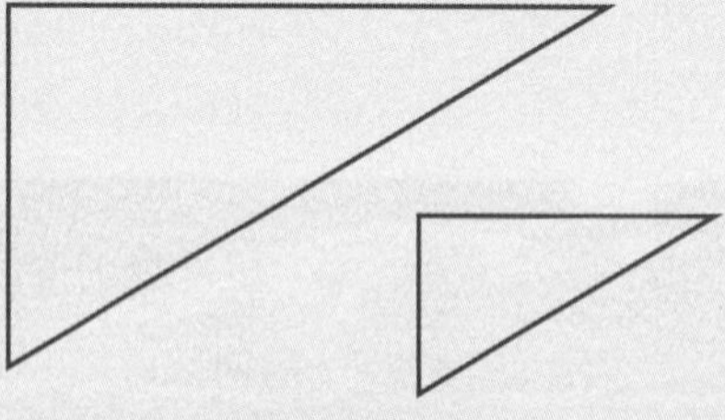

b

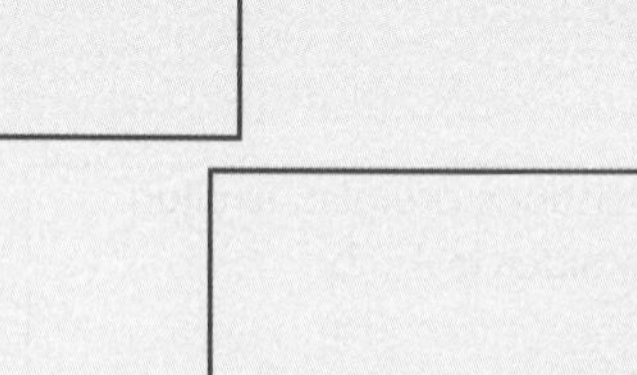

c

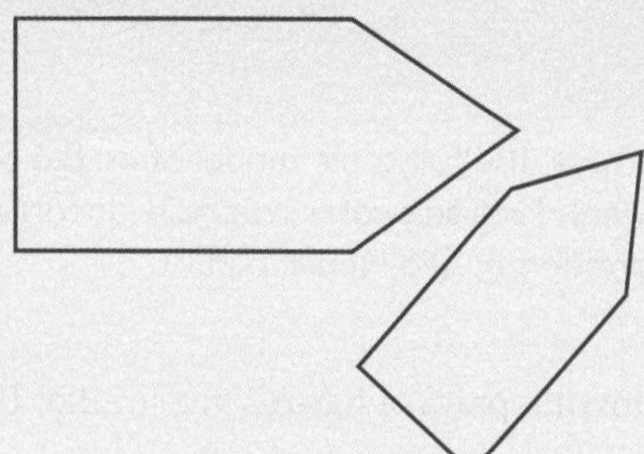

d

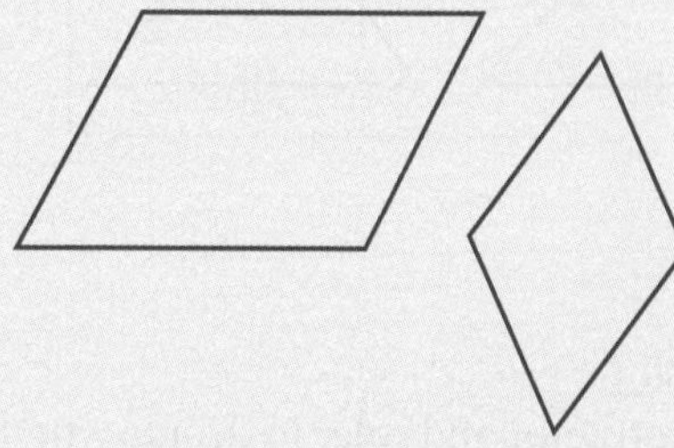

e

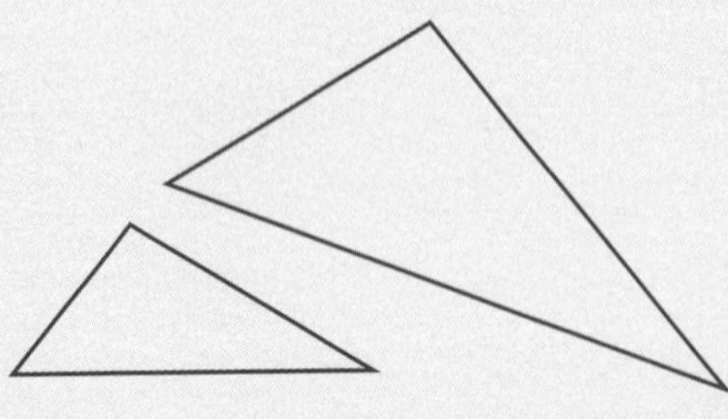

f

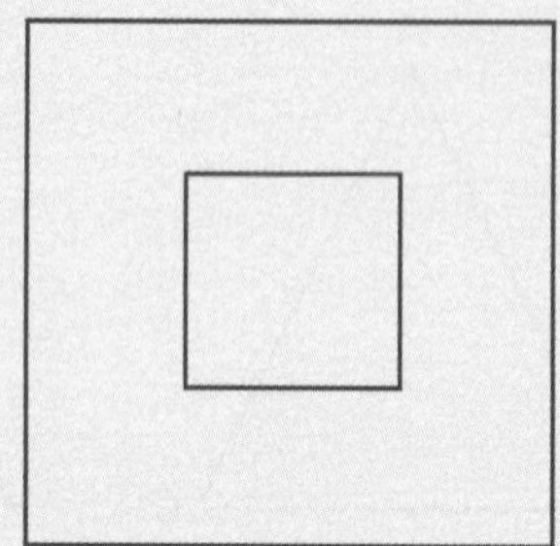

2 Find the size of the unknown angle for each pair of similar figures:

a

110° 70° 70° 110°

110° x° 70° 110°

b

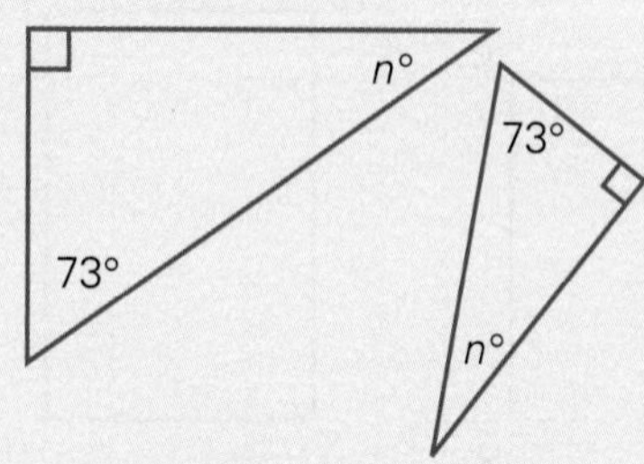

3 Enlarge the following shapes as shown:

a

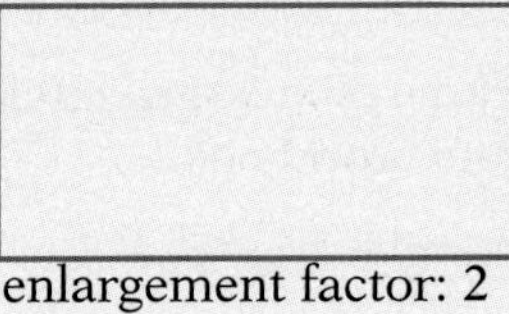

enlargement factor: 2

b

enlargement factor: 3

4 Use isometric paper to reduce the following shapes by the factor given:

a

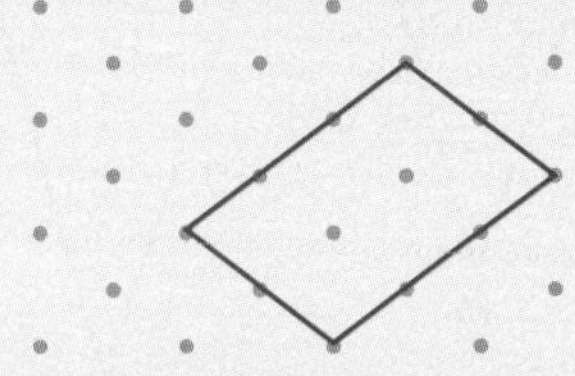

reduction factor: 2

b

reduction factor: 4

5 These two circles are similar.

i

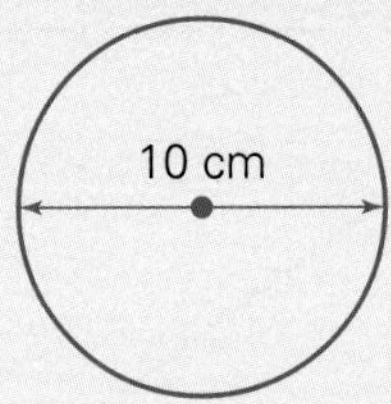

ii

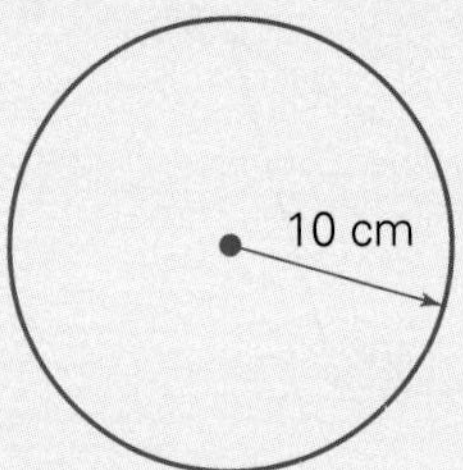

a Write the ratio of radius **i** to radius **ii**.

b What is the enlargement factor?

Level 2

6 Find the scale factor for the following pairs of similar figures:

a

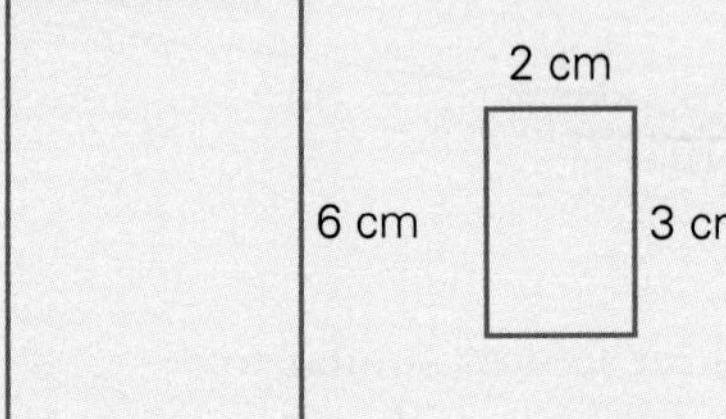

b

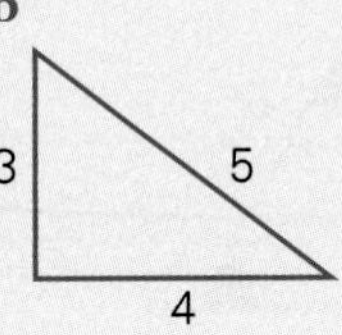

9 15 12

c

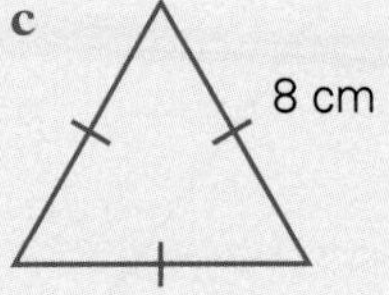

6 cm

d

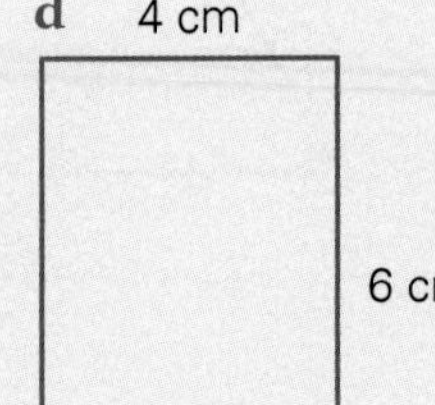

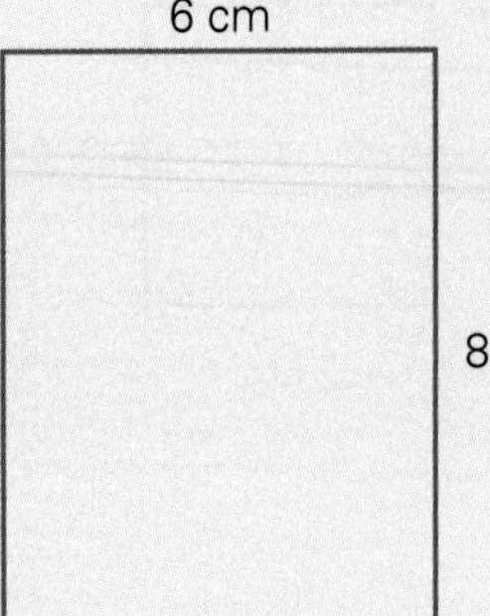

7 A circle of radius 3 cm is enlarged so that the ratio of the radii becomes 2 : 3.

a What is the radius of the larger circle?
b What is the ratio of the area of the initial circle to the area of enlarged circle?
c Construct both circles, using a pair of compasses in your workbook.

8 Write a statement of similarity for the following pairs of similar triangles:

a

ΔABC ||| Δ______

b

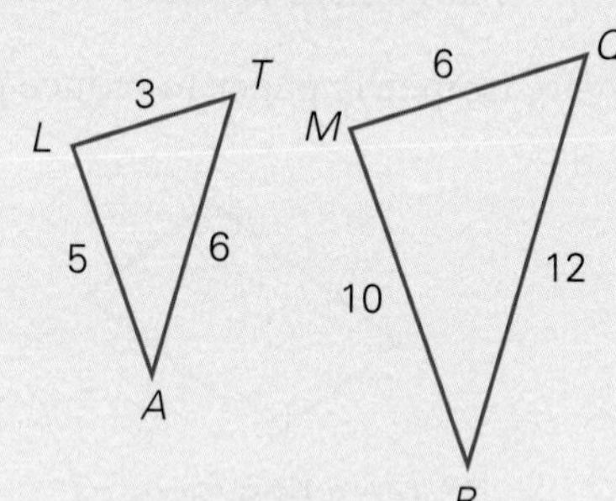

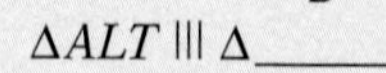

ΔALT ||| Δ______

c

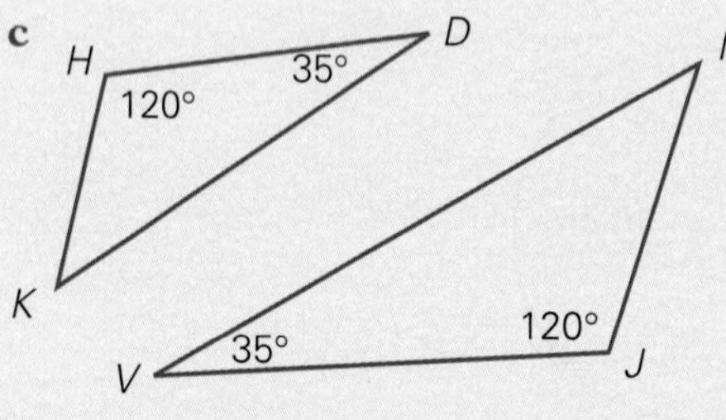

d

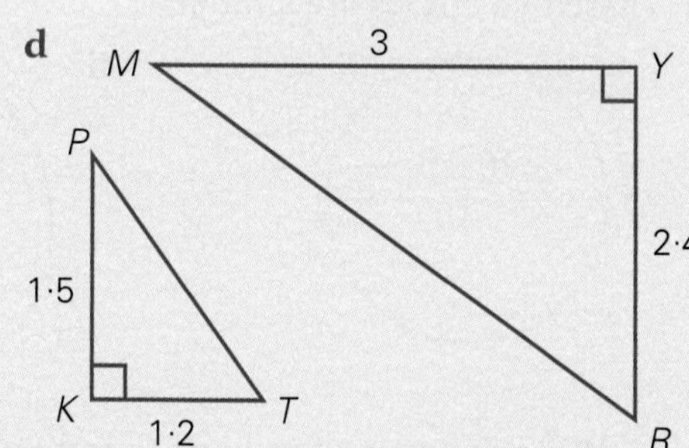

e

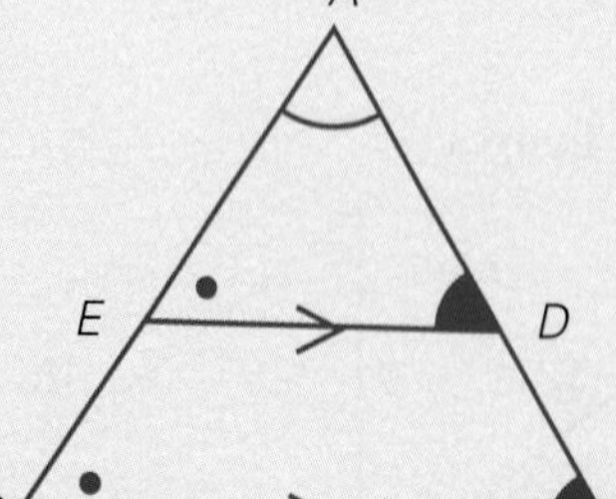

f

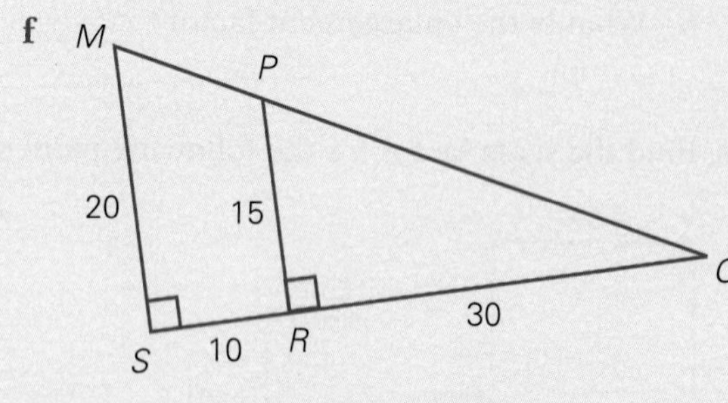

9 Construct, using a protractor, ruler and compass, three figures similar to:

a

b

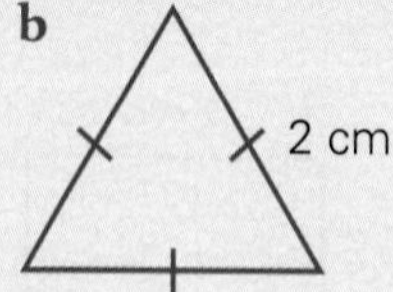

c

Investigation Similar figures

A Special features of similar figures

You will need: protractor, ruler, tracing paper, scissors

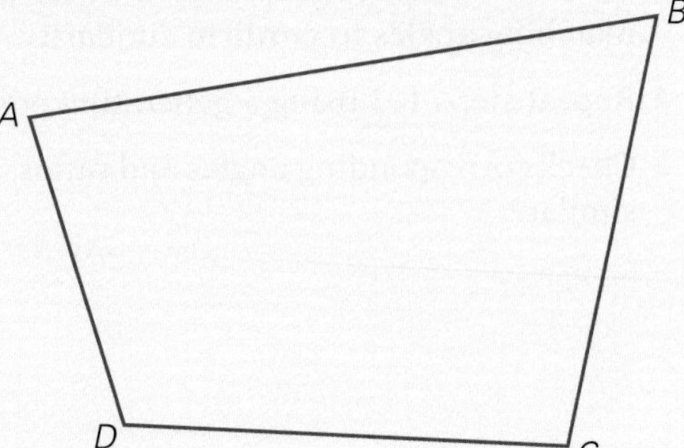

1 *ABCD* and *MPQR* are similar quadrilaterals. Trace *MPQR*, and cut out, and match angles. (Or use a protractor to measure them.)
Check ∠*A* and ∠*M*. What do you notice? Likewise, check ∠*B* and ∠*P*, ∠*C* and ∠*Q*, ∠*D* and ∠*R*. Summarise your findings.

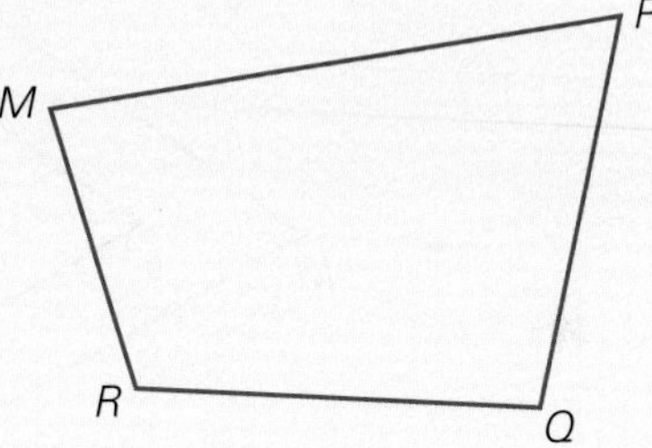

2 Measure all pairs of corresponding sides, and simplify the ratios:
$\frac{AB}{MP}$, $\frac{BC}{PQ}$, $\frac{CD}{QR}$, and $\frac{DA}{RM}$.
Summarise your findings.

B Generating similar figures

You will need: ruler, coloured pencils, protractor

1 Draw any polygon. Select a point *C* *outside* the figure.

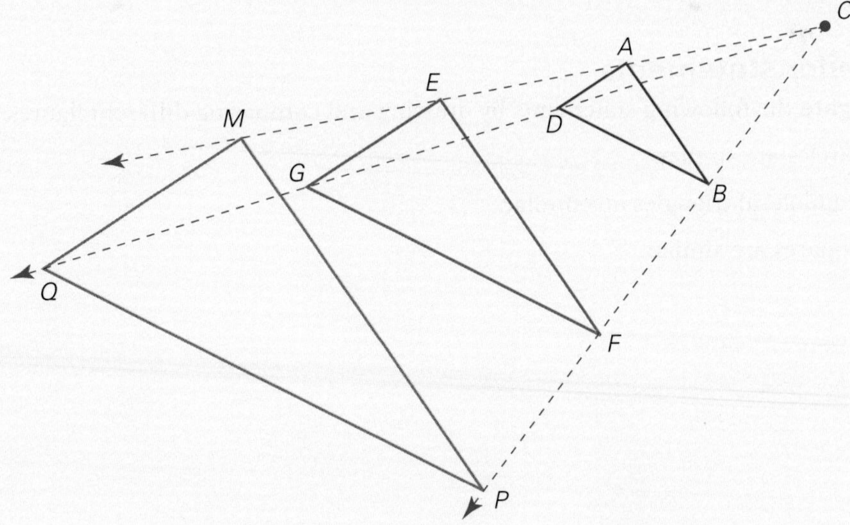

2 Draw a series of rays from *C* through the vertices of the polygon.

3 Select any point on one of the rays as a starter and draw lines parallel to the sides of the original figure (for example, ΔEFG and ΔMPQ above). All such polygons will be **similar** to the original.
Measure **i** all sides **ii** all angles. Check ratios of corresponding sides and sizes of matching angles to **confirm** similarity.

4 Repeat steps 1–3 using a generating point C *within* the original polygon.

5 Check corresponding angles and ratios of corresponding sides. Are the shapes created similar?

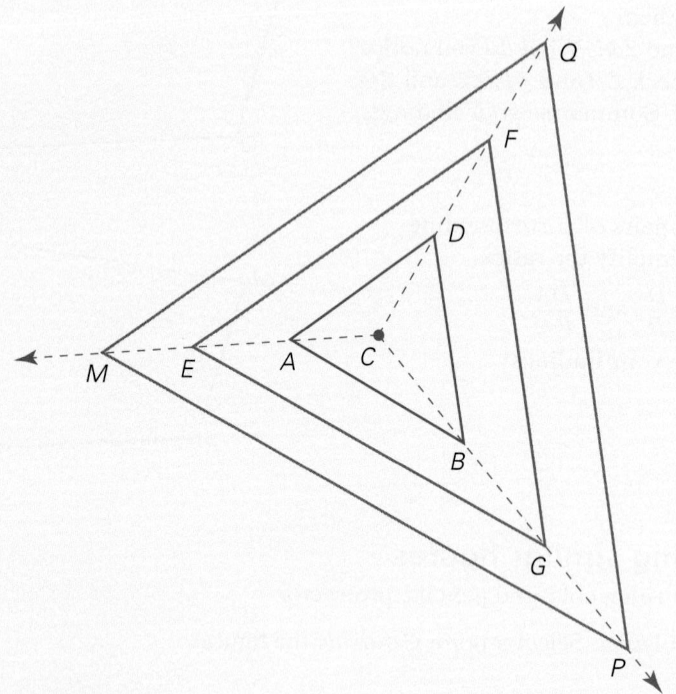

C Similar statements

Investigate the following statements by drawing and comparing different figures:

1 All circles are similar.

2 All equilateral triangles are similar.

3 All squares are similar.

14.4 Finding side lengths in similar figures

If two shapes are similar, then the corresponding sides are in proportion.

EXAMPLE 1

ABCD ||| *PQRS*. Find the value of x.

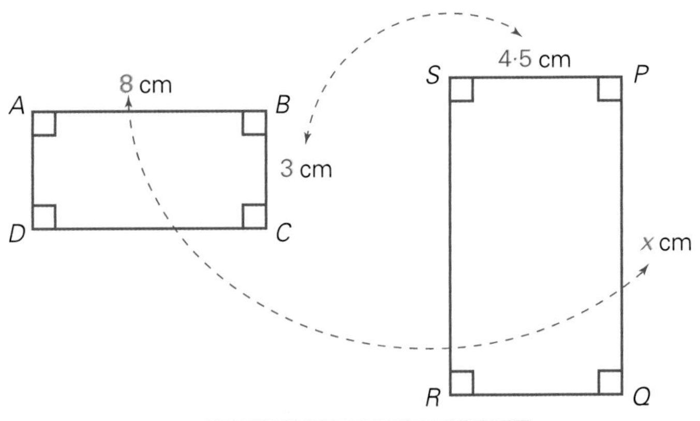

Match smallest to smallest.

Solution

$x : 8$

$4{\cdot}5 : 3$

Ratio of corresponding/matching sides $PQ \leftrightarrow AB$ and $SP \leftrightarrow BC$

i.e. $\frac{x}{8} = \frac{4{\cdot}5}{3}$ — Set as fractions.

$\therefore x = \frac{4{\cdot}5}{3} \times 8$ — Solve.

$= 12$

EXAMPLE 2

ΔABC ||| ΔMPQ. Find *MP*.

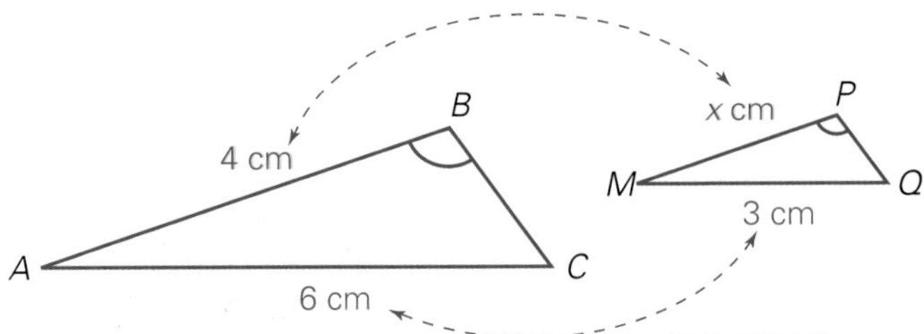

Solution

Ratio of corresponding sides *MPQ* 1st, $MP \leftrightarrow AB$ and $MQ \leftrightarrow AC$

$x : 4$

$3 : 6$

i.e. $\frac{x}{4} = \frac{3}{6}$ — Set as fractions.

$x = \frac{3}{6} \times 4$ — Solve.

$= 2$

$\therefore$ *MP* is 2 cm

EXERCISE 14D

Remember your ratio work from chapter 9.

1 Find x for the following ratios:

a $x : 2 = 4 : 8$ **b** $x : 7 = 5 : 8$ **c** $32 : 48 = x : 3$
d $1 : 5 = 4 : x$ **e** $3 : 5 = x : 7{\cdot}5$ **f** $9 : 5 = x : 2{\cdot}5$

2 Given $ABCD$ ||| $PQRS$, find the value of x.

a

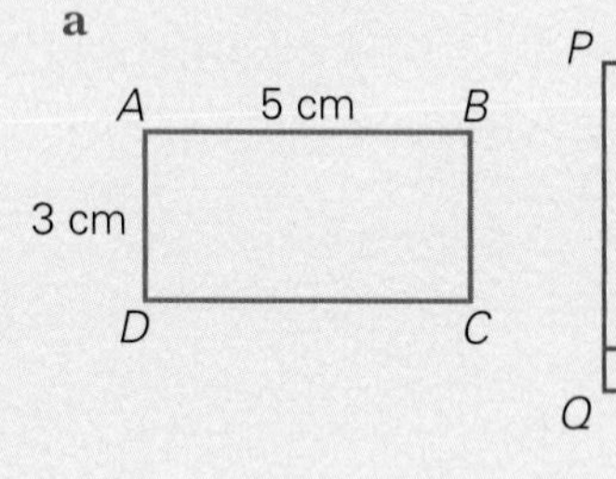

b

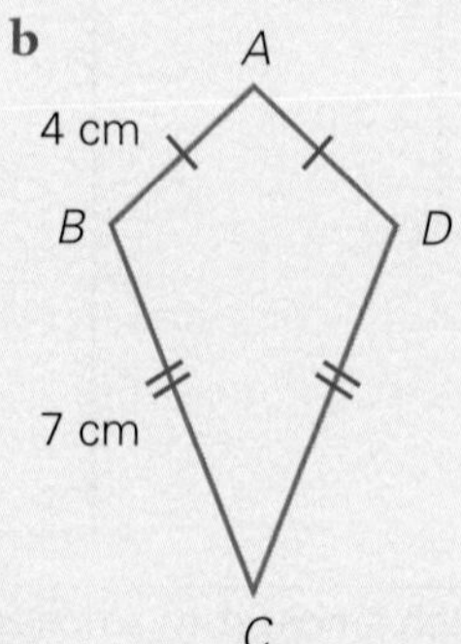

3 Given ΔABC ||| ΔXYZ, find the value of x.

a

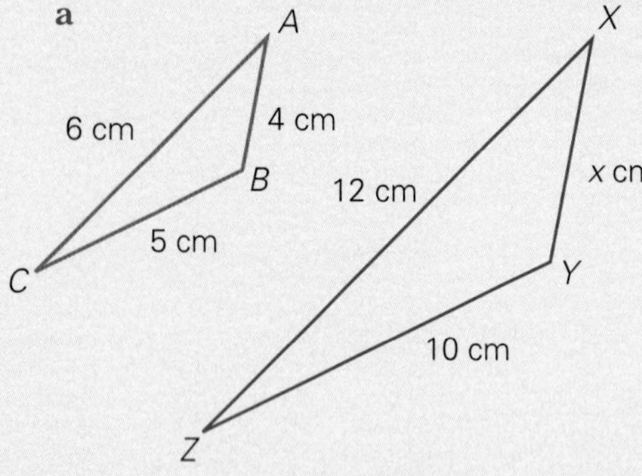

b

A
8 cm
10 cm
B
6 cm
C
X
25 cm
x cm
Z
Y

4 Sam is 1·5 metres tall and casts a 3-metre shadow. At the same time, a tree's shadow is 10 metes long. How high is the tree? (Note that similar triangles result.)

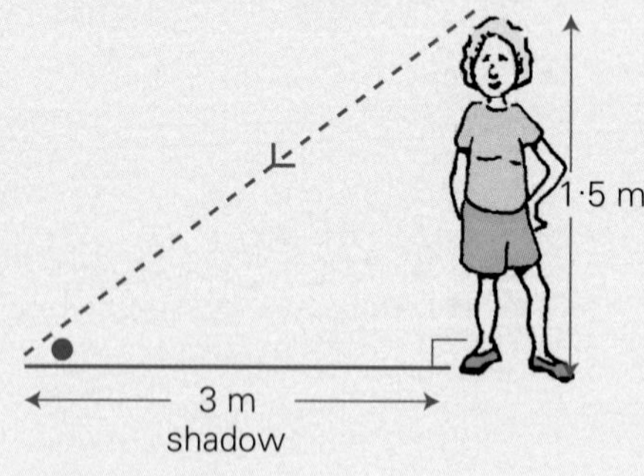

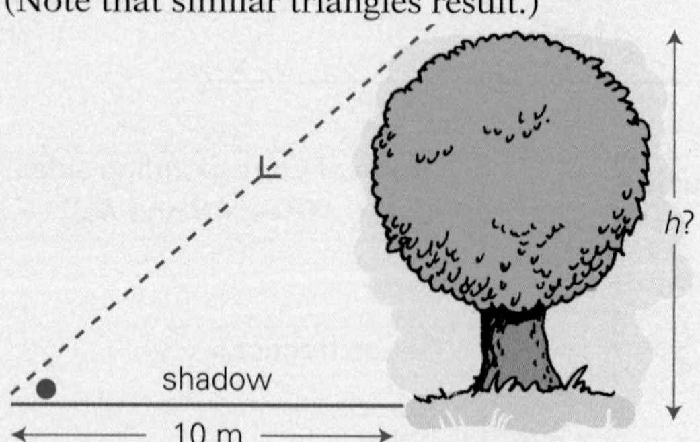

Level 2

5 Find the length of each of each unknown side in the following pairs of similar triangles:

a

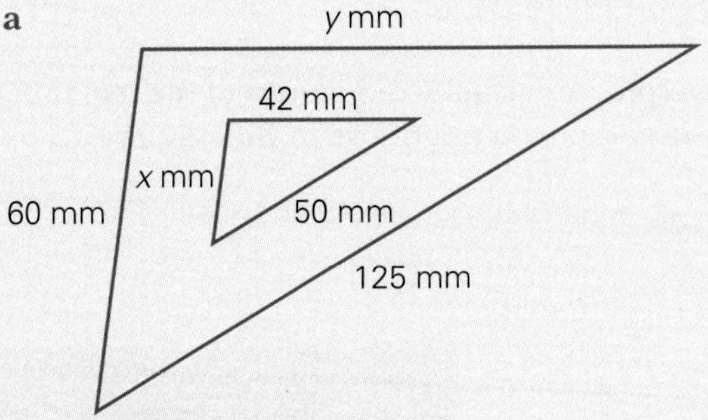

y mm
42 mm
x mm
60 mm
50 mm
125 mm

b

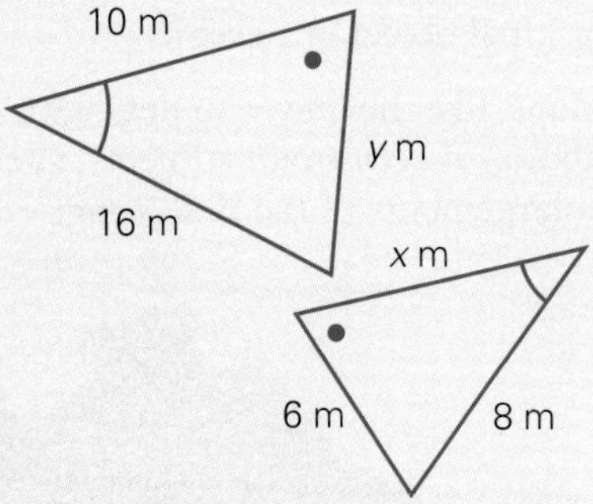

10 m
y m
16 m
x m
6 m
8 m

c

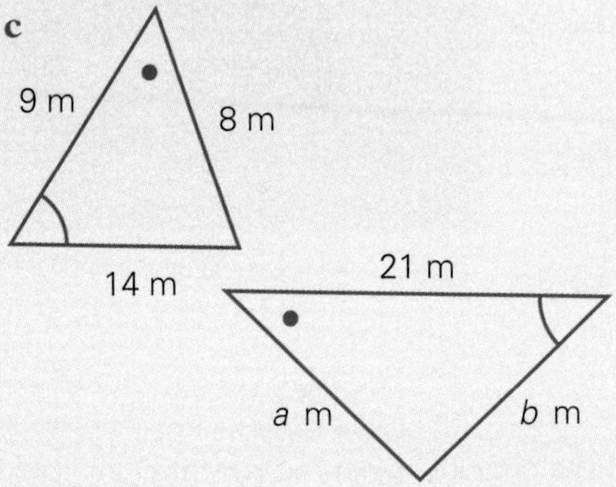

9 m
8 m
14 m
21 m
a m
b m

d

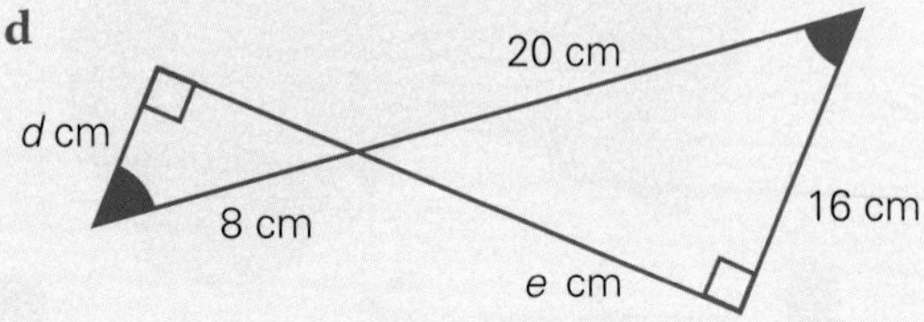

20 cm
d cm
8 cm
16 cm
e cm

e

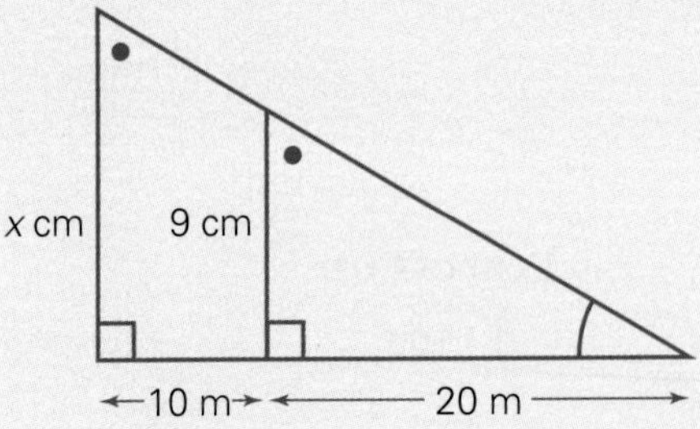

x cm
9 cm
10 m
20 m

f

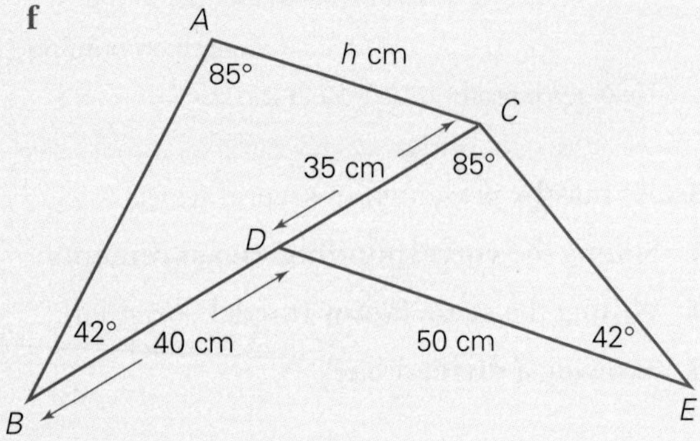

A
h cm
85°
C
35 cm
85°
D
42°
40 cm
50 cm
42°
B
E

6

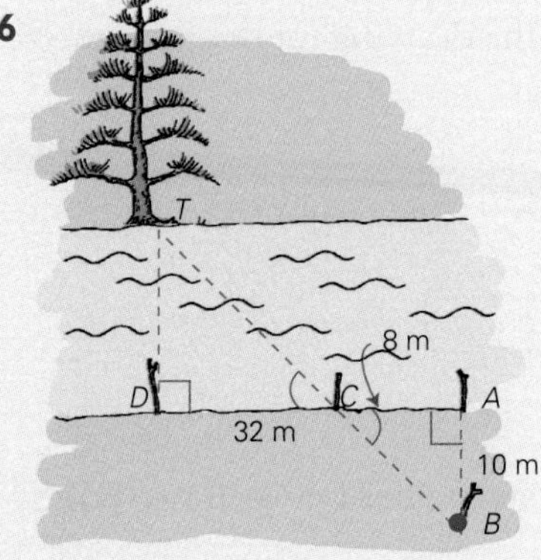

Megan's aim was to measure the width of the river. She placed a stick at A, walked back 10 metres at right angles to the river bank and set another stick at B.
She then sighted across the river to a prominent tree (T). She placed a third stick at C on the bank along that line of sight.
Finally, she located a stick at D, directly opposite the tree. She then took the other measurements shown on the diagram.
Use the fact that $\Delta TDC \ ||| \ \Delta BAC$ to find the width of the river (TD).

14.5 Scale drawings

Scale drawings are used by a great many professions including engineers, architects, builders and car designers.

A scale diagram is a drawing or picture of a real object or person. They are **similar** to the original object or person.

Maps, like the ones you deal with in Geography, are usually reductions of the real thing and photos of very small subjects, such as insects that you might use in Biology, are enlargements of the real thing.

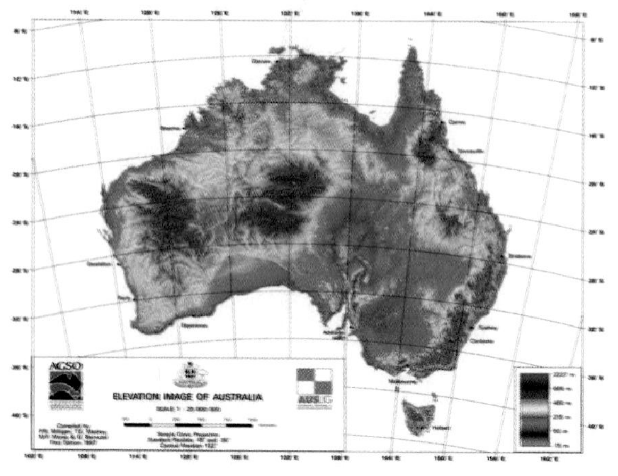

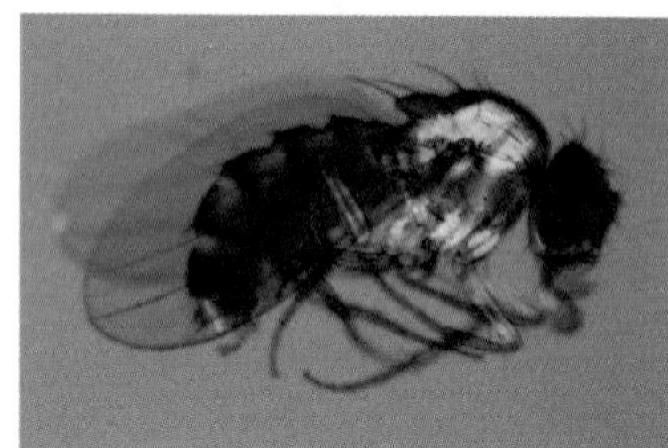

A scale drawing has exactly the same shape as the original object but is a different size.

The **ratio** is known as the **scale** factor and is:

length on drawing : real length

A scale factor is usually in the form of 1 : x.

Scales may be presented in several ways:

1 Stating the **corresponding measurements**: 1 cm to 1 m

2 Writing the **scale factor** (model : original): 1 : 100

3 Showing a **divided bar**:

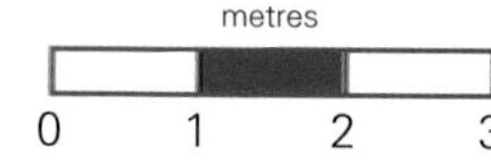

All methods give the same information: 1 centimetre on the drawing represents 1 metre on the real object.

EXAMPLE 1

Express the scale 1 cm to 2 m as a ratio in simplest form.

Solution

1 cm to 2 m = 1 cm : 200 cm — Convert 2 m to 200 cm.

= 1 : 200 — Simplify.

EXAMPLE 2

If the scale factor of a drawing is 1 : 10 000, what real distance lies between two points 4·8 cm apart on the drawing?

Solution

$$\begin{array}{rcl} & \text{scale} : \text{real} & \\ \times 4{\cdot}8 & 1 : 10\,000 & \times 4{\cdot}8 \\ & 4{\cdot}8 : x & \end{array}$$

$\therefore$ real distance $= 4{\cdot}8 \times 10\,000$
$= 48\,000$ cm
$= 480$ m

Express with a realistic unit.

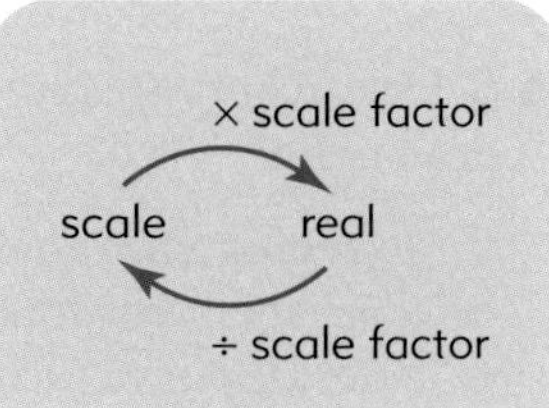

EXAMPLE 3

This is a scale drawing of a small car. Use a ruler and the given scale to calculate:

a total length
b total height

scale 1 : 60

Solution

a Diagram length = 6·4 cm
Real length = 6·4 cm × 60
= 384 cm
= 3·84 m

Measure length.

scale : real
1 : 60
6·4 : 60 × 6·4

b Diagram height = 2·3 cm
Actual height = 2·3 cm × 60
= 138 cm
= 1·38 m

Measure length.

Use ratio 1 : 60.

EXERCISE 14E

1 Express each scale in simplest ratio form:

a 1 cm represents 2 m
b 1 cm stands for 10 m
c 1 cm to 5 km
d 1 mm to 10 km
e 5 mm to 1 m
f 2 cm to 1 km

2 If the scale is 1 : 100, what actual length is represented by:

a 2 cm? b 5 cm? c 6·8 cm? d 8 mm?

3 If the scale is 1 : 80, what real length is represented by:

a 3 cm? b 7 mm? c 5·2 cm? d 8·9 mm?

4 For a scale of 1 cm to 10 m, find the *drawing* lengths which show real distances of:

a 20 m b 85 m c 5 m d 7680 mm

5 Use the scale drawing of the car in example 3 to find:

a length of the wheelbase
b diameter of a wheel
c horizontal distance from the top of the steering wheel to the driver's headrest

TEACHER

6 Select an appropriate scale and make a scale drawing of:

a a rectangular tennis court 24 m × 11 m
b a baseball diamond with sides of 18 m
c a desktop measuring 1200 mm by 600 mm
d a triangular sail measuring 5·8 m by 6·5 m by 3·2 m

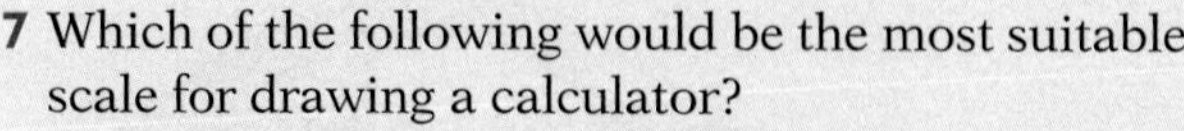

7 Which of the following would be the most suitable scale for drawing a calculator?

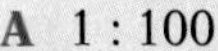

A 1 : 100 **B** 1 : 10 **C** 1 : 2 **D** 1 : 1

8 A 1 : 50 scale model of a 75-metre boat will be how long?

9 A 1 : 100 000 map shows two creeks 3·6 cm apart. How far apart are they really?

Level 2

10 If the real car is 5·95 m long and the scale drawing shows the corresponding length as 119 mm, what scale has been used.

11 This tree is drawn $\frac{1}{120}$ of actual size.

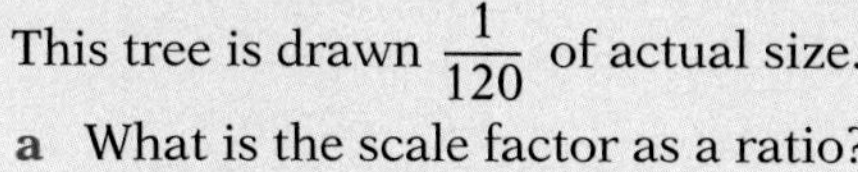

a What is the scale factor as a ratio?
b Use your ruler to find:
  i the real height of the tree
  ii the diameter of a trunk at its base

12 This house plan has been drawn to a scale of 1 : 250. Find:

a the dimensions of the double carport
b the cost of carpeting bedrooms 2 and 3 at \$126/m² fully laid (not including 'robes')
c the cost of tiling all the shaded areas shown at \$87.50/m² fully laid

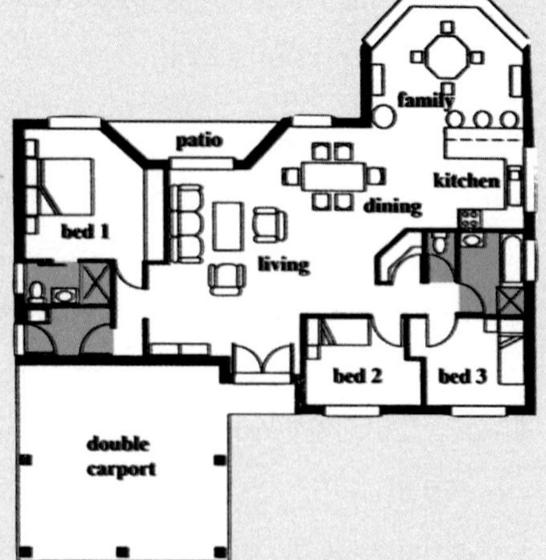

13 A scale is given as 3 : 2000.

First write the scale as 1 : *x*.

a Find the scale length given a real distance of:
  i 400 m **ii** 1 km **iii** 6 km **iv** 3000 m
b Find the real length represented by a length on the scale drawing of:
  i 6 cm **ii** 6 mm **iii** 1·2 cm **iv** 9 cm

Chapter Review

Language Links

congruent	enlargement	original	scale
corresponding	image	ratio	similar
drawing	matching	reduction	

1 Use every word in the list above to complete this paragraph:

A _____ drawing is an _____ which is either an _____ or a _____ of the _____. Figures which are _____ have _____ angles equal and corresponding sides in the same _____ .

If the _____ is one to one, then the _____ figures are also _____ and the _____ sides and angles are equal.

2 Explain the meaning of corresponding sides and angles. A diagram may help in your explanation.

3 **Explain** the difference in meaning between these two symbols: ||| and ≡.

4 List three or more areas where similar figures and scale drawings are used. **Explain** the benefits.

5 Write out the **procedure** required for each of the following shapes to be drawn.

A **procedure** should have a statement of goal followed by a series of steps; it should also include technical language.

a

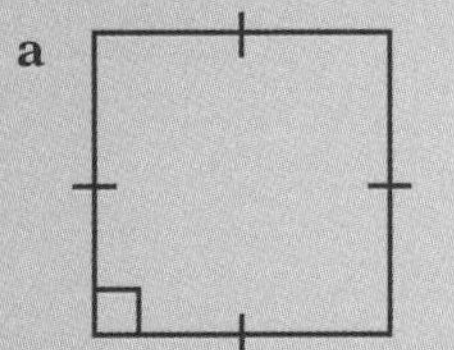

b

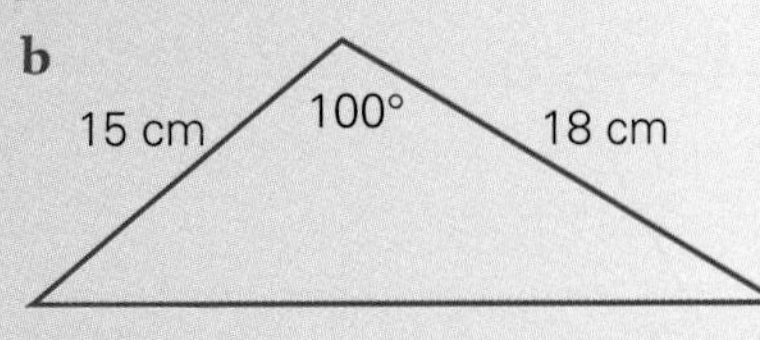

6 a Make up a set of instructions for drawing any pair of similar figures.
b Give these instructions to a classmate to complete and check that their diagrams are correct.

Chapter Review Exercises

1 Construct the following triangles: 14.1

a

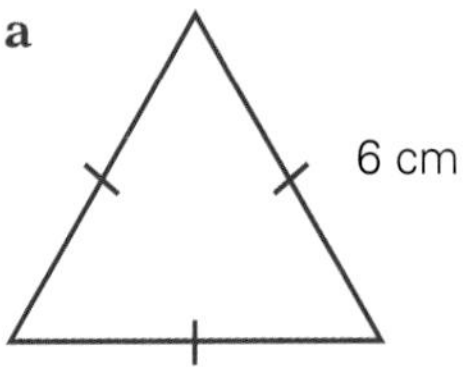

b

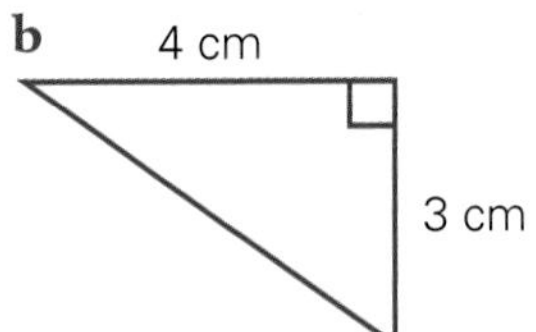

c

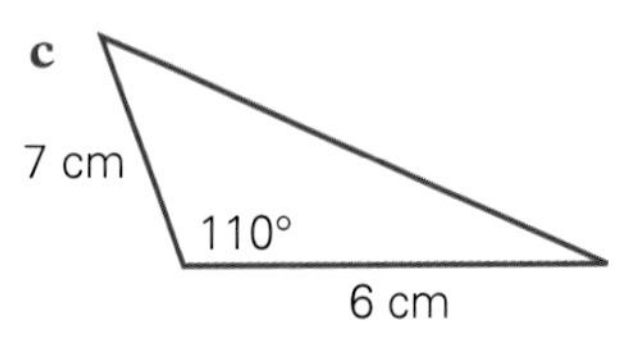

14.1 **2** Is it possible to construct a triangle with sides 4 cm, 7 cm and 12 cm?

14.2 **3 a** Use measurement to show that these two triangles are congruent.

b Complete: $\Delta ABC \equiv \Delta$ ___ .

c Name the side equal in length to PR.

d Name the angle equal in size to $\angle BCA$.

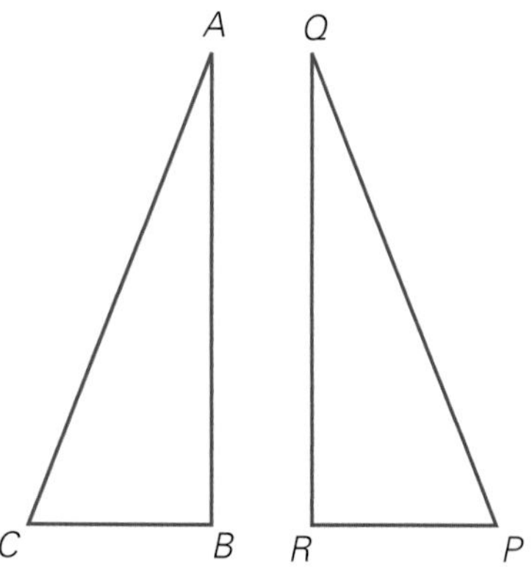

14.2 **4** The following two triangles are congruent. Copy and complete these statements:

a $DT =$ ______ **b** $PY =$ ______

c $TE =$ ______ **d** $\angle TDE =$ ______

e $\angle TED =$ ______ **f** $\angle DTE =$ ______

g ΔDTE ______ Δ ______

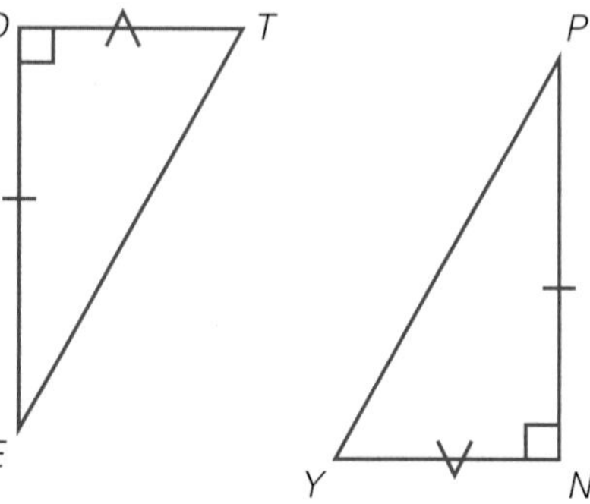

14.2 **5** Given $\Delta MPR \equiv \Delta QPR$, find the value of each pronumeral.

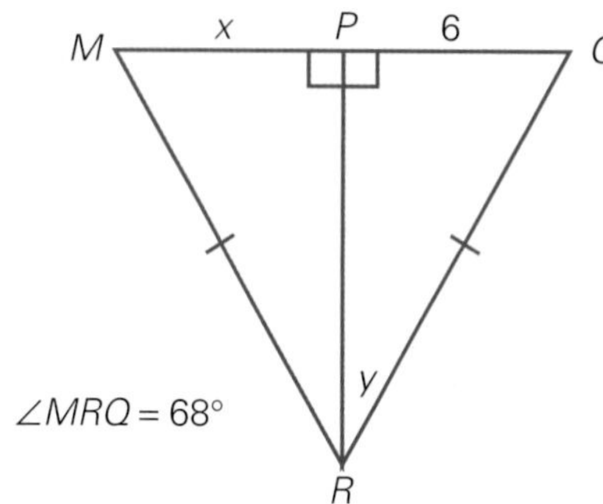

14.3 **6** Name the pairs of corresponding sides and angles in each set of similar figures:

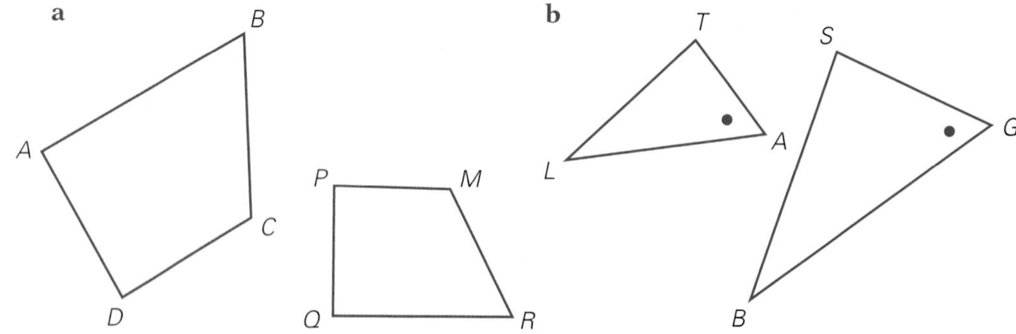

7 Construct a pair of similar parallelograms. 14.3

8 Find the scale factor for the following pairs of similar figures: 14.4

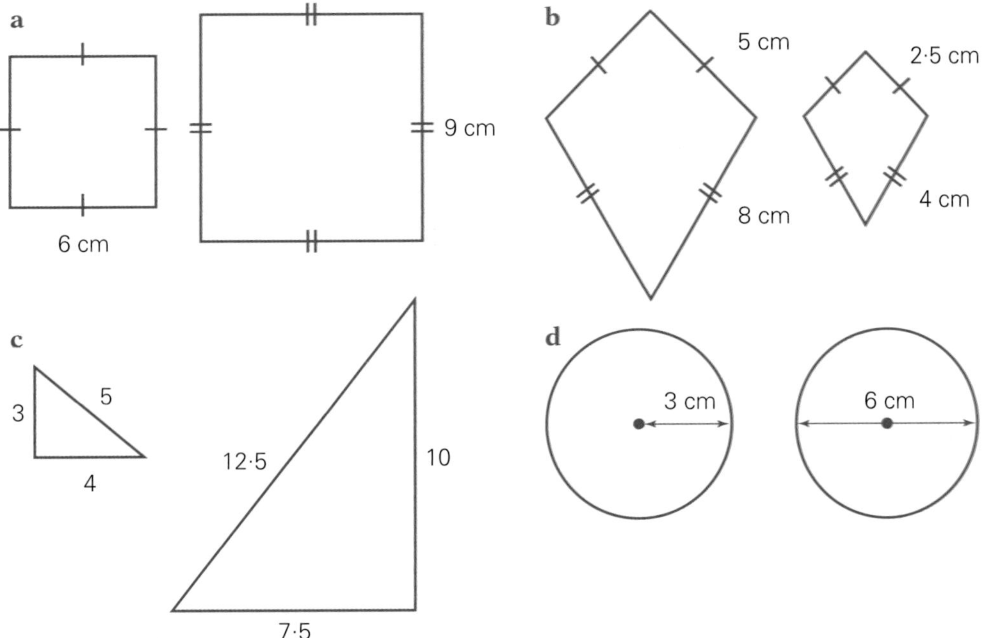

9 Find the value of the pronumerals, given $\Delta ABC \parallel\!| \Delta PQR$. 14.4

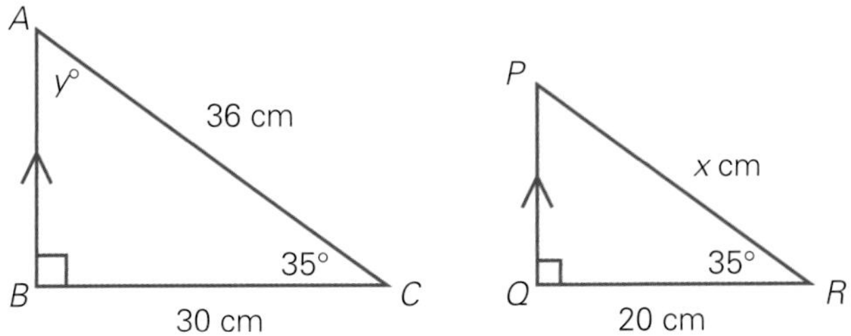

10 Write each scale in simplest ratio form: 14.5

a 1 mm to 2 m

b kilometres

0 10 20 30

11 The scale factor for a drawing is 1 : 50. 14.5

a What is the real length of a line that is shown in the drawing as 3·5 cm long?
b What length in the drawing would stand for a real length of 4 metres?

12 

The real shark represented in this scale diagram is 5·6 m long. Find: 14.5

a the scale used
b the actual height of the dorsal fin

Keeping Mathematically Fit

Part A: Non-calculator

1 Write down the value of $\frac{3}{10\,000} + \frac{5}{100}$.

2 Write 0·07, 0·007, $0{\cdot}0\dot{7}$ and $0{\cdot}00\dot{7}$ in ascending order.

3 Write $11\frac{3}{4}\%$ as a decimal.

4 Estimate the size of $\angle ABC$.

A

B C

5 At a family dinner there are five adults, four girls and six boys.
If one person is chosen at random to carve the roast, find the probability that the person chosen is an adult.

6 What are the coordinates of the point where the line $2x + y = 8$ cuts the x-axis?

7 Simplify $^{-}x - (x - 6)$.

8 N

2 500 000 2 700 000 What is the value of N?

9 Increase 180 by 10%.

10 Write an expression for the perimeter of ΔABC.

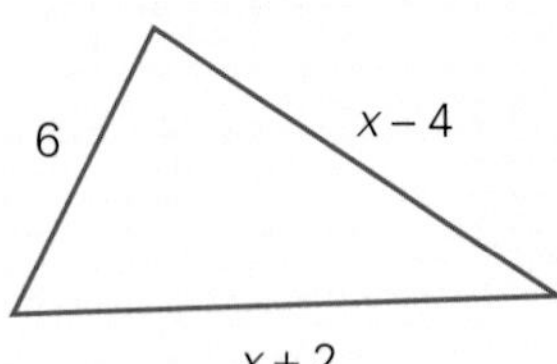

Part B: Calculator

1

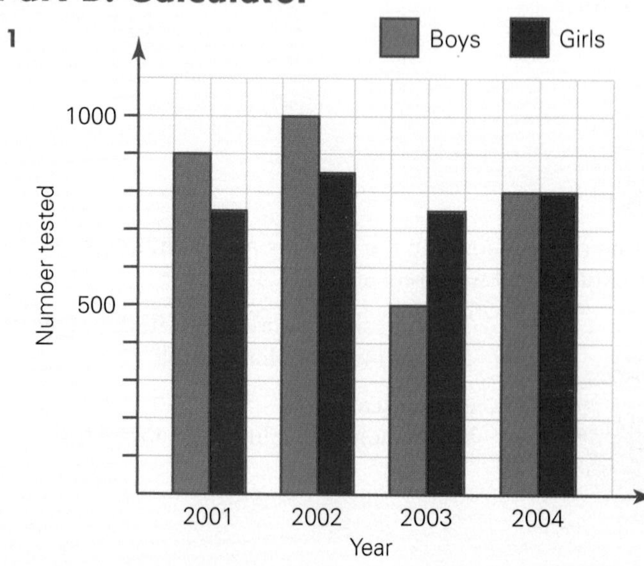

a How many boys were tested in 2003?

b What was the total number of students tested in 2002?

c In which year/s did the number of boys tested equal two-thirds the number of girls tested?

d How many students, altogether, were tested in the 4 years shown?

2 Find the area (answer to 2 decimal places):

a

4 cm

b

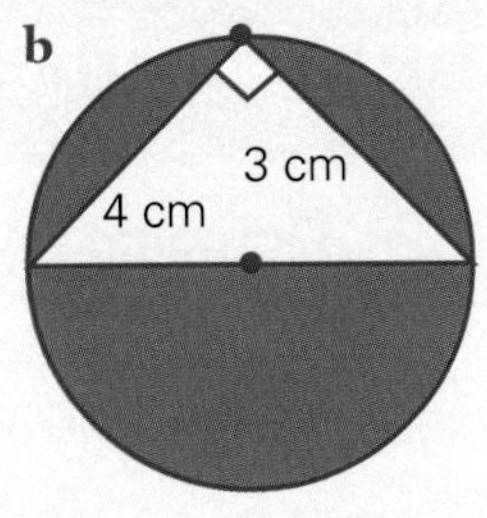

c

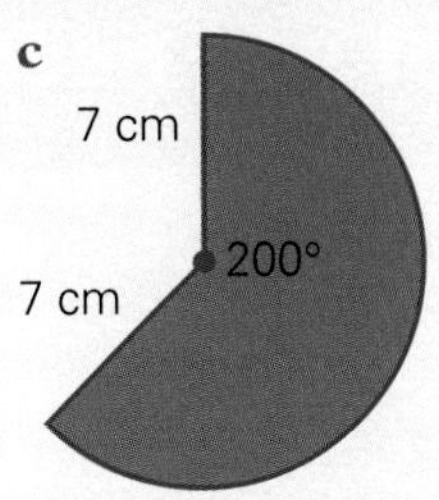

3

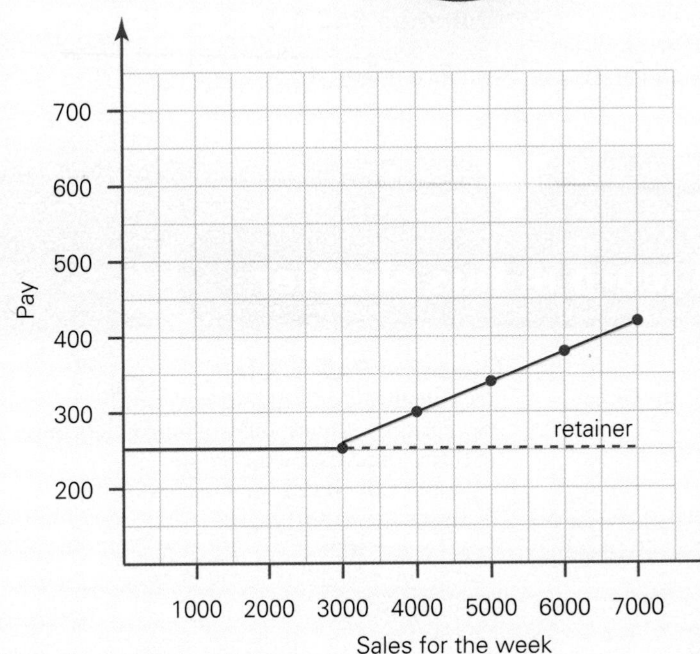

The graph shows the income for a given week related to the total sales generated.

a In a week where no sales are generated, what does the pay equal?

b How much is earned in a week when sales totalled \$6000?

c Extend the graph to find out the value of the sales required to earn \$500 p.w.

4 If $A = \pi r^2$, find the value of r, given $A = 34{\cdot}6\ \text{cm}^2$ (answer to 2 decimal places).

5 Solve:

a $3 - 4x = x + 9$ **b** $3(x + 7) = 2x - 3$ **c** $5x + 2 \geq 32$

6 A car travelling a constant 72 km/h leaves Sydney at 11:52 am. If the trip is 342 km, at what time does it arrive?

7 Give the coordinates of the point where the curve $y = 4 - 2x$ cuts the line $y = 6$.

8 **a** Find a set of six scores which have a mode of 9, a median of 6 and a range of 6.
b Find the mean of your six scores.
c Is there more than one set of scores? Are their means different?

TEACHER

1 What you need to know and revise

Outcome PAS 4.1:
Uses letters to represent numbers and translates between words and algebraic symbols

Outcome PAS 4.2:
Creates, records, analyses and generalises number patterns using words and algebraic symbols in a variety of ways

Outcome PAS 4.3:
Uses the algebraic symbol system to implify, expand and factorise simple algebraic expressions

Outcome PAS 4.4:
Uses algebraic techniques to solve linear equations and simple inequalities

2 What you will learn in this chapter

Outcome PAS 5.1.1:
Applies index laws to simplify algebraic expressions

Outcome PAS 5.2.1:
Simplifies, expands and factorises alebraic expressions involving fractions and negative and fractional indices

Outcome PAS 5.2.2:
Solves linear and simple quadratic equations, solves linear inequalities and solves simultaneous equations using graphical and analytical methods

Outcome PAS 5.3.1:
Uses algebraic techniques to simplify expressions, expand binomial products and factorise quadratic expressions

- applies the index laws using algebraic bases
  $a^m \times a^n = a^{m+n}$
  $a^m \div a^n = a^{m-n}$
  $(a^m)^n = a^{m \times n}$
  $a^0 = 1$
- expands binomial products
- solves equations involving fractions
- uses function notation

Working Mathematically outcomes WMS 4.1–4.5:
Students will be asked to ***question***, ***apply strategies***, ***communicate***, ***reason*** and ***reflect*** in the sections of this chapter.

Algebra II

Key mathematical terms you will encounter

algebraic expressions	factorise	index
	function notation	reciprocal

MathsCheck
Algebra II

1 Simplify:

a $3a + 3b - a$	**b** $5x + 8y - 12x$	**c** $a + b + a + b$	**d** $4a \times 5b$
e $24a - 30a$	**f** $3 + 9w + 10$	**g** $3a \div 3$	**h** $3 \div 3a$
i $4^2 \div 4$	**j** $6^5 \times 6^2$	**k** $7^8 \div 7^5$	**l** $(4^3)^5$
m $(2^3)^2 \div 2^4$	**n** $3^4 \times 2^7 \times 3^3$	**o** $24ab \div 12a$	**p** $12a \div 24ab$

2 Expand and simplify where necessary:

a $3(a + b)$	**b** $5(a - b)$	**c** $4(2w + 3)$
d $5(a - 4)$	**e** $3a + 2(a + 3)$	**f** $4(4d - 1) + 2d$
g $10 - (4 - 2x)$	**h** $6(4 - 2x) + 2x - 5$	**i** $5(a + 4b - 5)$
j $2(x + 3) + 4(x + 1)$	**k** $12p - (p^2 + 3p)$	**l** $5(x - 9) - 2(3x + 1)$

3 Factorise each of the following:

a $xy + 5x$	**b** $xy - 5y$	**c** $3ab + 12$	**d** $3ab + 12a$
e $24 - 36a$	**f** $16ab + 32a^2$	**g** $^-ab + a$	**h** $^-15xy^2 - 20xy$

4 Solve:

a $\frac{x}{3} = 9$	**b** $2x + 1 = 19$	**c** $x - 3 = 7$	**d** $3 - x = 7$
e $4 - 2x = 9$	**f** $\frac{a}{5} - 5 = 9$	**g** $5(2x + 7) = 35$	**h** $\frac{2x}{3} - 11 = 9$
i $3x = x + 9$	**j** $5s + 12 = 8 - 2s$	**k** $4 - \frac{x}{5} = 0$	**l** $5 - 2x = 10 + 6x$

5 Find the value of x in:

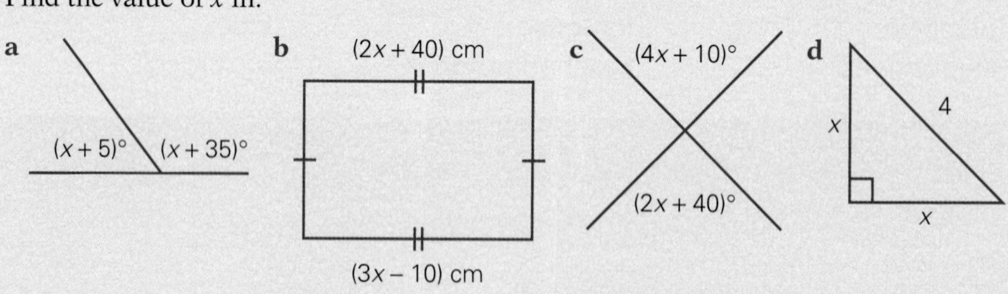

6 Without a calculator, evaluate:

a $1\frac{1}{2} + \frac{3}{4}$	**b** $\frac{3}{5} - \frac{1}{2}$	**c** $1\frac{2}{3} + 1\frac{1}{4}$	**d** $\frac{3}{5}$ of 20
e $\frac{3}{4} \times \frac{8}{9}$	**f** $1\frac{1}{2} \times \frac{4}{5}$	**g** $\frac{3}{5} \div \frac{1}{6}$	**h** $1\frac{1}{2} \div \frac{3}{8}$

7 Simplify:

a $\frac{4x}{5} - \frac{x}{5}$	**b** $\frac{7a}{10} - \frac{a}{5}$	**c** $\frac{x}{8} + \frac{3x}{4}$	**d** $\frac{7x}{10} - \frac{x}{3}$

8 Simplify:

a $\frac{4}{x} \times \frac{x}{5}$	**b** $\frac{ab}{7} \times \frac{2}{a}$	**c** $\frac{3a + 9}{3}$	**d** $\frac{a^2}{6} \div \frac{a}{12}$

TEACHER

Exploring New Ideas

15.1 Algebraic expressions and the index laws

Index notation can be used when the factor repeated is a pronumeral or a number.

$$4^5 = 4 \times 4 \times 4 \times 4 \times 4$$
$$x^5 = x \times x \times x \times x \times x$$
$$a^2b^4 = a \times a \times b \times b \times b \times b$$

WEB RESEARCH

The index laws looked at in chapter 1 can now be used to help simplify algebraic expressions.

THE FIRST AND SECOND INDEX LAWS

The first index law:

$$a^m \times a^n = a^{m+n}$$

When two numbers in index form with the same base are multiplied, the powers are added.

$$x^2 = x \times x$$
$$x^3 = x \times x \times x$$
$$\therefore x^2 \times x^3 = x \times x \times x \times x \times x$$
$$= x^5 \ (x^{2+3} = x^5)$$

The second index law:

$$a^m \div a^n = a^{m-n}$$

When numbers in index form with the same base are divided, the powers are subtracted.

$$a^5 = a \times a \times a \times a \times a$$
$$a^3 = a \times a \times a$$
$$\therefore a^5 \div a^3 = \frac{a^5}{a^3}$$
$$= \frac{a \times a \times a \times a \times a}{a \times a \times a}$$
$$= a \times a$$
$$= a^2 \ (a^{5-3} = a^2)$$

EXAMPLE 1

Write in index form:

a $x \times x \times x \times x \times x$ **b** $2 \times a \times a \times 3 \times b \times b \times b \times b$

Solution

a $x \times x \times x \times x \times x = x^5$

b $2 \times a \times a \times 3 \times b \times b \times b \times b = 2 \times 3 \times a \times a \times b \times b \times b \times b$
$= 6a^2b^4$

EXAMPLE 2

Simplify:

a $5^7 \times 5^3$ **b** $x^4 \times x^3$ **c** $4x^6y^6 \times 3x^2y$

Solution

a $5^7 \times 5^3 = 5^{7+3}$
$= 5^{10}$

b $x^4 \times x^3 = x^{4+3}$
$= x^7$

c $4x^6y^5 \times 3x^2y = 4 \times 3 \times x^6 \times x^2 \times y^5 \times y$
$= 12x^{6+2}y^{5+1}$
$= 12x^8y^6$

EXAMPLE 3

Simplify:

a $7^5 \div 7^4$ **b** $a^9 \div a^2$ **c** $10x^7y^4 \div 12x^3y^2$

Solution

a $7^5 \div 7^4 = 7^{5-4}$
$= 7^1$
$= 7$

b $a^9 \div a^2 = a^{9-2}$
$= a^7$

c $10x^7y^4 \div 12x^3y^2 = \dfrac{10x^7y^4}{12x^3y^2}$
$= \dfrac{5x^{7-3}y^{4-2}}{6}$
$= \dfrac{5x^4y^2}{6}$

WEB RESEARCH

SPREADSHEET ACTIVITY

EXERCISE 15A

1 Write the following in index form:

a $a \times a \times a \times a \times a \times a$ **b** $x \times x \times x \times x \times x \times y$ **c** $2x \times 2x \times 2x$
d $a \times b \times b \times a \times b \times b \times b$ **e** $5 \times w \times w \times p \times w \times 2 \times p$ **f** $(pq) \times (pq) \times (pq) \times (pq)$

2 Write the following in expanded form, without the use of indices:

a a^5 **b** p^3q^5 **c** $4a^7$ **d** $(3a)^5$

3 Simplify by writing each of the following in index notation:

a $3^2 \times 3^2$ **b** $2^3 \times 2^5$ **c** $7^3 \times 7^4$
d 5×5^{12} **e** $10^4 \times 10^5 \times 10^2$ **f** $3^4 \times 5^2 \times 3^2 \times 5^4$

4 Simplify these products:

a $g^4 \times g^3$ **b** $x^3 \times x^4$ **c** $a^5 \times a^4$ **d** $d^7 \times d^5$ **e** $x^5 \times x^8$
f $a^4 \times a^6$ **g** $s^7 \times s^2$ **h** $a^4 \times a$ **i** $n^6 \times n^2$ **j** $m^4 \times m^2$
k $n^3 \times n^2 \times n^3$ **l** $a^5 \times a^5$ **m** $d^5 \times d$ **n** $p^4 \times q^2$ **o** $n^6 \times n \times m$

5 Simplify by writing the following in index form:

a $6^3 \div 6^2$ **b** $5^{10} \div 5^4$ **c** $3^6 \div 3^2$ **d** $10^7 \div 10^4$ **e** $5^{12} \div 5^7 \div 5$

6 Simplify where possible:

a $x^{10} \div x^3$ **b** $a^2 \div a^2$ **c** $m^6 \div m^4$ **d** $m^3 \div m^2$ **e** $p^5 \div p^2$
f $p^5 \div p$ **g** $t^5 \div t^4$ **h** $t^5 \div t^5$ **i** $x^{15} \div x^8$ **j** $a^5 \div a^5$
k $w^{12} \div w^3 \div w^2$ **l** $n^6 \div n^5 \div n$ **m** $y^{12} \div y^{11}$ **n** $w^{10} \div w^4 \div w^3$ **o** $h^4 \div h \div h$

Level 2

7 Simplify:

a $3n^4 \times n^5$ **b** $7a^3a^7$ **c** $b^3 \times 6b$
d $3c^2 \times 5c^4$ **e** $6p^4 \times 2p^3$ **f** $8k^2 \times 3k$
g $3t \times 3t$ **h** $a^2b^3 \times a^4$ **i** $mp^2 \times m^3p^2$

8 Simplify:

a $\frac{a^5}{a^2}$ b $\frac{4x^7}{x^3}$ c $\frac{9c^6}{3c^2}$

d $\frac{8d^9}{12d^4}$ e $\frac{h^3 \times h^2}{h^4}$ f $\frac{a^2b^3}{ab}$

g $n^4r^2 \div n^3r$ h $\frac{24x^7a^3}{8a^2x^4}$ i $42b^3t^5 \div 12tb^5$

9 Copy and complete:

a $12n^2 = 6n \times$ ______ b $18p^7 = 2p^3 \times$ ______ c $56x^{12} = 8x^5 \times$ ______

10 True or false?

a $2^3 \div 2^3 = 2^0 = 1$ b $0 \div 2p^4 = 0$ c $(^-9)^3 \div (^-9)^2 = {^-9}$

Would this be true for *any* base?

15.2 Third and fourth index laws

The third index law:

$$(a^m)^n = a^{m \times n}$$

When a number in index form is raised to a power, the powers are multiplied.

$$\begin{aligned}(x^4)^5 &= x^4 \times x^4 \times x^4 \times x^4 \times x^4 \\ &= x^{4+4+4+4+4} \\ &= x^{4 \times 5} \\ &= x^{20}\end{aligned}$$

The fourth index law:

$$a^0 = 1$$

Any number (except zero) raised to the power of zero is 1.

$$\begin{aligned}a^4 \div a^4 &= a^{4-4} \\ &= a^0\end{aligned} \qquad \text{also} \qquad \begin{aligned}a^4 \div a^4 &= \frac{a^4}{a^4} \\ &= 1\end{aligned}$$

$$\therefore a^0 = 1$$

EXAMPLE 1

Remove the grouping symbols in:

a $(5^3)^2$ b $(x^7)^3$ c $(x^6y^2)^5$ d $(4x^9)^2$

Solution

a $(5^3)^2 = 5^{3 \times 2} = 5^6$

b $(x^7)^3 = x^{7 \times 3} = x^{21}$

c $(x^6y^2)^5 = (x^6)^5 \times (y^2)^5 = x^{30} \times y^{10} = x^{30}y^{10}$

d $(4x^9)^2 = 4^2 \times (x^9)^2 = 16x^{18}$

EXAMPLE 2

Simplify each of the following:

a 5^0 b p^0 c $4 + 3a^0$ d $(a + b - c)^0$

Solution

a $5^0 = 1$

b $p^0 = 1$

c $4 + 3a^0 = 4 + 3 \times a^0$
$= 4 + 3 \times 1$
$= 4 + 3$
$= 7$

d $(a + b - c)^0 = 1$

EXERCISE 15B

1 Apply the index laws to simplify each expression:

a $(2^3)^4$ b $(4^3)^3$ c $(a^2)^4$
d $(m^5)^2$ e $(p^3)^6$ f $(n^0)^5$
g $(a^3)^6$ h $(x^4)^7$ i $(n^6)^4$
j $(a^5)^5$ k $(p^6)^2$ l $(p^2)^6$

2 Simplify:

a 4^0 b 8^0 c a^0
d $(xy)^0$ e $2 + x^0$ f $a^0 + b$
g $5 + n^0$ h $5 + 3n^0$ i $(3 + 2m)^0$

Level 2

3 Simplify each of the following by removing the grouping symbols for:

a $(a^7)^5$ b $(3a^2)^2$ c $(4x^3)^3$
d $(3d)^4$ e $(5x^3y^2)^2$ f $({}^-2x^4)^2$
g $(5a^5b^3)^3$ h $(a^2b^3c)^4$ i $(5x^7y)^6$
j $(3 \times 5)^4$ k $(7 \times 6)^3$ l $(4 \times 7)^2$
m $(4m^6n)^2$ n $(7a^3b^4)^4$ o $(9a^5b^7)^2$

4 Simplify each of the following:

a $(3^2)^0 \times 3^4$ b $(8^0)^4 \times 8^3$ c $(4^5)^3 \div 4^{10}$
d $(4^3)^2 \div (4^2)^3$ e $(a^3)^4 \div (a^2)^5$ f $(a^5)^6 \times a^3$
g $p^4 \times (p^5)^2$ h $(w^3)^6 \div (w^2)^0$ i $(3x^4)^4 \div 9x^4$
j $5(b^2)^5$ k $a(b^3)^2$ l $(7d)^5 \div (d^5)^0$

5 Rewrite each of the following without brackets:

a $\left(\frac{x}{y}\right)^4$ b $\left(\frac{2}{3}\right)^2$ c $\left(\frac{4}{5}\right)^2$
d $\left(\frac{1}{3}\right)^3$ e $\left(\frac{a^2}{3}\right)^3$ f $\left(\frac{a^3}{b}\right)^3$
g $\left(\frac{xy^2}{2}\right)^2$ h $4\left(\frac{x}{y^2}\right)^2$ i $\left(\frac{4m}{2n}\right)^6$

6 Simplify each of the following:

a $(a^4b^2)^2 \div ab$ b $3m^2n^3 \times (2mn^3)^2$
c $(a^2b^4)^3 \times 5a^3b$ d $(2a^3b^2)^3 \div (b^2a)^2$
e $\dfrac{x^4y^6 \times (xy)^2}{2x^3y^2}$ f $\dfrac{(2a^3b^2)^3 \times a^4b}{(a^5b^3)^2}$

g $(w^4y^2)^6 \div w^{10}$ **h** $(m^3n^2)^4 \div (m^{12}n)^0$

i $(5x^4)^3 \div 25x^6$ **j** $\dfrac{3x^4y^2 \times (2x^3y)^2}{4x^6y^2}$

7 By using an appropriate index law, expand each of the following:

a $a^2(a^3 - 3a)$ **b** $4p(3p^5 + 2p^2)$ **c** $n^4(5 - 2n^3)$
d $x^3(7 - x^5)$ **e** $^-4p^5(p + 3p^3)$ **f** $^-w^3(2w^3 - 5w^4)$
g $4a^2b(a - b)$ **h** $3ab(a^3 - a^2b^5)$ **i** $a^3b^5(a^2b + ab^2)$
j $a^3 + a(a^2 - 3a)$ **k** $n^2(n^3 - 4n) + 6(n^3 + n^5)$ **l** $5x(2x^2 + 3x) - (x^2 + 4x)2x$

8 Fully factorise each of the following expressions:

a $x^5 - x^2$ **b** $a^3b - a^2b$ **c** $m^7 - 3m^4$
d $5p^7 - 25p^3$ **e** $x^4y^2 + x^2y^4$ **f** $p^5q^5 - p^3q$
g $4a^3 - 2a^4$ **h** $10a^4b^3 + 15a^3b^2$ **i** $m^8n - m^6n^2$

TEACHER

15.3 Binomial products

A **binomial** expression contains two terms.

$x + 2$, $4a - 9$, $ab - c$ are binomial expressions.

A **binomial product** is the multiplication of two factors, each of which are binomial expressions.

Examples are $(x + 2)(x + 5)$, $(4a - 9)(a - b)$ and $(ab - c)(ab + c)$.

Let us examine expanding $(x + 1)(x + 7)$ where $A = (x + 1)$.

$$(x + 1)(x + 7)$$
$$A(x + 7)$$
$$= A \times x + A \times 7$$
$$= x \times A + 7 \times A$$

Now we replace A with $(x + 1)$.

$$= x(x + 1) + 7(x + 1)$$

Expand.

$$= x \times x + x \times 1 + 7 \times x + 7 \times 1$$
$$= x^2 + x + 7x + 7$$

Simplify.

$$= x^2 + 8x + 7$$

We can see from the above that each term in the first bracket is multiplied by each term in the second bracket.

To expand binomial products:

$(a + b)(x + y) = a(x + y) + b(x + y)$
$= ax + ay + ba + by$

$(a + b)(x - y) = a(x - y) + b(x - y)$
$= ax - ay + bx - by$

pos × neg = neg

$(a - b)(x + y) = a(x + y) - b(x + y)$
$= ax + ay - bx - by$

neg × pos = neg

$(a - b)(x - y) = a(x - y) - b(x - y)$
$= ax - ay - bx + by$

neg × neg = pos

EXAMPLE 1

Expand each of the following binomial products:

a $(a + 5)(a + 7)$ **b** $(a - 5)(2a + 3)$ **c** $(7w - 6)(2w + 1)$

Solution

a $(a + 5)(a + 7)$
$= a(a + 7) + 5(a + 7)$
$= a^2 + 7a + 5a + 35$
$= a^2 + 12a + 35$

b $(a - 5)(2a + 3)$
$= a(2a + 3) - 5(2a + 3)$
$= 2a^2 + 3a - 10a - 15$
$= 2a^2 - 7a - 15$

c $(7w - 6)(2w + 1)$
$= 7w(2w + 1) - 6(2w + 1)$
$= 14w^2 + 7w - 12w - 6$
$= 14w^2 - 5w - 6$

EXAMPLE 2

Expand and simplify:

a $(5b - 9a)(5b + 9a)$ **b** $(3x + 4)^2$ **c** $(x - 7y)^2$

Solution

a $(5b - 9a)(5b + 9a)$
$= 5b(5b + 9a) - 9a(5b + 9a)$
$= 25b^2 + 45ab - 45ab - 81b^2$
$= 25b^2 - 81b^2$

b $(3x + 4)^2$
$= (3x + 4)(3x + 4)$
$= 3x(3x + 4) + 4(3x + 4)$
$= 9x^2 + 12x + 12x + 16$
$= 9x^2 + 24x + 16$

c $(x - 7y)^2$
$= (x - 7y)(x - 7y)$
$= x(x - 7y) - 7y(x - 7y)$
$= x^2 - 7xy - 7xy + 49y^2$
$= x^2 - 14xy + 49y^2$

EXERCISE 15C

1 Copy and complete:

a $(a + b)(c + d)$
$= a(c + d)$ ______ $(c + d)$
$= ac + ad$ ______
$=$ ____________

b $(x + 8)(x + 3)$
$= x($______$) + 8\ ($______$)$
$=$ ______ $+$ ______
$=$ ____________

c $(3a - 1)(a + 2)$
$= 3a(a + 2) - 1($______$)$
$=$ ______ $+ 6a$ ______
$=$ ____________

d $(3a - 5)(a + 7)$
$= 3a($______$) - 5($______$)$
$=$ ____________
$=$ ____________

e $(2a + 5)(3a + 1)$
$= 2a(3a + 1) +$ ___($______$)$
$= 6a^2 +$ ______ $+$ ______
$=$ ____________

f $(y - 2)(y - 3)$
$= y(y - 3)$ ______
$= y^2 - 3y$ ______
$=$ ____________

2 Expand the following:

a $(n + 6)(m + 2)$ **b** $(a + 7)(b - 1)$ **c** $(a + 5)(2b - 3)$
d $(4n + 1)(m + 2)$ **e** $(2m - 5)(n - 1)$ **f** $(3a - 10)(3b - 4)$

Level 2

3 Expand and simplify:

a $(a + 1)(a + 2)$ **b** $(x + 2)(x + 3)$ **c** $(x + 5)(x + 6)$
d $(w + 1)(w + 7)$ **e** $(p + 5)(p + 2)$ **f** $(a + 3)(a + 1)$
g $(q + 10)(q + 3)$ **h** $(m + 8)(m + 2)$ **i** $(a + 7)(a + 10)$

4 Expand and simplify the following binomial products:

a $(a + 3)(a - 4)$ **b** $(m + 5)(m - 1)$ **c** $(m - 5)(m - 1)$
d $(x - 4)(x + 3)$ **e** $(n - 3)(n + 4)$ **f** $(n - 3)(n - 4)$
g $(p - 8)(p - 1)$ **h** $(q + 3)(q - 5)$ **i** $(x - 3)(x - 3)$
j $(a + 4)(a - 6)$ **k** $(m - n)(m + n)$ **l** $(7 - p)(7 + p)$
m $(m + 8)(m - 8)$ **n** $(p - 5)(p + 5)$ **o** $(10 + w)(10 - w)$

5 Expand and simplify:

a $(2x + 1)(x + 3)$ **b** $(2x + 3)(2x + 1)$ **c** $(3m + 4)(m - 2)$
d $(x + 2)(3x - 1)$ **e** $(3x + 1)(2x + 1)$ **f** $(5n + 3)(2n - 3)$
g $(4n + 3)(3n - 4)$ **h** $(2n - 3)(2n + 7)$ **i** $(3 - 2n)(2 + n)$

6 Expand and simplify:

a $(a + 4)^2$ **b** $(m + 9)^2$ **c** $(n - 7)^2$
d $(x + 2)^2$ **e** $(n - 5)^2$ **f** $(2x + 1)^2$
g $(3x + y)^2$ **h** $(7m + n)^2$ **i** $(ab + 1)^2$

7 By using Pythagoras' theorem, find the value of x in:

a
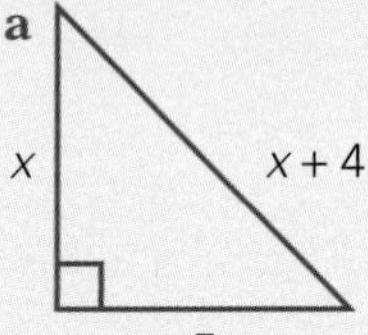

b
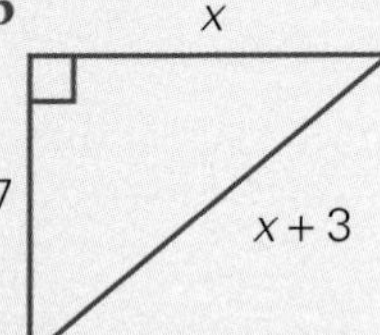

c
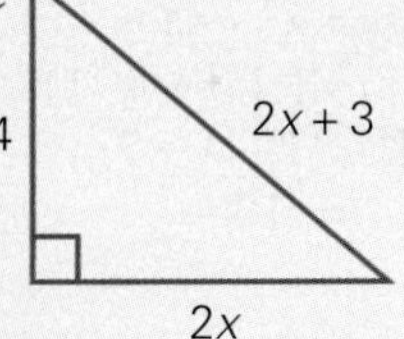

8 Expand and then simplify each of the following expressions:

a $(x + 3)(x + 2) + x^2 - 1$ **b** $(x + 4)^2 + 2x - 3$
c $10x + (x + 3)^2$ **d** $(2x + 3)(x - 5) - 6x - 10$
e $(2x + 1)^2 + 5 - 2x$ **f** $(x + 1)(x + 2) - (x - 1)(x + 1)$
g $(x + 4)(x + 1) - (x + 1)$ **h** $(x + 1)(x - 1) - x^2$
i $x^2 + (x + 2)(x + 3)$ **j** $(x + 2)^2 + x^2 + 2x - 3$
k $x^2 - (x - 1)^2$ **l** $x(x + 2)(x + 3)$

Try this

Pythagorean triads or triplets, as we know from chapter 6, are a set of three numbers that satisfy Pythagoras' theorem.

Plato's formula for a Pythagorean triplet was:

$(n^2 + 1)^2 = (n^2 - 1)^2 + (2n)^2$

By expanding the left-hand side and then the right-hand side, show that Plato's formula is true where n is a positive number.

15.4 Function notation

Function notation is a way of expressing a rule where each x-value inputted produces a unique answer. It is commonly used in the senior courses.

In this section we will look at substitutions into function notation.

EXAMPLE 1

If $f(x) = 3x - 4$ find the value of:

a $f(0)$ **b** $f(5)$ **c** $f(^-2)$

Solution

a
$$\begin{aligned} f(x) &= 3x - 4 \\ \therefore f(0) &= 3(0) - 4 \\ &= 3 \times 0 - 4 \\ &= {}^-4 \end{aligned}$$

b
$$\begin{aligned} f(x) &= 3x - 4 \\ \therefore f(5) &= 3(5) - 4 \\ &= 3 \times 5 - 4 \\ &= 11 \end{aligned}$$

c
$$\begin{aligned} f(x) &= 3x - 4 \\ \therefore f(^-2) &= 3(^-2) - 4 \\ &= 3 \times {}^-2 - 4 \\ &= {}^-10 \end{aligned}$$

EXAMPLE 2

Given $f(x) = x^2 + 4x - 7$, find the value of $f(^-1)$.

Solution

$$\begin{aligned} f(x) &= x^2 + 4x - 7 \\ \therefore f(^-1) &= (^-1)^2 + 4(^-1) - 7 \\ &= 1 - 4 - 7 \\ &= {}^-10 \end{aligned}$$

EXERCISE 15D

1 If $f(x) = 4x$, find the value of:

a $f(1)$ **b** $f(2)$ **c** $f(5)$ **d** $f(0)$ **e** $f(\frac{1}{2})$

2 If $f(x) = 3x + 3$, find the value of:

a $f(1)$ **b** $f(2)$ **c** $f(5)$ **d** $f(0)$ **e** $f(\frac{1}{2})$

3 If $f(x) = 5 - 3x$, find the value of:

a $f(1)$ **b** $f(4)$ **c** $f(0)$ **d** $f(^-1)$ **e** $f(\frac{2}{3})$

4 If $f(x) = \frac{x + 1}{2}$, find the value of:

a $f(1)$ **b** $f(3)$ **c** $f(0)$ **d** $f(^-1)$ **e** $f(^-3)$

5 If $f(x) = 2x^2 - 3x + 1$, find the value of:

a $f(0)$ **b** $f(1)$ **c** $f(2)$ **d** $f(^-3)$ **e** $f(10)$

Level 2

6 If $f(x) = 8x - 5$:

a Find the value of:
i $f(0)$ **ii** $f(2)$ **iii** $f(0) + f(2)$ **iv** $f(4) - f(^-1)$
b For what value of x does $f(x) = 19$?
c For what value of x does $f(x) = {}^-11$?

7 Given $f(x) = x^2 - x$ and $g(x) = x^2 + x$:

a Find $f(0)$ and $g(0)$.
b What is the significance of the answer to part **a**?
c Find the value of $f(3) + g(^-3)$.

8 Given $f(x) = 5x + 4$ and $g(x) = 2x - 7$:

a Find the value of:
i $f(0)$ **ii** $f(2)$ **iii** $f(^-2)$ **iv** $f(5) - f(1)$
b Find the value of:
i $g(5)$ **ii** $g(0)$ **iii** $g(^-7)$ **iv** $g(2) + g(^-1)$
c Find the value of $g(4) + f(^-1)$.
d Find the value of $f(x) + g(x)$.
e Find the value of $f(x) - g(x)$.
f Find the value of x for which $f(x) = g(x)$.

SPREADSHEET ACTIVITY

Try this

$$f(x) = \begin{cases} 2x \text{ when } x \geq 0 \\ 1 - x \text{ when } x < 0 \end{cases}$$

Find the value of:

a $f(2)$ **b** $f(^-4)$ **c** $f(0)$

Investigation Guess who?

1 I was America's greatest originator of puzzles. I lived between 1841 and 1911. You can still obtain copies of my puzzles today.

Can you guess who I am?

Match the binomial products to their expanded version to unlock the puzzle code. Not all letters are used.

D $(x+7)(x+9)$
I $(2x+3)(3-2x)$
A $(4-x)(2+x)$
K $(x-y)(x+y)$
M $(x-3y)(x-2y)$
Y $(y-x)(y+x)$
E $(x+1)^2$
N $(4+x)(2-x)$
L $(1-x)^2$
U $(x+3y)^2$
O $(2x+3)(x-3)$
S $(2x+3)(2x-3)$

$4x^2-9$	$8+2x-x^2$	$x^2-5xy+6y^2$

x^2-2x+1	$2x^2-3x-9$	y^2-x^2	$x^2-16x+63$

2 Now that you have worked out my identity, here is your next task.

Find three of my puzzles and present them, and their solutions, to your class. You may like to do this by way of a poster or as a live presentation in class. Be creative and remember to always quote your references.

15.5 Equations involving more than one fraction

We have solved many types of equations so far this year. It is now time to look at ways to solve equations resulting from fractions with different denominators. One method is shown in the example below.

EXAMPLE

Solve:

a $\frac{x}{3}+\frac{x}{5}=1$

b $\frac{3x}{4}-\frac{x}{2}=3$

Solution

a $\frac{x}{3}+\frac{x}{5}=1$ — Express $\frac{x}{3}$ and $\frac{x}{5}$ with a common

$\frac{5x}{15}+\frac{3x}{15}=1$ — Simplify.

$\frac{8x}{15}=1$ — Solve.

$8x=15$

$x=\frac{15}{8}$ — Don't forget to check your answer works!

b $\frac{3x}{4}-\frac{x}{2}=3$ — Express $\frac{3x}{4}$ and $\frac{x}{2}$ with a common

$\frac{3x}{4}-\frac{2x}{4}=3$ — Simplify.

$\frac{x}{4}=3$ — Solve.

$x=12$ — Check your solution.

EXERCISE 15E

1 Copy and complete each of the following:

a $\frac{x}{5} - \frac{x}{2} = 1$

$\frac{\square}{10} - \frac{5x}{10} = 1$

$\frac{\square}{10} = 1$

$\square = 10$

$x = \square$

b $\frac{3a}{5} + \frac{a}{4} = 2$

$\frac{\square}{20} + \frac{\square}{20} = 2$

$\frac{\square}{20} = 2$

$\square = 40$

$a = \square$

c $\frac{5a}{8} - \frac{a}{2} = 5$

$\square - \square = 5$

$\frac{\square}{8} = 5$

$\square = 40$

$a = \square$

Level 2

2 Solve each of the following equations:

a $\frac{3x}{5} + \frac{x}{2} = 1$ **b** $\frac{x}{2} + \frac{x}{5} = 4$ **c** $\frac{5a}{6} - \frac{a}{3} = {}^{-}1$

d $\frac{7x}{5} - \frac{x}{2} = 3$ **e** $\frac{w}{9} + \frac{w}{3} = 2$ **f** $\frac{5m}{4} - \frac{m}{2} = 7$

g $\frac{3x}{5} - \frac{x}{4} = 0$ **h** $\frac{5n}{6} + \frac{2n}{3} = 5$ **i** $\frac{x}{8} + \frac{x}{4} = {}^{-}2$

3 Solve:

a $\frac{x+1}{4} + \frac{x}{2} = 1$ **b** $\frac{2x}{5} + \frac{x+1}{2} = 3$ **c** $\frac{(x+1)}{2} + \frac{(x+3)}{4} = 2$ **d** $\frac{2x+1}{5} - \frac{(x+1)}{4} = 4$

TEACHER

4 Is $x = 1$ a solution to:

a $\frac{x}{4} + \frac{x}{5} = x - \frac{11}{20}$? **b** $\frac{x}{x-2} = 1$? **c** $\frac{x}{2-x} = 1$?

Investigation Discover the symbol

1 What does each symbol represent?

✪ + ✪ + ✪ + ♣ + ♠ = **31**

♣ + ♣ + ♣ + ♣ + ♣ = **20**

♦ + ♣ + ♥ + ♥ + ♥ = **36**

♠ + ♠ + ♣ + ♥ + ♥ = **30**

♣ + ♣ + ✪ + ♣ + ♣ = **24**

♥ + ♣ + ♣ + ✪ + ♠ = **29**

2 The product of a number and its square is 216. What is the number?

Chapter Review

Language Links

algebraic expressions | factorise | function notation | index | reciprocal

1 Explain the difference between an equation and expression.

2 Write the four index laws in words.

3 Explain the procedure required to expand binomial products.

4 What is meant by the reciprocal of a fraction?

Chapter Review Exercises

15.1 **1** Simplify:

a $a^{13} \times a^2$ b $w^{12} \div w^8$ c $4x^7 \times 6x^3$ d $7x^4 \times 3x^5$
e $x^{10} \div x^8$ f $16x^6 \div 2x^2$ g $10a^4 \div 5a^3$ h $a^4b^2 \times ab^3$

15.1 **2** Simplify:

a $(a^2)^5$ b $(x^5)^4$ c $(a^3)^7$ d $x^0 + 5y^0$
e $(5x^4)^2$ f $\left(\frac{x^2}{y}\right)^3$ g $(3a^4b^3)^4$ h $(7x^2y)^0$

15.1 **3** Fully simplify each of the following:

a $\frac{a^4b^6}{a^2b}$ b $9x^4 \div 12x^4$ c $(3x^4)^2 \times 5x^3$ d $4a^3b^2 \times {}^-3a^4b^6$
e $\frac{12a^5b^3}{6a^2b}$ f $\frac{(3a^4b)^2}{6a^3b}$ g $\frac{3ab^3 \times 2a^4b}{12a^3b^2}$ h $(5a^4b^3)^2 \times 3a^0$

15.2 **4** Expand and simplify:

a $(m + 6)(m + 3)$ b $(a + 7)(a - 2)$ c $(a - 3)(a + 4)$
d $(x - 8)(x - 2)$ e $(2m + 7)(3m + 2)$ f $(2x - 3)(x - 2)$

15.2 **5** Expand and simplify:

a $(x + 11)^2$ b $(2x + 5)^2$ c $(m - 3)^2$
d $(5 + w)(5 - w)$ e $(2p + 3)(2p - 3)$ f $(x - y)(x + y)$

15.3 **6** Given $f(x) = 5 + 4x - x^2$, find the value of:

a $f(0)$ b $f(1)$ c $f(5)$ d $f({}^-3)$

7 Solve each of the following equations:

a $\frac{x}{8} + \frac{3x}{4} = 1$ **b** $\frac{a}{8} + \frac{a}{6} = 1$ **c** $\frac{7a}{10} - \frac{a}{5} = 2$

8 Find the value of x in:

a

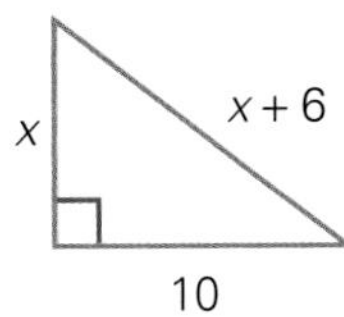

b

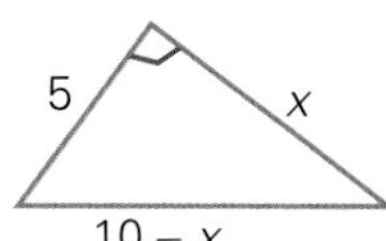

15.4
15.2
TEACHER

Keeping Mathematically Fit

Part A: Non-calculator

1 Write $\frac{52}{20}$ as a mixed number.

2 Write a whole number whose square root is between 5 and 6.

3

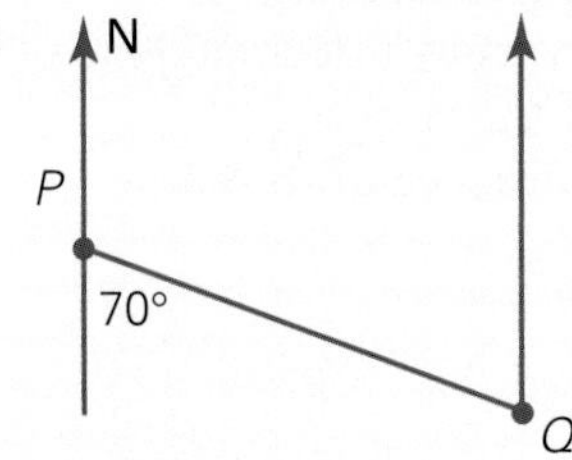

What is the bearing of P from Q?

4 Find $\frac{4}{5}$ of \$8.

5 A 5-litre container of sunscreen is used to fill tubes each holding 125 mL. How many tubes can be filled?

6 Find the value of $\frac{4}{100} + \frac{27}{1000}$.

7 The ratio of boys to girls in a class is 5 : 7. If there are 15 boys, how many students are in the class?

8 Find the value of $^{-}10 + (12 - 5 \times 6)$.

9 Simplify the ratio of 20 minutes to 2 hours.

10

A water tank is $\frac{2}{5}$ full with 12 000 litres in it. What is the tank's capacity?

Part B: Calculator

1 Find the value of $\dfrac{3}{4.6 \times \sqrt{9.8}}$, correct to 3 decimal places.

2 Simplify:

a $6^4 \times 6^3 \div 36$ **b** $3(x^2 + x - 1) + 2x^2 - 3x$ **c** $\dfrac{12ab}{18a}$

3 Find the mean age for the data displayed in the frequency histogram.

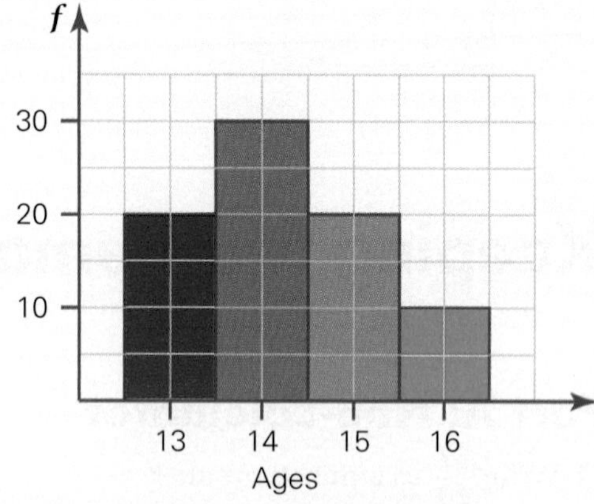

4

y
L
A
(2,0)
M
B
(8,0)
x

a Write down the coordinates of M.
b Write down the coordinates of L.
c Calculate, correct to 1 decimal place:
i the area of the circle
ii the circumference of the circle

5

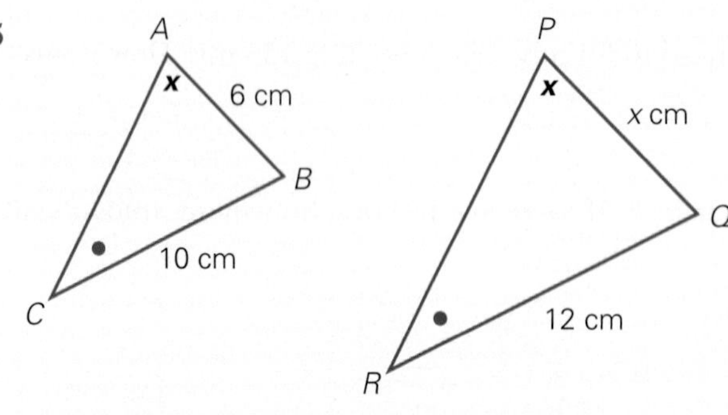

Given $\Delta ABC \parallel\!| \Delta PQR$, find the value of x.

6 If $f(x) = \dfrac{2x+1}{3}$, find the value of $f(^-2)$.

7 Calculate the simple interest earned on \$7000 at $6\frac{3}{4}\%$ p.a. for 30 months.

8 The sides of a rectangle increase by 10%.

a What is the percentage increase in the:
i perimeter?
ii area?

b Is the percentage increase the same for all rectangles?

TEACHER

Cumulative Review 5

Part A: Multiple-choice questions

Write the letter that corresponds to the correct answer in each of the following:

1 $10^3 \times 1000$ equals:

A 10^4 **B** 10^6 **C** $10\,000^3$ **D** 100^6

2 $4 - (2x - 3)$ equals:

A $1 - 2x$ **B** $8x - 12$ **C** $7 - 2x$ **D** $2x + 1$

3 The value of x is approximately:

A 15·6 **B** 15·7
C 12·1 **D** 7

4 If $x > 0$ and $y < 0$, which of the following is always true?

A $x + y > 0$ **B** $xy > 0$ **C** $\frac{x}{y} > 0$ **D** $x - y > 0$

5 The volume of the triangular prism is:

A 450 cm^3 **B** 3000 cm^3
C 1500 cm^3 **D** 1800 cm^3

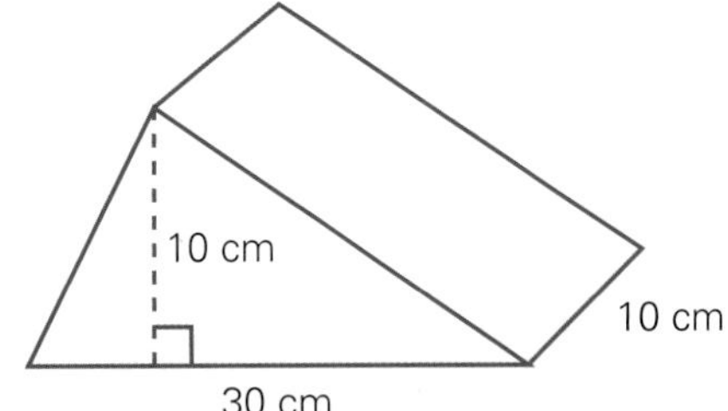

6 Which of the following angles is obtuse?

A 30° **B** 90° **C** 179° **D** 215°

7 The value of x is:

A 60 **B** 30
C 45 **D** 90

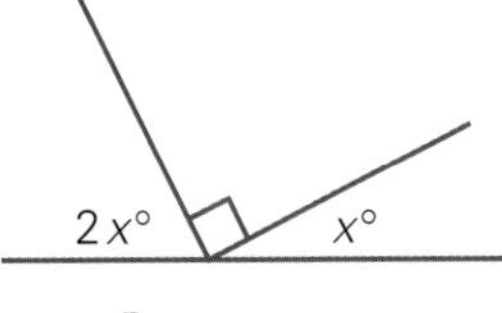

8 $\angle ABC$ is:

A 170° **B** 60°
C 130° **D** 70°

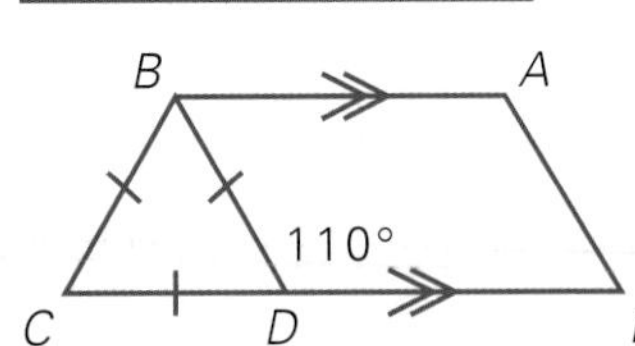

9

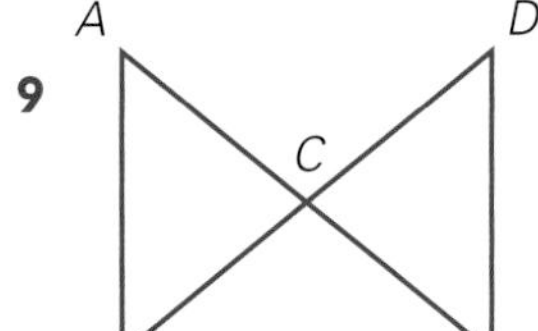

$\angle CAB$ and $\angle CED$ are:

A vertically opposite **B** corresponding
C co-interior **D** alternate

10 2 3 5 12 2 4 8 12 The mean is:

A 12 **B** 6 **C** 48 **D** 2 and 12

11 The circumference of a circle with diameter 4 cm is closest to:

A 25 cm **B** 12 cm **C** 50 cm **D** 13 cm

12 Which of the following has a capacity of 1 litre?

A

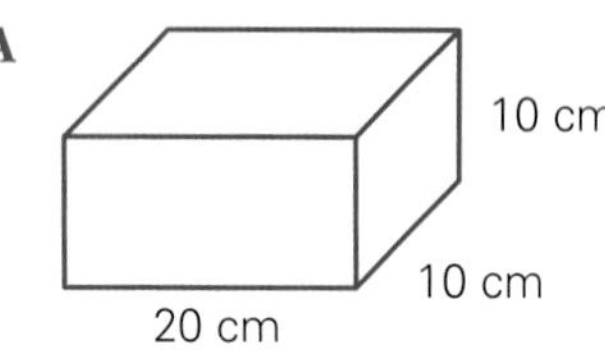

B

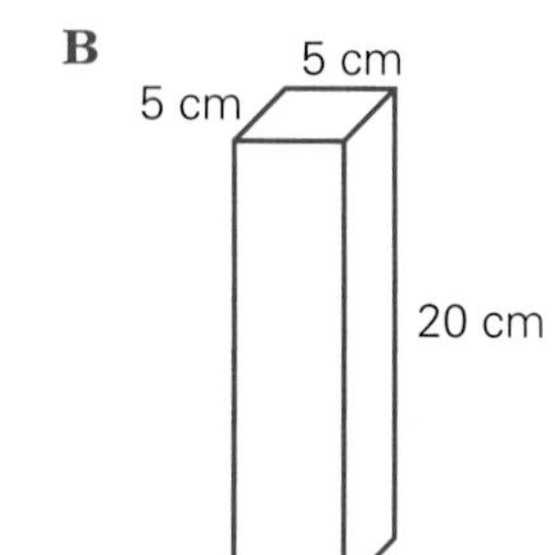

C

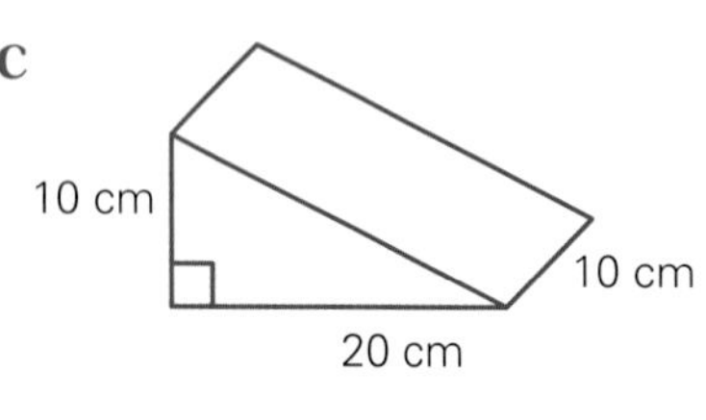

D

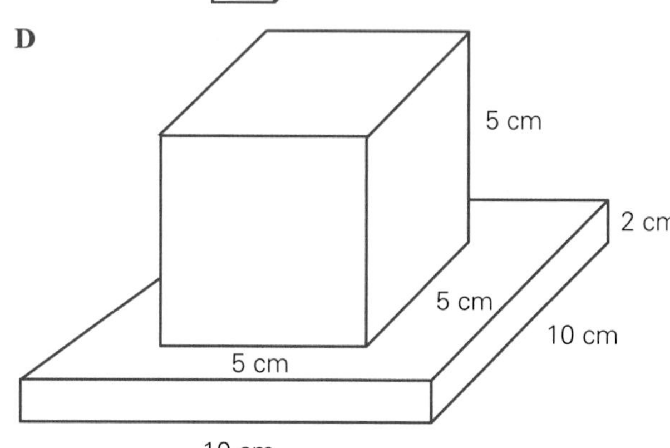

13 Which of the following straight lines pass through the point (1, 1)?

A $y = 2x - 1$ **B** $x + y + 2 = 0$ **C** $y = 2x$ **D** $x - y + 2 = 0$

14 In which quadrant does the point ($^-4$, $^-6$) lie?

A 1st **B** 2nd **C** 3rd **D** 4th

15 Which of the following has a mean of 4 and a mode of 3?

A 4, 4, 4, 4 **B** 3, 3, 4, 6 **C** 3, 3, 3, 4 **D** 2, 6, 8, 0

16 A bag of jellybeans has 42 red, 14 black, 28 green and 30 yellow jellybeans. If one jellybean is chosen at random to be eaten, the probability that it is black is:

A $\frac{50}{57}$ **B** $\frac{7}{57}$ **C** $\frac{7}{50}$ **D** $\frac{5}{19}$

17 By what factor has rectangle 1 been enlarged to produce rectangle 2?

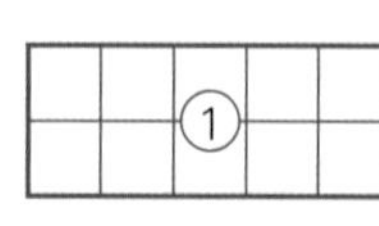

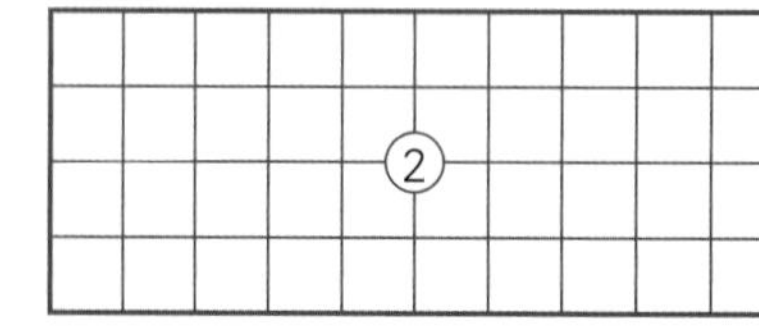

A $\frac{1}{2}$ **B** 2

C 30 **D** 4

18 The scale 1 mm : 1 m as a ratio is:

A 1 : 1000 **B** 1 : 100 **C** 100 : 1 **D** 10 000 : 1

19 Using a scale of 1 : 100 000, a measurement of 3·8 cm on a drawing represents what length on the real thing?

A 38 km **B** 3·8 km **C** 380 m **D** 3 800 000 cm

20 |————————————————————————|

This line represents 5 km. The scale used is:

A 1 : 50 000 **B** 1 : 500 000 **C** 2 : 1 **D** 50 000 : 1

21 Which of the following statements is true?

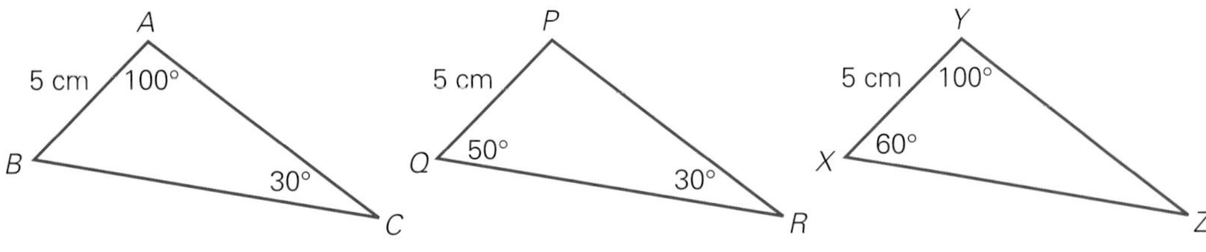

A $\Delta ABC \equiv \Delta PQR$ **B** $\Delta ABC \equiv \Delta YXZ$ **C** $\Delta PQR \equiv \Delta YXZ$ **D** none of these

22 $4x^6 \times 3x^4$ is equivalent to:

A $7x^{10}$ **B** $7x^{24}$ **C** $12x^{24}$ **D** $12x^{10}$

23 $2 - 3x^0 = ?$

A $^-1$ **B** 0 **C** 1 **D** 2

24 $x + \frac{x}{5} = ?$

A $\frac{2x}{5}$ **B** $\frac{6x}{5}$ **C** $\frac{x}{6}$ **D** $6x$

25 $\frac{1}{x} \div \frac{1}{y} = ?$

A $\frac{1}{xy}$ **B** $\frac{y}{x}$ **C** $\frac{x}{y}$ **D** xy

Part B: Short-answer questions

Show full working for each of the following.

1 Solve:

a $11(x + 6) = 10x + 12$ **b** $\frac{5 - 2x}{4} = x$ **c** $1 - x > 6$

2 Factorise:

a $5w - 10p$ **b** $^-ab - 3a$ **c** $12xy + 48x$

3

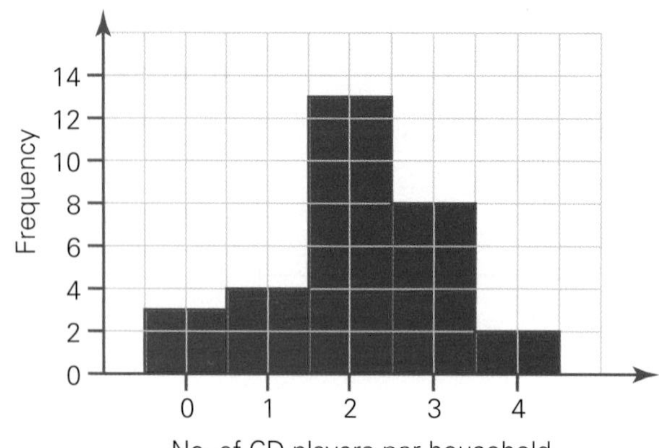

a How many households were surveyed?

b What fractionof those surveyed had two CD players?

c What percentage of those surveyed had four CD players?

4 Find $5\frac{1}{2}\%$ of:

a \$792 b \$81 000

5 Find the discounted price on a bike marked at \$990 during a 25%-off sale.

6 If 4% of an amount is 32, find the amount.

7 Find the value of each pronumeral, and give a reason:

a

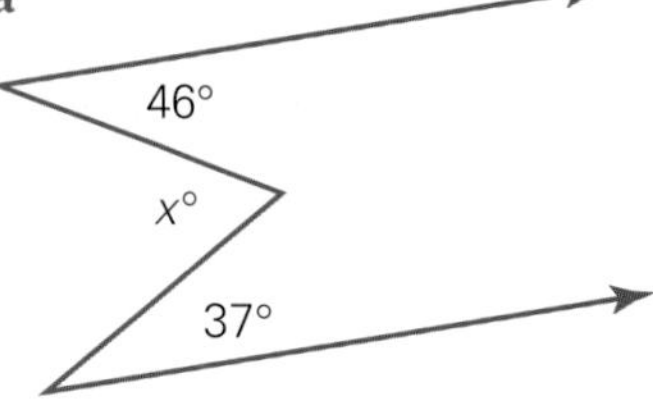

b

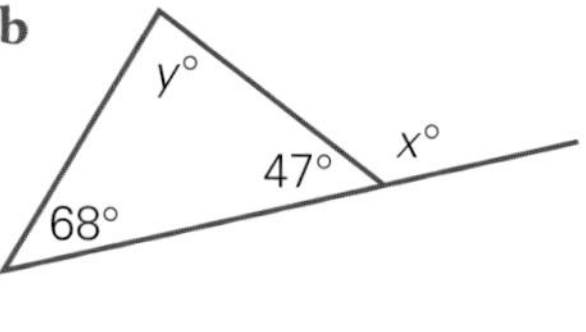

8 The angles in a quadrilateral are in the ratio of 2 : 2 : 3 : 5. Find the size of the largest angle.

9

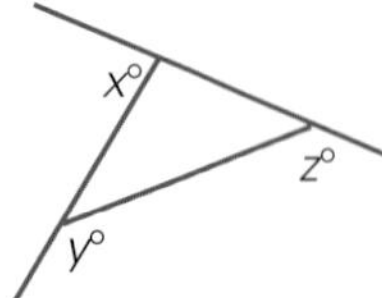

Find the value of $x + y + z$.

10 A play group has a ratio of males to females of 3 : 5. There are originally 30 girls in the group before two leave to join another group. What does the ratio of males to females become?

11 Find the surface area.

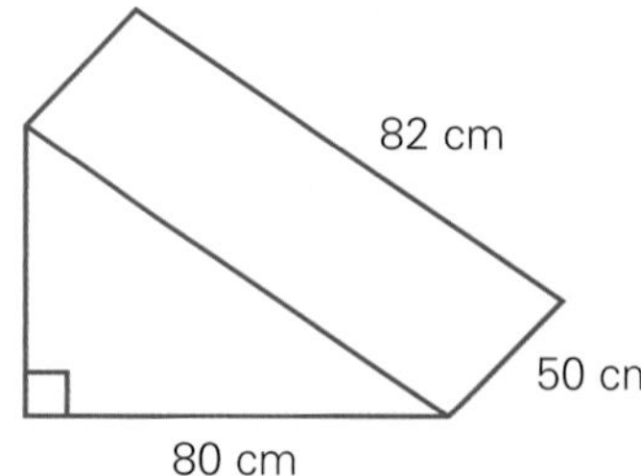

12 Find the volume.

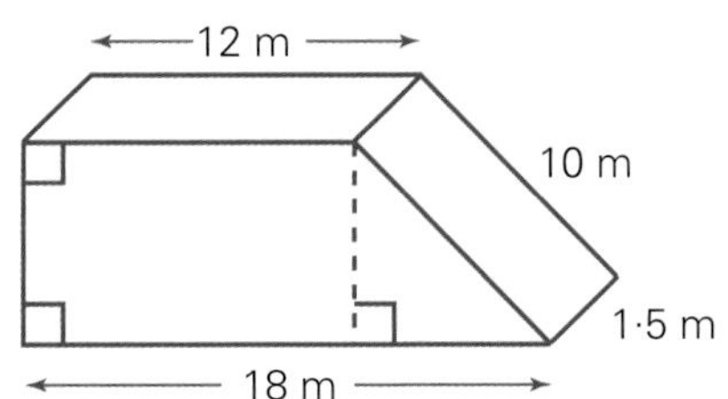

13 If $y = \frac{x}{3} + 1$, complete:

x	3	4	5	6
y				

14 Complete:

x	$^-2$	$^-1$	0	1	2
y					

for each of the following and graph them on the same number plane.

a $y = 2x$ **b** $y = 4x + 1$ **c** $y = 6 - x$

15 Sketch $y = x^2$ by first completing:

x	$^-2$	$^-1$	0	1	2
y					

16

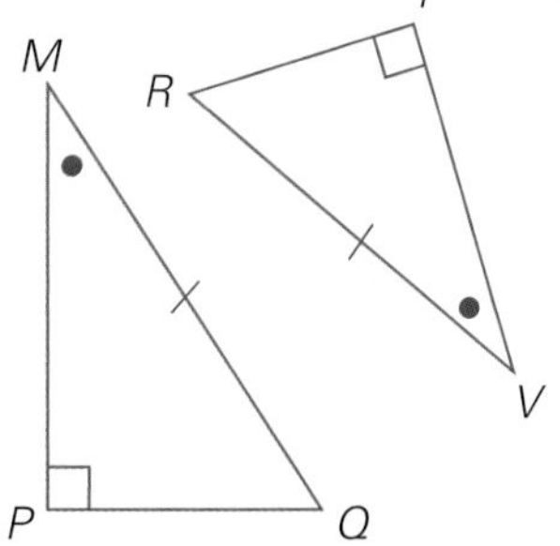

Write a statement of congruency about the two triangles.

17 Consider the triangles at right:

a State a congruency relationship, giving a reason.

b Name a side congruent to PK.

c If $PK = 13$ cm and $PX = 14$ cm, find the length of BC.

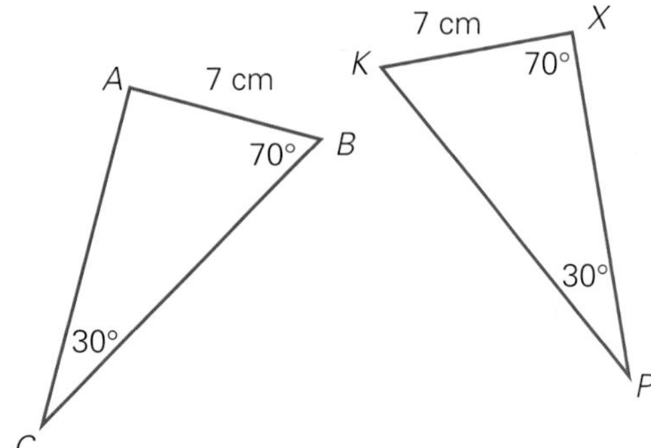

18 Which pairs of figures are similar?

a

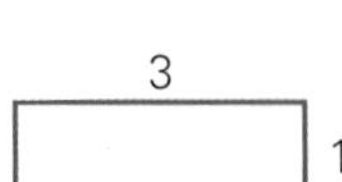

6

2

b

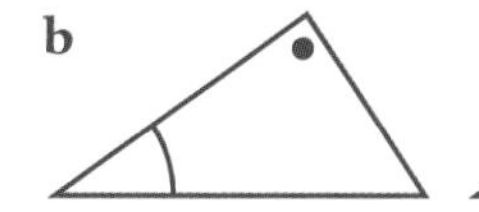

c

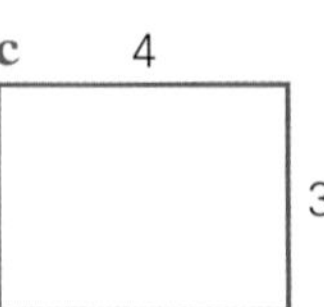

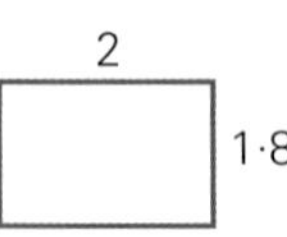

d

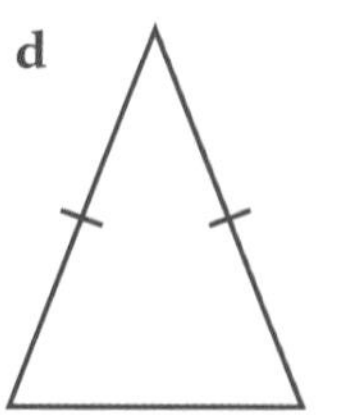

19 Find the value of the pronumeral/s in each of the following similar triangles.

a

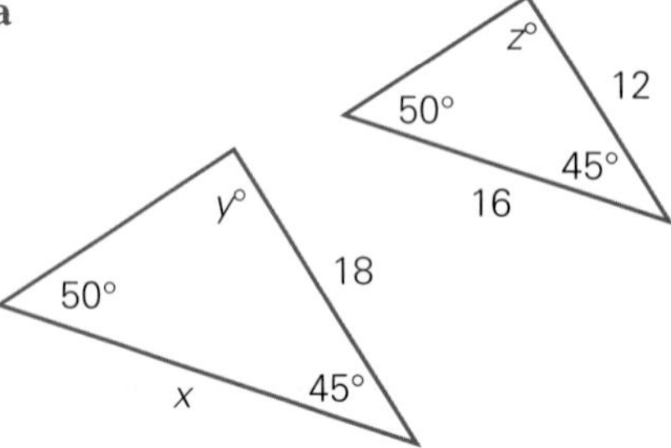

b

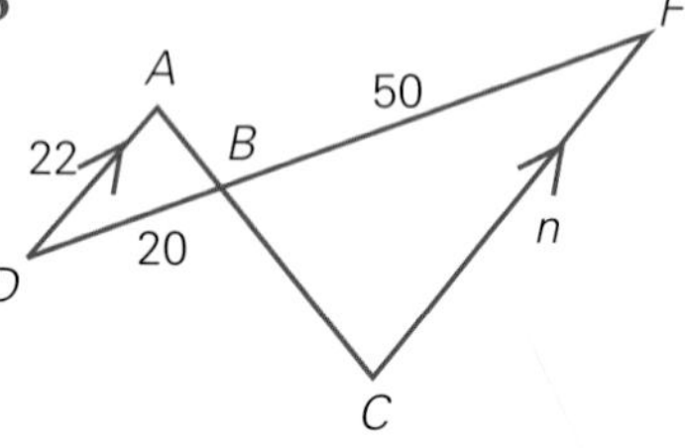

20 A letter is chosen at random from the word MATHEMATICS. What is the probability that the letter chosen is:

a an M? **b** a vowel? **c** not a vowel?
d an E or I? **e** an S? **f** a P?

21 Simplify:

a $\dfrac{9a - 18}{27}$ **b** $\dfrac{x + y}{3x + 3y}$

22 Simplify by using an appropriate index law:

a $a^{12} \times a^6$ **b** $x^{12} \div x^9$ **c** $(2a^7)^3$
d $5 + 3a^0$ **e** $7a^3b^2 \times 2ab^2$ **f** $(7a^4b^3)^2$

23 Simplify:

a $\dfrac{x}{7} + \dfrac{2x}{3}$ **b** $\dfrac{5x}{9} - \dfrac{x}{3}$ **c** $\dfrac{ab}{10} \times \dfrac{2a}{b}$ **d** $\dfrac{x}{4} \div \dfrac{x}{5}$

24 If $f(x) = 5x^3 - x^2$, find the value of:

a $f(0)$ **b** $f(1)$ **c** $f(^-1)$ **d** $f(2)$

25 Solve:

a $\dfrac{2x}{7} + \dfrac{x}{2} = 3$ **b** $\dfrac{x + 4}{5} - \dfrac{x}{2} = 1$

Answers

Chapter 1

MathsCheck

1 a 60 b 340 c 23 d 16 490

2 a $7 \times 10^4 + 4 \times 10^3 + 8 \times 10$
b $6 \times \frac{1}{10} + 8 \times \frac{1}{100}$
c $6 \times 100 + 5 \times 1 + 8 \times \frac{1}{1000}$

3 a 80 b $\frac{8}{100}$ c $\frac{8}{1000}$

4 a > b < c < d =

5 a 28 b 45 c 70 d 49 e 87 f 64

6 a $(17 - 4) \times 2 = 26$ b $18 \div (2 + 4) = 3$
c $15 \times (9 - 3) \div 3 = 30$

7 a F b T c T d F e F f T

8 a $45 = 3^2 \times 5$ b $75 = 5^2 \times 3$

9 a 15 b 225

10 a $^-14$ b $^-4$ c 7 d 0 e $^-5$ f $^-11$
g 15 h 6 i $^-12$ j $^-8$ k $^-13$ l 14

11 a $^-14$ b $^-14$ c 14 d $^-72$ e $^-45$ f $^-6$
g 3 h 0 i 25

12 a $^-4$ b $^-4$ c 4 d $^-2$ e $^-4$ f $^-6$
g 111 h $^-8$ i 0 j undefined k $^-1$
l $\frac{-1}{2}$

13 a $^-12$ b $^-12$ c 50 d 50

14 a 0·28 b 1·68 c 1·14 d 78
e 2700 f 64·2 g 0·008 h 0·36
i 0·0144 j 0·2 k 0·038 l 902
m 2·765 n 0·00935 o 0·0048 p 2·125
q 0·2 r 0·3

15 a $\frac{5}{12}$ b $\frac{8}{15}$ c $\frac{14}{15}$ d $\frac{8}{11}$

16 a $\frac{5}{6}$ b $\frac{1}{2}$ c $\frac{7}{10}$ d $\frac{7}{6}$ e $\frac{1}{8}$ f $\frac{7}{6}$
g $\frac{4}{15}$ h $\frac{7}{12}$ i $1\frac{11}{24}$ j $3\frac{3}{4}$ k $1\frac{1}{6}$ l $9\frac{5}{12}$

17 a $\frac{1}{12}$ b $\frac{1}{5}$ c $\frac{3}{10}$ d 25 e $\frac{3}{5}$ f $1\frac{1}{15}$
g $26\frac{3}{5}$ h $4\frac{2}{5}$

18 a $\frac{5}{4}$ b $\frac{7}{8}$ c $\frac{1}{6}$ d $\frac{3}{16}$

19 a 45 b 10 c 50 d $\frac{4}{5}$ e $\frac{7}{16}$ f 2
g $\frac{99}{200}$ h $\frac{9}{16}$

20 a $\frac{31}{42}$ b $\frac{1}{16}$ c $\frac{-2}{5}$ d $\frac{19}{252}$ e $\frac{2}{3}$ f $2\frac{2}{3}$

EXERCISE 1A

1 a 7^3 b 10^5 c 8^1 d $6^2 \times 5^4$

2 a $8 \times 8 \times 8$ b $1 \times 1 \times 1 \times 1 \times 1$
c 12×12 d $7 \times 7 \times 7 \times 7 \times 9 \times 9$

3 a 0 b 1000 c 32 d 72

4 a $2^3 \times 3^2$ b $3^3 \times 5^2 \times 2^2$ c $2^4 \times 3^2$
d $3^5 \times 7$

5 a 9 b 7 c 21 000 000

6 0^7, 1^{100}, 4^3, 3^4, 10^2

7 a 5^4 b 3^5 c 3^7 d $5^3 \times 7^2$

8 a 10^2 b 10^4 c 10^5 d 10^1 e 10^3
f 10^6 g 10^4 h 10^4

9 a 2 b 7 c 1 d 6

10 a $5^2 \times 7^2$ b $3^2 \times 11^2$
c $2^2 \times 5^2 \times 13^2$ d $2^3 \times 3^3 \times 7^3$

11 a 35 b 33 c 130 d 42

12 a 9 b 3 c 1 d 12 e 3 f 4

13 a 2 b 3 c 2 d 2 e 6 f 3

EXERCISE 1B

1 a $2 \times 2 \times 2 \times 2 \times 2 = 2^5$
b $10 \times 10 \times 10 \times 10 = 10^4$
c $\frac{3 \times 3 \times 3 \times 3}{3 \times 3} = 3^2$ d $\frac{7 \times 7 \times 7 \times 7 \times 7}{7 \times 7 \times 7 \times 7} = 7$

2 a 3^6 b 2^8 c 4^5 d 7^5

3 a 6^6 b 9^{10} c 2^8 d 2^4 e 4^8 f 12^4

4 a 2^2 b 8^3 c 3^1 d 5^4

5 a 6^2 b 8^4 c 7^2 d 4^2 e 11^3 f 9^1
g 7^2 h 16^2 i 2^0

6 a 2^{11} b 3^9 c 6^3 d 10^5 e 5^{11} f 7^{23}

7 a 2^4 b 5^5 c 7^6 d 17^6 e 9^1

8 a 5^2 b 5^{11} c 8^1 d 6^3 e 3^6 f 2^{16}

EXERCISE 1C

1 a 2^6 b 2^{20} c 3^{14} d 7^4 e 4^{15} f 7^{10}

2 a 1 b 1 c 1 d 1 e 1 f 1

3 a $2^{20} \times 3^4$ b $7^6 \times 5^4$ c $9^4 \times 3^{12}$
d $9^2 \times 5^4 \times 3^6$

4 a 8 b 7 c 9 d 3^8
e 7^{20} f 7^{20} g 1 h 10^{30}
i 9^7 j $8^2 = 64$ k $6^2 = 36$ l $\frac{9}{25}$
m $5^6 \times 8^3$ n 1 o $5^2 = 25$ p 5^1
q $3^2 = 9$ r 4^6

EXERCISE 1D

1 a 17·07 b 3·6 c 5 d $2\frac{1}{5}$ e $1\frac{1}{8}$
f 0·63 g $^-3·4$ h $^-18$ i 1600 j 25·251
k $1\frac{1}{7}$ l \$67

2 a 37·21 b 6859 c 78 125 d 110 592
e 4 f $^-8$ g 21·16 h 900

3 a 3·162 b 1·149 c 1·627 d 2·061
e 2 f $^-2$ g 0·5
h undefined

4 a 4·5 b 1·5 c 1·1 d 0·12

5 a 5 h 12 min b 1 h 15 min
c 7 h 24 min d 10 h 13 min 48 s
e 9 h 3 min f 1 h 7 min 30 s
g 6 h 30 min h 45 min
i 3 h 37 min 30 s

6 a 1·94 b 2·75 c 15·35 d $^-0·9$
e 1·37 f 0·69

7 **a** 3 h 20 min **b** 1 h 18 min
c 9 h 10 min **d** 1 h 40 min 48 s
e 4 h 5 min 30 s **f** 11 h 30 min 22·5 s
8 **a** 3 h, 3 h 20 min **b** 1 h, 1 h 18 min
c 9 h, 9 h 10 min **d** 2 h, 1 h 41 min
e 4 h, 4 h 6 min **f** 12 h, 11 h 30 min
9 **a** 6 h 45 min **b** 3 h 26 min
c 5 h 54 min 41 s **d** 25 min 57 s
10 **a** 1 **b** 9^7 **c** 49 (7^2)
d 10^3 (1000) **e** 7^5 **f** $5^7 \times 6^2$
g 16 384 (2^{14}) **h** 9^4 (6561) **i** 7^2 (49)

EXERCISE 1E

1 2688 **2** $1101.60
3 **a** 3724 **b** 49 557
4 $64 285.71
5 **a** $78 **b** $438
6 **a** 101 154·5 **b** 4214·8
c 70·2 **d** 3512·3
7 answers vary **8** 58·79 cm (2 dp)
9 40 books **10** $12.41 **11** ±9·65 (2 dp)
12 5 242 800
13 **a** 45 **b** 49 **c** 19 131 876
14 **a** approx: 0·0035 g **b** 0·0059 mg
c 0·00195 g
15 20·59
16 **a** 0·2, ⁻0·2, 0·64, ⁻0·64
b 0·016 384 **c** 11·8

EXERCISE 1F

1 $12\frac{1}{2}$ h **2** 13 **3** 21 **4** 10
5 $2 \times \$1 + 3 \times 50c + 3 \times 20c$
6 $2970 + 2184 = 5154$
7 **a** 10 **b** 100
8 **a** 6 **b** 9 **c** 18
9 **a** **i** 10 **ii** 100 **iii** 100 **b** no, discuss
10

11 answers vary
12 **a** 9 **b** 11, 22, 33, 44, 55, 66, 77, 88, 99
c 99

Chapter Review Exercises

1 **a** 970 **b** 1908 **c** 98 **d** 30 **e** 6 **f** 0·72
g 24 **h** 9 **i** 500
2 **a** 8^2 **b** 6^7 **c** $3^2 \times 5^3$
3 **a** $5 \times 5 \times 5 \times 5 \times 5 \times 5 \times 5$
b $2 \times 2 \times 2 \times 2 \times 2$
c 11×11
d $5 \times 5 \times 5 \times 2 \times 2 \times 2 \times 2$
4 **a** 2 **b** 4 **c** 2 **d** 1
5 **a** 9^{12} **b** 7^9 **c** 5^8 **d** $3^5 \times 7^{10}$
6 **a** 9^8 **b** 7^5 **c** 2^4 **d** 5^5
7 **a** 2^{15} **b** 5^8 **c** 7^9 **d** $2^{16} \times 3^{28}$
8 **a** 1 **b** 1 **c** 1 **d** 2^6
e 5 **f** 1 **g** 28 **h** 1

9 **a** 31·36 **b** 40·36 **c** 4·87 **d** 2·74
e 6·03 **f** 645·42
10 **a** 6 h 46 min **b** 1 h 48 min
c 5 h 27 min 36 s **d** 12 h 4 min
11 16, 32 or ⁻16, ⁻32 **12** $326 233
13 **a** $8{\cdot}39\dot{3}$ **b** 13·536 **c** 126

Keeping Mathematically Fit

PART A

1 205 000 **2** $4600 **3** $96 **4** 29·6 cm
5 0·0812 **6** 42 **7** 18·0 **8** octagon
9 5·5 cm
10 **a** 225 **b** 1 **c** $2\frac{1}{4}$ **d** 20

PART B

1 912·099 **2** 3·45 **3** $2\frac{1}{45}$ **4** 45
5 2·25 **6** 7 cm by 10 cm
7 12:35 pm, 11:52 pm **8** discuss

Chapter 2

MathsCheck

1 **a** $M = S + 2$ **b** $p = 2 \times h$ **c** $V = n - 6$
d $n = \frac{3}{4} \times p$
2 **a** 3, 5, 7, 9 **b** $3\frac{1}{2}$, 4, $4\frac{1}{2}$, 5
3 **a** 15, 21 **b** ⁻10, ⁻12 **c** 8, 4 **d** 13, 21
4 **a** $y = 5 \times x$ **b** $V = 3 \times n - 1$
5 **a**

n	1	2	3
V	4	8	12

$V = 4 \times n$

b

n	1	2	3
V	3	5	7

$V = 2 \times n + 1$

c

n	1	2	3	4
V	$\frac{1}{4}$	$\frac{1}{3}$	$\frac{3}{4}$	1

$V = \frac{1}{4} \times n$

d

n	1	2	3
V	2	6	10

$V = 4 \times n - 2$

6 **a**

n	1	2	3	4	5	6
V	⁻3	⁻6	⁻9	⁻12	⁻15	⁻18

b $V = {}^-3 \times n$ **c** ⁻72 **d** 16th
7 **a**

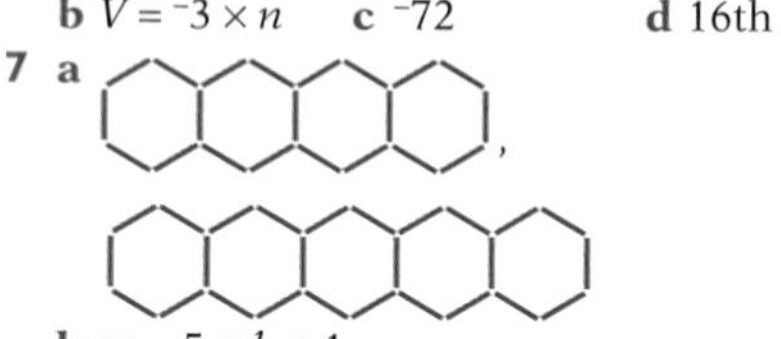

b $m = 5 \times h + 1$
c **i** 61 **ii** 9
8 **a** $T = 4 \times n + 1$ **b** **i** 77 **ii** 21
c no as n needs to be a positive integer
d 230

EXERCISE 2A

1 **a** $3 + p$ **b** $x + 7$ **c** $x - 8$ **d** $x + y + z$
e $m + n + 6$ **f** $x - y$
2 $(40 + x)m$ **3** $4a + 2b$
4 **a** $5x$ **b** x **c** $6p$ **d** $4m$ **e** $16w$ **f** $16w$
g 0 **h** $20p$ **i** $12w$ **j** f **k** $7a^2$ **l** $11p$
m $26m$ **n** $26m$ **o** $4pq$ **p** $2x^3$

5 a T b F c F d T e T f T
g T h F

6 a N b Y c Y d Y

7 a $10x$ b $8a$ c $6a$ d $4x$
e $14p$ f $3x$ g p h $3x^2$
i $11mn$ j $6x$ k $4a + 3b$ l $7g^2 + 4g$
m $6ab$ n 0 o $11w + w^2$

8 a ^-4x b ^-4a c ^-10x d ^-2p e $4xy$
f $^-4a^2$ g $9n$ h ^-6a i ^-5mn j $6m$
k ^-5n l ^-3xy m p n $^-10x^2y$ o $^-8a^2$
p ^-6xy

9 a $9h + 10$ b $6 + 8n$
c $3x^2 + 6x$ d $6x^2 + xy + 4y^2$
e $3a^2b + 7ab^2$ f $^-2j^3$
g $8 + 4x$ h $17x + 4$
i $4p + q + 6$ j $15 - 4x$
k $2p$ l $4b - a$
m $4x + 2y$ n $6y - x$
o $6y - 8x$ p $4a + 3b - a^2$
q $4a - 5a^2$ r $^-7q - 2p$
s $^-2x^2 - 3x$ t $9x - y$
u $2m - 5q - p$ v $5xy + 4x$
w $^-6ab - a^2 + a$ x $6a^2b - 3ab^2$

10 a $P = 4n$ b $P = 2x + 2y$ c $P = 8m$
d $P = 3a + 5x$ e $P = 10a - 8b$
f $P = 6x + 2y^2 + 2y$

EXERCISE 2B

1 a $5a$ b $\frac{m}{4}$ c pq d $\frac{xn}{5}$ e $\frac{x+y}{5}$

2 a $5n + 3$ b $2n + 8$ c $\frac{n}{8} - 2$ d $\frac{n+1}{4}$ e $\frac{n^2}{3n}$

3 a \$7.60 b $8p$ cents c np cents

4 a $\$3x$ b $\$52x$ c $\$nx$ d $\$\frac{x}{7}$

5 $\$\frac{x}{5}$

6 a \$11 b \$34 c $\$(3 + 2n)$

7 a $6a$ b $12a$ c $15ab$ d $12n^2$ e $10a^2b$
f $24xy$ g $36ab$ h $6nt$ i $30a^2$

8 a $\frac{m}{4}$ b $\frac{m}{n}$ c $\frac{x}{5}$ d $\frac{^-a}{3}$ e $\frac{^-x}{4}$ f $\frac{2}{m}$
g $\frac{13}{w}$ h $\frac{m}{n}$ i $\frac{ab}{c}$ j $\frac{3m}{n}$ k $\frac{5y}{^-x}$ l $\frac{6n}{m}$

9 a $3a$ b $4b$ c $4n$ d $2ab$ e 3 f $3e$

10 a $2x$ b $6c$ c $\frac{n}{2}$ d $\frac{d}{3}$ e $\frac{1}{2x}$ f 3
g $4x$ h $\frac{2}{3}$

11 a ^-6ab b ^-20c c $10ab$ d $^-16t^2$ e $\frac{mp}{4}$
f $0{\cdot}16nt$ g $2ab$ h $^-3a^2b^2$ i $5p$

12 a $3x + 1$ b $ab - 3$ c $5 + 4p$
d $7 + 4p$ e $8x - 6$ f $2 - 5x$
g $6 - 3x$ h $^-xy - 3$ i $12p - 6$
j $12xy + 2x$ k $20a - 3b$ l $25ab$
m $18a^2 - 18a$ n $10 - 21a^2$ o $50a^2$

13 a xy b y c y d $4p$ e 4 f $\frac{1}{b}$
g $25x$ h $16b$ i $4a^2$ j $5x$ k $\frac{1}{3a}$ l $2mt$
m $\frac{^-2}{3}$ n $\frac{1}{3n}$ o $\frac{^-2}{3}$ p $\frac{^-2n}{5}$ q $\frac{3bc}{2}$

14 a $9s^2$ b $40n^2$ c $\frac{15ab^2}{2}$

EXERCISE 2C

1 a $5(x + y)$ b $2(x + 5)$ c $4(a - 3)$
d $2(m - n)$ e $^-3(a + 5b)$

2 $5(x + y)$ **3** $25(m + n)$ dollars

4 a $2m + 2n$ b $2m - 2n$ c $3a + 12$
d $10 - 5p$ e $4p + 4q$ f $7x - 7y$
g $25 + 5x$ h $8 - 4a$ i $24 - 8w$
j $10a - 20$ k $4n + 4a$ l $3x + 6$
m $2a - 2t$ n $5n - 10$ o $12 - 6p$
p $3c + 3d$

5 a $^-4x + 4y$ b $^-3x - 3y$ c $^-5a + 5b$
d $^-6 + 3a$ e $^-9a - 9b$ f $^-10a + 70$
g $^-10a - 70$ h $^-12 + 4m$ i $^-4 - 2p$
j $^-6a + 36$ k $^-ab - 2a$ l $^-a - 3b$
m $^-x - y$ n $^-x + y$ o $^-5 + 2w$
p $^-x + 3y - 7$

6 a $10x + 5$ b $6x - 2$ c $12x + 8y$
d $12 - 8a$ e $6d + 9$ f $30 + 15x$
g $8n - 6$ h $12x + 16$ i $10x - 40$
j $14a + 21$ k $24a - 12b$ l $15a + 12b$
m $^-15m + 6$ n $^-3ab - 3ac$ o $6a - 3b + 3c$
p $^-6b + 8 - 2a$

7 a $a^2 + a$ b $a^2 + ab$ c $2a^2 - 3a$
d $m^2 + m$ e $p^2 - 3p$ f $n^2 + 5n$
g $^-a^2 + 2a$ h $^-12p + 3p^2$ i $2x^2 + 10x$
j $15p - 6p^2$ k $^-5nm - 5n^2$ l $^-6a^2 + 14a$
m $a^2 + ab - ac$ n $3a^2 - 2ab$ o $6a^2 + 15a$
p $^-18a^2 + 30ab$

8 a $2x + 9$ b $17 - 3a$ c $7x + 5y$
d $6x - 18$ e $6x - 24$ f $7n - 6$
g $8 + 15a$ h $35 - p$ i $4a + 5x$
j $^-2p - 3$ k $18 - 15x$ l ^-4b
m $6a + 10b$ n $x^2 - 2x$ o $13 - 10x$

9 a $3a + 8$ b $16 + 5a$ c $14 + 4a$
d $3a$ e $11a - 10$ f $5m - 12$
g $5 - x$ h $15 - x$ i $a - 3$
j $9a - 12$ k $3a^2 + 2$ l $3x + 10$
m $14x$ n $13a - 10$ o $16x - 15$

10 a $5a + 9$ b $5x + 10$ c $8x + 7$
d $7g - 26$ e $6x + 17$ f $3b^2 + 5b$
g ^-2x h $2a^2 - 7$ i $14a + 68$
j $13 - a$ k $2a^2 - 6a$ l 30
m ^-17h n $3 - 5x$ o $4b - ab - 6a^2$
p $7p^2 - 13p + 10$

EXERCISE 2D

1 a 3 b 4 c 12 d p e 3 f a
g 5 h 4 i 4 j x k $12a$ l $5m$
m 2 n $2x$ o $2xy$

2 a $2(t + 3)$ b $a(b - x)$ c $3(d + 2)$
d $2(2e - 3)$ e $h(h - 7)$ f $3k(2m + 3)$
g $^-5(n + p)$ h $6(w - 2)$ i $3(3c - 1)$
j $3s(1 + 2s)$

3 a $p(2 + p)$ b $3(2x + 1)$ c $a(b - 1)$
d $5(x + 1)$ e $4(3a - 1)$ f $4(5p + 3q)$

4 a $3(x + 1)$ b $4(p - 1)$ c $10(a - 1)$
d $6(v - 1)$ e $7(p + 1)$ f $4(1 + x)$

g $2(x-2)$ **h** $9(1-a)$ **i** $2(2x-1)$
j $5(2a-1)$ **k** $2(5+2p)$ **l** $4(3+2w)$
m $4(4+3a)$ **n** $2(9-5h)$ **o** $6(4x+3)$
p $4(4ab-3)$

5 **a** $2(a+b)$ **b** $3(a-b)$ **c** $3(a-2b)$
d $4(x-y)$ **e** $5(xy-1)$ **f** $6(a-2b)$
g $3(p-3q)$ **h** $7(x+2y)$ **i** $4(2m-3n)$
j $5(5p-q)$ **k** $4(a-2b)$ **l** $6(3p-4q)$

6 **a** $b(a+1)$ **b** $b(4a-3)$ **c** $6(2a-b)$
d $a(a+1)$ **e** $2b(2a-1)$ **f** $x(x-3)$
g $5d(1-d)$ **h** $4(pq-4)$ **i** $4(q-4p)$
j $m(m+4)$ **k** $3(x^2-2)$ **l** $w(1-w)$
m $p(1+p)$ **n** $c(c-1)$ **o** $p(5-p)$
p $n(1+4n)$ **q** $d(4d+1)$ **r** $15(y-x^2)$
s $w(w-x)$ **t** $p(7p-6)$

7 **a** $3a(3b-4)$ **b** $12x(2x-1)$ **c** $5(3-p)$
d $5y(y-4)$ **e** $4b(3a+4)$ **f** $4y(x-3)$
g $7x(2x-1)$ **h** $3t(3-4t)$ **i** $2a(5a+2b)$
j $10x(10-y)$ **k** $5(4a^2+3b^2)$ **l** $2x(8-15x)$
m $4q(3p-4)$ **n** $36(1-m^2)$ **o** $2(p+q-2)$
p $x(x+3-y)$

8 **a** $^{-}2(x-2)$ **b** $^{-}5(1+a)$ **c** $^{-}6(m-2)$
d $^{-}4(1+3x)$ **e** $^{-}2(p+4)$ **f** $^{-}3(x-2)$
g $^{-}3(x+2)$ **h** $^{-}5(2h-3)$ **i** $^{-}9(y+2)$
j $^{-}7(m+1)$ **k** $^{-}5(3-4p)$ **l** $^{-}6(1-3x)$
m $^{-}11(1+q)$ **n** $^{-}4(3x-4y)$ **o** $^{-}15(3-a)$
p $^{-}10(4a-3b)$

9 **a** $^{-}x(x-1)$ **b** $^{-}x(x+1)$ **c** $^{-}p(p-1)$
d $^{-}p(p+1)$ **e** $^{-}2p(q+5)$ **f** $^{-}3x(x-2)$
g $^{-}4a(2b-1)$ **h** $^{-}4a(3a+4)$ **i** $^{-}4a(2+a)$
j $^{-}5x(1-x)$ **k** $^{-}xy(x+1)$ **l** $^{-}4pq(1+4p)$
m $^{-}5(3x^2+4y)$ **n** $^{-}x(y-5x)$ **o** $^{-}p(q-p)$
p $^{-}4(2ab-3a+2b)$

10 **a** $^{-}4p(q+4p)$ **b** $4(2x+3y)$ **c** can't
d $r(r-1+a)$ **e** can't **f** can't

TRY THESE

a $(a+3)(2+b)$ **b** $(w+4)(w+2)$
c $(a+b)(a-b)$ **d** $4(x+y+k)$
e $p(p-q+3)$ **f** $a(a+b)(a+1)$

EXERCISE 2E

1 **a** 12 **b** $^{-}4$ **c** 0 **d** 10 **e** 1
2 **a** $^{-}1$ **b** 7 **c** $^{-}6$ **d** 22 **e** $^{-}75$
3 **a** 4 **b** 16 **c** 2 **d** 29 **e** $\frac{1}{2}$
4 **a** 24 **b** 27 **c** 33 **d** 32 **e** 160
f 5 **g** 7 **h** 18 **i** 16 **j** 12
k 2 **l** $1\frac{3}{4}$

5

7	1	12	0·75	13	18
9	9	0	0	9	0
0·06	0·02	0·0008	0·5	0·0404	0·0408
$\frac{3}{4}$	$\frac{^{-}1}{4}$	$\frac{1}{8}$	2	$\frac{1}{2}$	$1\frac{1}{8}$

6 **a** 51 **b** 28 **c** 56
7 **a** $^{-}6$ **b** $^{-}7$ **c** 16
8 **a** 8 **b** $^{-}8$ **c** $^{-}84$ **d** 1764
e $\frac{^{-}1}{18}$ **f** $^{-}18$ **g** 0 **h** 2·45 (2 dp)

9 9, 16, 36, 144, 16, 2·25

EXERCISE 2F

1 **a** $2a$ **b** $\frac{a}{2}$ **c** $\frac{b}{2}$ **d** $\frac{5a}{b}$
e y **f** $\frac{y}{x}$ **g** $\frac{2a}{3}$ **h** $\frac{2}{3}$

2 **a** $x+2$ **b** $\frac{x+2}{2}$ **c** $\frac{x+2}{3}$ **d** $\frac{3}{4}$
e $a+1$ **f** $\frac{a-b}{b}$ **g** 3 **h** $a(a+b)$

3 **a** $\frac{a+b}{2}$ **b** $\frac{x-2}{2}$ **c** $\frac{2a-b}{2}$ **d** $\frac{x+3}{3}$
e $a-1$ **f** $\frac{a+1}{2}$ **g** $\frac{1}{2}$ **h** $\frac{a+1}{2}$
i $\frac{a-2}{2}$ **j** $\frac{p-2}{4}$ **k** $\frac{x}{5}$ **l** $\frac{^{-}y}{x}$

EXERCISE 2G

1 **a** $\frac{2a}{3}$ **b** $\frac{x}{3}$ **c** $\frac{6a}{7}$ **d** $\frac{4w}{7}$
e $\frac{3x}{5}$ **f** $\frac{2x}{5}$ **g** $\frac{7x}{10}$ **h** $\frac{11x}{10}$
i $\frac{6a}{11}$ **j** $\frac{3x}{9}=\frac{x}{3}$ **k** $\frac{2a}{4}=\frac{a}{2}$ **l** 0

2 **a** $\frac{5a}{6}$ **b** $\frac{8x}{15}$ **c** $\frac{10x}{21}$ **d** $\frac{14x}{45}$
e $\frac{w}{56}$ **f** $\frac{17m}{72}$ **g** $\frac{2x}{5}$ **h** $\frac{2a}{9}$
i $\frac{5x}{12}$ **j** $\frac{x}{12}$ **k** $\frac{4w}{9}$ **l** $\frac{3m}{10}$
m $\frac{a}{4}$ **n** $\frac{5w}{9}$ **o** $\frac{^{-}11p}{35}$ **p** $\frac{x}{14}$

3 **a** $\frac{2x+1}{4}$ **b** $\frac{7x+11}{10}$ **c** $\frac{1-x}{4}$
d $\frac{3-k}{4}$ **e** $\frac{5-x}{12}$ **f** $\frac{w-2}{4}$
g $\frac{7x-13}{10}$ **h** $\frac{x+24}{30}$ **i** $\frac{x-17}{10}$

EXERCISE 2H

1 **a** $\frac{5}{4}$ **b** $\frac{3}{7}$ **c** $\frac{1}{8}$ **d** $\frac{y}{x}$ **e** $\frac{3}{a}$ **f** $\frac{b}{a}$
g $\frac{1}{x}$ **h** $\frac{1}{x^2}$

2 **a** $\frac{3a}{20}$ **b** $\frac{3x}{7}$ **c** $\frac{x}{4}$ **d** $\frac{a^2}{20}$ **e** $\frac{x^2}{8}$ **f** $\frac{4}{5}$
g $\frac{5w}{9}$ **h** $\frac{2a}{5}$ **i** $\frac{a}{7}$ **j** $\frac{2w}{11}$ **k** $\frac{3w}{5}$ **l** b

3 **a** a **b** $\frac{w}{2}$ **c** $\frac{a}{3}$ **d** $\frac{m}{4}$ **e** $\frac{2}{3}$ **f** $\frac{1}{2}$
g 6 **h** $\frac{4}{n}$ **i** $\frac{7}{8}$ **j** $\frac{15}{14}$ **k** $\frac{x}{y}$ **l** $\frac{b}{2}$

4 a $\frac{x+1}{5}$ b b c $\frac{x^2}{6}$ d $\frac{2b}{7}$

e $\frac{4x}{15}$ f $\frac{a^2}{15}$ g $\frac{x}{3}$ h $\frac{2b}{a}$

i $\frac{3x}{10}$ j a^3 k a^2 l 2

Chapter Review Exercises

1 a $a+2$ b $2a$ c $2m$ d $\frac{x}{2}$

e $2(x+y)$ f $\frac{m}{5x}$

2 a 4 b 8 c 1

3 a $12k$ b $30nt$ c $2a$
d 1 e $3b$ f x^2
g 1 h $10h$ i $^-24b^2$
j $5x$ k $7ab+2+b$ l $4x+9$

4 $1\frac{1}{4}, 1\frac{1}{2}, 1\frac{3}{4}, \ldots$

5 a $V=\frac{1}{n}$ b $V=n^2+1$

6 a 9 b $^-24$ c 46

7 a ^-3x b $^-10n^2$ c ^-60xy
d ^-15a e 10 f $5a-4b$

8 a $3k+21$ b $5a-20$
c $2b+3b^2$ d $4n^3-8n^2+12n$
e $^-6m+18$ f $^-28a^2+21ab$

9 a $17n+21$ b $x+6$ c $15p-8p^2$

10 a $P=7x+9$ b $A=3(3x+5)$

11 a $m(t-p)$ b $5(x+2)$ c $3c(2+3d)$
d $4x(3y-4x)$ e $x(x-5)$ f $^-7(h-5)$
g $8k(2-3k)$ h $^-2p(5p+6q)$

12 a 0, 3, 4, 3, 0, $^-5$
b 27, 1, 0, $^-1$, $^-27$, $^-343$
c 8, 4, 2, 2, 4, 8

13 a xy b $x+y$ c $x-y$ d $\frac{3}{4}$ e $\frac{x+3}{3x}$

14 a $\frac{2x}{7}$ b $\frac{4a}{3}$ c $\frac{7w}{10}$ d $\frac{5a}{4}$

15 a $\frac{5}{3}$ b $\frac{1}{4}$ c $\frac{8}{x}$ d $\frac{y}{x}$ e $\frac{1}{4a}$

16 a $\frac{a}{6}$ b $\frac{1}{6}$ c $\frac{a}{2}$ d $4a$

17 a 2 b ab c $\frac{2a}{b}$ d $\frac{y^2}{8x}$

18 a $\frac{1}{2}$ b $\frac{4}{p}$

Keeping Mathematically Fit

PART A

1 100 2 $25ab$ 3 \$2400 4 346·10
5 $x=6$ 6 11 600 7 92 8 560 000
9 375 10 1

PART B

1 3·815 2 16 791 3 $^-17·5$
4 a 52 b 73 c 48 d 115 e 45 f 140

5 a $5\frac{5}{6}$ b $\frac{13}{20}$ c $8\frac{1}{28}$

6 a 24 cm b 3·8 m c 158 cm
d 13·8 cm e 34 cm f 7·2 cm

7 a $8a$ b $\frac{y}{2}$ c $2x+8$ d $21a^2bc$

e $5x+12$ f $\frac{a}{2}$

8 $\frac{x}{2}$ and x, x^2 and $\frac{1}{2}$. Discuss other pairs.

Chapter 3

MathsCheck

1 a 0·15 b 0·27 c 0·56 d 0·19 e 0·95 f 0·82
g 0·99 h 0·07 i 0·01 j 0·08 k 0·1 l 0·9

2 a 75% b 140% c 95% d 56% e 27%
f 3% g 99% h 5% i 30% j 3%
k 104% l 12·25%

3 a $\frac{19}{100}$ b $\frac{23}{100}$ c $\frac{99}{100}$

d $\frac{24}{100}=\frac{6}{25}$ e $\frac{50}{100}=\frac{1}{2}$ f $\frac{25}{100}=\frac{1}{4}$

g $\frac{74}{100}=\frac{37}{50}$ h $\frac{60}{100}=\frac{3}{5}$ i $\frac{5}{100}=\frac{1}{20}$

4 a i 0·125 ii $\frac{1}{8}$ b i 0·087 ii $\frac{87}{1000}$

c i $0·\dot{3}$ ii $\frac{1}{3}$ d i $0·\dot{6}$ ii $\frac{2}{3}$

e i 1·9 ii $\frac{19}{10}$

5 a i 0·2 ii 20% b i 0·8 ii 80%
c i 0·8 ii 80% d i 0·3 ii 30%
e i 1 ii 100% f i 0·25 ii 25%
g i 0·75 ii 75% h i 0·3 ii 30%
i i 0·125 ii 12·5% j i 0·6 ii 60%
k i 0·0125 ii 1·25% l i $0·\dot{7}1428\dot{5}$
ii 71·4% (1 dp)

6 a $11\frac{1}{9}\%$ b $22\frac{2}{9}\%$ c $33\frac{1}{3}\%$ d $44\frac{4}{9}\%$

e $55\frac{5}{9}\%$ f $66\frac{2}{3}\%$ g $77\frac{7}{9}\%$ h $88\frac{8}{9}\%$

7 a $0·\dot{1}$ b $0·\dot{2}$ c $0·\dot{3}$ d $0·\dot{4}$
e $0·\dot{5}$ f $0·\dot{6}$ g $0·\dot{7}$ h $0·\dot{8}$

8 a 1·7% b 840% c 800% d $8\frac{1}{2}\%$

e 37·5% f 56·47% g 9·5% h $17\frac{1}{2}\%$

9 a 65% b 72% c 0·66 d $0·\dot{6}$ e $\frac{4}{3}$

10 $56\frac{2}{3}\%$ 11 $46\frac{56}{89}\%$

EXERCISE 3A

1 a 50% b $66\frac{2}{3}\%$ c 60% d $33\frac{1}{3}\%$
e $42\frac{6}{7}\%$ f 150%

2

$\frac{1}{5}$	20%
$\frac{3}{20}$	15%
$\frac{7}{20}$	35%
$\frac{3}{10}$	30%

3 a 25% b 20% c 60% d 50% e 80%
f 3% g 14% h 10% i 50% j 4%
k 9·5% l $33\frac{1}{3}$% m 45% n 5% o $16\frac{2}{3}$%
p 30%

4 80% 5 80%

6 Lydia = $83\frac{1}{3}$%, Hayley = $86\frac{2}{3}$%
∴ Hayley did better

7 a

Year	7	8	9	10	11	12	
Total	160	175	140	170	110	97	852

Total girls = 395
Total boys = 457
b 46·4% (1 dp) c 48·6% (1 dp)
d 24·3% (1 dp) e 10·7% (1 dp)
f 47·4% (1 dp)

8 a 4% b $66\frac{2}{3}$% c $66\frac{2}{3}$% d 60% e 29%
f 2% g 80% h 34·7% i 23·8% j 52·2%
k 35% l 51·3% m 1·7% n 1·6% o 25%
p 1·0%

9 a 75% b 30% c 167%

10 a 75% b 125% c $87\frac{1}{2}$% d 3·5%

11 a 75% b 42% c 4·5%

12 NSW = 10·4% (1 dp), VIC = 3·0%, TAS = 0·9%, QLD = 22·5%, SA = 12·8%, NT = 17·6%, WA = 32·8%

13 a 2·8% b 40% c 10·3% (1 dp) d 31·5%

14 tub B

15 a

%	$33\frac{1}{3}$	25	$11\frac{1}{9}$	50	$66\frac{2}{3}$	$83\frac{1}{3}$	$41\frac{2}{3}$

b No. The average occupancy for the week is only $44\frac{4}{9}$%

EXERCISE 3B

1 a 2 b 1 c 18 d 4 e 74 f 7·4
g 14·8 h 3·7 i 28 j 45 k 42 l 21

2 a 20 b 30 c 40 d 180
e $1 f 60 cm g 160 L h 990
i 2·1 m j 41·8 km k 15·12 g l $10·50

3 a $0.58 b 11·25 m c 215·28 d 25·55
e $2.36 f 0·09 g 31·84 h 0·06
i $1.13 j $3.26 k 2 m l 4·5

4 45 5 1400 6 5 h 46 min

7 70·56 8 $16 146 9 24

10 a 50 b 15 c 210 d 28·8 e 20

11 $238 12 10 out of 500

13 $5574.82 14 $205.97

15 a $7012.50 b 12·5%

16 2 h 18 min 14·4 s, $x = 2$, $y = 18$

17 y

EXERCISE 3C

1 a 105% b 107% c 112% d 101%
e 110% f 127% g $112\frac{1}{2}$% h $133\frac{1}{3}$%

2 a 92% b 88% c 90% d 50%
e 95% f 75% g 92·5% h $66\frac{2}{3}$%

3 a 1018·5 b 1037·9 c 1086·4 d 979·7
e 1067 f 1231·9 g 1091·25 h $1293\frac{1}{3}$

4 a 4600 b 4400 c 4500 d 2500
e 4750 f 3750 g 4625 h $3333\frac{1}{3}$

5 a 96·3 b 58·14 c 105·84 d 916·75
e 69 f 99 g 121 h 17·64
i 8·9012 j 10·8 k 88 l 632·7

6 $26 964

7 a $20.60 b $65.92 c $51.45

8 a $40.50 b $33.75 c $172.15

9 a 88% b 792 mangoes

10 $17 821.05

11 a $66.28 b $103.67

12 83 740 13 $448.20

14 a $55.64. No, as you decrease a larger amount by 8% than the $56 you started with
b $97.16
c $97.16. The order of increase and decrease doesn't matter—you achieve the same answer
d no

15 a $18 009 b $30 058.50 c $6951.30

16 $692.64 17 $22 050 18 $32 300

19 a $580 500 b $624 037.50
c $670 840.31

EXERCISE 3D

1 a 10% b 10% c 40% d 25% e 50%

2 a 50% b 25% c 40% d 10% e $33\frac{1}{3}$%

3 a

24	43%
50	50%
6	25%
1·44	16%
9 mm	18%

b

3	3%
50	33%
1	11%
1·474	19%
700 m	8%

4 21% (nearest %) 5 40% (nearest %)

6 20% 7 2·1% (1 dp)

8 a 12% b 88% c $902

9 21% 10 29% 11 336·4%

12 70·8% (1 dp)

13 a 24·3% (1 dp) b 6·1% p.a.

EXERCISE 3E

1 a

5	50%
6	25%
50	50%
5	2%
2·50	14·3%

b

5	50%
8	50%
50	50%
15	44·1%
4·25	4·5%

2 a $9 b 60%

3 a $3900 b 15·7%

4 39·8% **5** 9% **6** 39·8%

7

Cost price	Selling price	P/L	% P/L
\$470	\$400	L \$70	L 14·9%
\$520	\$470	L \$50	L 9·6%
\$50	\$75	P \$25	P 50%
\$87.27	\$96	P \$8.73	P 10%
\$500	\$585	P \$85	P 17%
\$450	\$600	P \$150	P $33\frac{1}{3}$%

8 **a** \$170 **b** 46·1% **c** 85·4%
9 25% **10** \$30 **11** 75·4%

EXERCISE 3F

1 **a** \$35 **b** \$70 **c** \$84 **d** \$332.50 **e** \$2450
2 **a** \$37.50 **b** \$75 **c** \$90 **d** \$356.25 **e** \$2625
3 **a** \$960 **b** \$14.27
4 **a** \$8400 **b** \$271 600

5

Fee	Saving
11 000	6 000
13 750	8 750
16 500	11 500
19 250	14 250
20 625	15 625

6 **a** \$370 **b** \$379 **c** \$460 **d** \$590
7 \$25 600 **8** \$1 090 000
9 **a** \$7500 **b** \$10 000 **c** \$13 000 **d** \$16 000 **e** \$15 250 **f** \$16 750 **g** \$19 295 **h** \$22 000
10 **a i** 0 **ii** 0 **iii** \$45 **b** \$2167
11 **a i** \$600 **ii** \$515 **b** \$11 250

EXERCISE 3G

1 **a** \$300 **b** \$900 **c** \$10.80 **d** \$450 **e** \$1600 **f** \$1000
2 **a** \$29.17 **b** \$29.17 **c** \$40 **d** \$210.94 **e** \$7487.50 **f** \$21.92
3 **a** 2·5% **b** 2·5% **c** 2·2% **d** 2·6% **e** 1%
4 12% **5** 5% **6** 2 years

EXERCISE 3H

1 **a** 450 **b** 50 **c** 200 **d** 500 **e** 90 **f** 5000 **g** 800 **h** 18 **i** 6000 **j** 550 **k** 288 **l** 360
2 **a** 1125 **b** 450 **c** 126·7 **d** 1800 **e** 2000 **f** 200 **g** 1340 **h** 300 **i** 2666·7 **j** 10·8 **k** 84 **l** 717·6
3 \$21 527.78 **4** \$1.64 **5** \$62.50

Chapter Review Exercises

1 **a** $\frac{1}{2}$ **b** $\frac{14}{25}$ **c** $\frac{3}{4}$ **d** $\frac{1}{3}$
2 **a** 0·77 **b** 0·07 **c** 0·25 **d** 0·125
3 **a** 5% **b** 20% **c** 37·5% **d** $66\frac{2}{3}$%
4 **a** $66\frac{2}{3}$% **b** 30% **c** 87·5% **d** $82{\cdot}\dot{2}$% **e** 112%
5 **a** 38·85 **b** 54 **c** 14 cents **d** 112·5 **e** 700 grams **f** 44 **g** \$22.50 **h** 5000
6 **a** 106·7 **b** 87·3 **c** 702·46 **d** 10·08 km **e** \$17.01 **f** \$787.50
7 \$7309.44
8 **a** \$33.96 **b** \$5.99
9 **a** \$530 **b** 61·3%
10 \$17 700
11 \$288.96
12 **a** \$744.60 **b** \$49.64 **c** 2·8% (1 dp)
13 **a** 7·5 **b** 150 **c** 600 **d** 37·5 **e** 375 **f** 750
14 **a** 4800 **b** 75
15 \$6117.24
16 **a** \$3400 **b** \$2890

Keeping Mathematically Fit

PART A

1 37·5 cm **2** 94 300 **3** 5 **4** ^-5x
5 $^-47$ **6** 18 **7** 4·6 m **8** 80·2
9 1·2 **10** $x = 2$

PART B

1 5·52 **2** $^-50$
3 **a** $4x - 2$ **b** $18x^2y$ **c** $\frac{ab}{2}$
4 **a** $12(ab - 1)$ **b** $12a(b - 2)$ **c** $^-x(x + 4)$
5 **a** 98 **b** 100 **c** 160
6 **a** 107·5 **b** \$6.75 **c** 0·05
7 80 **8** $49m^2$ (7×7)

Cumulative Review 1

Part A

1 B **2** A **3** D **4** C **5** B **6** B
7 A **8** C **9** A **10** B **11** B **12** C
13 C **14** C **15** A **16** A **17** D **18** C
19 D **20** B **21** D **22** D **23** A **24** B

Part B

1 **a** 32 **b** 1 **c** 1 **d** $\frac{1}{4}$ **e** $\frac{1}{81}$ **f** 0·008 **g** $^-64$ **h** 729
2 **a** 72% **b** 2% **c** 105·3% **d** 37·5%
3 60% **4** $87\frac{1}{2}$% **5** 108% **6** \$1440
7 \$472.50
8 **a** \$3000 **b** 18 000
9 profit of 95·6% **10** 0
11 **a** $3n + 3$ **b** $6xy$ **c** $2y - 6x$ **d** $x + 12$ **e** 2 **f** $5p^3$ **g** $\frac{3b}{5}$ **h** $2m^2 + 5m$ **i** $2x + 12$
12 **a** $3b + 3$ **b** $18 - 12k$ **c** $3p^2 + 12ap$ **d** $^-2x + 3y$ **e** $21 - 28a$ **f** $x^2 + 5x$
13 **a** $5(x + 2)$ **b** $6(y - 1)$ **c** $2x(4y - 3)$ **d** $7(p - 2q)$ **e** $^-x(x + 1)$ **f** $p(p + q + 2)$
14 **a** $\frac{23x}{14}$ **b** $\frac{3a}{10}$ **c** $\frac{x}{8}$ **d** $\frac{3a}{2}$ **e** $\frac{w}{14}$ **f** $\frac{3x + 2}{6}$

15 a $\frac{x^2}{4y}$ b $\frac{a}{7}$ c $2x$ d $\frac{3}{a}$ e 4 f $\frac{x}{y}$

Chapter 4

MathsCheck

1 a love triangle; 6000 b nearest 500
c Eq.blues, Rh.rock, D.C.O.M. d the first 4
e ⊙⊙ f 76·6%

2 a hamburgers; 23% b pies 2%
c 8% d 100; total no. surveyed
e 6 times
f 'don't know column'; or 'none of these'

3 a artists' fee b insurance
c prod. costs d 10 cm; 2 cm
e $\frac{1}{5}$ f \$20 000, \$8000
g 4000 h 1 cm = \$10 000

4 a blue and green b pink, largest sector
c pink, most popular colour
d 3 or 4 students

EXERCISE 4A

1 a

Score	Tally	Freq.
6	|	1
7	𝍸	5
8	|	1
9	||	2
10	|	1
	Total	10

b 7 c 4
d 3 times
e 10

2 a

Score	Tally	Freq.
7:00	|	1
7:30	||	2
8:00	||||	4
8:30	𝍸	5
9:00	𝍸	5
9:30	𝍸 |	6
10:00	|||	3
NST	||||	4
	Total	30

b 30 c 9:30
d 12 e 18
f $\frac{2}{15}$

3 a

Score	Tally	Freq.
0	𝍸	5
1	𝍸 𝍸	11
2	𝍸 |||	8
3	𝍸	5
4	|	1
	Total	30

b 1 c 4
d 4 e 0
f 16 g 6
h $\frac{4}{15}$ i 80%

EXERCISE 4B

1 a

No. of CD players per household	Freq.
0	◖
1	⊙ ◖
2	⊙ ⊙ ⊙ ⊙
3	⊙ ⊙
4	⊙
over 4	◖

Key: ⊙ = 4 households

b 2
c 8
d $\frac{2}{38}$
e 38

2 Title: __________

Brisbane $3\frac{1}{2}$ symbols
Sydney 5 symbols
Melbourne $4\frac{1}{2}$ symbols
Hobart $1\frac{1}{2}$ symbols
Adelaide 2 symbols
Perth 3 symbols
Darwin $\frac{1}{2}$ symbol

a Sydney b Darwin c 75 d 3rd last
e 300 000

3 a A: 15 B: 28 C: 34
b D: 6 E: 13 F: 22

4 a 10
b A: 30 B: 90 C: 120
c i first stroke above 50
ii midway bet. 50 and 100

5 a, b

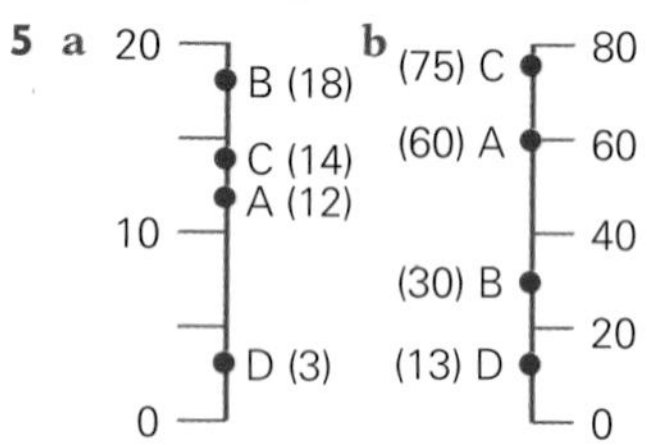

c, d

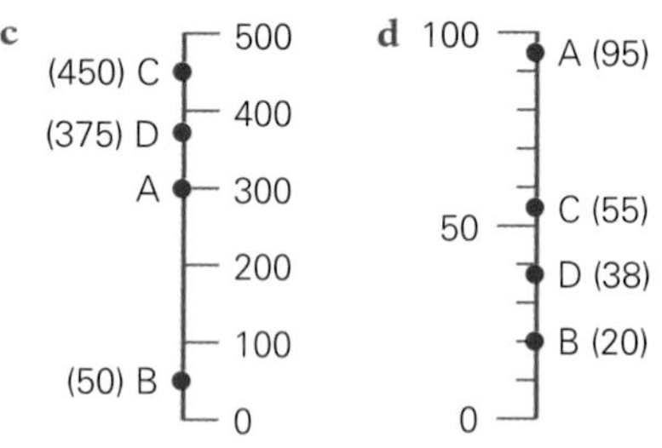

6

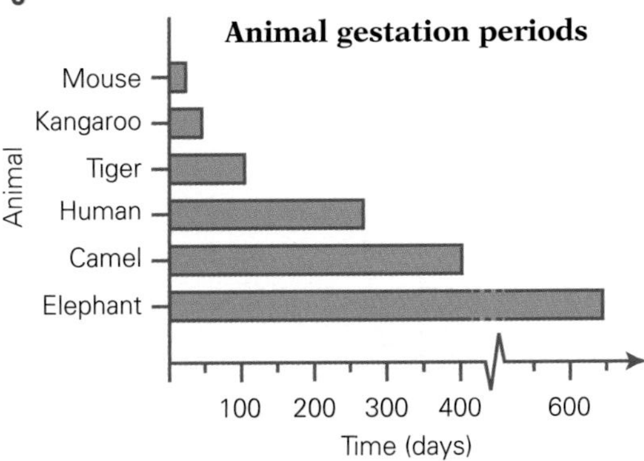

7 a

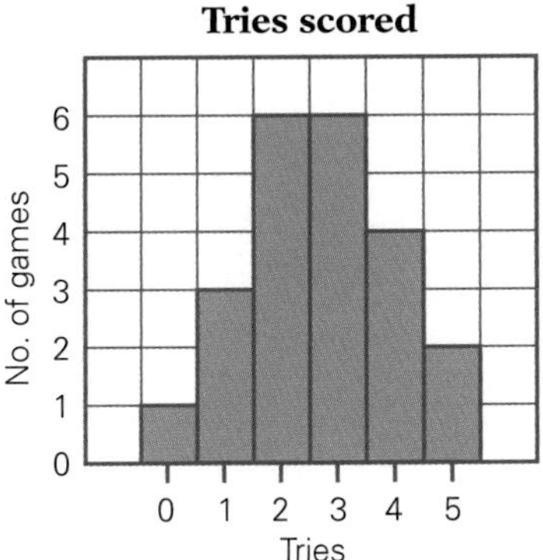

b 22 games

c i $\frac{1}{22}$ ii $\frac{3}{11}$ iii $\frac{6}{11}$

d 2·7

8 a

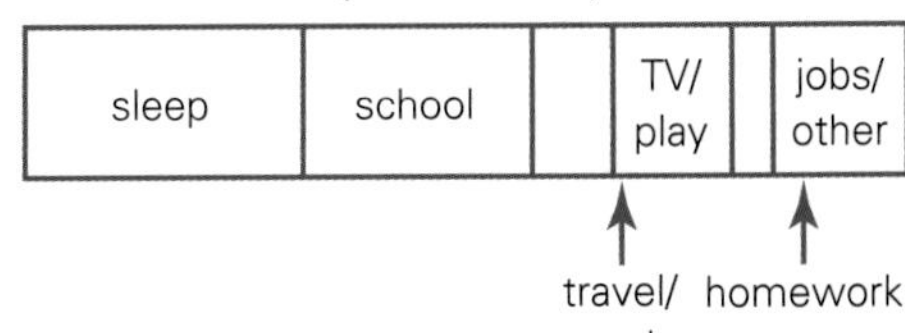

b sleep
school
jobs
TV
travel
homework

c school d $\frac{1}{8}$ e sleep

f discuss and compare

EXERCISE 4C

1 $\frac{120}{360} = \frac{1}{3}$ $\frac{90}{360} = \frac{1}{4}$ $\frac{60}{360} = \frac{1}{6}$ $\frac{45}{360} = \frac{1}{8}$

$\frac{40}{360} = \frac{1}{9}$ $\frac{72}{360} = \frac{1}{5}$ $\frac{270}{360} = \frac{3}{4}$

2 a movies b quiz shows

c news/current affairs and comedy

3 a 360 min b 90°, $\frac{1}{4}$ c $\frac{1}{6}$ d 120 min

4 a poker machines b catering. Discuss

c 240°, $\frac{2}{3}$ d 72°, $\frac{1}{5}$

e $576 000 f discuss

5 a C: 180° S: 90° V: 60° B: 30°

b

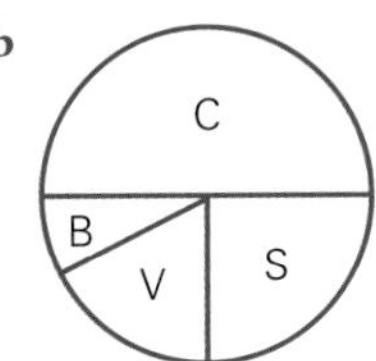

6 blue: 135° brown: 108° green: 72°
hazel: 45°

7 discuss

8 a 160° b $\frac{1}{4}$

c i $20 250 ii $30 375 iii $81 000

EXERCISE 4D

1 a 18 kg b 8 kg c 2 kg d 6th week
e 3rd week; horizontal interval
f 1 kg/week
g consistent or steady weight loss
h 15 kg i week 5

2 $1.50, $2, $3, $4.25, $2.50

3 a being filled with water
b 72 L
c 2 L water lost. Jason hopping into full bath
d stays the same
e 8 min, 5 min
f i 30 L ii 70 L iii 30 L
g 30 min h 7:04 pm
i water flowing in faster
j water out at steady rate
k 14 L/min

4 a

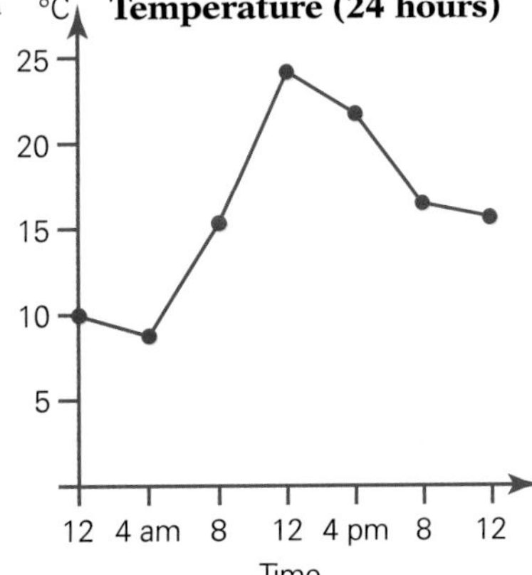

b

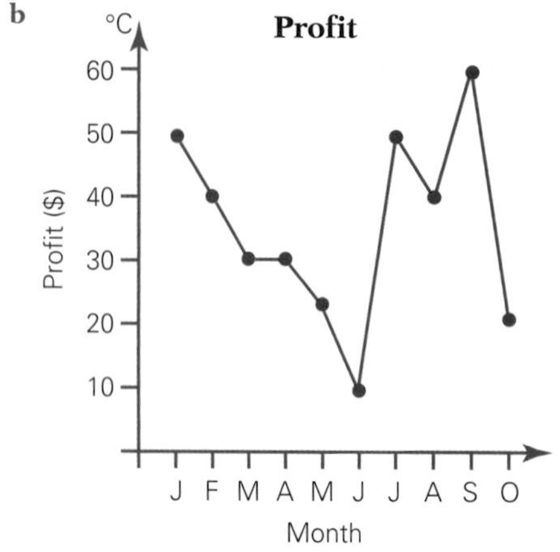

5

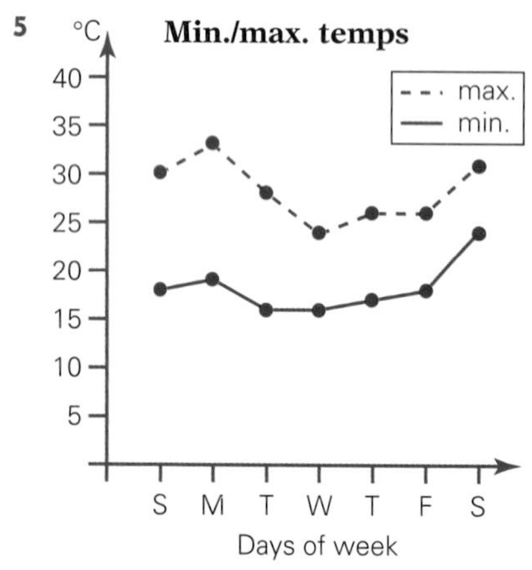

6 **a** 10 m **b** 10 m **c** 11 am **d** 6 m
e 7:30 am to 1:30 pm

EXERCISE 4E

1 **a** (iii) **b** (i) **c** (ii) steepest graph
2 **a** F **b** B **c** D **d** A **e** B, D & E
f G
3 discuss **4** C **5** B
6 **a** 11 am–11:30 am **b** 7 am
c turned straight around and returned home
d 5 hours **e** 240 km
f 4 hours **g** 20 km
h 1 hour **i** 4–4:30 pm
j a lift in a car **k** 60 km/h
l same distance from starting point at the same time

7 **a**

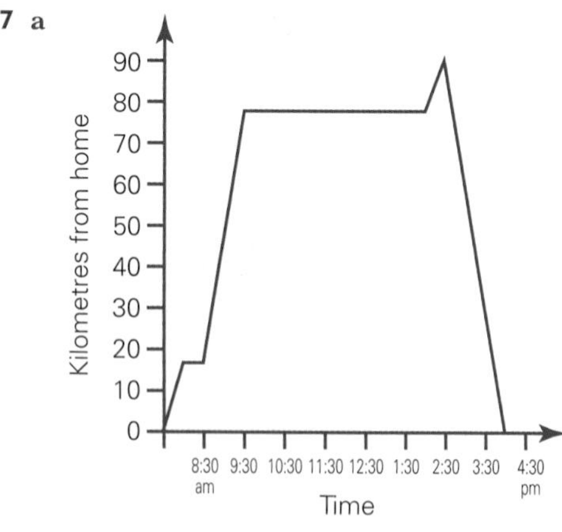

b 60 km/h **c** $8\frac{1}{2}$ h **d** discuss

EXERCISE 4F

1 **a** 1 cm = \$10 **b** \$2
c **i** US\$30 **ii** US\$30 **d** **i** A\$120 **ii** A\$20
e US\$350
2 **a** 120°F **b** approx. 40°C **c** approx. 30°F
3 **a** 50 miles **b** 25 miles **c** 32 km
d $37\frac{1}{2}$ miles/h **e** $\frac{5}{8}$
4 **a** horizontal axis
b **i** US\$20 **ii** NZ\$55
c A\$60 **d** A\$600
e **i** A\$100 **ii** NZ\$137.50
f \$200 **g** \$80
h **i** $\text{US} = \text{A} \times \frac{1}{2}$ **ii** $\text{NZ} = \frac{11\text{A}}{8}$
i US\$1.38, NZ\$3.78

5

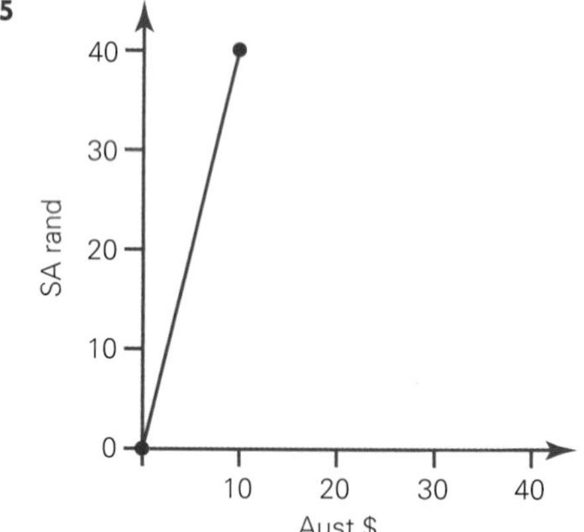

EXERCISE 4G

1 **a** \$1.50 **b** \$2.50 **c** **i** \$4 **ii** \$5
d less than 3·5 kg
2 **a** \$1.50 **b** \$2.50 **c** **i** \$4.00 **ii** \$5.00
d less than 3·5 kg **e** 50c **f** combined

3 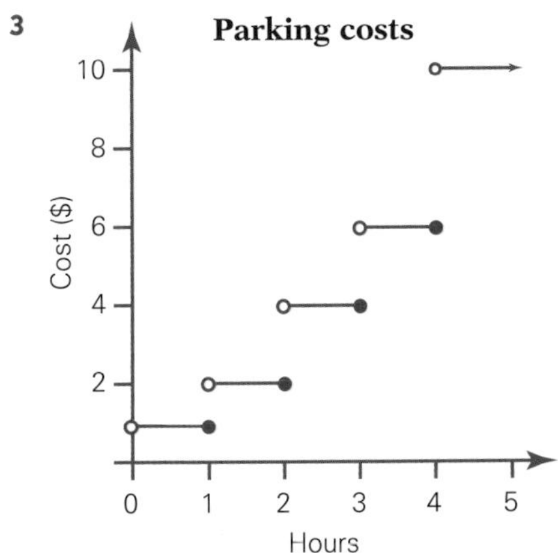

EXERCISE 4H

1 a vertical axis doesn't start at zero
 b 3D enlargement used to magnify effect
 c vertical axis broken, doesn't start at zero
 d columns not of equal width
 Discuss new graphs
2 a 1600, yes
 b A and B same scale, C has a smaller scale
 c A has a smaller scale than B or C and covers more years
 d discuss
3 a vertical axis broken
 b different symbols which vary in size and shape used to represent the same number of people
4 discuss and compare
5 discuss and compare

EXERCISE 4I

1 a \$5.80 b \$8.76
2 a Perth b Canberra
 c distance from Sydney
 d Canberra, Melbourne, Brisbane, Adelaide, Hobart, Darwin, Perth
3 a \$2.18 b \$1.73 c \$18.75 d \$30.28
 e 5 kg
4 a \$1.45/kg b \$4.38/kg
5 a 35c b 14c
 c 50c/min: longest distance; day rate—busiest time
 d 10c/min: shortest distance; least used time
 e over 745 km
 f 9c/min
 g i 20c ii \$1.08 iii \$1.70 iv \$1.35
 h \$2.88
 i They are charged at local call rates, not STD

Chapter Review Exercises

1 a 240 cars b
2 a discuss, i.e. favourite soft drinks
 b discuss; one example:
 = 4 students
 c COLA
 OTHER 5D
 d $\frac{1}{20}$

3 a

Score	Tally	Freq.
18	II	2
19	IIII	4
20	~~IIII~~ III	8
21	III	3
22	II	2
23	I	1
	Total	20

 b 20 days
 c 6 times
 d 6 times
4 a 19 b 4 c 289 d 28
 e win: 2, draw: 1 f 2 draws = 1 win
 g Wests, Illawarra, Gold Coast
 h win
5 a 1 kg b 1988 c 41 625 t
 d coffee is increasing to lead over tea; already more than double
 e coffee ≈ 2·5 kg/person
 tea ≈ 0·75 kg/person
6 a food b savings c \$6
 d soccer, swap cards e $\frac{1}{3}$ f 7 weeks
7 a government b retailer c $\approx \frac{1}{10}$
 d 16c e \$2 397 000 000 f 139°
8 a 1 cm = 5°C b 24 min ($\frac{2}{5}$ h)
 c i 11°C ii 27°C
 d 12 noon, 4:24 pm e 19°C
 f 15·5°C
9 a 60 km b 20 min c 20 km d 120 km
 e 80 min
10 a \$1 b \$1.50 c \$2 d < 6 h e < 5 h
11 a line b sector c column d column

Keeping Mathematically Fit

PART A

1 $2ab$ 2 3^6 3 $2^3 \times 5^2 \times 11$
4 $2\frac{5}{12}$ 5 $^-46$ 6 $\frac{9}{40}$ 7 12·5
8 4 9 5250 g 10 36

PART B

1 \$36.75 2 24 3 $^-6{\cdot}18$ 4 7%
5 a $^-72$ b 5 c $^-80$ d 144
6 a $2 - 2a$ b $\frac{2x}{y}$ c $5x + 7$
7 $y = x + 7$
8 a 2 or 3 b 2 values

Chapter 5

MathsCheck

1 a $12k$ b $20ap$ c $12m^2$ d $10n$ e $5d$

f $2ab$ g $8x + 13$ h ^-2c i $8p - 3q^2$

2 a $3k + 24$ b $5x - 15$ c $^-2x + 8$
d $2l + 2b$ e $4an + 6bn$ f $x^2 + 3xy$

3 a $m + 26$ b $6 - 5p$

4 $^-18$ **5** 10

6 16 → 25 → 75 → 25 → 16

7 a switch off b fill c surrender
d subtract 7 e halve f add 8

8 a 0 b 0 c 1 d 1 e 1 f 1
g 0 h 0 i 1

9 a a b x c p d $8n$ e t f w

10 a $^-6$ b $\frac{1}{8}$ c 4 d 4 e $^-2$ f 3
g 3 h 7 i 7 j $^-1$ k 1 l 7

EXERCISE 5A

1 b, f, g

2 a T b T c F d T e F f T
g F h T

3 a $+4$ b -3 c $\times 5$ d $\div 7$ e $\times 4, +8$
f $\times 2, -3$ g $\times {}^-6, \div 5$ h $+5, \div 3$ i $^-2, \times 3$

4 a -4 b $+3$ c $\div 5$ d $\times 7$ e $-8, \div 4$
f $+3, \div 2$ g $\times 5, \div {}^-6$ h $\times 3, -5$ i $\div 3, +2$

5 a

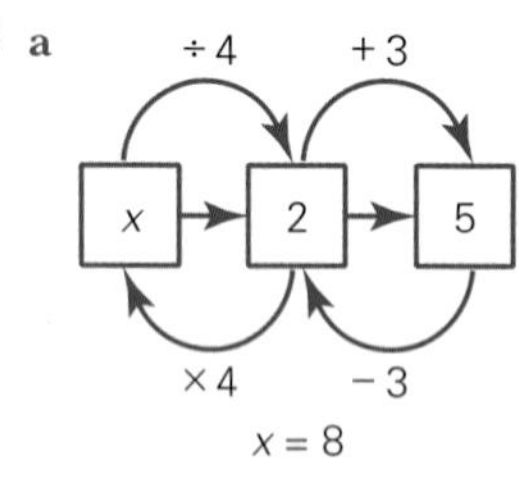

b

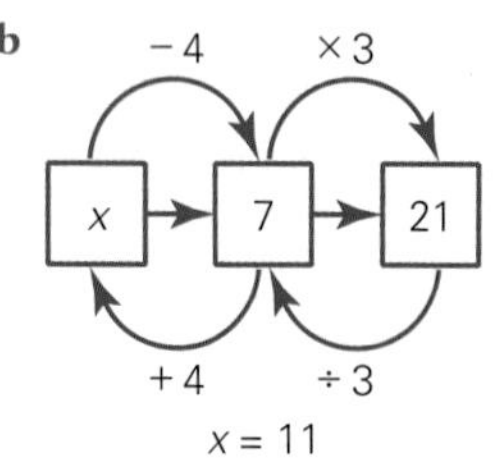

c

m = 2
↓ × 3 ↑ ÷ 3
$3m$ 3
↓ + 1 ↑ − 1
$3m + 1$ = 7

d

a = 3
↓ × 5 ↑ ÷ 5
$5a$ 15
↓ − 3 ↑ + 3
$5a - 3$ = 12

6 a T b F c F d T e F

7 a $a = b = 5$; $a = 1, b = 9$ are two solutions discuss others

8 a T b F c F d T e T f T

9 a A b B c A d A e B f B
g B h B i B

10 a $x = 3$ b $x = 2$ c $x = 2$ d $x = 5$
e $x = 5$ f $x = 24$

11 a 37 b 3 c discuss

EXERCISE 5B

1 a $t + 5 = 8$
$-5 \quad -5$
$t = 3$

b $n - 3 = 12$
$+3 \quad +3$
$n = 15$

c $8p = 24$
$\div 8 \quad \div 8$
$p = 3$

d $\frac{x}{5} = 10$
$\times 5 \quad \times 5$
$x = 50$

e $m - 5 = {}^-7$
$+5 \quad +5$
$m = {}^-2$

f $^-5c = 30$
$\div {}^-5 \quad \div {}^-5$
$c = {}^-6$

2 a $t = 5$ b $m = 7$ c $x = 33$ d $n = 4$
e $a = 24$ f $d = 9$ g $x = {}^-4$ h $a = {}^-3$
i $l = 3$ j $r = {}^-21$ k $p = {}^-3$ l $c = {}^-23$
m $b = \frac{1}{6}$ n $x = \frac{3}{4}$ o $a = 1\frac{1}{2}$ p $h = 70$
q $p = {}^-1$ r $x = {}^-5$

3 a $m = 2$ b $a = 2$ c $x = 2$ d $h = 6$
e $c = 19$ f $t = 1$

4 a $5p = 40$; $p = 8$ b $t - 9 = 32$; $t = 41$
c $x + 7 = 50$; $x = 43$ d $8x = 56$; $x = 7$
e $\frac{x}{9} = 8$; $x = 72$ f $\frac{2}{3}x = {}^-12$; $x = {}^-18$

5 a $x = {}^-7$ b $x = {}^-9$ c $p = 5$ d $x = \frac{^-1}{2}$
e $x = {}^-4$ f $x = 12$

6 a $x = 4$ b $x = 5$ c $x = 3$ d $x = 2$ e $p = 3$
f $a = 4$ g $a = 2$ h $d = 1$ i $x = 1$ j $p = 2$
k $p = 2$ l $s = 1$ m $d = 3$ n $d = 1\frac{1}{2}$ o $a = 1$
p $h = 3$ q $h = 3$ r $t = 2$ s $x = {}^-1$ t $p = {}^-2$
u $w = {}^-5$

7 a $x = 4$ b $a = 25$ c $a = 4$ d $d = 40$
e $s = 0$ f $a = 33$ g $w = 25$ h $q = 7$
i $r = 50$ j $x = 24$ k $x = 0$ l $x = 9$
m $x = 0$ n $a = 8$ o $w = {}^-9$

8 a $a = 13$ b $a = 15$ c $x = 1$ d $n = 12$
e $w = 2$ f $w = {}^-7$ g $x = 4$ h $x = \frac{5}{3}$
i $a = 9$ j $a = 14$ k $p = 7$ l $y = {}^-\frac{4}{5}$

9 a $x = 0{\cdot}5$ b $x = 0{\cdot}9$ c $h = 3\frac{1}{2}$ d $x = 7{\cdot}5$
e $p = 2{\cdot}8$ f $x = \frac{1}{4}$ g $x = 1{\cdot}2$ h $x = 0{\cdot}3$
i $p = 1{\cdot}9$ j $p = 0{\cdot}1$ k $w = 3\frac{1}{2}$ l $w = \frac{1}{2}$

10 **a** $x = {}^-9$ **b** $x = \frac{3}{4}$ **c** $x = 9$ **d** $x = {}^-100$ **e** $x = 0$

11 **a** $a = 5922$ **b** $x = {}^-90$ **c** $k = {}^-42\,320$ **d** $a = 1{\cdot}16$ **e** $x = {}^-45{\cdot}12$ **f** $x = 4311$

12 **a** $x = 3$ **b** $x = 4$ **c** $a = {}^-2$

13 **a** $p = 3$ **b** $k = 15$ **c** $a = 2$ **d** \$35

EXERCISE 5C

1 **a** $x + 10 = 19$, $x = 9$ **b** $7x = 35$, $x = 5$ **c** $x - 8 = 25$, $x = 33$ **d** $\frac{x}{4} = 9$, $x = 36$ **e** $2x + 7 = 27$, $x = 10$ **f** $\frac{x}{2} - 4 = 12$, $x = 32$ **g** $5x = 17$, $x = 3{\cdot}4$ **h** $x - 3 = {}^-3$, $x = 0$ **i** $4x - 9 = 23$, $x = 8$ **j** $3n + 5 = {}^-13$, $x = {}^-6$ **k** $x + x + 1 = 69$, $x = 34$

2 D **3** B **4** B **5** B

6 **a** 50 **b** 30 **c** 40, 41, 42 **d** 40, 42, 44

7 \$640

8 **a** $x = 8$ **b** $x = 4$ **c** $x = 36$ **d** $x = 15{\cdot}5$ **e** $x = 9\frac{2}{3}$ **f** $x = 10$

9 sides = 12 cm and 36 cm

10 8 m by 15 m; $A = 120\text{ m}^2$

11 28 and 33 **12** 120

EXERCISE 5D

1 **a** $x = 2$ **b** $a = 3$ **c** $x = 0$ **d** $a = \frac{1}{2}$ **e** $x = 2$ **f** $x = 14$ **g** $s = {}^-2$ **h** $a = {}^-1$ **i** $a = 11$ **j** $a = {}^-1$ **k** $a = 0$ **l** $a = 6$

2 **a** $x = 2$ **b** $x = \frac{5}{2}$ **c** $x = \frac{12}{5}$ **d** $x = 2$ **e** $x = \frac{5}{2}$ **f** $p = 0$ **g** $a = 1$ **h** $d = 0$ **i** $p = {}^-3$ **j** $n = 3$ **k** $k = 2$ **l** $w = 4$

3 **a** $x = 2$ **b** $a = 2$ **c** $x = \frac{-3}{2}$ **d** $a = 8$ **e** $x = {}^-4$ **f** $x = 4$ **g** $a = 5$ **h** $x = 2$ **i** $w = 5$ **j** $p = {}^-3$ **k** $b = {}^-3$ **l** $s = 1$

4 **a** $x = {}^-1$ **b** $p = \frac{16}{21}$ **c** $x = \frac{31}{12}$ **d** $x = \frac{21}{40}$ **e** $p = \frac{-3}{2}$ **f** $p = \frac{4}{3}$

5 **a** $x = 1$ **b** $x = 1$ **c** $x = 2$ **d** $d = 3$ **e** $p = 3$ **f** $x = \frac{19}{6}$ **g** $x = {}^-31$ **h** $p = {}^-2$ **i** $x = {}^-2$ **j** $x = 0$ **k** $x = {}^-2$ **l** $x = 1$

6 **a** $x = {}^-2$ **b** $a = \frac{17}{10}$ **c** $d = \frac{3}{5}$ **d** $t = 19$ **e** $x = 2$ **f** $f = {}^-34$ **g** $p = \frac{-3}{5}$ **h** $w = \frac{-5}{12}$

7 **a** 0 **b** $\frac{54}{7}$ **c** 2

8 **i** **a** $P = 2(x + 2) + 8$ **b** $x = 1$ **ii** **a** $P = 2(2x - 5) + 2x$ **b** $x = 4$ **iii** **a** $P = 4(20 - 3x)$ **b** $x = 5\frac{1}{2}$

9 \$21 **10** \$20

EXERCISE 5E

1 **a** $n = 1$ **b** $p = 7$ **c** $a = 2$

2 **a** $x = 3$ **b** $t = 2$ **c** $p = 4$ **d** $n = \frac{1}{2}$ **e** $c = 4$ **f** $n = {}^-8$ **g** $x = 2$ **h** $p = 5$ **i** $k = {}^-3$ **j** $x = 1$ **k** $t = 2$ **l** $x = 3$ **m** $y = 2$ **n** $x = {}^-1$ **o** $w = 3$

3 **a** $h = 6$ **b** $l = 7$ **c** $a = 3$ **d** $b = 4$ **e** $y = 1$ **f** $t = 0$ **g** $t = {}^-5$ **h** $v = 1$ **i** $x = 2$ **j** $y = 2$ **k** $p = 7$ **l** $x = {}^-2$ **m** $a = \frac{1}{2}$ **n** $p = {}^-2$ **o** $a = 5$

4 **a** $z = 6$ **b** $p = 2$ **c** $n = 4$ **d** $n = 1$ **e** $x = 2$ **f** $k = 13$ **g** $m = 1$ **h** $p = 2$

5 \$20 **6** 9 **7** $x = 30$ **8** \$80

9 **a** $a = 11, b = 18$ **b** $a = 20, b = 4$ **c** $a = 40, b = 28$ **d** $a = 32, b = 18$

10 **a** $\frac{8}{3}$ **b** 4 **c** 2, 4, 6 **d** $x = {}^-3$ **e** \$2.50

EXERCISE 5F

1 **a** $P = 22$ **b** $P = 22$ **c** $P = 30$ **d** $P = 44$ **e** $P = 32$ **f** $P = 1{\cdot}8$ **g** $P = 30{\cdot}8$ **h** $P = 60$

2 **a** $A = 24$ **b** $A = 24$ **c** $A = 54$ **d** $A = 120$ **e** $A = 63$ **f** $A = 0{\cdot}2$ **g** $A = 54{\cdot}88$ **h** $A = 225$

3 **a** $S = 50$ **b** $S = 90$ **c** $S = 36$

4 **a** $V = 12$ **b** $V = 25$ **c** $V = 55$ **d** $V = 153$ **e** $V = 5{\cdot}2$ **f** $V = 8{\cdot}8$ **g** $V = 109$ **h** $V = 32{\cdot}65$ **i** $V = 32{\cdot}5$

5 **a** $y = 13$ **b** $y = {}^-7$ **c** $y = 13$ **d** $y = 0$ **e** $y = 5$ **f** $y = {}^-5$ **g** $y = 11$ **h** $y = 27$ **i** $y = 3$

6 **a** $E = 36$ **b** $E = 10$ **c** $E = 0{\cdot}1056$ **d** $E = \frac{1}{32}$ **e** $E = 60$ **f** $E = 1$ **g** $E = 38{\cdot}808$ **h** $E = 2{\cdot}754$

7 **a** $C = {}^-17\frac{7}{9}$ **b** $C = {}^-12\frac{2}{9}$ **c** $C = 0$ **d** $C = 4\frac{4}{9}$ **e** $C = 37\frac{7}{9}$

8 **a** $I = 2400$ **b** $I = 270$ **c** $I = 1023{\cdot}75$ **d** $I = 33$

9 **a** $c = 5$ **b** $c = 10{\cdot}8$ (1 dp) **c** $c = 1$ **d** $c = 17$

10 **a** $x = 4$ **b** $a = 3$ **c** $x = \frac{17}{4}$

11 **a** $l = 3$ **b** $l = 5$ **c** $l = 20$ **d** $l = 9{\cdot}2$

12 **a** $b = 10$ **b** $b = 6$ **c** $b = 20\frac{2}{3}$ **d** $b = 40$

13 **a** $b = 3$ **b** $b = 7$ **c** $b = 1{\cdot}19$ **d** $b = 27{\cdot}25$

14 **a** $t = 3$ **b** $t = 5$ **c** $t = 384$ **d** $t = 1{\cdot}06$

15 **a** $a = 9$ **b** $a = 18{\cdot}42$

16 **a** $x = 4$ **b** $b = 5$ **c** $x = 8$ **d** $x = 0{\cdot}7$ **e** $x = 100$

17 **a** $F = 32$ **b** $F = 89{\cdot}6$ **c** $F = 212$ **d** $F = 77$ **e** $F = 50$

18 $h = 5{\cdot}32$ (2 dp) **19** $r = \frac{1}{2}$ **20** $x = 10$

21 $a = \pm 7{\cdot}9$ (1 dp) **22** $u = \pm 9{\cdot}4$ (1 dp)

EXERCISE 5G

1 **a** 3 **b** 2 **c** 1 **d** 5 **e** 4

2 **a** $x > 100\,000$ **b** $n \geq 50$ **c** boxes > 3
d $p > 100$ **e** $w \leq 30$

3 **a** $x < 0$; $^-6$, $^-2$, $\frac{-1}{2}$
b $x > 10$; 15, 24
c $x \geq 10$; 10, 15, 24
d $x \leq 0$; $^-6$, $^-2$, $\frac{-1}{2}$, 0
e $x \geq {}^-1$; $\frac{-1}{2}$, 0, 2, 5, 7, 10, 15, 24
f $x < 10$; $^-6$, $^-2$, $\frac{-1}{2}$, 0, 2, 5, 7

4 **a** 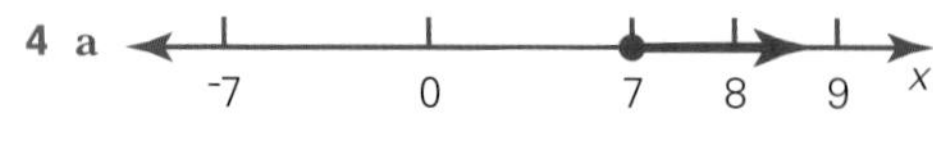

b 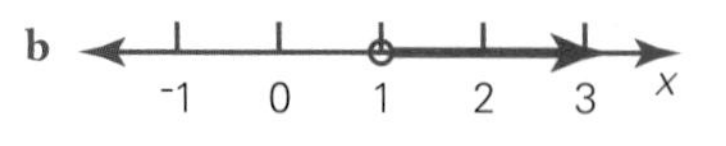

c 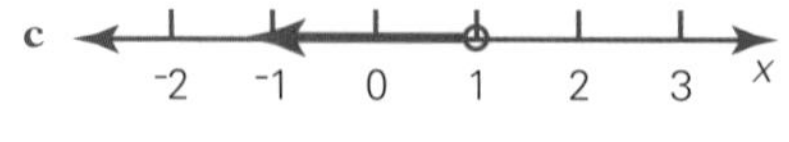

d

e

f

g

h

i

5 **a** $x \geq 1$ **b** $x < 1$ **c** $x < 6$ **d** $x \geq {}^-2$
e $x \leq 5$ **f** $x \geq {}^-2$

EXERCISE 5H

1 **a** $x > 5$ **b** $x < 5$ **c** $x > 5$ **d** $x \geq 16$
e $x > 7$ **f** $x < 7$ **g** $p \leq {}^-8$ **h** $a > 0$
i $x < 7$ **j** $x \leq 7$ **k** $m > {}^-3$ **l** $d > 5{\cdot}4$
m $x \leq 0{\cdot}7$ **n** $x \leq 12$ **o** $x > 4$

2 **a** 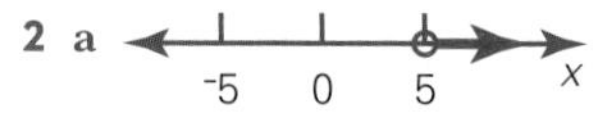

b

c

d

e

f

g

h

i

j

k

l

m

n

o

3 **a** $a \leq 2$ **b** $y > 3$ **c** $p > 5$ **d** $x \geq 4$
e $x < 1$ **f** $w \geq \frac{3}{2}$ **g** $x < 1$ **h** $x \geq 1$
i $p < 1$

4 **a** $x \leq 2$ **b** $a \leq 1$ **c** $x \geq 28$ **d** $x > {}^-15$
e $x < 4$ **f** $x < 30$ **g** $x > \frac{19}{2}$ **h** $x < \frac{-1}{2}$
i $x \geq \frac{-8}{3}$ **j** $x > 1$ **k** $x \geq \frac{3}{2}$ **l** $x < {}^-4\frac{1}{2}$

5 **a** $3n < 9$ **b** $\frac{3n}{4} < 6$ **c** $2n + 15 > 20$
$\therefore n < 3$ $\therefore n < 8$ $\therefore n > \frac{5}{2}$
d $x + x + 4 < 24$ **e** $4x + 7 \leq 27$
$\therefore x < 10$ $\therefore x \leq 5$

6 **a** $x > {}^-1$ **b** $x \leq {}^-3$ **c** $p \geq {}^-7$ **d** $a > 19$
e $w \leq {}^-6$ **f** $p > {}^-6$ **g** $w < {}^-5$ **h** $p \geq {}^-2$
i $a < \frac{5}{4}$

Chapter Review Exercises

1 c **2** b, e **3** a, b, c, d

4 **a** $p = 9$ **b** $m = 72$ **c** $x = 1$
d $w = {}^-2$ **e** $q = 12$ **f** $x = {}^-9$

5 **a** $t = 5$ **b** $c = 10$ **c** $d = 20$
d $w = \frac{8}{5}$ **e** $x = {}^-1$ **f** $w = 32$
g $w = {}^-4$ **h** $t = 8\frac{1}{2}$ **i** $x = {}^-8$

6 **a** $n = {}^-1$ **b** $h = 4$ **c** $k = 2$
d $v = 2$ **e** $p = 3$ **f** $w = {}^-1$

7 **a** $x = 1$ **b** $w = {}^-1$ **c** $b = 2$
d $c = 10$ **e** $d = \frac{8}{7}$ **f** $x = \frac{23}{15}$

8 **a** $P = 40$ **b** $l = 31$ **c** $a = 4\frac{1}{4}$

d $c = \pm30$ e $a = \pm4$

9 a $300 + 50x = 750$, $x = 9$
b $x + (x + 6) = 150$, larger is 78
c $3x + 6 = 177$; 57, 59, 61
d $140 + 45h = 500$; 8 hours
e $n + 14 = 2n - 4$; $n = 18$

10 a 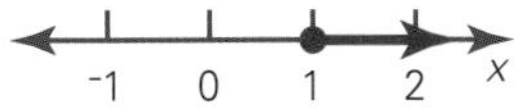

b 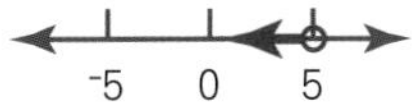

c 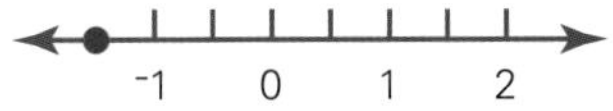

11 a $x > 3$ b $x < 3$ c $x > 54$
d $x \le 2$ e $x \ge 10$ f $x < \frac{^-3}{2}$

12 a an integer ≤ 12 b an integer $\le {}^-20$
c an integer ≤ 0

13 a $w > {}^-4$ b $p < {}^-1$ c $a \ge {}^-2$

14 $a \le 2$

Keeping Mathematically Fit

PART A

1 $\frac{1}{2}$ **2** 47 **3** $40 \div 10 = 32 \div 8$
4 121 **5** \$4.80 **6** 0·8 **7** 2
8 1 **9** $T = 20 - 4n$ **10** 1·1856

PART B

1 3·5 **2** \$207 **3** $3\frac{1}{14}$ **4** \$8.39 **5** $1\frac{3}{5}$
6 a 3^{13} b $2x + 13$ c $\frac{12y}{5}$
7 75·25 cm² **8**

4	9	2
3	5	7
8	1	6

, yes

Chapter 6

MathsCheck

1 a i ΔABC ii right-angle triangle
b i ΔPQR ii equilateral triangle
c i ΔXYZ ii isosceles triangle
d i ΔCDE ii scalene triangle

2 a BC b AB c XY d BC

3 a 44 cm b 22 cm c 24 cm d 29·2 cm

4 a 25 b 11·56 c 225 d 1681
e 25 f 56

5 a 12·81 b 6·40 c 9·64
d 9 e 7·42 f 10

EXERCISE 6A

constructions

EXERCISE 6B

1 a y b q c RQ d XY

2 a C b A c B

3 a $b^2 = a^2 + c^2$ b $r^2 = p^2 + q^2$
c $q^2 = r^2 + s^2$ d $z^2 = x^2 + y^2$

4 a $(AC)^2 = (BC)^2 + (AB)^2$ or $b^2 = a^2 + c^2$
b $(DE)^2 = (CD)^2 + (CE)^2$ or $c^2 = e^2 + d^2$

5 a $x^2 = 3^3 + 4^2$ b $x^2 = 5^2 + 12^2$
c $x^2 = 9^2 + 12^2$ d $x^2 = 8^2 + 15^2$
e $x^2 = 7^2 + 24^2$ f $x^2 = 60^2 + 80^2$

6 a $x = 5$ b $x = 13$ c $x = 15$ d $x = 17$
e $x = 25$ f $x = 100$

7 a $x = 6{\cdot}40$ b $w = 25{\cdot}61$ c $a = 4{\cdot}78$
d $p = 12{\cdot}81$ e $v = 9{\cdot}84$ f $a = 15{\cdot}58$

8 50 m **9** 75 m **10** 68 cm **11** $\sqrt{74}$

12 a 12·2 cm b 11·3 cm c 16·0 cm

13 a 5 b 7·2 c 5

14 2·12 m (2 dp) **15** $\sqrt{57}$ **16** 289 cm

17 a $a = 10{\cdot}6$, $P = 31{\cdot}6$ m b $w = 10{\cdot}8$, $P = 42{\cdot}8$ m
c $x = 5$, $P = 20$ cm

18 a 80·62 m b 78·10 m c 348·72 m
d \$8607.55

EXERCISE 6C

1 a $10^2 = b^2 + 8^2$
$b^2 = 10^2 - 8^2$
$= 36$
$b = \sqrt{36}$
$b = 6$

b $17^2 = x^2 + 8^2$
$x^2 = 17^2 - 8^2$
$x^2 = 225$
$x = \sqrt{225}$
$x = 15$

c $13^2 = p^2 + 12^2$
$p^2 = 13^2 - 12^2$
$= 25$
$p = \sqrt{25}$
$p = 5$

2 a $x = 9$ b $x = 48$ c $x = 28$
d $x = 40$ e $n = 9$ f $n = 4{\cdot}8$

3 a $p = 3{\cdot}5$ b $p = 6{\cdot}6$ c $p = 8{\cdot}0$
d $p = 8{\cdot}1$ e $p = 15$ f $p = 39{\cdot}2$

4 3·2 m **5** 90 m **6** 2·498 km (2498 m)

7 5·23 m (523 cm) **8** 9·2 m

9 a 8·02 km b 0·67 h (40 min)

10 a $a = 2{\cdot}6$, $b = 6{\cdot}3$, $P = 19{\cdot}9$ cm
b $x = 15$, $y = 13{\cdot}2$, $P = 54{\cdot}2$ m
c $x = 16$, $y = 46{\cdot}9$, $P = 112{\cdot}9$ m
d $x = 6{\cdot}6$, $y = 3{\cdot}3$, $P = 58{\cdot}3$ m

EXERCISE 6D

1 a

a	b	c
3	4	5
6	8	10
9	12	15
12	16	20
1·5	2·0	2·5

2 a yes b yes c yes d yes e yes f yes

3 a $(5x)^2 = (3x)^2 + (4x)^2$
b i 3, 4, 5 ii 21, 28, 35 iii 27, 36, 45
iv 105, 140, 175
c 95

4 a (5, 12, 13) (10, 24, 26) (15, 36, 39)
b (8, 15, 17) (16, 30, 34) (24, 45, 51)

5 a

a	b	c
3	4	5
5	12	13
7	24	25
8	15	17
9	40	41

6 a i 3, 4, 5 **ii** 15, 112, 113
b i 27, 364, 365 **ii** 31, 480, 481
7 $x = 30$

EXERCISE 6E

1 a no **b** no **c** yes **d** yes **e** yes **f** yes
2 a yes **b** no

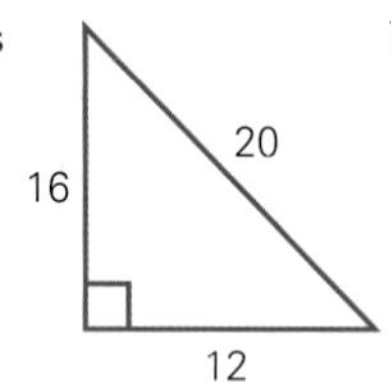

c yes

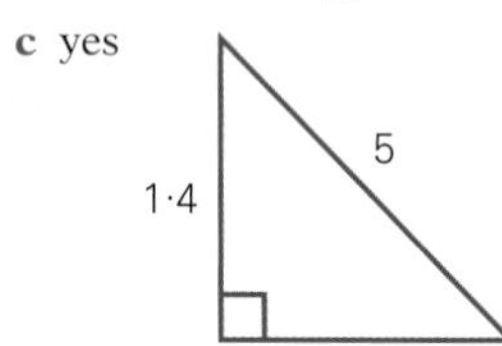

3 3500 mm
4 a 21, 20, 29 **b** 5, 12, 13
c 36, 27, 45 **d** 3·6, 4·8, 6
e 41, 40, 9 **f** 3·3, 5·6, 6·5

EXERCISE 6F

1 6·88 km (2 dp) **2** 2343 mm (0 dp)
3 5·9 m (1 dp) **4** 14·7 m (1 dp)
5 20·1 m **6** 26 km
7 7·155m (7155 mm) **8** 16 cm^2
9 31·22 cm^2 **10** 88·6 cm (1 dp)
11 a 8·66 **b** 8·60 **c** 14·70 **d** 15·81
12 a $\sqrt{20} \doteqdot 4{\cdot}47$ **b** 4·04 (2 dp)
13 4 cm by 8 cm **14** 22·45 m
15 right, isosceles triangle, $P = 34{\cdot}14$ cm
$A = 50$ cm^2

Chapter Review Exercises

1 a m **b** AC **c** 13
2 a $m^2 = a^2 + b^2$ **b** $25^2 = 7^2 + 24^2$
c $(DE)^2 = (DF)^2 + (FE)^2$
3 a 19·21 **b** 2·89 **c** 48·99
4 8·72 cm (2 dp) **5** no ($7{\cdot}4^2 \neq 6{\cdot}2^2 + 3{\cdot}1^2$)
6 4·6 cm (1 dp) **7** 10·8 m
8 a $x = \sqrt{113} \doteqdot 10{\cdot}6$ **b** $x = 13$
$y = \sqrt{138} \doteqdot 11{\cdot}7$ $y = \sqrt{56} \doteqdot 7{\cdot}5$
9 10·97 km (2 dp)
10 a 10·39 (2 dp) **d** 15·62 (2 dp)
11 $\sqrt{45}$

Keeping Mathematically Fit

PART A

1 497 000 **2** 1·296 **3** 32 **4** 12
5 9 and 10 **6** ⁻9 **7** 8, 13 **8** 140
9 624·8 **10** $\frac{3}{8}$

PART B

1 a pink **b** 35 **c** $\frac{2}{7}$
2 189·101 **3** $5\frac{1}{4}$ **4** $x = 0{\cdot}625$
5 7·04% (2 dp) **6** \$0.64
7 a $x = 60$ **b** $x = 11$ **c** $x = 3$
8 \$26.98, \$26.99, \$27, \$27.01, \$27.02

Cumulative Review 2

Part A

1 D **2** C **3** B **4** D **5** D
6 B **7** A **8** A **9** B **10** B
11 C **12** B **13** B **14** D **15** D

Part B

1 a 0·59 **b** 1·26 **c** 4·73
2 a 2^{12} **b** 4^4 **c** 2^{15}
d $4ab + 7a$ **e** $5a + 13$ **f** $x^2 + 2x + 5$
g $\frac{a}{3b}$ **h** $\frac{1}{3}$ **i** $\frac{1}{4}$
3 a \$34.72 **b** 6·12
4 \$890.65 **5** \$483
6 a Flinders **b** 2004 **c** 151
d Cook—2002, 2003 **e** 2002
f 144 **g** 26 **h** 500 pts
7 a $\frac{160}{360} = \frac{4}{9}$
b brain, fat, blood, bone, muscle
c 2·8% **d** 10·8 kg **e** 72 kg
8 a 28·3 **b** 1·8 **c** 12·78
9 a $n = 11$ **b** $n = {}^-10$ **c** $n = 6$
d $n = 4$ **e** $n = 5$ **f** $n = 11$
g $n = 1$ **h** $n = 1$ **i** $n = 10$
10 9
11 a $x > {}^-2$ **b** $x \leq 2\frac{1}{2}$
12 a $x \geq 2$ **b** $x < 30$

Chapter 7

MathsCheck

1 a $\angle ABC$ or $\angle CBA$ **b** acute **c** right
d obtuse **e** straight **f** reflex
g revolution **h** vertically opposite
i complementary **j** supplementary
k 360° **l** adjacent **m** right
n equilateral **o** isosceles **p** scalene
2 a right, isosceles triangle
b obtuse angle, isosceles triangle
3 a $\angle EBD$ or $\angle DBC$ **b** $\angle ABE$ or $\angle CBE$
c $\angle ABC$ **d** $\angle ABD$
e $\angle EBD$ and $\angle DBC$ or $\angle ABE$ and $\angle EBD$

f $\angle EBD$ and $\angle DBC$ g $\angle ABE$ and $\angle EBC$
h i 165° ii 15°

4 a $x = 33$ (D) b $w = 57$ (A) c $x = 63$ (E)
d $a = 90$ (B) e $x = 55$ (C) f $w = 145$ (A)
g $a = 120$ (A) h $\alpha = 104$ (D) i $\beta = 47$ (A)
j $x = 30$ (E) k $a = 50$ (B), $x = 145$ (C)
l $x = 34$ (C)

EXERCISE 7A

1 a corresponding b co-interior
c alternate d corresponding
2 i vertically opposite ii straight line
3 C 4 A
5 a b b a c m
6 a EG b AB and CD c M and P
d $\angle EPD$ e $\angle DPE$ f $\angle BMP$
g $\angle CPE$
7 $\angle AQS$ and $\angle PRP$, $\angle BQS$ and $\angle CRP$
8 $\angle ABP$ and $\angle XCA$, $\angle PBD$ and $\angle XCD$, $\angle ABQ$ and $\angle ACY$, $\angle QBD$ and $\angle YCD$

EXERCISE 7B

1 a $n = 108$ (corresponding $\angle$s on || lines)
b $n = 119$ (corresponding $\angle$s on || lines)
c $x = 70$ (co-interior $\angle$s on || lines)
d $n = 110$ (corresponding $\angle$s on || lines)
e $n = 100$ (corresponding $\angle$s on || lines)
f $n = 78$ (alternate $\angle$s on || lines)
g $n = 47$ (alternate $\angle$s on || lines)
h $n = 75$ (corresponding $\angle$s on || lines)
i $n = 72$ (alternate $\angle$s on || lines)
j $n = 85$ (co-interior $\angle$s on || lines)
k $n = 130$ (corresponding $\angle$s on || lines)
l $n = 95$ (corresponding $\angle$s on || lines)

2

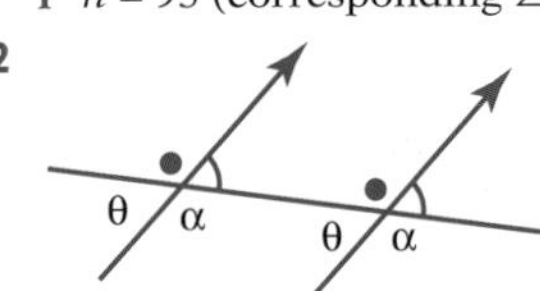

3

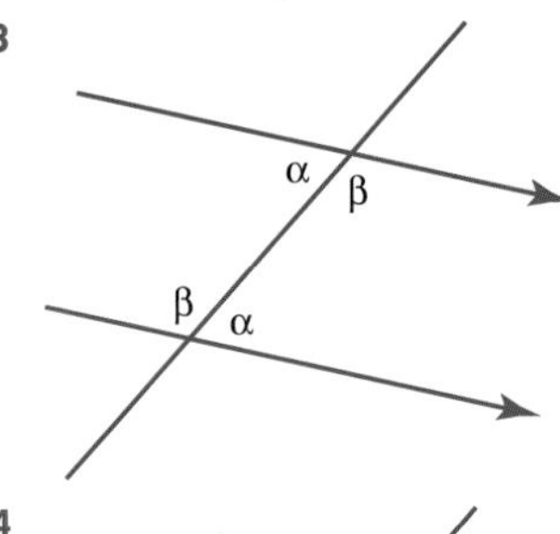

4

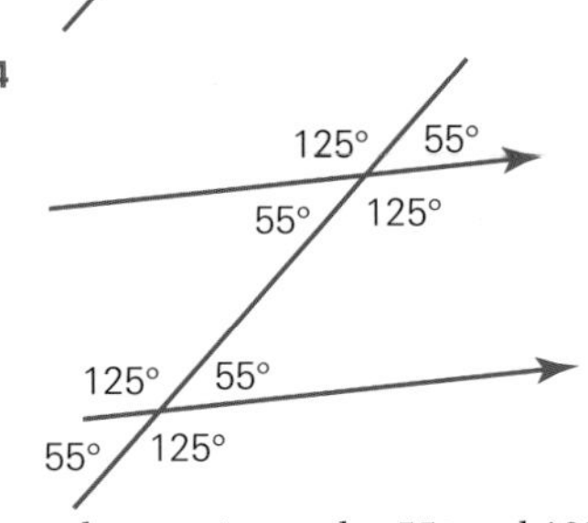

only two size angles 55° and 125°

5 a $x + 83 = 180$ (co-interior $\angle$s on || lines)
$x = 97$
$y = 83$
b $x = 90$ (alternate $\angle$s on || lines)
$x = y$ (vertically opposite $\angle$s)
$y = 90$
c $x + 70 = 180$ (co-interior $\angle$s on || lines)
$x = 110$
$x = y$ (vertically opposite)
$y = 110$

6 a as the alternate $\angle$s are equal
b as corresponding $\angle$s are equal
c as the co-interiors $\angle$s add to 180°

7 a $x = 120$ (co-interior $\angle$s on || lines) $y = 120$ (co-interior $\angle$s on || lines)
b $x = 46$ (co-interior $\angle$s on || lines) $y = 90$ (co-interior $\angle$s on || lines)
c $m = 48$ (alternate $\angle$s on || lines) $n = 54$ (alternate $\angle$s on || lines)
d $a = 67$ ($\angle$s on straight line) $b = 113$ (alternate $\angle$s on || lines)
e $\alpha = 108$ (co-interior $\angle$s on || lines) $y = 66$ (corresponding $\angle$s on || lines)
f $x = 51$ (alternate $\angle$s on || lines) $\beta = 39$ (adjacent complementary $\angle$s)
g $x = 94$ (corresponding $\angle$s on || lines) $y = 86$ ($\angle$s on a straight line)
h $z = 120$ (co-interior $\angle$s on || lines)
i $x = 70$ (vertically opposite $\angle$s) $y = 60$ (alternate $\angle$s on || lines)
j $\alpha = 64$ (corresponding $\angle$s on || lines) $\beta = 116$ ($\angle$s on a straight line)
k $x = 40$ (alternate $\angle$s on || lines) $y = 110$ ($\angle$s on a straight line)
l $y = 70$ (co-interior $\angle$s on || lines) $x = 125$ (alternate $\angle$s on || lines)

8 a no—co-interior $\angle$s aren't supplementary
b yes—corresponding $\angle$s are equal
c no—co-interior $\angle$s aren't supplementary

9 $a = 98$, $m = 82$, $c = 60$, $t = 120$
10 $a = 68$, $b = 112$, $c = 57$, $d = 123$
11 $a = 90$, $b = 54$, $c = 126$

EXERCISE 7C

1 a $n = 30$ ($\angle$ sum of Δ) b $x = 40$ ($\angle$ sum of Δ)
c $n = 19$ ($\angle$ sum of Δ) d $p = 35$ ($\angle$ sum of Δ)
e $c = 28$ ($\angle$ sum of Δ) f $a = 60$ ($\angle$ sum of Δ)
g $b = 90$ ($\angle$ sum of Δ) h $d = 70$ ($\angle$ sum of Δ)
i $n = 30$($\angle$ sum of Δ)

2 a $x = 105$ (exterior $\angle$ equals sum of 2 opposite interior $\angle$s)
b $n = 142$ (exterior $\angle$ equals sum of 2 opposite interior $\angle$s)
c $m + 20 = 130$ (exterior $\angle$ equals sum of 2 opposite interior $\angle$s)
$m = 110$
d $2r = 126$ (exterior $\angle$ equals sum of 2 opposite interior $\angle$s)
$r = 63$

e $4x = 84$ (exterior ∠ equals sum of 2 opposite interior ∠s)
$x = 21$
f $p = 50$ (∠s on a straight line) *or* (exterior ∠ equals sum of 2 opposite interior ∠s)

3 **a** $x = 56$
$y = 60$ (∠sum of Δ)
b $x = 72$ (alternate ∠s on || lines)
$140 + y = 180$
$y = 40$
c $x = 70$ (alternate ∠s on || lines)
$90 + 70 + y = 180$ (∠ sum of Δ)
$y = 20$

4 **a** $x = 40$ (vertically opposite ∠s), $y = 60$ (∠ sum of Δ)
b $x = 50$ (∠s on a straight line), $y = 60$ (∠ sum of Δ)
c $x = 50$ (∠s on a straight line), $y = 30$ (∠s on a straight line) $z = 100$ (∠ sum of Δ)
d $x = 115$ (co-interior ∠s on || lines), $y = 57$ (∠ sum of Δ) or (co-int ∠s || lines)
e $x = 95$ (∠ sum of Δ), $y = 70$ (∠ sum of Δ)
f $x = 50$ (vertically opposite ∠s), $y = 72$ (vertically opposite ∠s), $z = 58$ (∠ sum of Δ)

5 **a** $\angle BDC = 65°$ (∠ sum of ΔBCD)
$\angle DBF = 65°$ (alternate to $\angle BDC$, $BF \parallel CE$)
$\therefore x = 65$
b $\angle ACB = 58°$ (vertically opposite ∠s)
$\angle CAB = 80°$ (∠sum of ΔABC)
$\therefore x = 80$
c $\angle PSQ = 44°$ (∠s on a straight line)
$\angle PQS = 33$ (∠s on a straight line)
$\angle SPQ = 103$
$\therefore x = 103$

6 **a** no, ∠ sum ≠ 180° **b** no, ∠ sum ≠ 180°
c yes, ∠ sum = 180°

7 **a** $x = 25$ **b** $x = 30$ **c** $x = 20$

8 **a** $a = 100, b = 60$
b $a = 55, b = 35$, c = 100
c $a = 28, b = 78$

EXERCISE 7D

1 **a** $n = 80$ (angles opposite equal sides in an isosceles Δ)
b $x = 72$ (angles opposite equal sides in an isosceles Δ)
c $c = 45$ (angles opposite equal sides in an isosceles Δ)
d $x = 12$ (equal sides opposite equal in an isosceles Δ)
e $a = 64$ (∠ sum of Δ), $y = 3{\cdot}2$ (equal sides opposite equal ∠s isosceles Δ)
f $n = 60$ (angles of equilateral Δ)
g $d = 120$ (exterior angle of equilateral Δ)
h $a = 60$ (∠sum of Δ), $x = y = 35$ (equilateral Δ)
i $x = 72$ (alternate ∠s on || lines), $y = 72$ (angles opposite equal sides in an isosceles)

2 **a** $P = 70$ **b** $a = 100$ **c** $c = 45$
d $n = 25$ **e** $x = y = 60$ **f** $x = 80, y = 50$
g $x = 10$ **h** $a = 60, b = 120, c = 30$
i $a = 20, b = 140, c = 70, d = 90$
j $x = 25, y = 25, z = 130°$
k $x = y = 84$
l $n = 56, y = 68, z = 56$
m $x = 58, y = 32$
n $x = 42$ **o** $x = 30$

EXERCISE 7E

1 **a** $a = 90$ (∠ sum of quad)
b $a = 60$ (∠ sum of quad)
c $c = 90$ (∠ sum of quad)
d $a = 68$ (∠ sum of quad), $b = 112$ (∠s on a straight line)
e $h = 70$ (∠ sum of quad)
f $x = 25$ (∠ sum of quad)
g $x = 146$ (∠ sum of quad)
h $x = 40$ (∠ sum of quad)
i $x = 90$ (∠ sum of quad)
j $x = 110$(∠ sum of quad)
k $x = 97$ (∠s on a straight line), $y = 83$ (∠ sum of quad)
l $x = 69$ (vertically opposite ∠s), $y = 81$ (∠ sum of quad)

2 **a** $n = 110$ **b** $n = 30$ **c** $a = 35$
d $x = 40$ **e** $a = 120, b = 30, c = 70, d = 20$
f $a = 60, b = 30, c = 90$

3 **a** $m = 58$ (∠s on a straight line), $n = 58$(∠ sum of quad)
b $x = 246$ (∠ sum of quad), $y = 114$ (revolution)
c $\alpha = 82$(alternate ∠s on || lines), $\beta = 98$ (co-interior ∠s on || lines)
d $x = 65$ (∠s on a straight line) $y = 75$ (co-interior on || lines), $z = 115$ (∠ sum of quad)
e $x = 78$ (∠sum of isosceles Δ) $y = 70$ (∠ sum of isosceles ∠)
f $x = 45$ (∠ sum of quad), $y = 90$ (∠s on a straight line)
g $x = 82$ (co-interior ∠s on || lines), $b = 82$ (co-interior ∠s on || lines)
h $a = 70$ (co-interior ∠s on || lines), $b = 110$ (co-interior ∠s on || lines), $c = 70$ (∠ sum of quad)

EXERCISE 7F

1 **a** $\angle ABE = 82°$ (given)
$\angle PEF = 82°$ (alternate ∠s, $AC \parallel DF$)
$x + 82 = 180$
$x = 98$
b $\angle ABC = 73°$ (given)
$\angle BCD = 107°$
$\angle CDE = 107°$ (alternate ∠s on $AE \parallel BC$)
$\therefore x = 107$
c $\angle CBD = 48$ (given)
$\angle BCD = \angle BDC$
$\angle BCD = \angle BDC + 48° = 180°$
$\therefore \angle BCD = 66°$

ΔABC is equilateral
$\therefore \angle ACB = 60°$
$x = \angle ACB + \angle BCD$
$= 60 + 66$
$= 126$

2 Suggested solutions—setting out may vary.

a $\angle BAC = 31°$ (given)
$\angle BAC = \angle ABC$ ($\angle$s opposite equal sides, $AC = CB$)
$\angle ABC = 31°$
$a = 31$
$\angle ACD = \angle BAC + \angle ABC$ (exterior angle of ΔABC)
$= 62°$

b ΔAEF is isosceles, $AE = AF$
$\angle AEF = 70°$ (given)
$\angle AEF = \angle AFE$ ($\angle$s opposite equal sides)
$\angle AFE = 70°$
$\therefore a = 70$
$\angle AFE = \angle ACB$ (corresponding $\angle$s $EF \parallel BC$)
$\angle ACB = 70°$
$\therefore b = 70$
$\angle ACB + \angle ACD = 180$ ($\angle$s on a straight line)
$70 + c = 180$
$c = 110$

c $\angle ABF = 58$ (given)
$\angle ABF = \angle DBC$ (vertically opposite $\angle$s)
$\therefore a = 58$
$\angle DBC = \angle EGC$ (corresponding $\angle$s $DB \parallel EG$)
$\therefore b = 58$
$\angle CGE = \angle DEG$ (alternate $\angle$s $AC \parallel DE$)
$\therefore c = 58$

d $\angle ABE = 98°$, $\angle EDC = 112°$ (given)
$\angle ABE + \angle EBC = 180°$ ($\angle$s on a straight line)
$98 + a = 180$
$\therefore a = 82$
$\angle CBE + \angle BED = 180°$ (co-interior $\angle$s $AC \parallel ED$)
$a + b = 180$
$\therefore b = 98$
$a + b + c + 112 = 360$ ($\angle$ sum of quad $BCDE$)
$\therefore c = 68$

e $\angle ABD = 47°$, $\angle CBD = 55°$ (given)
$\angle ABD + \angle BDE = 180°$ (co-interior $\angle$s $AB \parallel CE$)
$47 + a = 180$
$\therefore a = 133$
$\angle ABC + \angle BCD = 180°$ (co-interior $\angle$s $AB \parallel CE$)
$47 + 55 + b = 180$
$\therefore b = 68$

f $\angle DEB = 126°$, $\angle EBF = 90°$ (given)
$\angle EBF + \angle EFB = 126°$ (exterior $\angle$ of ΔBEF)
$90 + a = 126$
$\therefore a = 36$
$\angle CBF = \angle EFB$ (alternate $\angle$s $AC \parallel DG$)
$\therefore b = 36$

3 (values of pronumerals only supplied)
a $a = 70, b = 70$ **b** $a = 33, b = 69, c = 78$
c $a = 60°, b = 60, c = 60, d = 60$

4 (values of pronumerals only supplied)
a $x = 40$ **b** $x = 94$ **c** $x = 80$ **d** $x = 50$
e $x = 73$ **f** $x = 126$

5 **a** $\alpha + \beta + \gamma = 360$
b $a + b + c + d = 360$

6 **a** $x = 28$ **b** $x = 72$ **c** $x = 34$ **d** $x = 70$
e $x = 120$ **f** $x = 30$ **g** $x = 33$ **h** $x = 155$
i $x = 100$

EXERCISE 7G
constructions

Chapter Review Exercises

1 **a** $n = 97$ ($\angle$ sum revolution)
b $m = 55$ ($\angle$ sum of Δ)
c $x = 20$ ($\angle$s on a straight line)
d $x = 30$ (adjacent complementary $\angle$s)
e $x = 82$ (corresponding $\angle$s on $\parallel$ lines), $y = 82$ (alternate $\angle$s on $\parallel$ lines)
f $x = 65$ ($\angle$ sum of isosceles Δ), $y = 115°$ (exterior $\angle$ of Δ equals the sum of the 2 opposite interior angles)

2 **a** $n = 80$ ($\angle$ sum of Δ)
b $x = 123$ (exterior $\angle$ equals sum of 2 opposite interior $\angle$s)
c $b = 53$ ($\angle$s on a straight line), $a = 37$ ($\angle$ sum of Δ)
d $x = 130$ ($\angle$ sum of quad)
e $x = 22{\cdot}5$ ($\angle$ sum of Δ)
f $c = 50$ ($\angle$ sum of isosceles Δ), $d = 12$ (equal sides opposite equal $\angle$s)

3 **a** isosceles **b** scalene
c right, isosceles Δ

4 **a** $AD \parallel BC$ as the alternate angles are equal
b AD is not $\parallel$ to CB as the alternate $\angle$s aren't equal

5 **a** $a = 120$ **b** $a = 90$ **c** $a = 88$ **d** $a = 75$
e $a = 67$ **f** $a = 47{\cdot}5$

Keeping Mathematically Fit

PART A

1 160 **2** 12 **3** $^-16$ **4** 7 and 8
5 $\frac{1}{3}$ **6** 97 **7** $15 \times 6 = 100 - 10$
8 $a = 10$ **9** west **10** 20

PART B

1

$24	$15	$80	$72	$90.91	$1.50
$32	$20	$88	$81	$100	$2.10
$33\frac{1}{3}\%$	$33\frac{1}{3}\%$	10%	$12\frac{1}{2}\%$	10%	40%

2 **a** $x = 15$ **b** $x = 9$ **c** $x = 26$
3 $115.50 **4** 1·6 **5** $\frac{89}{90}$
6 **a** $8(2a - 3)$ **b** $2(8a + b)$ **c** $^-16(a + 1)$
7 57·13 mm (2 dp) **8** 7 or $^-7$

Chapter 8

MathsCheck

1 **a** 5000 m **b** 1·8 m **c** 2 m **d** 2800 m **e** 0·8 m **f** 0·75 m **g** 3060 m **h** 904·5 m
2 **a** 2000 kg **b** 3700 kg **c** 1·2 kg **d** 0·454 kg **e** 0·08 kg **f** 1030 kg **g** 5 kg **h** 0·0175 kg
3 **a** 6000 L **b** 1540 L **c** 1·25 L **d** 2 400 000 L
4 **a** 1·4 **b** 45 **c** 0·8 **d** 0·01 **e** 0·006 **f** 0·02 **g** 50 **h** 50 **i** 0·02
5 **a** 290 cm **b** 1·2 kL **c** 1200 mg **d** 26 mm **e** 820 m **f** 500 kg
6 **a** 2·3 cm **b** 9·2 cm **c** 46 cm
7 **a** 30 cm **b** 119 mm **c** 5300 mm **d** 5·2 km **e** 9·4 m **f** 506 mm
8 **a** 7200 mm **b** 60 cm **c** 15 km **d** 10·6 m **e** 6000 mm **f** 6500 m (6·5 km)
9 115 m **10** 200 m
11 **a** 188 cm **b** 15 **c** 180 cm
12 **a** **i** rectangle
ii rectangular prism
iii

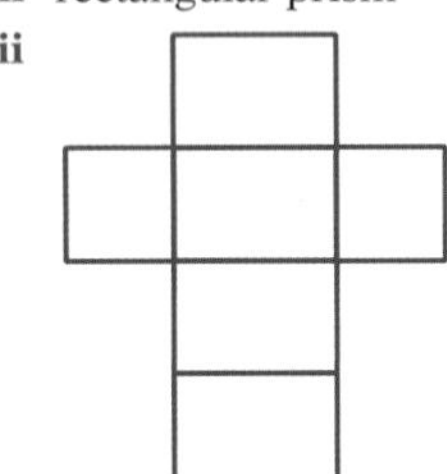

b **i** triangle
ii triangular prism
iii

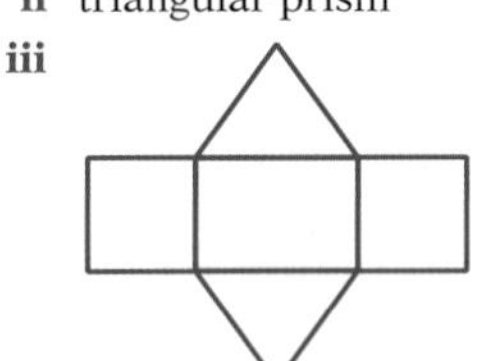

c **i** trapezium
ii trapezoidal prism
iii

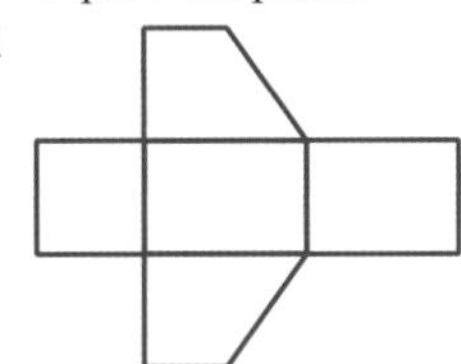

EXERCISE 8A

1 **a** 4·5 m to 5·5 m **b** 7·5 cm to 8·5 cm **c** 77·5 m to 78·5 m **d** 0·5 cm to 1·5 cm
2 **a** 22·5 cm to 23·5 cm **b** 3·85 kg to 3·95 kg **c** 19·45 m to 19·55 m **d** 8·65 m to 8·75 m
3 **a** 18·555 cm to 18·565 cm **b** 18·95 km to 19·05 km **c** 0·05 km to 0·15 km **d** 4·195 t to 4·205 t
4 **a** 149·5 cm to 150·5 cm **b** 149·95 cm to 150·05 cm
5 **a** 30 m **b** 145 g **c** 9 m
6 **a** 392 cm **b** 399·2 cm
7 **a** 9·15 cm **b** 9·25 cm **c** 36·6 cm to 37 cm
8 **a** 9·5 cm to 10·5 cm **b** 7·5 cm to 8·5 cm **c** 34 cm to 38 cm
9 5·75 kg to 5·855 kg **10** discuss

EXERCISE 8B

1 **a** 5 ha **b** $\frac{1}{2}$ ha **c** $\frac{4}{5}$ ha
2 **a** 30 000 m^2 **b** 17 000 ha **c** 3 m^2 **d** 0·6 m^2
3 **a** 0·5 **b** 4 **c** 0·407
4 **a** 96 m^2 **b** 900 mm^2 **c** 3·8 cm^2 **d** 800 cm^2 **e** 66 m^2 **f** 12 m^2 **g** 92·16 cm^2 **h** 240 mm^2 **i** 12 m^2
5 4750 m^2
6 **a** 1152 m^2 **b** 330 m^2 **c** 71·4% **d** $4068.90
7 240 mm or 24 cm
8 11 by 11; 121 cm^2; 12 by 10; 120 cm^2; 20 by 2; 40 cm^2 (discuss others)
9 1000 m^2
10 **a** 0·01 **b** 100 000 000 cm^2
11 **a** discuss **b** discuss
12 $h = \sqrt{19}$ **13** 130 cm^2
14 **a** 4·5 m and 2·5 m **b** 5·5 m and 3·5 m **c** 11·25 m^2 and 19·25 m^2
15 **a** 48 m^2 **b** 216 m^2 **c** 84 m^2 **d** 4 950 000 mm^2 **e** 106 m^2 **f** 34 m^2
16 **a** 64 m^2 **b** 35 m^2 **c** 36 m^2

EXERCISE 8C

1 **a** 16 m^2 **b** 1 m^2 **c** 25 m^2 **d** 17·64 m^2
2 **a** 8 m^2 **b** 13·5 cm^2 **c** 1550 cm^2 **d** 980 cm^2
3 **a** 1024 m^2 **b** 5628·48 cm^2
4 **a** 142 m^2 **b** 204 m^2
5 **a** 168 m^2 **b** 336 m^2
6 **a** 2400 cm^2 **b** 20 m^2 **c** 5·58 m^2
7 **a** 360 m^2 **b** 96 m^2
8 **a** 22 cm^2 **b** 23·16 m^2 **c** 10 240 cm^2
9 2850 cm^2
10 **a** 29·3 m^2 **b** 9·76 ≑ 10 L of paint

EXERCISE 8D

1 **a** 80 cm^3 **b** 300 cm^3 **c** 3600 cm^3 **d** 50 m^3 **e** 972 m^3 **f** 16 cm^3 **g** 480 cm^3 **h** 684 mm^3 **i** 5 m^3
2 **a** 30 m^3 **b** 6000 cm^3 **c** 14 m^3 **d** 72 m^3 **e** 9·72 cm^3 **f** 96 m^3 **g** 4 m^3 **h** 6 m^3 **i** 54 m^3
3 **a** 6000 mL **b** 15 mL **c** 8000 mL **d** 750 mL
4 **a** 0·36 L **b** 2 L **c** 500 L
5 **a** $V = x^3$
b **i** 8 m^3 **ii** 8000 cm^3 **iii** 0·125 m^3
6 **a** $V = lbh$
b **i** 1·5 m^3 **ii** 3·24 m^3 **iii** 0·55 m^3
c **i** 1·5 kL **ii** 3·24 kL **iii** 0·55 kL
7 **a** 800 cm^3 **b** 240 cm^3 **c** 160 cm^3

d 216 m^3 e 42 cm^3 f 1260 m^3
8 a 24 m^3 b 800 cm^3 c 22 500 cm^3
d 496 cm^3 e 560 m^3 f 10 080 cm^3
9 a 24 m^3 b 26·4 m^3 c 3·4 ≈ 4 truckloads
10 a 36 cm^2 b 6 cm
11 a 2 L b 48 600 L c 0·252 L
12 a 2 000 000 b 1 c 8
d 200 000 e 700 000 000 f 5
13 discuss 14 343 cm^3
15 a 726 m^2 b discuss
16 112 m^3

Chapter Review Exercises

1 a 3·6 b 1750 c 2040 d 0·51 e 4·25
f 2·08 g 2300 h 1·5 i 20 j 10 000
k 4000 l 70
2 a i 11·2 m ii 7·84 m^2
b i 6·5 m ii 1·12 m^2
c i 6000 mm ii 2 160 000 mm^2
d i 8m ii 2·56 m^2
e i 147 cm ii 1104 cm^2
f i 9·54 km ii 3.72 km^2
3 a 80 m^2 b 43 m^2 c 28 m^2
4 a 6·5 m to 7·5 m b 7·35 m to 7·45 m
c 7·435 m to 7·445 m
5 a 541·5 m^2 b 222 m^2 c 1536 m^2
6 a 160 m^3 b 288 m^3 c 756 000 cm^3
7 $1{\cdot}1\dot{6}$ ≈ 1 cm
8 a 0·672 m^3 b 0·672 kL c 4·04 m^2

Keeping Mathematically Fit

PART A

1 0·28 2 $10 + x$ 3 7·395 4 9800 m
5 $\$\frac{xy}{10}$ 6 $1\frac{4}{5}$ 7 $3 8 ⁻192
9 66 10 1575

PART B

1 a $x = 78$ b $x = 67$ c $x = 49$ d $x = 45$
2 a 3·332 m^2 b 82·28 m^2 c 54 m^2
3 ⁻1·15 4 24
5 a $10x$ b $2x + 4$ c $4x + 10$
6 15 7 $x \geq 3$
7 from 12·15 cm to 12·25 cm

Chapter 9

MathsCheck

1 a $\frac{1}{2}$ b $\frac{2}{3}$ c $\frac{2}{3}$ d $\frac{3}{4}$ e $\frac{3}{4}$
2 a 3 b 10 c 1 d 4
3 a $\frac{1}{2}$ b $\frac{1}{4}$ c $\frac{3}{5}$
4 a $20 b 60 kg c 60 m
5 a $3.99 b $11.97 c $29.93
6 a 1 pm b 2 pm, 1 hour
c no distance travelled
d 50 km e 325 km
f no, the graph doesn't end on the horizontal axis

EXERCISE 9A

1 a 2 : 3 b 2 : 5 c 5 : 3
2 a 5 : 3 b 1 : 3
3 a 8 : 4 = 2 : 1 b 1 : 2 c 8 : 12 = 2 : 3
4 a 4 : 7 b 2 : 4 = 1 : 2
5 a 1 b 6 c 4 d 16 e 9
f 12, 16 g 30 h 3 i 7 : 3
6 a $\frac{1}{2}$ b 2 : 3 c 7 : 4 d 8 : 15
e 7 : 2 f 5 : 6 g 6 : 1 h 3 : 7
i 3 : 4 j 3 : 4 k 1 : 4 l 3 : 1
m 19 : 10 n 3 : 5 : 6 o 3 : 5 : 6
7 1 : 2
8 a 10 : 1 b 3 : 8 c 1 : 2 d 1 : 100
9 a 2 : 3 b 7 : 50 c 3 : 8 d 7 : 50
e 3 : 73 f 18 : 5
10 a 1 : 6 b 1 : 6 c 1 : 5 d 4 : 3
e 5 : 2 f 54 : 70 : 9
11 a 1 : 6 b 4 : 5 c 20 : 1 d 1 : 5
e 51 : 10 f 21 : 30 : 7
12 a 25 : 64 b 14 : 9 c 1 : 1
13 a 1 : 1 b 1 : 2
14 Souths $1{\cdot}\dot{3}$: 1, Easts 1·25 : 1, ∴ Souths is better

EXERCISE 9B

1 $3, $7 2 $10, $90
3 a 8, 12 b 140, 40
c $36, $48 d 210 m, 240 m
e 150, 250 f 84, 12
4 a 4 cm, 6 cm b 4 cm, 3 cm
5 a 210 kg, 150 kg b 3 days, 1 day
c 600 kg, 400 kg d 0·35 m, 0·35 m
e $0.90, $2.70, $3.60
6 400 mL 7 200 mL 8 15 500 teenagers
9 $535 714, $714 286 10 24 kg
11 300 t 12 4·6875 m^2 13 67.5°
14 $29{\cdot}1\dot{6}$ cm 15 $49 000 16 $3\frac{1}{3}$ mL
17 1 : 1
18 a 2 : 15 b 7 : 15
19 9 : 25 20 20 : 63

EXERCISE 9C

1 $300 2 $2000 3 3 L 4 125 mL
5 a 75 min b 105 min
6 20 shovels 7 36 cm
8 62·5, ∴ 62 students 9 $400
10 a 17·5 m (17 500 mm) b 27 cm
11 144 mm 12 106·5 cents per litre
13 1050 mm 14 5 pm 15 43·5 cm
16 30° 17 x = 15 mL, z = 30 mL
18 164·4 (1 dp) 19 $94 090.91 20 $1900

EXERCISE 9D

1 a $10/h b $1.30/can c 43 people/bus
d 50c/kg e 30 km/h f 4 runs/over
g 8 km/L h 60 L/min i 25 emus/ha

j $30/h k 12·5 km/L l 18·2 pts/game
m $96/m n 2·5g/min o $2400/day

2 a 95c/kg b $7.95/m c 1.25c/g
d 78.9c/L e $156/m^2

3 a i double each quantity
ii halve each quantity
iii $\frac{1}{4}$ of each quantity
iv $1\frac{1}{2}$ times each quantity
b i double each quantity
ii halve each quantity
iii $\frac{5}{2}$ times each quantity

4 a 20 b $48 c 750 d $24 e 130
f 140 g 14 h 1·8

5 a $140 b 24 h c $22 050

6 a 40 sheep/h b 200 sheep c 320 sheep
d $7\frac{1}{2}$ hours

7 a 20 cents b $1.40 c $6.40

8 $110.36

9 a $7.40 b $55.50 c $1.39

10 a 8·4 km/L b 168 km
c 11·9 L/100 km d 416·5 L

11 a i 2880 ii 12
b i 2 min ii 10 s

12 a $3600 b $16\frac{2}{3}$ months

13 a 3 t b i 1000 m^2 ii 60 m^2

14 a 302·4 L b 23 d 3·6 h

15 Aust.: 2 per 125, India: 1 per 25 ⇒ 5 per 125
∴ India's rate is 2·5 times Australia's

16 Robert's council

17 a 2700 b 10 c 3600 d 540

18 a 12·5 km L b 32 L c $29.44 d 27·17 L

EXERCISE 9E

1 a 50 km/h b 70 km/h c 62·5 km/h
d 50 km/h e 30 km/h f 150 km/h

2 a 75 km/h b 100 km/h c 625 km/h
d 4 km/h e 60 km/h f 120 km/h
g 80 km/h h 45 km/h i 60 km/h

3 a 130 km b 325 km c 227·5 km
d 520 km e 32·5 km

4 a 3 h b 48 min c 20 min

5 a

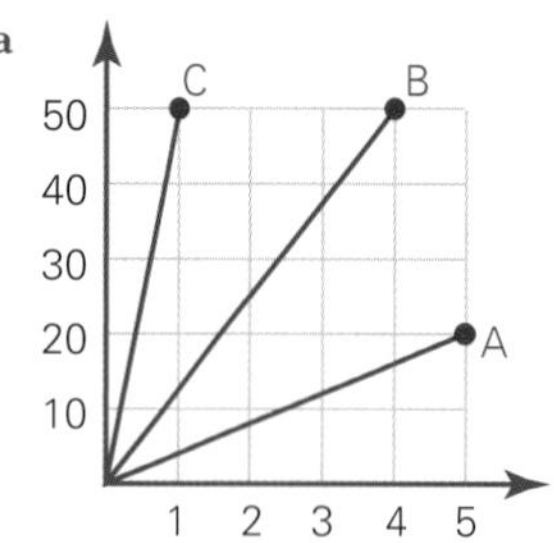

b car—steepest line
c car $\frac{50}{1} = 50$
cyclist $\frac{50}{4} = 12·5$
walker $\frac{20}{5} = 4$
the larger the gradient the faster the person, and the steeper the graph

6 a 1 cm to 1 h b 1 cm to 20 km
c 1 pm for 2 h d 160 km
e 28·6 km/h

7 a 200 km/h b 200 m/s c 8·1 m/min

8 a 13·3 b 28 s c 9·8 s

9 a 45 km b 7·5 km c $2\frac{1}{2}$ km
d 4·17 m

10 a 1 080 000 000 km/h b 36 000 000 km
c 497 s = 8·28 min d $0·01\dot{3}$ s

11 a 1224 km/h b 10·2 km c 14·7 s
d 13·8 s—aircraft flies overhead 0·9 s before its sound is heard

12 a 55 km/h b 1 pm, 1:30 pm
c $73·\dot{3}$ km/h ≈ 73 km/h d 75 min

13 a 5000 b 10 c 20 d 64·8

14 a 9·97 m/s b 35·89 km

15 a 1·755 m/s b 14·25 min c 6·318 km/h

16 a Lewis 10·14 m/s, Morcel 7·18 m/s, ∴ Lewis faster rate
b 2 min 28 s

Chapter Review Exercises

1 a 2 : 1 b 2 : 3

2 a 7 : 6 b 4 : 1 c 3 : 26

3 a 4 : 9 b 4 : 1 c 1 : 3
d 3 : 1 e 1 : 3 f 7 : 3

4 a 21 b 7 c 1·26 d 12, 20, 32
e 2·1 f 1 : 5 : 4 = 3 : 15 : 12

5 team A

6 a 16, 24 b 50, 250 c 480 cm, 120 cm

7 $132, $198, $330 **8** $220

9 $14.90 **10** 150 mL

11 a 25 L/h b 1·82 g/cm^3

12 a 48 km/h b 56 km/h

13 a 50 km/h b 2:27 pm

14 a $\frac{1}{2}$ km/min b $833\frac{1}{3}$ kg/h c 3000 m/h
d 3 km/h e 45 km/h f 500 m/min

Keeping Mathematically Fit

PART A

1 3 **2** 5 **3** 10 000 **4** 20
5 SE **6** 10% **7** ⁻3 **8** $0·\dot{3}$ to 0·5
9 720 cm^3 **10** 52 500

PART B

1 10 am
2 11 am to 11:36 am and 1 pm to 2 pm
3 225 km **4** 62·5 km/h **5** 6
6 $1008.32 **7** ⁻20
8 10 by 10 by 10 is one example

Cumulative Review 3

Part A

1 B	**2** C	**3** B	**4** B	**5** A
6 C	**7** D	**8** C	**9** A	**10** C
11 B	**12** D	**13** B	**14** C	**15** C

Part B

1 a 460·8 b 31·9225 c 2700

2 a 5^{10} b 5^2 (25) c $2^{20} \times 3^{10}$ d 1
e $7p + 9q$ f $^-21a^2b$ g $\frac{9}{10b}$ h $29a + 10b$

3 a $3a(5b - 1)$ b $^-x(y + 9)$ c $a(ab + 5)$

4 a 11·1 b 16·7 c 14·1

5 a i 13° ii 29° b 8 am
c yes, approx. 23·5 d 9° e 9°

6 a $w = {}^-6$ b $a = {}^-6\frac{1}{3}$ c $x = 9$ d $p \leq 4\frac{1}{3}$

7 a $\angle BAC$ b $\angle CMP$ c $\angle BCM$ d $\angle ACD$
e $\angle CBA$ or $\angle AMQ$ f $\angle QMC$ g $\angle CMP$

8 a $x = 30$ b $n = 50$ c $c = 26$

9 constructions

10 a i 6000 cm³ ii 6 L
b i 1·5 m³ ii 1500 L
c i 460 800 cm³ ii 460·8 L

11 a 51·36 m² b 1440 m²

12 a 1 : 3 b 1 : 4 c 1 : 12 d 12 : 17

13 a 5 : 3 b 7 : 15 c 3 : 2

14 a 20 b 49 c 64

15 $1\frac{3}{4}$ **16** \$336, \$504

17 160 kg, 240 kg, 320 kg

18 60

19 $4\frac{1}{2}$ cups flour, $\frac{3}{4}$ cup sugar

20 a 5 g/c b 7·5 m/s c 32 c/pencil

Chapter 10

MathsCheck

1 column graph

2 yes: states what the graph is about

3 yes: equally spaced

4 Sunday **5** 26 hours

6 $20\frac{5}{6}\%$ or $20{\cdot}8\dot{3}\%$

7 continuous: all values possible

8 10°C **9** $22\frac{1}{2}$°C **10** 10 am

11

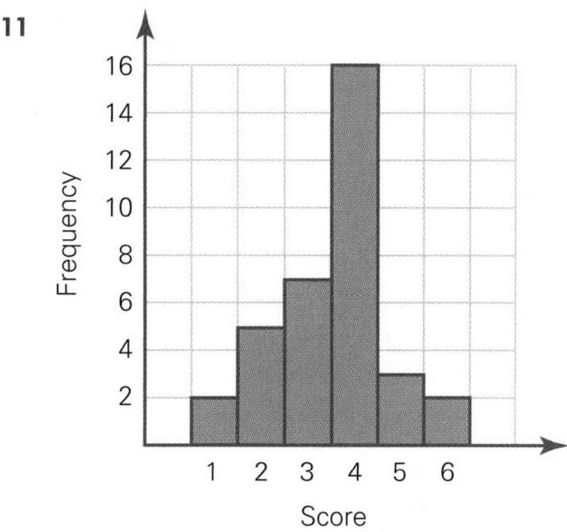

EXERCISE 10A

1 a quant. discrete b quant. continuous
c quant. continuous d quant. continuous
e quant. discrete f categorical

2 discuss **3** E discuss

4 a children b homeless
c country residents d people who work
e people in a hurry, people who don't shop themselves
f people who don't own, watch TV
g people who don't buy magazines

5 discuss

6 a B, A, D, C b B or A c discuss

7 discuss

8 discuss

9 discuss

10 discuss

EXERCISE 10B

1 a TV sets per family b 1 set
c 0 and 5 sets d 15 e 4
f 9 g quant. discrete
h

x	f
0	1
1	5
2	3
3	2
4	3
5	1
	15

2 a black b 15 c 100 d 10%
e categorical
f

Score	Freq.
Blue	10
Green	15
Yellow	15
Pink	25
Black	30
Orange	5
	100

3 a

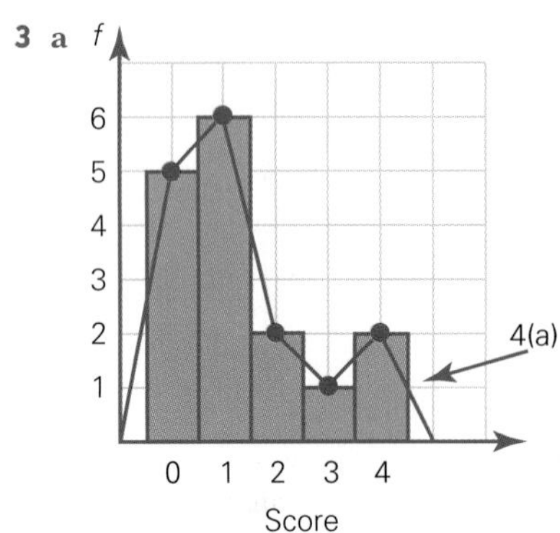

b

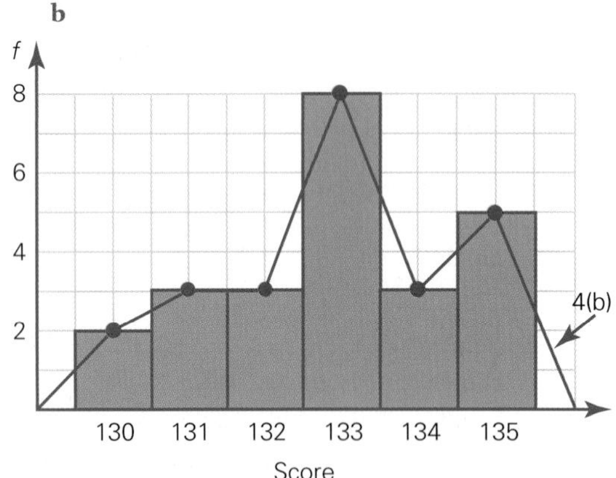

4 see quest **3**

5 a

Score	Tally	Freq.
0	ǂǂǂǂ	4
1	ǂǂǂǂ IIII	9
2	ǂǂǂǂ ǂǂǂǂ III	13
3	ǂǂǂǂ II	7
4	ǂǂǂǂ I	6
5	ǂǂǂǂ	5
6	II	2
7	III	3
8	I	1
	Total	50

b $\frac{26}{50} = \frac{13}{25}$

c 14%

d

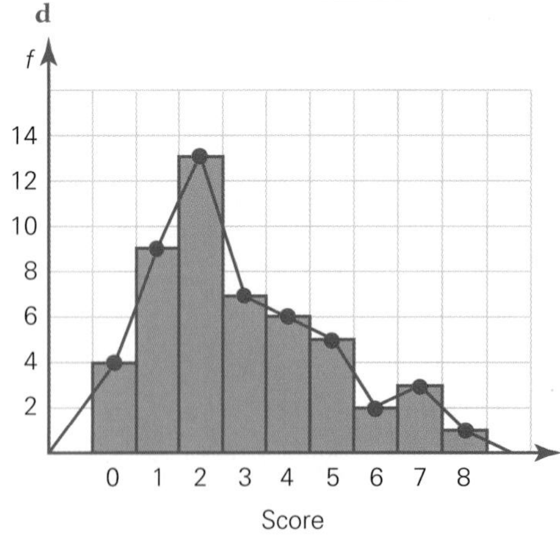

e 2·92

6 a

x	Tally	f
1	III	3
2	ǂǂǂǂ	5
3	ǂǂǂǂ II	7
4	ǂǂǂǂ I	6
5	ǂǂǂǂ IIII	9
6	ǂǂǂǂ ǂǂǂǂ	10
		40

b

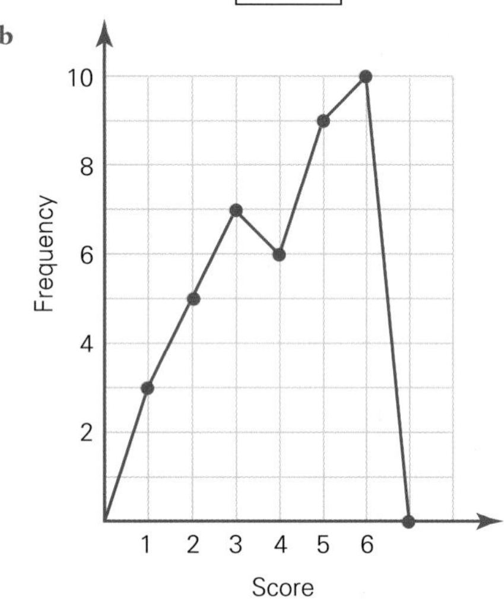

c 6

d 1, yes, as (1, 1) is the least likely pair

e 40 **f** $\frac{13}{40}$ **g** experiment; discuss

7 a

Vowel	Freq.
a	12
e	10
i	7
o	10
u	1
	40

b

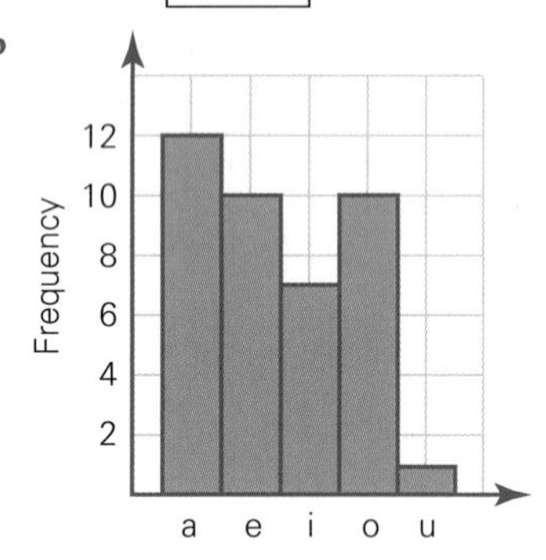

c *a* **d** *u* **e** 7 times

f 40 vowels **g** 25%

EXERCISE 10C

1 a 24 **b** 10/30 **c** 20/30 **d** 9

2 a 15 **b** 0 and 7 **c** $\frac{4}{15}$ **d** 7 **e** 2

f 2 per household

3 a

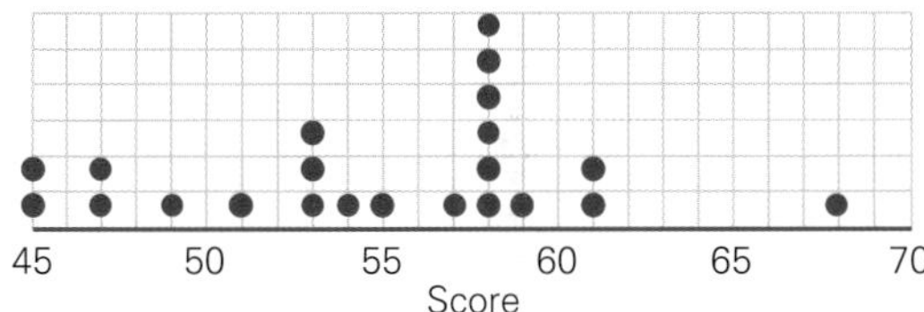

b clusters at 57–59 **c** 45 min
d traffic, red lights
e By 8:32 am (it usually takes 58 min or less)
f discuss

EXERCISE 10D

1 a 21 **b** 619 **c** 52 **d** 11

2 a

Stem	Leaf
0	2 3 3 5 7
1	1 3 5 5 6 7
2	2 3 5 6 8 8
3	0 3 3 5 5 9

b

Stem	Leaf
2	0 2 4 6 7 8
3	0 1 4 6 6 8 9
4	0 1 2 2 6 7 7 8 9 9

3 a 10, 10, 15, 21, 23, 25, 28, 33, 37, 38, 39, 39, 57, 57, 59
b 102, 103, 106, 115, 118, 119, 119, 120, 133, 135, 137, 140
c 245·1, 245·3, 245·8, 245·9, 246·0, 246·4, 246·6, 247·5, 247·5, 248·0, 248·4, 248·7

4 a

Stem	Leaf
1	8 9
2	0 3 3 4 4 8 9
3	0 3 5 8
4	2

b

Stem	Leaf
0	1 2 5 6
1	2 4 7
2	2 5
3	0 4
4	0 7
5	0

c

Stem	Leaf
10	0 2 5
11	1
12	0 4 5 5 6 9
13	0 5 8
14	2 5
15	0 5 6
16	0 2

5 a 3 or 4 **b** 7, 8 or 9 **c** 2
6 a i 18 **ii** 18
b 65, 65, 66, 67, 67, 70, 72, 72, 74, 76, 78, 81, 81, 81, 82, 82, 84, 86
c i group A **ii** group A
d 4
e group B—more students within the 'healthy' range

EXERCISE 10E

1 a 7·5 **b** 86 **c** 1·05 **d** 47·56 **e** 13·5
2 $1\frac{1}{9}$ goals **3** 4·4 mm **4** $(2x + 2y + 2) \div 5$
5 288 **6** 408 runs
7 a 718·75 cm **b** 888·75 cm **c** 148·125 cm
8 $12{\cdot}\dot{2}\dot{7}$ **9** $x = 37$ **10** $218\frac{2}{3}$ **11** 96%
12 a 4 **b** 5 **c** 7·2, 5, 8·3, 6·5

EXERCISE 10F

1 a mode = 2, median = 2·5, range = 3
b mode (none), median = 57, range = 57
c mode = 0, median = 2, range = 5
d mode (none), median = 99, range = 76
e mode = 7, median = 7, range = 6
f mode (none), median = 1·35, range = 3·1
2 a 3 **b** 58·4 **c** 2
d 79·8 **e** 7·3 (1 dp) **f** 1·46375
3 a 60 cents **b** \$1.20 **c** \$1.20 **d** 6
4 a mean = 39·375, mode = 0, median = 15·5, range = 192
b mean = 139·6, mode = 200, median = 101, range = 103
5 a 5 **b** none **c** increases the mean
d mode = 3, 5; median = 4
changes mode to only a score of 5 and increases the median to 5
6 a mean = 6·86 (2 dp), mode = 7, median = 7, range = 10
b mean = 13·71 (2 dp), mode = 14, median = 14, range = 20
c mean = 7·86 (2 dp), mode = 8, median = 8, range = 10
d mean = 5·86 (2 dp), mode = 6, median = 6, range = 10
7 2, 2, 3, 4, 4 or 2, 2, 3, 3, 5
8 Justin: equal means, modes Justin's median 90; range 35; Leigh's median 87; range 27
9 a \$524 000 **b** \$397 500
c median, as the \$1 500 000 distorts the true mean as it considered an outlier
d median
10 a team A: $\bar{x} = 1{\cdot}375$, range = 4; team B: $\bar{x} = 1{\cdot}5$, range = 3
b team B—lower range, less variation in points per game
11 a 10, 20, 30 **b** 10, 14, 16, 20 **c** 4, 4, 6, 9
12 e.g. 3, 4, 5, 5, 8

EXERCISE 10G

1 $\bar{x} = 14{\cdot}375$
2 a 3·5 **b** 134·5 **c** 8·6 **d** 17·1
3 a 4 **b** 134 **c** 9 **d** 18
4 a 8 **b** 5 **c** 3 **d** 4
5 a 2 **b** 12
6 a 14 **b** 118·3 **c** 86·8
7 mean = 34·49
mode = 30
median = 32
range = 52

8 a

x	f
36	4
37	5
38	15
39	5
40	4
41	3
42	5
43	5
44	2
45	2
	50

b 38 **c** 9 **d** 39·6 lollies
e 39 lollies
f i 38 lollies—most common number (mode)
ii mode does as it shows the score/s that **most** bags contain

9 a median = 11, mode = 11 **b** median = 138, mode = 121
c median = 16, mode = 12, 18

10 a median = 2, mode = 1 **b** median = 7, mode = 7
c median = 14, mode = 14

11 $x = 7$

12 a 4 **b** 1 goal
c

x	f
0	10
1	16
2	10
3	4
4	10
	50

d 1·76 **e** 1

13 a NSW, as more people travel or SA, to encourage more people to travel
b $26\frac{2}{3}\%$
c

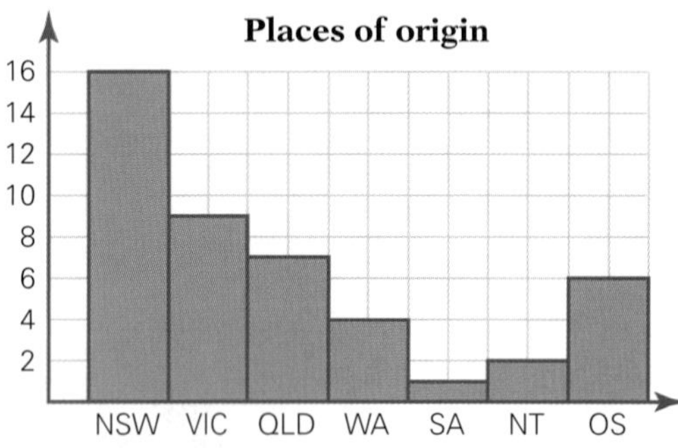

d 24·6 years of age **e** NSW

14 a

Colour	Freq.
W	9
R	5
B	6
G	4
S	4
O	2
	30

b white **c** 70% **d** white **e** 600 000
f discuss

15 a 5·8 **b** 68

16 a

Stem	Team A
2	4 9 9
3	0 4 5 6 6 7
4	0 0 0 4
5	0 2
6	

b team A: mean = 37·07, mode = 40, median = 36, range = 28
c team B: mean = 37·07, mode = 30, median = 38, range = 17
d team A is a better team (it has a higher mode) where as team B can be considered more consistent as it has a smaller range

17 discuss

EXERCISE 10H

1 a 12 **b** 40%

2 a i $\frac{3}{50}$ **ii** 6% **b** 1200

3 120 **4** 3 760 000 **5** 60

6 a 17 100 000 **b** 14 060 000

7 a discuss **b** C
c that you are more likely to have an accident than older drivers—you pay higher premiums

8 a, b

Science %
Maths %

c generally, yes **d** ≈ 70%

EXERCISE 10I

1

	Mean	Median	Min.	Max.	Range
a	5.3	5	2	9	7
b	6	6	3	10	7
c	16	15	12	21	9
d	72·9	73·5	52	95	43
e	442·83	445	401	476	75
f	3	3·1	1·5	4·8	3·3

2

	Mean	Median	Min.	Max.	Range
a	3	3	1	5	4
b	23·719	23·5	21	27	6
c	35·057	30	10	60	50

Chapter Review Exercises

1 a

x	f
9	1
10	3
11	2
12	2
13	6
14	3
15	3
	20

b

x	f
0	3
1	5
2	2
3	3
4	4
5	3
	20

2 a

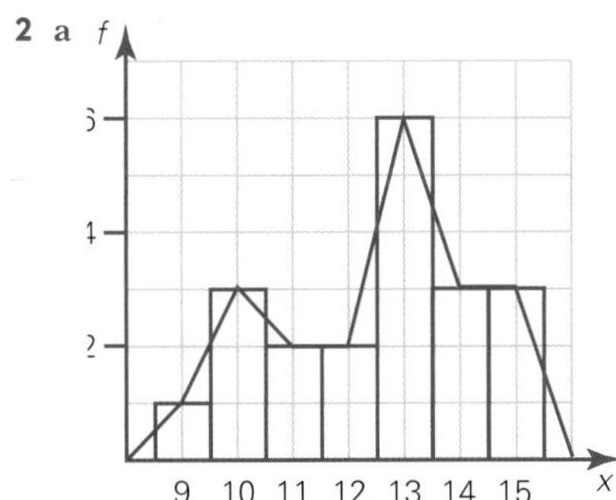

b

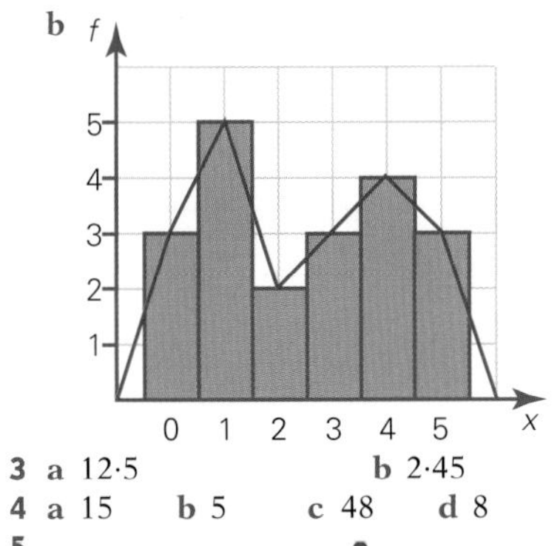

3 a 12·5 b 2·45

4 a 15 b 5 c 48 d 8 e 334

5

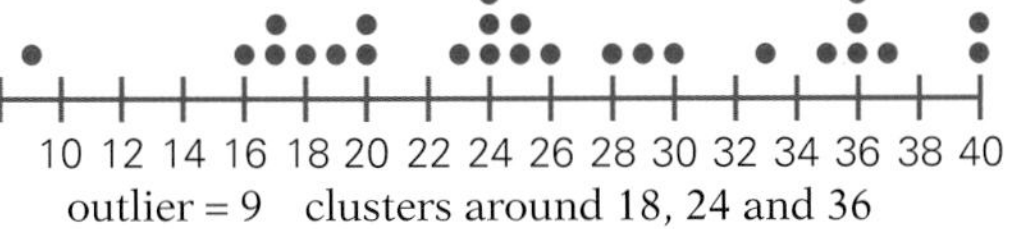

outlier = 9 clusters around 18, 24 and 36

6 a mean: 1·8
modes: 1, 3
median: 1·5
range: 4

b mean: 192·3
mode: 192
median: 192
range: 261

c mean: 8
mode: none
median: 8
range: 6

d mean: 0·776
mode: 0·75
median: 0·75
range: 0·25

7 a

x	f	fx
9	7	63
10	15	150
11	6	66
12	5	60
13	10	130
14	2	28
15	6	90
	51	587

b mean = 11·5 c 10 d 6 e 11
f individual scores only

8

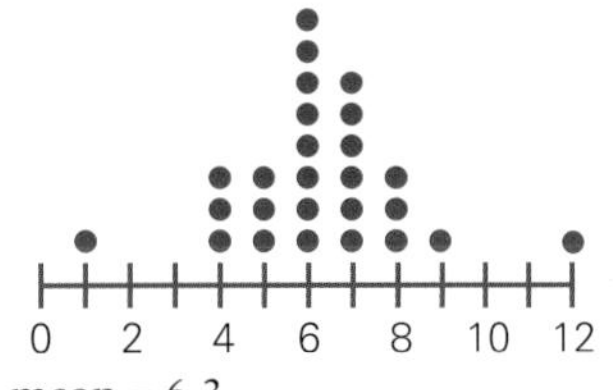

mean = 6·3
mode = 6
median = 6
range = 11

9 a mean = 63·8
median = 61
b mean = 136·9
median = 133·5

10 a \$2 b 45c c \$2.20 d \$2

11 a 20
b i 13 ii 14 iii 13·5 iv 4
c i 13 ii 14 iii 13 iv 4

12 1, 1, 1, 2, 2, 3, 4

Keeping Mathematically Fit

PART A

1 D 2 80 3 $a = 7$ $b = 3$ 4 64, 32
5 a 9 b $2n + 1$ 6 108° 7 11:05 am
8 $2\frac{2}{5}$ 9 $6 - 5x$ 10 8000 m²

PART B

1 55·2% (1 dp) 2 $x \geq 21$ 3 279·5m² (1 dp)
4 9·681 5 36° 6 90 km/h 7 39
8 a 3, 3, 4, 6 b discuss

Chapter 11

MathsCheck

1 a x-axis
b i (2, 2) ii (0, 3) iii (5, 3)
iv (4, 0) v $(1\frac{1}{2}, 0)$ vi (3, 5)
vii (5, 4) viii $(3\frac{1}{2}, 1\frac{1}{2})$
c i E ii K iii I iv H v O vi L
d B, H and D e O, B and L

2 **a** New York **b** London **c** Cairo **d** Rio **e** Tokyo **f** Mexico city

3 **a** 8°S, 110°E **b** 18°N, 75°E **c** 32°N, 115°W **d** 42°N, 15°E **e** 35°N, 120°E **f** 55°N, 40°E **g** 15°N, 125°E **h** 32°S, 20°E **i** 60°N, 155°W

4 **a** (⁻3, 2) **b** (⁻2, 0) **c** (0, ⁻1) **d** (3, ⁻1) **e** (⁻2, ⁻2)

5 **a** T **b** M **c** G **d** P

6 **a** x-axis **b** y-axis

7 **a** 3rd **b** 1st **c** 2nd

8 P and E, B and G, C and F; same vertical line

9 B and M, C and P, G and E; same horizontal line

10 (1, ⁻1) **11** (1, ⁻2)

12 BM = 4 units, BG = 3 units

13 **a**

x	0	1	2	3	4
$3x$	0	3	6	9	12

b

n	3	4	5	6	7
$10 - n$	7	6	5	4	3

c

x	0	1	2	3	4
$4x + 5$	5	9	13	17	21

d

n	⁻1	0	1	2
$3 - 2n$	5	3	1	⁻1

EXERCISE 11A

1 **a** (1, 4) (2, 8) (3, 12) (4, 16)

b

s	0	1	2	3	4
p	0	4	8	12	16

c a square of side 3 cm has a perimeter of 12 cm

d **i** 32 cm **ii** 40 cm

e as in between are included

2 **a**

x	0	2	4	8	11
y	3	5	7	11	14

b

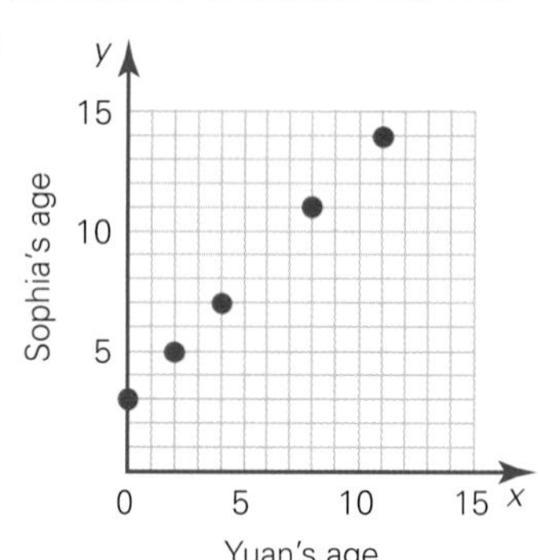

c yes, as it is possible to have an age that is not an integer

d **i** $6\frac{1}{2}$ **ii** 3 years 9 months

3 **a**

x	0	1	2	3	4
A	0	6	24	54	96

b

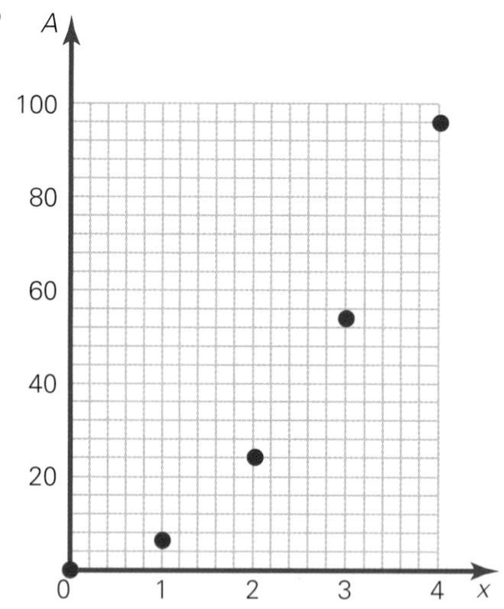

c since side lengths are not just whole numbers

4 **a**

x	1	2	3	4	5
y	3	5	7	9	11

b discrete, only use 'whole' matches

c

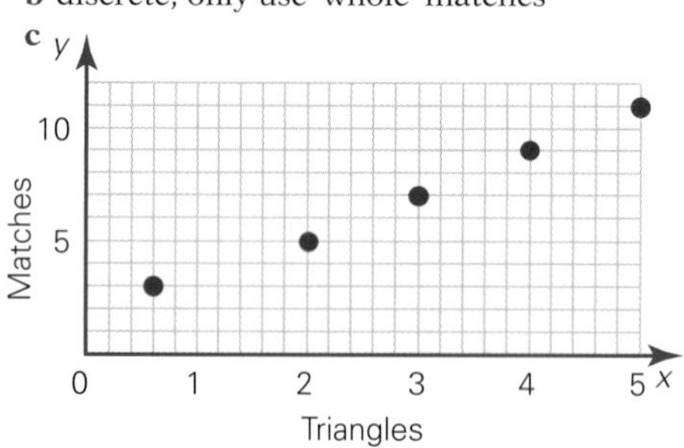

d **i** 17 **ii** 21

5 **a**

x	⁻3	⁻1	0	1	3
y	⁻3	⁻1	1	1	3

b

x	⁻3	⁻1	0	1	3
y	⁻1	1	2	3	5

c

x	⁻3	⁻1	0	1	3
y	⁻5	⁻3	⁻2	⁻1	1

d

x	⁻3	⁻1	0	1	3
y	⁻9	⁻3	0	3	9

e

x	⁻2	⁻1	0	1	2
y	⁻1	1	3	5	7

6 **a** (0, 4) (1, 5)

b

x	⁻2	⁻1	0	1	2
y	2	3	4	5	6

c $y = x + 4$

d (3, 7), (4, 8), etc.

7 **a** (⁻2 ,⁻5), (⁻1, ⁻4), (0, ⁻3), (1, ⁻2), (2, ⁻1)

b

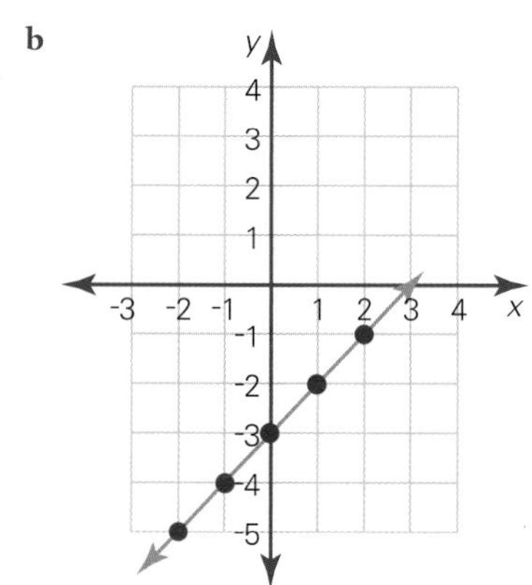

c $y = x - 3$

8 a

x	1	2	3	4	5
y	8	16	24	32	40

b

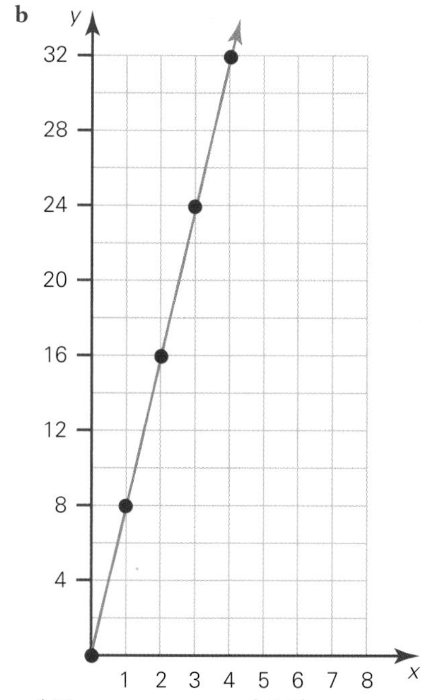

c \$48 **d** 5·5h

9

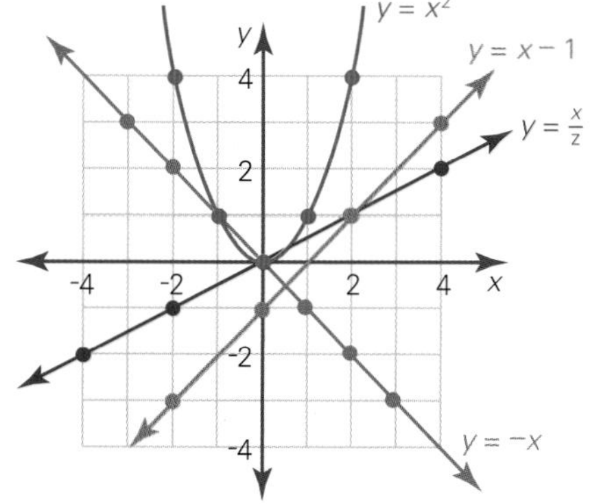

10 a a, b and d

b powers other than one on x

c **i** yes **ii** yes **iii** no **iv** no
v yes **vi** no **vii** yes **viii** yes

EXERCISE 11B

1 **a** A **b** A **c** B **d** A
e B **f** A **g** A **h** A

2 **a** $y = x + 3$ **b** $y = x - 3$
c $y = 3x$ **d** $y = x + 5$

3 **a** $y = {}^{-}x$ **b** $y = x - 4$
c $y = 3x + 1$ **d** $y = x^2$

4 **a** $y = x + 2$ **b** $y = x - 1$ **c** $y = \frac{x}{2}$ **d** $y = {}^{-}2x$

5 **a** (0, 0), (1, 5), (2, 10), (3, 15), (4, 20)
b (0, 10), (1, 9), (2, 8), (3, 7), (4, 6)
c (0, 0), (1, ⁻3), (2, ⁻6), (3, ⁻9), (4, ⁻12)

6 **a** $y = 3x$ **b** $y = 3x$
c $y = 5 - x$ **d** $y = x + 1$

7 a

x	⁻3	⁻2	⁻1
y	4	5	6

$y = x + 7$

b

x	⁻1	1	2	3
y	11	9	8	7

$y = 10 - x$

c

x	⁻3	⁻1	2	3
y	⁻15	⁻7	5	9

$y = 4x - 3$

d

x	⁻3	⁻2	2	3
y	14	12	4	2

$y = 8 - 2x$

8 $C = 5x + 5$ or $C = 5(x + 1)$

9 $C = 0{\cdot}95n$

10 $C = 2{\cdot}5 + 1{\cdot}5n$

EXERCISE 11C

1 a, b, e, g, h

2 a

x	⁻3	0	3
y	0	3	6

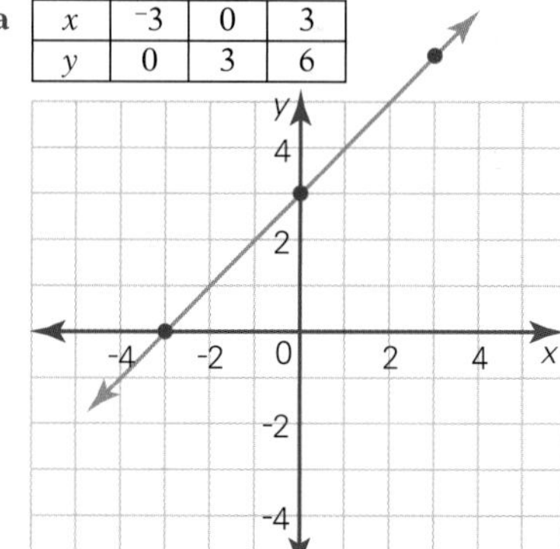

b

x	3	0	$^-1$
y	$^-1$	2	3

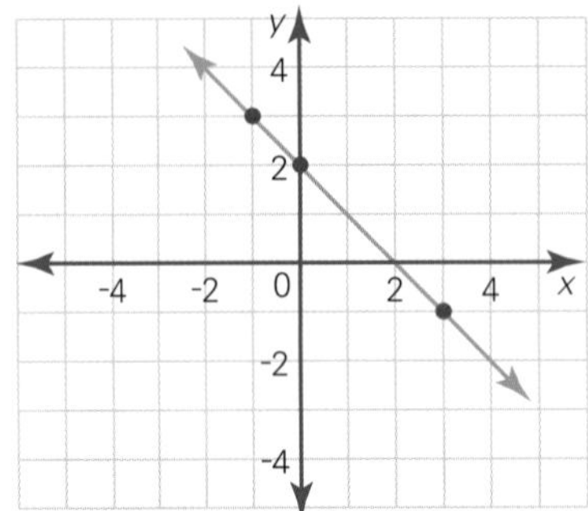

c

x	$^-1$	0	1
y	$^-5$	$^-1$	3

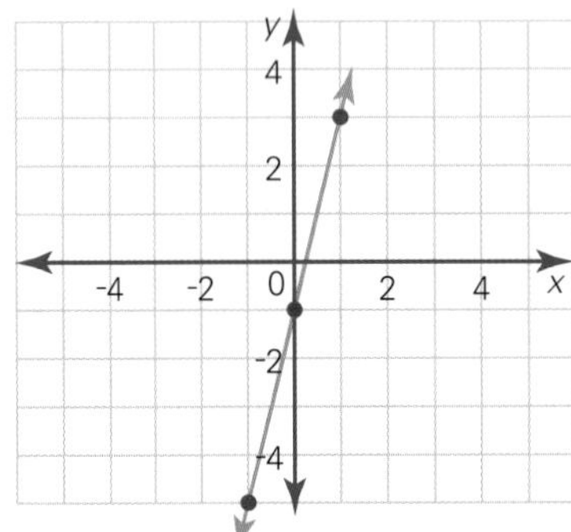

3

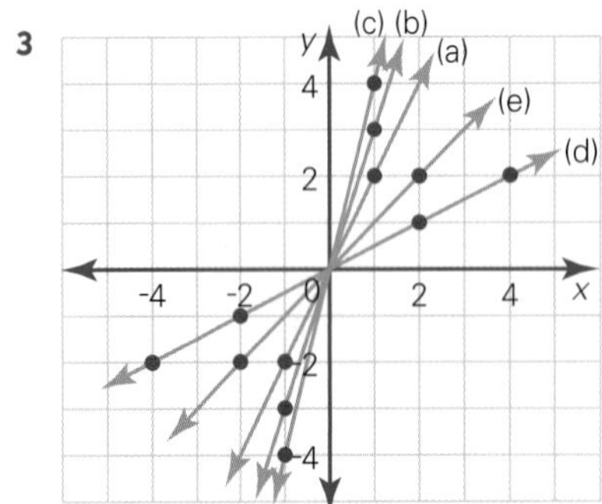

4 **a** (0, 0)
b the larger the number, the steeper the line

5

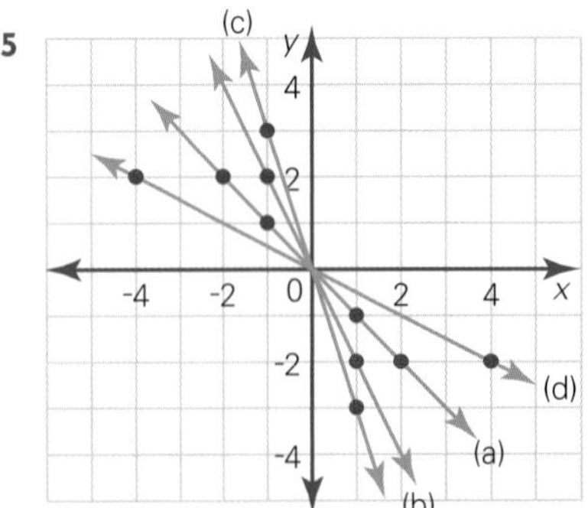

6 **a** (0, 0)
b the larger the 'negative' number, the steeper the line
c changes the direction (becomes a decreasing line)
d increasing (positive gradient)
e decreasing (negative gradient)

7

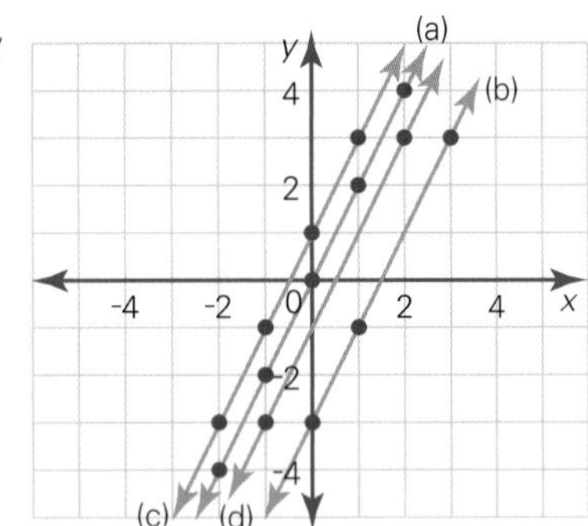

8 **a** yes
b the gradient, all $y = 2x$ …
c where the graph crosses the y-axis (the y-intercept)

9 yes, they hold true

10 **a**

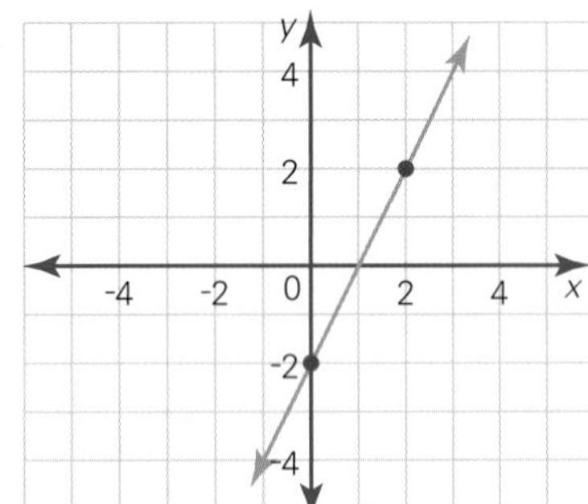

b

c

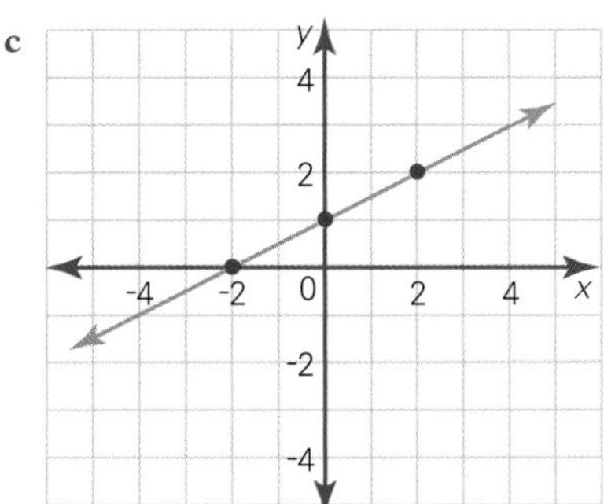

11 **a** (0, 5) **b** (0, 1) **c** (0, $^-2$)

12 **a** B **b** A **c** A **d** B **e** A

13 b, c, e **14** a, c

15 **a** $y = 2x - 4$ **b** $y = 4 - x$ **c** $y = \frac{x}{2} + 1$

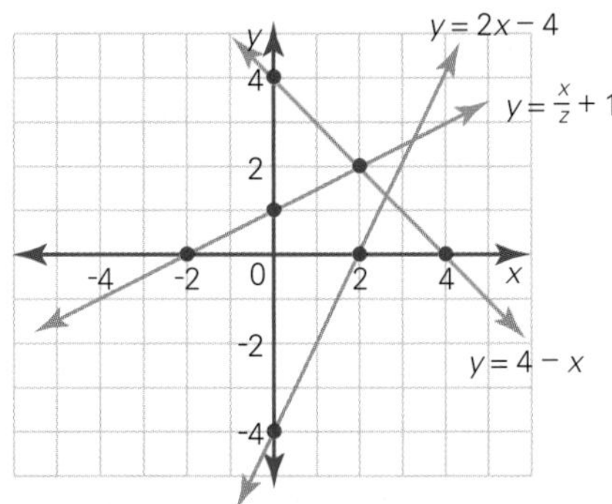

16 Increasing line with gradient of 1, crossing the y-axis at the point (0, 3)

EXERCISE 11D

1 (2, 1), (3, 1), ($^-4$, 1), (12, 1), ($^-6$, 1)

2 ($^-3$, 1), ($^-3$, 4), ($^-3$, 10), ($^-3$, 3)

3 **A** v **B** i **C** ii

D iii **E** vi **F** iv

4

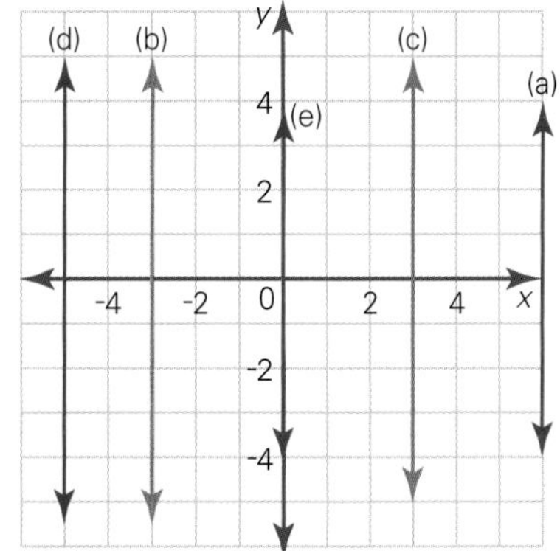

5

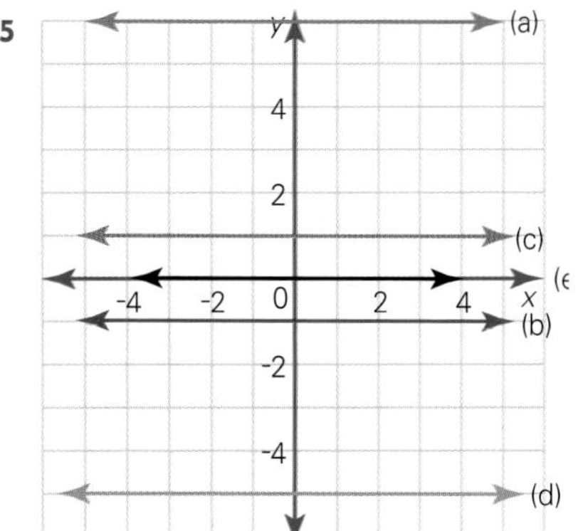

6 **a** **i** $y = 1$ **ii** $y = 3$ **iii** $y = {}^-3$

b **i** $x = 2$ **ii** $x = 3$ **iii** $x = {}^-4$

7 **a** $y = a$ (a is any number)

b $x = b$ (b is any number)

8 **a** $y = 6$ **b** $y = 1$ **c** $x = 3$ **d** $x = {}^-4$ **e** $y = 0$

EXERCISE 11E

1 **a** (1, 1) **b** ($^-1$, 2) **c** (4, $^-2$)

d (1, 2) **e** (0, $^-2$) **f** (4, 0)

2 **a** (1, 1) **b** **i** ($^-1$, $^-1$) **ii** (0, 2)

3 **a** (0, 5) **b** (2, 4) **c** (0, 2) **d** ($^-1$, 3)

e ($^-3$, $^-1$)

4 **a** (4, 2) **b** (0, 0) **c** ($^-2$, 0)

d ($^-1$, 1) **e** ($^-4$, $^-5$) **f** ($\frac{-1}{2}$, 3)

5 **a** T **b** F **c** F **d** T **e** T

6 graphs intersect when $n = 10$

7 **a** $R = 6n$

b breakeven when $n = 200$

c \$1200 profit

Chapter Review Exercises

1 **a** ($^-4$, 3) **b** (3, $^-2$)

2 **a** Q **b** P

3 3rd **4** B (3, $^-2$)

5 **a** 4 units **b** 6 units **c** 3 units

6 a, c, d

7

x	$^-3$	$^-2$	$^-1$	0	1	2	3
y	8	3	0	$^-1$	0	3	8

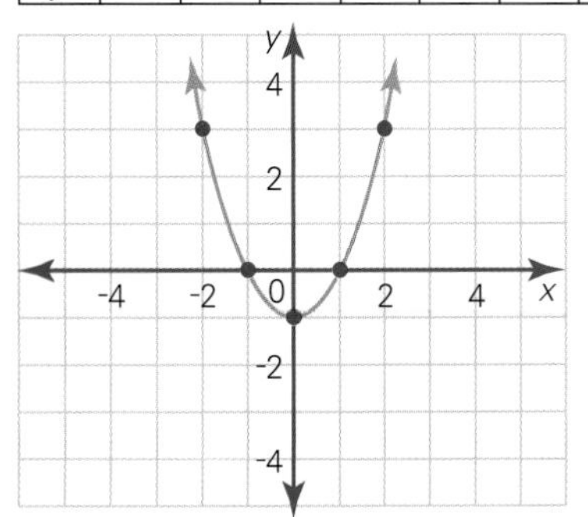

8 a

x	$^-2$	$^-1$	0	1	2
y	1	2	3	4	5

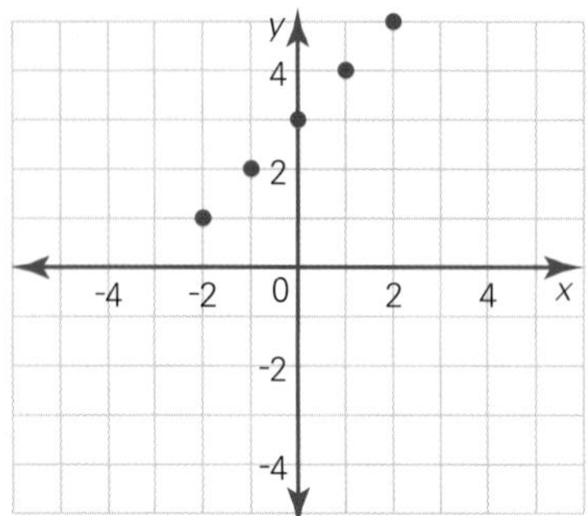

b

x	$^-3$	$^-1$	0	1	3
y	$^-11$	$^-5$	$^-2$	1	7

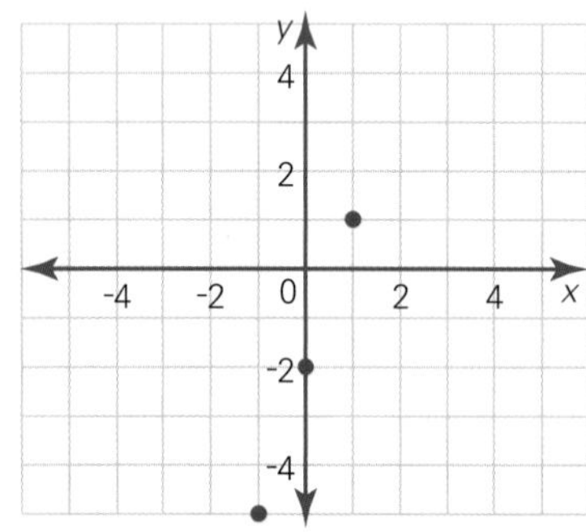

9 a

b decreasing **c** (0, 2)

10 a i

x	$^-1$	0	1
y	$^-2$	0	2

ii

x	$^-1$	0	1
y	2	0	$^-2$

iii

x	$^-1$	0	1
y	10	8	[illegible]

b

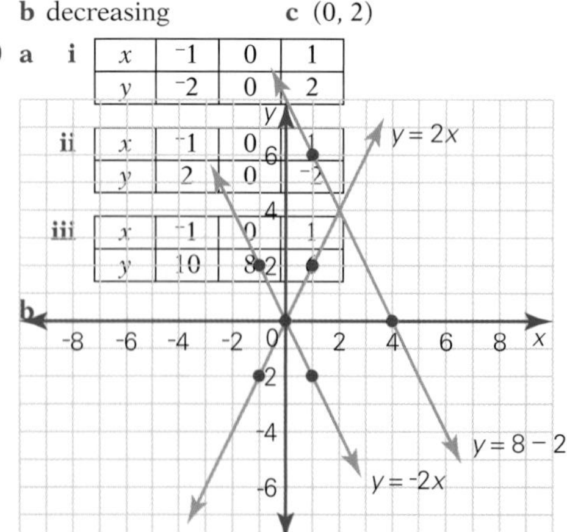

c $y = 2x$ and $y = {}^-2x$ **d** (0, 8)

e $y = 8 - 2x$ and $y = {}^-2x$

f (2, 4) **g** $y = x + 2$

11 a

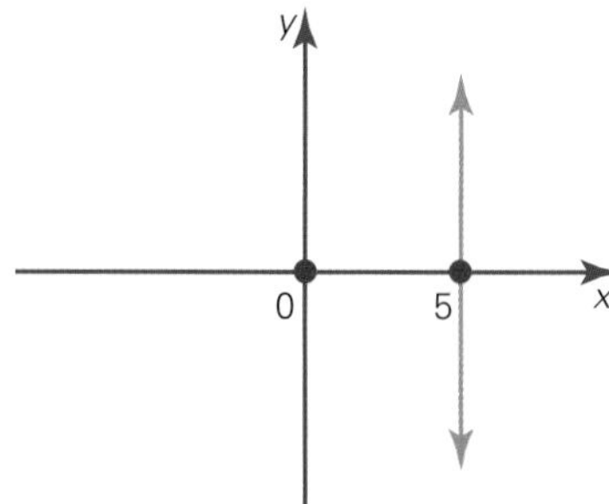

b

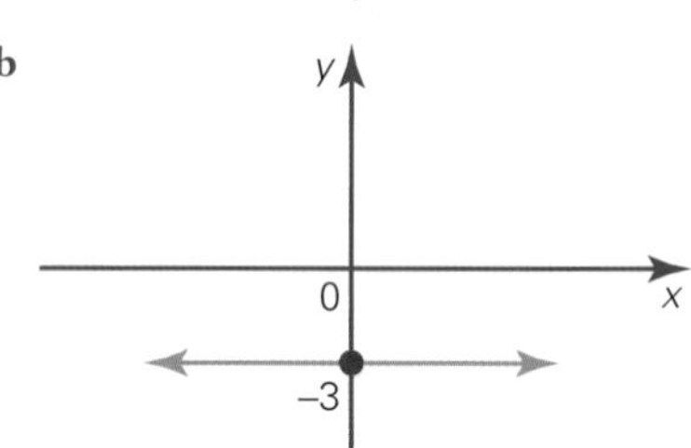

c

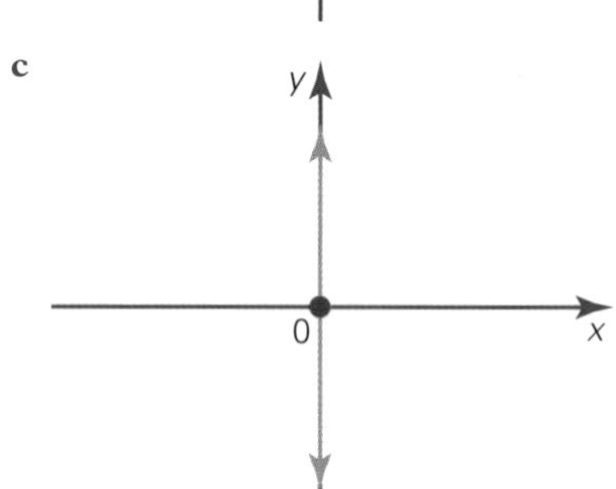

d

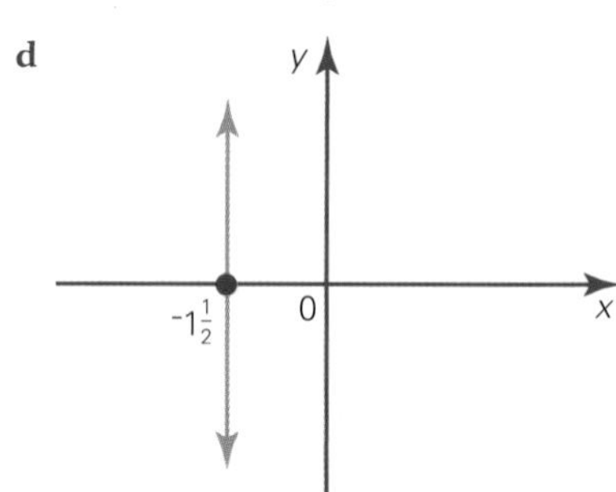

e

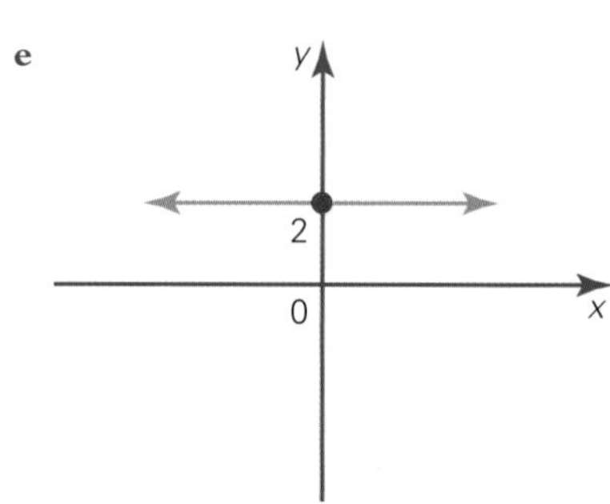

12 a (4, 8) **b** (1, $^-2$) **c** ($^-2$, 0) **d** (0, 0) **e** (2, 2)

13 a $y = 3x$ b $y = 5 - 2x$

14 a $y = \frac{x}{2}$ b $y = x + 4$ c $y = {}^{-}3x$ d $y = 2x - 4$

Keeping Mathematically Fit

PART A

1 $p + 2$ **2** 0·45 **3** 7·88 **4** 20 min
5 135° **6** 338, 507 **7** 12 **8** 94·2
9 12 **10** $21 600

PART B

1 0·57
2 a 5·14 (2 dp) b 1 c 4 d 11
3 $x - 13$ **4** 11 200 cm^2 **5** 67·2 L
6 a $8x + 12$ b $2a + 7b$ c $x + 7y + 4z$
7 $P = 32{\cdot}1$ cm, $A = 44{\cdot}18$ cm^2
8 m can take any value; lines in the form of $y = mx$ all pass through the origin

Chapter 12

MathsCheck

1 a 4 m b 12 m c 3 m d 0·45 m e 500 m
2 a 50 mm b 120 mm c 2000 mm d 500 mm e 340 mm
3 a 20 000 b 30 000 c 0·4 d 3
4 a $A = 65{\cdot}4481$ m^2, $P = 32{\cdot}36$ m
b $A = 22{\cdot}4$ m^2, $P = 19{\cdot}2$ m
c $A = 180$ m^2, $P = 90$ m
5 a $V = 3375$ cm^3 b $V = 144$ cm^3 c $V = 240$cm^3
6 a 5 L b 5 L c 500 L d 500 L e $\frac{3}{4}$ L f 1·8 L g 1500 L h 1500 L
7 a 1000 b 500 c 70 d 1 000 000 e 250 000 f 5
8 a 1 mL b $\frac{1}{2}$ mL c 70 mL d 1 000 000 mL e 250 000 mL f 5 000 000 mL

EXERCISE 12A

1 a sector b circumference c arc d semicircle
2 a diameter b radius c sector d chord e segment f semicircle g central angle
3 (Reduced diagram)

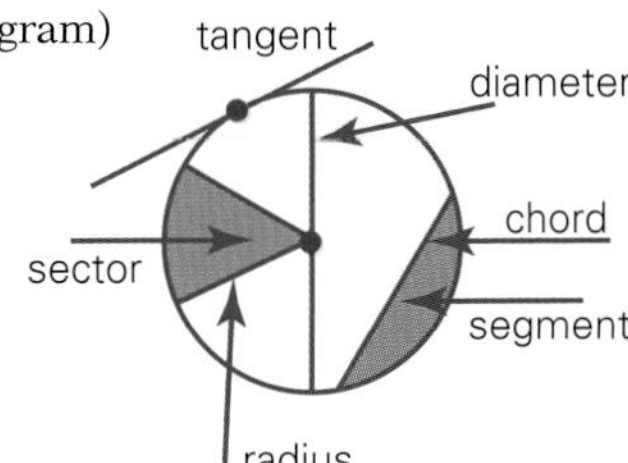

4 diameter = 2 × radius
5 a 2 radii and an arc b chord and an arc
6 constructions
7 chord AB and arc AB
8 a $\frac{1}{4}$ b $\frac{1}{6}$ c $\frac{1}{8}$ d $\frac{1}{3}$ e $\frac{1}{5}$ f $\frac{3}{4}$
9 2 radii equal, ∴ 2 sides of Δ are equal and the Δ is isosceles
10 construct and discuss **11** construction
12 a isosceles as $MC = CQ$ b 36°

EXERCISE 12B

1 a 30 cm b 60 mm c 114 m d 186 cm e 117 m f 7·5 km
2 a 62·8 cm b 251·2 mm c 94·2 cm d 314·0 m
3 a 44 m b 66 m
4 a 257·61 mm b 13·41 km c 54·04 m d 43·98 cm e 213·63 cm f 38·33 m g 9·42 cm h 13·82 cm i 28·27cm
5 40 200 km **6** 45·55 mm
7 a 182 cm b 1820 m c just over 824 revolutions
8 a 8·45 cm to 8·55 cm b 53·09 … cm to 53·72 … cm
9 a 8·5 m b 53·4 m ≑ 54 m c $864 d 27
10 a 7·9 m b 31·4 m c 24·4 m
11 a 18 m b 51 m c 38 m d 36 m e 28 m f 33 m
12 a 158·3 m b 552·7 m c 10 969·9 mm d 25·1 mm e 101·1 cm f 29·7 m
13 a $r = 5{\cdot}52$cm (2 dp)
b the circumference is accurate to the nearest mm and the radius should have more decimal places, ∴ 2 dec. pl. are required
14 a 199·9 m b 50·0 cm
15 3600 m **16** 153 m
17 yes, as the diameter of the pipe (4·77 cm) is less than the length and width of the brick
18 1 : 2 **19** 1 min 34 s
20 25·7 cm (1 dp) **21** 20π metres

EXERCISE 12C

1 a 300 m^2 b 1200 m^2 c 45·63 m^2 d 48 m^2 e 13·8675 m^2 f 3072 m^2
2 a 1256 cm^2 b 2826 cm^2 c 12·56 m^2
3 a 154 cm^2 b 38·5 m^2 c 9·625 cm^2
4 a 141·03 cm^2 b 475·29 m^2 c 1170·21 m^2 d 907·92 mm^2 e 3·14 cm^2 f 2375·83 mm^2
5 a 100π mm^2 b 144π mm^2 c 4π cm^2 d 16π cm^2 e 49π cm^2 f $\frac{\pi}{16}$ m^2
6 44·2m^2 (1 dp) **7** 1257 km^2 **8** $2734.44
9 a 11 310 cm^2 b 15 394 cm^2
10 28·3 cm^2 (1 dp) **11** 12·89% (2 dp)
12 a 10 691 m^2 b 48 m^2 c 1 m^2 d 236 m^2
13 a 4·0 cm b 6·3 mm c 6·0 cm
14 7896 m^2
15 a 4 b 3 c $3{\cdot}14 \times 4^2$ or $\frac{22}{7} \times 4^2$ d 25·12 units

EXERCISE 12D

1 a 30 cm b 900 cm^2 c 353·4 cm^2 d 1253·4 cm^2
2 a 1500 m^2 b 982 m^2 c 2482 m^2
3 193·4 m^2 4 123·6 m^2 5 175·9 m^2
6 a 263·9 cm^2 b 48·7 m^2 c 47 242·3 m^2
7 a 19·63 m^2 b 34·36 cm^2 c 142·47 mm^2
8 50π cm^2 9 7696·9 mm^2
10 four times (ratio = 1 : 4)
11 a 120 cm by 60 cm b 1548 cm^2

EXERCISE 12E

1

Volume	402·12	15·71	0·60	370 431·47	0·03

2 a 1 131 cm^3 b 44 277·6 cm^3 c 186·0 cm^3 d 11 938 cm^3 e 2 513 cm^3 f 12 566 cm^3
3 a 192·42 cm^3 b 73·89 cm^3 c 365·21 cm^3
4 *A*: 196 cm^3, *B*: 392·7 cm^3, ∴ *B* is larger
5 a 503 mL b 1005 mL c 31 416 mL
6 a 300π cm^3 (942·5 cm^3) b 7200 cm^3 c 1545 cm^3
7 60 cm
8 a 144π cm^2 b 1008π cm^3
9 a 23 562 cm^3 b 226 195 cm^3 c 50 800 cm^3
10 3·9 L
11 a 8189 m^3 b 3 480
12 812 015 coins
13 a 22·167 m^3 b 22 167 L c $22 100 d 554 cars
14 5·05 m 15 14 cm
16 a 9·95 cm b 150·8 cm^3
17 $V = \dfrac{\pi d^2 h}{4}$

Chapter Review Exercises

1 3, 3·14, $\frac{22}{7}$
2 a diameter b sector/quadrant c semicircle d radius e $\frac{1}{4}$ f $\frac{1}{12}$
3 11·4 cm
4 ① segment (minor) ② tangent
5 a $C = 103·7$ mm, $A = 855·3$ mm^2
b $C = 7·5$ cm, $A = 4·5$ cm^2
c $C = 182·2$ cm, $A = 2642·1$ cm^2
6 a $P = 32·0$ cm, $A = 68·2$ cm^2
b $P = 31·4$ cm, $A = 64·8$ cm^2
c $P = 48·3$ cm, $A = 137·4$ cm^2
7 a 2513·3 cm^2 b 442·9 m^2 c 85·8 cm^2 d 114·2 cm^2
8 a 288π cm^3 b 20π cm^3 c 28π m^3
9 a 3054 cm^3 b 1 130 973·4 cm^3
10 26·8 cm^3 (1 dp) 11 13 cm
12 1·3 cm

Keeping Mathematically Fit

PART A

1 1000 m 2 76% 3 $x = 3$
4 $3x + 6$ 5 60 km 6 (1, 7)
7 8 8 $\frac{2}{3}$ 9 $807.50
10 1

PART B

1 a 62 cm^2 b 31·74 cm^2 c 48 m^2
2 25·2 3 $13\frac{1}{8}$
4 a $x > 5$ b $x = 3$ c $x = {}^-8$
5 88
6

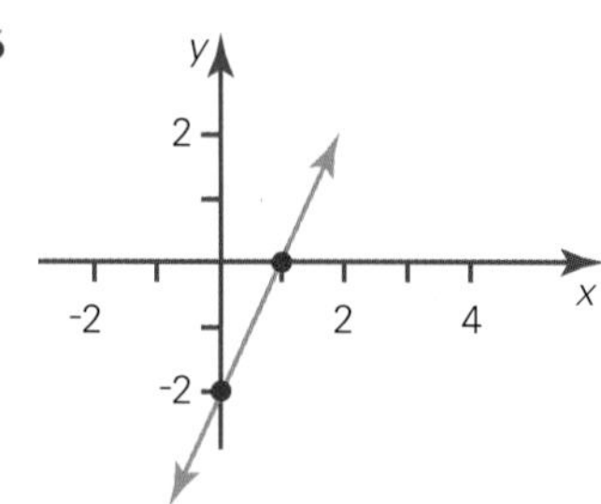

7 25·2 L
8 square, rhombus, parallelogram, rectangle

Cumulative Review 4

Part A

1 D	2 C	3 C	4 D	5 C	6 A
7 D	8 B	9 B	10 B	11 A	12 A
13 D	14 C	15 C	16 B	17 D	18 C
19 B	20 D	21 D	22 C		

Part B

1 1·40
2 a $4a$ b $2x$ c $3^6 = 729$ d $2a$
3 $10 185 4 30% 5 $5750
6 a $x = {}^-1$ b $x = {}^-4$ c $x > 1$ d $x \le {}^-1$
7 a $x = 0·9$ b $x = 24$ c $x = \sqrt{61} \doteqdot 7·8$
8 a $x = 68$ (vertically opposite angles)
$y = 68$ (alternate angles on || lines)
$z = 112$ (∠s on a straight line)
b $x = 66$ (∠ sum of isosceles Δ)
$y = 66$ (alternate angles on || lines)
9 a *ABC*: 22%, Com: 50%
b 16%
c children, information and education 6%
d 20%
10 a mean = 7·53 (2 dp) mode = 6, 9 median = 8 range = 15
b mean = 20·96 (2 dp) mode = 21 median = 21 range = 3
c mean = 18·21 (2 dp) mode = 12, 19 median = 19 range = 35

11 a

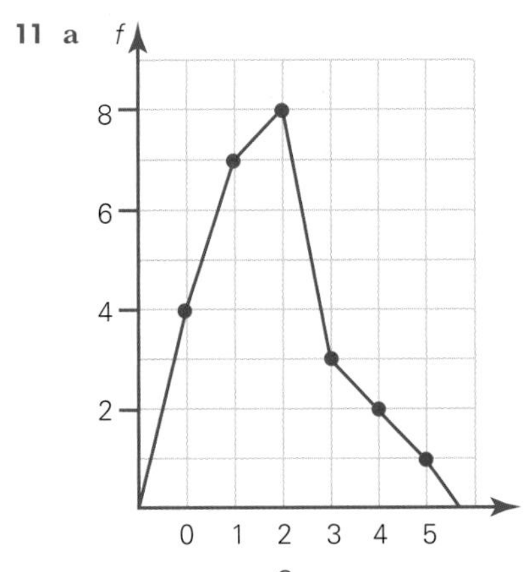

b i 25 **ii** 5 **iii** 2 **iv** 1·96

12 a $75{\cdot}4\dot{3}$ **b** 47·2 **c** 84·25 **d** 48·9

13 a $y = 2x$ **b** $y = 2x - 1$

14 13·14 (2 dp) **15** (1, 5)

16 a i 94·2 cm **ii** 706·9 cm^2

b i 182·2 m **ii** 2642·1 m^2

c i 45·9 cm **ii** 167·4 cm^2

17 a i 103 cm **ii** 628 cm^2

b i 132 m **ii** 1043 cm^2

c i 228 cm **ii** 1885 cm^2

18 a 8482 m^3 **b** 58 974 m^3

c i 44 m^3 **ii** 44 000 L

Chapter 13

MathsCheck

1 a $\frac{3}{10}$ **b** $\frac{3}{4}$ **c** $\frac{3}{25}$ **d** $\frac{17}{25}$ **e** $\frac{19}{25}$

2 a 40 **b** 800 **c** 300 **d** 150 **e** 2400 **f** 200

3 a $\frac{1}{3}$ **b** $\frac{4}{15}$ **c** $\frac{1}{6}$ **d** $\frac{5}{6}$ **e** $\frac{3}{5}$

f $\frac{7}{30}$ **g** $\frac{23}{30}$ **h** 0

4 a

x	f
3	1
4	1
5	6
6	3
7	5
8	3
9	1

b 3

1 c i $\frac{3}{10}$ **ii** $\frac{3}{20}$ **iii** $\frac{1}{10}$ **iv** $\frac{3}{5}$

v 0 **vi** $\frac{9}{20}$

2 a 40 **b** 5 **c** $\frac{1}{8}$

d i $\frac{1}{10}$ **ii** $\frac{3}{8}$ **iii** $\frac{1}{2}$

EXERCISE 13A

1

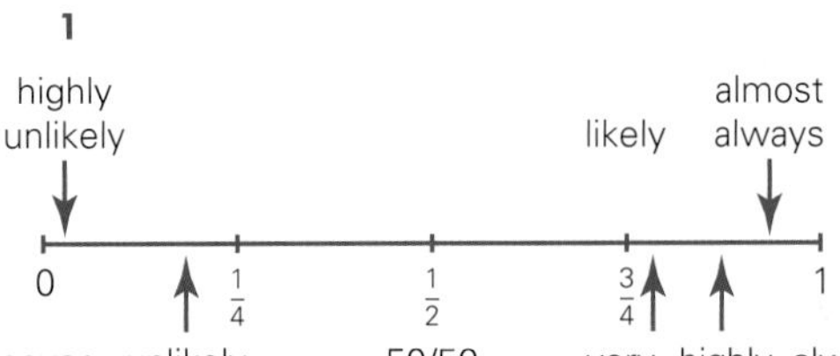

2 a 0 **b** 0 **c** 1 **d** 0 **e** 0 **f** 0

g 1 **h** 0 **i** 1 **j** 0 **k** 0

3 discuss

4 a 0 **b** 50-50 **c** highly likely

d highly unlikely **e** certain

f certain (discuss other answers)

5 a unlikely **b** even chance **c** even chance

d even chance **e** unlikely **f** unlikely

g unlikely **h** even chance

6 a 0 **b** 0·5 **c** discuss

d 0·7 **e** 0·5

7 unlikely a, d, i; even chance, b, c, e; very likely g, h; certain f

8 A, D, C, B **9** D, C, B, A

EXERCISE 13B

1 a 4

b i $\frac{1}{4}$ **ii** $\frac{1}{4}$ **iii** $\frac{1}{2}$ **iv** $\frac{1}{4}$

v $\frac{3}{4}$ **vi** $\frac{1}{4}$ **c** 1

2 a 4

b i $\frac{1}{4}$ **ii** $\frac{1}{4}$ **iii** $\frac{3}{4}$ **iv** 0

v $\frac{1}{4}$ **vi** $\frac{1}{2}$ **vii** $\frac{1}{2}$ **viii** $\frac{1}{2}$

c triangular pyramid

3 a $\frac{1}{30}$ **b** $\frac{1}{30}$ **c** $\frac{1}{30}$ **d** $\frac{1}{2}$ **e** $\frac{1}{3}$ **f** $\frac{9}{30}$

4 a $\frac{1}{9}$ **b** $\frac{1}{9}$ **c** $\frac{1}{3}$ **d** $\frac{2}{3}$ **e** 0

5 a $\frac{3}{10}$ **b** $\frac{3}{10}$ **c** $\frac{7}{10}$ **d** $\frac{1}{10}$ **e** $\frac{2}{5}$

6 a $\frac{1}{8}$ **b** $\frac{1}{8}$ **c** $\frac{1}{8}$ **d** $\frac{1}{2}$ **e** $\frac{1}{4}$ **f** $\frac{7}{8}$

7 a $\frac{1}{5}$ **b** $\frac{1}{5}$ **c** $\frac{2}{5}$ **d** $\frac{3}{5}$ **e** $\frac{2}{5}$ **f** $\frac{3}{5}$

g $\frac{1}{5}$ **h** 1

8 a $\frac{2}{5}$ **b** $\frac{4}{15}$ **c** 0 **d** 1

9 a 26 **b** 26 **c** 13 **d** 13 **e** 13 **f** 13

g 4 **h** 4 **i** 2 **j** 2 **k** 4 **l** 4

m 2 **n** 1

10 a $\frac{1}{2}$ **b** $\frac{1}{2}$ **c** $\frac{1}{4}$ **d** $\frac{1}{4}$ **e** $\frac{1}{4}$ **f** $\frac{1}{4}$

g $\frac{1}{13}$ **h** $\frac{1}{13}$ **i** $\frac{1}{26}$ **j** $\frac{1}{26}$ **k** $\frac{1}{13}$ **l** $\frac{1}{13}$

m $\frac{1}{26}$ n $\frac{1}{52}$ o $\frac{1}{13}$ p $\frac{2}{13}$ q $\frac{12}{13}$ r $\frac{3}{4}$

s $\frac{1}{26}$ t $\frac{1}{52}$

11 **a** 123, 132, 213, 231, 312, 321

b i $\frac{1}{6}$ ii $\frac{1}{3}$ iii $\frac{1}{3}$ iv $\frac{1}{3}$ v $\frac{1}{3}$

vi 1 vii $\frac{2}{3}$ viii 0

12 **a** $\frac{1}{6}$ **b** $\frac{4}{15}$ **c** $\frac{1}{3}$ **d** $\frac{2}{15}$

13 **a** 40% ($\frac{2}{5}$) **b** 35% ($\frac{7}{20}$) **c** 17% ($\frac{17}{100}$)

d 3% ($\frac{3}{100}$)

b 170

14 **a** WAS, WSA, AWS, ASW, SAW, SWA

b i $\frac{1}{3}$ ii $\frac{2}{3}$ iii $\frac{1}{3}$ iv $\frac{1}{6}$

EXERCISE 13C

1 **a** obtaining a tail
b choosing a black card
c obtaining an even number
d losing the raffle (ie. not winning)
e testing negative
f choosing a red card

2 $\frac{5}{9}$

3 **a** $\frac{8}{9}$ **b** $\frac{2}{5}$ **c** $\frac{1}{4}$ **d** 0·3 **e** 0·56

4 **a** 1, 2: $\frac{1}{3}$ **b** 3, 4, 5, 6 : $\frac{2}{3}$

5 0·16 **6** $\frac{99}{100}$ **7** $\frac{4}{7}$

8 **a** 2 **b** $\frac{4}{5}$

9 **a** $\frac{2}{9}$ **b** $\frac{7}{9}$ **c** $\frac{2}{9}$ **d** $\frac{7}{9}$ **e** $\frac{4}{9}$

f $\frac{5}{9}$ **g** $\frac{1}{9}$ **h** $\frac{8}{9}$ **i** 1

10 **a** $\frac{4}{7}$ **b** $\frac{3}{5}$ **c** $\frac{6}{35}$ **d** $\frac{29}{35}$ **e** 1 **f** 1

11 **a** $\frac{1}{13}$ **b** $\frac{1}{13}$ **c** $\frac{1}{2}$ **d** $\frac{1}{2}$ **e** $\frac{12}{13}$ **f** 1

Chapter Review Exercises

1 **a** certain **b** certain **c** impossible
d impossible **e** even chance

2 **a** 1, 2, 3, 4, 5, 6 **b** H, T **c** B, G

3 **a** $\frac{1}{6}$ **b** $\frac{1}{2}$ **c** $\frac{1}{6}$ **d** $\frac{1}{4}$ **e** $\frac{4}{7}$

4 **a** $\frac{1}{2}$ **b** $\frac{1}{13}$ **c** $\frac{1}{26}$ **d** $\frac{1}{13}$ **e** $\frac{1}{26}$ **f** $\frac{3}{13}$

g $\frac{1}{52}$ **h** $\frac{1}{4}$ **i** $\frac{3}{4}$ **j** $\frac{3}{52}$

5 **a** $\frac{1}{3}$ **b** $\frac{5}{12}$ **c** $\frac{3}{4}$ **d** $\frac{3}{4}$ **e** $\frac{7}{12}$

6 **a** $\frac{4}{11}$ **b** $\frac{1}{11}$ **c** $\frac{7}{11}$ **d** $\frac{13}{22}$ **e** $\frac{1}{22}$

7 0·22 (22%)

8 **a** $\frac{4}{7}$ **b** 9 **c** 0

Keeping Mathematically Fit

PART A

1 560, 320 **2** $\frac{3}{10}$ **3** 0·071 **4** 24

5 120 **6** 16 000

7 $4x = 7$ is incorrect

solution is $\frac{4x-5}{2} = 6$

$4x - 5 = 12$

$4x = 17$

$x = \frac{17}{4}$

8 10 and 25

9 **a** (4, 8) **b** $40u^2$

10 $\frac{3a}{b}$

PART B

1 **a** 6·57 (2 dp) **b** 9 **c** 7

2 245 cm^2 **3** 90 km/h

4 **a** $x = 11$ **b** $x = {}^-10$ **c** $x = \frac{5}{6}$

5 $550 000 **6** $690 **7** (2, 2)

8 **a** $a = 32$ $b = 16$ **b** $^-32, {}^-16$

Chapter 14

MathsCheck

1 **a** isosceles triangle
b isosceles, right triangle
c equilateral triangle

2 **a** $x = 56$, alternate ∠s on || lines
b $p = 72$, vertically opposite angles
c $a = 74$, corresponding angles on || lines;
$b = 106$, angles on a straight line
d $x = 60$, angles in an equilateral triangle
e $x = 86$, angle sum of an isosceles triangle
f $x = 106$, angle sum of a quad

3 **a** 3 : 100 **b** 1 : 100 **c** 1 : 200
d 1 : 10 **e** 7 : 2000 **f** 1 : 2500

4 **a** 1 : 250 **b** 1 : 1000 **c** 1 : 75
d 1 : 200 **e** 1 : 300 **f** 1 : 2 500 000

5 **a** rotation **b** translation

EXERCISE 14A

1 constructions **3** construction

2 construction

4 **a** T **b** F **c** T

5 constructions **6** constructions

7 **a** no **b** yes **c** no **d** no

8 no as $8 > 4{\cdot}5 + 3$

9 **a** sides 3 cm, 4 cm and the angle between 60°
c two angles of 60° and 50° and the side opposite the 60° = 7 cm
d a right angle and the two short sides 10 cm and 8 cm

10 construction

EXERCISE 14B

1 a, b, e, f

2

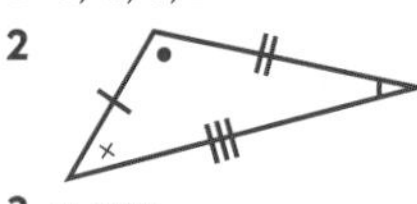

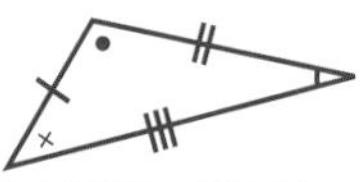

3 **c** yes **d** $ABCD \equiv PQRS$

4 **a** PB **b** PT **c** TB **d** $\angle PBT$
e $\angle CAM$ **f** $\angle MCA$ **g** $2{\cdot}25\ \text{cm}^2$

5 **b** $\Delta LKF \equiv \Delta NGH$
c **i** HG **ii** LF **iii** $\angle NHG$
iv $\angle KLF$

6 **a** B **b** A, C **c** A, B, C

7 **a** ME **b** XZ **c** $\angle AME$ **d** $\angle XTZ$

8 **a** **i** ΔTQB **ii** $\angle Q$ **iii** BQ
b **i** 30° **ii** 70° **iii** 6 cm
iv 13 cm **v** 15 cm

9 **a** 35 cm **b** 28 cm **c** 51° **d** 92°
e 20 cm **f** 37°

10 **a**

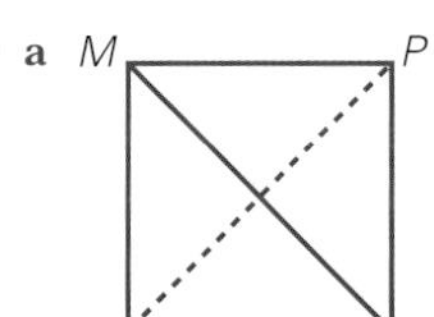

$\Delta PMQ \equiv \Delta RQM$ or $\Delta MPR \equiv \Delta QPR$

b

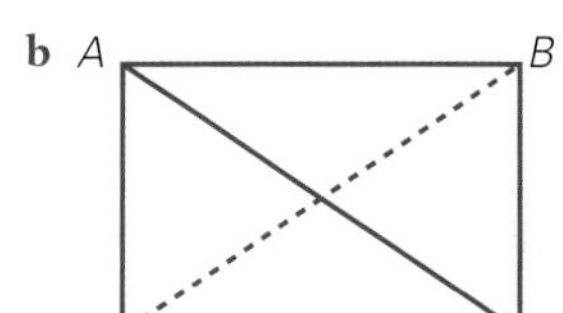

$\Delta ABC \equiv \Delta CDA$ or $\Delta ABD \equiv \Delta CDE$

c

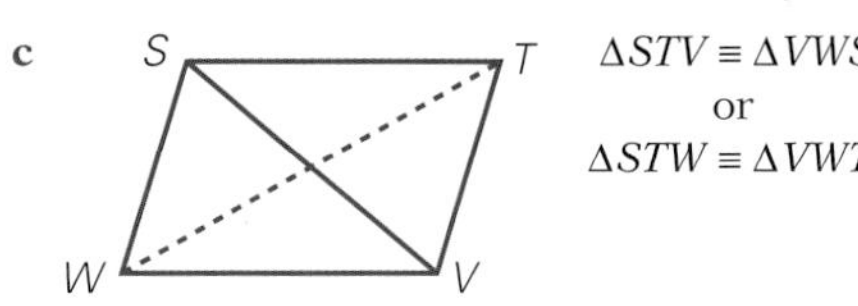

$\Delta STV \equiv \Delta VWS$ or $\Delta STW \equiv \Delta VWT$

d

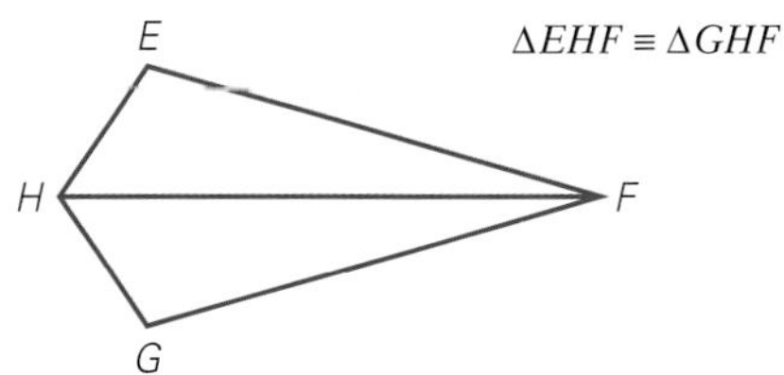

$\Delta EHF \equiv \Delta GHF$

e

J
I
K
L

$\Delta IJL \equiv \Delta KLJ$
$\Delta IJK \equiv \Delta KLJ$

f

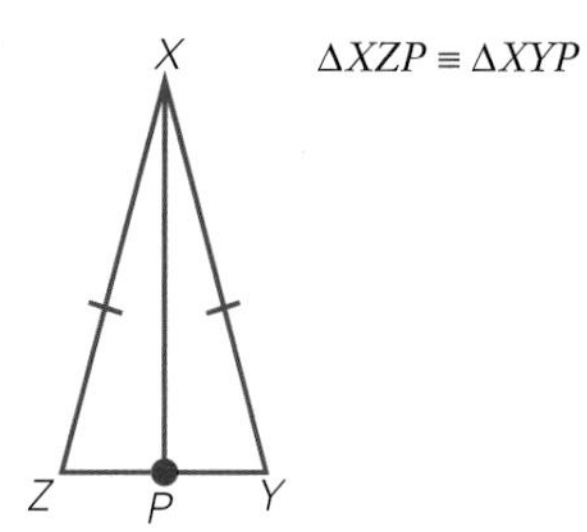

$\Delta XZP \equiv \Delta XYP$

11 **a** $\Delta AEB \equiv \Delta DEC$, $\Delta ACB \equiv \Delta DBC$

b

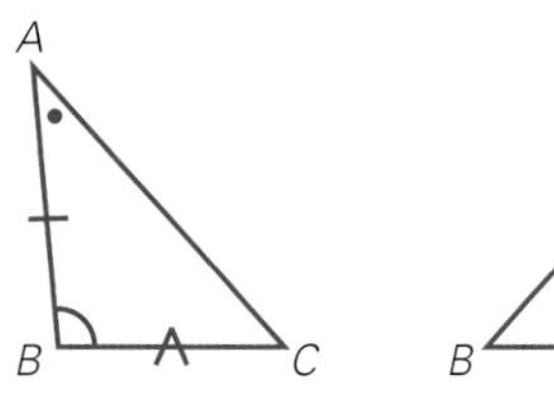

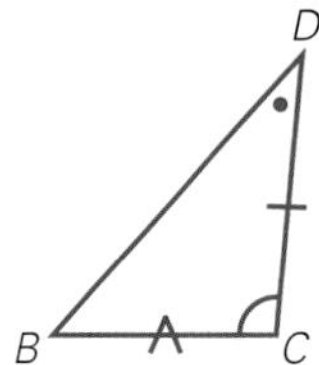

12 **a** Y **b** N **c** Y

13 **a** $x = 80, y = 60$ **b** $n = 57$
c $c = 30, b = 1{\cdot}5, x = 2{\cdot}5$
d $x = 30, y = 15, p = 25$

EXERCISE 14C

1 **a** yes, 2 : 1 **b** no **c** yes, 3 : 2
d no **e** yes, 3 : 5 **f** yes, 5 : 2

2 **a** $x = 70$ **b** $n = 17$

3 **a** rectangle 7 cm by 3 cm
b kite 4·5 cm and 7·5 cm

4 **a** sides 1 by 1·5 **b** all sides 1

5 **a** 1 : 2 **b** 2

6 **a** reduction factor 2 (2 : 1)
b enlargement factor 3 (1 : 3)
c reduction factor $\frac{3}{4}$ (4 : 3)
d enlargement factor $\frac{3}{2}$ (2 : 3)

7 **a** $4\frac{1}{2}$ cm **b** 4 : 9

8 **a** $\Delta ABC \mathbin{|||} \Delta DEF$ **b** $\Delta ALT \mathbin{|||} \Delta BMQ$
c $\Delta HDK \mathbin{|||} \Delta JVM$ **d** $\Delta PKT \mathbin{|||} \Delta MYR$
e $\Delta AED \mathbin{|||} \Delta ACB$ **f** $\Delta QPR \mathbin{|||} \Delta QMS$

9 constructions

EXERCISE 14D

1 **a** $x = 1$ **b** $x = \frac{35}{8}$ **c** $x = 2$ **d** $x = 20$
e $x = 4{\cdot}5$ **f** $x = 4{\cdot}5$

2 **a** $x = 9$ **b** $x = 2$

3 **a** $x = 8$ **b** $x = 20$

4 $x = 5$ metres

5 **a** $x = 24, y = 105$ **b** $x = 5, y = 12$
c $a = 18\frac{2}{3}, b = 32\frac{2}{3}$ **d** $d = 6{\cdot}4$, e = 12

e $x = 13{\cdot}5$ f $h = 52{\cdot}5$
6 40 m

EXERCISE 14E

1 a 1 : 200 b 1 : 1000
c 1 : 500 000 d 1 : 10 000 000
e 1 : 200 f 1 : 50 000
2 a 2 m b 5 m c 6·8 m d 0·8 m
3 a 2·4 m b 56 cm c 4·16 m d 71·2 cm
4 a 2 cm b 8·5 cm c 5 mm d 7·68 cm
5 a 2460 mm b 570 mm c 720 mm
6 scale drawings 7 C
8 150 cm (1·5 m) 9 3·6 km
10 1 : 50
11 a 1 : 120 b 6 m c 840 mm
12 a 5.75 m × 5.25 m b $1685 c $1034
13 a i 60 cm ii 1·5 m iii 9 m iv 4·5 m
b i 40 m ii 4 m iii 8 m iv 60 m

Chapter Review Exercises

1 constructions 2 no, as $12 > 4 + 7$
3 a yes b $\Delta ABC \equiv \Delta QRP$ c CB d $\angle RPQ$
4 a NY b ET c YP d $\angle YNP$
e $\angle YPN$ f $\angle NYP$ g $\Delta DTE \equiv \Delta NYP$
5 $x = 6, y = 34°$
6 a $ABCD \parallel\!| QRMP$, $AB \leftrightarrow QR$ $\angle A \leftrightarrow \angle Q$
$BC \leftrightarrow RM$ $\angle B \leftrightarrow \angle R$
$CD \leftrightarrow MP$ $\angle C \leftrightarrow \angle M$
$DA \leftrightarrow PQ$ $\angle D \leftrightarrow \angle P$
b $\Delta LTA \parallel\!| \Delta BSG$ $LT \leftrightarrow BS$ $\angle T \leftrightarrow \angle S$
$TA \leftrightarrow SG$ $\angle L \leftrightarrow \angle B$
$AL \leftrightarrow GB$ $\angle A \leftrightarrow \angle G$
7 construction
8 a enlargement factor $1\frac{1}{2}$ (2 : 3)
b reduction factor 2 (2 : 1)
c enlargement factor 2·5 (2 : 5)
d enlargement factor 1 (1 : 1)
9 a $x = 24, y = 55$
10 a 1 : 2 000 b 1 : 1 000 000
11 a 1·75 m b 8 cm
12 a 1 : 100 b 63 cm

KeepIng Mathematically Fit

PART A

1 0·0503 2 0·007, $0{\cdot}00\dot{7}$, 0·07, $0{\cdot}0\dot{7}$
3 0·1175 4 30° 5 $\frac{1}{3}$ 6 $x = 4$
7 $^-2x + 6$ 8 2 650 000 9 198 10 $P = 2x + 4$

PART B

1 a 500 b 1850 c 2003 d 6350
2 a 50·27 cm^2 b 13·63 cm^2 c 85·52 cm^2
3 a $250 b $400 c $8000
4 3·32cm
5 a $x = \frac{^-6}{5}$ b $x = {}^-24$ c $x \geq 6$
6 4:37 pm 7 ($^-1$, 6)
8 a one set 3, 4, 7, 8, 9, 9
b mean of set (a) = $6\frac{2}{3}$
c yes, discuss other sets such as 3, 5, 7, 8, 9, 9

Chapter 15

MathsCheck

1 a $2a + 3b$ b $8y - 7x$ c $2a + 2b$ d $20ab$
e ^-6a f $13 + 9w$ g a h $\frac{1}{a}$
i 4 j 6^7 k 7^3 (343) l 4^{15}
m $2^2 = 4$ n $3^7 \times 2^7$ o $2b$ p $\frac{1}{2b}$
2 a $3a + 3b$ b $5a - 5b$ c $8w + 12$
d $5a - 20$ e $5a + 6$ f $18d - 4$
g $6 + 2x$ h $19 - 10x$ i $5a + 20b - 25$
j $6x + 10$ k $9p - p^2$ l $^-x - 47$
3 a $x(y + 5)$ b $y(x - 5)$ c $3(ab + 4)$
d $3a(b + 4)$ e $12(2 - 3a)$ f $16a(b + 2a)$
g $^-a(b - 1)$ h $^-5xy(3y + 4)$
4 a $x = 27$ b $x = 9$ c $x = 10$ d $x = {}^-4$
e $x = \frac{^-5}{2}$ f $a = 70$ g $x = 0$ h $x = 30$
i $x = 4{\cdot}5$ j $s = \frac{^-4}{7}$ k $x = 20$ l $x = \frac{^-5}{8}$
5 a $x = 70$ b $x = 50$ c $x = 15$ d $x = \sqrt{8}$
6 a $2\frac{1}{4}$ b $\frac{1}{10}$ c $2\frac{11}{12}$ d 12
e $\frac{2}{3}$ f $1\frac{1}{5}$ g $3\frac{3}{5}$ h 4
7 a $\frac{3x}{5}$ b $\frac{a}{2}$ c $\frac{7x}{8}$ d $\frac{11x}{30}$
8 a $\frac{4}{5}$ b $\frac{2b}{7}$ c $a + 3$ d $2a$

EXERCISE 15A

1 a a^6 b x^4y c $8x^3$ d a^2b^5
e $10w^3p^2$ f $(pq)^4 = p^4q^4$
2 a $a \times a \times a \times a \times a$
b $p \times p \times p \times q \times q \times q \times q \times q$
c $4 \times a \times a \times a \times a \times a \times a \times a$
d $(3a) \times (3a) \times (3a) \times (3a) \times (3a)$
3 a 3^4 b 2^8 c 7^7 d 5^{13} e 10^{11}
f $3^6 \times 5^6$
4 a g^7 b x^7 c a^9 d d^{12} e x^{13}
f a^{10} g s^9 h a^5 i n^8 j m^6
k n^8 l a^{10} m d^6 n p^4q^2 o n^7m
5 a 6 b 5^6 c 3^4 d 10^3 e 5^4
6 a x^7 b $1 (a^0)$ c m^2 d m e p^3
f p^4 g t h 1 i x^7 j 1
k w^7 l 1 m y n w^3 o h^2
7 a $3n^9$ b $7a^{10}$ c $6b^4$ d $15c^6$ e $12p^7$
f $24k^3$ g $9t^2$ h a^6b^3 i m^4p^4
8 a a^3 b $4x^4$ c $3c^4$ d $\frac{2d^5}{3}$ e h
f ab^2 g nr h $3x^3a$ i $\frac{7t^4}{2b^2}$
9 a $2n$ b $9p^4$ c $7x^7$
10 a T b T c T

EXERCISE 15B

1 **a** 2^{12} **b** 4^9 **c** a^8 **d** m^{10} **e** p^{18} **f** n^0
g a^{18} **h** x^{28} **i** n^{24} **j** a^{25} **k** p^{12} **l** p^{12}

2 **a** 1 **b** 1 **c** 1 **d** 1 **e** 3 **f** $1 + b$
g 6 **h** 8 **i** 1

3 **a** a^{35} **b** $9a^4$ **c** $64x^9$
d $81d^4$ **e** $25x^6y^4$ **f** $4x^8$
g $125a^{15}b^9$ **h** $a^8b^{12}c^4$ **i** $15\,625x^{42}y^6$
j $3^4 \times 5^4$ **k** $7^3 \times 6^3$ **l** $4^2 \times 7^2$
m $16m^{12}n^2$ **n** $2401a^{12}b^{16}$ **o** $81a^{10}b^{14}$

4 **a** 3^4 **b** 8^3 **c** 4^5 **d** 1 **e** a^2 **f** a^{33}
g p^{14} **h** w^{18} **i** $9x^{12}$ **j** $5b^{10}$ **k** ab^6
l $16\,807d^5$

5 **a** $\frac{x^4}{y^4}$ **b** $\frac{4}{9}$ **c** $\frac{16}{25}$ **d** $\frac{1}{27}$ **e** $\frac{a^6}{27}$ **f** $\frac{a^9}{b^3}$
g $\frac{x^2y^4}{4}$ **h** $\frac{4x^2}{y^4}$ **i** $\frac{64m^6}{n^6}$

6 **a** a^7b^3 **b** $12m^4n^9$ **c** $5a^9b^{13}$ **d** $8a^7b^2$
e $\frac{x^3y^6}{2}$ **f** $8a^3b$ **g** $w^{14}y^{12}$ **h** $m^{12}n^8$
i $5x^6$ **j** $3x^4y^2$

7 **a** $a^5 - 3a^3$ **b** $12p^6 + 8p^3$ **c** $5n^4 - 2n^7$
d $7x^3 - x^8$ **e** $^-4p^6 - 12p^8$ **f** $^-2w^6 + 5w^7$
g $4a^3b - 4a^2b^2$ **h** $3a^4b - 3a^3b^6$ **i** $a^5b^6 + a^4b^7$
j $2a^3 - 3a^2$ **k** $7n^5 + 2n^3$ **l** $8x^3 + 7x^2$

8 **a** $x^2(x^3 - 1)$ **b** $a^2b(a - 1)$
c $m^4(m^3 - 3)$ **d** $5p^3(p^4 - 5)$
e $x^2y^2(x^2 + y^2)$ **f** $p^3q(p^2q^4 - 1)$
g $2a^3(2 - a)$ **h** $5a^3b^2(2ab + 3)$
i $m^6n(m^2 - n)$

EXERCISE 15C

1 **a** $ac + ad + cb + bd$ **b** $x^2 + 11x + 24$
c $3a^2 + 5a - 2$ **d** $3a^2 + 16a - 35$
e $6a^2 + 17a + 5$ **f** $y^2 - 5y + 6$

2 **a** $nm + 2n + 6m + 12$ **b** $ab - a + 7b - 7$
c $2ab - 3a + 10b - 15$ **d** $4nm + 8n + m + 2$
e $2mn - 2m - 5n + 5$ **f** $9ab - 12a - 30b + 40$

3 **a** $a^2 + 3a + 2$ **b** $x^2 + 5x + 6$
c $x^2 + 11x + 30$ **d** $w^2 + 8w + 7$
e $p^2 + 7p + 10$ **f** $a^2 + 4a + 3$
g $q^2 + 13q + 30$ **h** $m^2 + 10m + 16$
i $a^2 + 17a + 70$

4 **a** $a^2 - 1 - 12$ **b** $m^2 + 4m - 5$ **c** $m^2 - 6m + 5$
d $x^2 - x - 12$ **e** $n^2 + n - 12$ **f** $n^2 - 7n + 12$
g $p^2 - 9p + 8$ **h** $q^2 - 2q - 15$ **i** $x^2 - 6x + 9$
j $a^2 - 2a - 24$ **k** $m^2 - n^2$ **l** $49 - p^2$
m $m^2 - 64$ **n** $p^2 - 25$ **o** $100 - w^2$

5 **a** $2x^2 + 7x + 3$ **b** $4x^2 + 8x + 3$
c $3m^2 - 2m - 8$ **d** $3x^2 + 5x - 2$
e $6x^2 + 5x + 1$ **f** $10n^2 - 9n - 9$
g $12n^2 - 7n - 12$ **h** $4n^2 + 8n - 21$
i $6 - n - 2n^2$

6 **a** $a^2 + 8a + 16$ **b** $m^2 + 18m + 81$
c $n^2 - 14n + 49$ **d** $x^2 + 4x + 4$
e $n^2 - 10n + 25$ **f** $4x^2 + 4x + 1$
g $9x^2 + 6xy + y^2$ **h** $49m^2 + 14mn + n^2$
i $a^2b^2 + 2ab + 1$

7 **a** $x = \frac{9}{8}$ **b** $x = 6\frac{2}{3}$ **c** $x = \frac{7}{12}$

8 **a** $2x^2 + 5x + 5$ **b** $x^2 + 10x + 13$
c $x^2 + 16x + 9$ **d** $2x^2 - 13x - 25$
e $4x^2 + 2x + 6$ **f** $3x + 3$
g $x^2 + 4x + 3$ **h** $^-1$
i $2x^2 + 5x + 6$ **j** $2x^2 + 6x + 1$
k $2x - 1$ **l** $x^3 + 5x^2 + 6x$

EXERCISE 15D

1 **a** 4 **b** 8 **c** 20 **d** 0 **e** 2
2 **a** 6 **b** 9 **c** 18 **d** 3 **e** 4·5
3 **a** 2 **b** $^-7$ **c** 5 **d** 8 **e** 3
4 **a** 1 **b** 2 **c** $\frac{1}{2}$ **d** 0 **e** $^-1$
5 **a** 1 **b** 0 **c** 3 **d** 28 **e** 171
6 **a** **i** $^-5$ **ii** 11 **iii** 6 **iv** 40
b $x = 3$ **c** $x = \frac{^-3}{4}$
7 **a** $f(0) = 0, g(0) = 0$ **b** where $f(x) = g(x)$
c 12
8 **a** **i** 4 **ii** 14 **iii** $^-6$ **iv** 20
b **i** 3 **ii** $^-7$ **iii** $^-21$ **iv** $^-12$
c 0 **d** $7x - 3$ **e** $3x + 11$ **f** $x = \frac{^-11}{3}$

EXERCISE 15E

1 **a** $x = \frac{^-10}{3}$ **b** $a = \frac{40}{17}$ **c** $a = 40$
2 **a** $x = \frac{10}{11}$ **b** $x = 5\frac{5}{7}$ **c** $a = {}^-2$
d $x = 3\frac{1}{3}$ **e** $w = 4\frac{1}{2}$ **f** $m = 9\frac{1}{3}$
g $x = 0$ **h** $n = 3\frac{1}{3}$ **i** $x = {}^-5\frac{1}{3}$
3 **a** $x = 1$ **b** $x = 2\frac{7}{9}$ **c** $x = 1$ **d** $x = 27$
4 **a** yes **b** no **c** yes

Chapter Review Exercises

1 **a** a^{15} **b** w^4 **c** $24x^{10}$ **d** $21x^9$ **e** x^2
f $8x^4$ **g** $2a$ **h** a^5b^5
2 **a** a^{10} **b** x^{20} **c** a^{21} **d** 6 **e** $25x^8$
f $\frac{x^6}{y^3}$ **g** $81a^{16}b^{12}$ **h** 1
3 **a** a^2b^5 **b** $\frac{3}{4}$ **c** $45x^{11}$ **d** $^-12a^7b^8$
e $2a^3b^2$ **f** $\frac{3a^5b}{2}$ **g** $\frac{a^2b^2}{2}$ **h** $75a^8b^6$
4 **a** $m^2 + 9m + 18$ **b** $a^2 + 5a - 14$
c $a^2 + a - 12$ **d** $x^2 - 10x + 16$
e $6m^2 + 25m + 14$ **f** $2x^2 - 7x + 6$
5 **a** $x^2 + 22x + 121$ **b** $4x^2 + 20x + 25$
c $m^2 - 6m + 9$ **d** $25 - w^2$
e $4p^2 - 9$ **f** $x^2 - y^2$
6 **a** 5 **b** 8 **c** 0 **d** $^-16$
7 **a** $x = 1\frac{1}{7}$ **b** $a = 3\frac{3}{7}$ **c** $a = 4$
8 **a** $x = 5\frac{1}{3}$ **b** $x = 3\frac{3}{4}$

Keeping Mathematically Fit

PART A

1 $2\frac{3}{5}$ 2 $\sqrt{26}$ to $\sqrt{35}$ 3 290°
4 \$6.40 5 40 6 0·047 7 36
8 $^-28$ 9 1 : 6 10 30 000 L

PART B

1 0·208
2 a 6^5 (7776) b $5x^2 - 3$ c $\frac{2b}{3}$
3 14·25
4 a (5, 0) b (5, 3)
c i 28·3 units^2 ii 18·8 units
5 $x = 7·2$ 6 $^-1$ 7 \$1181.25
8 a i 10% ii 21% b yes

Cumulative Review 5

Part A

1 B 2 C 3 C 4 D 5 C 6 C
7 B 8 C 9 D 10 B 11 D 12 C
13 A 14 C 15 B 16 B 17 B 18 A
19 B 20 A 21 A 22 D 23 A 24 B
25 B

Part B

1 a $x = {}^-54$ b $x = \frac{5}{6}$ c $x < {}^-5$
2 a $5(w - 2p)$ b $^-a(b + 3)$ c $12x(y + 4)$
3 a 30 b $\frac{13}{30}$ c $6\frac{2}{3}\%$
4 a \$43.56 b \$4455
5 \$742.50 6 800
7 a $x = 83$ b $x = 133, y = 65$
8 150° 9 $x + y + z = 360$
10 9 : 14 11 10 440 cm^2 12 180 m^3

13

x	3	4	5	6
y	2	$2\frac{1}{3}$	$2\frac{2}{3}$	3

14 a

x	$^-2$	$^-1$	0	1	2
y	$^-4$	$^-2$	0	2	4

b

x	$^-2$	$^-1$	0	1	2
y	$^-7$	$^-3$	1	5	9

c

x	$^-2$	$^-1$	0	1	2
y	8	7	6	5	4

15

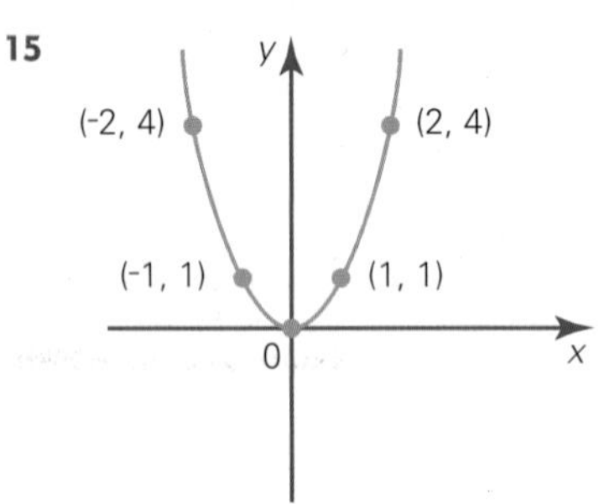

16 $\Delta MPQ \equiv \Delta VTR$
17 a $\Delta ABC \equiv \Delta KXP$ b AC c 14 cm
18 a Y b Y c N d N
19 a $x = 24, y = z = 85$ b $n = 55$
20 a $\frac{2}{11}$ b $\frac{4}{11}$ c $\frac{7}{11}$ d $\frac{2}{11}$ e $\frac{1}{11}$ f 0
21 a $\frac{3a - 6}{9}$ b $\frac{1}{3}$
22 a a^{18} b x^3 c $8a^{21}$ d 8
e $14a^4b^4$ f $49a^8b^6$
23 a $\frac{17x}{21}$ b $\frac{2x}{9}$ c $\frac{a^2}{5}$ d $\frac{5}{4}$
24 a 0 b 4 c $^-6$ d 36
25 a $x = 3\frac{9}{11}$ b $x = \frac{^-2}{3}$